# Managing Salt Tolerance in Plants

## Molecular and Genomic Perspectives

T0136177

# Managing Salt Tolerance in Plants

## Molecular and Genomic Perspectives

Edited by

**Shabir Hussain Wani**
**Mohammad Anwar Hossain**

CRC Press
Taylor & Francis Group
Boca Raton London New York

CRC Press is an imprint of the
Taylor & Francis Group, an **informa** business

CRC Press
Taylor & Francis Group
6000 Broken Sound Parkway NW, Suite 300
Boca Raton, FL 33487-2742

First issued in paperback 2020

© 2016 by Taylor & Francis Group, LLC
CRC Press is an imprint of Taylor & Francis Group, an Informa business

No claim to original U.S. Government works

ISBN-13: 978-1-4822-4513-4 (hbk)
ISBN-13: 978-0-367-65875-5 (pbk)

---

**Library of Congress Cataloging-in-Publication Data**

---

Managing salt tolerance in plants : molecular and genomic perspectives / edited by Shabir Hussain
    Wani, Mohammad Anwar Hossain.
      pages cm
    Includes bibliographical references and index.
    ISBN 978-1-4822-4513-4
    1. Plants--Effect of salt on. 2. Salt-tolerant crops. I. Wani, Shabir Hussain, editor. II. Hossain,
Mohammad Anwar, editor.

    QK753.S32M36 2015
    581.7'88--dc23                                      2015016372

---

**Visit the Taylor & Francis Web site at**
**http://www.taylorandfrancis.com**

**and the CRC Press Web site at**
**http://www.crcpress.com**

# CONTENTS

# PREFACE

Abiotic stress has become one of the decisive factors dwindling crop productivity worldwide. Salinity stress is one of the major abiotic stresses affecting agriculture, with more than 80 million hectares of irrigated land affected worldwide. This book describes salinity stress in plants and its effects on plant growth and productivity. It also addresses the management aspect of salinity stress in crops through molecular and genomic approaches. This edited book attempts to bring together all the biochemical, physiological, and molecular techniques exploited to develop crop plants with increased salinity tolerance. Through this book, an attempt has been made to integrate the most recent findings about the key biological determinants of salinity stress tolerance with contemporary crop improvement approaches. Twenty-two chapters written by leading scientists involved in plant salinity stress research worldwide provide an ample coverage of the key factors impacting world crop production.

Chapter 1 discusses the understanding of salt stress response from the gene to the whole plant level. Chapter 2 explains the mechanisms of salt stress tolerance in halophytes. Chapter 3 is concerned with vacuolar sodium sequestration in plant breeding for salinity tolerance. Chapter 4 discusses salt stress signaling pathways. Chapter 5 discusses the physiological and biochemical approaches for salinity tolerance. Chapter 6 describes the plant cell organelle proteomics and salinity tolerance. Chapter 7 depicts the function of heat shock proteins in salt tolerance. Chapter 8 deals with the transcription factors involved in salt tolerance. Chapter 9 discusses the role of the glyoxalase system and salinity stress tolerance, and Chapter 10 depicts ROS metabolism and salt stress tolerance in plants. Chapter 11 describes the insights of hydrogen peroxide–induced salt stress tolerance, Chapter 12 deals with the roles of ascorbate-glutathione cycle in salt stress tolerance of plants, and Chapter 13 discusses polyamine metabolism and salinity stress tolerance. In Chapter 14, metabolomics and salt stress tolerance are reviewed. Chapter 15 discusses plant–microbe interaction and salt stress tolerance, Chapter 16 summarizes molecular breeding for salt stress tolerance, and Chapter 17 is about mutation breeding for salt stress tolerance. Chapter 18 is concerned with the present status and future prospects of transgenic approaches for salinity tolerance, Chapter 19 discusses proline engineering for enhanced salt stress tolerance, Chapter 20 describes salt stress tolerance in plants in relation to ion transporters and genetic engineering, and Chapter 21 reviews transgenic plants for higher antioxidant contents and salt stress tolerance. The final Chapter 22 summarizes the insights of salinity tolerance–based transcriptomic studies.

The facts presented in this book call attention to primary genetic, physiological, and biochemical acquaintance of plant salinity stress, which may lead to both conventional and biotechnological applications that finally lead to enhanced crop productivity in stressful environments. We hope that the book will be very helpful for plant abiotic stress researchers, graduate students, and university teachers. It will also be of interest to environmental scientists, biochemists, and policy makers.

We give special thanks to all the authors for their stupendous and sensible work in producing such fine chapters. We also thank Dr. C. R. Crumly (senior acquisitions editor, CRC Press), Stephanie Morkert (project coordinator, CRC Press), and other members of CRC Press, Taylor & Francis Group, for their guidance and support during the progress of this important book. Thanks also to all the well-wishers, teachers, colleagues, research students, and family members. Without their unending support, motivation, and encouragement, the grueling task of writing this book would not have been possible.

Finally, we bow in reverence to Allah, who blessed us with the favor of plentiful academic work.

# EDITORS

**Dr. Shabir Hussain Wani** received his PhD in plant breeding and genetics through his thesis "Transgenic rice for abiotic stress tolerance" from Punjab Agricultural University, Ludhiana, India, in 2009. Then he worked as research associate in the Biotechnology Laboratory, Central Institute of Temperate Horticulture (ICAR), Rangreth, Srinagar, India, for two years, up to October 2011. In November 2011, he joined the Krishi Vigyan Kendra (Farm Science Centre) as programme coordinator (i/c) at Senapati, Manipur, India. Since May 2013, he has been assistant professor/scientist at the Division of Genetics and Plant Breeding, SKUAST-K, Srinagar, Kashmir. Dr. Wani teaches courses related to plant breeding, seed science and technology, and stress breeding. He has published more than 75 papers in journals and books of national and international repute. He has also edited several books on current topics in crop improvement. His PhD research fetched the first prize in the North Zone at the national level competition in India. He received the Young Scientist Award from the Society for the Promotion of Plant Science Research, Jaipur, India, in 2009. He is a fellow of the Society for Plant Research, India. He has attended several national and international conferences and presented his research.

**Dr. Mohammad Anwar Hossain** received his PhD in agriculture, with specialization in plant abiotic stress physiology and molecular biology from Ehime University, Japan, in 2011. He received his bachelor's and master's degrees in agriculture and genetics and plant breeding from Bangladesh Agricultural University, Bangladesh. He also earned his master of science in agriculture from Kagawa University, Japan. He has been an associate professor at the Department of Genetics and Plant Breeding, Bangladesh Agricultural University since September 2011. Dr. Hossain teaches plant breeding, advanced plant breeding, molecular plant breeding, and stress breeding. He has published 17 research articles, 6 book chapters, and 4 review articles on the important aspects of plant physiology and breeding, plant stress responses, environmental problems in relation to plant species, and the role of micronutrients on plant defense system. Presently, he is working on elucidating the molecular mechanisms of plant abiotic stress tolerance in plants, supervising graduate students, and serving as editorial board member of several plant science journals. He attended and presented abstracts and posters in national and international conferences in different countries. He is a professional member of the Indian Society of Genetics and Plant Breeding, Bangladesh Association for Plant Tissue Culture and Biotechnology, and the Seed Science Society of Bangladesh.

# CONTRIBUTORS

GETNET DINO ADEM
School of Land and Food
University of Tasmania
Hobart, Tasmania, Australia

L.V. ARUNA
Department of Biochemistry and Bioinformatics
GITAM University
Visakhapatnam, India

MUHAMMAD ARSLAN ASHRAF
Department of Botany
Government College University
Faisalabad, Pakistan

S.N. BEGUM
Plant Breeding Division
Bangladesh Institute of Nuclear Agriculture
Mymensingh, Bangladesh

SOUMEN BHATTACHARJEE
Department of Botany
University of Burdwan
West Bengal, India

SUJIT KUMAR BISHI
Division of Crop Improvement
ICAR-Central Soil Salinity Research Institute
Haryana, India

JAYAKUMAR BOSE
School of Land and Food
University of Tasmania
Hobart, Tasmania, Australia

ABHISHEK BOHRA
Crop Improvement Division
Indian Institute of Pulses Research (IIPR)
Kanpur, India

FAIÇAL BRINI
Centre of Biotechnology of Sfax
University of Sfax
Sfax, Tunisia

DAVID J. BURRITT
Department of Botany
University of Otago
Dunedin, New Zealand

I. CAÇADOR
Centre of Oceanography
University of Lisbon
Lisbon, Portugal

ANANYA CHAKRABARTY
Department of Botany
University of Burdwan
West Bengal, India

MINGNA CHEN
Department of Genetics and Plant Breeding
Shandong Peanut Research Institute
Qingdao, Shandong, People's Republic of China

NA CHEN
Department of Genetics and Plant Breeding
Shandong Peanut Research Institute
Qingdao, Shandong, People's Republic of China

XIAOYUAN CHI
Department of Genetics and Plant Breeding
Shandong Peanut Research Institute
Qingdao, Shandong, People's Republic of China

VISHWANATHAN CHINNUSAMY
Division of Plant Physiology
Indian Agricultural Research Institute
New Delhi, India

K.G.K. DEEPAK
Department of Biochemistry and Bioinformatics
GITAM University
Visakhapatnam, India

NEETI DHAKA
Department of Botany
University of Delhi
New Delhi, India

B. DUARTE
Centre of Oceanography
University of Lisbon
Lisbon, Portugal

HAI FAN
College of Life Science
Shandong Normal University
Jinan, Shandong, People's Republic of China

YUN FAN
School of Land and Food
University of Tasmania
Hobart, Tasmania, Australia

KAOUTHAR FEKI
Centre of Biotechnology of Sfax
University of Sfax
Sfax, Tunisia

BRIAN V. FORD-LLOYD
School of Biosciences
University of Birmingham
Birmingham, United Kingdom

MASAYUKI FUJITA
Department of Applied Biological Science
Kagawa University
Kagawa, Japan

N. HOQUE
Biotechnology Division
Bangladesh Institute of Nuclear Agriculture
Mymensingh, Bangladesh

MOHAMMAD ANWAR HOSSAIN
Department of Genetics and Plant Breeding
Bangladesh Agricultural University
Mymensingh, Bangladesh

MOHAMMAD RASHED HOSSAIN
Department of Genetics and Plant Breeding
Bangladesh Agricultural University
Mymensingh, Bangladesh

and

School of Biosciences
University of Birmingham
Birmingham, United Kingdom

IQBAL HUSSAIN
Department of Botany
Government College University
Faisalabad, Pakistan

MUHAMMAD IQBAL
Department of Botany
GC University
Faisalabad, Pakistan

MIRZAMOFAZZAL ISLAM
Plant Breeding Division
and
Biotechnology Division
Bangladesh Institute of Nuclear Agriculture
Mymensingh, Bangladesh

ROHIT JOSHI
Plant Molecular Biology Group
International Center for Genetic Engineering and
    Biotechnology
New Delhi, India

P.B. KAVI KISHOR
Department of Genetics
Osmania University
Andra Pradesh, India

HYUN-JEE KIM
School of Applied Biosciences
Kyungpook National University
Daegu, Republic of Korea

ASHWANI KUMAR
Central Soil Salinity Research Institute
Haryana, India

KUNDAN KUMAR
Department of Biological Sciences
Birla Institute of Technology and Science
Goa, India

VINAY KUMAR
Department of Biotechnology
University of Pune
Pune, India

JEONG-DONG LEE
Delta Research Center
University of Missouri
Portageville, Missouri

TERESA LIVERMORE
School of Biosciences
University of Birmingham
Birmingham, United Kingdom

MAHESH KUMAR MAHATMA
Directorate of Groundnut Research
Gujarat, India

ANITA MANN
Division of Crop Improvement
ICAR-Central Soil Salinity Research Institute
Haryana, India

KAREEM A. MOSA
Department of Biotechnology
Al-Azhar University
Cairo, Egypt

NAGESWARA RAO REDDY NEELAPU
Department of Biochemistry and Bioinformatics
GITAM University
Visakhapatnam, India

LIJUAN PAN
Shandong Peanut Research Institute
Qingdao, People's Republic of China

MALABIKA ROY PATHAK
Department of Life Sciences
Arabian Gulf University
Manama, Kingdom of Bahrain

JEREMY PRITCHARD
School of Biosciences
University of Birmingham
Birmingham, United Kingdom

RIZWAN RASHEED
Department of Botany
Government College University
Faisalabad, Pakistan

M.K. SAHA
Department of Biotechnology
Bangladesh Agricultural University
Mymensingh, Bangladesh

WALID SAIBI
Centre of Biotechnology of Sfax
University of Sfax
Sfax, Tunisia

SERGEY SHABALA
School of Agricultural Science
University of Tasmania
Hobart, Tasmania, Australia

J. GROVER SHANNON
Delta Research Center
University of Missouri
Portageville, Missouri

GARIMA SHARMA
School of Biosciences
University of Birmingham
Birmingham, United Kingdom

VARSHA SHRIRAM
Department of Botany
University of Pune
Pune, India

BAHAA T. SHAWKY
Department of Microbial Chemistry
National Research Center
Giza, Egypt

BALWANT SINGH
Division of Molecular Biology and Biotechnology
National Research Centre on Plant
    Biotechnology
New Delhi, India

PREETI SINGH
Faculty of Biology
Technion-Israel Institute of Technology
Haifa, Israel

CHALLA SUREKHA
Department of Biochemistry and Bioinformatics
GITAM University
Visakhapatnam, India

NEVEEN B. TALAAT
Department of Plant Physiology
Cairo University
Giza, Egypt

THEODORE W. THANNHAUSER
Agricultural Research Service, USDA
Cornell University
Ithaca, New York

LAURA VICKERS
Crop and Environment Sciences
Harper Adams University
Newport, United Kingdom

BAO-SHAN WANG
College of Life Science
Shandong Normal University
Jinan, Shandong, People's Republic of China

CHUANTANG WANG
Department of Biotechnology
Shandong Peanut Research Institute
Qingdao, Shandong, People's Republic of China

MIAN WANG
Department of Genetics and Plant Breeding
Shandong Peanut Research Institute
Qingdao, Shandong, People's Republic of China

TONG WANG
Department of Genetics and Plant Breeding
Shandong Peanut Research Institute
Qingdao, Shandong, People's Republic of China

SHABIR HUSSAIN WANI
Division of Genetics and Plant Breeding
Sher-E-Kashmir University of Agricultural Sciences
    and Technology of Kashmir
Kashmir, India

ZHEN YANG
Department of Genetics and Plant Breeding
Shandong Peanut Research Institute
Qingdao, Shandong, People's Republic of China

SHANLIN YU
Department of Genetics and Plant Breeding
Shandong Peanut Research Institute
Qingdao, Shandong, People's Republic of China

MEIXUE ZHOU
School of Land and Food
University of Tasmania
Hobart, Tasmania, Australia

SUPING ZHOU
Department of Agricultural and Environmental
    Sciences
Tennessee State University
Nashville, Tennessee

CHAPTER ONE

# Understanding Plant Stress Response and Tolerance to Salinity from Gene to Whole Plant

*Kaouthar Feki, Walid Saibi, and Faiçal Brini*

## CONTENTS

*Abstract.* Salinity is a major environmental stress that limits agriculture production. Hence, it is essential to produce salt-tolerant crops for sustaining food production. Understanding the molecular basis of salt-stress signaling and tolerance mechanisms is essential for breeding and genetic engineering of salt tolerance in crop plants. Plant adaptation or tolerance to salinity stress involves complex physiological traits, metabolic pathways, and molecular or gene networks. In many plants, the salt tolerance is associated with the ability to exclude sodium from the shoot, to prevent its accumulation in photosynthetic tissues. Salinity stress involves changes in various physiological and metabolic processes, depending on severity and duration of the stress, and ultimately inhibits crop production. In this chapter, we mainly discuss about the effect of salinity on plants and tolerance mechanisms that permit the plants to withstand stress.

*Keywords:* Functional genomics, Ion homeostasis, Salt stress, Salinity tolerance

## 1.1 INTRODUCTION

Salinity is one of the most serious factors limiting food production, because it limits crop yield with adverse effects on germination and restricts use of land previously uncultivated (Munns and Tester 2008). High salinity causes water stress, ion toxicity, nutritional disorders, oxidative stress, alteration of metabolic processes, membrane disorganization, reduction of cell division, and genotoxicity (Zhu 2002, 2007; Munns 2002). The complex "plant response to abiotic stress" involves many genes and biochemical molecular mechanisms. The analysis of the functions of stress-inducible genes is an important tool not only to understand the molecular mechanisms of stress tolerance and the responses of higher plants but also to improve the stress tolerance of plants by genomic strategies. The susceptibility or tolerance to high salinity stress in plants is a coordinated action of multiple stress-responsive genes, which also cross talk with other components of stress signal transduction pathways. Several types of gene belonging to different metabolic functions have been identified and used for over-expression into glycophytic plants to enhance salinity stress tolerance. The stress-related genes are generally classified into two major groups. The first one is involved in signaling cascades, transcriptional control, and the degradation of transcripts or proteins. The member of the second group functions in membrane protection and osmoprotection as antioxidants and as reactive oxygen species (ROS) scavengers (Pardo 2010). Plant responses to salinity and mechanisms conferring plant salinity tolerance have been studied for a long time. Using modern genetic approaches like genome sequencing, reverse genetics methods, and identification and characterization of key genes involved in salt-stress signaling, the understanding of salt tolerance mechanisms is substantially in progress especially salt ion signaling and transport (Hasegawa et al. 2000; Flowers 2004; Kosova et al. 2013). This chapter provides an overview of our current understanding of the mechanisms contributing to salt-stress tolerance in plants and the contribution of the genomic, transcriptomic, proteomic, and metabolic investigations to understand plant salinity tolerance.

## 1.2 EFFECTS OF SALINITY ON PLANTS

Salt stress imposes a major environmental threat to agriculture, and its adverse impacts are serious problems in regions where saline water is used for irrigation. It is estimated that about 7% of world agricultural land is affected by salinity and that this number could increase up to 20% in the future due to land salinization as a consequence of artificial irrigation and unsuitable land management. Regarding irrigated soils that contribute to roughly one-third of the global food production, it is estimated that nearly one-half of the total area of irrigated land could be adversely affected by salinization (Munns 2002; Munns and Tester 2008). High salinity causes hyperosmotic stress and ion disequilibrium that produce secondary effects (Hasegawa et al. 2000; Zhu 2001). Indeed, salts dissolved in the soil solution reduce the water potential (i.e., diminish water availability to the plant), and water uptake by roots is thermodynamically hampered. Thus, plants have to cope with osmotic effect by the mechanisms of osmotic adjustment. The stomatal closure often observed in salt-treated plants ameliorates tissue dehydration by limiting water losses (Fricke et al. 2004). Due to a reduced stomatal conductance, the rate of photosynthetic $CO_2$ assimilation is generally reduced by salinity. Salinity may cause nutrient deficiencies or imbalances, due to the competition of $Na^+$ and $Cl^-$ with nutrients such as $K^+$ and $Ca^{2+}$. Plants cope with increased ion concentrations either via salt ion exclusion from the cells or salt ion compartmentation in the intracellular compartments. Plants are generally classified as glycophytes or halophytes referring to their capacity to grow on highly saline environments (Flowers et al. 1977). Under salt stress, halophytes accumulate salts and have a capacity to growth on salinized soils in coastal and arid regions due to specific mechanisms of salt tolerance developed during their phylogenetic adaptation. However, glycophytes, including most crop plants, tend to exclude salt, and they are severely inhibited or even killed by 100–200 mM NaCl (Zhu 2007). Halophytes represent an ideal target for understanding the genetic and molecular basis for their adaptation in saline conditions (Subudhi and Baisakh 2011). Some halophytes have evolved unique adaptations such as salt glands and bladders, succulence, life cycle avoidance, and salt-induced facultative metabolism to cope with salinity (Flowers et al. 1977, 1986, 2010; Shabala and MacKay 2011).

Salt stress has various effects on plant physiological processes such as increased respiration rate and ion toxicity, changes in plant growth, mineral distribution, membrane permeability (Gupta et al. 2002), and decreased efficiency of photosynthesis (Boyer 1976; Downton 1977; Hasegawa et al. 2000;

Munns 2002; Kao et al. 2003). The chlorophyll content decreases in many salt-susceptible plants such as tomato (Lapina and Popov 1970), potato (Abdullah and Ahmed 1990), and pea (Hamada and El-Enany 1994). The derived reduction in the photosynthetic rate of salt-sensitive plants can increase the production of ROS. Salinity is well established to induce oxidative stress, which occurs due to the production of ROS such as superoxide radical ($O^-$), hydrogen peroxide ($H_2O_2$), and hydroxyl radical ($OH^-$). These species of oxygen are highly cytotoxic and can seriously react with vital biomolecules such as lipids, proteins, and nucleic acid, causing lipid peroxidation, protein denaturation, and DNA mutation, respectively (Scandalios 1993; McCord 2000; Breusegem et al. 2001; Mittler 2002). ROS are rapidly removed by antioxidative mechanisms, which include specific ROS-scavenging antioxidative enzymes and small nonenzymatic molecules that act as ROS scavenger such as ascorbate, glutathione (GSH), α-tocopherol, flavonoids, anthocyanines, polyphenolic compound, and carotenoids.

## 1.3 SALT TOLERANCE MECHANISM

The plant response to salinity consists of numerous processes that must function in coordination to alleviate both cellular hyperosmolarity and ion disequilibrium.

### 1.3.1 Ion Homeostasis: Transport Determinants and Their Regulation

Elevated salts lead to a passive salt ion penetration via plasma membrane and to an accumulation of salt ions in cell cytoplasm, which can lead to inhibition of intracellular enzyme activity (Munns and Tester 2008). Some halophytes are able to reduce shoot $Na^+$ accumulation through an intracellular sensing mechanism that indirectly regulates inward $K^+$ conductance (Robinson et al. 1997; Véry et al. 1998). Many salt-tolerant halophytes accumulate higher shoot $Na^+$ concentrations than less salt-tolerant halophytes or glycophytes, which is indicative of greater $Na^+$ homeostasis capacity (Rus et al. 2006; Flowers and Colmer 2008; Munns and Tester 2008; Baxter et al. 2010). In general, to prevent the accumulation of $Na^+$ in the cytoplasm, plants have developed three mechanisms that function in a cooperative manner, which are (1) reduction of $Na^+$ entry into the cell, (2) activation of $Na^+$ extrusion at the root–soil interface, and (3) compartmentalization of $Na^+$ in the vacuole (Tester and Davenport 2003).

### 1.3.1.1 Na+ Influx

Under salinity, sodium enters into root cell through cation channels or selective or nonselective transporters or into the root xylem stream via an apoplastic pathway (Chinnusamy et al. 2005). One of the key responses to salt stress is to maintain cellular ion homeostasis by restricting the accumulation of toxic sodium (Clarkson and Hanson 1980; Tester and Davenport 2003). The major pathway for passive $Na^+$ entry into roots at high soil NaCl concentrations is the voltage-dependent nonselective cation channels (NSCCs) (Tester and Davenport 2003; Demidchik and Maathuis 2007). Many studies have demonstrated that NSCCs are directly involved in a multitude of stress responses, growth and development, uptake of nutrients, and calcium signaling. NSCCs can also function in the perception of external stimuli and as signal transducers for ROS, pathogen elicitors, cyclic nucleotides, membrane stretch, amino acids, and purines (Demidchik and Maathuis 2007). Due to the similarity between $Na^+$ and $K^+$, voltage-dependent $K^+$ inward rectifiers or outward rectifiers appear to be one path for $Na^+$ entry into root cells (Blumwald et al. 2000). The members of the HKT gene family are $Na^+$-specific transporters, although they are initially described as high-affinity $K^+$ transporters. Generally, the HKT members of subfamily 1, which has a highly conserved serine in the first pore loop of the protein, have a relatively higher $Na^+$ to $K^+$ selectivity than subfamily 2 HKT transporters (Horie et al. 2009; Pardo 2010; Yao et al. 2010). Whereas the HKT family is comprised of a single gene in *Arabidopsis thaliana*, encoding a $Na^+$-selective transporter (Uozumi et al. 2000; Mäser et al. 2001), it is much larger in cereals: nine HKT genes are present in rice (*Oryza sativa*), and 6–24 HKT are expected to be present in barley (*Hordeum vulgare*) and wheat species (Garciadeblás et al. 2003; Huang et al. 2008; Ben Amar et al. 2013). In *Arabidopsis*, AtHKT1 has been shown to control $Na^+$ accumulation in the shoots in salt-stress conditions by mediating $Na^+$ retrieval from the ascending xylem sap in the roots (Sunarpi et al. 2005; Davenport et al. 2007) and $Na^+$ recirculation from shoots to roots via phloem sap loading (Berthomieu et al. 2003). Many studies in cereals have shown that natural variation in the activity or expression of HKT transporters may be a genetic resource for enhanced NaCl tolerance. Indeed, the inactivation or suppression of low affinity Na⁺ transporter can improve plant salt tolerance. For example, the *hkt1* mutation suppresses the salt hypersensitivity and K⁺-deficient phenotype of the *Arabidopsis sos3* mutant (Rus et al. 2001). In addition, transgenic wheat expressing antisense *HKT1* showed less Na⁺ uptake and significant growth under salinity compared with control plants (Laurie et al. 2002).

### 1.3.1.2 Na⁺ Exclusion

A critical factor of salinity tolerance in plants is the ability to exclude Na⁺ from the shoot, and the modification of specific Na⁺ transport processes has yielded enhanced salinity tolerance. In shoots, high concentrations of Na⁺ can cause a range of osmotic and metabolic problems for plants. Sodium exclusion is one of the major mechanisms conferring salt tolerance in cereal crops including rice, wheat, and barley (Gorham et al. 1990; Munns et al. 2006). Bread wheat (*Triticum aestivum*, AABBDD) is, in general, a better Na⁺ "excluder" than durum wheat (*Triticum turgidum* ssp. *durum*, AABB), a trait controlled by the *Kna1* locus on chromosome 4D, which corresponds to an *HKT1;5*-like gene (Dvorák et al. 1994; Dubcovsky et al. 1996; Byrt et al. 2007). However, an unusual durum wheat named Line 149 has a salt-tolerant phenotype similar to the bread wheat. This is due to the presence of two major genes for Na⁺ exclusion, named *Nax1* and *Nax2* (Munns et al. 2003). The *Nax1* locus is associated with the exclusion of Na⁺ from leaf blades only upon salt stress, both by retrieval of Na⁺ from the xylem in roots and leaf sheaths and by recirculation of Na⁺ from the shoots to the roots via the phloem. Concerning

the *Nax2* gene, it also reduces the Na⁺ transport from root to shoot and has a higher rate of K⁺ transport, resulting in enhanced K⁺ versus Na⁺ discrimination in the leaf (James et al. 2006). High-resolution mapping and sequencing analyses of known Na⁺ transporter genes have suggested that the *Nax1* and *Nax2* loci are attributable to polymorphisms in wheat *HKT* genes encoding protein of the subfamily 1 with preferred Na⁺ transport (Huang et al. 2006; Byrt et al. 2007). *Nax2* was shown to be homologous to *Kna1* in *T. aestivum*, namely, *TaHKT8* (Byrt et al. 2007).

In *Arabidopsis*, the plasma membrane Na⁺/H⁺ antiporter SOS1 plays a crucial role in sodium extrusion from root epidermal cells at the root-soil interface under salinity (Shi et al. 2000, 2002). In salt-stress conditions, SOS1 protein controls long-distance Na⁺ transport since this ion is transported from the root to the shoot via the xylem (Shi et al. 2002). This critical function was demonstrated also in the halophytic *Arabidopsis* relative *Thellungiella salsuginea* (Oh et al. 2009) and in tomato (Olías et al. 2009). SOS1, SOS2, and SOS3 proteins are three essential components of SOS signaling pathway, which mediate cellular signaling under salt stress, to maintain ion homoeostasis (Figure 1.1). SOS1 protein is the direct target of SOS signaling pathway, and it is regulated through protein phosphorylation by the alternative SOS2/SOS3 and SOS2/CBL10 protein kinase complexes (Qiu et al. 2002; Quintero et al. 2002, 2011; Quan et al. 2007). ABI2 is the negative regulatory of this pathway, through the inhibition of SOS2 kinase activity or the activity of SOS2 targets, suggesting a cross talk between the ABA pathway and SOS pathway (Ohta et al. 2003).

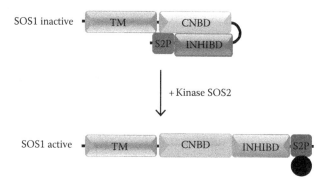

**Figure 1.1.** The different steps of SOS signaling pathway for salt-stress adaptation. Salt stress induces an accumulation of sodium in the cytosol, producing Ca²⁺ signal that activates the SOS3/SOS2 protein kinase complex. SOS2 activates the antiporter SOS1 by phosphorylation, the tonoplast Na⁺/H⁺ antiporter NHX and CAX1 (H⁺/Ca⁺ antiporter), restoring the ionic homeostasis.

Many other SOS1 proteins are also regulated by *Arabidopsis* SOS2/SOS3 complex, such as rice OsSOS1 (Atienza et al. 2007), tomato SlSOS1 (Olías et al. 2009), bread wheat (Xu et al. 2008), and durum wheat TdSOS1 (Feki et al. 2011). The activation mechanism of *Arabidopsis* and durum wheat SOS1 proteins involves the phosphorylation by the kinase SOS2 and inactivation of an autoinhibitory domain located at the C-terminal end of these transporters. Indeed, SOS1 is maintained in a resting state by the autoinhibitory domain, which is the target of SOS2/SOS3 complex and interacts intramolecularly with an adjacent domain that is essential for SOS1 activity. Upon salinity stress, the $Ca^{2+}$-dependent SOS2/SOS3 protein kinase complex phosphorylates SOS1 at the phosphorylation sites and relieves SOS1 from autoinhibition, presumably by displacing the autoinhibitory domain (Figure 1.2) (Quintero et al. 2011; Feki et al. 2011). SOS3 is a calcium-binding protein capable of sensing calcium transients elicited by salt stress (Liu and Zhu 1998; Ishitani et al. 2000). SOS2 is a serine/threonine protein kinase whose catalytic domain is evolutionarily related to that of the yeast protein SNF1 and animal AMP-activated kinases (Liu et al. 2000). It was shown that SOS3 recruits SOS2 to the plasma membrane to achieve efficient interaction with SOS1 (Quintero et al. 2002).

In addition to the direct target SOS1 protein, CAX1 ($H^+/Ca^{2+}$ antiporter) and NHX proteins are an additional target of SOS2 activity (Quintero et al. 2002; Cheng et al. 2004).

It has been reported that SOS proteins may have novel roles in addition to their functions in sodium homeostasis. For example, these proteins play a role in the dynamics of cytoskeleton under stress. Indeed, SOS3 plays a key role in mediating $Ca^{2+}$-dependent reorganization of actin filaments during salt stress (Ye et al. 2013). In addition, SOS1 is phosphorylated by MPK6, which is implicated in the organization and dynamics of mitotic and cortical microtubules (Müller et al. 2010; Yu et al. 2010).

### 1.3.1.3 Na⁺ Sequestration

Under salt stress, plants have evolved an osmotic adjustment mechanism that maintains water uptake and turgor under osmotic stress conditions. For osmotic adjustment, plants use some organic compatible solutes such as proline, betaine, and soluble sugars and also inorganic ions like $Na^+$ and $K^+$. Proteins of the NHX family function in the sequestration of $Na^+$ in the vacuole, endosomal transporter, luminal pH control, and vesicle trafficking (Pardo et al. 2006). Plant NHX exchangers have 10–12 transmembrane domains

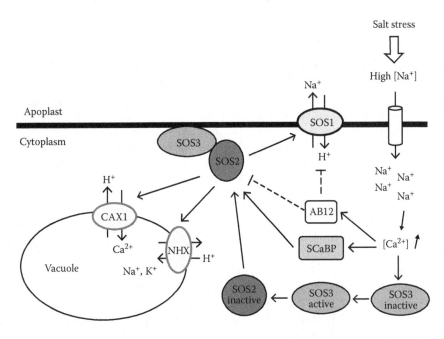

Figure 1.2. Model illustrating the activation mechanism of AtSOS1. (Adapted from Quintero, F.J. et al., *Proc. Natl. Acad. Sci. USA*, 108, 2611, 2011.)

with a hydrophilic C-terminal tail, which has vacuolar localization (Yamaguchi et al. 2003). The Na$^+$/H$^+$ exchange was electroneutral and driven by the vacuolar proton gradient established by the activity of the proton pumps V-ATPase and V-PPase (Blumwald 1987). The first Na$^+$/H$^+$ antiporter exchanger identified in plant was AtNHX in *Arabidopsis* (Gaxiola et al. 1999). The NHX proteins belong to the NHX/NHE subfamily of the cation proton antiporter CPA1 family (Saier 2000). The NHX proteins were subdivided into two classes. In plants, the class-I NHX proteins are localized in the tonoplast and function as (Na$^+$, K$^+$)/H$^+$ exchanger that accumulate Na$^+$ and K$^+$ into vacuoles, thereby contributing to osmotic regulation and the generation of turgor essential for cell expansion. However, the localization of class-II NHX proteins is in the endosomal compartments, and they function as K$^+$/H$^+$ exchange to prevent the accumulation of potentially toxic Na$^+$ into the endosomal lumen (Pardo et al. 2006). The microarray analysis showed that class-I AtNHX antiporters are expressed in leaves (AtNHX1, 2 and 4) or roots (AtNHX3) under the application of salt or osmotic stress. AtNHX1, 2, and 5 are expressed especially in guard cells compared to surrounding mesophyll cells (Shi and Zhu 2002; Rodriguez-Rosales et al. 2009). Although the first suggestion that AtNHX1 is specific to Na$^+$ transport, later studies have shown that AtNHX1 expressed in plants also catalyzes K$^+$/H$^+$ antiporter, albeit with lower affinity (Apse et al. 1999, 2003; Yokoi et al. 2002). Tomato LeNHX2 protein was purified and reconstituted into liposome showing that this protein catalyses relatively specific K$^+$/H$^+$ antiport (Venema et al. 2003). In addition to ionic homeostasis regulation, plant NHX proteins are implicated in endosomal and vacuolar pH regulation. Involvement of plant NHX genes in vacuolar pH regulation was most clearly demonstrated analyzing the dependence of flower color on vacuolar pH (Yoshida et al. 2005; Fukada-Tanaka et al. 2000). The involvement of ScNHX1 in pH regulation was demonstrated in yeast by the elimination of this protein, producing an acidification of the vacuolar and cytoplasmic pH (Brett el al. 2005). Both class-I and class-II plant NHX isoforms complement NaCl, KCl, and hygromycin sensitivity of the yeast ScNHX1 disruption mutant (Quintero et al. 2000; Yokoi et al. 2002; Venema et al. 2003). It is thus tempting to suggest a role for plant NHX proteins in endosomal pH regulation and protein trafficking as well.

The overexpression of NHX antiporters isolated from different plant species induced tolerance not only to salt but also to drought stress (Apse et al. 1999; Zhang and Blumwald 2001; Xue et al. 2004; Liu et al. 2010; Brini et al. 2007). Both class-I and class-II NHX antiporter from glycophytes or halophytes seem to have a similar effect on salt tolerance (Rodriguez-Rosaled et al. 2008; Shi et al. 2008).

### 1.3.1.4 Sodium Transport in the Whole Plant

Sodium is transported from soil solution through symplastic, apoplastic, or transcellular pathways up to the endodermis where a hydrophobic barrier that includes the Casparian strip restricts apoplastic movement (Schreiber et al. 2005; Plett and Moller 2010). The movement of Na$^+$ from root to shoot in xylem vessels occurs by bulk flow, driven primarily by xylem vessel tension caused by the vapor pressure gradient (Epstein and Bloom 2005).

NSCCs, and possibly other cation transporters like HKT and KUP/HAK transporters, are thought to mediate Na$^+$ influx in root cells. HKT mediates Na$^+$ entry in roots under low external concentrations of Na$^+$ and K$^+$, yet they are not likely to play a significant role in the salinity tolerance of plants (Horie et al. 2009). In *Arabidopsis*, SOS1 protein mediates sodium exclusion at the root–soil interface thereby reducing the net uptake of Na$^+$. Under moderate stress, SOS1 functions to load Na$^+$ into the xylem in roots for its transfer and storage in leaf mesophyll vacuoles. Whereas under severe salt stress, SOS1 is proposed to function in unloading Na$^+$ from the root xylem to reduce Na$^+$ damage of leaves that might be caused by exceeding the capacity of Na$^+$ sequestration in leaf cell vacuoles (Shi et al. 2002). SOS1 protein and AtHKT1;1 protein work in concert to regulate the Na$^+$ distribution between roots and shoots (Figure 1.3). Indeed, AtHKT1;1 is preferentially expressed in the vasculature, and it unloads Na$^+$ from xylem vessels to xylem parenchyma cells (Sunarpi et al. 2005). To translocate Na$^+$ back to roots, Na$^+$ is loaded into shoot phloem cells via symplastic diffusion (Sunarpi et al. 2005) or facilitated by HKT1-like proteins, preventing Na$^+$ overaccumulation in shoots (Berthomieu et al. 2003). However, the retranslocation of Na+ from leaf via phloem is little compared to the amount imported in the transpiration stream via the xylem (Tester and Davenport 2003; James et al. 2006).

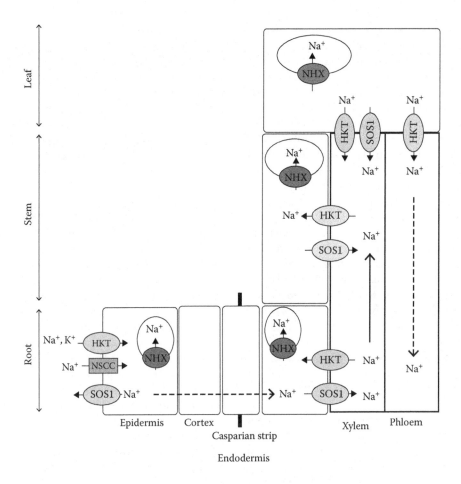

**Figure 1.3.** Sodium transport in the whole plant subjected to salinity stress. Thick arrows in the xylem and phloem indicate the flow of $Na^+$ in the sap. The dashed arrow symbolizes the reduced basipetal flow of $Na^+$ via phloem compared to the acropetal flux via xylem. (Adapted from Pardo, J.M., *Curr. Opin. Biotechnol.*, 21, 185, 2010.)

### 1.3.2 Osmotic Tolerance

Under salinity conditions, salt ions induce a decrease of osmotic potential with a passive salt ion penetration into plant cells. Osmotic stress decreases water availability for the cells, and it leads to a decreased water uptake resulting in cellular dehydration. The decrease in cellular turgor is sensed by the osmosensor histidine kinase at the cell plasma membrane (Urao et al. 1999). The osmotic stress is then transduced via a series of phosphorylation and calcium signaling to nucleus producing changes in gene expression. Plasma membrane phospholipids like phospholipases C and D are also implicated in osmotic stress signaling, and they lead to formation of small signaling molecules such as inositol-1, 4, 5-triphosphate ($IP_3$), diacylglycerol, and phosphatidic acid. These molecules induce $Ca^{2+}$ signaling events leading to signal transduction to nucleus.

To regulate the osmotic balance within the cells, plants produce compatible osmolytes known as osmoprotectants. They are organic compounds of uncharged polarity and do not interfere with the cellular metabolism even at high concentrations (Wyn Jones et al. 1977). They mainly include proline, glycine betaine, sugar, and polyols, and they are implicated in the protection of the structure and in the regulation of osmotic balance within the cell via continuous water influx (Hasegawa et al. 2000). Contrary to the other amino acids, proline concentration rises in salt-stress condition (El-Shintinawy and El-Shourbagy 2001).

Proline is synthesized either from glutamic acid or ornithine, and the biosynthetic pathway comprises two major enzymes, which are the pyrroline carboxylic acid synthetase and pyrroline carboxylic acid reductase. It has been demonstrated that the addition of proline enhanced salt tolerance in many plants like olive (*Olea europaea*), tobacco (*Nicotiana tabacum*), and *A. thaliana*, by increasing the activity of some antioxidative enzyme activities (Abraham et al. 2003; Hoque et al. 2008; Ben Ahmed et al. 2010).

Glycine betaine is an amphoteric quaternary ammonium compound, and it stabilizes protein and protects the cell by osmotic adjustment and the photosynthetic apparatus from stress damages and reduction of ROS (Gadallah 1999; Makela et al. 2000; Ashraf and Foolad 2007; Cha-Um and Kirdmanee 2010). It has been showed that the treatment with glycine betaine ameliorates the ultrastructure of *O. sativa* seedlings when exposed to salt stress (Rahman et al. 2002). In addition, the application of glycine betaine to stressed plants increases the photosynthetic rate and growth level (Cha-Um and Kirdmanee 2010; Ahmad et al. 2013).

Under salt stress, some carbohydrates like sugars are accumulated within the cell in a number of plants belonging to different species to assure osmoprotection, carbon storage, and scavenging of ROS (Gupta and Huang 2014). ABA may regulate osmolyte biosynthesis in plants under salt stress. For example, Xiong et al. (2001) demonstrated that the production of ABA under osmotic stress regulates the *P5CS* gene involved in proline biosynthesis.

### 1.3.3 Antioxidant Regulation of Salinity Tolerance

Salt stress induces an accumulation of ROS that are detrimental to cells at high concentrations because they cause oxidative damage to membrane lipids, proteins, and nucleic acids (Gómez et al. 1999; McCord 2000; Breusegem et al. 2001; Mittler 2002). Despite the potential of ROS for causing harmful oxidations, it is now well established that they are also implicated in the control of plant growth and development as well as priming acclimatory responses to stress stimuli (Foyer and Noctor 2005, 2009). To cope with ROS, living organisms evolved antioxidant defense systems, comprised of enzymatic and nonenzymatic components. Several enzymes are involved in the detoxification of the activated oxygen species like superoxide dismutase (SOD), ascorbate peroxidase (APX), catalase (CAT), glutathione reductase (GR), and glutathione peroxidase (GPX). Transgenic plants overexpressing ROS-scavenging enzymes showed increased tolerance to oxidative stress (Wang et al. 1999; Roxas et al. 1997, 2000; Foyer et al. 1995). SOD is one of the most important enzymes used against oxidative stress in the plant defense system, and it occurs ubiquitously in every cell of all types of plants (Ashraf 2009). With a few exceptions, Cu/Zn-SODs are generally found in the cytosol of eukaryotic cells and chloroplasts; the Mn-SODs are found in the matrix of mitochondria and in prokaryotes; the Fe-SODs are generally found in prokaryotes and have been reported to exist in some plants (Duke and Salin 1985). A membrane-associated Mn-SOD has been reported in chloroplasts of some plants (Hayakawa et al. 1984). CATs are one of the $H_2O_2$-metabolizing proteins in plants, and they are highly active enzymes that do not require cellular reductants as they primarily catalyze a dismutase reaction (Mhamdi et al. 2010). CATs have a very fast turnover rate, but a much lower affinity for $H_2O_2$ than APX and PRX (peroxiredoxins), which have KM values below 100 μM (Mittler and Zilinskas 1991; Konig et al. 2002). Three CAT genes are present in the genome of *Arabidopsis*, in which two are located on chromosome 1 (*CAT1* and *CAT3*) and one on chromosome 4 (*CAT2*) (Frugoli et al. 1996). The *CAT1* gene is mainly expressed in pollen and seeds, *CAT2* in photosynthetic tissues but also in roots and seeds, while *CAT3* is associated with vascular tissues but also leaves (Mhamdi et al. 2010). APXs are thought to be the most important $H_2O_2$ scavengers operating both in the cytosol and chloroplasts. They use ascorbic acid (AsA) as a reducing substrate and form part of a cycle, known as the ascorbate–GSH or Halliwell–Asada cycle.

The nonenzymatic antioxidant system includes AsA, GSH, α-tocopherols (vitamin E), flavonoids, anthocyanines, polyphenolic, and carotenoids compounds (Noctor and Foyer 1998; Schafer et al. 2002). AsA and GSH, the most abundant soluble antioxidants in plants, play a key role in plant defense against oxidative stress (Foyer and Noctor 2011). It has been demonstrated that salinity tolerance is positively correlated with the activity of antioxidant enzymes, such as SOD, CAT, GPX, APX, and GR, and with the accumulation of non-enzymatic antioxidant compounds (Sairam et al. 2002, 2005; Mandhania et al. 2006; Koca et al. 2007; Khosravinejad et al. 2008; Gapinska et al. 2008; Turhan et al. 2008).

## 1.4 PLANT ADAPTATION TO SALINITY: GENOMIC, TRANSCRIPTOMIC, PROTEOMIC, AND METABOLIC REGULATIONS

In general, four major factors could determine the tolerance of plants to abiotic stress at the molecular level, which are genomic, transcriptomic, proteomic, and metabolic levels.

### 1.4.1 Genomic Regulation

Various mechanisms of gene regulation have been identified from transcriptional initiation, to RNA processing, and to the posttranslational modification of a protein. Great advances in the comparison of genomes and the transcriptomes of different organisms have contributed to the development of comparative genomics as one of the most promising fields in the area (Caicedo and Purugganan 2005). Despite the genomes of *A. thaliana* and Thellungiella parvula having very similar size and gene number, there are significant differences in gene copy number in certain functional categories important for stress tolerance. In fact, the T. parvula genome reveals a higher gene copy number of several genes involved in transport like *AVP1*, *HKT1*, and *NHX8* than *A. thaliana* genome. In contrast, the T. parvula genome contains lower gene copy numbers of several genes involved in signal transduction with respect to *A. thaliana* (Dassanayake et al. 2011). Among monocotyledonous plant species, rice has a high degree of synteny with genomes of others cereals plants like wheat, barley, and maize (Caicedo and Purugganan 2005; Paterson et al. 2005).

### 1.4.2 Transcriptomic Regulation

Transcriptomic analysis provides detailed knowledge about gene expression at the mRNA level, which is widely used to screen candidate genes involved in stress responses. The availability of the complete genome sequence of some model plants like *A. thaliana* and *O. sativa* has allowed the development of whole genome tiling microarrays. This constitutes a new powerful technology that has already made possible the identification of several unannotated transcripts responsive to abiotic stress (Gregory et al. 2008; Matsui et al. 2008). Many genes induced by salinity have been identified by the analysis of gene expression profile (Kreps et al. 2002; Oono et al. 2006; Jianping and Suleiman 2007). It has been reported that in *A. thaliana* and rice, the transcript profile changes under different abiotic stresses like high salinity, cold, and drought (Rabbani et al. 2003; Gong et al. 2005).

Salinity produces an upregulation of some genes and transcription factor in different plant species. These genes can be classified into the following functional categories: ion transport (SOS, AtNHX, H+-ATPase genes); ROS-scavenging; molecular chaperones; and dehydration-related transcription factors (Table 1.1). SOS1 is a component of the SOS signaling pathway, and it plays an important role in ion homoeostasis. SOS1 gene is upregulated by salinity in monocotyledonous and dicotyledonous species like in *A. thaliana* and in bread wheat (Shi et al. 2002; Xu et al. 2008). In ROS-scavenging genes, we cited the case of *Apx* gene, in which its expression is rapidly induced by various stress conditions, such as paraquat, abscisic acid, ethylene, drought, and heat shock, suggesting an important role in stress tolerance (Mittler and Zilinskas 1991). Molecular chaperones play an important role in salt-stress response, like heat-shock proteins (HSP). In rice, the application of high salt stress induced an upregulation of OsHsp17.0 and OsHsp23.7 genes (Zou et al. 2009). Genes involved in osmoprotectant biosynthesis are also upregulated under salt stress (Zhu 2002). In response to salinity, many transcription factors have been identified, and they are capable of controlling the expression of a broad range of target genes by binding to the specific cis-acting element. The expression of the transcription factor bZIP genes was upregulated in the salt-sensitive wheat cultivar, compared to the salt-tolerant cultivar (Johnson et al. 2002). Thus, in response to salinity, the expression profile is different between the salt-tolerant and salt-sensitive cultivars. This suggestion was observed in other plants like in Arabidopsis and Arabidopsis-related halophyte (Thellungiella halophyla). Indeed, contrary to

TABLE 1.1

*Examples of upregulated genes in response to salinity stress.*

| Gene functions | Gene name | Species | References |
|---|---|---|---|
| Ion transport | SOS1 | *Arabidopsis thaliana* | Shi et al. (2002) |
| | | *Triticum aestivum* | Xu et al. (2008) |
| | | *Oryza sativa* | Atienza et al. (2007) |
| | | *Solanum lycopersicum* | Olías et al. (2009); Wang et al. (2010) |
| | | *Puccinellia tenuiflora* | |
| | | *Populus trichocarpa* | Tang et al. (2010) |
| ROS scavenging | Fe-SOD | *Thellungiella halophyla* | Taji et al. (2004) |
| | Apx | *Spinacia oleracea* | Yoshimura et al. (2000) |
| | | *Pisum sativum* | Mittler and Zilinskas (1992) |
| Heat-shock proteins | HSP | *Oryza sativa* | Zou et al. (2009) |
| | | *Daucus carota* | Song and Ahn et al. (2011) |
| Transcription factor | DREB | *Arabidopsis thaliana* | Tang et al. (2011) |
| | AlSAP | *Aeluropus littoralis* | Ben Saad et al. (2010) |

*Arabidopsis*, a large number of known abiotic and biotic stress-inducible genes, such as *Fe-SOD, P5CS, PDF1.2, AtNCED, P-protein, β-glucosidase,* and *SOS1*, were expressed in salt cress at high levels even in the absence of stress. Thus, it is possible that the salt tolerance of salt cress is due to constitutive overexpression of many genes that function in stress tolerance and that are stress inducible in *Arabidopsis* (Taji et al. 2004). Comparative analysis of salt-sensitive rice (line IR64) and salt-tolerant rice (Pokkali) has led to identification of some salinity-responsive genes, displaying a higher expression in Pokkali than in IR64 (Kumari et al. 2009). These two examples showed that salt-tolerant plants are able to cope with stress due to the expression induction of some genes implicated in salt-stress response and as a consequence the production of more proteins and response efficiently to stress.

In addition to the upregulation of some genes, downregulated genes are emerging now as essential components of the response to salinity. For example, downregulation of β-carotene hydroxylase increases β-carotene and total carotenoids enhancing salt-stress tolerance in transgenic cultured cells of sweet potato (Kim et al. 2012). Therefore, it is possible that in plants there is a mutual regulation mechanism between different genes and proteins and signals underlying different processes of plant adaptation to abiotic stress.

### 1.4.3 Proteomic and Metabolic Regulations

Several common stress-responsive proteins are expressed in response to various abiotic stresses in different plants species, which are either upregulated or downregulated by salinity stress (Zhang et al. 2012). Several functional groups of proteins affected by salt stress include proteins involved in ion transport, signaling, energy metabolism (photosynthesis, respiration, ATP production), protein and lipid metabolism, metabolism of osmolytes and phytohormones, and stress-related proteins. Comparative proteome responses to salt stress have been analyzed in some related plant species with contrasting salinity tolerance like *A. thaliana* and *T. salsuginea* (Pang et al. 2010), rice and salt-tolerant wild rice *Porteresia coarctata* (Sengupta and Majumder 2009), and common wheat (*T. aestivum*) cv. Jinan 177 and *T. aestivum*/*Thinopyrum ponticum* (Wang et al. 2008). These studies give information about differential protein abundance but not about protein function under salinity, and therefore, validation of comparative proteomics should be done by protein functional analysis. Other approaches like posttranslation modifications, protein–protein interactions, tissue and subcellular localization, and phenotype influence by silencing or overexpressing the gene coding for a protein interest have to be employed to unravel the role of the proteins in acquisition and development of salinity tolerance in plants. Consequently, large-scale high-throughput proteome analyses must be integrated with transcriptomic and metabolomic analyses to improve our understanding of the stress response.

Another significant research approach in plant system biology is metabolomics, which involves the study of metabolome. Plant metabolites implicated in salinity tolerance include polyols like mannitol; dimethylsulfonium compounds; glycine betaine; sugars such as sucrose, trehalose, and fructans; or amino acids like proline. The concentration of these osmolytes increases in plant subjected to salt stress, suggesting their importance in salt tolerance.

## 1.5 CONCLUSIONS AND FUTURE RESEARCH PERSPECTIVES

Salinity tolerance involves a complex of responses at molecular, cellular, and whole plant levels. An understanding of how single-cell responses to salt are coordinated with an organism and whole-plant responses to maintain an optimal balance between salt uptake and compartmentation is fundamental to our knowledge of how plants successfully adapted to salt stress. Despite the significant advancements in the fields of genomic, transcriptomic, proteomic, and metabolomic techniques, there is lack of the integration of information among these four regulation levels. Therefore, the combined approach is essential for the determination of the key pathways or processes controlling salinity tolerance.

## ACKNOWLEDGMENT

This work was supported by grants from the Ministry of Higher Education, Scientific Research and Information Technology and Communication, Tunisia.

## REFERENCES

Abdullah, Z. and R. Ahmed. 1990. Effect of pre and post kinetin treatment on salt tolerance of different potato cultivars growing on saline soils. *Journal of Agronomy and Crop Science* 165:94–102.

Abraham, E., G. Rigo, G. Szekely, R. Nagy, C. Koncz, and L. Szabados. 2003. Light dependent induction of proline biosynthesis by abscisic acid and salt stress is inhibited by brassinosteroid in *Arabidopsis*. *Plant Molecular Biology* 51:363–372.

Ahmad, R., C.J. Lim, and S.Y. Kwon. 2013. Glycine betaine: A versatile compound with great potential for gene pyramiding to improve crop plant performance against environmental stresses. *Plant Biotechnology Reports* 7:49–57.

Apse, M., G. Aharon, W. Sneddon, and E. Blumwald. 1999. Salt tolerance conferred by overexpression of a vacuolar $Na^+/H^+$ antiport in *Arabidopsis*. *Science* 285:1256–1258.

Apse, M.P., J.B. Sottosanto, and E. Blumwald. 2003. Vacuolar cation/H+ exchange, ion homeostasis, and leaf development are altered in a T-DNA insertional mutant of AtNHX1, the *Arabidopsis* vacuolar $Na^+/H^+$ antiporter. *Plant Journal* 36:229–239.

Ashraf, M. 2009. Biotechnological approach of improving plant salt tolerance using antioxidants as markers. *Biotechnology Advances* 27:84–93.

Ashraf, M. and M.R. Foolad. 2007. Roles of glycine betaine and proline in improving plant abiotic stress resistance. *Environmental and Experimental Botany* 59:206–216.

Atienza M.J., X. Jiang, B. Garciadeblas, I. Mendoza, J.K. Zhu, J.M. Pardo, and F.J. Quintero. 2007. Conservation of salt overly sensitive pathway in rice. *Plant Physiology* 143:1001–1012.

Baxter, I., J.N. Brazelton, D. Yu, Y.S. Huang, B. Lahner, E. Yakubova, Y. Li et al. 2010. A coastal cline in sodium accumulation in *Arabidopsis thaliana* is driven by natural variation of the sodium transporter AtHKT1;1. *PLOS Genetics* 6:e1001193.

Ben Ahmed, C., B. Ben Rouina, S. Sensoy, M. Boukhriss, and F. Ben Abdullah. 2010. Exogenous proline effects on photosynthetic performance and antioxidant defense system of young olive tree. *Journal of Agricultural and Food Chemistry* 58:4216–4222.

Ben Amar, S., F. Brini, H. Sentenac, K. Masmoudi, and A.A. Very. 2013. Functional characterization in *Xenopus* oocytes of $Na^+$ transport systems from durum wheat reveals diversity among two HKT1;4 transporters. *Journal of Experimental Botany* doi:10.1093/jxb/ert361.

Ben Saad, R., N. Zouari, W. Ben Ramdhan, J. Azaza, D. Meynard, E. Guiderdoni, and A. Hassairi. 2010. Improved drought and salt stress tolerance in transgenic tobacco overexpressing a novel A20/AN1 zinc-finger "AlSAP" gene isolated from the halophyte grass *Aeluropus littoralis*. *Plant Molecular Biology* 72:171–190.

Berthomieu, P., G. Conéjéro, A. Nublat, W.J. Brackenbury, C. Lambert, C. Savio, N. Uozumi et al. 2003. Functional analysis of *AtHKT1* in *Arabidopsis* shows that $Na^+$ recirculation by the phloem is crucial for salt tolerance. *EMBO Journal* 22:2004–2014.

Blumwald, E. 1987. Tonoplast vesicles as a tool in the study of ion transport at the plant vacuole. *Physiologia Plantarum* 69:731–734.

Blumwald, E. 2000. Sodium transport and salt tolerance in plants. *Current Opinion in Cell Biology* 12:431–434.

Brett, C.L., D.N. Tukaye, S. Mukherjee, and R. Rao. 2005. The yeast endosomal $Na^+(K^+)/H^+$ exchanger Nhx1 regulates cellular pH to control vesicle trafficking. *Molecular Biology of the Cell* 16:1396–1405.

Breusegem, F.V., E. Vranova, J.F. Dat, and D. Inze. 2001. The role of active oxygen species in plant signal transduction. *Plant Science* 161:405–414.

Brini, F., M. Hanin, I. Mezghanni, G. Berkowitz, K. Masmoudi. 2007. Overexpression of wheat $Na^+/H^+$ antiporter TNHX1 and $H^+$-pyrophosphatase TVP1 improve salt and drought stress tolerance in *Arabidopsis thaliana* plants. *Journal of Experimental Botany* 58(2):301–308.

Byrt, C.S., J.D. Platten, W. Spielmeyer, R.A. James, E.S. Lagudah, E.S. Dennis, M. Tester, and R. Munns. 2007. HKT1;5-like cation transporters linked to $Na^+$ exclusion loci in wheat, Nax2 and Kna1. *Plant Physiology* 143:1918–1928.

Caicedo, A.L. and M.D. Purugganan. 2005. Comparative plant genomics. Frontiers and prospects. *Plant Physiology* 138:545–547.

Cha-Um, S. and C. Kirdmanee. 2010. Effect of glycinebetaine on proline, water use, and photosynthetic efficiencies, and growth of rice seedlings under salt stress. *Turkish Journal of Agriculture and Forestry* 34:517–527.

Cheng, N.H., J.K. Pittman, J.K. Zhu, and K.D. Hirschi. 2004. The protein kinase SOS2 activates the *Arabidopsis* $H^+/Ca^{2+}$ antiporter CAX1 to integrate calcium transport and salt tolerance. *The Journal of Biological Chemistry* 23:2922–2926.

Chinnusamy, V., A. Jagendorf, and J. Zhu. 2005. Understanding and improving salt tolerance in plants. *Crop Science* 45:437–448.

Clarkson, D.T. and J.B. Hanson. 1980. The mineral nutrition of higher plants. *Annual Review of Plant Physiology* 31:239–298.

Dassanayake, M., D.H. Oh, J.S. Haas, A. Hernandez, H. Hong, S. Ali, D.J. Yun et al. 2011. The genome of the extremophile crucifer *Thellungiella. parvula*. *Nature Genetics* 43:913–918.

Davenport, R.J., A. Munoz-Mayor, D. Jha, P.A. Essah, A. Rus, and M. Tester. 2007. The $Na^+$ transporter AtHKT1;1 controls retrieval of $Na^+$ from the xylem in *Arabidopsis*. *Plant Cell and Environment* 30:497–507.

Demidchik, V. and F.J. Maathuis. 2007. Physiological roles of nonselective cation channels in plants: From salt stress to signalling and development. *The New Phytologist* 175:387–404.

Downton, W.J.S. 1977. Photosynthesis in salt-stressed grapevines. *Australian Journal of Plant Physiology* 4:183–192.

Dubcovsky, J., G. Santa Maria, E. Epstein, M.C. Luo, and J. Dovorak. 1996. Mapping of the $K^+/Na^+$ discrimination locus Kna1 in wheat. *Theoretical and Applied Genetics* 92:448–454.

Duke, M.V. and M.L. Salin. 1985. Purification and characterization of an iron-containing superoxide dismutase from a eukaryote, *Ginkgo biloba*. *Archives of Biochemistry and Biophysics* 243:305–314.

Dvorák, J., M.M. Noaman, S. Goyal, and J. Gorham. 1994. Enhancement of the salt tolerance of *Triticum turgidum* L by the Kna1 locus transferred from *Triticum aestivum* L. chromosome 4D by homoeologous recombination. *Theoretical and Applied Genetics* 87:872–877.

El-Shintinawy, F. and M.N. El-Shourbagy. 2001. Alleviation of changes in protein metabolism in NaCl-stressed wheat seedlings by thiamine. *Biologia Plantarum* 44:541–545.

Epstein, E. and A.J. Bloom. 2005. *Mineral Nutrition of Plants: Principles and Perspectives*, 2nd edn., Sinauer Associates, Sunderland, MA.

Feki, K., F.J. Quintero, J.M. Pardo, and K. Masmoudi. 2011. Regulation of durum wheat $Na^+/H^+$ exchanger TdSOS1 by phosphorylation. *Plant Molecular Biology* 76:545–556.

Flowers, T.J. 2004. Improving crop salt tolerance. *Journal of Experimental Botany* 55:307–319.

Flowers, T.J. and T.D. Colmer. 2008. Salinity tolerance in halophytes. *New Phytology* 179:945–963.

Flowers, T.J., H.K. Galal, and L. Bromham. 2010. Evolution of halophytes: Multiple origins of salt tolerance in land plants. *Functional Plant Biology* 37:604–612.

Flowers, T.J., M.A. Hajibagheri, and N.J.W. Clipson. 1986. Halophytes. *Quarterly. Review of Biology* 61:313–337.

Flowers, T.J., P.F. Troke, and A.R. Yeo. 1977. The mechanism of salt tolerance in halophytes. *Annual Review of Plant Physiology Plant Molecular Biology* 28:89–121.

Foyer, C.H. and G. Noctor. 2005. Oxidant and anti-oxidant signalling in plants: A re-evaluation in the concept of oxidative stress in a physiological context. *Plant, Cell and Environment* 28:1056–1071.

Foyer, C.H. and G. Noctor. 2009. Redox regulation in photosynthetic organisms: Signaling, acclimation, and practical implications. *Antioxidants and Redox Signaling* 11:861–905.

Foyer, C.H. and G. Noctor. 2011. Ascorbate and glutathione: The heart of the redox hub. *Plant Physiology* 155:2–18.

Foyer, C.H., N. Souriau, S. Perret, M. Lelandais, K.J. Kunert, C. Pruvost, and L. Jouanin. 1995. Overexpression of glutathione reductase but not glutathione synthetase leads to increases in antioxidant capacity and resistance to photoinhibition in poplar trees. *Plant Physiology* 109:1047–1057.

Fricke, W., G. Akhiyarova, D. Veselov, and G. Kudoyarova. 2004. Rapid and tissue-specific changes in ABA and in growth rate in response to salinity in barley leaves. *Journal of Experimental Botany* 55:1115–1123.

Frugoli, J.A., H.H. Zhong, M.L. Nuccio, P. McCourt, M.A. McPeek, T.L. Thomas, and C.R. McClung. 1996. Catalase is encoded by a multigene family in Arabidopsis thaliana (L.) Heynh. *Plant Physiology* 112:327–336.

Fukada-Tanaka, S., Y. Inagaki, T. Yamaguchi, N. Saito, and S. Iida. 2000. Colour-enhancing protein in blue petals. *Nature* 407:581.

Gadallah, M.A.A. 1999. Effects of proline and glycine betaine on Vicia faba responses to salt stress. *Biologia Plantarum* 42:249–257.

Gapinska, M., M. Sklodowska, and B. Gabara. 2008. Effect of short and long-term salinity on the activities of antioxidative enzymes and lipid peroxidation in tomato roots. *Acta Physiologia Plantarum* 30:11–18.

Garciadeblás, B., M.E. Senn, M.A. Bañuelos, and A. Rodríguez-Navarro. 2003. Sodium transport and HKT transporters: The rice model. *The Plant Journal* 34:788–801.

Gaxiola, R.A., R. Rao, A. Sherman, P. Grisafi, S.L. Alper, and G.R. Fink. 1999. The *Arabidopsis thaliana* proton transporters, AtNhx1 and Avp1, can function in cation detoxification in yeast. *Proceedings of the National Academy of Sciences USA* 96:1480–1485.

Gómez, J.M., J.A. Hernández, A. Jiménez, L.A. del Rio, and F. Sevilla. 1999. Differential response of antioxidative enzymes of chloroplasts and mitochondria to long-term NaCl stress of pea plants. *Free Radical Research* 31:11–18.

Gong, Q., P. Li, S. Ma, S. Indu Rupassara, and H.J. Bohnert. 2005. Salinity stress adaptation competence in the extremophile Thellungiella halophila in comparison with its relative *Arabidopsis thaliana*. *Plant Journal* 44:826–839.

Gorham, J., R.G. Wyn Jones, and A. Bristol. 1990. Partial characterization of the trait for enhanced $K^+$-$Na^+$ discrimination in the D genome of wheat. *Planta* 180:590–597.

Gregory, B.D., J. Yazaki, and J.R. Ecker. 2008. Utilizing tilling microarray for whole-genome analysis in plants. *Plant Journal* 53:636–644.

Gupta, B. and B. Huang. 2014. Mechanism of salinity tolerance in plants: Physiological, biochemical, and molecular characterization. *International Journal of Genomics* 2014:18–37.

Gupta, N.K., S.K. Meena, S. Gupta, and S.K. Khandelwal. 2002. Gas exchange, membrane permeability, and ion uptake in two species of Indian jujube differing in salt tolerance. *Photosynthetica* 40:535–539.

Hamada, A.M. and A.E. El-Enany. 1994. Effect of NaCl salinity on growth, pigment and mineral element contents, and gas exchange of broad bean and pea plants. *Biologia Plantarum* 36:75–81.

Hasegawa, P.M., R.A. Bressan, J.K. Zhu, and H.J. Bohnert. 2000. Plant cellular and molecular responses to high salinity. *Annual Review of Plant Physiology and Plant Molecular Biology* 51:463–499.

Hayakawa, T., S. Kanematsu, and K. Asada. 1984. Occurrence of Cu,Zn-superoxide dismutase in the intrathylakoid space of spinach chloroplasts. *Plant Cell Physiology* 25:883–889.

Hoque, M.A., M.N.A. Banu, Y. Nakamura, Y. Shimoishi, and Y. Murata. 2008. Proline and glycinebetaine enhance antioxidant defense and methylglyoxal detoxification systems and reduce NaCl-induced damage in cultured tobacco cells. *Journal of Plant Physiology* 165:813–824.

Horie, T., F. Hauser, and J.I. Schroeder. 2009. HKT transporter-mediated salinity resistance mechanisms in *Arabidopsis* and monocot crop plants. *Trends in Plant Science* 14:660–668.

Huang, S., W. Spielmeyer, E.S. Lagudah, and R. Munns. 2008. Comparative mapping of HKT genes in wheat, barley and rice, key determinants of $Na^+$ transport and salt tolerance. *Journal of Experimental Botany* 59:927–937.

Huang, S.H., W. Spielmeyer, E.S. Lagudah, R. James, D.J. Platten, E.S. Dennis, and R. Munns. 2006. A sodium transporter (HKT7) is a candidate for Nax1, a gene for salt tolerance in durum wheat. *Plant Physiology* 142:1718–1727.

Ishitani, M., J. Liu, U. Halfter, C.S. Kim, M. Wei, and J.K. Zhu. 2000. SOS3 function in plant salt tolerance requires myristoylation and calcium binding. *Plant Cell* 12:1667–1677.

James, R.A., R.J. Davenport, and R. Munns. 2006. Physiological characterization of two genes for $Na^+$ exclusion in durum wheat, Nax1 and Nax2. *Plant Physiology* 142:1537–1547.

Johnson, R.R., R.L. Wagner, S.D. Verhey, and M.K. Walker-Simmons. 2002. The abscisic acid-responsive kinase PKABA1 interacts with a seed-specific abscisic acid response element-binding factor, TaABF, and phosphorylates TaABF peptide sequences. *Plant Physiology* 130:837–846.

Jianping, P.W. and S.B. Suleiman. 2007. Monitoring of gene expression profiles and identification of candidate genes involved in drought responses in Festuca mairei. *Molecular Genetics and Genomics* 277:571–587.

Kao, W.Y., T.T. Tsai, and C.N. Shih. 2003. Photosynthetic gas exchange and chlorophyll *a* fluorescence of three wild soybean species in response to NaCl treatments. *Photosynthetica* 41:415–419.

Khosravinejad, F., R. Heydari, and T. Farboodnia.2008. Antioxidant responses of two barley varieties to saline stress. *Pakistan Journal of Biological Science* 11:905–909.

Kim, S.H., Y.O. Ahn, M.J. Ahn, H.S. Lee, and S.S. Kwak. 2012. Down-regulation of β-carotene hydroxylase increases β-carotene and total carotenoids enhancing salt stress tolerance in transgenic cultured cells of sweetpotato. *Phytochemistry* 74:69–78.

Koca, H., M. Bor, F. Ozdemir, and I. Turkan. 2007. Effect of salt stress on lipid peroxidation, antioxidative enzymes and proline content of sesame cultivars. *Environmental and Experimental Botany* 60:344–351.

Konig, J., M. Baier, F. Horling, U. Kahmann, G. Harris, P. Schurmann, and K.J. Dietz. 2002. The plant-specific function of 2-Cys peroxiredoxin-mediated detoxification of peroxides in the redoxhierarchy of photosynthetic electron flux. *Proceedings of the National Academy of Sciences USA* 99:5738–5743.

Kosova, K., I.T. Prasil, and P. Vitamvas. 2013. Protein contribution to plant salinity response and tolerance acquisition. *International Journal of Molecular Sciences* 14:6757–6789.

Kreps, J.A., Y. Wu, H.S. Chang, T. Zhu, X. Wang, and J.F. Harper. 2002. Transcriptome changes for Arabidopsis in response to salt, osmotic, and cold stress. *Plant Physiology* 130:2129–2141.

Kumari, S., V. Panjabi, H.R. Kushwaha, S.K. Sopory, S.L. Singla-Pareek, and A. Pareek. 2009. Transcriptome map for seedling stage specific salinity stress response indicates a specific set of genes as candidate for saline tolerance in *Oryza sativa* L. *Functional and Integrative Genomics* 9:109–123.

Lapina, L.P. and B.A. Popov. 1970. Effect of sodium chloride on photosynthetic apparatus of tomatoes. *Fiziologiya Rastenii* 17:580–584.

Laurie, S., K.A. Feeney, F.J.M. Maathuis, P.J. Heard, S.J. Brown, and R.A. Leigh. 2002. A role for HKT1 in sodium uptake by wheat roots. *Plant Journal* 32:139–149.

Liu, J. and J.K. Zhu. 1998. A calcium sensor homolog required for plant salt tolerance. *Science* 280:1943–1945.

Liu, P., G.D. Yang, H. Li, C.C. Zheng, and C.A. Wu. 2010. Overexpression of NHX1s in transgenic Arabidopsis enhances photoprotection capacity in high salinity and drought conditions. *Acta Physiologia Plantarum* 32:81–90.

Makela, P., J. Karkkainen, and S. Somersalo. 2000. Effect of glycinebetaine on chloroplast ultrastructure, chlorophyll and protein content, and RuBPCO activities in tomato grown under drought or salinity. *Biologia Plantarum* 43:471–475.

Mandhania, S., S. Madan, and V. Sawhney. 2006. Antioxidant defense mechanism under salt stress in wheat seedlings. *Biologia Plantarum* 227:227–231.

Mäser, P., S. Thomine, J.I. Schroeder, J.M. Ward, K. Hirschi, H. Sze, I.N. Talke et al. 2001. Phylogenetic relationships within cation transporter families of *Arabidopsis*. *Plant Physiology* 126:1646–1667.

Matsui, A., J. Ishida, T. Morosawa, Y. Mochizuki, E. Kaminuma, T.A. Endo, M. Okamoto et al. 2008. *Arabidopsis* transcriptome analysis under drought, cold, high-salinity and ABA treatment conditions using a tiling array. *Plant Cell Physiology* 49:1135–1149.

McCord, J.M. 2000. The evolution of free radicals and oxidative stress. *The American Journal of Medicine* 108:652–659.

Mhamdi, A., G. Queval, S. Chaouch, S. Vanderauwera, F.V. Breusegem, and G. Noctor. 2010. Catalase function in plants: A focus on *Arabidopsis* mutants as stress-mimic models. *Journal of Experimental Botany* 61:4197–4220.

Mittler, R. 2002. Oxidative stress, antioxidant and stress tolerance. *Trends in Plant Science* 7:405–410.

Mittler, R. and B.A. Zilinskas. 1991. Purification and characterization of pea cytosolic ascorbate peroxidase. *Plant Physiology* 97:962–968.

Müller, J., M. Beck, U. Mettbach, G. Komis, G. Hause, D. Menzel, and J. Šamaj. 2010. *Arabidopsis* MPK6 is involved in cell division plane control during early root development, and localizes to the pre-prophase band, phragmoplast, *trans*-Golgi network and plasma membrane. *Plant Journal* 61:234–248.

Munns, R. 2002. Comparative physiology of salt and water stress. *Plant, Cell & Environment* 25:239–250.

Munns, R., R.A. James, and A. Läuchli. 2006. Approaches to increasing the salt tolerance of wheat and other cereals. *Journal of Experimental Botany* 57:1025–1043.

Munns, R., G.J. Rebetzke, S. Husain, R.A. James, and R.A. Hare 2003. Genetic control of sodium exclusion in durum wheat. *Australian Journal of Agricultural Research* 54:627–635.

Munns, R. and M. Tester. 2008. Mechanisms of salinity tolerance. *Annual Review of Plant Biology* 59:651–681.

Noctor, G. and C.H. Foyer. 1998. Ascorbate and glutathione: Keeping active oxygen under control. *Annual Review of Plant Physiology and Plant Molecular Biology* 49:249–279.

Oh, D.H., E. Leidi, Q. Zhang, S.M. Hwang, Y. Li, F.J. Quintero, X. Jiang et al. 2009. Loss of halophytism by interference with SOS1 expression. *Plant Physiology* 151:210–222.

Ohta, M., Y. Guo, U. Halfter, and J.K. Zhu. 2003. A novel domain in the protein kinase SOS2 mediates interaction with the protein phosphatase 2C ABI2. *Proceedings of the National Academy of Sciences USA* 100:11771–11776.

Olías, R., E. Zakia, L. Jun, D. Pazalvarez, M.C. Marin-Manzano, M. Mari Carmen, J.M. Pardo, and B. Andrés. 2009. The plasma membrane Na+/H+ antiporter SOS1 is essential for salt tolerance in tomato and affects the partitioning of Na+ between plant organs. *Plant Cell Environment* 32:904–916.

Oono, Y., M. Seki, M. Satou, K. Lida, K. Akiyama, T. Sakurai, M. Fujita, K. Yamaguchi-Shinozaki, and K. Shinozaki. 2006. Monitoring expression profile of Arabidopsis genes during cold acclimation and deacclimation using DNA microarrays. *Functional and Integrative Genomics* 6:212–234.

Pang, Q., S. Chen, S. Dai, Y. Chen, Y. Wang, and X. Yan. 2010. Comparative proteomics of salt tolerance in *Arabidopsis thaliana* and *Thellungiella halophila*. *Journal of Proteome Research* 9:2584–2599.

Pardo, J.M. 2010. Biotechnology of water and salinity stress tolerance. *Current Opinion Biotechnology* 21:185–196.

Pardo, J.M., B. Cubero, E.O. Leidi, and F.J. Quintero. 2006. Alkali cation exchangers: Roles in cellular homeostasis and stress tolerance. *Journal of Experimental Botany* 57:1181–1199.

Paterson, A.H., M. Freeling, and T. Sasaki. 2005. Grains of knowledge: Genomics of model cereals. *Genome Research* 15:1643–1650.

Plett, D.C. and I.S. Moller. 2010. Na+ transport in glycophytic plants: What we know and would like to know. *Plant Cell and Environment* 33:612–626.

Qiu, Q., Y. Guo, M.A. Dietrich, K.S. Schumaker, and J.K. Zhu. 2002. Regulation of SOS1, a plasma membrane Na+/H+ exchanger in *Arabidopsis thaliana*, by SOS2 and SOS3. *Proceedings of the National Academy of Sciences USA* 99:8436–8441.

Quan, R., H. Lin, I. Mendoza, Y. Zhang, W. Cao, Y. Yang, M. Shang, S. Chen, J.M. Pardo, and Y. Guo. 2007. SCABP8/CBL10, a putative calcium sensor, interacts with the protein kinase SOS2 to protect *Arabidopsis* shoots from salt stress. *Plant Cell* 19:1415–1431.

Quintero, F.J., M.R. Blatt, and J.M. Pardo. 2000. Functional conservation between yeast and plant endosomal Na+/H+ antiporters. *FEBS Letters* 471:224–228.

Quintero, F.J., J. Martinez-Atienza, I. Villalta, X. Jiang, W.Y. Kim, Z. Ali, H. Fujii et al. 2011. Activation of the plasma membrane Na+/H+ antiporter salt-overly-sensitive 1 (SOS1) by phosphorylation of an auto-inhibitory C-terminal domain. *Proceedings of the National Academy of Sciences USA* 108:2611–2616.

Rabbani, M.A., K. Maruyama, H. Abe, M. Ayub Khan, K. Katsura, Y. Ito, K. Yoshiwara, M. Seki, K. Shinozaki, and K. Yamaguchi-Shinozaki. 2003. Monitoring expression profiles of rice genes under cold, drought, and high-salinity stresses and abscisic acid application using cDNA microarray and RNA gel-blot analyses. *Plant Physiology* 133:1755–1767.

Rahman, S., H. Miyake, and Y. Takeoka. 2002. Effects of exogenous glycinebetaine on growth and ultrastructure of salt-stressed rice seedlings (*Oryza sativa* L.). *Plant Production Science* 5:33–44.

Robinson, M.F., A.A. Very, D. Sanders, and T.A. Mansfield. 1997. How can stomata contribute to salt tolerance? *Annals of Botany* 80:387–393.

Rodríguez-Rosales, M.P., F.J. Galvez, R. Huertas, M.N. Aranda, M. Baghour, O. Cagnac, and K. Venema. 2009. Plant NHX cation/proton antiporters. *Plant signaling & Behavior* 4:265–276.

Rodríguez-Rosales, M.P., X.J. Jiang, F.J. Gálvez, M.N. Aranda, B. Cubero, and K. Venema. 2008. Overexpression of the tomato K+/H+ antiporter LeNHX2 confers salt tolerance by improving potassium compartmentalization. *New Phytologist* 179:366–377.

Roxas, V.P., S.A. Lodhi, D.K. Garrett, J.R. Mahan, and R.D. Allen. 2000. Stress tolerance in transgenic tobacco seedlings that overexpress glutathione S-transferase/glutathione peroxidase. *Plant Cell Physiology* 41:1229–1234.

Roxas, V.P., R.K. Smith, Jr, E.R. Allen, and R.D. Allen. 1997. Overexpression of glutathione S-transferase/glutathione peroxidase enhances the growth of transgenic tobacco seedlings during stress. *Nature Biotechnology* 15:988–991.

Rus, A., I. Baxter, B. Muthukumar, J. Gustin, B. Lahner, E. Yakubova, and D.E. Salt. 2006. Natural variants of AtHKT1 enhance Na+ accumulation in two wild populations of Arabidopsis. *PLOS Genetics* 2:1964–1973.

Rus, A., S. Yokoi, A. Sharkhuu, M. Reddy, B.H. Lee, T.K. Matsumoto, H. Koiwa, J.K. Zhu, R.A. Bressan, and P.M. Hasegawa. 2001. AtHKT1 is a salt tolerance determinant that controls Na⁺ entry into plant roots. *Proceedings of the National Academy of Sciences USA* 98: 14150–14155.

Saier Jr., M.H. 2000. A functional-phylogenetic classification system for transmembrane solute transporters. *Microbiology and Molecular Biology Reviews* 64:354–411.

Sairam, R.K., K.V. Rao, and G.C. Srivastava. 2002. Differential response of wheat genotypes to long-term salinity stress in relation to oxidative stress, antioxidant activity and osmolyte concentration. *Plant Science* 163:1037–1046.

Sairam, R.K., G.C. Srivastava, S. Agarwal, and R.C. Meena. 2005. Differences in antioxidant activity in response to salinity stress in tolerant and susceptible wheat genotypes. *Biologia Plantarum* 49: 85–91.

Scandalios, J.G. 1993. Oxygen stress and superoxide dismutases. *Plant Physiology* 101:7–12.

Schafer, R.Q., H.P. Wang, E.E. Kelley, K.L. Cueno, S.M. Martin, and G.R. Buettner. 2002. Comparing carotene, vitamin E and nitric oxide as membrane antioxidants. *Biological Chemistry* 383:671–681.

Schreiber, L., R. Franke, K.D. Hartman, K. Ranathunge, and E. Steudle. 2005. The chemical composition of suberin in apoplastic barriers affects radial hydraulic conductivity differently in the roots of rice (*Oryza sativa* L cv. IR64) and corn (*Zea mays* L. cv. Helix). *Journal of Experimental Botany* 56:1427–1436.

Sengupta, S. and A.L. Majumder. 2009. Insight into the salt tolerance factors of a wild halophytic rice, *Porteresia. coarctata*: A physiological and proteomic approach. *Planta* 229:911–929.

Shabala, S. and A. MacKay. 2011. Ion transport in halophytes. *Advances in Botanical Research* 57:151–199.

Shi, H., M. Ishitani, C. Kim, and J.K. Zhu. 2000. The *Arabidopsis thaliana* salt tolerance gene SOS1 encodes a putative Na⁺/H⁺ antiporter. *Proceedings of the National Academy of Sciences USA* 97:6896–6901.

Shi, H., F.J. Quintero, J.M. Pardo, and J.K. Zhu. 2002. The putative plasma membrane Na⁺/H⁺ antiporter SOS1 controls long-distance Na⁺ transport in plants. *Plant Cell* 14:465–477.

Shi, H. and J.K. Zhu. 2002. Regulation of expression of the vacuolar Na⁺/H⁺ antiporter gene AtNHX1 by salt stress and abscisic acid. *Plant Molecular Biology* 50:543–550.

Shi, L.Y., H.Q. Li, X.P. Pan, G.J. Wu, and M.R. Li. 2008. Improvement of *Torenia fournieri* salinity tolerance by expression of *Arabidopsis* AtNHX5. *Functional Plant Biol* 35:185–192.

Song, N.H. and Y.J. Ahn. 2011. DcHsp17.7, a small heat shock protein in carrot, is tissue-specifically expressed under salt stress and confers tolerance to salinity. *New Biotechnology* 28:698–704.

Subudhi, P.K. and N. Baisakh. 2011. *Spartina alterniflora* Loisel., a halophyte grass model to dissect salt stress tolerance. In vitro *Cellular and Developmental Biology-Plant* 47:441–457.

Sunarpi, H., T. Horie, J. Motoda, M. Kubo, H. Yang, K. Yoda, R. Horie et al. 2005. Enhanced salt tolerance mediated by AtHKT1 transporter induced Na⁺ unloading from xylem vessels to xylem parenchyma cells. *The Plant Journal* 44:928–938.

Taji, T., M. Seki, M. Satou, T. Sakurai, M. Kobayashi, K. Ishiyama, Y. Narusaka, M. Narusaka, J.K. Zhu, and K. Shinozaki. 2004. Comparative genomics in salt tolerance between *Arabidopsis* and *Arabidopsis*-related halophyte salt cress using *Arabidopsis* microarray. *Plant Physiology* 135:1697–1709.

Tang, R.J., H. Liu, Y. Bao, Q.D. Lv, L. Yang, and H.X. Zhang. 2010. The woody plant poplar has a functionally conserved salt overly sensitive pathway in response to salinity stress. *Plant Molecular Biology* 74:367–380.

Tester, M. and R.J. Davenport. 2003. Na⁺ transport and Na⁺ tolerance in higher plants. *Annals of Botany* 91:503–527.

Turhan, E., H. Gulen, and A. Eris. 2008. The activity of antioxidative enzymes in three strawberry cultivars related to salt-stress tolerance. *Acta Physiologia Plantarum* 30:201–208.

Uozumi, N., J. Kim, F. Rubio, T. Yamaguchi, S. Muto, A. Tsuboi, E.P. Bakker, T. Nakamura, and J.I. Schroeder. 2000. The *Arabidopsis* HKT1 gene homolog mediates inward Na⁺ currents in *Xenopus laevis* oocytes and Na⁺ uptake in *Saccharomyces cerevisiae*. *Plant Physiology* 122:1249–1259.

Urao, T., B. Yakubov, R. Satoh, K. Yamaguchi-Shinozaki, M. Seki, T. Hirayama, and K. Shinozaki. 1999. A transmembrane hybrid-type histidine kinase in *Arabidopsis* functions as an osmosensor. *Plant Cell* 11:1743–1754.

Venema, K., A. Belver, M.C. Marín-Manzano, M.P. Rodríguez-Rosales, and J.P. Donaire. 2003. A novel intracellular K⁺/H⁺ antiporter related to Na⁺/H⁺ antiporters is important for K⁺ ion homeostasis in plants. *The Journal of Biological Chemistry* 278:22453–22459.

Very, A.A., M.F. Robinson, T.A. Mansfield, and D. Sanders. 1998. Guard cell cation channels are involved in Na⁺-induced stomatal closure in a halophyte. *Plant Journal* 14:509–521.

Wang, J., H. Zhang, and R.D. Allen. 1999. Overexpression of an *Arabidopsis* peroxisomal ascorbate peroxidase gene in tobacco increases protection against oxidative stress. *Plant Cell Physiology* 40:725–732.

Wang, M.C., Z.Y. Peng, C.L. Li, F. Li, C. Liu, and G.M. Xia. 2008. Proteomic analysis on a high salt tolerance introgression strain of *Triticum aestivum/Thinopyrum ponticum*. *Proteomics* 8:1470–1489.

Wang, X., R. Yang, B. Wang, G. Liu, C. Yang, and Y. Cheng. 2010. Functional characterization of a plasma membrane Na+/H+ antiporter from alkali grass (*Puccinellia tenuiflora*). *Molecular Biology Reports* 38:4813–4822.

Wyn Jones, R.G., R. Storey, R.A. Leigh, N. Ahmad, and A.Pollard. 1977. A hypothesis on cytoplasmic osmoregulation. In: Marre, E., Cifferi, O. (Eds.), *Regulation of Cell Membrane Activities in Plants*. Elsevier, Amsterdam, the Netherlands, pp. 121–136.

Xiong, L., Z. Gong, C.D. Rock, S. Subramanian, Y. Guo, W. Xu, D. Galbraith, and J.K. Zhu. 2001. Modulation of abscisic acid signal transduction and biosynthesis by an Sm-like protein in *Arabidopsis*. *Developmental Cell* 1:771–781.

Xu, H., X. Jiang, K. Zhan, X. Cheng, X. Chen, J.M. Pardo, and D. Cui. 2008. Functional characterization of a wheat plasma membrane Na+/H+ antiporter in yeast. *Archives of Biochemistry and Biophysics* 473:8–15.

Xue, Z.Y., D. Zhi, G. Xue, H. Zhang, Y. Zhao, and G. Xia. 2004. Enhanced salt tolerance of transgenic wheat (*Tritivum aestivum* L.) expressing a vacuolar Na+/H+ antiporter gene with improved grain yields in saline soils in the field and a reduced level of leaf Na+. *Plant Science* 167:849–859.

Yamaguchi, T., M.P. Apse, H. Shi, and E. Blumwald. 2003. Topological analysis of a plant vacuolar Na+/H+ antiporter reveals a luminal C terminus that regulates antiporter cation selectivity. *Proceedings of the National Academy of Sciences USA* 100:12510–12515.

Yao, X., T. Horie, S. Xue, H.Y. Leung, M. Katsuhara, D.E. Brodsky, Y. Wu, and J.I. Schroeder. 2010. Differential sodium and potassium transport selectivities of the rice OsHKT2;1 and OsHKT2;2 transporters in plant cells. *Plant Physiology* 152:341–355.

Ye, J., W. Zhang, and Y. Guo. 2013. *Arabidopsis* SOS3 plays an important role in salt tolerance by mediating calcium-dependent microfilament reorganization. *Plant Cell Reports* 32:139–148.

Yoshida, K., M. Kawachi, M. Mori, M. Maeshima, M. Kondo, M. Nishimura, and T. Kondo. 2005. The involvement of tonoplast proton pumps and Na+(K+)/H+ exchangers in the change of petal colour during flower opening of morning glory, *Ipomea tricolor* cv. Heavenly Blue. *Plant Cell Physiology* 46:407–415.

Yoshimura, K., Y. Yabuta, T. Ishikawa, and S. Shigeoka. 2000. Expression of spinach ascorbate peroxidases isoenzymes in response to oxidative stresses. *Plant Physiology* 123:223–234.

Yokoi, S., F.J. Quintero, B. Cubero, T. Ruiz, R.A. Bressan, P.M. Hasegawa, and J.M. Pardo. 2002. Differential expression and function of *Arabidopsis thaliana* NHX Na+/H+ antiporters in the salt stress response. *The Plant Journal* 30:1–12.

Yu, L., J. Nie, C. Cao, Y. Jin, M. Yan, F. Wang, J. Liu, Y. Xiao, Y. Liang, and W. Zhang. 2010. Phosphatidic acid mediates salt stress response by regulation of MPK6 in *Arabidopsis thaliana*. *New Phytologist* 188:762–773.

Zhang, H. and E. Blumwald. 2001. Transgenic salt-tolerant tomato plants accumulate salt in foliage but not in fruit. *Nature Biotechnology* 19:765–768.

Zhang, H., B. Han, T. Wang, S. Chen, H. Li, Y. Zhang, and S. Dai. 2012. Mechanisms of plant salt response: Insights from proteomics. *Journal of Proteome Research* 11:49–67.

Zhu, J.K. 2001. Plant salt tolerance. *Trends in Plant Science* 6:66–71.

Zhu, J.K. 2002. Salt and drought stress signal transduction in plants. *Annual Review of Plant Biology* 53:247–273.

Zou J., A. Liu, X. Chen, X. Zhou, G. Gao, W. Wang, and X. Zhang. 2009. Expression analysis of nine rice heat shock protein genes under abiotic stresses and ABA treatment. *Journal of Plant Physiology* 166:851–861.

# CHAPTER TWO

# Mechanisms of Salt Stress Tolerance in Halophytes

## BIOPHYSICAL AND BIOCHEMICAL ADAPTATIONS

*I. Caçador and B. Duarte*

## CONTENTS

*Abstract.* The salinization of soils is one of the most important factors impacting plant productivity. About 3.6 billion of the world's 5.2 billion ha of agricultural dryland have already suffered erosion, degradation, and salinization. This arises the need to arrange solutions to overcome the stress imposed by salinity to the typical glycophytic crops, such as the improvement of these crops or the use of halophytes in their substitution. Halophytes typically are considered to be plants able to complete their life cycle in environments where the salt concentration is around 200 mM NaCl or higher, representing 1% of the world flora. Different strategies are identified to overcome salt stress as adaptation mechanism from these type of plants. The adjustment to salinity is a complex phenomenon characterized by a high degree of ecological complexity, structural changes, and physiological adjustments both at the biochemical and biophysical levels. These adaptations have naturally evolved in halophytes as responses to their colonization of saline ecosystems, and therefore making halophytes good model plants to study the tolerance mechanisms underlying these salinity constrains.

As photosynthesis is a prerequisite for biomass production, halophytes adapted their electronic transduction pathways and the entire energetic metabolism to overcome the stress imposed by the excessive ionic concentration in their cells. The maintenance of the homeostasis between the $Na+$, $K+$, and $Ca+$ concentrations is in the basis of all cellular stress in particular in terms of redox potential and energy transduction. A salt-stressed cell is unable to process the electronic energy fluxes leading to the accumulation of lethal excessive energy. In the present work, the biophysical mechanisms underlying energy capture and transduction in halophytes are discussed and their relation to the biochemical mechanisms (osmocompatible solute production, pigment profile alteration, antioxidant enzymatic and nonenzymatic defenses), integrating data from the photosystem light harvesting complexes (LHC), passing through electronic transport chains (ETCs) to the quinone pools and the carbon harvest and energy dissipation metabolism and the inevitable antioxidant processes, in order to draw a map of some of the diversity of metabolic mechanisms of salt stress tolerance.

## 2.1 INTRODUCTION

Earth is in fact a salt planet. Seventy percent of its surface is covered by salt water, the oceans, with concentrations of Na+ around 500 mM in contrast with the low K+ concentrations of 9 mM (Flowers, 2004). Also, the remaining 30% of Earth's surface is being severely affected by an increased salinization phenomenon, mostly due to the increased soil use to agriculture and its irrigation procedures (Zhang and Shi, 2014). Aggravating this situation are the ongoing climate changes increasing drought, air temperature, and salt water intrusion in coastal soils (Duarte et al., 2013a). This will have severe impacts in the planet's terrestrial primary production with a special emphasis on the crop production. Salinity-induced damage in plants include reduction of leaf expansion, stomata closure, reduced primary production, biomass losses due to water deficit, and deficiency in essential nutrients like K+ (Mahajan and Tuteja, 2005; Rahnama et al., 2011; James et al., 2011). Although this is true for most of Earth's flora, halophytes are the exception, being highly productive under saline conditions.

Halophytes are defined as plant species that can survive and reproduce under growth conditions with more than 200 mM NaCl, comprising only 1% of world flora (Flowers and Colmer, 2008). Some of these species are what may be called "obligate halophytes," like *Suaeda maritima* and *Mesembryanthemum crystallinum*, requiring saline environments for optimal growth, while other species like *Puccinellia maritima* and *Thellungiella halophila* are "facultative halophytes" with optimal growth without salt in their substrate but tolerating high NaCl concentrations (Flowers, 1972; Gong et al., 2005; Gao et al., 2006, 2012; Agarie et al., 2007; Wang et al., 2007, 2009). Salt tolerance results from a complex network of mechanisms involving multiple biochemical and physiological traits. Over the last decades, this issue attracted several investigation groups as the global

soil salinization problem intensified, and the need to understand these mechanisms increased with the main objective of applying this knowledge to economically important crops. On the other hand, another source of interest arises as some halophytes were identified as potential food sources with high nutritional value and with possibilities to be cultured in arid environments of the poorer regions of the planet, such as the African desert countries. Several halophytes like *Aster tripolium* (Ventura et al., 2013), *Chenopodium quinoa* (Eisa et al., 2012), and *Salicornia sp.* (Ventura and Sagi, 2013) are already identified and commercially used as food sources in some countries.

## 2.2 MORPHOLOGICAL ADAPTATIONS

Some of the more evident adaptations to arid salt environments are immediately detected while observing halophyte morphology. There are typically two mechanisms that halophytes undergo in order to overcome high salinities: secretion and exclusion. The secretion-based strategy implies the existence of specialized salt glands, normally located at the leaf surface, which excrete the excess salt and thus avoiding its potential negative effects on cell metabolism (Figure 2.1). This tolerance mechanism is probably one of the most well studied in halophytes (Rozema et al., 1981; Waisel et al., 1986). The accumulated salt crystals on the leaf surface are then washed by rain or, since most halophytes inhabit coastal areas, are washed away by tidal waters (Balsamo et al., 1995).

On the other hand, *Suaeda fruticosa* (Amaranthaceae), for example, is a typical excluder, retaining higher amounts of K+ and Ca²+ inside its cells, thus avoiding the entrance of Na+ (Figure 2.2). This exclusion strategy is often accompanied by a dilution strategy implying that the halophyte increases its intracellular water content in order to decrease

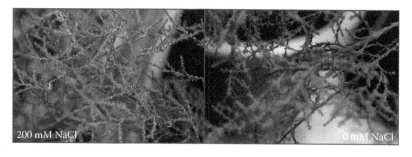

Figure 2.1. *Tamarix gallica* leaves of individuals subjected to 200 and 0 mM NaCl. (Photo by B. Duarte.)

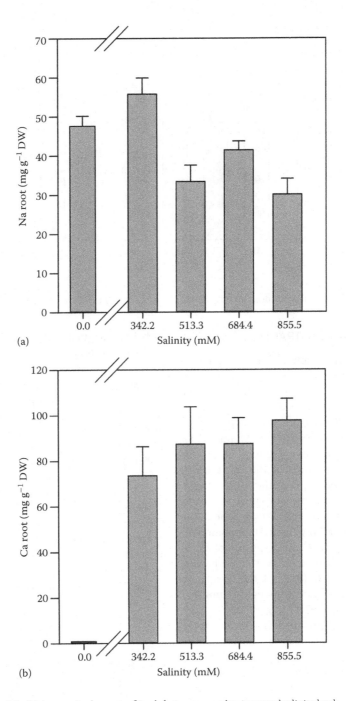

Figure 2.2. Na (a) and Ca (b) ionome in the roots of *Suaeda fruticosa* exposed to increased salinity levels.

the ionic concentration inside its cells (Figure 2.3). Nevertheless, all these morphological adaptations have implications at the biophysical and biochemical levels. This was also observed for T. *halophila* retaining higher potassium and accumulating less sodium, while increasing its transpiration rate resulting in a high water uptake (Volkov and Amtmann, 2006).

On the basis of this differential ionic absorption are specific protein-like ionic channels. A total of 32 salt-induced differentially expressed proteins were identified in T. *halophila* (Pang et al., 2010). In stress situations, $K^+$ transporter proteins are preferentially expressed counterbalancing the extracellular $Na^+$ concentration.

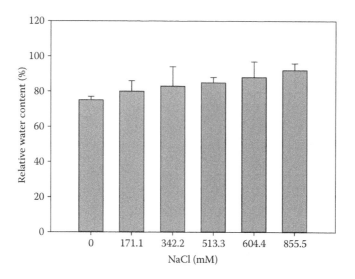

Figure 2.3. *Suaeda fruticosa* photosynthetic stems relative water content exposed to increased salinity levels.

## 2.3 BIOPHYSICAL CONSTRAINS

As all other excessive ionic accumulation, excessive salinity also has its redox implications at the cellular level, unbalancing the electronic fluxes inside the cell. A decrease in photosynthesis capacity is very common in salt-stressed plants (Munns and Termaat, 1986; Munns, 1993; Qiu et al., 2003; Jaleel et al., 2007), mostly due to a low osmotic potential of the soil solution (osmotic stress), specific ion effects (salt stress), nutritional imbalances, or, more usually, a combination of these factors (Zhu, 2003). One of the consequences of salinity-induced limitation of photosynthetic capacity is the exposure of plants to excess energy with inevitable consequences on the photosystem II (PSII), if the dissipation mechanisms are not efficient enough (Demming-Adams and Adams, 1992; Qiu et al., 2003), since plants under salt stress use less photon energy for photosynthesis (Megdiche et al., 2008). The effects of salinity on photosynthesis include several other consequences besides the damage on PSII. Also the photosynthetic carbon harvesting is affected by disturbances on leaf water relations and osmotic potential (Munns, 2002; Zhao et al., 2007) on the chloroplast membrane systems and on the pigment composition (carotenoids and chlorophyll). To avoid damages to the PSII, plants have developed several strategies to dissipate excessive energy, protecting the photosynthetic apparatus. Comparing a glycophyte species with a halophyte one, the differences in a global examination of the PSII activity are evident (Figure 2.4). Both real (operational) and maximum activities of PSII suffer drastic decreases in activity due to salt stress in glycophyte species. On the other hand, a halophytic species very well adapted to salt environments shows almost no differences along a salinity gradient even at oceanic salt concentrations.

PSII quantum yield gives rapid and valuable informations on the overall ongoing processes in the PSII, but in order to understand the causes of these changes as well as the mechanisms that allow halophytes to overcome salt stress, one has to delve deeper into the biophysics and energetics of the chloroplast. PSII efficiency relies essentially on two major processes: (1) photon harvesting, entrapment, and transport throughout the transport chain and (2) excessive electronic energy dissipation. Examining the first one and specially the electronic transport depending on the tolerance and defense mechanism, two behaviors can be observed (Figure 2.5). Observing the rapid light curves obtained for both species at different stress levels, it is possible to observe that there are evident differences either between species or, in particular for *Halimione portulacoides*, between stress levels. In *S. fruticosa*, the maximum electron transport rate (ETR), photosynthetic efficiency, and the onset of light saturation are very similar among healthy and stressed individuals, only with small differences also regarding the rETR at different light exposures. On the other hand,

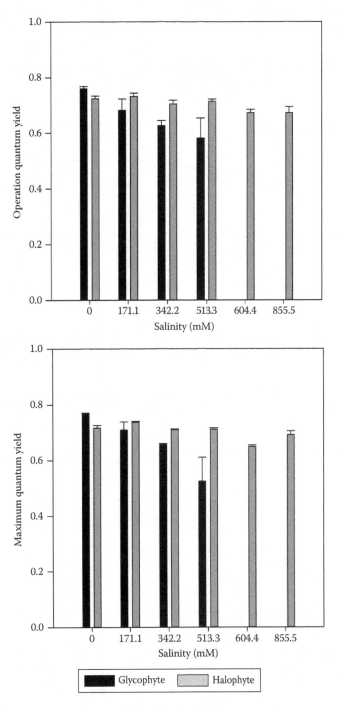

Figure 2.4. Operational and maximum photosystem II efficiency in a glycophyte and in a halophyte along a salinity gradient.

H. *portulacoides* stressed and healthy individuals exhibited very distinct photosynthetic parameters. Not only the photosynthetic efficiency and the onset of light saturation were reduced to zero, but also the maximum ETR was lower in stressed individuals. Observing S. *fruticosa* healthy and stressed individuals, there are no major differences neither between the ETR nor in the onset of light saturation, indicating a normal functioning in the ETC. As for H. *portulacoides*, not only the

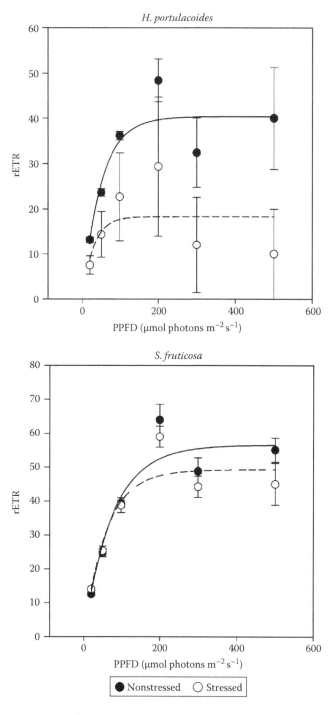

Figure 2.5. Electron transport rate at different light intensities in *Halimione portulacoides* and *Suaeda fruticosa* stressed and nonstressed individuals. (From Duarte, B. et al., *Plant Physiol. Biochem.*, 67, 178, 2013b.)

ETR is rather decreased in stressed individuals, but also these individuals have a smaller onset for light saturation, indicating incapacity to use the absorbed photons into primary photochemistry. This inevitably leads to an accumulation of large amounts of lethal energy that, as stated before, can destroy the D1 protein, impairing the photochemical apparatus. Again two tolerance mechanisms are evidenced between these two Amaranthaceae species. S. fruticosa has salinity tolerance mechanisms that allow the photosystems to absorb light even at high Na concentrations, while in H. portulacoides these mechanisms appear to be absent or inactivated, leading to lower light and carbon harvesting efficiencies. In fact S. fruticosa exhibits a common feature among halophytes: elevated salt concentrations improve some energetic mechanisms. Delving even deeper into the electronic processes, one can distinguish how the energy fluxes that, in sum result in the PSII activity, are affected by salt stress.

Looking deeper into the photochemical mechanisms (Figure 2.6), it can be observed that in S. fruticosa the major factor responsible for the decrease in the photosynthetic rate is due to salinity adverse effects in the quinone pools. Both the electron flow from the ETC to the quinone pool (Sm) and the quinone reduction turnover rate were rather decreased, leading to an excess of energy accumulated at this level (Kalaji et al., 2011). In H. portulacoides, the negative effects driven by salt stress leads to higher amounts of energy dissipated rather than trapped in the photochemical reactions, ultimately having as a consequence the destruction of the D1 protein (Rintamäki et al., 1995). In these individuals, there is a small probability that an incident photon can move an electron throughout the ETC and also a reduced efficiency of a trapped electron to move further than the oxidized quinone, reducing this way the maximum yield of primary photochemistry (Kalaji et al., 2011). Although excessive salt acts at different levels in the two different analyzed species (in the photon reception in H. portulacoides and in the reduction of the quinone pool in S. fruticosa), all these effects are well summarized overlooking the reduced performance index in stressed individuals, due to its dependency on the efficiency, yield of energy transfer, and primary photochemistry (Figure 2.7). The behavior exhibited by S. fruticosa is very similar to the one found in

T. gallica when supplied with 200 mM NaCl and can be easily detected using a rapid induction Kautsky curve (Figure 2.8). This type of analysis is very quick and allows a rapid interpretation of the overall energetic fluxes underlying the PSII activity. In this assessment, two phases can be distinguished: O-J phase or photochemical phase and the J-I-P phase or thermal phase. The first one is considered to be a good proxy of the photochemical energy production work ongoing inside the chloroplasts, while the second one reflects the ability to dissipate excessive amounts of energy throughout thermal dissipation. It is possible to observe that T. gallica individuals have similar photochemical activity both with and without salt, but the individuals supplemented with 200 mM NaCl have a higher ability to dissipate excessive energy throughout heat formation. This is one of the mechanisms that halophytes exhibit to overcome excessive energy absorbed to the photosystems while under stress, avoiding this way the photodestruction of the photosynthetic apparatus (Duarte et al., 2013b).

## 2.4 BIOCHEMICAL IMPLICATIONS

Beyond the biophysical processes, halophytes have also a battery of biochemical adjustments to counteract, at the molecular level, the cellular stress imposed by excessive ionic concentrations, namely, Na. Still in the chapter of the photosynthetic light harvesting mechanisms, also the pigment profiles are often affected by elevated salt concentrations. On the other hand, under favorable conditions, the increase in efficiency of the photosystems, consequence, for example, of optimal salt concentrations, a frequently observed strategy is the decrease of the antenna size since there is no need to have large LHC as it would be in stressful conditions for maximization of light harvesting (Rabhi et al., 2012), as it can be evaluated by its pigmentar proxy, the chl a/b ratio (Figure 2.9). The increase of the chl a/b ratio is directly related to an increase in the number of active light harvesting reaction centers and is commonly used as indicator of plant photochemical capacity enhancement, leading to an increase in processing the absorbed light, even at normal light conditions. On the other hand, when the halophyte is away from its optimum salt

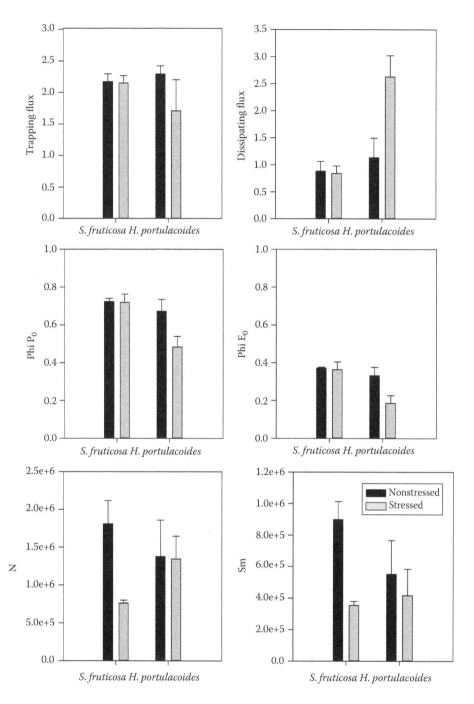

Figure 2.6. Rapid transient O-J-I-P curve calculated parameters in stressed and nonstressed individuals of *Suaeda fruticosa* and *Halimione portulacoides* species. (From Duarte, B. et al., *Plant Physiol. Biochem.*, 67, 178, 2013b.)

conditions, the excessive energy reaching the photosystems must be dissipated (Duarte et al., 2013b). *H. portulacoides* appears to have a physiological optimum at median NaCl concentrations (513.3 mM) similar to those observed in its natural habitat at the estuaries.

On the other hand, when this increase in LHC is not enough to process all the incoming solar radiation, the plant needs to dissipate its energy, either by fluorescence quenching or by a pigment cycle involving a class of carotenoids called xanthophylls (Demming-Adams and Adams, 1992).

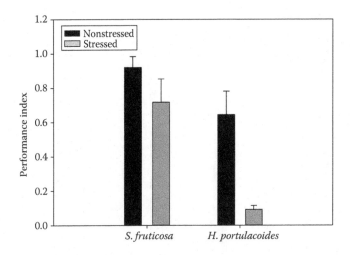

Figure 2.7. Performance index derived from the J-I-P test, in stressed and nonstressed individuals of *Suaeda fruticosa* and *Halimione portulacoides* species. (From Duarte, B. et al., *Plant Physiol. Biochem.*, 67, 178, 2013b.)

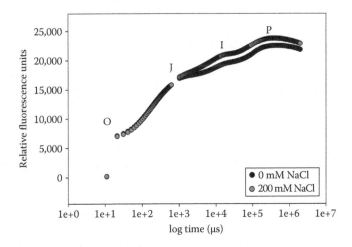

Figure 2.8. Kautsky curve from *Tamarix gallica* individuals exposed to 0 and 200 mM NaCl.

Additionally in extreme stress conditions, an unhealthy plant cannot withstand what in normal cases would be considered a normal dose of light, and thus even at low solar radiances, it undergoes photoinhibition and thus needs to dissipate energy. Another evident signal of environmental stress is the xanthophyll cycle malfunctioning, as revealed by the increase in the de-epoxidation state index (Figure 2.10). When the absorbed light exceeds the plant photochemical capacity (as revealed previously by the decrease in the chi a/b ratio), even in normal light conditions, this excess energy may be transferred to the ever-present oxygen. In this context, the conversion of violaxanthin to zeaxanthin throughout the xanthophyll cycle is considered to be one of the most effective energy dissipation mechanisms (Demmig-Adams and Adams, 1992). Also the total-chlorophyll-to-total-carotenoid ratio points in the same direction. An increase in this ratio occurred in stressed individuals of both species, indicating that, although all pigments suffer a drastic decrease under stress, chlorophylls decrease in a smaller proportion than carotenoids, enhancing the light harvesting efficiency and counteracting stress (Figure 2.10).

Although this turnover toward the carotenoid production is not evident to the naked eye, sometimes another phenomenon can be seen overlooking halophytic extensions, especially during summer. During warm seasons, sediment water evaporates, greatly increasing the salinity in the

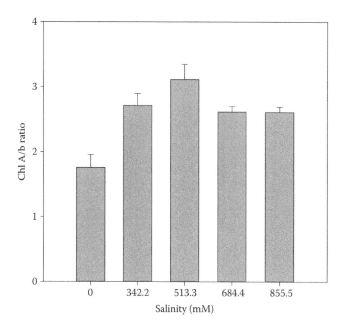

Figure 2.9. Chlorophyll a/b ratio in *Halimione portulacoides* leaves from individuals exposed to a salinity gradient.

sediments sometimes to values twice those observed in seawater. In these conditions, it is often observable in Amaranthaceae salt marshes large extensions of plants exhibiting a strong red coloration (Figure 2.11). This coloration is due to a presence of water-soluble pigments from the betacyanin class, normally produced as a response to salinity, anoxia, or thermal stresses (Wang et al., 2006). Betacyanins play an important role in scavenging reactive oxygen species (ROS) (Stintzing and Carle, 2004), generated under environmental stress conditions. Wang et al. (2006) found similar results for other Amaranthaceae species (*Suaeda salsa*), suggesting that this betacyanin production is part of a common defense mechanism against environmental stresses, namely, salinity. Commonly, these pigments can also be indicators of a high betaine production, a quaternary ammonium compound, mainly accumulated in the chloroplast in order to counteract high Na concentrations in this compartment (Rhodes and Hanson, 1993; McNeil et al., 1999). As for the cytosol, the plant tends to accumulate proline, an amino acid but also a quaternary ammonium compound, in this compartment as an effective osmoregulator of the ionic pressure exerted by excessive salt concentrations. Comparing a glycophyte with a halophyte species, the differences are evident (Figure 2.12), with the halophyte species highly adapted to salinity, with

an enormous production of betain in order to balance and regulate the osmotic potential inside its photosynthetic compartments.

Halophytes are often classified as extremophyte species inhabiting extremely arid environments under extreme abiotic conditions adverse to life support, particularly high salinity levels. Another interesting adaptation developed by this group of plants was the acquisition and development of a highly efficient battery of antioxidant enzymes. As any other excessive ionic concentration, high Na concentrations generate ROS due to its reactions with the cellular biological elements (Duarte et al., 2013c). Halophytes evidence a highly efficient enzymatic system of rapid response to salinity changes that are rapidly activated when the medium conditions are shifted aside from the halophyte optimum (Figure 2.13).

This battery has its higher expression while overlooking its first line of defense, superoxide dismutase. This enzyme catalyses the conversion of the highly toxic superoxide anions to hydrogen peroxide. As a second line of defense, intrinsically connected to the first one are the peroxidase class of enzymes such as catalase, ascorbate peroxidase, and guaiacol peroxidase. All three enzymes have as major function in the hydrogen peroxide detoxification and thus in the reduction of ROS to nondamaging concentrations.

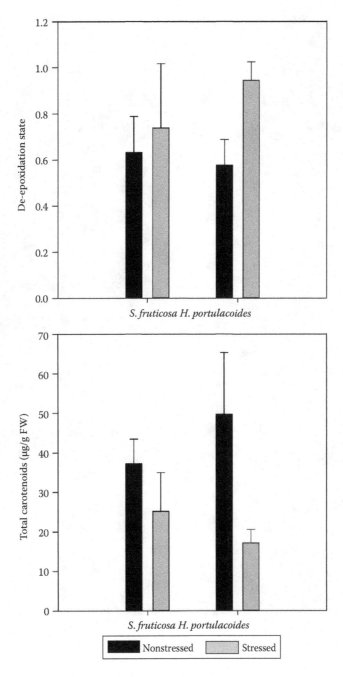

Figure 2.10. De-epoxidation state and total carotenoids in *Suaeda fruticosa* and *Halimione portulacoides* stressed and nonstressed individuals. (From Duarte, B. et al., *Plant Physiol. Biochem.*, 67, 178, 2013b.)

It is possible to observe that these enzymes are activated both at very low concentrations of Na (below the physiological optimum) and at seawater Na concentrations (considered excessive), pointing out to a physiological Na dependence in certain halophytes, like *H. portulacoides* (Figure 2.13).

## 2.5 FINAL REMARKS

Halophytes are extremely plastic species with a high degree of adaptation to saline habitats being therefore excellent models to study salt resistance and tolerance mechanisms. Nevertheless, some halophytes have recently been pointed

Figure 2.11. *Suaeda fruticosa* exhibiting a red coloration due to excessive salt concentrations. (Photo by B. Duarte.)

out as potential cash crops for replacing usual crops that cannot tolerate the excess salt in an increasingly salinized world. Their tolerance to salt goes from simple morphological adjustments like increasing turgescence or specific salt glands to efficient energy dissipation mechanisms based on the electronic flux adjustment inside the chloroplast or to the production of specific molecules with the main objective of counteracting the osmotic unbalance driven by excessive salt. Nowadays, the metabolic biophysical and biochemical mechanisms underlying these processes are relatively well described for several halophytes. This opens a new door for which physiology can be allied to biotechnology identifying the key genes under these processes, which can be introduced into nontolerant crops allowing their cultivation in arid and saline lands maintaining the food supply in some of the poorer regions of the planet.

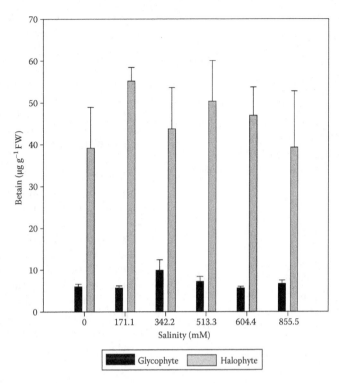

Figure 2.12. Betain concentration in the leaves of a glycophyte and of a halophyte along a salinity gradient.

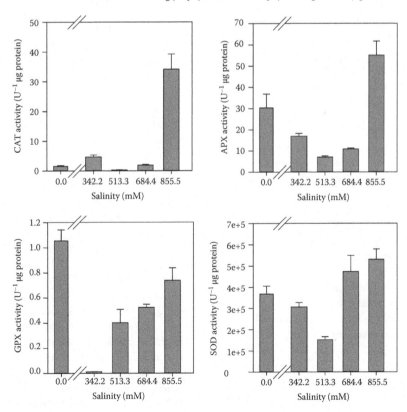

Figure 2.13. Antioxidant enzymatic activities in the leaves of *Halimione portulacoides* exposed to a salinity gradient.

## REFERENCES

Agarie, S., Shimoda, T., Shimizu, Y., Baumann, K., Sunagawa, H., Kondo, A., Ueno, O., Nakahara, T., Nose, A., and Cushman, J.C. Salt tolerance, salt accumulation, and ionic homeostasis in an epidermal bladder-cell-less mutant of the common ice plant Mesembryanthemum crystallinum. Journal of Experimental Botany 58 (2007): 1957–1967.

Balsamo, R.A., Adams, M.E., and Thomson, W.W. Electrophysiology of the salt glands of Avicennia germinans. International Journal of Plant Science 156 (1995): 658–667.

Demmig-Adams, B. and Adams II, W.W. Photoprotection and other responses of plants to light stress. Annual Review of Plant Physiology and Plant Molecular Biology 43 (1992): 599–626.

Duarte, B., Caçador, I., Marques, J.C. and Croudace, I. Tagus Estuary salt marshes feedback to sea level rise over a 40-year period: Insights from the application of geochemical indices. Ecological Indicators 34 (2013a): 268–276.

Duarte, B., Santos, D., and Caçador, I. Halophyte antioxidant feedback seasonality in two salt marshes with different degrees of metal contamination: Search for an efficient biomarker. Functional Plant Biology 40 (2013c): 922–930.

Duarte, B., Santos, D., Marques, J.C., and Caçador, I. Ecophysiological adaptations of two halophytes to salt stress: Photosynthesis, PS II photochemistry and antioxidant feedback—Implications for resilience in climate change. Plant Physiology and Biochemistry 67 (2013b): 178–188.

Eisa, S., Hussein, S., Geissler, N., and Koyro, H.-W. Effect of NaCl salinity on water relations, photosynthesis and chemical composition of Quinoa (Chenopodium quinoa Willd.) as a potential cash crop halophyte. Australian Journal of Crop Science 6 (2012): 357–368.

Flowers, T.J. Salt tolerance in Suaeda maritima (L.) Dum: The effect of sodium chloride on growth, respiration, and soluble enzymes in a comparative study with Pisum sativum L. Journal of Experimental Botany 23 (1972): 310–321.

Flowers, T.J. Improving crop salt tolerance. Journal of Experimental Botany 55 (2004): 307–319.

Flowers, T.J. and Colmer, T.D. Salinity tolerance in halophytes. New Phytologist 179 (2008): 945–963.

Gao, F., Gao, Q., Duan, X.G., Yue, G.D., Yang, A., and Zhang, J. Cloning of an H+-PPase gene from Thellungiella halophila and its heterologous expression to improve tobacco salt tolerance. Journal of Experimental Botany 57 (2006): 3259–3270.

Gong, Q., Li, P., Ma, S., Rupassara, S.I., and Bohnert, Hans J. Salinity stress adaptation competence in the extremophile Thellungiella halophila in comparison with its relative Arabidopsis thaliana. Plant Journal 44 (2005): 826–839.

Guo, Q., Wang, P., Ma, Q., Zhang, J.-L, Bao, A.-K., and Wang, S.-M. Selective transport capacity for K+ over Na+ is linked to the expression levels of PtSOS1 in halophyte Puccinellia tenuiflora. Functional Plant Biology 39 (2012): 1047–1057.

Jaleel, C.A., Gopi, R., Manivannan, P., and Panneerselvam, R. Antioxidative potentials as a protective mechanism in Catharanthus roseus (L.) G. Don. plants under salinity stress. Turkish Journal of Botany 31 (2007): 245–251.

James, R.A., Blake, C., Byrt, C.S., and Munns, R. Major genes for Na+ exclusion, Nax1 and Nax2 (wheat HKT1;4 and HKT1;5), decrease Na+ accumulation in bread wheat leaves under saline and waterlogged conditions. Journal of Experimental Botany 62 (2011): 2939–2947.

Kalaji, H.M., Govindjee, Bosa, K., Koscielniak, J., and Zuk-Golaszewska, K. Effects of salt stress on photosystem II efficiency and CO$_2$ assimilation of two Syrian barley landraces. Environmental and Experimental Botany 73 (2011): 64–72.

Mahajan, S. and Tuteja, N. Cold, salinity and drought stresses: An overview. Archives of Biochemistry and Biophysics 444 (2005): 139–158.

McNeil, S.D., Nuccio, M.L., and Hanson, A.D. Betaines and related osmoprotectants. Targets for metabolic engineering of stress resistance. Plant Physiology 120 (1999): 945–949.

Megdiche, W., Hessini, K., Gharbi, F., Jaleel, C., Ksouri, R., and Abdelly, C. Photosynthesis and photosystem II efficiency of two salt-adapted halophytic seashore Cakile maritima ecotypes. Photosynthetica 46 (2008): 410–419.

Munns, R. Physiological processes limiting plant growth in saline soil. Some dogmas and hypothesis. Plant Cell and Environment 16 (1993): 15–24.

Munns, R. Comparative physiology of salt and water stress. Plant, Cell and Environment 25 (2002): 239–250.

Munns, R. and Termaat, A. Whole-plant responses to salinity. Australian Journal of Plant Physiology 13 (1986): 143–160.

Pang, Q., Chen, S., Dai, S., Chen, Y., Wang, Y. and Yan, X. Comparative proteomics of salt tolerance in Arabidopsis thaliana and Thellungiella halophila. Journal of Proteome Research 9 (2010): 2584–2599.

Qiu, N., Lu, Q., and Lu, C. Photosynthesis, photosystem II efficiency and the xanthophyll cycle in the salt-adapted halophyte Atriplex centralasiatica. New Phytologist 159 (2003): 479–486.

Rabhi, M., Castagna, A., Remorini, D., Scattino, C., Smaoui, A., Ranieri, A., and Abdelly, C. Photosynthetic responses to salinity in two obligate halophytes: *Sesuvium portulacastrum* and *Tecticornia indica*. *South African Journal of Botany* 79 (2012): 39–47.

Rahnama, A., James, R.A., Poutini, K., and Munns, R., Stomatal conductance as a screen for osmotic stress tolerance in durum wheat growing in saline soil. *Functional Plant Biology* 37 (2011): 255–263.

Rhodes, D. and Hanson, A.D. Quaternary ammonium and tertiary sulfonium compounds in higher plants. *Annual Review in Plant Physiology and Plant Molecular Biology* 44 (1993): 357–384.

Rintamäki, E., Salo, R., Lehtonen, E., and Aro, E.-M. Regulation of D1 protein degradation during photoinhibition of photosystem II in vivo: Phosphorylation of the D1 protein in various plant groups. *Planta* 195 (1995): 379–386.

Rozema, J., Gude, H., and Pollack, G. An ecophysiological study of the salt secretion of four halophytes. *New Phytologist* 89 (1981): 207–217.

Stintzing, F.C. and Carle, R. Functional properties of anthocyanins and betalains in plants, food, and in human nutrition. *Trends in Food Science and Technology* 15 (2004): 19–38.

Ventura, Y., Myrzabayeva, M., Alikulov, Z., Cohen, S., Shemer, Z., and Sagi, M. The importance of iron supply during repetitive harvesting of *Aster tripolium*. *Functiona Plant Biology* 40 (2013): 968–976.

Ventura, Y. and Sagi, M. Halophyte crop cultivation: The case for *Salicornia* and *Sarcocornia*. *Environmental and Experimental Botany* 92 (2013): 144–153.

Volkov, V. and Amtmann, A. *Thellungiella halophila*, a salt tolerant relative of *Arabidopsis thaliana*, has specific root ion-channel features supporting $K^+/Na^+$ homeostasis under salinity stress. *Plant Journal* 48 (2006): 342–353.

Waisel, Y., Eshel, A., and Agami, M. Salt balance of leaves pf the mangrove *Avicennia marina*. *Physiologia Plantarum* 67 (1986): 67–72.

Wang, C.-M., Zhang, J.-L., Liu, X.-S., Li, Z., Wu, G.-Q., Cai, J.-Y., Flowers, T.J., and Wang, S.-M. *Puccinellia tenuiflora* maintains a low $Na^+$ level under salinity by limiting unidirectional $Na^+$ influx resulting in a high selectivity for $K^+$ over $Na^+$. *Plant Cell and Environment* 32 (2009): 486–496.

Wang, C.Q., Zhao, J.-Q., Chen, M., and Wang, B.-S. Identification of betacyanin and effects of environmental factors on its accumulation in halophyte *Suaeda salsa*. *Journal of Plant Physiology and Molecular Biology* 32 (2006): 195–201.

Wang, S.-M., Zhang, J.-L., and Flowers, T.J. Low affinity $Na^+$ uptake in the halophyte *Suaeda maritima*. *Plant Physiology* 145 (2007): 559–571.

Zhang, J.-L. and Shi, Huazhong. Physiological and molecular mechanisms of plant salt tolerance. *Photosynthesis Research* 115 (2014): 1–22.

Zhao, G.Q., Ma, B.L., and Ren, C.Z. Growth, gas exchange, chlorophyll fluorescence, and ion content of naked oat in response to salinity. *Crop Science* 47 (2007): 123–131.

Zhu, J.-K. Regulation of ion homeostasis under salt stress. *Current Opinion in Plant Biology* 6 (2003): 441–445.

CHAPTER THREE

# Targeting Vacuolar Sodium Sequestration in Plant Breeding for Salinity Tolerance

*Getnet Dino Adem, Jayakumar Bose, Meixue Zhou, and Sergey Shabala*

## CONTENTS

*Abstract.* Salinity is a major environmental issue affecting crop production around the globe, and creating a salt-tolerant germplasm is absolutely essential meeting the 2050 challenge of feeding a 9.3 billion population. While most efforts of plant breeders were focused around genes and mechanisms responsible for exclusion of cytotoxic Na$^+$ from uptake, this is not the strategy naturally salt-tolerant halophyte species use. One of the hallmarks of halophytes is their ability to safely deposit large volumes of salt in their vacuoles, in the process termed vacuolar sodium sequestration. This chapter reviews molecular and physiological mechanisms mediating this process and prospects of their targeting in breeding programs.

*Keywords*: Na$^+$ sequestration, SV and FV channels, Vacuolar H$^+$-ATPase, Vacuolar H$^+$-PPiase

## 3.1 INTRODUCTION

Salinity is a major environmental issue affecting crop production around the globe. Up to 7% of the total land surface is saline (Flowers and Yeo, 1995;

Munns, 2002), and about one-third of the world's irrigated land suffers from secondary-induced salinization (Flowers and Yeo, 1995). The global cost of irrigation-induced salinity is equivalent to an estimated US$12 billion per year (Qadir et al., 2008), and it is predicted that up to 50% of currently arable land may be lost by the year 2050 (Wang et al., 2003). This trend may have very serious economic, social, and political implications. Taking rice as an example, it is the second most produced cereal in the world and the most important staple food in Asia and the West Indies. The USDA forecast for 2013–2014 global rice production is 473.2 million tons (milled basis) (http://oryza.com/news/usda) that showed a steady growth of about 1.6%–1.8% p.a. Unfortunately, rice is rather sensitive to salinity (Yeo et al., 1999) and generally cannot grow in soils with EC levels above 5 dS/m. At the same time, most of the major rice-producing countries such as China, India, Indonesia, and Pakistan experience severe and increasing salinity problems, with most soils far exceeding the 5 dS/m limit. Thus, meeting the 2050 challenge of feeding a 9.3 billion population requires a qualitative breakthrough in breeders' efforts to create a salt-tolerant germplasm. It was argued (Shabala, 2013) that this qualitative leap can be achieved by learning from halophytes, understanding mechanisms employed by these naturally salt-tolerant species to deal with excessive salt, and incorporating those mechanisms into conventional crops. One of the hallmarks of halophytes is their ability to safely deposit large volumes of salt in their vacuoles, in the process termed *vacuolar sodium sequestration*. This chapter reviews molecular and physiological mechanisms mediating this process and prospects of their targeting in breeding programs.

## 3.2 ESSENTIALITY OF VACUOLAR Na$^+$ SEQUESTRATION FOR PLANT SALINITY TOLERANCE

Halophyte plants are evolutionarily adapted to thrive in seawater and can sometimes tolerate NaCl concentration in excess of 1 M (Flowers et al., 1977). On the contrary, there is a substantial yield reduction in glycophytes when grown at much modest (e.g., around 100 mM) salinities. However, there is some growth response overlap between these two adaptations; for example, sugar beet can grow in up to 260 mM salt, as it descended from a halophyte ancestor (Greenway and Munns, 1980). A common physiological mechanism preserved in both halophytes and glycophytes is a maintenance of optimal (low)

Na$^+$/K$^+$ ratio in the metabolically active cytoplasm (Greenway and Munns, 1980; Jeschke et al., 1983; Gorham et al., 1990). To achieve this goal, most glycophytes exclude Na$^+$ from the shoot by either restricting it from loading into the shoot, its extrusion to the external medium, or restricting its entry into the root (Gorham et al., 1990; Tester and Devenport, 2003; Colmer et al., 2005; Munns and Tester, 2008). The unidirectional influx of Na$^+$ in glycophytes is thermodynamically passive and poorly controlled (Tester and Devenport, 2003). Taking the case in wheat genotypes, the study using $^{22}$Na$^+$ trace element showed that there is little difference among the genotypes in Na$^+$ accumulation (Davenport et al., 1997); similar results were reported for barley cultivars contrasting in salinity tolerance (Chen et al., 2007a). Hence, the two main mechanisms for glycophytes like wheat would be restriction of Na$^+$ loading to the xylem and active extrusion of Na$^+$ to the external growth media (Cuin et al., 2011). The extrusion of Na$^+$ out of the plant cell into the rhizosphere is a thermodynamically active process that consumes energy. Unlike animals that use Na$^+$-ATPases to pump Na$^+$ out of the cell, plants rely on the plasma membrane Na$^+$/H$^+$ exchanger to actively expel Na$^+$ out of the root cells (Zhu, 2003). This electroneutral exchange activity is the only transport activity that is used for Na$^+$ export under physiological conditions (Zhu, 2003). The active transport of Na$^+$ by the Na$^+$/H$^+$ exchanger protein is coupled with the H$^+$ gradient that is developed across plasma membrane as an electrochemical potential generated by proton pumps at the plasma membrane (Blumwald et al., 2000; Hasegawa et al., 2000; Shi et al., 2000). The other mechanism to reduce Na$^+$ accumulation in the shoot in glycophytes is the control of Na$^+$ loading into the xylem (Moller et al., 2009).

In a stark contrast to glycophytes, halophytes accumulate very substantial quantities of Na$^+$ in their aboveground tissues. Here, optimal cytosolic K$^+$/Na$^+$ ratio is maintained by efficient Na$^+$ sequestration in leaf vacuoles. This sequestration is made possible by the presence of tonoplast Na$^+$/H$^+$ exchange proteins (Apse et al., 1999; Blumwald, 2000). Because of this, halophytes are classified as salt includers and use Na$^+$ as a cheap osmoticum for maintaining cell turgor and hence shoot expansion growth under saline conditions (Flowers and Colmer, 2008).

## 3.3 MOLECULAR AND PHYSIOLOGICAL MECHANISMS FOR Na⁺ LOADING INTO VACUOLE

### 3.3.1 Molecular Identity of NHX Proteins

As a pioneering work, a gene coding for a Na$^+$/H$^+$ exchanger (NHX1) was identified by in silico analysis of the yeast genome (Nass et al., 1997). It was then shown that under salt stress, NHX1 was responsible for prevacuolar/vacuolar compartmentalization of Na$^+$ ions (Nass and Rao, 1998), revealing the functional role of this exchanger. Na$^+$/H$^+$ antiporter exchanges protons for Na$^+$ ions across membranes and is particularly active in the vacuoles in plants, algae, and fungi. However, this exchanger (antiporter) is also known to exist in animals, yeasts, bacteria, and plants (Blumwald, 2000). This antiporter situated in the vacuole in plants removes Na$^+$ ions from the cytoplasm energized by the electrochemical potential created by the pumping of H$^+$ into the vacuole by two proton pumps, vacuolar H$^+$-inorganic pyrophosphatase (V-inorganic pyrophosphatase [V-PPase}, E.C. 3.6.1.1), and vacuolar H$^+$-ATPase (V-adenosine triphosphatase [V-ATPase], E.C. 3.6.1.3). Under saline stress, plants must maintain high K$^+$/Na$^+$ ratios in the cytoplasm, and for doing so, Na$^+$ ions need to be accumulated in the vacuole, away from the cytoplasm. For achieving this, plant vacuolar Na$^+$/H$^+$ antiporter is of paramount importance (Fukuda et al., 2004a). The *Arabidopsis thaliana* Na$^+$/H$^+$ antiporter gene (AtNHX1) was the first plant NHX homologue to be cloned (Gaxiola et al., 1999). There are six isoforms of AtNHX in *Arabidopsis* identified, and AtNHX1 and AtNHX2 have widespread transcript distribution and are highly expressed. Nevertheless, AtNHX3 and AtNHX4 transcript are exclusively found in flowers and roots (Yokoi et al., 2002; Aharon et al., 2003).

### 3.3.2 NHX Proteins and Their Role in Salt Tolerance

Apse et al. (1999) showed that by overexpressing AtNHX1, it was possible to improve salt tolerance. Similarly, Bayat et al. (2011) showed that overexpression of the barley (*Hordeum vulgare*) HvNHX2 gene in *Arabidopsis* showed that the plant has substantially increased salt tolerance and grows normally with 200 mM NaCl, while the wild-type *Arabidopsis* plant exhibits necrosis. Furthermore, transgenic tomato overexpressing Na$^+$/H$^+$ antiporter was able to thrive at 200 mM NaCl, with the concentration of Na$^+$ in

the fruit very low, which is a very important consideration for fruit quality (Zhang and Blumwald, 2001). Transgenic *Brassica napus* overexpressing AtNHX1 was able to grow, flower, and produce seeds in the presence of 200 mM NaCl, with seed yield and oil quality unaffected (Zhang et al., 2001). Similar improvement in salt tolerance was obtained by overexpressing NHX antiporter from cotton into tobacco (Wu et al., 2004), and salt tolerance in rice was also achieved by overexpressing OsNHX1 (Fukuda et al., 2004b). This finding indicates that it is possible to improve salt tolerance of plants via compartmentalization of Na$^+$ in the vacuoles. In addition to overexpression experiments, gene expression studies on both glycophytes (*A. thaliana* and *Oryza sativa*) and halophytes (*Mesembryanthemum crystallinum* and *Atriplex gmelinii*) showed that the antiporter gene expression increases during salt stress conditions, substantiating the role of the antiporter gene in salt tolerance. The antiporter is also more active in salt-tolerant plant barley (Garbarino and Dupont, 1988, 1989; Fukuda et al., 1998). Overall, it has been reported that mainly in dicot plants and in few monocots, Na$^+$/H$^+$ antiporters play an important role in salt tolerance of various types of plants.

A vacuolar Na$^+$/H$^+$ antiporter (HvNHX1) cDNA clone was identified in barley, and the transcript level of the gene was studied under both salt stress and osmotic stress. It was indicated that the transcript level of HvNHX1 was increased five- and twofold as compared to the control when roots were treated with 200 mM NaCl after 5 and 24 h, respectively. Similarly, exposure of barley plants to 400 mM mannitol, which simulates osmotic stress, increased the transcript level of HvNHX1 (Fukuda et al., 2004a). The HvNHX1 amino acid sequence had 70% identity with the HvNHX2. HvNHX1 has a conserved motif [86]LFFIYLLPPI[95] that is found in plant, yeast, and mammalian Na$^+$/H$^+$ antiporters (Fukuda et al., 2004a). Furthermore, cross talk was observed between osmotic stress and accumulation of Na$^+$ in the vacuole where osmotic stress increased abscisic acid synthesis, which in turn upregulates AtNHX1. This suggests a possible role of the AtNHX1 in osmotic stress (Shi and Zhu, 2002).

### 3.3.3 Regulation of NHX Proteins

Under saline conditions, haplotype plants use inorganic ions such as Na$^+$, K$^+$, and Cl$^-$ to maintain osmotic and turgor pressure (Flowers et al., 1977; Glenn et al., 1999). However, glycophytes prefer the de novo synthesis of compatible solutes to effect the same. Osmolytes have a substantial carbon cost (it takes 50–70 moles of ATP for producing 1 mole of osmolytes). This carbon cost has a yield penalty exhausting the energy currency that it would have been used for other cellular processes (Shabala and Shabala, 2011). The case is different in halophytes, and it is much more efficient than glycophytes. The halophytes produce compatible solutes only in their cytosol where it covers only 10% of the cell volume as a function of osmotic adjustment. The excess toxic sodium will be efficiently sequestered away from the metabolic pathways in the vacuole (Shabala, 2013). The improvement of glycophyte plants using overexpression of NHX was reported by many authors. The partial account of these efforts has been reviewed in earlier paragraphs. Nevertheless, these efforts have not produced improved crop plants that can be used by farmers as salt-tolerant crops (Flowers, 2004; Shabala, 2013). This may be plausibly explained as follows. The NHX $Na^+/H^+$ exchanger is energized by vacuolar $H^+$ pumps. Hence, the constitutive activity of NHX as a result of its overexpression should be supported by increased activity of vacuolar $H^+$-ATPase and $H^+$-PPase. This will be jeopardized by two reasons. First, the ATP pool will be exhausted by the de novo synthesis of compatible solutes. Second, further ATP will also be consumed to maintain the depolarized membrane to help $K^+$ retention (Shabala and Cuin, 2008). For these two reasons, the activity of vacuolar $H^+$-ATPase will be hampered leaving the possibility of energizing the $Na^+/H^+$ exchange process to the $H^+$-PPases. Nonetheless, the activity of $H^+$-PPases is dependent upon the availability of $K^+$ in the cytosol (Rea and Poole, 1993), and the $K^+$ pool in the cytosol is substantially reduced under saline condition. This $K^+$ loss is resulted from depolarization activated (Chen et al., 2007b; Shabala and Cuin, 2008), and ROS induced $K^+$ efflux channels located in the plasma membrane (Cuin and Shabala, 2007; Demidchik et al., 2010). In order for the transgenic plants overexpressing NHX to function properly and provide a higher output, most of the regulation and other factors involved, discussed later, has to be met by the plant.

Furthermore, there is a protein–protein interaction regulation via the interaction of CBL10 with NHX to extrude sodium to the vacuole (Apse et al., 1999; Weinl and Kudla, 2009). In addition, further insight on the regulation of NHX exists, that is, the activity of the NHX protein is regulated by binding calmodulin-like protein (CaM15) at the C-terminus. NHX-calmodulin interaction reduces the $V_{max}$ of the $Na^+/H^+$ exchange activity but not the $K^+/H^+$ exchange activity (Yamaguchi et al., 2005). This not only regulates the accumulation of $Na^+$ but also affects the cytoplasmic $K^+/Na^+$ ratio. Calmodulin itself needs $Ca^{2+}$ to be activated and interact with other proteins. This regulation happens in the vacuolar lumen ruling out that this calcium signaling has a direct link with salinity. CIPK 24 (SOS2) has also been attributed to the regulation of NHX, in which it interacts with CBL10 and phosphorylates the NHX C-terminus, although direct evidence for this is yet to be reported. In this line, Yamaguchi et al.(2005) reported that the NHX1 protein regulation is performed by calmodulin-like protein at the C-terminus in a $Ca^{2+}$ and pH-dependent manner, and this regulatory protein decreases the $V_{max}$ of the $Na^+/H^+$ exchange but not for $K^+/H^+$ exchange. Overall, different calcium signals are involved in either up- or downregulation of NHX (Maathuis, 2014).

### 3.3.4 Selectivity of NHX Proteins and Impact of Their Activity of $K^+$ Homeostasis

Previous reports have indicated that NHX1 protein tonoplast $Na^+/H^+$ exchange causes $Na^+$ to be sequestered in the vacuole, thereby rendering salt tolerance (Apse et al., 1999; Gaxiola et al., 1999; Quintero et al., 2000). Nevertheless, despite the presence of many reports explaining the enhancement for salt tolerance by overexpressing the NHX1 protein, the mechanism behind this transporter toward salt tolerance is yet to be unravelled (Tester and Devenport, 2003; Pardo et al., 2006). Many have failed to show direct correlation between salt tolerance and $Na^+$ accumulation by NHX protein (Ohta et al., 2002; Fukuda et al., 2004b; Wu et al., 2009). On the contrary, some reported greater $K^+$ content rather than $Na^+$ content in the tissue of the transgenic plant expressing NHX protein (Xue et al., 2004; Wu et al., 2005; Zhao et al., 2006; Rodriguez-Rosales et al., 2008). Biochemical studies showed that AtNHX1, a class-1 protein, catalyzes the exchange of both $Na^+$ and $K^+$ as antiport for $H^+$ with similar affinity (Venema et al., 2002; Apse et al., 2003; Yamaguchi et al., 2005). In a similar study, it was shown that the Km of AtNHX1 protein was estimated to be 22 ± 12.8 mM for $Na^+$ and 20.9 ± 14.7 mM for $K^+$. $V_{max}$ was also similar for both cations (116 ± 20 for $Na^+$ and

96.5 ± 19.9 for $K^+$, measured as the relative change in fluorescence [$\Delta F/F_{max}$] $mg^{-1}$ $min^{-1}$) indicating that the AtNHX1 protein has similar selectivity for both ions, strengthening the report cited earlier that under normal growth conditions, AtNHX1 mediates the uptake of $K^+$ into vacuoles, whereas, under salinity stress, $Na^+$ is also transported due to the rising concentration of this ion (Leidi et al., 2010). Salt-induced repression of HAK5 has been reported (Nieves-Cordoneset al., 2008) raising an argument that the $K^+$ supply will be hampered by the low $K^+$ influx into the cytosol affecting salt tolerance thereby creating another node (HAK) to be considered for salt tolerance. Sodium is transported across the tonoplast in a bidirectional manner (Demidchik et al., 2002; Tester and Devenport, 2003). For sustained $Na^+$ compartmentation, the $Na^+/H^+$ has to exchange the influx of $Na^+$ continuously, which is challenged by the backflow of the ion from vacuole to the cytosol through ion channels (Demidchik et al., 2002; Bonales-Alatorre et al., 2013b). Even if the $K^+/Na^+$ ratio in the cytosol is high and the NHX proteins do not discriminate between the two ions, they still can contribute to the compartmentation of $Na^+$ provided that the back leak from the vacuole to the cytosol is low (Leidi et al., 2010). When the LeNHX2 was overexpressed in A. thaliana plants, the transgenic plants overexpressing this gene grew equally well in normal growth conditions as compared to the control plants; however, growth was inhibited under $K^+$ limiting growth conditions (Rodriguez-Rosales et al., 2008). It is long reported that in the case of limiting $K^+$ supply, the vacuolar $K^+$ concentration tends to deplete in favor of cytoplasmic $K^+$ concentration (Walker et al., 1996). It is, therefore, that the increased accumulation of $K^+$ in the internal compartment of LeNHX2-RGS(H)10 overexpressing plants causes growth inhibition in such conditions, as it negatively affects the cytoplasmic $K^+$ pool thereby causing the observed growth inhibition (Rodriguez-Rosales et al., 2008). The knowledge of $K^+$ transport process at the vacuole is much more limited compared to the plasma membrane $K^+$ transport. Eventually, the $K^+$ transport mechanisms and the proteins involved therein at the vacuole are starting to be unravelled. The tonoplast is energized by two types of proton pumps, namely, V-ATPase that uses ATP as energy source and V-PPase that uses inorganic pyrophosphate

to generate energy. In the majority of species and cell types, these proton pumps generate pH gradient of 1–2 units (acidic inside) and electric charge (membrane potential) of 20–40 mV causing a more positive membrane potential in the vacuole lumen than the cytosol. Thus, $K^+$ is excluded from the vacuole in $K^+$-rich cells unless the transport is coupled with energy; however, the backflow can be effected by $K^+$-permeable ion channels as it is a downhill. The fact that when the cytoplasmic $K^+$ is decreasing, the slow-activating vacuolar (SV) channels release $K^+$ from the vacuole to the cytoplasm to keep the homeostasis of the ion by partitioning it between the vacuole and the cytoplasm (Isayenkov et al., 2010; Hedrich and Marten, 2011).

## 3.4 FUELLING NHX ACTIVITY

### 3.4.1 Essentiality of $H^+$-Pumping for the Sequestration Process

The maintenance of proton gradient between cytoplasm (pH ≈ 7.2) and vacuole (pH ≈ 5.5) is the main driving force for the NHX-mediated $Na^+$ transport into the vacuole (Apse and Blumwald, 2007). As NHX functioning will increase the vacuolar pH by leaking $H^+$ back into the cytoplasm (Pittman, 2012), there is a need for $H^+$-translocating transporters in the tonoplast membrane to pump $H^+$ back into the vacuole. The combined action of vacuolar $H^+$-translocating enzymes, namely, V-ATPase (EC 3.6.1.3) and V-PPase (EC 3.6.1.1), has been suggested to perform this function (Gaxiola et al., 2007). Indeed, increased pH gradient generated by both V-ATPases and V-PPases is responsible for increased NHX activity in salt-treated potato cell cultures (Queirós et al., 2009).

The V-ATPase and V-PPase are redundant in pumping $H^+$ into the lumen of the vacuole. Hence, questions arise about what are the specific roles and relative contribution of each of this vacuolar $H^+$-pumps in the vacuolar acidification. Double knockouts of vacuolar specific Arabidopsis V-ATPases (VHA-a2 and VHA-a3) increased the vacuolar pH from 0.5 to 6.4 (Krebs et al., 2010), whereas V-PPase (AVP1) knockout mutant increased the vacuolar pH by only 0.2–0.3 pH units (Li et al., 2005; Ferjani et al., 2011). Thus, it appears that V-ATPase plays the major role in comparison with V-PPase, at least under control conditions. Under stressed conditions, the demand for ATP increases along with low ATP turnover rate (discussed in other sections in this chapter), which may make the V-PPase the major player for vacuolar pH homeostasis maintenance. The aforementioned notion is possible, because *Arabidopsis* V-PPase (*AVP1*) has been shown to recover vacuolar acidification in a yeast V-ATPase (*VMA1*) knockout mutant (Pérez-Castiñeira et al., 2011).

### 3.4.2 Regulation of Vacuolar H[+]-ATPases under Saline Conditions

The V-ATPases are large (nanometer scale), abundant, multisubunit proteins highly conserved throughout all life Kingdoms (Sze et al., 2002 Schumacher and Krebs, 2010). These H[+] pumps have two subcomplexes, namely, a peripheral $V_1$ complex with eight subunits (VHA-A to -H) performing ATP hydrolysis and a membrane-integral $V_0$ complex with six subunits (VHA-a, -c, -c′, -c″, -d, and -e) performing the translocation of H[+] from the cytosol into the lumen of endomembrane compartments (Schumacher and Krebs, 2010). The total number of genes encoding VHA can vary from 14 in *Chlamydomonas* up to 54 in soybean (Schumacher and Krebs, 2010). Among these members, only VHA-a2 and VHA-a3 have been shown to be vacuolar specific, while others are located in the *trans*-Golgi network (TGN; e.g., AHA-a1), endoplasmic reticulum, and secretory vesicles (Sze et al., 2002; Schumacher and Krebs, 2010). The V-ATPase plays a central role in plant growth and development. For example, decreased V-ATPases have been shown to cause gametophytic and/or embryolethality (Dettmer et al., 2005) and growth defects (Brüx et al., 2008).

Enhanced V-ATPase activity is believed to drive Na[+] sequestration into the vacuole because (1) salt stress increases both the V-ATPase transcripts and protein levels, as well as pump activity, both in glycophyte and halophyte species (Dietz et al., 2001; Wang et al., 2001; Schumacher and Krebs, 2010; Zhou et al., 2011), (2) *Arabidopsis* mutants such as *de-etiolated 3* (*det3*) and *vha-c3* that have decreased V-ATPase activity show increased salt sensitivity (Padmanaban et al., 2004; Batelli et al., 2007), (3) overexpression of V-ATPase subunit c1 from a halophyte grass *Spartina alterniflora* (SaVHAc1) increased salt tolerance of rice (Baisakh et al., 2012), and (4) inhibition of the V-ATPase activity abolished the NHX transport in isolated vacuoles (Zhou et al., 2011). By contrast, salt tolerance and Na[+] accumulation was not affected in an *Arabidopsis vha-a2 vha-a3* double mutant lacking two tonoplast-specific isoforms of the membrane-integral V-ATPase subunit VHA-a. Moreover, VHA-a2- and VHA-a3-mediated V-ATPase activity found to be essential for efficient nutrient (e.g., Zn) storage rather than Na[+] sequestration (Krebs et al., 2010). Endosomal V-ATPase mode of Na[+] detoxification has been proposed (Schumacher and Krebs, 2010) to explain the aforementioned controversy because (1) RNAi-lines of TGN-localized VHA-a1 were found to be salt sensitive (Krebs et al., 2010); (2) inhibition of *VHA-a1* alone restricts cell expansion, suggesting TGN eventually merges with vacuoles to effect cell expansion (Brüx et al., 2008); and (3) colocalization of NHX5 and NHX6 with *VHA-a1* in TGN along with TGN's involvement in vesicular trafficking suggests a possible functional relationship between the NHXs and endosomal V-ATPase in Na[+] sequestration inside the vacuole (Bassil et al., 2011).

The V-ATPase is a key player that consumes the bulk of the ATP produced in the cell, to regulate the transport of cytosolic and vacuolar pools of major metabolites and ions in accordance with the metabolic demands of salt-stressed plants (Barkla et al., 2009). The ATP demand for salt-stressed plants is usually higher than the unstressed plants to survive and maintain growth (Munns and Termaat, 1986). In reality, the available ATP decreases about 40%–50% in comparison with control plants within 3 days of salt stress (Bose et al., 2014). Hence, increased activity of V-ATPase is probably a transient response and is least likely to contribute for Na[+] sequestration into the vacuole during long-term salt exposure.

### 3.4.3 Regulation of Vacuolar H[+]-PPases under Saline Conditions

The PPases are single subunit proteins containing a homodimer of a single polypeptide. Unlike V-ATPase, the PPases are found only in plants and their algal antecedents and a few species of phototrophic bacteria (Drozdowicz and Rea, 2001). Plant V-PPase are highly conserved, sharing at least 85% of amino acid sequences (Zhen et al., 1997; Drozdowicz and Rea, 2001). Two homologues of V-PPases are found in plants and are classified as type I and II. The type I V-PPases (AVP1) are strongly dependent on cytosolic K[+] for their activity and moderately sensitive to inhibition by Ca[2+], whereas the type II V-PPases (AVP2) are K[+] insensitive but extremely

Ca$^{2+}$ sensitive (Drozdowicz and Rea, 2001; Gaxiola et al., 2012). Furthermore, the type I V-PPases are localized at the tonoplast (Maeshima, 2000), while the type II V-PPases are less abundant (~0.2% of the type I V-PPase) and found mainly in the Golgi apparatus and the *trans*-Golgi network (Segami et al., 2010). Vacuolar PPase has been shown to participate in postgermination growth (Ferjani et al., 2011) and auxin-mediated organ development (Li et al., 2005).

About 200 biosynthetic reactions have been known to release pyrophosphate (PPi) as a by-product, and the concentration of PPi is usually in the mM range (Heinonen, 2001). The V-PPases use this PPi instead of ATP to drive H$^+$ transport across the endomembranes. Thus, V-PPase has been suggested as the predominant proton pump in the vacuoles of growing cells (Nakanishi and Maeshima, 1998) and may function as a backup system for the V-ATPase under ATP limiting conditions (Maeshima, 2000). Salt stress has been shown to decrease the available ATP in barley roots (Bose et al., 2014), and pharmacological evidence suggested that the plant might switch to using V-PPase instead of V-ATPase to sequester Na$^+$ into the vacuole. This notion is well supported by the following observations in barley roots: (1) V-PPase activity found to be stimulated rapidly (within 1 h of salt stress) in the salt-tolerant cultivar and remained higher than control plants up to 1 day after salt stress (Fukuda et al., 2004a); (2) ATP utilization exceeded the ATP synthesizing capacity in the sensitive cultivar, but the two processes were well balanced in the tolerant cultivar (Fan et al., 1989); (3) expression of AVP1 and NHX was synchronized in the tolerant cultivar (Fukuda et al., 2004a); and (4) fluctuations in vacuolar pH was accompanied by the Na$^+$ influx into the vacuole (Fan et al., 1989).

Contribution of the V-PPase activity to the Na$^+$ sequestration through NHX has been demonstrated in two ways. First, an *Arabidopsis* mutant solely relies on V-PPase for vacuolar H$^+$ pumping found to sequester Na$^+$ (Krebs et al., 2010). Second, coexpression of *Suaeda salsa* (a halophyte), Na$^+$/H$^+$ antiporter (SsNHX1), and *Arabidopsis* AVP1 improved salt tolerance in rice than the expression of SsNHX1 alone (Zhao et al., 2006). In addition, numerous examples can be derived from the literature that overexpression of type I V-PPase can improve the salt tolerance of diverse plant species, including *Arabidopsis*, rice, barley, maize, alfalfa, creeping bent grass, peanut, cotton, and tobacco (Gaxiola et al.,

2012; Schilling et al., 2014). Glasshouse and field trails also proved that transgenic crops overexpressing V-PPase can produce greater biomass and yield than respective wild types under salt stress (Gaxiola et al., 2012; Schilling et al., 2014). Increased root and shoot growth in V-PPase overexpressing plants is attributed to the increased cell numbers (Li et al., 2005; Gonzalez et al., 2010), suggesting these plants will have greater tonoplast membrane surface-to-volume ratio than wild type; this trait is considered essential for efficient Na$^+$ transport and sequestration into the vacuole. Interestingly, transgenic plants overexpressing V-PPase were also found to be drought tolerant, and this drought tolerance was achieved by the increased osmolyte transport into the vacuole (Gaxiola et al., 2012). Since salt stress also imposes osmotic stress (Munns and Tester, 2008), V-PPase overexpressing plants will transport and sequester Na$^+$ into the vacuole in order to use it as a cheap osmoticum. This notion has to be confirmed through additional experiments involving Na$^+$-selective fluorescent dyes.

## 3.5 RETAINING Na$^+$ IN THE VACUOLE

### 3.5.1 Pathways of Na$^+$ Back Leak into the Cytosol

While efficient Na$^+$ loading into the vacuole mediated by NHX exchangers is essential for intracellular Na$^+$ sequestration, by itself it will be not sufficient to confer tissue tolerance in plants. Indeed, the reported values for tonoplast potential are typically within −20 to +20 mV range (Bonales-Alatorre et al., 2013a,b), while the concentration Na$^+$ gradient between the vacuole and the cytosol is at least four- to fivefold (Shabala and Mackay, 2011). If not properly controlled, vacuolar Na$^+$ may simply back leak into the cytosol under such conditions. Thus, the passive tonoplast Na$^+$ conductance should be reduced to avoid energy-consuming futile Na$^+$ cycling between the cytosol and vacuole. Electrophysiological and molecular experiments have identified two major types of tonoplast channels that may mediate Na$^+$ flux from vacuole into cytosol, namely, SV and FV channels. Both are nonselective cation channels that are ubiquitously and abundantly expressed in higher plant vacuoles (Hedrich and Marten, 2011). Their molecular identity and control modes are discussed in the following sections.

### 3.5.2 Molecular Identity and Regulation of SV and FV Vacuolar Channels

The SV channel is the most abundant and best described as a vacuolar ion channel. It is permeable to both mono- and divalent cations and is activated by cytosolic $Ca^{2+}$ and positive vacuolar voltage (Hedrich and Neher, 1987; Ward and Schroeder, 1994; Pottosin et al., 1997, 2001). In *Arabidopsis*, the SV channel is encoded by TPC1 gene (Peiter et al., 2005) and represents a two-pore channel that belongs to a family of voltage-gated cation channels consisting of two homologous domains with six transmembrane helices and one pore domain each (Hedrich and Marten, 2011). The SV channel seems to be ubiquitous among terrestrial plants (embryophytes), including ferns and liverworts (Hedrich et al., 1988). Only one member of the TPC family was identified so far in both *Arabidopsis* (Furuichi et al., 2001) and rice (Kurusu et al., 2004), indicating that the SV channel might be formed by a TPC1 homodimer. An *Arabidopsis* knockout mutant lacking TPC1 (*tpc1–2*) does not show any SV channel activity, and TPC1-overexpressing lines have increased SV channel activity (Peiter et al., 2005).

Two major factors activate SV channel: positive shifts in tonoplast potentials and elevated cytosolic $Ca^{2+}$ (Hedrich and Neher, 1987). The open probability of the SV channel is increased by some reducing agents such as dithiothreitol or glutathione (Carpaneto et al., 1999; Scholz-Starke et al., 2004), and SV currents are decreased by 14–3–3 proteins (van den Wijngaard et al., 2001). SV currents are also strongly blocked by physiologically relevant concentrations of polyamines (Dobrovinskaya et al., 1999), resulting from the direct block of the channel's pore. SV channel activity is also decreased under low vacuolar pH conditions and is also affected by vacuolar $Ca^{2+}$ and $Mg^{2+}$ (Schulzlessdorf and Hedrich, 1995; Pottosin et al., 1997, 2004). SV channel voltage dependence is shifted toward more positive potentials by the presence of high amounts of $Na^+$ in the vacuole (Ivashikina and Hedrich, 2005).

In a stark contrast to SV channels, the molecular identity of FV channels remains elusive. At the physiological level, FV channel is permeable for monovalent cations only (Pottosin and Dobrovinskaya, 2014). FV vacuolar currents carried by the FV channel dominate the tonoplast electrical characteristics at physiological cytosolic free $Ca^{2+}$ concentrations ($<1$ μM) (Hedrich and Neher, 1987) and are inhibited by divalent cations from either side of the membrane (Tikhonova et al., 1997; Bruggemann et al., 1999). FV channels only weakly differentiate between $K^+$ and $Na^+$, both in glycophytes (Maathuis and Prins, 1990) and halophyte (Bonales-Alatorre et al., 2013a) species.

### 3.5.3 Role of SV and FV Channels in Salinity Tolerance in Plants

Over the last 20 years, only a handful of papers dealt with the issue of "locking" $Na^+$ in plant vacuoles, and the conclusions from these studies were rather controversial. Ivashikina and Hedrich (2005) found that an increase in luminal $Na^+$/$K^+$ ratio, mimicking the accumulation of $Na^+$ in vacuoles during salt stress, shifted the threshold for SV activation to positive potentials, so reducing SV channel open probability under saline conditions. At the same time, Maathuis et al. (1992) reported significant SV channel activity in mesophyll cell vacuoles of high (200 mM)-salt-grown plants of a halophyte species of *Suaeda maritima*. The same group has compared properties of SV channels from roots of *Plantago media* (salt sensitive) and *Plantago maritima* (salt tolerant). While SV channel activity diminished in salt-grown plants, no difference in any of the studied SV channel characteristics (voltage gating, unitary conductance, and $Na^+$/$K^+$ selectivity) was found (Maathuis and Prins, 1990).

Recently, a comprehensive assessment of the role of tonoplast SV and FV channels in vacuolar $Na^+$ sequestration was undertaken in our laboratory using quinoa (*Chenopodium quinoa*) species. Quinoa is a facultative halophyte of a high nutritional value and agricultural importance (Jacobsen et al., 2003; Adolf et al., 2012) in which optimal growth is observed at external salinities of around 150 mM (Hariadi et al., 2011; Shabala, 2013). Being a halophyte species, quinoa sequesters significant amounts of $Na^+$ in mesophyll cell vacuoles; the extent of sequestration ability, however, depends strongly on the leaf position and the presence of salt bladders on the leaf surface. While young leaves rely heavily on $Na^+$ exclusion to salt bladders, old ones possessing far fewer salt bladders depend almost exclusively on $Na^+$ sequestration to mesophyll vacuoles (Bonales-Alatorre et al., 2013a). It was found that although old leaves accumulate more $Na^+$, this does not compromise their leaf photochemistry, as old leaves had an intrinsically lower density of the FV current compared with young ones. Moreover, FV channel density decreases about twofold in plants grown under high salinity. Also, the intrinsic activity of SV channels in old leaves of salt-grown plants was sevenfold lower compared with young ones (Bonales-Alatorre et al., 2013a). It was concluded that the negative control of SV and FV tonoplast channels activity in old leaves reduces $Na^+$ leak thus enabling efficient sequestration of $Na^+$ to their vacuoles, protecting the leaf photochemistry and conferring to salinity tolerance in this species.

A further support for this conclusion was obtained by studying the properties of tonoplast FV and SV channels in two quinoa genotypes contrasting in their salinity tolerance (Bonales-Alatorre et al., 2013b). The FV conductance at physiological (near zero) tonoplast potentials in the salt-sensitive Q5206 genotype was three- to fivefold higher than the salt-tolerant Q16 and were further reduced in salt-grown plants. Salinity also caused a significant shift in the voltage dependence of SV channels, resulting in a 10-fold decrease of mean activity of SV channels at physiologically attainable transtonoplast potentials (Bonales-Alatorre et al., 2013b).

Another interesting aspect of that study was related to the fact that in a tolerant genotype, the tonoplast channel activity is intrinsically lower, while in sensitive genotypes, this reduction is stress inducible (Bonales-Alatorre et al., 2013b). This mirrors the expression of tonoplast NHX $Na^+/H^+$ antiporters (constitutive in halophytes [Matoh et al., 1989; Yokoi et al., 2002], inducible in glycophytes [Garbarino and Dupont, 1988; Adler et al., 2010]). It remains to be answered as to whether these changes in SV and FV activity originate from changes at the transcript level (expression patterns) of these transporters under saline conditions, posttranslational regulation, or the existence of alternative splicing forms or whether the observed difference is attributed to channel regulation by some cytosolic or luminal compound(s).

## 3.6 TARGETING VACUOLAR SEQUESTRATION IN BREEDING PROGRAMS

Plant breeders have been trying to develop stress-tolerant crops using genetic variation in crops, at intraspecific, interspecific, and intergeneric levels. Two main approaches were employed:

traditional breeding and transgenic approach. Traditional breeding relies largely on the natural intraspecific genetic variations. When introducing a gene from different species, there is not much success due to reproductive barrier and/or the risk of other undesirable traits transferred with the target traits. One of the examples is the introduction of BYDV resistance to wheat from *Thinopyrum intermedium* in which nearly 20% of the distal portion of 7DL was replaced by a chromosome fragment from *T. intermedium* (Banks et al., 1995). The replacement of a segment transferred many other genes to wheat at the same time (Ayala-Navarrete et al., 2013). To avoid this problem, genetic engineering strategy is more preferred, as it only deals with the specific gene(s) transferred. Many abiotic stress tolerance genes have been successfully expressed in intergeneric species. These included expressing barley MATE gene in wheat (Zhou et al., 2013), sorghum *SbMATE* gene in barley (Zhou et al., 2014), and *Arabidopsis AtNHX1* gene in tomato (Zhang and Blumwald, 2001). Without any doubt, transgenic technology will continue to aid the search for the cellular mechanisms that underlie tolerance. However, the public acceptance of transgenic plants may face consumer backlash, which limits the use of transgenic approaches in improving abiotic stress tolerance. The effectiveness of transgenic strategy can also be greatly reduced when targeting salt tolerance, which is controlled by many different mechanisms (Munns and Tester, 2008; Shabala, 2013). Thus it is not surprising that some transgenic lines with a specific mechanism did not show much better salt tolerance.

As shown previously, vacuolar $Na^+$ sequestration is one of the key mechanisms for plants to tolerate salt stress and, hence, should be a target of breeding programs (Shabala, 2013). Significant (e.g., 10- to 20-fold) differences in this trait were found among varieties of the same species (e.g., wheat [Wu et al., 2014]). This makes it possible to breed for $Na^+$ sequestration through the traditional breeding approach. However, the direct selection of this trait in a breeding program is not practical due to the complexity of the screening process and the limitation of the amount of genotypes to be screened. The development of molecular biology techniques enabled the development of molecular markers that link to the specific gene(s) controlling this trait. QTL mapping of the trait is the first step toward marker-assisted selection. QTL are

DNA segments in the genome of an organism linked with a particular trait. QTL mapping of the trait not only will provide plant breeders with a better understanding of the basis for the genetic correlation between different traits but also has potential to facilitate a more efficient incremental improvement of the trait. The use of QTL has improved the efficiency of selection, in particular for those traits that are controlled by several genes and are highly influenced by environmental factors (Flowers, 2004) and those traits with a very complicated screening process such as $Na^+$ sequestration. Once DNA markers have been identified closely linked to the QTL involved in the expression of $Na^+$ sequestration, they can be efficiently used in future marker-assisted selection.

When breeding for $Na^+$ sequestration, many undesirable or unwanted genes will be inevitably transferred during the breeding process due to the close linkage between the gene for $Na^+$ sequestration and the gene(s) for other unwanted traits. Two approaches can help to avoid this: (1) fine mapping of the QTL to find a marker cosegregating with the gene or even gene-specific markers and (2) more backcrosses and selecting from larger populations using closely linked or gene-specific markers for $Na^+$ sequestration and those for other unwanted genes. Gene-specific markers based on single-nucleotide polymorphisms or small indels (insertion/deletions) are direct markers because sequence information provides the exact nature of the allelic variants. They are far more prevalent than other markers and may provide a high density of markers near a locus of interest (Bian et al., 2013). The gene-specific markers can also be used to detect the genes from different populations.

Since salt tolerance was controlled by many different mechanisms, we should not expect miraculous consequences from just transferring the NHX gene into a variety. To make transgenic lines overexpressing $Na^+/H^+$ NHX exchangers fully functional under saline stress, several other conditions should be met (Shabala, 2013): (1) $Na^+$ back leak into the vacuole should be prevented by effective control of SV and FV channels conductance; (2) plants should not heavily invest into the production of compatible solutes, so leaving a greater part of the ATP pool available for fueling tonoplast $H^+$-ATPase; and (3) plants should possess mechanisms for efficient $K^+$ retention to enable tonoplast $H^+$-PPases to function. Thus, the progress in breeding for efficient $Na^+$ sequestration may be achieved only by pyramiding several set of genes conferring the aforementioned traits.

## 3.7 CONCLUSIONS

Both glycophytes and halophytes need to keep an optimal $Na^+/K^+$ ratio in a metabolically active cytosolic compartment. This may be achieved via effective compartmentalization of $Na^+$ in the vacuole, a mechanism potentially existing in all species but that is dominant in the halophytes, naturally salt-loving plants. Effective $Na^+$ sequestration, however, cannot be achieved by merely overexpressing the amount of NHX protein. Several other conditions have to be met including ensuring that NHX exchangers are properly energized and that futile $Na^+$ cycling between the vacuole and the cytosol is prevented by effective control of tonoplast $Na^+$ conductance. As far as we know, this aspect has never been considered or done in conjunction with any attempt to improve salinity tolerance via the overexpression of tonoplast $Na^+/H^+$ NHX exchangers in crops. This may explain the slow progress and the lack of truly tolerant cultivars in farm fields despite over 15 years of genetic manipulation with the NHX gene.

## REFERENCES

Adler, G., Blumwald, E., Bar-Zvi, D. 2010. The sugar beet gene encoding the sodium/proton exchanger 1 (BvNHX1) is regulated by a MYB transcription factor. *Planta*, 232: 187–195.

Adolf, V. I., Shabala, S., Andersen, M. N., Razzaghi, F., Jacobsen, S. E. 2012. Varietal differences of quinoa's tolerance to saline conditions. *Plant and Soil*, 357: 117–129.

Aharon, G. S., Apse, M. P., Duan, S. L., Hua, X. J., Blumwald, E. 2003. Characterization of a family of vacuolar $Na^+/H^+$ antiporters in *Arabidopsis thaliana*. *Plant and Soil*, 253: 245–256.

Apse, M. P., Aharon, G. S., Snedden, W. A., Blumwald, E. 1999. Salt tolerance conferred by overexpression of a vacuolar $Na^+/H^+$ antiport in *Arabidopsis*. *Science*, 285: 1256–1258.

Apse, M. P., Blumwald, E. 2007. $Na^+$ transport in plants. *FEBS Letters*, 581: 2247–2254.

Apse, M. P., Sottosanto, J. B., Blumwald, E. 2003. Vacuolar cation/$H^+$ exchange, ion homeostasis, and leaf development are altered in a T-DNA insertional mutant of *AtNHX1*, the *Arabidopsis* vacuolar $Na^+/H^+$ antiporter. *Plant Journal*, 36: 229–239.

Ayala-Navarrete, L. I., Mechanicos, A. A., Gibson, J. M., Singh, D., Bariana, H. S., Fletcher, J., Shorter, S., Larkin, P. J. 2013. The Pontin series of recombinant alien translocations in bread wheat: Single translocations integrating combinations of Bdv2, Lr19 and Sr25 disease-resistance genes from *Thinopyrum intermedium* and *Th. ponticum*. *Theoretical and Applied Genetics*, 126: 2467–2475.

Baisakh, N., RamanaRao, M. V., Rajasekaran, K., Subudhi, P., Janda, J., Galbraith, D., Vanier, C., Pereira, A. 2012. Enhanced salt stress tolerance of rice plants expressing a vacuolar $H^+$-ATPase subunit c1 (SaVHAc1) gene from the halophyte grass *Spartina alterniflora* Löisel. *Plant Biotechnology Journal*, 10: 453–464.

Banks, P. M., Larkin, P. J., Bariana, H. S., Lagudah, E. S., Appels, R., Waterhouse, P. M., Brettell, R. I. S. et al. 1995. The use of cell-culture for subchromosomal introgressions of barley yellow dwarf virus resistance from *Thinopyrum intermedium* to wheat. *Genome*, 38: 395–405.

Barkla, B. J., Vera-Estrella, R., Hernández-Coronado, M., Pantoja, O. 2009. Quantitative proteomics of the tonoplast reveals a role for glycolytic enzymes in salt tolerance. *The Plant Cell Online*, 21: 4044–4058.

Bassil, E., Ohto, M.-a., Esumi, T., Tajima, H., Zhu, Z., Cagnac, O., Belmonte, M., Peleg, Z., Yamaguchi, T., Blumwald, E. 2011. The *Arabidopsis* intracellular $Na^+/H^+$ antiporters NHX5 and NHX6 are endosome associated and necessary for plant growth and development. *The Plant Cell Online*, 23: 224–239.

Batelli, G., Verslues, P. E., Agius, F., Qiu, Q., Fujii, H., Pan, S., Schumaker, K. S., Grillo, S., Zhu, J.-K. 2007. SOS2 promotes salt tolerance in part by interacting with the vacuolar $H^+$-ATPase and upregulating its transport activity. *Molecular and Cellular Biology*, 27: 7781–7790.

Bayat, F., Shiran, B., Belyaev, D. V. 2011. Overexpression of HvNHX2, a vacuolar $Na^+/H^+$ antiporter gene from barley, improves salt tolerance in *Arabidopsis thaliana*. *Australian Journal of Crop Science*, 5: 428–432.

Bian, M., Waters, I., Broughton, S., Zhang, X. Q., Zhou, M. X., Lance, R., Sun, D. F., Li, C. D. 2013. Development of gene-specific markers for acid soil/aluminium tolerance in barley (*Hordeum vulgare* L.). *Molecular Breeding*, 32: 155–164.

Blumwald, E. 2000. Sodium transport and salt tolerance in plants. *Current Opinion in Cell Biology*, 12: 431–434.

Blumwald, E., Aharon, G. S., Apse, M. P. 2000. Sodium transport in plant cells. *Biochimica Et Biophysica Acta-Biomembranes*, 1465: 140–151.

Bonales-Alatorre, E., Pottosin, I., Shabala, L., Chen, Z.-H., Zeng, F., Jacobsen, S.-E., Shabala, S. 2013a. Differential activity of plasma and vacuolar membrane transporters contributes to genotypic differences in salinity tolerance in halophyte species, *Chenopodium quinoa*. *International Journal of Molecular Sciences*, 14: 9267–9285.

Bonales-Alatorre, E., Shabala, S., Chen, Z.-H., Pottosin, I. 2013b. Reduced tonoplast fast-activating and slow-activating channel activity is essential for conferring salinity tolerance in a facultative halophyte, Quinoa. *Plant Physiology*, 162: 1–13.

Bose, J., Shabala, L., Pottosin, I., Zeng, F., Velarde-BuendíA, A.-M., Massart, A., Poschenrieder, C., Hariadi, Y., Shabala, S. 2014. Kinetics of xylem loading, membrane potential maintenance, and sensitivity of $K^+$-permeable channels to reactive oxygen species: Physiological traits that differentiate salinity tolerance between pea and barley. *Plant, Cell and Environment*, 37: 589–600.

Bruggemann, L. I., Pottosin, I I, Schonknecht, G. 1999. Selectivity of the fast activating vacuolar cation channel. *Journal of Experimental Botany*, 50: 873–876.

Brüx, A., Liu, T.-Y., Krebs, M., Stierhof, Y.-D., Lohmann, J. U., Miersch, O., Wasternack, C., Schumacher, K. 2008. Reduced V-ATPase activity in the *trans*-Golgi network causes oxylipin-dependent hypocotyl growth inhibition in *Arabidopsis*. *The Plant Cell Online*, 20: 1088–1100.

Carpaneto, A., Cantu, A. M., Gambale, F. 1999. Redox agents regulate ion channel activity in vacuoles from higher plant cells. *FEBS Letters*, 442: 129–132.

Chen, Z., Zhou, M., Newman, I. A., Mendham, N. J., Zhang, G., Shabala, S. 2007a. Potassium and sodium relations in salinised barley tissues as a basis of differential salt tolerance. *Functional Plant Biology*, 34: 150–162.

Chen, Z. H., Pottosin, I I, Cuin, T. A., Fuglsang, A. T., Tester, M., Jha, D., Zepeda-Jazo, I. et al. 2007b. Root plasma membrane transporters controlling $K^+/Na^+$ homeostasis in salt-stressed barley. *Plant Physiology*, 145: 1714–1725.

Colmer, T. D., Munns, R., Flowers, T. J. 2005. Improving salt tolerance of wheat and barley: Future prospects. *Australian Journal of Experimental Agriculture*, 45: 1425–1443.

Cuin, T. A., Bose, J., Stefano, G., Jha, D., Tester, M., Mancuso, S., Shabala, S. 2011. Assessing the role of root plasma membrane and tonoplast $Na^+/H^+$ exchangers in salinity tolerance in wheat: In planta quantification methods. *Plant Cell and Environment*, 34: 947–961.

Cuin, T. A., Shabala, S. 2007. Compatible solutes reduce ROS-induced potassium efflux in *Arabidopsis* roots. *Plant Cell and Environment*, 30: 875–885.

Davenport, R. J., Reid, R. J., Smith, F. A. 1997. Sodium-calcium interactions in two wheat species differing in salinity tolerance. *Physiologia Plantarum*, 99: 323–327.

Demidchik, V., Cuin, T. A., Svistunenko, D., Smith, S. J., Miller, A. J., Shabala, S., Sokolik, A., Yurin, V. 2010. Arabidopsis root $K^+$-efflux conductance activated by hydroxyl radicals: Single-channel properties, genetic basis and involvement in stress-induced cell death. *Journal of Cell Science*, 123: 1468–1479.

Demidchik, V., Davenport, R. J., Tester, M. 2002. Nonselective cation channels in plants. *Annual Review of Plant Biology*, 53: 67–107.

Dettmer, J., Schubert, D., Calvo-Weimar, O., Stierhof, Y. D., Schmidt, R., Schumacher, K. 2005. Essential role of the V-ATPase in male gametophyte development. *The Plant Journal*, 41: 117–124.

Dietz, K. J., Tavakoli, N., Kluge, C., Mimura, T., Sharma, S. S., Harris, G. C., Chardonnens, A. N., Golldack, D. 2001. Significance of the V-type ATPase for the adaptation to stressful growth conditions and its regulation on the molecular and biochemical level. *Journal of Experimental Botany*, 52: 1969–1980.

Dobrovinskaya, O. R., Muniz, J., Pottosin, II. 1999. Inhibition of vacuolar ion channels by polyamines. *Journal of Membrane Biology*, 167: 127–140.

Drozdowicz, Y. M., Rea, P. A. 2001. Vacuolar $H^+$ pyrophosphatases: From the evolutionary backwaters into the mainstream. *Trends in Plant Science*, 6: 206–211.

Fan, T., Higashi, R. M., Norlyn, J., Epstein, E. 1989. *In vivo* $^{23}$Na and $^{31}$P NMR measurement of a tonoplast $Na^+/H^+$ exchange process and its characteristics in two barley cultivars. *Proceedings of the National Academy of Sciences of the United States of America*, 86: 9856–9860.

Ferjani, A., Segami, S., Horiguchi, G., Muto, Y., Maeshima, M., Tsukaya, H. 2011. Keep an eye on PPi: The vacuolar-type $H^+$-pyrophosphatase regulates postgerminative development in *Arabidopsis*. *The Plant Cell Online*, 23: 2895–2908.

Flowers, T. J. 2004. Improving crop salt tolerance. *Journal of Experimental Botany*, 55: 307–319.

Flowers, T. J., Colmer, T. D. 2008. Salinity tolerance in halophytes. *New Phytologist*, 179: 945–963.

Flowers, T. J., Troke, P. F., Yeo, A. R. 1977. Mechanism of salt tolerance in halophytes. *Annual Review of Plant Physiology and Plant Molecular Biology*, 28: 89–121.

Flowers, T. J., Yeo, A. R. 1995. Breeding for salinity resistance in crop plants: Where next? *Australian Journal of Plant Physiology*, 22: 875–884.

Fukuda, A., Chiba, K., Maeda, M., Nakamura, A., Maeshima, M., Tanaka, Y. 2004a. Effect of salt and osmotic stresses on the expression of genes for the vacuolar $H^+$-pyrophosphatase, $H^+$-ATPase subunit A, and $Na^+/H^+$ antiporter from barley. *Journal of Experimental Botany*, 55: 585–594.

Fukuda, A., Nakamura, A., Tagiri, A., Tanaka, H., Miyao, A., Hirochika, H., Tanaka, Y. 2004b. Function, intracellular localization and the importance in salt tolerance of a vacuolar $Na^+/H^+$ antiporter from rice. *Plant and Cell Physiology*, 45: 146–159.

Fukuda, A., Yazaki, Y., Ishikawa, T., Koike, S., Tanaka, Y. 1998. $Na^+/H^+$ antiporter in tonoplast vesicles from rice roots. *Plant and Cell Physiology*, 39: 196–201.

Furuichi, T., Cunningham, K. W., Muto, S. 2001. A putative two pore channel *AtTPC1* mediates $Ca^{2+}$ flux in *Arabidopsis* leaf cells. *Plant and Cell Physiology*, 42: 900–905.

Garbarino, J., Dupont, F. M. 1988. NaCl induces a $Na^+/H^+$ antiporter in tonoplast vesicles from barley roots. *Plant Physiology*, 86: 0231–0236.

Garbarino, J., Dupont, F. M. 1989. Rapid induction of $Na^+/H^+$ exchange activity in barley root tonoplast. *Plant Physiology*, 89: 1–4.

Gaxiola, R. A., Palmgren, M. G., Schumacher, K. 2007. Plant proton pumps. *FEBS Letters*, 581: 2204–2214.

Gaxiola, R. A., Rao, R., Sherman, A., Grisafi, P., Alper, S. L., Fink, G. R. 1999. The *Arabidopsis thaliana* proton transporters, AtNhx1 and Avp1, can function in cation detoxification in yeast. *Proceedings of the National Academy of Sciences of the United States of America* 96: 1480–1485.

Gaxiola, R. A., Sanchez, C. A., Paez-Valencia, J., Ayre, B. G., Elser, J. J. 2012. Genetic manipulation of a "vacuolar" $H^+$-PPase: From salt tolerance to yield enhancement under phosphorus-deficient soils. *Plant Physiology*, 159: 3–11.

Glenn, E. P., Brown, J. J., Blumwald, E. 1999. Salt tolerance and crop potential of halophytes. *Critical Reviews in Plant Sciences*, 18: 227–255.

Gonzalez, N., De Bodt, S., Sulpice, R., Jikumaru, Y., Chae, E., Dhondt, S., Van Daele, T., De Milde, L., Weigel, D., Kamiya, Y. 2010. Increased leaf size: Different means to an end. *Plant Physiology*, 153: 1261–1279.

Gorham, J., Jones, R. G. W., Bristol, A. 1990. Partial characterization of the trait for enhanced $K^+/Na^+$ discrimination in the D-genome of wheat. *Planta*, 180: 590–597.

Greenway, H., Munns, R. 1980. Mechanism of salt tolerance in nonhalophytes. *Annual Review of Plant Physiology*, 31: 149–190.

Hariadi, Y., Marandon, K., Tian, Y., Jacobsen, S. E., Shabala, S. 2011. Ionic and osmotic relations in quinoa (*Chenopodium quinoa* Willd.) plants grown at various salinity levels. *Journal of Experimental Botany*, 62: 185–193.

Hasegawa, P. M., Bressan, R. A., Zhu, J. K., Bohnert, H. J. 2000. Plant cellular and molecular responses to high salinity. *Annual Review of Plant Physiology and Plant Molecular Biology*, 51: 463–499.

Hedrich, R., Barbierbrygoo, H., Felle, H., Flugge, U. I., Luttge, U., Maathuis, F. J. M., Marx, S. et al. 1988. General mechanisms for solute transport across the tonoplast of plant vacuoles-A patch clamp survey of ion channels and proton pumps. *Botanica Acta*, 101: 7–13.

Hedrich, R., Marten, I. 2011. TPC1-SV channels gain shape. *Molecular Plant*, 4: 428–441.

Hedrich, R., Neher, E. 1987. Cytoplasmic calcium regulates voltage-dependent ion channels in plant vacuoles. *Nature*, 329: 833–836.

Heinonen, J. K. 2001. *Biological Role of Inorganic Pyrophosphate.* London, U.K.: Kluwer Academic Publishers.

Isayenkov, S., Isner, J. C., Maathuis, F. J. M. 2010. Vacuolar ion channels: Roles in plant nutrition and signalling. *FEBS Letters*, 584: 1982–1988.

Ivashikina, N., Hedrich, R. 2005. $K^+$ currents through SV-type vacuolar channels are sensitive to elevated luminal sodium levels. *Plant Journal*, 41: 606–614.

Jacobsen, S. E., Mujica, A., Jensen, C. R. 2003. The resistance of quinoa (*Chenopodium quinoa* Willd.) to adverse abiotic factors. *Food Reviews International*, 19: 99–109.

Jeschke, W. D., Stelter, W., Reising, B., Behl, R. 1983. Vacuolar $Na^+/K^+$ exchange, its occurrence in root-cells of *Hordeum*, *Atriplex* and *Zea* and its significance for K+/Na+ discrimination in roots. *Journal of Experimental Botany*, 34: 964–979.

Krebs, M., Beyhl, D., Görlich, E., Al-Rasheid, K. A. S., Marten, I., Stierhof, Y.-D., Hedrich, R., Schumacher, K. 2010. *Arabidopsis* V-ATPase activity at the tonoplast is required for efficient nutrient storage but not for sodium accumulation. *Proceedings of the National Academy of Sciences of the United States of America*, 107: 3251–3256.

Kurusu, T., Sakurai, Y., Miyao, A., Hirochika, H., Kuchitsu, K. 2004. Identification of a putative voltage-gated $Ca^{2+}$-permeable channel (OsTPC1) involved in $Ca^{2+}$ influx and regulation of growth and development in rice. *Plant and Cell Physiology*, 45: 693–702.

Leidi, O. E., Barragan, V., Rubio, L., El-Hamdaoui, A., Ruiz, T. M., Cubero, B., Fernandez, A. J. et al. 2010. The AtNHX1 exchanger mediates potassium compartmentation in vacuoles of transgenic tomato. *The Plant Journal*, 61: 495–506.

Li, J., Yang, H., Ann Peer, W., Richter, G., Blakeslee, J., Bandyopadhyay, A., Titapiwantakun, B. et al. 2005. *Arabidopsis* $H^+$-PPase *AVP1* regulates auxin-mediated organ development. *Science*, 310: 121–125.

Maathuis, F. J. M. 2014. Sodium in plants: Perception, signalling, and regulation of sodium fluxes. *Journal of Experimental Botany*, 65: 849–858.

Maathuis, F. J. M., Flowers, T. J., Yeo, A. R. 1992. Sodium-chloride compartmentation in leaf vacuoles of the halophyte *Suaeda maritima* (L) Dum and its relation to tonoplast permeability. *Journal of Experimental Botany*, 43: 1219–1223.

Maathuis, F. J. M., Prins, H. B. A. 1990. Patch clamp studies on root cell vacuoles of salt tolerant and salt sensitive *Plantago* species—Regulation of channel activity by salt stress. *Plant Physiology*, 92: 23–28.

Maeshima, M. 2000. Vacuolar $H^+$-pyrophosphatase. *Biochimica et Biophysica Acta (BBA)-Biomembranes*, 1465: 37–51.

Matoh, T., Ishikawa, T., Takahashi, E. 1989. Collapse of ATP-induced pH gradient by sodium ions in microsomal membrane vesicles prepared from *Atriplex gemlinii* leaves—Possibility of $Na^+/H^+$ antiport. *Plant Physiology*, 89: 180–183.

Moller, I. S., Gilliham, M., Jha, D., Mayo, G. M., Roy, S. J., Coates, J. C., Haseloff, J., Tester, M. 2009. Shoot $Na^+$ exclusion and increased salinity tolerance engineered by cell type-specific alteration of $Na^+$ transport in Arabidopsis. *Plant Cell*, 21: 2163–2178.

Munns, R. 2002. Comparative physiology of salt and water stress. *Plant, Cell & Environment*, 25: 239–250.

Munns, R., Termaat, A. 1986. Whole-plant responses to salinity. *Australian Journal of Plant Physiology*, 13: 143–160.

Munns, R., Tester, M. 2008. Mechanisms of salinity tolerance. *Annual Review of Plant Biology*, 59: 651–681.

Nakanishi, Y., Maeshima, M. 1998. Molecular cloning of vacuolar $H^+$-pyrophosphatase and its developmental expression in growing hypocotyl of mung bean. *Plant Physiology*, 116: 589–597.

Nass, R., Cunningham, K. W., Rao, R. 1997. Intracellular sequestration of sodium by a novel $Na^+/H^+$ exchanger in yeast is enhanced by mutations in the plasma membrane $H^+$-ATPase—Insights into mechanisms of sodium tolerance. *Journal of Biological Chemistry*, 272: 26145–26152.

Nass, R., Rao, R. 1998. Novel localization of a Na⁺/H⁺ exchanger in a late endosomal compartment of yeast—Implications for vacuole biogenesis. *Journal of Biological Chemistry*, 273: 21054–21060.

Nieves-Cordones, M., Miller, A. J., Aleman, F., Martinez, V., Rubio, F. 2008. A putative role for the plasma membrane potential in the control of the expression of the gene encoding the tomato high-affinity potassium transporter HAK5. *Plant Molecular Biology*, 68: 521–532.

Ohta, M., Hayashi, Y., Nakashima, A., Hamada, A., Tanaka, A., Nakamura, T., Hayakawa, T. 2002. Introduction of a Na⁺/H⁺ antiporter gene from *Atriplex gmelini* confers salt tolerance to rice. *FEBS Letters*, 532: 279–282.

Padmanaban, S., Lin, X., Perera, I., Kawamura, Y., Sze, H. 2004. Differential expression of vacuolar H⁺-ATPase subunit c genes in tissues active in membrane trafficking and their roles in plant growth as revealed by RNAi. *Plant Physiology*, 134: 1514–1526.

Pardo, J. M., Cubero, B., Leidi, E. O., Quintero, F. J. 2006. Alkali cation exchangers: Roles in cellular homeostasis and stress tolerance. *Journal of Experimental Botany*, 57: 1181–1199.

Peiter, E., Maathuis, F. J. M., Mills, L. N., Knight, H., Hetherington, A. M., Pelloux, J., Sanders, D. 2005. TPC1, a vacuolar Ca²⁺-activated channel controlling germination and stomatal movement. *Comparative Biochemistry and Physiology A—Molecular & Integrative Physiology*, 141: S345–S345.

Pérez-Castiñeira, J. R., Hernández, A., Drake, R., Serrano, A. 2011. A plant proton-pumping inorganic pyrophosphatase functionally complements the vacuolar ATPase transport activity and confers bafilomycin resistance in yeast. *Biochemical Journal*, 437: 269–278.

Pittman, J. K. 2012. Multiple transport pathways for mediating intracellular pH homeostasis: The contribution of H⁺/ion exchangers. *Frontiers in Plant Science*, 3: 1–8.

Pottosin, II, Dobrovinskaya, O. R., Muniz, J. 2001. Conduction of monovalent and divalent cations in the slow vacuolar channel. *Journal of Membrane Biology*, 181: 55–65.

Pottosin, II, Tikhonova, L. I., Hedrich, R., Schonknecht, G. 1997. Slowly activating vacuolar channels can not mediate Ca²⁺-induced Ca²⁺ release. *Plant Journal*, 12: 1387–1398.

Pottosin, I., Dobrovinskaya, O. 2014. Non-selective cation channels in plasma and vacuolar membranes and their contribution to K⁺ transport *Journal of Plant Physiology*, 171: 732–742.

Pottosin, I., Martinez-Estevez, M., Dobrovinskaya, O., Muniz, J., Schonknecht, G. 2004. Mechanism of luminal Ca²⁺ and Mg²⁺ action on the vacuolar slowly activating channels. *Planta*, 219: 1057–1070.

Qadir, M., Tubeileh, A., Akhtar, J., Larbi, A., Minhas, P., Khan, M. 2008. Productivity enhancement of salt-affected environments through crop diversification. *Land Degradation and Development*, 19: 429–453.

Queirós, F., Fontes, N., Silva, P., Almeida, D., Maeshima, M., Gerós, H., Fidalgo, F. 2009. Activity of tonoplast proton pumps and Na⁺/H⁺ exchange in potato cell cultures is modulated by salt. *Journal of Experimental Botany*, 60: 1363–1374.

Quintero, F. J., Blatt, M. R., Pardo, J. M. 2000. Functional conservation between yeast and plant endosomal Na⁺/H⁺ antiporters. *FEBS Letters*, 471: 224–228.

Rea, P. A., Poole, R. J. 1993. Vacuolar H⁺-translocating pyrophosphatase. *Annual Review of Plant Physiology and Plant Molecular Biology*, 44: 157–180.

Rodriguez-Rosales, M. P., Jiang, X. Y., Galvez, F. J., Aranda, M. N., Cubero, B., Venema, K. 2008. Overexpression of the tomato K⁺/H⁺ antiporter LeNHX2 confers salt tolerance by improving potassium compartmentalization. *New Phytologist*, 179: 366–377.

Schilling, R. K., Marschner, P., Shavrukov, Y., Berger, B., Tester, M., Roy, S. J., Plett, D. C. 2014. Expression of the *Arabidopsis* vacuolar H⁺-pyrophosphatase gene (AVP1) improves the shoot biomass of transgenic barley and increases grain yield in a saline field. *Plant Biotechnology Journal*, 12: 378–386.

Scholz-Starke, J., De Angeli, A., Ferraretto, C., Paluzzi, S., Gambale, F., Carpaneto, A. 2004. Redox-dependent modulation of the carrot SV channel by cytosolic pH. *FEBS Letters*, 576: 449–454.

Schulzlessdorf, B., Hedrich, R. 1995. Protons and calcium modulate SV-type channels in the vacuolar-lysosomal compartment- channel interaction with calmodulin inhibitors. *Planta*, 197: 655–671.

Schumacher, K., Krebs, M. 2010. The V-ATPase: Small cargo, large effects. *Current Opinion in Plant Biology*, 13: 724–730.

Segami, S., Nakanishi, Y., Sato, M. H., Maeshima, M. 2010. Quantification, organ-specific accumulation and intracellular localization of type II H⁺-pyrophosphatase in *Arabidopsis thaliana*. *Plant and Cell Physiology*, 51: 1350–1360.

Shabala, S. 2013. Learning from halophytes: Physiological basis and strategies to improve abiotic stress tolerance in crops. *Annals of Botany*, 112: 1209–1221.

Shabala, S., Cuin, T. A. 2008. Potassium transport and plant salt tolerance. *Physiologia Plantarum*, 133: 651–669.

Shabala, S., Mackay, A. 2011. Ion transport in halophytes. In: *Plant Responses to Drought and Salinity Stress: Developments in a Post-Genomic Era*, ed. I. Turkan. pp. 151–199. New York: Elsevier Ltd.

Shabala, S., Shabala, L. 2011. Ion transport and osmotic adjustment in plants and bacteria. *Biomolecular Concept*, 2: 407–419.

Shi, H. Z., Ishitani, M., Kim, C. S., Zhu, J. K. 2000. The *Arabidopsis thaliana* salt tolerance gene SOS1 encodes a putative $Na^+/H^+$ antiporter. *Proceedings of the National Academy of Sciences of the United States of America*, 97: 6896–6901.

Shi, H. Z., Zhu, J. K. 2002. Regulation of expression of the vacuolar $Na^+/H^+$ antiporter gene *AtNHX1* by salt stress and abscisic acid. *Plant Molecular Biology*, 50: 543–550.

Sze, H., Schumacher, K., Müller, M. L., Padmanaban, S., Taiz, L. 2002. A simple nomenclature for a complex proton pump: *VHA* genes encode the vacuolar $H^+$-ATPase. *Trends in Plant Science*, 7: 157–161.

Tester, M., Devenport, R. 2003. Mechanism of salinity tolerance: $Na^+$ tolerance and $Na^+$ transport in higher plants. *Annals of Botany*, 91: 503–527.

Tikhonova, L. I., Pottosin, II, Dietz, K. J., Schonknecht, G. 1997. Fast-activating cation channel in barley mesophyll vacuoles: Inhibition by calcium. *Plant Journal*, 11: 1059–1070.

Venema, K., Quintero, F. J., Pardo, J. M., Donaire, J. P. 2002. The *Arabidopsis* $Na^+/H^+$ exchanger *AtNHX1* catalyzes low affinity $Na^+$ and $K^+$ transport in reconstituted liposomes. *Journal of Biological Chemistry*, 277: 2413–2418.

Walker, D. J., Leigh, R. A., Miller, A. J. 1996. Potassium homeostasis in vacuolate plant cells. *Proceedings of the National Academy of Sciences of the United States of America*, 93: 10510–10514.

Wang, B. S., Luttge, U., Ratajczak, R. 2001. Effects of salt treatment and osmotic stress on V-ATPase and V-PPase in leaves of the halophyte Suaeda salsa. *Journal of Experimental Botany*, 52: 2355–2365.

Wang, W. X., Vinocur, B., Altman, A. 2003. Plant responses to drought, salinity and extreme temperatures: Towards genetic engineering for stress tolerance. *Planta*, 218: 1–14.

Ward, J. M., Schroeder, J. I. 1994. Calcium-activated $K^+$ channels and calcium-induced calcium release by slow vacuolar ion channels in guard cell vacuoles implicated in the control of stomatal closure. *Plant Cell*, 6: 669–683.

Weinl, S., Kudla, J. 2009. The CBL-CIPK $Ca^{2+}$-decoding signaling network: Function and perspectives. *New Phytologist*, 184: 517–528.

van den Wijngaard, P. W. J., Bunney, T. D., Roobeek, I., Schonknecht, G., de Boer, A. H. 2001. Slow vacuolar channels from barley mesophyll cells are regulated by 14–3-3 proteins. *FEBS Letters*, 488: 100–104.

Wu, C., Gao, X., Kong, X., Zhao, Y., Zhang, H. 2009. Molecular cloning and functional analysis of a $Na^+/H^+$ antiporter gene ThNHX1 from a halophytic plant *Thellungiella halophila*. *Plant Molecular Biology Reporter*, 27: 1–12.

Wu, C. A., Yang, G. D., Meng, Q. W., Zheng, C. C. 2004. The cotton GhNHX1 gene encoding a novel putative tonoplast $Na^+/H^+$ antiporter plays an important role in salt stress. *Plant and Cell Physiology*, 45: 600–607.

Wu, H., Shabala, L., Liu, X., Azzarello, E., Zhou, M., Pandolfi, C., Chen, Z. H., Bose, J., Mancuso, S., Shabala, S. 2015. Linking salinity stress tolerance with tissue-specific $Na^+$ sequestration in wheat roots. *Frontiers in Plant Science*, 6: 1–13.

Wu, Y. Y., Chen, Q. J., Chen, M., Chen, J., Wang, X. C. 2005. Salt-tolerant transgenic perennial ryegrass (Lolium perenne L.) obtained by *Agrobacterium tumefaciens*-mediated transformation of the vacuolar $Na^+/H^+$ antiporter gene. *Plant Science*, 169: 65–73.

Xue, Z. Y., Zhi, D. Y., Xue, G. P., Zhang, H., Zhao, Y. X., Xia, G. M. 2004. Enhanced salt tolerance of transgenic wheat (Tritivum aestivum L.) expressing a vacuolar $Na^+/H^+$ antiporter gene with improved grain yields in saline soils in the field and a reduced level of leaf $Na^+$. *Plant Science*, 167: 849–859.

Yamaguchi, T., Aharon, G. S., Sottosanto, J. B., Blumwald, E. 2005. Vacuolar $Na^+/H^+$ antiporter cation selectivity is regulated by calmodulin from within the vacuole in a $Ca^{2+}$- and pH-dependent manner. *Proceedings of the National Academy of Sciences of the United States of America*, 102: 16107–16112.

Yeo, A. R., Flowers, S. A., Rao, G., Welfare, K., Senanayake, N., Flowers, T. J. 1999. Silicon reduces sodium uptake in rice (Oryza sativa L.) in saline conditions and this is accounted for by a reduction in the transpirational bypass flow. *Plant Cell and Environment*, 22: 559–565.

Yokoi, S., Quintero, F. J., Cubero, B., Ruiz, M. T., Bressan, R. A., Hasegawa, P. M., Pardo, J. M. 2002. Differential expression and function of *Arabidopsis thaliana* NHX $Na^+/H^+$ antiporters in the salt stress response. *Plant Journal*, 30: 529–539.

Zhang, H. X., Blumwald, E. 2001. Transgenic salt-tolerant tomato plants accumulate salt in foliage but not in fruit. *Nature Biotechnology*, 19: 765–768.

Zhang, H. X., Hodson, J. N., Williams, J. P., Blumwald, E. 2001. Engineering salt-tolerant Brassica plants: Characterization of yield and seed oil quality in transgenic plants with increased vacuolar sodium accumulation. *Proceedings of the National Academy of Sciences of the United States of America*, 98: 12832–12836.

Zhao, F. Y., Zhang, X. J., Li, P. H., Zhao, Y. X., Zhang, H. 2006. Co-expression of the *Suaeda salsaSsNHX1* and Arabidopsis *AVP1* confer greater salt tolerance to transgenic rice than the single *SsNHX1*. *Molecular Breeding*, 17: 341–353.

Zhen, R.-G., Kim, E., Rea, P. 1997. The molecular and biochemical basis of pyrophosphate-energized proton translocation at the vacuolar membrane. *Advances in Botanical Research*, 25: 297–337.

Zhou, G. F., Delhaize, E., Zhou, M. X., Ryan, P. R. 2013. The barley *MATE* gene, *HvAACT1*, increases citrate efflux and $Al^{+3}$ tolerance when expressed in wheat and barley. *Annals of Botany*, 112: 603–612.

Zhou, G. F., Delhaize, E., Zhou, M. X., Ryan, P. R. 2014. Enhancing the aluminium tolerance of barley by expressing the citrate transporter genes *SbMATE* and *Frd3*. *Journal of Experimental Botany*, 65: 2381–2390.

Zhou, S., Zhang, Z., Tang, Q., Lan, H., Li, Y., Luo, P. 2011. Enhanced V-ATPase activity contributes to the improved salt tolerance of transgenic tobacco plants overexpressing vacuolar $Na^+/H^+$ antiporter *AtNHX1*. *Biotechnology Letters*, 33: 375–380.

Zhu, J. K. 2003. Regulation of ion homeostasis under salt stress. *Current Opinion in Plant Biology*, 6: 441–445.

CHAPTER FOUR

# Salt Stress Signaling Pathways

## SPECIFICITY AND CROSS TALK

*Rohit Joshi, Balwant Singh, Abhishek Bohra, and Vishwanathan Chinnusamy*

CONTENTS

*Abstract.* Soil salinity is one of the major environmental constraints to crop production, restricting the efficient utilization of agricultural land worldwide. Most plants will come in contact with $Na^+$ at some stage during their life cycle, although their level of exposure may vary greatly. All plants tested so far have been shown to take up $Na^+$ in the low affinity range. Salt stress is a complex phenomenon in plants imposing hyperionic and hyperosmotic stress, ionic imbalance, disrupting several major metabolic activities and thus limiting crop productivity. The osmotic stress and ionic excess arise in sequence with two distinct phases, with genotypic differences in salt tolerance. The tolerance mechanism toward salt stress is to minimize ion disequilibrium or osmotic stress or alleviate the consequent secondary effects caused by these stresses. The activation of the SOS signaling pathway is well recognized as a key mechanism for $Na^+$ exclusion and ion homeostasis control at the cellular level. Similarly, polyamines also function as stress messengers during plant responses to salinity stress signaling. Polyamines bind to and regulate the activity and subcellular localization of target proteins implicated in salt and osmotic stress signaling and development. Salt stress promotes the production of reactive oxygen species such as hydrogen peroxide and the superoxide anion. Increasing evidence indicates that lipid signalling is an integrative part of the complex regulatory network in plant response toward salinity stress. Besides this, calcium has been recognized as an important secondary messenger in plants under various developmental cues, as well as salinity. Secondary messengers such as $Ca^{2+}$, cGMP, and reactive oxygen species show rapid increase after elevation in the salt concentration. Salt- and hyperosmosis-induced secondary messengers activate various members of the MAPK family, which play pivotal roles in the plant's responses. The knowledge of salinity stress was greatly enhanced by identifying the convergent and divergent pathways between salinity and other abiotic stress responses. This chapter gives an overview of different salt stress signaling pathways and discusses their specificity and cross talk during signaling responses.

*Keywords*: Abiotic stress, cross talk, osmoprotection, salinity, salt tolerance, signal transduction

## 4.1 INTRODUCTION

World agriculture is currently dealing with a challenging task of producing 70% more food for an extra 2.3 billion people by 2050 (Kumar et al. 2013). The major abiotic stresses that plants face most frequently and that adversely affect growth include high salinity, flooding drought, cold, and heat. They negatively influence the survival, biomass production, and crop productivity, leading to 60% yield losses of staple food crops, which is alarming for food security worldwide (Mantri et al. 2012). Soil salinity is one of the major environmental constraint to crop production, among other abiotic stresses, restricting the utilization of about 830 mha of agriculture land worldwide (Yadav et al. 2011). Globally more than 80 mha of irrigated land (i.e., 40% of total irrigated land) has already been deteriorated by salt (Xiong and Zhu 2001), and 1.5 mha is still adding to this each year as a result of increasing soil salinity levels (Munns and Tester 2008) and is expected to increase due to global climate changes and irrigation practices (Roy et al. 2014). With the current scenario of increasing salinity stress, it is estimated that by 2050, almost 50% of the current cultivated land will be lost (Wang et al. 2003). High salinity is reported to affect plants in several ways, that is, nutritional disorders, membrane disorganization, alteration of metabolic processes, water stress, ion toxicity, oxidative stress, reduction of cell division as well as expansion, and genotoxicity (Hasegawa et al. 2000; Munns 2002; Zhu 2007). Salinity stress extensively affects the agricultural yield, mainly because crops show evidence of reduced growth rates and tillering, and over months, reproductive development is significantly affected (Munns and Tester 2008). Salt stress is a complex phenomenon in plants imposing hyperionic and hyperosmotic stress and ionic imbalance, disrupting several major metabolic activities and thus limiting the crop productivity (Munns and Tester 2008). These effects are associated with the activation of salinity-induced molecular networks involved in protein expressions, regulations of stress-related genes, stress perception, signal transduction, and subsequently metabolic alterations. Recently, transcriptomic and proteomic analyses specified that salt stress induces complex plant biological modifications at the whole system level (Zhang et al. 2011).

The main solution to such a problem is to enhance the ability of plants to sustain growth and productivity in saline soils comparative to their growth in nonsaline soils, that is, to reduce the effect of salinity on growth and yield. However, significant progress has already been made to increase the crop yield using conventional breeding. But the target of improving abiotic stress tolerance in crops has limited success because of complex and multigenic nature of traits with narrow genetic variation in between the genetic pools of major crops (Pardo 2010). Various proteins and genes have been reported to affect the resistance toward different environmental stresses in a defined array in plant species, which together make a complex network with a myriad of individual elements and their cross talk during signal transduction pathways that are not easily incorporated into a holistic model. To employ the wild relatives of crop plants as a resource of genetic determinants of stress tolerance is labor intensive and time-consuming. Further, undesirable genes are transferred frequently along with the desirable ones. Classical breeding approaches have shown that these stress tolerance traits are located in different quantitative trait loci, which hamper during genetic selection of these traits, despite a few success stories.

Several physiological, molecular, and functional studies have given a wealth of information on salt-responsive genes and their products and shed light on species-related variation in mechanisms of salinity tolerance. The aim of this chapter is to compile the insights into salt stress signaling in plants attained by detailed analysis of cellular organelles, which are known to be major targets harboring salt stress defense components because of their significant role in receptor signaling, ion sequestration, and ion transport. We also summarize different cellular pathways functional during signal transduction under salt stress. We also address signaling components and recapitulate current knowledge on salt stress defense mechanisms. Finally, we have derived conclusions for future research directions during salinity stress.

## 4.2 OSMOTIC AND IONIC STRESS

Under normal conditions, the osmotic pressure in plant cells is greater than that in soil solution. This higher osmotic pressure is employed by plant cells for uptake of water and essential mineral nutrients in root cells from the soil solution. However, under salt stress, the osmotic pressure in the soil solution exceeds that of internal osmotic pressure in plant cells because of high salt concentration and, thus, decreases the ability of plants to uptake water and essential minerals like $K^+$ and $Ca^{2+}$ (Kader and Lindberg 2010). Growth is reduced by salinity via several quite distinct processes, which are related either to the accumulation of salt in the shoot or which are independent of shoot salt accumulation. Munns (1988) proposed a two-phase model of plant growth under salt stress, hypothesizing the dominance of water deficit in the first phase and those of ions in the second phase. They hypothesized that osmotic stress and ion excess are not alternatives, but arise in sequence with two distinct phases, with genotypic differences in salt tolerance being apparent only in the second phase. Osmotic stress occurs instantaneously when roots come in contact with solutions containing unfavorably high concentrations of salts in hydroponic systems or in soil, resulting in inhibited cell expansion and cell division, as well as stomatal closure (Munns 2002; Flowers 2004). At this stage, plants need to regulate osmotically for water potential and turgor to attain homeostasis. The reduction of growth during the first phase would be due to the osmotic strength of the salt solution external to the roots. It is perhaps better described as a "shoot salt accumulation–independent" effect. The first phase might be quite long (several weeks)

for some plants, whereas short (several days) for other plants. Tolerance to osmotic stress is a vital component of the plant reaction to salt stress. After salt application, plants generally attain osmotic homeostasis comparatively rapidly, either during several hours or at least within few minutes following salt stress (Munns 2002, 2005; Roshandel and Flowers 2009). Osmotic adjustment refers to the modification of osmotic pressure inside plant cells in response to osmotic stress and needs to be much stronger if plants are to endure the effects of salt (osmotic) shock. The second phase would start only after salt accumulated to toxic levels so as to cause injury and hence diminish the supply of assimilates to the growing regions. In this phase of plant responses to salinity, there is a slower onset of inhibition of growth occurring over several days to weeks. Further, ionic stress results in premature senescence of older leaves, leading to reduction in the total available photosynthetic area available (Cramer and Nowak 1992) and in toxicity symptoms (chlorosis, necrosis) in mature leaves because of high $Na^+$, which affects plants by disturbing protein synthesis and interfering with enzyme activity (Hasegawa et al. 2000; Munns 2002; Läuchli and Grattan 2007).

## 4.3 RESPONSE OF CROPS TO SALT STRESS

Sodium ($Na^+$) is a profuse element, which makes up approximately 3% of the earth's crust. Furthermore, it is abundantly found in almost all surface and subterranean water bodies including seas and oceans where it can reach over 5% (w/w). Thus, it is not surprising that most plants will come in contact with $Na^+$ at some stage during their life cycle although their level of exposure may vary greatly. At the extreme end, terrestrial and aquatic plants can encounter several levels of salinity (NaCl) that are two or three times higher than that of seawater (~540 mM), but very few plant species are capable of surviving such an onslaught. Less extreme, and more widespread, are low and intermediary levels of salinization that arose all over the world. Salinization can arise through natural sources, such as the local geology or proximity to coastal areas or through the use of high salt–containing irrigation water (Maathuis 2013).

For almost all terrestrial plants, $Na^+$ is not essential either for growth and development or for reproduction. An exception is a subgroup of

C$_4$ plants for which Na$^+$ has been reported to be crucial, for example, *Atriplex* spp., *Kochia childsii*, millet, and few other C$_4$ grasses. These C$_4$ species require Na$^+$ at trace levels to drive any particular transport process, the uptake of pyruvate into chloroplasts by a Na$^+$-pyruvate cotransporter (Furumoto et al. 2011). In contrast, in salt-tolerant (halophytic) plants, even high concentrations of Na$^+$ encourage growth, for example, *Salicornia* spp. and *Suaeda maritima*. However, at low levels, Na$^+$ not only is nontoxic but can also be very helpful particularly in low K$^+$ concentrations (Subbarao et al. 2003). This happens because, in hydrated form, Na$^+$ and K$^+$ are chemically and structurally very identical (Amtmann and Sanders 1999). Thus, K$^+$ plays several roles in plant cells, including few of the metabolic ones. In soils where K$^+$ is limited, increased Na$^+$ utilization could lower agriculture's dependence on costly potash fertilizer. Uptake of Na$^+$ at the root–soil boundary also help in the phytoremediation of moderately saline soils. Additionally, the existence of minute amounts of Na$^+$ in the growth medium has been reported to improve the taste of many crops including asparagus, barley, broccoli, and beet (Maathuis 2013).

Furthermore, low levels of Na$^+$ are reported to be beneficial in various conditions: moderate and high levels of salt are unfavorable to the majority of plants that are classified as glycophytes. Several reasons of salinity were reported that inhibit plant growth (Flowers and Colmer 2008). The first reason is the high concentrations of inorganic ions such as Na$^+$ and Cl$^-$ that lower the water potential and therefore induce osmotic stress. A reasonable response to counter this shock is the uptake of these ions themselves: a strategy that is primarily utilized by several salt-tolerant plant species. However, this response can lead to a second problem, that is, a significant increase in cellular ion contents, particularly of Na$^+$ that negatively affects cellular biochemistry. A third phase of salinity is the injurious effect that Na$^+$ and Cl$^-$ ions have on the acquisition and allocation of essential nutrients such as NO$_3^-$, Ca$^{2+}$, and K$^+$. These lethal effects occur at both the cellular and whole tissue levels and cause many terrestrial plants to struggle with even moderate salt levels (Maathuis 2013).

Slower growth is an adaptive feature for plant survival under stress, because it allows plants to rely on multiple resources to combat stress. One reason of reduction in growth rate during stress is insufficient photosynthesis owing to closure of stomata, resulting in inadequate carbon dioxide uptake. Some plants were shown to be highly responsive to stress as they panic and almost stop their growth even under mild stress. In contrast, few plants are probably irresponsive to stress and thus run the risk of dying by their continuous growth even under severe stress. Just as salinity has different effects on a plant, there are also many mechanisms for these plants to tolerate this stress. These mechanisms can be classified into three main categories: first, osmotic tolerance, which is regulated by long-distance signals that reduce shoot growth and is triggered before shoot Na$^+$ accumulation; second, ion exclusion, where Na$^+$ and Cl$^-$ transport processes in roots reduce the accumulation of toxic concentrations of Na$^+$ and Cl$^-$ within leaves; and finally, tissue tolerance, where high salt concentrations are found in leaves but are compartmentalized at the cellular and intracellular levels (Roy et al. 2014).

How plants deal with salinity and how much their strategies are successful against these stresses vary widely among species. Halophytes are naturally adapted to salt and prefer saline environments. Either they control net salt uptake in extremely well manner or they compartmentalize salt in different tissues to make them least harmful (Galvan-Ampudia and Testerink 2011). Glycophytes show limited approaches toward dealing with salt and do not grow well on saline soils. However, there is a significant difference in salt tolerance among glycophytes. Monocotyledonous crops, such as barley and wheat, are moderately tolerant, while several dicot crops are extremely sensitive toward salt stress (Galvan-Ampudia and Testerink 2011). Several morphological adaptations are evolved by halophytes in their leaves, such as succulence, wax deposition, and specialized organs for salt extrusion (e.g., salt glands or bladders), which can actively extrude salt as well as allow them to save water. Although glycophytes have not evolved any stable morphological features to deal with salt, they can adjust root system architecture and root growth in response to salinity stress. Salt stress inhibits primary root growth by decreasing cell division and elongation, resulting in loss of root apical meristem (West et al. 2004). Modifications in root system architecture during salt stress differentially regulate auxin distribution in the root, which determines both the postembryonic development of lateral roots and the direction of root growth (Miyazawa et al. 2009).

At the whole plant level, the detrimental effects of high concentrations of salt on plants can be observed in the form of increasing death of affected plants and/or decrease in their productivity. Reductions in plant growth due to salt stress are often associated with decreases in photosynthetic activities, such as the electron transport (Greenway and Munns 1980). Because of the important roles of several hormones in regulating cell elongation, it would not be surprising if stress inhibits cell expansion by reducing the concentration of growth-promoting hormones such as auxin, cytokinin, gibberellins, and brassinolides. Previous studies with transgenic plants overexpressing stress-tolerant components have also implied that there are connections between stress and growth regulation. Constitutive overexpression of transgenes in plants does not generally seem to compromise plant growth, suggesting that energy is not limiting for plant growth under normal conditions (Zhu 2001).

The effect of stress on flowering time can be attributed, in part, to induced changes in the epigenome. Epigenetics refers to heritable, self-perpetuating changes in gene activities that are not caused by changes in nucleotide sequence and are associated with chemical modifications of chromatin (Bonasio et al. 2010). These modifications take place in the nucleosome at different levels through reversible biochemical reactions that include DNA methylation and histone tail modifications. Eukaryotic cells respond to the environment by modifying their gene expression profiles, a process that usually involves specific chromatin modifications (Chinnusamy and Zhu 2009). Moreover, loss of DNA methylation reduces the ability of plants to tolerate salt stress conditions (Yaish et al. 2011).

Stress-inducible miRNAs and their predicted targets were detected in *Arabidopsis* and also reported to be conserved among other plant species (Sunkar and Zhu 2004). In rice, global expression analysis revealed crucial role for miRNAs in controlling gene expression when plants are exposed to saline conditions (Shen et al. 2010). The STRESS RESPONSE SUPPRESSOR1 and 2 (STRS1 and 2) genes code for DEAD-box RNA helicases that are suppressed when plants are exposed to salt stress conditions. Mutant lines for these genes display higher tolerance in comparison with the wild type (Kant et al. 2007). GENERAL CONTROL NON-REPRESSED PROTEIN5 (AtGCN5) is a major histone acetyltransferase in *Arabidopsis* and acts as a central point between histone modification and miRNA production (Stockinger et al. 2001). Mutation within the coding region of this gene causes pleiotropic effects on plant development, which leads to impaired floral production where petals are transformed into stamens and sepals into filamentous structures (Bertrand et al. 2003). Recently, it was shown that ADA2b (transcriptional coactivator of AtGCN5-containing complexes) positively regulates salt-induced genes by maintaining the required acetylation level of histones H4 and H3, with the ada2b-1 mutant being hypersensitive to salt and ABA (Hark et al. 2009; Kaldis et al. 2010). SGF29A-1 is another component of the AtGCN5 complex, but unlike ada2b-1 mutant, sgf29a-1 displays enhanced salt tolerance compared with the wild type (Kaldis et al. 2010). Salt stress–induced cytoskeletal dynamic changes are also been shown earlier. Previous studies on actin filaments and microtubule array response to salt stress in *Arabidopsis* show that salt stress can induce actin filament assembly and bundle formation (Wang et al. 2011). On the other hand, when seedlings were exposed to salt stress for long duration, the polymerized MFs gets depolymerized and then again repolymerize (Wang et al. 2010). The initial depolymerization of microtubules appears in both low and high NaCl concentration (Wang et al. 2007).

Plant molecular adaptations to salt stress involve the components of the salt overly sensitive (SOS) pathway, salt- and ABA-induced accumulation of reactive oxygen species (ROS), salt-inducible transcription factors, and cytoskeleton, peroxisomes, apoplastic proteins, and glycolytic enzymes (Ahuja et al. 2010). Further detailed studies show evidence on enolase as a multifunctional protein across species (Barkla et al. 2009), implying that it may also play a regulatory or sensory role in multiple stresses at diverse cellular locations. High salinity stress significantly modified the lipid composition of the plasma membrane in broccoli (*Brassica oleracea*) (López-Pérez et al. 2009). It was also suggested that this modification could influence membrane stability or the activity of plasma membrane proteins such as aquaporins or $H^+$-ATPase, to provide a mechanism controlling water permeability and acclimation to salinity stress (López-Pérez et al. 2009). Moreover, under salinity, polyamine oxidase activity was reported to provide ROS production in the apoplast, sustaining maize leaf elongation (Rodriguez et al. 2009). An apoplastic protein, OsRMC (root meander curling), showed drastic abundance in response to salt stress, highlighting the important role of apoplastic proteins in salt tolerance (Zhang et al. 2009). Cell type–specific expression of HKT1;1 in the mature root stele of *Arabidopsis* showed an efficient way for decreasing shoot $Na^+$ accumulation and increasing salinity tolerance (Møller et al. 2009).

Plant adaptive strategies to stress are coordinated and fine-tuned by adjusting growth, development, and cellular and molecular activities. Significant progress has been made earlier in understanding the physiological, cellular, and molecular mechanisms of plant responses to environmental stress factors. It is now clear that salinity tolerance can be complex and involves many genes, as has been pointed out for several decades, which has usually been attributed to the multigenic nature of salinity tolerance (Roy et al. 2014). It is therefore necessary not to study the molecular genetic basis of salinity tolerance as a particular trait in itself, but to study the mechanisms of traits that are hypothesized to contribute to salinity tolerance. Differences in traits must then be correlated with differences in tolerance, as measured by performance in the field. Responses to perturbations are usually accompanied by major changes in the plant transcriptome, proteome, and metabolome. Recent research has made efficient use of these "omic" approaches to identify transcriptional, proteomic, and metabolic networks linked to stress perception and response (Ahuja et al. 2010).

## 4.4 SALT UPTAKE, PARTITIONING, AND REMOBILIZATION

Based on previous research, we now have a reasonably comprehensive picture of how $Na^+$ enters plants and how it is distributed within cells and tissues. Understanding how $Na^+$ enters plants and how it is distributed between different tissues and within cell compartments is vital not only to improve crop resistance toward salt stress but also to increase the value of $Na^+$ as a functional nutrient. In both cases, we require insights into the identities and properties of membrane transporters that catalyze $Na^+$ movement (Maathuis 2013).

There may be several other, less characterized mechanisms that impact $Na^+$ uptake and,

therefore, participate in $Na^+$ sensing. These include amino acids and purines as both of these compounds activate $Na^+$-permeable cation channels (Demidchik et al. 2011). Another parameter is the membrane potential itself as the onset of salinity is typically followed by rapid depolarization of epidermal and cortical root cells (Maathuis 2006). By contrast, hyperosmotic shock generally causes cell hyperpolarization. Thus, changes in membrane potential could act as signals to relay different stimuli where salinity and osmotic shock are concerned. However, cell depolarizations are observed along with the increase in cation concentrations in the root medium. In fact, monovalent cations such as $Na^+$, $K^+$, $Rb^+$, and $Cs^+$ all give very similar depolarizations as a function of external concentration. This lack of specificity can make membrane potentials unsuitable to report changes in salinity and/or $Na^+$. Several glycophytes can be classified as salt excluders, that is, species that prevents large accumulation of salt in photosynthesizing tissues (Marschner 1995). Such species show a relatively high $K^+/Na^+$ selectivity where salt translocation is concerned, possibly via the reabsorption of salts in the basal parts of the root (Lessani and Marschner 1978). In addition, translocation of $Na^+$ from the shoot to the root has also been reported, and such mechanisms would also contribute to low shoot salt concentrations (Marschner 1995).

Long-distance transport of $Na^+$ to the shoot depends upon its loading into the xylem. The exact transport mechanisms for xylem loading of $Na^+$ are still unknown but thought to comprise passive loading (mediated by $Na^+$-permeable ion channels located at the xylem parenchyma interface and active loading mediated by $Na^+$/$H^+$ exchangers [Maathuis 2013]). In *Arabidopsis*, SOS1 is expressed not only in the root epidermis but also in xylem parenchyma during $Na^+$ loading into the xylem sap under moderate salt stress (Shi et al. 2000). However, its exact function depends on the severity of the salinity stress, which includes the $Na^+$ removal from the xylem stream under excessive salt stress. The cation antiporter CHX (cation/$H^+$ exchanger) has also been implicated as a mechanism to move $Na^+$ into the xylem (Hall et al. 2006). AtCHX is found to be mainly expressed in the root endodermis, and its loss of function reduced levels of $Na^+$ in the xylem sap without affecting phloem $Na^+$ concentrations; however, CHX21 and its homologue CHX23 are also reported to be involved in K+ homeostasis (Evans et al. 2012). In *Arabidopsis*, loss-of-function mutations in the HKT1 gene also result in the over-accumulation of Na+ in shoots and rendered plant hypersensitive to Na+ (Berthomieu et al. 2003; Moller et al. 2009). In other species too, HKT isoforms have been implicated in long-distance Na+ movement, especially in cereals. In rice, OsHKT is a plasma membrane Na+ transporter expressed in xylem parenchyma cells that retrieves Na+ from the xylem sap (Ren et al. 2005). The activity of OsHKT1;5 results in less Na+ load in shoot tissue and, therefore, a considerably higher K+/Na+ ratio in leaf tissue.

All plants tested so far have been shown to take up Na+ in the low-affinity range (Zhang et al. 2010), but it is also reported that many plants take up Na+ in the high-affinity range (Haro et al. 2010). At low levels, Na+ is nontoxic, whereas uptake in the high-affinity range may be completely passive process. Various transporters contribute to high-affinity uptake (Haro et al. 2010). Members of the HKT subfamily mediate Na+ transport, but some isoforms show a distinct inhibition whenever trace levels of K+ are present. For example, K+ inhibits barley HKT2 activity with a Kd of almost 30 μM, and hence, Na+ uptake only proceeds when K+ is very low or absent (Mian et al. 2011).

More complicated regulatory mechanisms that depend on secondary messengers may also be important in regulating Na+ uptake. Secondary messengers such as $Ca^{2+}$, cGMP, and ROS show rapid transient increase in cytosolic levels after elevation of the salt concentration (Donaldson et al. 2004). After the onset of salt and osmotic stress, an increase in cellular cGMP was observed (Donaldson et al. 2004), which affects unidirectional Na+ influx (Essah et al. 2003) and directly inhibits nonselective ion channels in root protoplasts (Maathuis and Sanders 2001). This could explain the inhibitory effect of externally applied cGMP on net Na+ uptake. During salinity, cGMP promotes K+ uptake and affects transcript levels of many genes, particularly those encoding membrane transporters (Maathuis 2013).

Growth inhibition by Na+ and/or Cl− toxicity is one of the major adverse effects of salt stress in plants. However, for Gramineae crop like rice, Na+ is the principal reason for causing damage (Kader and Lindberg 2010). High concentration of sodium ion (>10 mM) is toxic to plants mainly because of its adverse effects on ion homeostasis and cellular metabolism. Therefore, maintenance of low level of Na+ in cells is vital for plants. Plants remove Na+ from the cytoplasm by using vacuolar- and plasma membrane–localized Na+/H+ transporters. Na+/H+ transporters are membranous proteins that transport protons (H+) across the membrane in exchange for Na+. This exchange activity requires H+ electrochemical gradient across the membrane, which is generated by the H+ pumps such as H+-ATPase present on plasma membrane or vacuolar H+-pyrophosphatase. Similarly, as a cofactor in cytosol, K+ activates more than 50 enzymes, which are very susceptible to high cytosolic Na+ and high Na+/K+ ratios. Therefore, apart from low cytosolic Na+, maintenance of a low cytosolic Na+/K+ ratio is also crucial for the functioning of cells (Soni et al. 2013).

With the increase of ambient Na+, low-affinity fluxes occur that are mediated by HKT-type carriers (James et al. 2012), nonselective ion channels (Amtmann and Sanders 1999), K+-selective ion channels such as AKT1 (Zhang et al. 2010), and outward rectifying K+ channels (Amtmann and Sanders 1999). Typically, this causes considerable amounts of Na+ to accumulate in cells and tissues. There are several ways plants can perceive and transduce these powerful stimuli. Membrane receptors can detect the change of physical forces on membranes and cell walls that follow after change in turgor. One of the candidates is *Arabidopsis* histidine kinase AtHK1, which has periplasmic domains that record changes in turgor by measuring the distance between the membrane and the cell wall. HK1 activation, in turn, leads to MAPK signaling cascade and altered gene expression (Chefdor et al. 2006). Alternatively, distortion of cell wall membrane geometry can be relayed by mechanosensitive ion channels that open in response to membrane expansion. The fairly nonselective properties of these transporters means that channel opening would cause large membrane depolarizations and may induce increased cytoplasmic $Ca^{2+}$ levels, which is a potent signal to relay changes in environmental osmolarity.

In fact, salinity-motivated $Ca^{2+}$ transients were observed in both endodermis and pericycle (Kiegle et al. 2000). In contrast to K+, Na+ can easily accumulate to toxic levels in the cytosol of the cell, and to prevent this, several mechanisms are reported. Compartmentation of Na+ in the vacuole probably occurs in all tissues and is a principal strategy to detoxify Na+ while still retaining its contribution as an osmolyte to lower water potential. Antiport mechanisms, especially from the NHX family, were identified early on as $H^+$-driven exchangers at the tonoplast that load $Na^+$ into the vacuole. AtNHX1 overexpression significantly improved salinity tolerance in *Arabidopsis* (Apse et al. 1999), and subsequent manipulation of the expression of NHX1 orthologs in other species such as tomato (Zhang and Blumwald 2001), rice (Fukuda et al. 2004), and wheat (Xue et al. 2004) showed the fundamental role of this protein plays in salt tolerance and explains why it is a major focus for genetic engineering. Under low $Na^+$ growth conditions, NHX exchangers mainly mediate $K^+/H^+$ exchange rather than $Na^+/H^+$ exchange (Barragan et al. 2012). This dual selectivity means that the exact NHX function during salinity is sometimes difficult to differentiate, because it directly impacts on cytoplasmic $Na^+$ and $K^+$ and stomatal function (Barragan et al. 2012) and indirectly modifies $Na^+$ and $K^+$ translocation (Xue et al. 2004).

## 4.5 OSMOTIC ADJUSTMENT AND OSMOPROTECTION

High salinity causes ion disequilibrium and hyperosmotic stress that produce secondary effects or pathologies (Hasegawa et al. 2000; Zhu 2001). Fundamentally, plants cope by either avoiding or tolerating salt stress. Thus, either plants are dormant during the salt incident or there must be cellular adjustment to tolerate the saline environment. Tolerance mechanisms can be categorized as those that function to minimize ion disequilibrium or osmotic stress or alleviate the consequent secondary effects caused by these stresses. Cellular response to turgor reduction is osmotic adjustment. The chemical potential of the saline solution initially establishes a water potential imbalance between the apoplast and symplast that leads to decrease in turgor, which causes growth retardation (Bohnert et al. 2005). Growth retardation occurs when turgor is reduced below the yield threshold of the cell wall. Cellular dehydration begins when the water potential difference is greater than that can be compensated by turgor loss (Taiz et al. 2015). Cytosolic and organellar machinery of glycophytes and halophytes is equivalently sensitive toward $Na^+$ and $Cl^-$. Thus, osmotic adjustment is achieved in these compartments by accumulation of compatible osmoprotectants and osmolytes (Bohnert and Jensen 1996; Bohnert et al. 2005). However, $Na^+$ and $Cl^-$ are energetically efficient osmolytes for osmotic adjustment and are compartmentalized into the vacuole to minimize cytotoxicity (Blumwald et al. 2000; Niu et al. 2005). Since plant cell growth occurs mainly due to directional expansion mediated by an increase in vacuolar volume, compartmentalization of $Na^+$ and $Cl^-$ facilitates osmotic adjustment that is essential for cellular development (Hasegawa et al. 2000).

To counter osmotic stress imposed by high salinity, plants need to synthesize compatible organic solutes that accumulate in the cytosol and organelles where they function in osmotic adjustment and osmoprotection (Kader and Lindberg 2010). Compatible solute accumulation as a response to osmotic stress is a ubiquitous process in organisms ranging from bacteria to plants and animals. However, solutes that accumulate may vary with the organism and even between plant species. Majority of organic osmotic solutes consists of simple sugars (i.e., fructose and glucose), sugar alcohols (glycerol and methylated inositols), and complex sugars (trehalose, raffinose, and fructans) (Bohnert and Jensen 1996). Others include quaternary amino acid derivatives (proline, glycine betaine, β-alanine betaine, and proline betaine), tertiary amines (1,4,5,6-tetrahydro-2-mehyl-4-carboxyl pyrimidine), and sulfonium compounds (choline-o-sulfate, dimethyl sulfonium propironate) (Nuccio et al. 1999). Several organic osmolytes were assumed to be osmoprotectants, as their levels of accumulation are insufficient to facilitate osmotic adjustment. Glycine betaine prevents the integrity of plasma membrane and thylakoid after exposure to saline solutions (Rhodes and Hanson 1993).

On the other hand, $K^+$ and $Na^+$, if compartmentalized into the vacuole, were major compatible inorganic solutes used by the plant under salinity stress. For all these defense-response mechanisms to be active during osmotic stress and ionic toxicity, plants initially need to perceive the stress and then activate the whole signaling cascade, starting by an elevation of $[Ca^{2+}]_{cyt}$, either in coordination with changes in cytosolic pH $[pH_{cyt}]$ or individually. In addition, many of the osmoprotectants enhance stress tolerance of plants when expressed as transgene products (Bohnert and Jensen 1996; Zhu 2001). An adaptive biochemical function of osmoprotectants is ROS scavenging that are by-products of hyperosmotic and ionic stresses and cause membrane dysfunction and cell death (Bohnert and Jensen 1996). A common feature of compatible solutes is that these compounds can accumulate in high concentrations without disturbing intracellular biochemistry (Bohnert and Jensen 1996). Compatible solutes have the capacity to maintain the activity of enzymes that are in saline solutions. These compounds have minimal effect on pH or charge balance of the cytosol or luminal compartments of organelles. The synthesis of compatible osmolytes is often achieved by the diversion of basic intermediary metabolites into unique biochemical reactions.

## 4.6 SIGNALING PATHWAYS FOR ION HOMEOSTASIS

One of the key responses toward salt stress is to maintain cellular ion homeostasis by restricting the accumulation of toxic sodium ($Na^+$) (Clarkson and Hanson 1980; Tester and Davenport 2003). A well-defined signaling pathway known to be required for control of ion homeostasis is the SOS signaling pathway. Activation of the SOS signaling pathway has long been recognized as a key mechanism for $Na^+$ exclusion and ion homeostasis control at the cellular level (Zhu 2000). Recently, combinatorial approaches have been used to unravel the novel roles of the SOS pathway and the functions of other complex regulatory networks that regulate ion homeostasis in plants as discussed in the following text:

### 4.6.1 SOS Pathway of $Na^+$ Exclusion, Remobilization, and Partitioning

Mechanisms by which plants combat with salinity stress include responses that counteract the osmotic component of the stress and the exclusion of $Na^+$ from tissues. A principal mechanism leading to tissue tolerance engages cellular adjustments that sequester $Na^+$ into vacuoles (Munns and Tester 2008). The SOS pathway regulates $Na^+/K^+$ ion homeostasis when plants were grown at high salt concentrations and operates to maintain low cytoplasmic concentrations of sodium by sequestering $Na^+$ in vacuoles (Ahuja et al. 2010). After salt stress perception, $Ca^{2+}$ spike was recorded in the cytoplasm of root cells that activates the SOS signal transduction cascade to protect the cells from damage because of

excessive ion accumulation (Chinnusamy et al. 2005). Using forward genetics approach, several salt-oversensitive loci were identified (Shi et al. 2000). One of these was SOS1 that encodes a plasma membrane–located $Na^+/H^+$ antiporter that is found to be involved in the extrusion of $Na^+$ from the cytoplasm, an important aspect of salinity tolerance (Shi et al. 2000, 2002). SOS1 is expressed in several tissues, but its transcript levels were elevated particularly in the root epidermis and around the vascular tissue after several hours or days of salt stress. SOS1 activity is directly responsive to phosphorylation by the serine/threonine protein kinase SOS2, which belongs to the SnRK3 (sucrose nonfermenting-1-related protein kinase-3) family of protein kinases (Hrabak et al. 2003; Ji et al. 2013) or CIPK24, a member of the CIPK (calcium-induced protein kinase) family (Chinusamy et al. 2004).

Previously, SOS3-like calcium-binding protein 8 (SCaBP8, also known as calcineurin B-like [CBL10]) has been shown as an alternative regulator of SOS2 activity that functions primarily in the shoots (Quan et al. 2007). CIPK24 is activated when it associates with the calcineurin-B-like (CBL) calcium sensor CBL4 (SOS3), which is found to be more prominent in roots (Du et al. 2011). SOS3 encodes a myristoylated calcium-binding protein that appears to function as a primary calcium sensor to perceive the increase in cytosolic $Ca^{2+}$ triggered by excess $Na^+$ that has entered the cytoplasm. After binding of $Ca^{2+}$, CBL4 dimerizes, allowing its interaction with an NAF (asparagine, alanine, phenylalanine) domain on CIPK24. CBL binding releases the C-terminal autoinhibition domain of CIPK24, thereby activating this kinase. CIPK-CBL- or SCaBP8-SOS2-mediated phosphorylation at the SOS1 C-terminus further removes the SOS1 autoinhibitory domain, activating the antiporter (Quintero et al. 2011). Thus, a linear pathway can be hypothesized that starts with a salt-induced $Ca^{2+}$ transient and terminates with the increased activity of an antiporter that limits cytoplasmic $Na^+$ accumulation (Quintero et al. 2011).

Loss of function in any of the SOS genes not only leads to increased salt sensitivity but also changes homeostasis of other cations, particularly $K^+$, reported to be the most sensitive one identified till date (Wu et al. 1996). The severe salt sensitivity of sos1 mutants at the early stages of stress suggested its function in protecting root cells, which was recognizable by the high expression of SOS1 in the root tip epidermis (Shi et al. 2002). Interpretation of CBL4 (SOS3) and CIPK24 (SOS2) knockouts is even more complex as both are capable of interacting with various other proteins. For example, CBL4 also interacts with $K^+$ transport by initiating endoplasmic reticulum to plasma membrane trafficking of the $K^+$ channel AKT2 (Held et al. 2010). Binding of $Ca^{2+}$ to CBL4 can be expected to be instantaneous followed by dimerization and then binding to CIPK24. Several previous studies suggest that the $Ca^{2+}$ signal senses osmotic rather than ionic perturbation. In addition, salt-/osmotic-induced $Ca^{2+}$ transients are short lived, and cytoplasmic $Ca^{2+}$ is restored to normal levels within 1 or 2 min. The whole cascade is likely to be completed within minutes, when vacuolar and cytoplasmic $Na^+$ concentration may have risen but not up to the harmful levels.

Additionally, SOS2 may also affect the activity of other transporters such as HKT1, which is involved in $Na^+$ uptake (Rus et al. 2001; Laurie et al. 2002) through interaction with SCaBP8 and NHX1, which is responsible for the extrusion of $Na^+$ into the vacuole (Apse et al. 1999; Weinl and Kudla 2009). Compartmentalization of $Na^+$ that escapes export into vacuoles through SOS1 action is considered to be mediated by NHX1, a vacuolar membrane $Na^+/H^+$ antiporter (Apse et al. 1999, 2003). In a different $[Ca^{2+}]$- and pH-dependent pathway, vacuolar $Na^+/H^+$ transport activity and the selectivity of NHX1 are regulated by calmodulin (CaM)-like CaM15 (Yamaguchi et al. 2005). The activated SOS2/3 complex, in turn, also activates CAX1 (Cheng et al. 2004) that is found to be involved in vacuolar $Na^+$ transport (Qiu et al. 2004).

Recent studies indicated that the homeostatic regulation of root functions that occur through SOS1 is not as simple as it was though initially. Electrophysiological analysis of roots revealed that $K^+$ and $H^+$ flux accumulation in *sos* mutants in response to salt stress is highly different from the wild type. It appears that $Na^+$-induced activity of SOS1 does not exclusively depend on the SOS3–SOS2 complex (Shabala et al. 2005). Indeed, SOS1 is a target of the phospholipase D (PLD) signaling pathway in ion sensing and dynamic equilibrium adjustment under salt stress (Yu et al. 2010). Exposure to salt stress causes an increase in enzyme activity of PLDα1 in *Arabidopsis*, resulting in rapid and transient accumulation of the lipid secondary messenger phosphatidic acid (PA). PA in turn activates mitogen-activated protein kinase 6 (MPK6), which can directly phosphorylate SOS1 (Yu et al. 2010). Loss-of-function mutants of PLDα1 and MPK6 show altered $Na^+$ accumulation in the shoot and increased sensitivity to salt stress. Although the PLD signaling pathway mediates active ion exclusion and homeostatic maintenance through SOS1, it appears to operate in parallel to the SOS3–SOS2 complex. Thus, the regulatory mechanisms of $Na^+$ flux and ion homeostasis appear far more sophisticated than our current understanding, and the linear SOS3/SCaBP8-SOS2–SOS1 signaling pathway is not the only regulatory cellular process for $Na^+$ exclusion (Ji et al. 2013).

## 4.7 SIGNALING PATHWAYS FOR OSMOTIC ADJUSTMENT AND OSMOPROTECTION

### 4.7.1 Polyamine Accumulation and Stress Response

PAs are a group of phytohormone-like aliphatic amine compounds, and majority of them are triamine Spd $[NH_2(CH_2)_3NH(CH_2)_4NH_2]$, tetramine Spm $[NH_2(CH_2)_3NH(CH_2)_4NH(CH_2)_3NH_2]$, and their diamine obligate precursor Put $[NH_2(CH_2)_4NH_2]$ that are ubiquitous in all plant cells. They are now regarded as a new class of growth substances, which play a pivotal role in the regulation of plant developmental and physiological processes (Gill and Tuteja 2010) such as cell division, embryogenesis, germination of seeds and dormancy breaking in tubers, plant morphogenesis, development of flower buds, fruit set and growth, fruit ripening, and response toward biotic and abiotic stresses (Groppa and Benavides 2008). PAs are involved in the regulation of several basic cellular processes, including DNA replication, transcription, translation, cell proliferation, modulation of enzyme activities, cellular cation/anion balance, and membrane stability. Additionally, exceptional PAs, like 1,3-diaminopropane, cadaverine, canavalmine, and homospermidine, have been detected in a large number of biological systems, including bacteria, algae, plants, and animals. At the physiological pH, PAs are found as cations, and this polycationic nature of PAs is one of their important properties affecting their biological activities (Gill and Tuteja 2010). PAs may also function as stress messengers in plant responses to different stress signals (Liu et al. 2007) and are well known for their antisenescence and antistress abilities because of their acid-neutralizing and antioxidant properties as well as their membrane- and cell wall–stabilizing activities (Zhao and Yang 2008). It has already been reported that PAs play important roles in modulating the defense response of plants to diverse environmental stresses including metal toxicity, drought, salinity, chilling, and oxidative stress (Gill and Tuteja 2010).

Previous studies were made on the effect of salinity on plant growth, ethylene production, and polyamine levels in B. *oleracea*, *Beta vulgaris*, *Spinacia oleracea*, *Cucumis melo*, *Capsicum annuum*, *Lactuca sativa*, and *Lycopersicon esculentum* (Zapata et al. 2004). They found that PA levels changed with salinity, and mostly putrescine decreased while spermidine and spermine increased. The increase in (Spd + Spm)/Put ratio along with salinity in all species serves to increase the salinity tolerance. Duan et al. (2008) reported that salinity stress increased proline contents in the roots of tolerant cultivar Changchun mici than in susceptible cultivar Jinchun No. 2. They have also noted a marked increase in arginine decarboxylase, ornithine decarboxylase, S-adenosylmethionine decarboxylase, and diamine oxidase activities, as well as free spermidine and spermine, soluble conjugated and insoluble bound putrescine, spermidine, and spermine contents in the roots of tolerant cultivar in comparison to susceptible cultivar under salt stress. He et al. (2008) reported changes in enzymatic and nonenzymatic antioxidant capacity of the spermidine synthase–overexpressing transgenic apple in response to salt or mannitol stress. They found that before stress treatment, the spermidine contents and spermidine synthase activity were similar in transgenic plants than the wild type. However, under salt or mannitol stress, both the wild type and transgenic exhibited accumulated spermidine, with the latter accumulating more. Lefevre et al. (2001) studied the importance of ionic and osmotic components of salt stress on changes in free polyamine levels in the seedling of two sensitive (IKP) and tolerant (Pokkali) cultivars of rice under isoosmotic concentrations of salt or polyethylene glycol and found that putrescine has differential roles in nonphotosynthetic organs versus photosynthetic, because it accumulated to high amounts in the roots of Pokkali in comparison to IKP, whereas an opposite trend

was recorded in the shoots. They have shown that tyramine was also present at much higher concentrations in the roots of Pokkali and its level clearly increased in response to ionic stresses. Urano et al. (2004) studied the role of putrescine in salt tolerance in *Arabidopsis*. They found the induction of *ADC2* in response to salt stress causing the accumulation of free putrescine. Further, to analyze the roles of stress-inducible *ADC2* gene and endogenous putrescine in stress tolerance, they isolated a Ds insertion mutant of *ADC2* gene (*adc2-1*) and characterized its phenotypes under salt stress, where they reported 25% reduction in free putrescine content to that in the control plants, which did not increase under salt stress. These *adc2-1* mutants were more sensitive to salt stress than control plants, but stress sensitivity of *adc2-1* was recovered by the addition of exogenous putrescine. They concluded that endogenous putrescine plays an important role in salt tolerance in *Arabidopsis*, and *ADC2* is an important gene for the production of putrescine not only under salinity conditions but also under normal conditions.

It has been shown earlier that exogenous application of polyamines is also an effective approach for increasing stress tolerance in plants, leading to enhanced crop productivity (Yiu et al. 2009). Furthermore, it has been noted that genetic transformation with polyamine biosynthetic genes encoding arginine decarboxylase, ornithine decarboxylase, S-adenosylmethionine decarboxylase, or spermidine synthase improved environmental stress tolerance in various plant species (Liu et al. 2007). It is interesting to note that transgenic plants overexpressing ADC, SPDS, or SAMDC34 can tolerate multiple stresses including salinity, drought, and low and high temperature (Wi et al. 2006; Prabhavathi and Rajam 2007; Wen et al. 2008). During the last few years, several genes encoding biosynthetic enzymes of PA pathway such as ADC, ODC, SAMDC, or SPDS improved environmental stress tolerance in various plant species and allowed the molecular approach to become a useful tool for gaining new insights on the regulation of PA metabolism (Gill and Tuteja 2010).

### 4.7.2 ABA Signaling

The phytohormone ABA serves as an endogenous messenger in several aspects of plant development including seed development, desiccation tolerance of seeds, and seed dormancy and plays a crucial role in the plant's response to abiotic (drought, salinity, cold, and hypoxia) and biotic stresses. Upon perception of a stress signal, ABA formation is induced primarily in vascular tissues, and ABA is exported from the site of biosynthesis, and uptake is stimulated into other cells by specific ATP-dependent transporters. The mechanism allows the rapid distribution of ABA into neighboring tissues. Although the upregulation of ABA biosynthesis in response to osmotic stress is well known, the signaling pathways by which ABA biosynthetic genes are upregulated are still unknown. A $Ca^{2+}$-dependent signaling pathway appears to regulate the expression of ABA biosynthetic genes such as zeaxanthin epoxidase, 9-cis-epoxycarotenoid dioxygenase, ABA-aldehyde oxidase, and molybdenum cofactor sulfurase (Xiong et al. 2002). Genetic analysis of ABA-deficient mutants clarify the necessity of ABA signaling in stomatal control of water loss from plants (Schroeder et al. 2001). Stress-responsive genes have been proposed to be regulated by both ABA-dependent and ABA-independent signaling pathways (Shinozaki and Yamaguchi-Shinozaki 2000; Zhu 2002).

ABA-independent stress-responsive gene expression has been thought to be regulated through DRE cis-elements, while ABA-dependent pathways activate gene expression through ABRE cis-elements. In ABA-independent pathways during osmotic stress signaling, AP2-type transcription factors, DREB2A and DREB2B, transactivate the DRE cis-element of stress-responsive genes (Liu et al. 1998). However, cloning and transgenic analysis of a DREB1-related transcription factor, CBF4 in *Arabidopsis*, showed that regulation of DRE elements may also be mediated by an ABA-dependent pathway. Thus, stress-responsive proteins act as effectors of cold, drought, and salt tolerance, but different signaling pathways control the expression of stress-responsive genes during cold and osmotic stresses.

ABA response element–binding proteins are unique to plants, so its activity for sucrose non-fermenting-1-related protein kinases (SnRK-1) has evolved during plant evolution. Subfamilies of protein kinases related to SnRK-1 have also emerged during plant evolution: these are SnRK2 and SnRK3 (Halford and Hey 2009), which are large and relatively diverse compared with SnRK-1, with 10 and 25 members in *Arabidopsis*,

respectively. There is now convincing reports that members of the SnRK2 and SnRK3 subfamilies of protein kinases are involved in ABA-mediated and other signaling pathways that regulate plant responses to drought, cold, salt, and osmotic stresses (Umezawa et al. 2004). SnRKs are closely related to calcium-dependent protein kinases (CDPKs) and in fact, SnRK3-type protein kinases are likely to be calcium dependent, because they interact with CBL calcium-binding proteins (Guo et al. 2002). Due to this reason, SnRK3s are also known as CBL-interacting protein kinases (CIPKs). In Arabidopsis, of the total 25 SnRK3-encoding genes, SnRK3.11 (also known as CIPK24) and SOS2 are known to impart salt tolerance (Liu et al. 2000). Structurally, given the presence of conserved PPI domain, part of the regulatory domain of the SnRK3 remains similar to that of the DNA repair and replication checkpoint protein kinase i.e. CHK1 (Ohta et al. 2003). CHK1 is required for cell cycle arrest in response to DNA damage, and it has been reported that SnRK3 mutants show cell cycle defects at the root meristem when subjected to salt stress (Hey et al. 2010).

### 4.7.3 ROS Signaling

The ROS signaling network is an evolutionarily conserved signal transduction network found in all aerobic organisms (Suzuki et al. 2011). It uses ROS such as singlet oxygen ($_1O^2$), superoxide ($O^{2-}$), and/or hydrogen peroxide ($H_2O_2$) as signal transduction molecules to control a large array of biological processes, ranging from the regulation of development and growth to responses to biotic and/or abiotic stimuli (Suzuki et al. 2011). Low levels of ROS accumulate constantly as by-products of metabolism, but their concentration increases during stress conditions such as osmotic stress and salinity and causes ROS-associated injury (Abbasi et al. 2007; Zhu et al. 2007; Giraud et al. 2008). To minimize damage caused by these potentially harmful molecules, plants produce antioxidants and ROS-scavenging enzymes (Apel and Hirt 2004; Miller et al. 2010). Salt stress promotes the production of ROS such as hydrogen peroxide ($H_2O_2$) and the superoxide anion (Miller et al. 2010). Although ROS are toxic by nature and upregulation of antioxidant enzymes (e.g., superoxide dismutases, catalases, and peroxidases) often mitigates sensitivity, they can also serve as signaling molecules. Salt-induced

ROS emerge within minutes after the stress application (Hong et al. 2009), mostly in the form of $H_2O_2$, and this may depend on the activity of NADPH oxidases.

The rapid increase in ROS production, referred to as "the oxidative burst," was shown to be essential for many of these processes, and genomic studies have shown that respiratory burst oxidase homologue (Rboh) genes, encoding plasma membrane–associated NADPH oxidases, are the main producers of signal transduction–associated ROS in cells during these processes (Mittler et al. 2004; Torres and Dangl 2005). In addition to active generation of ROS signals by the plants, like NADPH oxidases, ROS are also generated due to metabolic imbalances during stress and therefore channeled by the plant to serve as a stress signal to activate acclimation and defense mechanisms that in turn counteract associated oxidative stress (Mittler et al. 2004; Davletova et al. 2005; Miller et al. 2008). Leshem et al. (2007) further studied that suppression of the salt-specific induction of NADPH oxidase–mediated ROS production within endosomes in the phosphatidylinositol 3 kinase mutants (pi3k) caused a reduction in oxidative stress but resulted in enhanced sensitivity toward salt. Similarly, Boursiac et al. (2008) shed light on the regulatory role and identified that ROS formation in roots plays in response to salt in inhibiting hydraulic conductivity Lpr. This led to a novel for ROS in regulating the opening of aquaporins in Arabidopsis roots during salt stress through cellular signaling mechanism (Boursiac et al. 2008). Elucidating the mechanisms that control ROS signaling in cells during drought and salt stresses could therefore provide a powerful strategy to increase the tolerance of crops to these environmental stress conditions.

Although relatively less is known about downstream targets of ROS during salt stress, it is generally accepted that MAP kinase cascades are involved (Miller et al. 2010). Several transcription factors also displayed a restricted or a predominant stress specificity, for example, CERK1 and WRKY70 for salt stress and CML38 (calcium-binding protein) and ANAC053 (Arabidopsis NAC domain–containing protein 53) for osmotic stress. Still very less is known about the function of these ROS-responsive regulators during osmotic stress or salinity, but they are reported to play an important role in the specific acclimation to stress in harmony with the oxidative

homoeostasis under each condition (Vanderbeld and Snedden 2007). NAC domain–containing proteins are plant-specific transcription factors that interact with a wide range of responses including development, hormone signaling, and abiotic stress responses, that is, salinity (Pinheiro et al. 2009). It is likely that WRKY70 and Zat7 are involved in rendering APX1-deficient plants more salt stress tolerant (Ciftci-Yilmaz et al. 2007; Miller et al. 2007). Previous studies have shown that the chloroplastic protein ENH1 (enhancer of sos3-1), a rubredoxin-containing domain, is found to be involved in the reduction of $O^{2-}$ in some bacteria and is also involved in both ROS and ion homeostasis. Deficiency in ENH1 led to increased ROS levels in response to NaCl, rendering the plant sensitive to salt and further enhancing the salt-sensitive phenotype of sos3 mutant (Zhu et al. 2007). Later it was found that ENH1 may function in a salt tolerance pathway that involves SOS2 (Zhu et al. 2007). One such transcription is ethylene response factor 1 (ERF1) in rice (Schmidt et al. 2013), which after activation binds to multiple promoters, including MAPKs. Increased ERF expression led to enhanced salt tolerance (Schmidt et al. 2013). ROS also directly impact on ion fluxes in Arabidopsis roots where outward rectifying $K^+$ channels were directly activated by ROS (Demidchik et al. 2010) and hence could explain the regularly observed $K^+$ loss in plant roots during salt stress (Verslues et al. 2007).

### 4.7.4 Phosphatidic Acid Signaling

Increasing evidence indicates that lipid signaling is an integrative part of the complex regulatory network in plant response toward salinity and drought (Hong et al. 2009; Li et al. 2009; Munnik and Testerink 2009). Modifications in membrane lipids produce different classes of signaling messengers, such as PA, diacylglycerol (DAG), DAG-pyrophosphate, lysophospholipids, free fatty acids, oxylipins, phosphoinositides, and inositol polyphosphates (Bargmann and Munnik 2006; Boss et al. 2008). The production of these mediators is regulated by different families of enzymes, particularly phospholipases, lipid kinases, and/or phosphatases. PA is a minor class of membrane lipids, constituting less than 1% of total phospholipids in most plant tissues. PLD, which hydrolyses membrane phospholipids to PA and a free head group, is a major family of phospholipases in plants. The production of the lipid mediator PA is a key mode of action by which PLDs modulate plant functions. Several studies indicate that PLD and PA play important and complex roles during plant drought and salt stress tolerance (Bargmann and Munnik 2006; Wang et al. 2006; Bargmann et al. 2009; Hong et al. 2009). Genetic and biochemical experiments have shown that PA produced by PLDs plays an important role in stomatal closure, root growth, and plant tolerance toward salinity and water deficits.

PA notably plays a role in plant stress signaling, and almost every environmental cue has been found to trigger a rapid PA response. These include salinity, drought, cold, heat, wounding, and pathogen attack, through activation of either PLD, the PLC/DGK pathway, or both (Arisz et al. 2009; Li et al. 2009; Testerink and Munnik 2011). In addition to this, several genomic studies supported the role for PA in stress responses and several PA target proteins have also been identified to be involved in biotic and abiotic stress signaling. Osmotic stress, in the form of salinity or drought, triggers the fast and transient formation of PA in both green algae and higher plants including *Chlamydomonas*, *Dunaliella*, *Arabidopsis*, tomato, tobacco, alfalfa, and rice (Einspahr et al. 1988; Munnik et al. 2000; Katagiri et al. 2001; Meijer et al. 2002; Arisz et al. 2009; Bargmann et al. 2009; Hong et al. 2010). Based on radiolabeling experiments, PLC/DGK pathways have also been shown to be activated by salinity (Munnik et al. 2000; Arisz et al. 2009; Testerink and Munnik 2011). Of these PI-PLCs, AtPLC1 was shown to be induced in response to salinity and drought (Hirayama et al. 1995) and is required for ABA-induced inhibition of germination and gene expression (Sanchez and Chua 2001). Previously, the NADPH oxidase isoforms RbohD and RbohF were found to bind PA (Zhang et al. 2009).

Additionally, it has been shown earlier that PA binds to and regulates the activity and subcellular localization of target proteins implicated in salt and osmotic stress signaling and development (Galvan-Ampudia and Testerink 2011). These include MPK6 (Yu et al. 2010), sucrose nonfermenting 1-related protein kinase 2 (SnRK2) (Testerink et al. 2004), the serine/threonine protein phosphatase type-C ABI1 (Mishra et al. 2006), 3-phosphoinositide-dependent kinase 1 (PDK1) (Anthony et al. 2004), and PINOID protein kinase (Zegzouti et al. 2006). MPK6 activation

in response to salinity stress was shown to be dependent on PLDα1-generated PA formation (Yu et al. 2010). It is previously been shown that ribosomal protein S6 kinase (S6K) has been implicated in hyperosmotic stress response in *Arabidopsis* (Mahfouz et al. 2006). Additionally, S6K is a substrate of the phosphoinositide-dependent protein kinase 1 (PDK1), and PA has been shown to bind to PDK1 from *Arabidopsis*. The PA–PDK1 interaction activates the PDK1 and AGC2-1 kinases to promote root hair growth (Anthony et al. 2004). In particular, PLDε-produced PA promotes root hair formation under nitrogen deprivation and salt stress (Hong et al. 2009). The activity of PLD increases under various hyperosmotic stresses, such as high salinity (Testerink and Munnik 2005; Bargmann et al. 2009). pldε KO mutant seedlings also exhibit reduced primary root growth under hyperosmotic stress conditions, but their phenotype is reported to be the result of PLDε in nutrient signaling (Hong et al. 2009).

PLDα1, the most predominant PLD found in plants, is activated by ABA to produce PA. PLDα1-deficient plants exhibited a higher transpirational rate than wild-type plants, whereas the overexpression of PLDα1 reduced transpirational water loss by rendering the plants more sensitive to ABA (Sang et al. 2001). The stomata in epidermal peels from PLDα1-deficient *Arabidopsis* plants fail to close in response to ABA, whereas externally applied PA promoted the stomatal closure in wild-type and PLDα1-deficient *Arabidopsis* plants (Sang et al. 2001; Mishra et al. 2006). The PLD-mediated production of PA has also been implicated in promoting stomatal closure in *Vicia faba* (Jacob et al. 1999). These results suggest that PLDα1 and PA plays a positive role in ABA effects on preventing water loss. PLDα1-produced PA binds to ABI1, a PP2C, which functions as a negative regulator in ABA signaling in stomata closure (Zhang et al. 2004). On the other hand, PLDα1 also interacts with Gα protein (a heterotrimeric Gα protein) in *Arabidopsis* to prevent opening of stomata (Zhao and Wang 2004; Mishra et al. 2006). Gα and two G protein–coupled receptor–like proteins GTG1 and GTG2 are involved in mediating ABA response in *Arabidopsis* (Pandey and Assmann 2004; Pandey et al. 2009). Previously, two ABI1 interactors were identified as ABA receptors that bind and inhibit PP2C activity (Ma et al. 2009; Park et al. 2009). Therefore, interactions of PLD/PA with ABI1 and the G protein place PLDα1 and PA in the initial events during

ABA sensing and signaling. Additionally, PLDα1 is also found to be involved in response to salt stress. KO lines of *Arabidopsis* PLDα1 rendered plants less tolerance to salt stress (Bargmann et al. 2009).

PLDδ differs from PLDα1 in subcellular association, gene expression pattern, activity requirements, and substrate preference (Wang and Wang 2001; Qin and Wang 2002; Wang et al. 2006). PLDδ is activated in response to high salinity and rapid dehydration (Katagiri et al. 2001; Bargmann et al. 2009). PLDδ mRNA is induced by dehydration and high salinity (Katagiri et al. 2001). PLDδ-antisense plants did not display clear phenotypic changes, but PLDδ-KO *Arabidopsis* plants were more susceptible to salt stress. PLDα1/PLDδ double mutants accumulated only 30% of the PA in wild type and were more susceptible to salt stress than PLDα1 and PLDδ single mutants (Bargmann et al. 2009). It has also been reported that *Arabidopsis* PLDδ is activated by $H_2O_2$ (Zhang et al. 2003). These results indicate that PLDα1 and PLDδ have distinctively different as well as overlapping functions under salt stress. The role of PLDδ in salt and freezing stresses suggests that PLDδ plays a positive role in hyperosmotic response, possibly through its role in mediating cellular response to ROS.

### 4.7.5 Calcium Signaling

Calcium is an essential nutrient for the growth and development of plants (Kader and Lindberg 2010). It plays important structural role in producing plant tissues and enables them to grow better. Calcium increases the resistance in plant tissues under various stresses including both biotic and abiotic stresses (Sanders et al. 1999; Knight 2000). Besides these fundamental roles, calcium has been recognized long back as an important secondary messenger in plants under various developmental cues as well as under different stresses, including salinity stress. In plant cells, under normal conditions, the cytosolic calcium concentration, $[Ca^{2+}]_{cyt}$, was maintained at nanomolar level in the range of 10–200 nM, whereas the concentration of $Ca^{2+}$ in the cell wall, vacuole, endoplasmic reticulum, and mitochondria is 1–10 mM (Reddy 2001). Also in rice protoplasts, as well as in quince protoplasts, different $[Ca^{2+}]_{cyt}$ changes were obtained under sodium stress and under osmotic stress (D'Onofrio and Lindberg 2009). Kiegle et al. (2000) reported significant quantitative differences in $[Ca^{2+}]_{cyt}$

elevation in different cells of *Arabidopsis* roots. Both under osmotic stress (440 mM mannitol) and salt stress (220 mM NaCl), the endodermis and pericycle cells displayed extended oscillation in $[Ca^{2+}]_{cyt}$ that were different from the responses of other cell types. These changes in $[Ca^{2+}]_{cyt}$ appear to be supplied from the apoplast or from the internal stores like ER, Golgi bodies, mitochondria, or vacuole (Sanders et al. 2002). Calcium-permeable channels in the plasma membrane, which are activated by membrane depolarization, are found to elevate $[Ca^{2+}]_{cyt}$ in many species after the perception of a range of stimuli (White 2000; D'Onofrio and Lindberg 2009). The generation of $Ca^{2+}$ increase in the cytosol further modulates other messengers like inositol phosphate, which induces a further $Ca^{2+}$ elevation in the cytosol through the opening of inositol-(1,4,5)-triphosphate ($IP_3$)-regulated $Ca^{2+}$ channels (Tracy et al. 2008).

The increase in $[Ca^{2+}]_{cyt}$ under salinity stress is detected by SOS3, a $Ca^{2+}$ sensor. Within seconds after sensing of salinity stress, a transient, stable, or oscillating change in $[Ca^{2+}]_{cyt}$ concentration is obtained. This change is required for the activation of downstream response mechanisms through either induction or downregulation of the responsive genes. The nature of $[Ca^{2+}]_{cyt}$ signal in terms of amplitude, frequency, and duration of the peak or signal is likely to have a specific role in encoding the particular information for plants under salinity stress. Specific $Ca^{2+}$ signatures are important for plant cells to sense for the subsequent events in the signaling processes. Such processes may change with the particular stress (Kiegle et al. 2000), the rate of stress development (Plieth et al. 1999), previous exposure to stress conditions (Knight et al. 1997), and the tissue type (Kader and Lindberg 2010). In a few studies with root tissues or root protoplasts, salt stress was reported to reduce $[Ca^{2+}]_{cyt}$. Within minutes of application of 100 mM NaCl to root cells of *Arabidopsis* (Halperin et al. 2003) or to corn root protoplast (Lynch and Läuchli 1988), there was a decrease in $[Ca^{2+}]_{cyt}$. In contrast, several studies revealed an increase in $[Ca^{2+}]_{cyt}$ after salt stress (Knight 2000; Kader et al. 2007; D'Onofrio and Lindberg 2009).

Plants possess three main families of calcium sensors: CaM, CBL, and CDPKs. Unlike CaM and CBL, which must relay the $Ca^{2+}$-induced conformational change to protein partners, CDPKs have the unique feature of both $Ca^{2+}$-sensing and $Ca^{2+}$-responding activities within a single protein to directly translate $Ca^{2+}$ signals into phosphorylation events (Boudsocq and Sheen 2013). CDPKs have been shown to function in many different aspects of plant biology, from environmental stress signaling upon the perception of abiotic and biotic stress stimuli to hormone-regulated processes in development (Asano et al. 2012).

Regulating $Ca^{2+}$ signals is also a key component of stress signaling to ensure that specific responses are induced by each stress. Autoinhibited calcium ATPase in *Arabidopsis* was reported earlier to play a crucial role in generating the appropriate $Ca^{2+}$ signal in yeast exposed to high salinity (Anil et al. 2008). Genetic analysis of the FIERY1 (FRY1/SAL1) locus of *Arabidopsis* suggested the involvement of $IP_3$ (which in turn generates cytosolic $Ca^{2+}$ oscillations) in ABA, salt, and cold stress signaling. ABA-induced transcriptional reprogramming through ABRE-binding factors is likely to be a key feature of CDPK signaling in both monocots and dicots. Drought and salinity trigger the production of ROS, which must be detoxified. In rice, OsCPK12 regulates ROS homeostasis under salt stress conditions by inducing the expression of ROS scavenger genes OsAPX2/OsAPX8 and repressing the NADPH oxidase gene OsRBOHI, leading to increased salt tolerance (Asano et al. 2012). Interestingly, unlike in defense responses (Boudsocq et al. 2010), AtCPK3 does not regulate gene expression in salt stress signaling but modulates the phosphoproteome independently of MAPK cascades (Mehlmer et al. 2010).

## 4.7.6 MAPK Signaling

In addition to several signaling pathways involved in abiotic stress response in plants, MAPK cascade is one of the major pathways. This signaling module links external stimuli with several cellular responses and is evolutionary conserved among eukaryotic organisms (Sinha et al. 2011). MAPK cascades are conserved signaling modules found in all eukaryotes, which transduce environmental and developmental cues into intracellular responses. A MAPK cascade is mainly composed of MAP kinase kinase kinases (MAP3Ks/MAPKKKs/MEKKs), MAP kinase kinases (MAP2Ks/MAPKKs/MEKs/MKKs), and MAP kinases (MAPKs/MPKs) (Rodriguez et al. 2010). During stress, stimulated plasma membrane activates MAP3Ks or MAP kinase kinase kinase kinases (MAP4Ks). MAP4Ks may act as adaptors, linking upstream signaling steps to the core MAPK cascades. MAPKs are serine/threonine kinases able to phosphorylate a wide range of substrates, including other kinases and transcription factors. In several species, MAPK cascades have been shown to be involved in signaling pathways in response to abiotic stresses such as cold, salt, touch, wounding, heat, UV, osmotic shock, and heavy metals (Sinha et al. 2011).

MAPK cascades are particularly important in controlling cross talk between stress responses, as many MAPKs are activated by more than one type of stress or plant hormone and are thus able to integrate different signals (Rohila and Yang 2007; Andreasson and Ellis 2010). For example, *Arabidopsis* MPK6 functions in response to ethylene synthesis, cold and salt stress, pathogen signaling, and stomatal control (Rodriguez et al. 2010). In rice, at least 5 of the 17 identified MAPK genes are inducible by both biotic and abiotic stresses (Rohila and Yang 2007), and certain MAPKs can directly influence both pathways. Of these, overexpression of OsMAPK5 in rice transgenic plants increased tolerance, while suppression led to hypersensitivity to various stresses including salt. Salt- and hyperosmosis-induced secondary messengers activate diverse signaling proteins including various members of the MAPK family, which play pivotal roles in the plant's responses (Kim et al. 2011; Persak and Pitzschke 2013). High salt concentrations, drought, and abscisic acid (ABA) all induce the activation of a Raf-like MAP3K known as drought hypersensitive mutant 1 in rice (Ning et al. 2010). In *Arabidopsis*, salt concentrations that are high enough to induce a hyperosmotic response in addition to sodium toxicity cause the activation of MKKK20, which lies upstream of MPK6 (Kim et al. 2012). Interestingly, *Arabidopsis* mekk1 mutants exhibit improved growth under high salinity conditions. This suggests that in contrast to MKKK20, MEKK1 may be a negative regulator of the salt response (Su et al. 2007).

In *Arabidopsis*, previous study has demonstrated that the MEKK1 (an MAPKKK) mRNA accumulated in response to high salinity (Mizoguchi et al. 1996). Under salt stress, MAPK pathway involves MEKK1 as an upstream activator of MKK2 and the downstream MAPKs MPK4 and MPK6 (Teige et al. 2004). MKP1 plays a negative role in salt stress signaling through MAPKs

(MPK6 and MPK4) (Smékalová et al. 2014). One, however, is the cold- and salt-responsive MEKK1-MKK1/2-MPK4/6 signaling cascade, which seems to have a bidirectional interaction with ROS: the MEKK1 protein was found to be activated and stabilized by $H_2O_2$, and also the MAPK components MPK4 and MPK6 were activated by ROS and abiotic stresses. During salt and cold stresses, MPK6 (and MPK4) was activated by MKK2 (Teige et al. 2004), but although MKK3 has also been shown to activate MPK6, it is not required for salt tolerance (Takahashi et al. 2007). In addition, the overexpression of MKK9, another MPK3/6-activating MAPKK, rendered transgenic plants sensitive to salt treatment (Xu et al. 2008). A 46 kDa SIMK (salt stress–induced MAPK) in alfalfa was reported to be activated by salt. Yeast-2-hybrid assay has shown an upstream activator kinase SIMKK that interacts specifically with SIMK and enhanced the salt-induced activation of SIMK in vivo, as well as in vitro. Osmotic stress reportedly activated the expression of AtMPK3, AtMPK4, and AtMPK6 in Arabidopsis (Droillard et al. 2004). Recently three salt stress–induced MAPKs, ZmMPK3, ZmMAPK5, and ZmSIMK1, have been identified in Zea mays (Wang et al. 2010).

## 4.8 CROSS TALK BETWEEN DIFFERENT SIGNALING PATHWAYS: SALINITY AND OTHER BIOTIC AND ABIOTIC STRESS SIGNALING CROSS TALK

The term "cross talk" is used generally to refer to situations where different signaling pathways share one or more intermediates/components or have some common outputs. Signal transduction pathways in plants are very well developed, while at the same time, they are extremely complex to reveal all the cross talks. The simple reason behind these complexities is that the plants are sessile and experience both biotic and abiotic stresses at single location. Several abiotic stresses lead to both general and specific effects on plant growth and development. Plants have evolved complex signaling pathways in response to many stimuli, such as salt, drought, cold, wounding, and pathogen invasion, which lead to acquired plasticity in metabolic functions and developmental switches to cope with changing environmental conditions (Genoud and Metraux 1999). Cross talk between different pathways appears to be the most common feature in plants, as exemplified by biotic defenses involving ethylene, salicylic acid,

and jasmonic acid (JA) (Ma et al. 2006) or by the DREB/CBF pathway on which several signals from abiotic stress conditions converge (Chinnusamy et al. 2004). The knowledge of salinity stress was greatly enhanced by identifying the convergent and divergent pathways between salinity and other abiotic stress responses and the nodes of signaling convergence. The investigation of plant molecular responses to multiple stresses has often focused on overlapping transcriptional patterns. To this effect, several studies have been carried out in which different groups of plants are exposed to either one stress or another in parallel, and their gene expression patterns compared. Overlapping sets of genes that are regulated by both stresses are then identified and proposed to represent generalized nodes of cross talk between signaling pathways (Swindell 2006; Kilian et al. 2007; Huang et al. 2008). It has been hypothesized that these genes are the targets for improving stress tolerance in crop plants (Denby and Gehring 2005; Swindell 2006). Genes were identified that were commonly induced by both biotic and abiotic stress and therefore are important in regulating cross talk between pathways (Narusaka et al. 2004; Atkinson and Urwin 2012).

Often the response to one type of stress renders plants more resistant to another type of stress, a phenomenon called cross-tolerance (Wurzinger et al. 2011). Both MAPKs and CDPKs have been implicated in cross-tolerance between biotic and abiotic stress responses. Overexpression of pathogen-induced MYB transcription factors increases salt tolerance in tomato, and CDPK activities were found to be responsible in this cross-tolerance (Abuqamar et al. 2009). Previous studies have revealed CDPKs as central regulators of $Ca^{2+}$-mediated immune and stress responses that are critical for plant survival. Besides some specificity, several CDPKs have shown functional redundancy in providing plant responsiveness and adaptability toward both abiotic and biotic stresses. Several complex interactions have been observed between CDPKs and MAPK cascades from synergism to independence and antagonism (Boudsocq and Sheen 2013). The multiple cross talks between CDPKs and SnRK2s or MAPKs provide additional regulatory points to fine-tune plant immune and stress responses. Analysis of the expression levels of several flg22 inducible genes in mesophyll protoplasts expressing deregulated AtCPK5 and AtMKK4 identified MAPK-specific, CDPK-specific, CDPK/MAPK synergistic, and MAPK dominant target genes.

Few CDPKs were reported to play opposite roles in distinct cell types. Different CDPK substrates are likely to be responsible for diverse physiological functions in diverse cellular processes.

Perhaps some of the strongest evidence for cross talk during abiotic stress signaling in plants comes from studies of MAPK cascades. The *Arabidopsis* genome encodes approximately 60 MAPKKKs, 10 MAPKKs, and 20 MAPKs. Signals perceived by the 60 MAPKKKs have to be transduced through 10 MAPKKs to 20 MAPKs, which offer scope for cross talk between different stress signals. MAPKs are implicated in developmental, hormonal, biotic, and abiotic stress signaling (Ligterink and Hirt 2000). Several members of MAPK cascades are activated by multiple stresses, which suggests that MAPK cascades act as points of convergence during stress signaling. Salt stress induces the expression and activity of AtMEKK1 (Ichimura et al. 2000) that activates AtMPK4 in vitro (Huang et al. 2000). AtMPK4 is activated by osmotic stress, cold, low humidity, touch, and wounding (Ichimura et al. 2000). AtMKK2 has been identified as a key regulator of cold and salt-stress response in *Arabidopsis*, but was also reported to be involved in response to *Pseudomonas syringae* infections (Wurzinger et al. 2011). AtCPK3 was shown to be involved during salt stress acclimation and also identified as part of the herbivore response signaling along with AtCPK13 (Kanchiswamy et al. 2010). AtCPK3 and AtCPK13 knockout mutants accumulated equal levels of the phytohormones JA, ABA, and ethylene as the wild types in response to infection, indicating that AtCPK3 and AtCPK13 are not upstream regulators of phytohormone biosynthesis. It was assumed that ABA-induced expression of CAT1 was mediated by *Arabidopsis* MKK1 (MAP2K) in response to $H_2O_2$ signaling. The next component that was found to be involved in this signaling pathway was MPK6. CAT1 expression was suppressed in an mpk6 mutant but was enhanced in an MPK6-overexpressing plant line. These results clearly suggest that MPK6 and MKK1 affect the ABA-mediated expression of CAT1 expression in *Arabidopsis*. It is therefore likely that the MKK1/MPK6 module is an important component of the ABA-dependent signaling pathway that is responsible for $H_2O_2$ production and stress responses (Xing et al. 2008). ABA is therefore likely to play a key role in fine-tuning of observed responses toward simultaneous biotic and abiotic stresses.

ABA treatment represses the systemic acquired resistance pathway both upstream and downstream of salicylic acid induction in *Arabidopsis* and tobacco, as well as inhibits the accumulation of defense compounds such as lignins and phenyl propanoids (Mohr and Cahill 2007; Yasuda et al. 2008; Kusajima et al. 2010). Salicylic acid, in turn, also interferes during abiotic stress signaling. ABA also has a positive effect on pathogen defense systems (Asselbergh et al. 2008; Ton et al. 2009). Now ABA is considered as a global regulator of stress response that predominantly suppresses pathogen defense pathways; thus, it acts as "controlling switch" between biotic and abiotic stress responses and allows plants to respond to multiple stresses (Asselbergh et al. 2008). ABA production is therefore a critical factor in determining how plants respond under multiple stresses. The ethylene receptors ETR1 and ETR2 are two-component histidine protein kinases that function upstream of constitutive triple response (CTR1), a Raf-like MAP3K with a unique N-terminal regulatory domain and a C-terminal Ser/Thr kinase domain (Kieber et al. 1993; Huang et al. 2003; Ouaked et al. 2003). Ethylene blocks CTR1 and thereby activates the MKK9/MPK3/MPK6 cascade, which can stabilize the production of positive regulators of ethylene signaling (Ju et al. 2012).

## 4.9 CONCLUSION AND FUTURE PROSPECTS

The devastating impact of salinization on plant growth and yield has aggravated the food and environmental issues. The most effective measure to solve this worldwide problem is to increase the salt tolerance in crop plants. The strategies of plants toward the adverse effect of salinity stress include ion regulation and compartmentation, biosynthesis of compatible solutes, and antioxidant enzyme formation. Several high-throughput techniques in molecular biology are now introduced, which enable us to understand the complex signaling network in plants. Recent studies revealed that the response of plants toward combinations of stress results in a high degree of complexity in plant response, which are controlled by the cross talk between their specific signaling pathways. Thus, future challenge should be directed toward combining our knowledge along with the genetic engineering approaches to utilize it for the improvement of crop yield.

## ACKNOWLEDGMENTS

RJ acknowledges Start-Up Research Grant (Young Scientists) from Science and Engineering Research Board, Department of Science and Technology, Government of India.

## REFERENCES

Abbasi, A.R., M. Hajirezaei, D. Hofius, U. Sonnewald and L.M. Voll. 2007. Specific roles of alpha- and gamma-tocopherol in abiotic stress responses of transgenic tobacco. *Plant Physiology* 143: 1720–1738.

Abuqamar, S., H. Luo, K. Laluk, M.V. Mickelbart and T. Mengiste. 2009. Crosstalk between biotic and abiotic stress responses in tomato is mediated by the AIM1 transcription factor. *The Plant Journal* 58: 347–360.

Ahuja, I., R.C.H. de Vos, A.M. Bones and R.D. Hall. 2010. Plant molecular stress responses face climate change. *Trends in Plant Science* 15: 664–674.

Amtmann, A. and D. Sanders. 1999. Mechanisms of Na$^{(+)}$ uptake by plant cells. *Advances in Botanical Research* 29: 75–112.

Andreasson, E. and B. Ellis. 2010. Convergence and specificity in the *Arabidopsis* MAPK nexus. *Trends in Plant Science* 15: 106–113.

Anil, V.S., P. Rajkumar, P. Kumar and M.K. Mathew. 2008. A plant $Ca^{2+}$ pump, ACA2, relieves salt hypersensitivity in yeast. Modulation of cytosolic calcium signature and activation of adaptive Na$^+$ homeostasis. *The Journal of Biological Chemistry* 283: 3497–3506.

Anthony, R.G., R. Henriques, A. Helfer, T. Meszaros, G. Rios, C. Testerink, T. Munnik, M. Deak, C. Koncz and L. Bogre. 2004. A protein kinase target of a PDK1 signalling pathway is involved in root hair growth in *Arabidopsis*. *EMBO Journal* 23: 572–581.

Apel, K. and H. Hirt. 2004. Reactive oxygen species: Metabolism, oxidative stress and signal transduction. *Annual Review of Plant Biology* 55: 373–399.

Apse, M.P., G.S. Aharon, W.A. Snedden and E. Blumwald. 1999. Salt tolerance conferred by overexpression of a vacuolar Na$^+$/H$^+$ antiport in *Arabidopsis*. *Science* 285: 1256–1258.

Apse, M.P., J.B. Sottosanto and E. Blumwald. 2003. Vacuolar cation/H$^+$ exchange, ion homeostasis, and leaf development are altered in a T-DNA insertional mutant of AtNHX1, the *Arabidopsis* vacuolar Na$^+$/H$^+$ antiporter. *The Plant Journal* 36: 229–239.

Arisz, S.A., C. Testerink and T. Munnik. 2009. Plant PA signaling via diacylglycerol kinase. *Biochimica et Biophysica Acta* 1791: 869–875.

Asano, T., N. Hayashi, M. Kobayashi, N. Aoki, A. Miyao, I. Mitsuhara, H. Ichikawa, S. Komatsu, H. Hirochika, S. Kikuchi and R. Ohsugi. 2012. A rice calcium-dependent protein kinase OsCPK12 oppositely modulates salt-stress tolerance and blast disease resistance. *The Plant Journal* 69: 26–36.

Asselbergh, B., D. De Vleesschauwer and M. Hofte. 2008. Global switches and fine-tuning-ABA modulates plant pathogen defense. *Molecular Plant-Microbe Interactions* 21: 709–719.

Atkinson, N.J. and P.E. Urwin. 2012. The interaction of plant biotic and abiotic stresses: From genes to the field. *Journal of Experimental Botany* 63: 3523–3544.

Bargmann, B.O. and T. Munnik. 2006. The role of phospholipase D in plant stress responses. *Current Opinion in Plant Biology* 9: 515–522.

Bargmann, B.O., A.M. Laxalt, B. ter Riet, B. van Schooten, E. Merquiol, C. Testerink, M.A. Haring, D. Bartels and T. Munnik. 2009. Multiple PLDs required for high salinity and water deficit tolerance in plants. *Plant Cell and Physiology* 50: 78–89.

Barkla, B.J., R. Vera-Estrella, M. Hernández-Coronado and O. Pantoja. 2009. Quantitative proteomics of the tonoplast reveals a role for glycolytic enzymes in salt tolerance. *The Plant Cell* 21: 4044–4058.

Barragan, V., E.O. Leidi, Z. Andres, L. Rubio, A. De Luca, J.A. Fernandez, B. Cubero and J.M. Pardo. 2012. Ion exchangers NHX1 and NHX2 mediate active potassium uptake into vacuoles to regulate cell turgor and stomatal function in *Arabidopsis*. *The Plant Cell* 24: 1127–1142.

Berthomieu, P., G. Conéjéro, A. Nublat, W.J. Brackenbury, C. Lambert, C. Savio, N. Uozumi, S. Oiki, K. Yamada, F. Cellier, F. Gosti, T. Simonneau, P.A. Essah, M. Tester, A.A. Véry, H. Sentenac and F. Casse. 2003. Functional analysis of AtHKT1 in *Arabidopsis* shows that Na$^{(+)}$ recirculation by the phloem is crucial for salt tolerance. *EMBO Journal* 22: 2004–2014.

Bertrand, C., C. Bergounioux, S. Domenichini, M. Delarue and D.X. Zhou. 2003. *Arabidopsis* histone acetyltransferase AtGCN5 regulates the floral meristem activity through the WUSCHEL/AGAMOUS pathway. *Journal of Biological Chemistry* 278: 28246–28251.

Blumwald, E., G.S. Aharon and M.P. Apse. 2000. Sodium transport in plant cells. *Biochemica et Biophysica Acta* 1465: 140–151.

Bohnert, H.J. and R.G. Jensen. 1996. Strategies for engineering water-stress tolerance in plants. *Trends in Biotechnology* 14: 89–97.

Bohnert, H.J., D.E. Nelson and R.G. Jensen. 2005. Adaptations to environmental stresses. *Plant Cell* 7: 1099–1111.

Bonasio, R., S. Tu and D. Reinberg. 2010. Molecular signals of epigenetic states. *Science* 330: 612–616.

Boss, W., D. Lynch and X. Wang. 2008. Lipid-mediated signaling. *Annual Plant Reviews* 33: 202–243.

Boudsocq, M. and J. Sheen. 2013. CDPKs in immune and stress signaling. *Trends in Plant Science* 18: 30–40.

Boudsocq, M., M.R. Willmann, M. McCormack, H. Lee, L. Shan, P. He, J. Bush, S.H. Cheng and J. Sheen. 2010. Differential innate immune signalling via Ca(2+) sensor protein kinases. *Nature* 464: 418–422.

Boursiac, Y., J. Boudet, O. Postaire, D.T. Luu, C. Tournaire-Roux and C. Maurel. 2008. Stimulus-induced downregulation of root water transport involves reactive oxygen species-activated cell signalling and plasma membrane intrinsic protein internalization. *The Plant Journal* 56: 207–218.

Chefdor, F., H. Benedetti, C. Depierreux, F. Delmotte, D. Morabito and S. Carpin. 2006. Osmotic stress sensing in *Populus*: Components identification of a phosphorelay system. *FEBS Letters* 580: 77–81.

Cheng, N.H., J.K. Pittman, J.K. Zhu and K.D. Hirschi. 2004. The protein kinase SOS2 activates the *Arabidopsis* H$^+$/Ca$^{2+}$ antiporter CAX1 to integrate calcium transport and salt tolerance. *Journal of Biological Chemistry* 279: 2922–2926.

Chinnusamy, V. and J.K. Zhu. 2009. Epigenetic regulation of stress responses in plants. *Current Opinion in Plant Biology* 12: 133–139.

Chinnusamy, V., K. Schumaker and J.K. Zhu. 2004. Molecular genetic perspectives on cross-talk and specificity in abiotic stress signalling in plants. *Journal of Experimental Botany* 55: 225–236.

Chinnusamy, V., A. Jagendorf and J.K. Zhu. 2005. Understanding and improving salt tolerance in plants. *Crop Science* 45: 437–448.

Ciftci-Yilmaz, S., M.R. Morsy, L. Song, A. Coutu, B.A. Krizek, M.W. Lewis, D. Warren, J. Cushman, E.L. Connolly and R. Mittler. 2007. The ear-motif of the C2H2 zinc-finger protein ZAT7 plays a key role in the defense response of *Arabidopsis* to salinity stress. *Journal of Biological Chemistry* 282: 9260–9268.

Clarkson, D.T. and J.B. Hanson. 1980. The mineral nutrition of higher-plants. *Annual Review of Plant Physiology* 31: 239–298.

Cramer, G.R. and R.S. Nowak. 1992. Supplemental manganese improves the relative growth, net assimilation and photosynthetic rates of salt-stressed barley. *Physiologia Plantarum* 84: 600–605.

D'Onofrio, C. and S. Lindberg. 2009. Sodium induces simultaneous changes in cytosolic calcium and pH in salt-tolerant quince protoplasts. *Journal of Plant Physiology* 166: 1755–1763.

Davletova, S., L. Rizhsky, H. Liang, Z. Shengqiang, D.J. Oliver, J. Coutu, V. Shulaev, K. Schlauch and R. Mittler. 2005. Cytosolic ascorbate peroxidase 1 is a central component of the reactive oxygen gene network of *Arabidopsis*. *The Plant Cell* 17: 268–281.

Demidchik, V., T.A. Cuin, D. Svistunenko, S.J. Smith, A.J. Miller, S. Shabala, A. Sokolik and V. Yurin. 2010. *Arabidopsis* root K$^+$-efflux conductance activated by hydroxyl radicals: Single-channel properties, genetic basis and involvement in stress-induced cell death. *Journal of Cell Science* 123: 1468–1479.

Demidchik, V., Z. Shang, R. Shin, R. Colaço, A. Laohavisit, S. Shabala and J.M. Davies. 2011. Receptor-like activity evoked by extracellular ADP in *Arabidopsis* root epidermal plasma membrane. *Plant Physiology* 156: 1375–1385.

Denby, K. and C. Gehring. 2005. Engineering drought and salinity tolerance in plants: Lessons from genome-wide expression profiling in *Arabidopsis*. *Trends in Biotechnology* 23: 547–552.

Donaldson, L., N. Ludidi, M.R. Knight, C. Gehring and K. Denby. 2004. Salt and osmotic stress cause rapid increases in *Arabidopsis thaliana* cGMP levels. *FEBS Letters* 569: 317–320.

Droillard, M.J., M. Boudsocq, H. Barbier-Brygoo and C. Lauriere. 2004. Involvement of MPK4 in osmotic stress response pathways in cell suspensions and plantlets of *A. thaliana*: Activation by hypoosmolarity and negative role in hyperosmolarity tolerance. *FEBS Letters* 574: 42–48.

Du, W., H. Lin, S. Chen, Y. Wu, J. Zhang, A.T. Fuglsang, M.G. Palmgren, W. Wu and Y. Guo. 2011. Phosphorylation of SOS3-like calcium-binding proteins by their interacting SOS2-like protein kinases is a common regulatory mechanism in *Arabidopsis*. *Plant Physiology* 156: 2235–2243.

Duan, J.J., J. Li, S.R. Guo and Y.Y. Kang. 2008. Exogenous spermidine affects polyamine metabolism in salinity-stressed *Cucumis sativus* roots and enhances short-term salinity tolerance. *Journal of Plant Physiology* 165: 1620–1635.

Einspahr, K.J., T.C. Peeler and G.A. Thompson Jr. 1988. Rapid changes in polyphosphoinositide metabolism associated with the response of *Dunaliella salina* to hypoosmotic shock. *Journal of Biological Chemistry* 263: 5775–5779.

Essah, P.A., R. Davenport and M. Tester. 2003. Sodium influx and accumulation in *Arabidopsis*. *Plant Physiology* 133: 307–318.

Evans, A.R., D. Hall, J. Pritchard and H.J. Newbury. 2012. The roles of the cation transporters CHX21 and CHX23 in the development of *Arabidopsis thaliana*. *Journal of Experimental Botany* 63: 59–67.

Flowers, T.J. 2004. Improving crop salt tolerance. *Journal of Experimental Botany* 55: 307–319.

Flowers, T.J. and T.D. Colmer. 2008. Salinity tolerance in halophytes. *New Phytologist* 179: 945–963.

Fukuda, A., A. Nakamura, A. Tagiri, H. Tanaka, A. Miyao, H. Hirochika and Y. Tanaka. 2004. Function, intracellular localization and the importance in salt tolerance of a vacuolar $Na^+/H^+$ antiporter from rice. *Plant and Cell Physiology* 45: 146–159.

Furumoto, T., T. Yamaguchi, Y. Ohshima-Ichie, M. Nakamura, Y. Tsuchida-Iwata, M. Shimamura, J. Ohnishi, S. Hata, U. Gowik, P. Westhoff, A. Bräutigam, A.P. Weber and K. Izui. 2011. A plastidial sodium-dependent pyruvate transporter. *Nature* 476: 472–475.

Galvan-Ampudia, C.S. and C. Testerink. 2011. Salt stress signals shape the plant root. *Current Opinion in Plant Biology.* 14: 296–302.

Genoud, T. and J.P. Metraux. 1999. Crosstalk in plant cell signaling: Structure and function of the genetic network. *Trends in Plant Science* 4: 503–507.

Gill, S.S. and N. Tuteja. 2010. Polyamines and abiotic stress tolerance in plants. *Plant Signaling and Behavior* 5: 26–33.

Giraud, E., L.H. Ho, R. Clifton, A. Carroll, G. Estavillo, Y.F. Tan, K.A. Howell, A. Ivanova, B.J. Pogson, A.H. Millar and J. Whelan. 2008. The absence of ALTERNATIVE OXIDASE1a in *Arabidopsis* results in acute sensitivity to combined light and drought stress. *Plant Physiology* 147: 595–610.

Greenway, H. and R. Munns. 1980. Mechanisms of salt tolerance in nonhalophytes. *Annual Review of Plant Physiology and Plant Molecular Biology* 31: 149–190.

Groppa, M.D. and M.P. Benavides. 2008. Polyamines and abiotic stress: Recent advances. *Amino Acids* 34: 35–45.

Guo, Y., L. Xiong, C.P. Song, D. Gong, U. Halfter and J.K. Zhu. 2002. A calcium sensor and its interacting protein kinase are global regulators of abscisic acid signaling in *Arabidopsis*. *Developmental Cell* 3: 233–244.

Halford, N.G. and S.J. Hey. 2009. SNF1-related protein kinases (SnRKs) act within an intricate network that links metabolic and stress signalling in plants. *Biochemical Journal* 419: 247–259.

Hall, D., A.R. Evans, H.J. Newbury and J. Pritchard. 2006. Functional analysis of CHX21: A putative sodium transporter in *Arabidopsis*. *Journal of Experimental Botany* 57: 1201–1210.

Halperin, S.J., S. Gilroy and J.P. Lynch. 2003. Sodium chloride reduces growth and cytosolic calcium, but does not affect cytosolic pH, in root hairs of *Arabidopsis thaliana* L. *Journal of Experimental Botany* 54: 1269–1280.

Hark, A.T., K.E. Vlachonasios, K.A. Pavangadkar, S. Rao, H. Gordon, I.D. Adamakis, A. Kaldis, M.F. Thomashow and S.J. Triezenberg. 2009. Two *Arabidopsis* orthologs of the transcriptional coactivator ADA2 have distinct biological functions. *Biochimica and Biophysica Acta* 1789: 117–124.

Haro, R., M.A. Banuelos and A. Rodriguez-Navarro. 2010. High affinity sodium uptake in land plants. *Plant and Cell Physiology* 51: 68–79.

Hasegawa, P.M., R.A. Bressan, J.K. Zhu and H.J. Bohnert. 2000. Plant cellular and molecular responses to high salinity. *Annual Review of Plant Physiology and Plant Molecular Biology* 51: 463–499.

He, L., Y. Ban, H. Inoue, N. Matsuda, J. Liu and T. Moriguchi. 2008. Enhancement of spermidine content and antioxidant capacity in transgenic pear shoots overexpressing apple spermidine synthase in response to salinity and hyperosmosis. *Phytochemistry* 69: 2133–2141.

Held, K., F. Pascaud, C. Eckert, P. Gajdanowicz, K. Hashimoto, C. Corratgé-Faillie, J.N. Offenborn, B. Lacombe, I. Dreyer, J.B. Thibaud and J. Kudla. 2010. Calcium-dependent modulation and plasma membrane targeting of the AKT2 potassium channel by the CBL4/CIPK6 calcium sensor/protein kinase complex. *Cell Research* 21: 1116–1130.

Hey, S.J., E. Byrne and N.G. Halford. 2010. The interface between metabolic and stress signaling. *Annals of Botany* 105: 197–203.

Hirayama, T., C. Ohto, T. Mizoguchi and K. Shinozaki. 1995. A gene encoding a phosphatidylinositol-specific phospholipase C is induced by dehydration and salt stress in *Arabidopsis thaliana*. *Proceedings of the National Academy of Sciences, USA* 92: 3903–3907.

Hong, Y., S.P. Devaiah, S.C. Bahn, B.N. Thamasandra, M. Li, R. Welti and X. Wang. 2009. Phospholipase D epsilon and phosphatidic acid enhance *Arabidopsis* nitrogen signaling and growth. *The Plant Journal* 58: 376–387.

Hong, Y., W. Zhang and X. Wang. 2010. Phospholipase D and phosphatidic acid signalling in plant response to drought and salinity. *Plant, Cell and Environment* 33: 627–635.

Hrabak, E.M., C.W. Chan, M. Gribskov, J.F. Harper, J.H. Choi, N. Halford, J. Kudla, S. Luan, H.G. Nimmo, M.R. Sussman, M. Thomas, K. Walker-Simmons, J.K. Zhu and A.C. Harmon. 2003. The *Arabidopsis* CDPK-SnRK superfamily of protein kinases. *Plant Physiology* 132: 666–680.

Huang, D.Q., W.R. Wu, S.R. Abrams and A.J. Cutler. 2008. The relationship of drought-related gene expression in *Arabidopsis thaliana* to hormonal and environmental factors. *Journal of Experimental Botany* 59: 2991–3007.

Huang, Y., H. Li, R. Gupta, P.C. Morris, S. Luan and J.J. Kieber. 2000. ATMPK4, an *Arabidopsis* homolog of mitogen-activated protein kinase, is activated in vitro by AtMEK1 through threonine phosphorylation. *Plant Physiology* 122: 1301–1310.

Huang, Y., H. Li, C.E. Hutchison, J. Laskey and J.J. Kieber. 2003. Biochemical and functional analysis of CTR1, a protein kinase that negatively regulates ethylene signaling in *Arabidopsis*. *The Plant Journal* 33: 221–233.

Ichimura, K., T. Mizoguchi, R. Yoshida, T. Yuasa and K. Shinozaki. 2000. Various abiotic stresses rapidly activate *Arabidopsis* MAP kinases ATMPK4 and ATMPK6. *The Plant Journal* 24: 655–665.

Jacob, T., S. Ritchie, S.M. Assmann and S. Gilroy. 1999. Abscisic acid signal transduction in guard cells is mediated by phospholipase D activity. *Proceedings of the National Academy of Sciences, USA* 96: 12192–12197.

James, R.A., C. Blake, A.B. Zwart, R.A. Hare, A.J. Rathjen and R. Munns. 2012. Impact of ancestral wheat sodium exclusion genes Nax1 and Nax2 on grain yield of durum wheat on saline soils. *Functional Plant Biology* 39: 609–618.

Ji, H., J.M. Pardo, G. Batelli, M.J. Van Oosten, R.A. Bressan and X. Li. 2013. The salt overly sensitive (SOS) pathway: Established and emerging roles. *Molecular Plant* 6: 275–286.

Ju, C., G.M. Yoon, J.M. Shemansky, D.Y. Lin, Z.I. Ying, J. Chang, W.M. Garrett, M. Kessenbrock, G. Groth, M.L. Tucker, B. Cooper, J.J. Kieber and C. Chang. 2012. CTR1 phosphorylates the central regulator EIN2 to control ethylene hormone signaling from the ER membrane to the nucleus in *Arabidopsis*. *Proceedings of the National Academy of Sciences, USA* 109: 19486–19491.

Kader, M.A. and S. Lindberg. 2010. Cytosolic calcium and pH signaling in plants under salinity stress. *Plant Signaling and Behaviour* 5: 233–238.

Kader, M.A., S. Lindberg, T. Seidel, D. Golldack and V. Yemelyanov. 2007. Sodium sensing induces different changes in free cytosolic calcium concentration and pH in salt-tolerant and salt-sensitive rice (*Oryza sativa* L.) cultivars. *Physiologia Plantarum* 130: 99–111.

Kaldis, A., D. Tsementzi, O. Tanriverdi and K.E. Vlachonasios. 2010. *Arabidopsis thaliana* transcriptional co-activators ADA2b and SGF29a are implicated in salt stress responses. *Planta* 233: 749–762.

Kanchiswamy, C.N., H. Takahashi, S. Quadro, M.E. Maffei, S. Bossi, C. Bertea, S.A. Zebelo, A. Muroi, N. Ishihama, H. Yoshioka, W. Boland, J. Takabayashi, Y. Endo, T. Sawasaki and G. Arimura. 2010. Regulation of *Arabidopsis* defense responses against *Spodoptera littoralis* by CPK-mediated calcium signaling. *BMC Plant Biology* 10: 97.

Kant, P., S. Kant, M. Gordon, R. Shaked and S. Barak. 2007. STRESS RESPONSE SUPPRESSOR1 and STRESS RESPONSE SUPPRESSOR2, two DEAD-box RNA helicases that attenuate *Arabidopsis* responses to multiple abiotic stresses. *Plant Physiology* 145: 814–830.

Katagiri, T., S. Takahashi and K. Shinozaki. 2001. Involvement of a novel *Arabidopsis* phospholipase D, AtPLDd, in dehydration-inducible accumulation of phosphatidic acid in stress signalling. *The Plant Journal* 26: 595–605.

Kieber, J.J., M. Rothenberg, G. Roman, K.A. Feldmann and J.R. Ecker. 1993. CTR1, a negative regulator of the ethylene response pathway in *Arabidopsis*, encodes a member of the raf family of protein kinases. *Cell* 72: 427–441.

Kiegle, E., C.A. Moore, J. Haseloff, M.A. Tester and M.R. Knight. 2000. Cell-type-specific calcium responses to drought, salt and cold in the *Arabidopsis* root. *The Plant Journal* 23: 267–278.

Kilian, J., D. Whitehead, J. Horak, D. Wanke, S. Weinl, O. Batistic, C. D'Angelo, E. Bornberg-Bauer, J. Kudla and K. Harter. 2007. The AtGenExpress global stress expression data set: Protocols, evaluation and model data analysis of UV-B light, drought and cold stress responses. *The Plant Journal* 50: 347–363.

Kim, S.H., D.H. Woo, J.M. Kim, S.Y. Lee, W.S. Chung and Y.H. Moon. 2011. *Arabidopsis* MKK4 mediates osmotic-stress response via its regulation of MPK3 activity. *Biochemical and Biophysical Research Communications* 412: 150–154.

Kim, J.M., D.H. Woo, S.H. Kim, S.Y. Lee, H.Y. Park, H.Y. Seok, W.S. Chung, Y.H. Moon. 2012. *Arabidopsis* MKKK20 is involved in osmotic stress response via regulation of MPK6 activity. *Plant Cell Reports* 31: 217–224.

Knight, H. 2000. Calcium signaling during abiotic stress in plants. *International Review of Cytology-A Survey of Cell Biology* 195: 269–324.

Knight, H., A.J. Trewavas and M.R. Knight. 1997. Calcium signaling in *Arabidopsis thaliana* responding to drought and salinity. *The Plant Journal* 12: 1067–1078.

Kumar, K., M. Kumar, S.R. Kim, H. Ryu and Y.G. Cho. 2013. Insights into genomics of salt stress response in rice. *Rice* 6: 27.

Kusajima, M., M. Yasuda, A. Kawashima, H. Nojiri, H. Yamane, M. Nakajima, K. Akutsu and H. Nakashita. 2010. Suppressive effect of abscisic acid on systemic acquired resistance in tobacco plants. *Journal of General Plant Pathology* 76: 161–167.

Läuchli, A. and S.R. Grattan. 2007. Plant growth and development under salinity stress, in M.A. Jenks et al. (eds.), *Advances in Molecular Breeding Toward Drought and Salt Tolerant Crops*, Springer, New York, pp. 1–32.

Laurie, S., K.A. Feeney, F.J.M. Maathuis, P.J. Heard, S.J. Brown and R.A. Leigh. 2002. A role for HKT1 in sodium uptake by wheat roots. *The Plant Journal* 32: 139–149.

Lefevre, I., E. Gratia and S. Lutts. 2001. Discrimination between the ionic and osmotic components of salt stress in relation to free polyamine level in rice (*Oryza sativa*). *Plant Science* 16: 943–952.

Leshem, Y., L. Seri and A. Levine. 2007. Induction of phosphatidylinositol 3-kinase-mediated endocytosis by salt stress leads to intracellular production of reactive oxygen species and salt tolerance. *The Plant Journal* 51: 185–197.

Lessani, H. and H. Marschner. 1978. Relation between salt tolerance and long-distance transport of sodium and chloride in various crop species. *Australian Journal of Plant Physiology* 5: 27–37.

Li, M., Y. Hong and X. Wang. 2009. Phospholipase D- and phosphatidic acid-mediated signaling in plants. *Biochimica et Biophysica Acta* 1791: 927–935.

Ligterink, W. and H. Hirt. 2000. MAP kinase pathways in plants: Versatile signalling tools. *International Review of Cytology* 201: 209–258.

Liu, J., M. Ishitani, U. Halfter, C.S. Kim and J.K. Shu. 2000. The *Arabidopsis thaliana* SOS2 gene encodes a protein kinase that is required for salt tolerance. *Proceedings of the National Academy of Sciences, USA* 97: 3730–3734.

Liu, J.H., H. Kitashiba, J. Wang, Y. Ban and T. Moriguchi. 2007. Polyamines and their ability to provide environmental stress tolerance to plants. *Plant Biotechnology* 24: 117–126.

Liu, Q., M. Kasuga, Y. Sakuma, H. Abe, S. Miura, K. Yamaguchi-Shinozaki and K. Shinozaki. 1998. Two transcription factors, DREB1 and DREB2, with an EREBP/AP2 DNA binding domain separate two cellular signal transduction pathways in drought- and low-temperature-responsive gene expression, respectively, in *Arabidopsis*. *Plant Cell* 10: 1391–1406.

López-Pérez, L., M.C. Martínez-Ballesta, C. Maurel and M. Carvajal. 2009. Changes in plasma membrane lipids, aquaporins and proton pump of broccoli roots, as an adaptation mechanism to salinity. *Phytochemistry* 70: 492–500.

Lynch, J. and A. Läuchli. 1988. Salinity affects intracellular calcium in corn root protoplasts. *Plant Physiology* 87: 351–356.

Ma, S., Q. Gong and H.J. Bohnert. 2006. Dissecting salt stress pathways. *Journal of Experimental Botany* 57: 1097–1107.

Ma, Y., I. Szostkiewicz, A. Korte, D. Moes, Y. Yang, A. Christmann and E. Grill. 2009. Regulators of PP2C phosphatase activity function as abscisic acid sensors. *Science* 324: 1064–1068.

Maathuis, F.J.M. 2006. cGMP modulates gene transcription and cation transport in *Arabidopsis* roots. *The Plant Journal* 45: 700–711.

Maathuis, F.J.M. 2013. Sodium in plants: Perception, signalling, and regulation of sodium fluxes. *Journal of Experimental Botany* 65: 849–858.

Maathuis, F.J.M. and D. Sanders. 2001. Sodium uptake in *Arabidopsis thaliana* roots is regulated by cyclic nucleotides. *Plant Physiology* 127: 1617–1625.

Mahfouz, M.M., S. Kim, A.J. Delauney and D.P. Verma. 2006. *Arabidopsis* TARGET OF RAPAMYCIN interacts with RAPTOR, which regulates the activity of S6 kinase in response to osmotic stress signals. *The Plant Cell* 18: 477–490.

Mantri, N., V. Patade, S. Penna, R. Ford and E. Pang. 2012. Abiotic stress responses in plants: Present and future, in P. Ahmad and M.N.V. Prasad (eds.), *Abiotic Stress Responses in Plants: Metabolism, Productivity and Sustainability*, Springer, New York, pp. 1–19.

Marschner, H. 1995. *Mineral Nutrition of Higher Plants*, 2nd edition, Academic Press, London, U.K., pp. 388–390.

Mehlmer, N., B. Wurzinger, S. Stael, D. Hofmann-Rodrigues, E. Csaszar, B. Pfister, R. Bayer and M. Teige. 2010. The Ca$^{(2+)}$-dependent protein kinase CPK3 is required for MAPK-independent salt-stress acclimation in *Arabidopsis*. *The Plant Journal* 63: 484–498.

Meijer, H.J., B. ter Riet, J.A. van Himbergen, A. Musgrave and T. Munnik. 2002. KCl activates phospholipase D at two different concentration ranges: Distinguishing between hyperosmotic stress and membrane depolarization. *The Plant Journal* 31: 51–59.

Mian, A., R.J.F.J. Oomen, S. Isayenkov, H. Sentenac, F.J.M. Maathuis and A.A. Véry. 2011. Overexpression of a Na$^+$ and K$^+$-permeable HKT transporter in barley improves salt tolerance. *The Plant Journal* 63: 468–479.

Miller, G., N. Suzuki, L. Rizhsky, A. Hegie, S. Koussevitzky and R. Mittler. 2007. Double mutants deficient in cytosolic and thylakoid ascorbate peroxidase reveal a complex mode of interaction between reactive oxygen species, plant development, and response to abiotic stresses. *Plant Physiology* 144: 1777–1785.

Miller, G., V. Shulaev and R. Mittler. 2008. Reactive oxygen signaling and abiotic stress. *Physiologia Plantarum* 133: 481–489.

Miller, G., N. Suzuki, S. Ciftci-Yilmaz and R. Mittler. 2010. Reactive oxygen species homeostasis and signaling during drought and salinity stresses. *Plant, Cell and Environment* 33: 453–467.

Mishra, G., W. Zhang, F. Deng, J. Zhao and X. Wang. 2006. A bifurcating pathway directs abscisic acid effects on stomatal closure and opening in *Arabidopsis*. *Science* 312: 264–266.

Mittler, R., S. Vanderauwera, M. Gollery and F. Van Breusegem. 2004. Reactive oxygen gene network of plants. *Trends in Plant Science* 9: 490–498.

Miyazawa, Y., A. Takahashi, A. Kobayashi, T. Kaneyasu, N. Fujii and H. Takahashi. 2009. GNOM-mediated vesicular trafficking plays an essential role in hydrotropism of *Arabidopsis* roots. *Plant Physiology* 149: 835–840.

Mizoguchi, T., K. Irie, T. Hirayama, N. Hayashida, K. Yamaguchi-Shinozaki, K. Matsumoto and K. Shinozaki. 1996. A gene encoding a mitogen-activated protein kinase kinase kinase is induced simultaneously with genes for a mitogen-activated protein kinase and an S6 ribosomal protein kinase by touch, cold, and water stress in *Arabidopsis thaliana*. *Proceedings of the National Academy of Sciences, USA* 93: 765–769.

Mohr, P.G. and D.M. Cahill. 2007. Suppression by ABA of salicylic acid and lignin accumulation and the expression of multiple genes in *Arabidopsis* infected with *Pseudomonas syringae* pv. tomato. *Functional and Integrative Genomics* 7: 181–191.

Møller, I.S., M. Gilliham, J. Deepa, G.M. Mayo, S.J. Roy, J.C. Coates, J. Haseloff and M. Tester. 2009. Shoot Na⁺ exclusion and increased salinity tolerance engineered by cell type-specific alteration of Na⁺ transport in *Arabidopsis*. *The Plant Cell* 21: 2163–2178.

Munnik, T. and C. Testerink. 2009. Plant phospholipid signaling: "In a nutshell." *Journal of Lipid Research* 50: S260–S265.

Munnik, T., H.J.G. Meijer, B. ter Riet, H. Hirt, W. Frank, D. Bartels and A. Musgrave. 2000. Hyperosmotic stress stimulates phospholipase D activity and elevates the levels of phosphatidic acid and diacylglycerol pyrophosphate. *The Plant Journal* 22: 147–154.

Munns, R. 1988. Causes of varietal differences in salt tolerance, in S.K. Sinha et al. (eds.), *Proceedings of the International Congress of Plant Physiology*, Vol. 2, Greated Kalla New Delhi, New Delhi, pp. 960–968.

Munns, R. 2002. Comparative physiology of salt and water stress. *Plant, Cell and Environment* 25: 239–250.

Munns, R. 2005. Genes and salt tolerance: Bringing them together. *New Phytologist* 167: 645–663.

Munns, R. and M. Tester. 2008. Mechanisms of salinity tolerance. *Annual Review of Plant Biology* 59: 651–681.

Narusaka, Y., M. Narusaka, M. Seki, T. Umezawa, J. Ishida, M. Nakajima, A. Enju and K. Shinozaki. 2004. Crosstalk in the responses to abiotic and biotic stresses in *Arabidopsis*: Analysis of gene expression in cytochrome P450 gene superfamily by cDNA microarray. *Plant Molecular Biology* 55: 327–342.

Ning, J., X. Li, L.M. Hicks and L. Xiong. 2010. A Raf-like MAPKKK gene DSM1 mediates drought resistance through reactive oxygen species scavenging in rice. *Plant Physiology* 152: 876–890.

Niu, X., R.A. Bressan, P.M. Hasegawa and J.M. Pardo. 2005. Ion homeostasis in NaCl stress environments. *Plant Physiology* 109: 735–742.

Nuccio, M.L., D. Rhodes, S.D. McNeil and A.D. Hanson. 1999. Metabolic engineering of plants for osmotic stress resistance. *Current Opinion in Plant Biology* 2: 128–134.

Ohta, M., Y. Guo, U. Halfter and J.K. Zhu. 2003. A novel domain in the protein kinase SOS2 mediates interaction with the protein phosphatase 2C ABI2. *Proceedings of the National Academy of Sciences, USA* 100: 11771–11776.

Ouaked, F., W. Rozhon, D. Lecourieux and H. Hirt. 2003. A MAPK pathway mediates ethylene signaling in plants. *EMBO Journal* 22: 1282–1288.

Pandey, S. and S.M. Assmann. 2004. The *Arabidopsis* putative G protein coupled receptor GCR1 interacts with the G protein a subunit GPA1 and regulates abscisic acid signaling. *The Plant Cell* 16: 1616–1632.

Pandey, S., D.C. Nelson and S.M. Assmann. 2009. Two novel GPCR type G proteins are abscisic acid receptors in *Arabidopsis*. *Cell* 136: 136–148.

Pardo, J.M. 2010. Biotechnology of water and salinity stress tolerance. *Current Opinion in Biotechnology* 21: 185–196.

Park, S.Y., P. Fung, N. Nishimura, D.R. Jensen, H. Fujii, Y. Zhao, S. Lumba, J. Santiago, A. Rodrigues, T.F. Chow, S.E. Alfred, D. Bonetta, R. Finkelstein, N.J. Provart, D. Desveaux, P.L. Rodriguez, P. McCourt, J.K. Zhu, J.I. Schroeder, B.F. Volkman and S.R. Cutler. 2009. Abscisic acid inhibits type 2C protein phosphatases via the PYR/PYL family of START proteins. *Science* 324: 1068–1071.

Persak, H. and A. Pitzschke. 2013. Tight interconnection and multi-level control of *Arabidopsis* MYB4 in MAPK cascade signalling. *PLoS One* 8: e57547.

Pinheiro, G.L., C.S. Marques, M.D. Costa, P.A. Reis, M.S. Alves, C.M. Carvalho, L.G. Fietto and E.P. Fontes. 2009. Complete inventory of soybean NAC transcription factors: Sequence conservation and expression analysis uncover their distinct roles in stress response. *Gene* 444: 10–23.

Plieth, C., U.P. Hansen, H. Knight and M.R. Knight. 1999. Temperature sensing by plants: the primary characteristics of signal perception and calcium response. *Plant Journal* 18: 491–497.

Prabhavathi, V.R. and M.V. Rajam. 2007. Polyamine accumulation in transgenic eggplant enhances tolerance to multiple abiotic stresses and fungal resistance. *Plant Biotechnology* 24: 273–282.

Qin, C. and X. Wang. 2002. The *Arabidopsis* phospholipase D family. Characterization of a calcium-independent and phosphatidylcholine-selective PLDz1 with distinct regulatory domains. *Plant Physiology* 128: 1057–1068.

Qiu, Q.S., Y. Guo, F.J. Quintero, J.M. Pardo, K.S. Schumaker and J.K. Zhu. 2004. Regulation of vacuolar $Na^+/H^+$ exchange in *Arabidopsis thaliana* by the salt-overly-sensitive (SOS) pathway. *Journal of Biological Chemistry* 279: 207–215.

Quan, R., H. Lin, I. Mendoza, Y. Zhang, W. Cao, Y. Yang, M. Shang, S. Chen, J.M. Pardo and Y. Guo. 2007. SCABP8/CBL10, a putative calcium sensor, interacts with the protein kinase SOS2 to protect *Arabidopsis* shoots from salt stress. *Plant Cell* 19: 1415–1431.

Quintero, F.J., J. Martinez-Atienza, I. Villalta, X. Jiang, W.Y. Kim, Z. Ali, H. Fujii, I. Mendoza, D.J. Yun and J.K. Zhu. 2011. Activation of the plasmamembrane Na/H antiporter salt-overly-sensitive 1 (SOS1) by phosphorylation of an auto-inhibitory C-terminal domain. *Proceedings of the National Academy of Sciences, USA* 108: 2611–2616.

Reddy, A.S.N. 2001. Calcium: Silver bullet in signaling. *Plant Science* 160: 381–404.

Ren, Z.H., J.P. Gao, L.G. Li, X.L. Cai, W. Huang, D.Y. Chao, M.Z. Zhu, Z.Y. Wang, S. Luan and H.X. Lin. 2005. A rice quantitative trait locus for salt tolerance encodes a sodium transporter. *Nature Genetics* 37: 1141–1146.

Rhodes, D. and A.D. Hanson. 1993. Quaternary ammonium and tertiary sulfonium compounds in higher plants. *Annual Review of Plant Physiology and Plant Molecular Biology* 44: 357–384.

Rodríguez, A.A., S.J. Maiale, A.B. Menéndez and O.A. Ruiz. 2009. Polyamine oxidase activity contributes to sustain maize leaf elongation under saline stress. *Journal of Experimental Botany* 60: 4249–4262.

Rodriguez, M.C., M. Petersen and J. Mundy. 2010. Mitogen activated protein kinase signaling in plants. *Annual Review of Plant Biology* 61: 621–649.

Rohila, J.S. and Y.N. Yang. 2007. Rice mitogen-activated protein kinase gene family and its role in biotic and abiotic stress response. *Journal of Integrative Plant Biology* 49: 751–759.

Roshandel, P. and T. Flowers. 2009. The ionic effects of NaCl on physiology and gene expression in rice genotypes differing in salt tolerance. *Plant and Soil* 315: 135–147.

Roy, S.J., S. Negrão and M. Tester. 2014. Salt resistant crop plants. *Current Opinion in Biotechnology* 26: 115–124.

Rus, A., B.H. Lee, A. Munoz-Mayor, A. Sharkhuu, K. Miura, J.K. Zhu, R.A. Bressan and P.M. Hasegawa. 2001. AtHKT1 facilitates $Na^+$ homeostasis and $K^+$ nutrition in planta. *Plant Physiology* 136: 2500–2511.

Sanchez, J.P. and N.H. Chua. 2001. *Arabidopsis* PLC1 is required for secondary responses to abscisic acid signals. *The Plant Cell* 13: 1143–1154.

Sanders, D., C. Brownlee and J.F. Harper. 1999. Communicating with calcium. *Plant Cell* 11: 691–706.

Sanders, D., J. Pelloux, C. Brownlee and J.F. Harper. 2002. Calcium at the cross-roads of signaling. *Plant Cell* 14: 401–417.

Sang, Y., D. Cui and X. Wang. 2001. Phospholipase D and phosphatidic acid-mediated generation of superoxide in *Arabidopsis*. *Plant Physiology* 126: 1449–1458.

Schmidt, R., D. Mieulet, H.M. Hubberten, T. Obata, R. Hoefgen, A.R. Fernie, J. Fisahn, B. San Segundo, E. Guiderdoni, J.H. Schippers and B. Mueller-Roeber. 2013. Salt-responsive ERF1 regulates reactive oxygen species-dependent signaling during the initial response to salt stress in rice. *Plant Cell* 25: 2115–2131.

Schroeder, J.I., J.M. Kwak and G.J. Allen. 2001. Guard cell abscisic acid signalling and engineering drought hardiness in plants. *Nature* 410: 327–330.

Shabala, L., T.A. Cuin, I.A. Newman and S. Shabala. 2005. Salinity induced ion flux patterns from the excised roots of *Arabidopsis* sos mutants. *Planta* 222: 1041–1050.

Shen, J., K. Xie and L. Xiong. 2010. Global expression profiling of rice microRNAs by one-tube stem–loop reverse transcription quantitative PCR revealed important roles of microRNAs in abiotic stress responses. *Molecular and General Genetics* 284: 477–488.

Shi, H., M. Ishitani, C. Kim and J.K. Zhu. 2000. The *Arabidopsis thaliana* salt tolerance gene SOS1 encodes a putative Na/H antiporter. *Proceedings of the National Academy of Sciences, USA* 97: 6896–6901.

Shi, H., F.J. Quintero, J.M. Pardo and J.K. Zhu. 2002. The putative plasmamembrane Na$^+$/H$^+$ antiporter SOS1 controls long distance Na$^+$ transport in plants. *Plant Cell* 14: 465–477.

Shinozaki, K. and K. Yamaguchi-Shinozaki. 2000. Molecular responses to dehydration and low temperature: Differences and cross-talk between two stress signaling pathways. *Current Opinion in Plant Biology* 3: 217–223.

Sinha, A.K., M. Jaggi, B. Raghuram and N. Tuteja. 2011. Mitogen-activated protein kinase signaling in plants under abiotic stress. *Plant Signaling and Behavior* 6: 196–203.

Smékalová, V., A. Doskočilová, G. Komis and J. Šamaj. 2014. Crosstalk between secondary messengers, hormones and MAPK modules during abiotic stress signalling in plants. *Biotechnology Advances* 32: 2–11.

Soni, P., G. Kumar, N. Soda, S.L. Singla-Pareek and A. Pareek. 2013. Salt overly sensitive pathway members are influenced by diurnal rhythm in rice. *Plant Signaling and Behavior* 8: e24738.

Stockinger, E.J., Y. Mao, M.K. Regier, S.J. Triezenberg and M.F. Thomashow. 2001. Transcriptional adaptor and histone acetyltransferase proteins in *Arabidopsis* and their interactions with CBF1, a transcriptional activator involved in cold-regulated gene expression. *Nucleic Acids Research* 29: 1524–1533.

Su, S.H., M.C. Suarez-Rodriguez and P. Krysan. 2007. Genetic interaction and phenotypic analysis of the *Arabidopsis* MAP kinase pathway mutations mekk1 and mpk4 suggests signaling pathway complexity. *FEBS Letters* 581: 3171–3177.

Subbarao, G.V., O. Ito, W.L. Berry and R.M. Wheeler. 2003. Sodium: A functional plant nutrient. *Critical Reviews in Plant Sciences* 22: 391–416.

Sunkar, R. and J.K. Zhu. 2004. Novel and stress-regulated microRNAs and other small RNAs from *Arabidopsis*. *The Plant Cell* 16: 2001–2019.

Suzuki, N., G. Miller, J. Morales, V. Shulaev, M.A. Torres and R. Mittler. 2011. Respiratory burst oxidases: The engines of ROS signaling. *Current Opinion in Plant Biology* 14: 691–699.

Swindell, W.R. 2006. The association among gene expression responses to nine abiotic stress treatments in *Arabidopsis thaliana*. *Genetics* 174: 1811–1824.

Taiz, L., E. Zeiger, I.M. Møller and A. Murphy. 2015. Plant Physiology and Development. 6th edn. Inc., Publishers, Sunderland, MA.

Takahashi, F., R. Yoshida, K. Ichimura, T. Mizoguchi, S. Seo, M. Yonezawa, K. Maruyama, K. Yamaguchi-Shinozaki and K. Shinozaki. 2007. The mitogen-activated protein kinase cascade MKK3-MPK6 is an important part of the jasmonate signal transduction pathway in *Arabidopsis*. *Plant Cell* 19: 805–818.

Teige, M., E. Scheikl, T. Eulgem, R. Dóczi, K. Ichimura, K. Shinozaki, J.L. Dangl and H. Hirt. 2004. The MKK2 pathway mediates cold and salt stress signaling in *Arabidopsis*. *Molecular Cell* 15: 141–152.

Tester, M. and R. Davenport. 2003. Na$^+$ tolerance and Na+ transport in higher plants. *Annals of Botany* 91: 503–527.

Testerink, C. and T. Munnik. 2005. Phosphatidic acid: A multifunctional stress signaling lipid in plants. *Trends in Plant Science* 10: 368–375.

Testerink, C. and T. Munnik. 2011. Molecular, cellular, and physiological responses to phosphatidic acid formation in plants. *Journal of Experimental Botany* 62: 2349–2361.

Testerink, C., H.L. Dekker, Z.Y. Lim, M.K. Johns, A.B. Holmes, C.G. Koster, N.T. Ktistakis and T. Munnik. 2004. Isolation and identification of phosphatidic acid targets from plants. *The Plant Journal* 39: 527–536.

Ton, J., V. Flors and B. Mauch-Mani. 2009. The multifaceted role of ABA in disease resistance. *Trends in Plant Science* 14: 310–317.

Torres, M.A. and J.L. Dangl. 2005. Functions of the respiratory burst oxidase in biotic interactions, abiotic stress and development. *Current Opinion in Plant Biology* 8: 397–403.

Tracy, F.E., M. Gilliham, A.N. Dodd, A.A. Webb and M. Tester. 2008. NaCl-induced changes in cytosolic free Ca2+ in *Arabidopsis thaliana* are heterogeneous and modified by external ionic composition. *Plant Cell and Environment* 31: 1063–1073.

Umezawa, T., R. Yoshida, K. Maruyama, K. Yamaguchi-Shinozaki and K. Shinozaki. 2004. SRK2C, a SNF1-related protein kinase 2, improves drought tolerance by controlling stress-responsive gene expression in *Arabidopsis thaliana*. *Proceedings of the National Academy of Sciences, USA* 101: 17306–17311.

Urano, K., Y. Yoshiba, T. Nanjo, T. Ito, K. Yamaguchi-Shinozaki and K. Shinozaki. 2004. *Arabidopsis* stress-inducible gene for arginine decarboxylase AtADC2 is required for accumulation of putrescine in salt tolerance. *Biochemical and Biophysical Research Communications* 313: 369–375.

Vanderbeld, B. and W.A. Snedden. 2007. Developmental and stimulus-induced expression patterns of *Arabidopsis* calmodulin-like genes CML37, CML38 and CML39. *Plant Molecular Biology* 64: 683–697.

Verslues, P.E., G. Batelli, S. Grillo, F. Agius, Y.S. Kim, J. Zhu, M. Agarwal, S. Katiyar-Agarwal and J.K. Zhu. 2007. Interaction of SOS2 with nucleoside diphosphate kinase 2 and catalases reveals a point of connection between salt stress and $H_2O_2$ signaling in *Arabidopsis thaliana*. *Molecular Cell Biology* 27: 7771–7780.

Wang, C. and X. Wang 2001. A novel phospholipase D of *Arabidopsis* that is activated by oleic acid and associated with the plasma membrane. *Plant Physiology* 127: 1102–1112.

Wang, C., J. Li and M. Yuan. 2007. Salt tolerance requires cortical microtubule reorganization in *Arabidopsis*. *Plant and Cell Physiology* 48: 1534–1547.

Wang, C., L.J. Zhang and R.D. Huang. 2011. Cytoskeleton and plant salt stress tolerance. *Plant Signaling and Behavior* 6: 29–31.

Wang, J., H. Ding, A. Zhang, F. Ma, J. Cao and M. Jiang. 2010. A novel mitogen-activated protein kinase gene in maize (*Zea mays*), ZmMPK3, is involved in response to diverse environmental cues. *Journal of Integrative Plant Biology* 52: 442–452.

Wang, W., B. Vinocur and A. Altman. 2003. Plant responses to drought, salinity and extreme temperatures: Towards genetic engineering for stress tolerance. *Planta* 218: 1–14.

Wang, X., S.P. Devaiah, W. Zhang and R. Welti. 2006. Signaling functions of phosphatidic acid. *Progress in Lipid Research* 45: 250–278.

Weinl, S. and J. Kudla. 2009. The CBL-CIPK $Ca^{2+}$-decoding signaling network: Function and perspectives. *New Phytologist* 184: 517–528.

Wen, X.P., X.M. Pang, N. Matsuda, M. Kita, H. Inoue, Y.J. Hao, C. Honda and T. Moriguchi. 2008. Overexpression of the apple spermidine synthase gene in pear confers multiple abiotic stress tolerance by altering polyamine titers. *Transgenic Research* 17: 251–263.

West, G., D. Inze and G.T. Beemster. 2004. Cell cycle modulation in the response of the primary root of *Arabidopsis* to salt stress. *Plant Physiology* 135: 1050–1058.

White, P.J. 2000. Calcium channels in higher plants. *Biochimica et Biophysica Acta* 1465: 171–189.

Wi, S.J., W.T. Kim and K.Y. Park. 2006. Overexpression of carnation S-adenosyl methionine decarboxylase gene generates a broad-spectrum tolerance to abiotic stresses in transgenic tobacco plants. *Plant Cell Reports* 25: 1111–1121.

Wu, S.J., D. Lei and J.K. Zhu. 1996. SOS1, a genetic locus essential for salt tolerance and potassium acquisition. *Plant Cell* 8: 617–627.

Wurzinger, B., A. Mair, B. Pfister and M. Teige. 2011. Cross-talk of calcium-dependent protein kinase and MAP kinase signaling. *Plant Signaling and Behavior* 6: 8–12.

Xing, Y., W. Jia and J. Zhang. 2008. AtMKK1 mediates ABA-induced CAT1 expression and $H_2O_2$ production via AtMPK6-coupled signaling in *Arabidopsis*. *The Plant Journal* 54: 440–451.

Xiong, L. and J.K. Zhu. 2001. Abiotic stress signal transduction in plants: Molecular and genetic perspectives. *Physiologia Plantarum* 112: 152–166.

Xiong, L.M., K.S. Schumaker and J.K. Zhu. 2002. Cell signaling during cold, drought and salt stress. *Plant Cell* 14: 165–183.

Xu, J., Y. Li, Y. Wang, H. Liu, L. Lei, H. Yang, G. Liu and D. Ren. 2008. Activation of MAPK kinase 9 induces ethylene and camalexin biosynthesis and enhances sensitivity to salt stress in *Arabidopsis*. *The Journal of Biological Chemistry* 283: 26996–27006.

Xue, Z.Y., D.Y. Zhi, G.P. Xue, H. Zhang, Y.X. Zhao and G.M. Xia. 2004. Enhanced salt tolerance of transgenic wheat (*Tritivum aestivum* L.) expressing a vacuolar $Na^+/H^+$ antiporter gene with improved grain yields in saline soils in the field and a reduced level of leaf $Na^+$. *Plant Science* 167: 849–859.

Yadav, S., M. Irfan, A. Ahmad and S. Hayat. 2011. Causes of salinity and plant manifestations to salt stress: A review. *Journal of Environmental Biology* 32: 667–685.

Yaish, M.W., J. Colasanti and S.J. Rothstein. 2011. The role of epigenetic processes in controlling flowering time in plants exposed to stress. *Journal of Experimental Botany* 62: 3727–3735.

Yamaguchi, T., G.S. Aharon, J.B. Sottosanto and E. Blumwald. 2005. Vacuolar $Na^+/H^+$ antiporter cation selectivity is regulated by calmodulin from within the vacuole in a $Ca^{2+}$- and pH-dependent manner. *Proceedings of the National Academy of Sciences, USA* 102: 16107–16112.

Yasuda, M., A. Ishikawa, Y. Jikumaru, M. Seki, T. Umezawa, T. Asami, A. Maruyama-Nakashita, T. Kudo, K. Shinozaki, S. Yoshida and H. Nakashita. 2008. Antagonistic interaction between systemic acquired resistance and the abscisic acid-mediated abiotic stress response in *Arabidopsis*. *Plant Cell* 20: 1678–1692.

Yiu, J.C., L.D. Juang, D.Y.T. Fang, C.W. Liu and S.J. Wu. 2009. Exogenous putrescine reduces flooding-induced oxidative damage by increasing the antioxidant properties of Welsh onion. *Scientia Horticulturae* 120: 306–314.

Yu, L., J. Nie, C. Cao, Y. Jin, M. Yan, F. Wang, J. Liu, Y. Xiao, Y. Liang and W. Zhang. 2010. Phosphatidic acid mediates salt stress response by regulation of MPK6 in *Arabidopsis thaliana*. *New Phytologist* 188: 762–773.

Zapata, P.J., M. Serrano, M.T. Pretel, A. Amoros and M.A. Botella. 2004. Polyamines and ethylene changes during germination of different plant species under salinity. *Plant Science* 167: 781–788.

Zegzouti, H., W. Li, T.C. Lorenz, M. Xie, C.T. Payne, K. Smith, S. Glenny, G.S. Payne and S.K. Christensen. 2006. Structural and functional insights into the regulation of *Arabidopsis* AGC VIIIa kinases. *Journal of Biological Chemistry* 281: 35520–35530.

Zhang, H.X. and E. Blumwald. 2001. Transgenic salt-tolerant tomato plants accumulate salt in foliage but not in fruit. *Nature Biotechnology* 19: 765–768.

Zhang, J., Y. Zhang, Y. Du, S. Chen and H. Tang. 2011. Dynamic metabonomic responses of tobacco (*Nicotiana tabacum*) plants to salt stress. *Journal of Proteome Research* 10: 1904–1914.

Zhang, L., L.H. Tian, J.F. Zhao, Y. Song, C.J. Zhang and Y. Guo. 2009. Identification of an apoplastic protein involved in the initial phase of salt stress response in rice root by two-dimensional electrophoresis. *Plant Physiology* 149: 916–928.

Zhang, W., C. Qin, J. Zhao and X. Wang. 2004. Phospholipase D$\alpha$1-derived phosphatidic acid interacts with ABI1 phosphatase 2C and regulates abscisic acid signaling. *Proceedings of the National Academy of Sciences, USA* 101: 9508–9513.

Zhang, W., C. Wang, C. Qin., T. Wood, G. Olafsdottir, R. Welti and X. Wang. 2003. The oleate-stimulated phospholipase D, PLD$\delta$, and phosphatidic acid decrease $H_2O_2$-induced cell death in *Arabidopsis*. *The Plant Cell* 15: 2285–2295.

Zhang, Z., A. Rosenhouse-Dantsker, Q.Y. Tang, S. Noskov and D.E. Logothetis. 2010. The RCK2 domain uses a coordination site present in Kir channels to confer sodium sensitivity to Slo2.2 channels. *Journal of Neuroscience* 30: 7554–7562.

Zhao, H. and H. Yang. 2008. Exogenous polyamines alleviate the lipid peroxidation induced by cadmium chloride stress in *Malus hupehensis* Rehd. *Scientia Horticulturae* 116: 442–447.

Zhao, J. and X. Wang. 2004. *Arabidopsis* phospholipase D$\alpha$1 interacts with the heterotrimeric G-protein a-subunit through a motif analogous to the DRY motif in G-protein-coupled receptors. *Journal of Biological Chemistry* 279: 1794–1800.

Zhu, J., X. Fu, Y.D. Koo, J.K. Zhu, F.E. Jenney Jr., M.W. Adams, Y. Zhu, H. Shi, D.J. Yun, P.M. Hasegawa and R.A. Bressan. 2007. An enhancer mutant of *Arabidopsis* salt overly sensitive 3 mediates both ion homeostasis and the oxidative stress response. *Molecular and Cellular Biology* 27: 5214–5224.

Zhu, J.K. 2000. Genetic analysis of plant salt tolerance using *Arabidopsis*. *Plant Physiology* 124: 941–948.

Zhu, J.K. 2001. Plant salt tolerance. *Trends in Plant Science* 6: 66–71.

Zhu, J.K. 2002. Salt and drought stress signal transduction in plants. *Annual Review of Plant Biology* 53: 247–273.

Zhu, J.K. 2007. Plant salt stress. In: O'Daly A (ed) *Encyclopedia of Life Sciences*, John Wiley and Sons Ltd, Chichester, UK, pp. 1–3.

CHAPTER FIVE

# Physiological and Biochemical Approaches for Salinity Tolerance

*Muhammad Arslan Ashraf, Muhammad Iqbal, Iqbal Hussain, and Rizwan Rasheed*

## CONTENTS

*Abstract.* Food insecurity is more prominent in developing countries, and the problem is further worsened by different abiotic stresses that cause a substantial decline in the production of major crops. Among various abiotic stresses, salt stress is one of the major environmental limitations that greatly inhibits plant growth and affect both quantity and quality of yield. Plants employ a number of adaptations to avoid inhibitory effects of salinity. These adaptations include cellular homeostasis mainly through reduced $Na^+$ uptake and subsequent compartmentalization in the vacuoles. Strategies such as exogenous application of osmoprotectants or induction of overproduction of endogenous osmoprotectants could help the plants to overcome osmotic shock at early stage of salt stress as well as to maintain growth at later growth stages. In this chapter, we have highlighted potential physiochemical traits directly linked with salt tolerance of plants. In addition, we have discussed various possibilities of manipulating the physiochemical traits for induction of salt tolerance in the plants.

*Keywords:* Antioxidative system, nutritional and hormonal imbalances, osmolytes synthesis, osmotic stress, physiological approaches, ROS, salt tolerance, specific ion toxicity

## 5.1 INTRODUCTION

There has been a rapid increase in human population. According to an estimate, it will reach 9.15 billion in 2050 (Alexandratos and Bruinsma, 2012). This alarming rise in population would hamper the production of food. Plants are of prime importance in food production as human nutrition is dependent on plants (Chrispeels and Sadava, 2003). Food insecurity is more prominent in developing countries. The problem is further worsened by different environmental hazards that are known to cause substantial decline in the production of major crops. Fluctuations in temperature, ionic imbalance, drought, and salinity are among major environmental adversities, limiting crop yield worldwide.

Among various abiotic stresses, salt stress is one of the major environmental limitations that greatly inhibits plant growth and yield quality. Soil is defined as saline if a saturated paste of soil possesses electrical conductivity (ECe) of 4 dS m$^{-1}$ or greater (U.S. Salinity Laboratory Staff, 1954). On the basis of inherent potential of plants to tolerate increasing levels of salt in the growth medium, plants are grouped as salt tolerant (halophytes) and salt sensitive (glycophytes). Halophytes can grow in a medium with elevated salt concentrations, whereas glycophytes cannot grow in a medium with high salt concentrations (Flowers et al., 1977). Glycophytes mainly include grain crops that show inhibited growth at soil ECe > 4 dS m$^{-1}$.

Long ago, Bernstein and Hayward (1958) reported that response of crop plants to salinity stress varied with developmental growth stages. For instance, plants at the vegetative stage are sensitive to salinity (Läuchli and Epstein, 1990; Maas and Grattan, 1999) and gradually become tolerant to salinity at later growth stages. During germination or emergence stage, salt tolerance is taken as percent survival of plants, while at later growth stages, reduction in relative growth rate is correlated with salinity tolerance of plants (Läuchli and Grattan, 2007). Early developmental stages are adversely affected by salinity, but salt-induced inhibitory effects vary depending on whether the harvested part is root, leaf, stem, shoot, fiber, grain, or fruit. Inhibitory effects of salinity are more evident on shoot growth as compared to root growth (Läuchli and Epstein, 1990). Salinity stress can increase sterility and affect maturity time in wheat and rice. In addition, salt stress causes a decrease in the number of florets per spike (Maas and Poss, 1989a,b; Khatun et al., 1995). Therefore, management strategies are required to reduce the adverse effects of salinity at critical developmental stages of plants so as to minimize salt-induced reduction in yield and yield quality. This could be accomplished by improving our understanding of plant salt tolerance at different developmental growth stages (Läuchli and Grattan, 2007). Crop plants are generally tolerant to salinity stress at germination stage. Salinity stress causes a significant delay in seed germination, but germination percentage remains unaffected (Maas and Poss, 1989a,b). This information leads to categorization of salinity tolerance potential of plants at distinct developmental growth stages. For instance, 10 dS m$^{-1}$ of NaCl salinity stimulated germination in *Limonium perezii* seeds (Carter et al., 2005). However, higher levels of salt generally reduced germination percentage (Badia and Meiri, 1994; Mauromicale and Licandro, 2002). Sugar beet plant categorized as salt tolerant was reported to be sensitive to salinity stress at germination stage (Läuchli and Epstein, 1990). In this context, Tajbakhsh et al. (2006) reported that screening of crop plants for salinity stress at germination stage is essential for determining salt tolerance potential of crops, since this stage largely determines crop establishment. Salinity stress severely affects plant growth at the seedling and early vegetative growth stages as compared to germination stage in a number of plants such as melon (Botia et al., 2005), tomato (Del Amor et al., 2001), red orach (Wilson et al., 2000), spinach (Wilson et al., 2000), cowpea (Maas and Poss, 1989b), sorghum (Maas et al., 1986), corn (Maas et al., 1983), barley (Ayers et al., 1952), rice (Pearson and Ayers, 1996), and cotton (Abul-Naas and Omran, 1974). Likewise, when corn and wheat were grown under greenhouse condition, a marked reduction in shoot biomass and yield was recorded (Maas et al., 1983; Maas and Poss, 1989a). A significant reduction in leaf area and dry mass was recorded in wheat plants when exposed to salinity stress at the reproductive stage (Zheng et al., 2010). Salt stress shortened reproductive cycle in wheat that greatly inhibited spike development and caused significant reduction in yield (Läuchli and Grattan, 2007). Plants of two genetically diverse spring wheat cultivars (cv. S-24 and MH-97) exhibited variable response to salinity and were more sensitive at early developmental growth stages than at later growth stages (Ashraf et al., 2012). Rice was found to be susceptible to salinity stress at the vegetative and reproductive stages (Zeng et al., 2001; Moradi and Ismail, 2007). It is evident from the literature that plant salt tolerance potential varies with the developmental growth stages of plants. Therefore, understanding of salinity tolerance mechanisms at different growth stages will help to highlight the most sensitive stage of plant. Then countermeasures could be taken to escape the inhibitory effects of salinity at the sensitive growth stages.

Elevated levels of NaCl in the growth medium markedly deteriorate yield and yield quality of crop plants. Plants employ a number of adaptations to avoid inhibitory effects of salinity. These adaptations mainly include reduced $Na^+$ uptake and subsequent compartmentalization in vacuoles. Plants grown under saline conditions passively uptake a greater amount of $Na^+$ that results in $Na^+$ buildup in the cytosol (Jacoby, 1999; Blumwald, 2000; Mansour et al., 2003; Ashraf et al., 2008). Plants are also reported to actively uptake $Na^+$ through $Na^+/H^+$ antiporters (Ratner and Jacoby, 1976; Niu et al., 1993; Shi et al., 2003). Uptake and accumulation of $Na^+$ in crop plants are also regulated through $Na^+/H^+$ antiporters and proton pumps. There are two types of $Na^+/H^+$ antiporters: cell membrane $Na^+/H^+$ antiporter that operates at the plasma membrane and vacuolar $Na^+/H^+$ antiporter that operates at tonoplast (Apse et al., 1999; Wu et al., 2004). All organisms are reported to possess $Na^+/H^+$ antiporters (Padan et al., 2001).

Researchers are now generating crop plants with enhanced expression of different antiporters through genetic engineering. Overexpression of antiporters checks the higher uptake of toxic ions, thereby inducing salinity tolerance in crop plants. Transgenic Arabidopsis lines were generated by introducing MsNHX1 and ZmOPR1 antiporter genes from alfalfa and maize, respectively. A significantly greater germination was recorded in these transgenic lines under saline conditions (Bao-Yan et al., 2008; Gu et al., 2008). Crop plants having genetically engineered antiporters exhibited significantly higher biomass under salinity regimes. For instance, when PgNHX1 antiporter gene of Pennisetum glaucum was introduced into the rice plant, significantly higher shoot and root lengths were recorded under salinity stress (Verma et al., 2007). Likewise, transgenic lines of tobacco also exhibited improved plant dry masses under saline conditions (Wu et al., 2004; Zhang et al., 2008). However, in general, plants use an integrative approach to tolerate salinity stress (Figure 5.1).

## 5.2 INHIBITORY EFFECTS OF SALINITY ON GROWTH AND PHYSIOCHEMICAL ATTRIBUTES

Salinity exerts inhibitory effects on plant growth and development through (1) osmotic stress, (2) ionic toxicity, (3) generation of reactive oxygen species (ROS), and (4) hormonal and nutritional imbalances.

### 5.2.1 Osmotic Stress

Elevated levels of salts in the soil result in rapid decline in soil water potential. This drop in soil water potential affects the permeability of cell membranes and water conductivity of roots, thereby resulting in reduced influx of water by the plant (Munns, 2002, 2009, 2011; Ashraf et al., 2012; Álvarez et al., 2012). Growth arrest in plants during the early stage of salt stress is largely due to salt-induced osmotic stress (Munns and Tester, 2008; Sobhanian et al., 2010; Munns, 2011; Carassay et al., 2012). Turgor potential is positively associated with plant growth, and thus, decline in turgor is the major cause of suppressed plant growth under saline conditions (Munns, 2002; Ashraf, 2004; Yadav et al., 2011). Osmotic component of salt stress exerts more adverse effects on plant growth than ionic toxicity. Decline in plant growth and development was recorded in Arabidopsis when exposed to mannitol-induced osmotic stress during the early phase of ontogeny (Skirycz et al., 2011). A marked decline in plant biomass was recorded in rice plants grown under NaCl- or PEG-induced osmotic stress at different growth stages (Castillo et al., 2005). Two hydroponically grown diverse wheat cultivars (salt-sensitive MH-97 and salt-tolerant S-24) showed a marked reduction in plant biomass at different growth stages (vegetative, boot, and reproductive) under varying concentrations of NaCl. It was shown that growth arrest was largely due to osmotic component of the salt stress (Ashraf et al., 2012). When 5-week old Phlomis purpurea plants were exposed to 26 weeks of water and salt stress, a significant decline in plant growth was recorded. This decline in plant growth was largely due to osmotic component of salt stress (Álvarez et al., 2012). Suppression of plant growth in response to salt stress varies according to the extent of stress (mild or severe). Mild osmotic stress causes rapid growth inhibition of stem and leaves, whereas

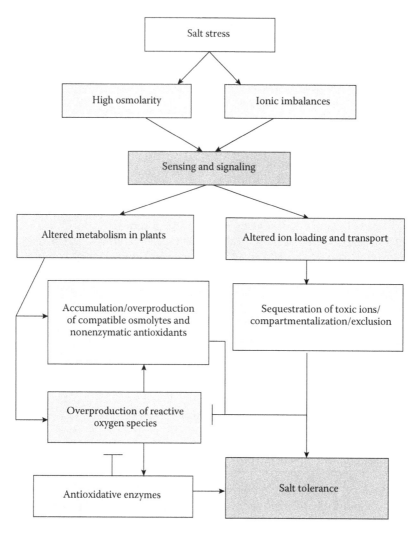

Figure 5.1. A schematic representation of major processes in plants involved in the tolerance to salinity.

the root may continue to grow (Spollen et al., 1993; Bartels and Sunkar, 2005). Several factors, including gradual or rapid imposition of stress, time scale of response, and plant species, are held responsible for growth inhibition in plants due to NaCl-induced osmotic stress. Exogenous application of osmoprotectants or induction of overproduction of endogenous osmoprotectants could help the plants to overcome osmotic shock at early stage of salt stress.

## 5.2.2 Specific Ion Toxicity

Crop productivity is greatly hampered in response to specific ion toxicity, thereby resulting in crop failures. Response of crop plants is variable among different plant species, but most crop plants are susceptible (Nawaz et al., 2010). Growth media containing elevated levels of toxic salts such as sodium chloride and sodium sulfate results in the buildup of lethal ions first in the old leaves. Under higher salt concentrations, plant cells are unable to compartmentalize ions into the vacuoles, resulting in cellular injury. This rapid accumulation of ions in the cytoplasm also inhibits enzyme activity. Salt deposition in the cell wall results in the dehydration of cells (Munns, 2005). However, in maize cultivars with differential response to salinity, such mechanism was not found (Mühling and Läuchli, 2002).

Plant tolerance to specific ion toxicity involves (1) minimized entry of salts into the roots, (2) minimized transport to the shoot, and (3) minimized cytosolic concentration of toxic salts. To avoid deposition of toxic ions in the shoot, plant roots tend to exclude toxic $Na^+$ and $Cl^-$ ions in the soil solution (Munns, 2005). However, using durum wheat cultivars with variable rates of $Na^+$ transport to leaf, Husain et al. (2003) observed that traits other than $Na^+$ exclusion at elevated NaCl salinity were important, because under such circumstances, specific ion toxicity played little role, whereas salt-induced osmotic stress was largely responsible for inhibition in plant growth and yield (Husain et al., 2003). Deposition of $Na^+$ in the cell wall displaces $Ca^{2+}$ that disrupts cell wall synthesis, thereby resulting in inhibited plant growth (Xue et al., 2004). Likewise, tissue concentration of $Cl^-$, greater than 80 mM alters plant morphology, reduces leaf thickness, and makes stomata less responsive to environmental changes. Chloride is readily available to the plant as it is not adsorbed by the soil. As a result, plants take up elevated levels of $Cl^-$, and subsequent buildup of $Cl^-$ in leaves results in leaf burn, death of leaf tissue, and early leaf drop (Marchner, 1995). It is evident from the available literature that buildup of toxic ions imposes deleterious effects on plant growth and development. However, the degree of these injurious effects is variable among plant species, and salt concentrations in the growth medium could also alter these effects.

### 5.2.3 Production of ROS

Molecular oxygen is prerequisite for life on the earth, but its reduction through any means results in the synthesis of ROS. Of various environmental constraints, salinity also induces the synthesis of ROS in the plants (Ali and Alqurainy, 2006). Hydroxyl radical ($\cdot$OH), superoxide ($\cdot O_2^-$), hydrogen peroxide ($H_2O_2$), and singlet oxygen ($^1O_2$) are among the lethal ROS. ROS are likely to react with a number of molecules to generate more free radicals so as to attain stable electronic configuration (Ashraf, 2009). For instance, electrons leaked from respiratory chains reduce molecular oxygen to superoxide radical (Mittler, 2002). Enzymatic dismutation of superoxide generates hydrogen peroxide, which is also produced during different cellular processes such as $\beta$-oxidation of fatty acids and photorespiration reactions (Parida and Das, 2005; Ashraf, 2009; Nawaz et al., 2010). Being reactive in nature, ROS induce substantial damage to a number of metabolites and macromolecules such as DNA, RNA, pigments, proteins, and

lipids that results in a cascade of vicious processes (Lamb and Dixon, 1997; Mittler, 2002). Plants also synthesize lower levels of ROS under normal conditions (Polle, 2001). However, exposure of plants to environmental hazards results in enhanced synthesis of lethal ROS (Desikan et al., 2001; Pastori and Foyer, 2002; Karpinski et al., 2003; Laloi et al., 2004; Ashraf, 2009). Furthermore, osmotic stress-induced stomatal closure resulted in low intercellular $CO_2$ concentration that inhibited photosynthesis. The inhibition in photosynthesis leads to the generation of superoxide radical in the chloroplasts that impairs plant growth in terms of photooxidation and photoinhibition. Superoxide does not react with proteins or lipids, but protonated form of superoxide induces lipid peroxidation (Asada and Takahashi, 1987; Ashraf, 2009). Activity of peroxidases and ribonucleotide reductase is markedly inhibited by superoxide (Ashraf, 2009). Higher cellular levels of superoxide result in the generation of lethal $H_2O_2$ and ($\cdot$OH) (Hideg, 1997). Hydrogen peroxide induces substantial damage to plant growth and development as it is a potential inhibitor of the Calvin cycle (Shen et al., 1997; Ashraf, 2009). In addition, interaction of metal ions or metal chelates with hydrogen peroxide results in the generation of lethal ($\cdot$OH) (Temple et al., 2005). Hydrogen peroxide is considered the most lethal ROS as 10 $\mu$M of this biomolecule in chloroplast could result in 50% reduction in photosynthesis (Kaiser, 1979). When four wheat cultivars (two tolerant cultivars, Kharchia 65, KRL 19, and two sensitive cultivars, HD 2009, HD 2687) were grown under salinity stress, tolerant cv. Kharchia 65 exhibited minimal oxidative damage in response to $H_2O_2$. This behavior of Kharchia 65 was associated with its potential to resist the excessive cellular accumulation of $H_2O_2$ (Sairam et al., 2005). Salt induced higher levels of $H_2O_2$ in tomato (Mittova et al., 2003), pea (Gomez et al., 2004), and potato (Fidalgo et al., 2004) cells. In addition, generation of ROS also varied with developmental growth stages of crop plants (Ashraf, 2009). Some ROS such as $H_2O_2$ also function as second messengers, thereby mediating stress-responsive signaling pathways and gene regulation.

### 5.2.4 Nutritional and Hormonal Imbalances

Higher levels of $Na^+$ and $Cl^-$ in saline soils often exceed those of micro- and macronutrients that ultimately cause nutritional imbalances in the plants. This, in turn, substantially reduces yield and yield quality (Grattan and Grieve, 1999). Higher level of soluble salts in the soil solution greatly lowers water potential of the soil that disturbs plant–water relations, and uptake and utilization of mineral nutrients. Plants under saline conditions are prone to nutrient imbalance and deficiencies as a result of interactions between salts and mineral nutrients (McCue and Hanson, 1990). Ionic imbalance due to elevated cellular levels of $Na^+$ and $Cl^-$ largely impairs uptake of other nutrients such as $Ca^{2+}$, $K^+$, and $Mg^{2+}$ (Karimi et al., 2005). Also excessive levels of $Na^+$ and $Cl^-$ in the root zone inhibits $K^+$ uptake and causes $K^+$ deficiency that leads to chlorosis and necrosis (Gopa and Dube, 2003). These nutritional imbalances inhibit a variety of metabolic processes including the activities of various enzymes (Munns, 2002; De Lacerda et al., 2003). Potassium is known to play a positive role in the osmoregulation, protein synthesis, maintenance of cell turgidity, and improvement of photosynthesis (Freitas et al., 2001; Ashraf, 2004; Ashraf et al., 2011). Potassium and calcium are also required to maintain the integrity of cell membranes (Wenxue et al., 2003). Under saline conditions, plant tissues require adequate level of $K^+$ maintained by selective uptake of potassium and selective cellular $K^+$ and $Na^+$ compartmentation, and subsequent distribution in the shoot (Munns et al., 2000; Carden et al., 2003; Nawaz et al., 2010). In addition, the presence of higher levels of NaCl in the root zone inhibits the uptake of other mineral nutrients (Parida and Das, 2005). Elevated levels of NaCl in the growth medium increase cellular levels of $Na^+$ and $Cl^-$ and decrease those of $Mg^{2+}$, $K^+$, and $Ca^{2+}$ in a variety of crop plants (Khan et al., 1999, 2000, 2001). For instance, genetically diverse canola cultivars exhibited a significant decrease in $Ca^{2+}$ and $K^+$ under saline conditions (Tunçtürk et al., 2011). Exposure of two strawberry cultivars (salt-sensitive cv. Elsanta and salt-tolerant cv. Korona) to 40 and 80 mM of NaCl salinity resulted in higher endogenous levels of $Zn^{2+}$, $K^+$, $Na^+$, $Cl^-$, P, and N. However, this salt-induced increase in mineral nutrients was greater in the salt-tolerant cultivar (Keutgen and Pawelzik, 2008). Salinity stress resulted in a marked increase in $Na^+$ and decrease in $K^+$ and $K^+/Na^+$ ratio in maize plants. (Carpici et al., 2010). Similarly, two differentially tolerant wheat cultivars (salt-sensitive cv. WH-542

and salt-tolerant cv. KRL-19) exhibited decline in $K^+/Na^+$ ratio under salinity stress. However, this decline was more in the salt-sensitive cultivar (Mandhania et al., 2006). Salinity stress caused a significant perturbation in the ionic composition of two rice cultivars (cv. Begunbitchi and cv. Lunishree) with variable salt tolerance potential. Higher levels of NaCl increased $Na^+$ and decreased $K^+$ contents in both rice cultivars. This salt-induced increase in $Na^+$ and decrease in $K^+$ was higher in cv. Lunishree (Khan and Panda, 2008). *Vicia faba* exhibited increase in $Cl^-$, $Na^+$, and $Ca^{2+}$ contents and decrease in $K^+/Na^+$ ratio in response to salinity stress (Gadallah, 1999). Likewise, leguminous plants exhibited 45% increase in $Na^+$ contents over control plants (Kurban et al., 1999). Overall, it can be inferred that plants suffer severe nutritional deficiency under elevated levels of NaCl in the growth medium. This imbalance could be attributed to the $Na^+$-induced inhibited uptake of mineral nutrients, particularly $K^+$.

Plant hormones play a pivotal role in growth and developmental processes as well as in signal transduction of plants under various environmental constraints (Pedranzani et al., 2003). Higher levels of NaCl in the growth medium cause increase in the endogenous levels of cytokinins and abscisic acid (ABA) in the crop plants (Thomas et al., 1992; Aldesuquy et al., 1998; Vaidyanathan et al., 1999). ABA alters the expression of a number of genes in plants grown under saline regimes (de Bruxelles et al., 1996). Alteration in gene expression due to ABA played an essential role in inducing tolerance in rice plants under saline stress (Gupta et al., 1998). *Citrus sinensis* exposed to NaCl salinity exhibited increase in endogenous levels of ethylene, ABA, and aminocycloprpane-1-carboxylic acid (Gómez-Cadenas et al., 1998). Ethylene exerts direct effects on the cell cycle by inhibiting the activity of cyclin-dependent kinase A (Skirycz et al., 2011). However, ABA is reported to mitigate the adverse effects of salt stress on photosynthesis, translocation of assimilates, and growth (Popova et al., 1995). ABA causes stomatal closure by altering ion fluxes in the guard cells when under salt stress. Furthermore, higher ABA levels improve calcium uptake that maintains cell membrane integrity in plants, thereby regulating the uptake and transport of mineral nutrients under saline stress (Chen et al., 2001). Salt-induced leaf abscission due to ethylene release was mitigated by ABA in citrus. This effect of ABA was ascribed to lower accumulation of Cl⁻ in the leaves (Gómez-Cadenas et al., 2002). Salinity tolerance of plants is also improved by jasmonates. Higher concentration of jasmonates was positively associated with salt tolerance in tomato (Hilda et al., 2003). Jasmonates mediate signal transduction in important mechanisms of plants such as defense responses, flowering, and senescence. Jasmonates also play a pivotal role in the osmoregulation possibly by inducing higher accumulation of organic osmolytes and decreasing ionic buildup. For instance, Fedina and Tsonev (1997) reported that exogenously applied jasmonates resulted in higher levels of proline and lower accumulation of toxic Na⁺ and Cl⁻ in pea.

## 5.3 IMPROVEMENT OF SALT TOLERANCE IN PLANTS

Researchers have produced salt-tolerant lines of crop plants by utilizing genetic variation in plant species at intergenic and inter- or intraspecific levels (Ashraf and Akram, 2009). Crop plants do not have a great degree of genetic variation in gene pool that is considered as the main hindrance for plant breeders. Conventional breeding has not been proven to be a successful approach for generating salt-tolerant plants. Therefore, we cannot expect a higher extent of improvement in salt tolerance potential of crop plants through conventional breeding. Wild relatives of crop plants are an important source of salinity tolerance genes, but these genes could not be transferred easily to crop plants largely due to the reproductive barrier (Ashraf et al., 2008). This makes the conventional approach time-consuming and labor intensive. Moreover, transfer of undesired genes along with salt-tolerant genes is often a problem associated with the conventional approach. Therefore, as an alternative approach, genetic engineering is being utilized for introducing salinity tolerance–related genes in crop plants.

Genetically modified plants possess foreign genes derived from other species. However, in most parts of the world, salt-tolerant cultivars raised through conventional breeding are preferred over salt-tolerant cultivars generated by genetic engineering. There are several hindrances in terms of different legal, social, and political barriers that restrain the release of genetically modified plants for field trials (Denby and Gehring, 2005; Parida and Das, 2005; Ashraf and

Akram, 2009), even though genetically modified plants with desired tolerance against environmental constraints have now been grown in different parts of the world. Brazil, China, Canada, and United States are among the countries where transgenic plants with tolerance against abiotic stress are widely grown (ISAAA, 2006). Salinity tolerance in plants is a multigenic trait that could not be easily improved by genetic engineering. Therefore, researchers have now focused on a number of genes (ion transport proteins, late embryogenesis abundant (LEA) proteins, antioxidants, heat shock, and organic osmolytes) to improve salinity tolerance of plants through genetic engineering (Ashraf et al., 2008). Under the circumstances, the use of various physiological and biochemical approaches to overcome salt intolerance could be an easy, low-cost, efficient, and thus attractive technique. A number of researchers have effectively used various physiological and biochemical approaches including exogenous use of different chemicals to induce salt tolerance in different crops. In view of the recent literature, promising biochemical traits directly linked with the salt tolerance of plants are shown in Table 5.1.

### 5.3.1 Physiological Approaches

#### 5.3.1.1 Water Relations

Plants exposed to saline stress undergo osmotic adjustments through accumulation of different organic and inorganic osmolytes (Hernández and Almansa, 2002; Chaparzadeh et al., 2003; Taiz and Zeiger, 2010). Salt-induced perturbations in turgor potential greatly determine plant growth (Taiz and Zeiger, 2010). Greater decline in leaf osmotic potential than soil water potential results turgor maintenance of plants under NaCl stress (Rajasekaran et al., 2001). A variety of crop plants exhibit significant variation in turgor potential under salinity regimes (Robinson et al., 1983; Lloyd et al., 1987; Thiel et al., 1988; Heuer and Plaut, 1989; Yang et al., 1990; Ashraf and O'Leary, 1996a,b; Romero-Aranda et al., 2001).

Water relation attributes (water and osmotic and turgor potentials) are variably affected at different developmental growth stages under salinity regimes (Millar et al., 1968; Ashraf and Shahbaz, 2003; de Azevedo Neto et al., 2004; Singh et al., 2010). Variation in salinity tolerance

TABLE 5.1

*Efficacy of different biochemical agents for the induction of salt tolerance in different plants.*

| Biochemical agents | Concentration used | Plant species | Mode of exogenous application | Plant response/outcome | References |
|---|---|---|---|---|---|
| *Plant Growth Regulators* | | | | | |
| Cytokinins | $10^{-6}$ M | Potato | Pre- and posttransplantation application | Kinetin pretreatment promoted growth and tuberization by increasing reducing sugars and sodium contents that played a critical part in osmotic adjustment. Authors further claimed induction of salinity tolerance in potato by prekinetin treatment. | Abdullah and Ahmad (1990) |
| | 150 mg $L^{-1}$ | Wheat | Seed priming | Kinetin priming showed beneficial effects on water use efficiency and photosynthetic rate under salt stress. | Iqbal and Ashraf (2005a) |
| | 25, 50, and 100 mg $L^{-1}$ | Wheat | Seed priming | Increased plant fresh and dry biomass. | Afzal et al. (2005) |
| | 100, 150, and 200 mg $L^{-1}$ | Wheat | Seed priming | Plants raised from seeds primed with kinetin had better growth and grain yield under saline conditions. | Iqbal et al. (2006a) |
| | 100, 150, and 200 mg $L^{-1}$ | Wheat | Seed priming | Plants raised from kinetin-primed (150 mg $L^{-1}$) seeds showed a consistent improvement in growth and grain yield in diverse cultivars that was positively correlated with leaf indoleacetic acid and negatively with ABA concentrations under both saline and nonsaline conditions. | Iqbal and Ashraf (2006) |
| | 10 μM | *Nigella sativa* | Foliar spray | Kinetin application prevented the plants from oxidative damage by modulating the antioxidant defense system under saline conditions. | Shah (2011) |
| Auxins | 5, 25, and 50 mg $L^{-1}$ | Wheat | Seed priming | Plants treated with auxins had better flag leaf area and fresh and dry masses. Furthermore, auxin-treated plants also exhibited greater pigment formation. | Aldesuquy (2000) |
| | 0.1, 1.0, and 10 μM | Mung bean | Foliar spray | Mitigated the effects of salt stress on the activities of SOD, POX, and CAT as well as MDA contents. | Chakrabarti and Mukherji (2003) |
| | | | | Auxin-priming-mediated hormonal homeostasis enhanced net photosynthetic rate, growth, and grain yield and thus conferred tolerance to salinity. | Iqbal and Ashraf (2013a) |
| | | | | Treatment with NAA (150 mg $L^{-1}$) altered hormonal balance (lowered leaf free ABA and putrescine while raised salicylic acid and spermidine concentrations) and ionic homeostasis under salt stress to induce salt tolerance. | Iqbal and Ashraf (2013b) |

*(Continued)*

TABLE 5.1 (*Continued*)

*Efficacy of different biochemical agents for the induction of salt tolerance in different plants.*

| Biochemical agents | Concentration used | Plant species | Mode of exogenous application | Plant response/outcome | References |
|---|---|---|---|---|---|
| Brassinosteroids | 3.0 μM | Rice | Seed priming | Brassinosteroids application elevated the levels of soluble proteins and nucleic acids and successfully mitigated the adverse effects of salinity on germination and seedling growth. | Anuradha and Rao (2001) |
| | 0, 0.052, 0.104, and 0.156 μM | Wheat | Rooting medium | Brassinosteroids application improved plant growth and yield under saline conditions possibly by increasing the translocation of photoassimilates to grain. | Ali et al. (2008) |
| | $10^{-11}, 10^{-9}$, and $10^{-7}$ M | Rice | Seed priming | Treatment with 24-epibrassinolide (EBL) enhanced growth, levels of protein and proline contents and antioxidant enzymes activities while decreased MDA contents under salt stress. | Sharma et al. (2013) |
| | $10^{-10}$ M | Oil seed rape (*Brassica napus* L.) | Rooting medium | The protective effect of EBL under salt stress was probably associated with EBL antioxidant effect, rather than the hormone-induced accumulation of proline and of low-molecular-weight phenolics, as well as with the ability to regulate water status by maintaining intracellular ion homeostasis. | Efimova et al. (2014) |
| | 0.1, 0.5, 1, and 2 mg $L^{-1}$ brassinolide | Wheat | Growth medium | Brassinolide (1 mg $L^{-1}$) counteracted the inhibitory effects of salinity and caused an increase in growth parameters, carbohydrate fractions, total soluble proteins, and amylase and protease activities both in the roots and shoots. | El-Feky and Abo-Hamad (2014) |
| Gibberellins | 0.1, 1.0, and 10 μM | Mung bean | Foliar spray | Salinity stress caused a significant rise in the activities of SOD, POX, and CAT as well as malondialdehyde contents. Application of the hormone mitigated the inhibitory effects of salt stress on the aforementioned attributes. | Chakrabarti and Mukherji (2003) |
| | 100 and 150 mg $L^{-1}$ | Wheat | Seed priming | Gibberellin-treated plants exhibited better growth and grain yield under saline conditions. Plants treated with gibberellin had lower levels of toxic $Na^+$ and greater contents of $Ca^{2+}$ and $K^+$ in different plant parts under saline conditions. | Iqbal and Ashraf (2013c) |
| TRIA | 10 mM | Soybean | Foliar spray | The treated salt-stressed plants had higher specific leaf area, relative water contents, chlorophyll pigments, nucleic acids, total soluble sugars and proteins contents, and lower proline contents. | Krishnan and Kumari (2008) |

(*Continued*)

## TABLE 5.1 (Continued)

*Efficacy of different biochemical agents for the induction of salt tolerance in different plants.*

| Biochemical agents | Concentration used | Plant species | Mode of exogenous application | Plant response/outcome | References |
|---|---|---|---|---|---|
| | 0, 10 and 20 μM | Wheat | Seed priming | Plants raised from seeds treated with TRIA had greater Na$^+$ and K$^+$ levels, and shoot Cl$^-$ use efficiency under saline conditions. | Perveen et al. (2012) |
| | 0, 0.5, and 1 mg L$^{-1}$ | Canola (*Brassica napus* L.) | Seed priming | TRIA increased shoot fresh weight, number of seeds per plant through better photosynthetic rate and higher shoot and root K$^+$ contents, and free proline and GB contents. | Shahbaz et al. (2013) |
| Salicylic acid | 0.5 and 1.0 mM | Cucumber | Foliar spray | SA promoted the activities of sucrose phosphate synthase, sucrose synthase, and amylases. The changes in various sugar contents resulted in the accumulation of soluble sugars in SA-treated seedlings, especially nonreducing sugars in the roots. The increased sugars could function as osmotic regulators and facilitate water uptake and retention in the plant cells, thereby conferring seedlings an enhanced tolerance to salinity stress. | Dong et al. (2011) |
| | 0, 0.10, 0.50, or 1.00 mM | *Pistacia vera* L | Foliar spray | SA (0.5 and 1.0 mM) ameliorated the salt stress injuries by inhibiting increases in proline content and the leaf electrolyte leakage. | Bastam et al. (2013) |
| | 10–500 μM SA | Arabidopsis | Root pretreatment | SA-pretreatment-ameliorated salinity stress by counteracting NaCl-induced membrane depolarization and by decreasing K$^+$ efflux. | Jayakannan et al. (2013) |
| ***Osmoprotectants*** | | | | | |
| Proline | 20 mM | Tobacco | Linsmaier and Skoog medium | Salinity stress caused a significant rise in the endogenous levels of ROS, lipid peroxidation, apoptosis-like cell death, and ATP content. Application of proline and GB markedly decreased the accumulation of ROS and lipid peroxidation but did not influence the apoptosis-like cell death and ATP content. Likewise, the expression of antioxidant genes like CAT and lignin-forming POX was greatly enhanced due to the application of osmoprotectants. However, mitigation of salinity stress was greater when proline was applied. | Banu et al. (2009) |

(Continued)

TABLE 5.1 (*Continued*)

*Efficacy of different biochemical agents for the induction of salt tolerance in different plants.*

| Biochemical agents | Concentration used | Plant species | Mode of exogenous application | Plant response/outcome | References |
|---|---|---|---|---|---|
| | 10 mM | Rice | Nutrient solution | Salinity stress caused a marked decline in plant growth and increase in $Na^+/K^+$ ratio, proline and $H_2O_2$ contents, and activities of various antioxidant enzymes (SOD, POD, CAT, and APX). Exogenous application of proline decreased $Na^+/K^+$ ratio and markedly increased endogenous levels of proline, while decreased the activities of the aforementioned antioxidant enzymes. Trehalose application significantly reduced $Na^+/K^+$ ratio and endogenous proline levels. Activities of SOD and POD decreased, while that of APX increased in response to treatment with trehalose under saline conditions. | Nounjan et al. (2012) |
| | 10 and 20 mM | Egg plant | Foliar spray | NaCl salinity reduced plant growth, water use efficiency, photosynthetic rate, photosystem II efficiency, and $K^+$ and $Ca^{2+}$ contents of plants. Proline application as foliar spray improved plant growth and water use efficiency of plants. In nutshell, proline application was not successful in inducing tolerance to plants with respect to morphological and physiological attributes. | Shahbaz et al. (2013) |
| | 1 mM | Barley | Seed priming | NaCl salinity markedly reduced plant growth, leaf water content, chlorophyll contents, and soluble sugars. In contrast, NaCl salinity caused a significant rise in the activities of different antioxidant enzymes (SOD, POD, CAT). Plants treated with proline showed better growth and improved the aforementioned attributes under saline conditions. Proline-treated plants exhibited marked changes in the activities of antioxidant enzymes in the presence or absence of NaCl salinity. | Agami (2014) |
| GB | 50 mM | Egg plant | Foliar application | Salt stress prominently suppressed plant growth, yield, and gas exchange characteristics. Endogenous levels of $K^+$ and $Ca^{2+}$ as well as $K^+/Na^+$ ratios decreased in response to salinity. In contrast, GB and proline levels also with toxic ions $Na^+$ and $Cl^-$ increased in plants subjected to salinity stress. Application of GB mitigated the adverse effects of salinity on growth, yield, and some important physiological characteristics of plants. | Abbas et al. (2010) |

(*Continued*)

TABLE 5.1 (*Continued*)

*Efficacy of different biochemical agents for the induction of salt tolerance in different plants.*

| Biochemical agents | Concentration used | Plant species | Mode of exogenous application | Plant response/outcome | References |
|---|---|---|---|---|---|
| | 50 mM | Okra | Foliar application | Salt stress caused a prominent reduction in plant biomass, yield, photosynthesis, and stomatal conductance. Endogenous levels of proline and GB also enhanced in leaves and roots, while the levels of $K^+$, $Ca^{2+}$, and $K^+/Na^+$ ratios decreased markedly under saline conditions. Foliar-applied GB markedly mitigated the negative effects of salinity on plant biomass production, yield, and gas exchange characteristics. GB application also reduced accumulation of toxic ions $Na^+$ and $Cl^-$, while increased the buildup of $K^+$ and $Ca^{2+}$ under saline conditions. | Habib et al. (2012) |
| | 0, 1, 5, and 25 mM | Pepper | Seed priming | Salt stress suppressed plant growth. However, GB-treated plants had better growth under saline conditions by enhancing chlorophyll and proline levels, and photosynthetic rate. GB application also improved relative water content, water potential, and activity of superoxide dismutase, while significantly decreased lipid peroxidation and membrane permeability. Moreover, GB treatment restricted the accumulation of toxic ions $Na^+$ and $Cl^-$ and stopped the leakage of $K^+$ that maintained a lower $Na^+/K^+$ ratio. | Korkmaz et al. (2012) |
| *Nonenzymatic Antioxidants* | | | | | |
| AsA | 0, 50, and 150 mg $L^{-1}$ | Wheat | Rooting medium | Salinity stress resulted in a prominent decline in growth, photosynthesis, and endogenous levels of AsA, while increase in $Na^+$ and $Cl^-$ and decrease in $K^+$ and $Ca^{2+}$ and $K^+/Na^+$ ratios were recorded. Application of AsA ameliorated the negative effects of salt stress on growth and photosynthesis by enhancing endogenous levels of AsA and activity of CAT enzyme. Plants treated with AsA had greater accumulation of $K^+$ and $Ca^+$ under saline conditions. | Athar et al. (2008) |
| | 0 and 30 mmol $L^{-1}$ | Canola | Rooting medium | Plant growth, photosynthesis, chlorophyll content, and nitrate reductase activity reduced in response to salinity. However, salt stress increased the accumulation of proline and activity of antioxidant enzymes. Exogenous application of AsA mitigated the adverse effects of salinity on plants growth, photosynthesis, and yield. | Bybordi (2012) |

(*Continued*)

TABLE 5.1 (*Continued*)

*Efficacy of different biochemical agents for the induction of salt tolerance in different plants.*

| Biochemical agents | Concentration used | Plant species | Mode of exogenous application | Plant response/outcome | References |
|---|---|---|---|---|---|
| **α-Tocopherols** | 0, 50, and 100 mg L$^{-1}$ | Faba bean | Foliar spray | Salt stress suppressed plant growth and yield and increased MDA and proline levels. However, application of α-tocopherols mitigated inhibitory effects of salt stress by improving antioxidant activity and proline levels and decreasing MDA contents. | Orabi and Abdelhamid (2014) |
| | 0, 0.25, 0.50, and 1.0 mM | Faba bean | Foliar spray | There was a prominent decline in different growth attributes, key physiological characteristics, and yield of faba bean under saline conditions. α-Tocopherol treatment improved plant growth, yield, relative water content, membrane stability index, and nutrient relation under salinity stress. Authors claimed that α-tocopherol treatment activated the antioxidant defense system of plants that prevented the faba bean plants from salt-induced oxidative damage. | Semida et al. (2014) |
| **Oxidants** | | | | | |
| Nitric oxide | 10 μM DETA/NO and 10 μM DETA | Soybean | Rooting medium | Salinity stress suppressed growth, nodule number and nodule weight, and nodule numbers. Application of nitric oxide (NO) alleviated salinity stress in terms of improved plant growth, reduced H$_2$O$_2$ levels, and enhanced activity of APX. Authors proposed that NO-induced alleviation of salinity stress could be associated with its direct scavenging of H$_2$O$_2$ or its potential to enhance APX activity. | Lin et al. (2012) |
| | 100 μM of sodium nitroprusside | Cucumber | Rooting medium | Salt stress suppressed plant growth mainly due to ROS-induced oxidative damage. Salinity stress also increased H$_2$O$_2$ contents and lipid peroxidation. Application of NO declined H$_2$O$_2$ levels and lipid peroxidation. Moreover, plant growth was also better in response to NO application under saline conditions. | Egbichi et al. (2014) |
| **Polyamines** | | | | | |
| Spermine | 2.5 mM | Wheat | Seed priming | Spermine-priming altered hormonal balance and induced salt tolerance. | Iqbal et al. (2006b) |

(*Continued*)

## TABLE 5.1 (*Continued*)
Efficacy of different biochemical agents for the induction of salt tolerance in different plants.

| Biochemical agents | Concentration used | Plant species | Mode of exogenous application | Plant response/outcome | References |
|---|---|---|---|---|---|
| | 0.3 mM | Cucumis sativus | Foliar spray | Salinity stress decreased plant growth, chlorophyll contents, and $F_v/F_m$ and increased malondialdehyde contents and generation of superoxide anion. Application of spermine improved plant growth, chlorophyll levels, and photosystem II efficiency. This effect of spermine was correlated with spermine-induced higher activity of antioxidant enzymes. Authors proposed that spermine prevented plant from salt-induced oxidative damage by enhancing the activities of antioxidant enzymes and photosystem II efficiency. | Shu et al. (2013) |
| | 0.3 mM | Wheat | Seed priming | Grain priming with spermine alleviated the adverse effects of seawater stress by stimulating leaf area expansion, pigment production, and photosynthetic activity. | Aldesuquy et al. (2014) |
| Spermidine | 5 mM | Wheat | Seed priming | Increased stomatal conductance and reduced shoot Cl⁻ concentrations and alleviated salt stress. | Iqbal and Ashraf (2005b) |
| | 1 mM | Cucumber | Foliar spray | Salt stress reduced plant growth, chlorophyll contents, and photosystem II efficiency. Application of spermidin improved the aforementioned attributes under saline conditions. Exogenously added spermidine improved plant salt tolerance mainly by increasing the endogenous levels of polyamines. | Shu et al. (2011) |

potential of ten safflower lines at the vegetative stage was determined by using water relations as potential selection criteria. Salt stress caused a marked decline in water and osmotic potentials and increase in turgor potential in all safflower lines (Siddiqi and Ashraf, 2008). Likewise, salt-induced perturbations in water relation attributes were also recorded in maize (Nawaz and Ashraf, 2007), jute (Chaudhuri and Choudhuri, 1997), halophyte perennial grass *Urochondra setulosa* (Gulzar et al., 2003), *Suaeda salsa* (Lu and Vonshak, 2002), brassica (Singh et al., 2010), and *Vigna radiata* (Win et al., 2011). It is evident from the literature that salinity stress differentially affects various water relation attributes of plants at different growth stages. Therefore, water relations of plants could be used as potential selection criteria to determine the most susceptible growth stage under saline regimes.

### 5.3.1.2 Photosynthetic Pigments

Salt stress affects photosynthetic pigments differently in young and old leaves. Older leaves exhibit toxic effects of salinity in terms of chlorosis, and continued saline stress may lead to fall of these older leaves (Hernández et al., 1995, 1999; Gadallah, 1999, Agastian et al., 2000; Ashraf, 2004). Salt levels in the growth medium affect chlorophyll contents differently (Winicov and Button, 1991; Locy et al., 1996). Salt stress is known to cause a significant decline in endogenous levels of 5-aminolevulinic acid (ALA) that act as a precursor for protochlorophyllide. Exposure of protochlorophyllide to light results in its conversion to chlorophyll (Santos, 2004). Moreover, salt stress also decreases glutamic acid that is an essential component for the synthesis of 5-ALA (Beale and Castelfranco, 1974; Santos and Caldeira, 1999; Santos et al., 2001; Santos, 2004). When *Grevillea* plants were exposed to salinity stress, a marked decline in carotenoids and chlorophyll and increase in anthocyanin were recorded. Tomato plants exhibited decline in chlorophyll and carotenoids in response to NaCl salinity (Khavarinejad and Mostofi, 1998). However, increase in chlorophyll content in response to saline stress has also been reported in *Amaranthus* (Wang and Nil, 2000). A significant reduction in photosynthetic pigments due to salinity was recorded in rice (Alamgir and Ali, 1999), blue panic grass (*Panicum antidotale* Retz.) (Ashraf and Shahbaz, 2003), wheat

(Sairam et al., 2005; Zheng et al., 2008), and pea (Hernández et al., 1999). Crop plants also showed variation in salt-induced decline in chlorophyll contents at different developmental growth stages (Sairam et al., 2002). Two wheat cultivars (HD 2009 and KRL1–4) exhibited variable degradation of chlorophyll contents in response to NaCl salinity at different developmental stages. These authors further reported that salt-induced decline in photosynthetic pigments at different growth stages was due to reduced *de novo* synthesis of chlorophyll or enhanced activity of chlorophyllase enzyme. *Sesbania grandiflora* showed decline in chlorophyll contents at seedling stage when exposed to saline stress (Dhanapackiam and Ilyas, 2010). These authors further observed that decline in chlorophyll contents was correlated with plant growth stage and type of salt used, for example, $Na_2SO_4$ and NaCl. Salt-induced degradation of chlorophyll was also recorded in *Brassica juncea* L. (cv. Kraniti) at different growth stages (vegetative, boot, and reproductive). Decline in the endogenous levels of chlorophyll due to salinity was greater in plants at the reproductive stage than those of other growth stages. Degradation of photosynthetic pigments was also observed in salt-tolerant (DK961) and salt-sensitive (JN17) wheat cultivars at the reproductive stage (Zheng et al., 2008). Exposure of *Momordica charantia* plants to salinity stress at different growth stages (preflowering, flowering, and postflowering) resulted in marked decline in photosynthetic pigments. This decline was observed at all three growth stages (Agarwal and Shaheen, 2007). When hydroponically grown pepper (*Capsicum annuum* L.) was exposed to salinity at the seedling stage, a significant degradation in chlorophyll contents was recorded (Chookhampaeng, 2011). Decline in chlorophyll contents at the different growth stages in wheat plants has also been reported by Sairam et al. (2002). Chlorophyll is among important nonstomatal factors that greatly determines photosynthetic rates under saline stress. Therefore, complete profiling of salinity effect on chlorophyll at different growth stages would contribute toward identifying the susceptible and tolerant growth stages of crop plants.

### 5.3.1.3 Soluble Proteins

Both increase and decrease in total soluble proteins has been recorded under salinity stress. Mild salt stress resulted in increase in total soluble proteins, but this attribute decreased markedly under severe salinity stress in mulberry (Agastian et al., 2000). Likewise, Doganlar et al. (2010) reported salt-induced decline in total soluble proteins in three tomato cultivars. When rice plants were exposed to salinity stress, a significant decline in soluble proteins was recorded (Amirjani, 2010a). In contrast, there are a number of reports available in the literature that depict the increase in soluble proteins under salinity stress. For instance, wheat plants exhibited increase in soluble proteins due to varying levels of NaCl salinity (Afzal et al., 2006). Salt-induced increase in soluble proteins was found in cowpea under different concentrations of salt in the growth medium (Lobato et al., 2009). Likewise, radish plants had markedly higher endogenous levels of total soluble proteins in response to mild salt concentrations in the growth medium (Turan et al., 2010). Moreover, salt-induced perturbation in total soluble proteins is also variable at different growth stages (Garg et al., 2006). For instance, increase in total soluble proteins at different growth stages was also recorded in wheat plants under NaCl salinity (Sairam et al., 2002).

### 5.3.1.4 Photosynthesis

Salt stress is known to cause a substantial decline in photosynthetic and transpiration rates and stomatal conductance in crop plants (Gibberd et al., 2002; Tezara et al., 2002; Burman et al., 2003; Ashraf, 2004). Inhibited photosynthesis in plants under salinity stress is largely attributed to stomatal closure (Rajesh et al., 1998; Stepien and Klobus, 2006). Excessive endogenous levels of ABA in response to salt-induced osmotic stress greatly reduce stomatal conductance, photosynthetic pigments, activity of Rubisco, and intercellular $CO_2$ concentration. Accumulation of sucrose and electron transport is also affected due to salt-induced elevated endogenous levels of ABA (Ashraf, 2004). Photosynthetic tissues with higher levels of salt exhibit shrinkage of thylakoids, stacking of adjacent membranes, inhibition of photosystem II (PSII), and significant decrease in chloroplast $K^+$ contents (Sharma and Hall, 1991; Murata et al., 2007). Moreover, under saline conditions, sugar consumption of growing tissues is impaired, thereby resulting in higher sugar concentration in mesophyll tissues. Feedback inhibition of higher levels of sugar in mesophyll cells

also reduces photosynthesis (Munns et al., 1982; Ashraf, 2004). Salinity tolerance of plants is a complex mechanism, entailing a number of factors including maintenance of net photosynthesis, stomatal conductance, and higher concentration of chlorophyll (Krishnaraj et al., 1993; Salama et al., 1994; Lakshmi et al., 1996; Khan et al., 2009). A number of studies indicated a positive association between photosynthesis and yield in different crop plants, for example, *Spinacia oleracea* (Robinson et al., 1983); *Phaseolus vulgaris* (Seemann and Critchley, 1985); *Zea mays* (Crosbie and Pearce, 1982); *Vigna mungo* (Chandra Babu et al., 1985); *Spartina alterniflora, Panicum hemitomon*, and *Spartina patens* (Hester et al., 2001); *Gossypium barbadense* (Cornish et al., 1991); *Gossypium hirsutum* (Pettigrew and Meredith, 1994); and *Brassica* (Ashraf, 2001). In contrast, some reports also exhibit no or a little correlation between yield and photosynthesis, for example, *Triticum aestivum* (Hawkins and Lewis, 1993; Ashraf and O'Leary, 1996b), *Hibiscus cannabinus* (Curtis and Läuchli, 1986), *Olea europaea* (Loreto et al., 2003), *Trifolium repens* (Rogers and Noble, 1992), and *Hordeum vulgare* (Rawson et al., 1988).

Degree of salinity-induced damage to photosynthesis in crop plants is variable at different growth stages (Ashraf and Parveen, 2002; Walia et al., 2005; Moradi and Ismail, 2007). For instance, rice was reported to exhibit significant variation in photosynthesis reduction at different growth stages under saline stress. Salt-induced decline in photosynthesis was greater in rice plants at the vegetative stage than that at the reproductive stage (Lutts et al., 1995; Walia et al., 2005; Moradi and Ismail, 2007). A marked decline in net photosynthesis was recorded in *Arabidopsis* when exposed to salinity stress at the vegetative stage (Ho et al., 2010). Likewise, wheat plants also exhibited salt-induced decline in photosynthesis at the vegetative and reproductive stages. However, this decline in photosynthesis in response to salinity stress was greater in salt-sensitive cv. Potohar as compared to that of salt-tolerant cv. SARC-I at all growth stages (Ashraf and Parveen, 2002). Likewise, exposure of *Linum usitatissimum* to saline stress at different growth stages exhibited a marked decline in photosynthesis, and this decline was apparent at all growth stages (Khan et al., 2007).

The available literature clearly exhibited the variation in salt-induced reduction in photosynthesis of crop plants. Since changes in photosynthesis are taken as important selection criteria of plants for salt tolerance, information of salt-induced inhibition in photosynthesis would help to improve the salt tolerance of plants by taking countermeasures.

### 5.3.1.5 Chlorophyll Fluorescence

Chlorophyll fluorescence is an effective, rapid, and nondestructive method to determine the degree of damage to photosynthetic apparatus in response to different environmental constraints (Palta, 1992; Sestak and Stiffel, 1997; Baker and Rosenqvist, 2004; Percival, 2004; Kauser et al., 2006; Baker, 2008). Researchers use this technique to identify the differential salt tolerance of crop plants in breeding experiments. Alteration in chlorophyll $a$ fluorescence due to inhibited PSII activity in response to salinity can be effectively determined with the help of this technique. There are contrasting reports available in the literature that exhibit variable response of crop plants to chlorophyll fluorescence attributes under saline conditions. Maize cultivars exhibited no alterations in $F_m$ and $F_v$ under salinity stress. This indicated that salt stress did not affect functioning of PSII (Shabala et al., 1998). Long ago, Smillie and Nott (1982) reported that salinity stress caused significant perturbations in chlorophyll fluorescence attributes in crop plants. These researchers were of the view that chlorophyll fluorescence technique could be used as potential selection criteria to identify the underlying cause of inhibited plant growth under saline stress (Smillie and Nott, 1982). Salinity stress resulted in a marked inhibition of electron transport chain in thylakoid membrane and integrity of PSII in $Brassica$ (Alia et al., 1993). Sorghum plants exhibited decrease in the activity of PSII, photochemical quenching coefficient (qP), and electron transport rate (ETR) under salinity stress (Netondo et al., 2004). Rice plants when exposed to salinity stress in the growth medium showed a marked rise in ETR and decline in $F_v/F_m$ (Amirjani, 2010b). Likewise, two rice cultivars with contrasting response to salinity stress exhibited no change in $F_v/F_m$, while there was significant rise in the values of qN in salt-sensitive rice cultivar (Dionisio-Sese and Tobita, 2000). Salinity stress caused variable alterations in a number of chlorophyll fluorescence attributes of four wheat cultivars. Among chlorophyll fluorescence parameters, rise

in nonphotochemical quenching (NPq), decline in quantum yield of dark adapted leaves ($F_v/F_m$), photochemical quenching (qP), and quantum yield of PSII of light-adapted leaf ($\Phi$PSII) were recorded in wheat plants under salinity regimes (Abdeshahian et al., 2010).

Plants also exhibit differential changes in chlorophyll $a$ fluorescence parameters at different growth stages under salinity stress (da Silva et al., 2011). For instance, rice plants were reported to exhibit variation in salt tolerance potential at the vegetative and reproductive stages in terms of decline in ETR and rise in NPq. A great variation in chlorophyll fluorescence attributes was determined in $Jatropha\ curcas$ plants at different harvest times (when plants were harvested after 3 and 7 days of salinity and after 3 days of stress recovery). Maximum quantum yield ($F_v/F_m$) and qP remained unaffected under saline stress, while 39% decline in quantum yield of primary photochemistry ($\Delta F/F_m$) was recorded in plants exposed to 14 days of salinity. NPq increased in plants harvested after 3 days of recovery and under salinity stress (da Silva et al., 2011). Exposure of 28-day-old tomato cultivars to salinity stress resulted in a marked perturbation in various chlorophyll fluorescence attributes. For instance, salinity stress caused a significant decline maximum quantum yield, qP, and integrity of PSII in tomato cultivars. The given information clearly indicated that changes in chlorophyll $a$ fluorescence greatly depend upon the type of species and developmental growth stage. Nonetheless, it is a potential technique that can be employed to screen crop plants for relative salt tolerance at different growth stages.

## 5.4 SALT-INDUCED VARIATION IN OSMOLYTE PRODUCTION IN PLANTS

Plants exhibit a number of mechanisms to tolerate salinity stress. Of these mechanisms, synthesis of organic osmolytes is one of the most important adaptation of plants (Ashraf and Akram, 2009; Nawaz et al., 2010). These are low-molecular-weight organic compounds with high solubility. Elevated cellular levels of these organic compounds are not toxic. Osmolytes are known to protect plants in a variety of ways, for example, osmotic adjustment, detoxification of ROS, stabilization of proteins/enzymes, and integrity of cell membranes (Bohnert and Jensen, 1996; Ashraf and Foolad, 2007). These organic osmolytes are also referred to as osmoprotectants owing to their involvement in protecting cellular components from dehydration damage (Ashraf and Foolad, 2007): trehalose, proline, polyols, and quaternary ammonium compounds like glycine betaine (GB), alaninebetaine, prolinebetaine, pipecolatebetaine, hydroxyprolinebetaine, and choline O-sulfate (Rhodes and Hanson, 1993). Despite the fact that transgenic plants are produced to enhance the cellular levels of different osmolytes, the desired levels of osmoprotectants have not yet been achieved. Researchers try to overcome this constraint by using exogenous application of organic osmolytes so as to induce tolerance in crop plants against various abiotic stresses. GB and proline have been extensively used by the researchers for the induction of tolerance in plants against abiotic stresses. The potential of different biochemical agents recently used for salt tolerance is summarized in Table 5.2.

## 5.4.1 Glycine Betaine

Among different quaternary ammonium compounds, plants tend to accumulate higher levels of GB in response to different abiotic stresses (Mansour, 2000; Mohanty et al., 2002; Yang et al., 2004). Higher concentration of GB mainly in chloroplast results in protection of thylakoid membranes and osmotic adjustment, thereby leading to higher net photosynthesis (Robinson and Jones, 1986; Genard et al., 1991). A variety of plant species exhibited accumulation of GB under stress conditions, for example, wheat (*T. aestivum*), sugar beet (*Beta vulgaris*), spinach (*S. oleracea*), barley (*H. vulgare*), and sorghum (*Sorghum bicolor*) (Weinberg et al., 1984; Fallon and Phillips, 1989; McCue and Hanson, 1990; Rhodes and Hanson, 1993; Yang et al., 2004). Salt-tolerant cultivars have the potential to accumulate greater levels of GB than those of salt-sensitive cultivars (Ashraf and Foolad, 2007). Higher endogenous levels of GB were also recorded in salt-sensitive and salt-tolerant cultivars of clover (*Trifolium alexandrinum*) (Varshney et al., 1988). In some studies, there was no correlation between accumulation of GB and salinity tolerance, for example, *Triticum*, *Agropyron*, and *Elymus* (Wyn Jones et al., 1984). *Citrus limon* did not exhibit significant accumulation of GB under saline conditions (Piqueras et al., 1996). These studies depict that involvement of GB in induction of salinity tolerance varies among plant species (Ashraf and Foolad, 2007). There are a number of plant species that lack the potential to synthesize GB naturally under normal or stress conditions, for example, *Arabidopsis* (*Arabidopsis thaliana*), tobacco (*Nicotiana tabacum*), rice (*Oryza sativa*), and mustard (*Brassica* spp.) (Rhodes and Hanson, 1993). Transgenic plants of these species made by the introduction of a gene involved in the biosynthesis of GB exhibited higher cellular levels of GB that resulted in salinity tolerance (Ashraf and Akram, 2009). However, these transgenic plants exhibited lower accumulation of GB than those of natural accumulators (Rhodes and Hanson, 1993). Lower levels of GB in transgenic plants are attributed to the limited supply of choline (a known substrate for GB production) or to the inhibited transport of choline in chloroplasts that are the sites for GB synthesis (Nuccio et al., 1998; Huang et al., 2000; McNeil et al., 2000). Plants also exhibit significant variation in GB production during distinct phases of ontogeny (Sariam et al., 2002). Different plant parts (root, leaf, and stem) of *Amaranthus tricolor* had shown markedly variable accumulation of GB under salinity stress. Accumulation of GB was also significantly variable in different plant parts during distinct phases of ontogeny. For instance, greater endogenous levels of GB were detected during the unfolding stage, while GB contents were minimal during the maturation and senescence stages (Yu-Mei et al., 2004). Exposure of two wheat cultivars with differential salinity tolerance resulted in higher accumulation of GB at different growth stages. Salt-tolerant cultivar Kharchia 65 had greater endogenous levels of GB as compared to salt-sensitive cv. KRL 19 at different growth stages (Sairam et al., 2002). Available literature showed that GB accumulation in plants depended on the type of species and also on the developmental growth stage.

## 5.4.2 Proline

Proline is considered to be an essential amino acid synthesized by higher plants. It plays a number of vital functions such as protection of subcellular structures, osmotic adjustment, buffer in cellular redox potential, and detoxification of ROS. After stress recovery, plants are also reported to extract nitrogen from proline (Ashraf and Foolad, 2007). There are numerous reports in the literature that indicated the higher production of

## TABLE 5.2
*Biochemical traits aassociated with salt tolerance in different plants.*

| Plant species | Carbohydrates | GB | Proline | Total amino acids | Total polyamines | References |
|---|---|---|---|---|---|---|
| Arachis hypogaea L. | Increased | Increased | Increased | Increased | Increased | Ranganayakulu et al. (2013) |
| Oryza sativa L. | Increased | — | — | — | — | Nemati et al. (2011) |
| O. sativa L. | Increase in soluble sugars, insoluble starch, sucrose, glucose, fructose | — | — | — | — | Boriboonkaset et al. (2013) |
| Lycopersicon esculentum Mill | Increase in soluble carbohydrates | — | Increased | — | — | Mohamed and Ismail (2011) |
| Sorghum bicolor L. | Increase in total sugars and reducing sugars | — | — | — | — | Gill et al. (2002) |
| Lygeum spartum L. | Increased | — | | Increased | — | Nedjimi (2011) |
| Matricaria chamomilla L. | Increased | — | Increased | — | — | Heidari and Sarani (2012) |
| Melissa officinalis L. | Increased | — | Increased | — | — | Khalid and Cai (2011) |
| Atriplex portulacoides | Increase in soluble sugars up to 800 mM NaCl while further increase in salinity decreased | — | Increased with increase in the salinity levels | — | — | Benzarti et al. (2014) |
| Salsola dendroides | Increased | — | Increased | — | — | Heidari-Sharifabad and Mirzaie-Nodoushan (2006) |
| Triticum aestivum L. | Increased | Increased | Increased | — | — | Sairam et al. (2002) |
| Phragmites karka | Total soluble sugars remained unchanged | — | Proline contents remained unchanged | — | — | Abideen et al. (2014) |
| Phaseolus vulgaris L. | Increase in total soluble sugars | Increased | Increased | — | — | Rady et al. (2013) |
| Zea mays L. | Decrease in total soluble sugars | — | Increased | — | — | Agami et al. (2013) |
| Vicia faba | Increase in soluble carbohydrates while decrease in total carbohydrates | — | Increased | Increased | — | Dawood et al. (2015) |
| Cucumis sativus L. | — | — | Increased | — | Increased | Duan et al. (2008) |

proline by crop plants under different environmental stresses (Kavi-Kishore et al., 2005; Ashraf and Foolad, 2007; Summart et al., 2010; Ashraf et al., 2012). Maize plants exhibited greater cytosolic levels of proline under salinity stress (Turan et al., 2009). Hydroponically grown 20-day-old wheat and sorghum cultivars produced higher cellular levels of proline in response to salt stress (Heidari, 2009). Higher accumulation of this amino acid was also detected in different plant parts (root, stem, and leaf) of two tomato cultivars (Isfahani and Shirazy) grown under varying concentrations of salinity (Amini and Ehsanpour, 2005). Exposure of four rice genotypes (Pokkali, Nonabokra, BR 9, IRATOM-24) to salt stress at the seedling stage resulted in a marked increase in cellular levels of proline (Alamgir et al., 2007). When six wheat cultivars (Pasban-90, Lu-26, E-30, E-2, E-37, E-38) were subjected to drought or salinity stress, a significant rise in endogenous content of proline was recorded (Bano and Aziz, 2003). Crop plants are shown to possess greater cellular levels of proline at the early growth stages as compared to that of later growth stages under abiotic stress. The underlying cause of this decline in proline contents at later growth stages is higher activity of proline-degrading enzyme "proline oxidase" (Madan et al., 1995). Two genetically diverse wheat cultivars (salt-tolerant cv. S-24 and salt-sensitive cv-MH-97) were studied for relative accumulation of proline at different growth stages under saline conditions. It was observed that both wheat cultivars showed higher levels of proline at the vegetative stage as compared to the boot and reproductive growth stages (Ashraf et al., 2012). Exposure of two *Brassica* cultivars (SR2P1–2 and SRP6–2) to salinity stress resulted in higher cellular levels of proline at different growth stages (Madan et al., 1995). Higher levels of proline were also recorded in 60-day-old wheat plants (Sairam et al., 2002). Salt-induced increase in proline was observed in maize at the seedling stage (Cha-um and Kirdmanee, 2009). The given information shows that proline plays a vital role in the induction of tolerance to plants against abiotic stresses. Furthermore, plants tend to exhibit higher levels of proline at the early developmental stage as compared to later growth stages. Therefore, proline could be used as a biochemical selection criterion to characterize the crop plants for salt tolerance at different growth stages.

### 5.4.3 Enzymatic and Nonenzymatic Antioxidants

Salt-induced higher cellular levels of ROS are considered lethal for plant growth and development. Being highly reactive, ROS induce substantial damage to lipids, proteins, DNA, and several other metabolites (Ashraf, 2009). Crop plants possess a defensive strategy to overcome ROS-induced oxidative damage. This strategy entails enzymatic and nonenzymatic components of antioxidant system. Dehydroascorbate reductase, glutathione reductase (GR), monodehydroascorbate reductase (MDHAR), catalase (CAT), superoxide dismutase (SOD), and ascorbate peroxidase (APX) comprise enzymatic component of antioxidant system, while carotenoids, ascorbic acid, glutathione, and tocopherols are among important nonenzymatic compounds in plants (Mateo et al., 2004; Gupta et al., 2005; Ashraf, 2009). Endogenous levels of superoxide and $H_2O_2$ are substantially reduced by the activity of several antioxidant enzymes such as SOD, POD, CAT, and APX (Ali and Alqurainy, 2006). SOD is considered as the most vital enzymatic component of the antioxidant system as it governs the conversion of lethal superoxide to $H_2O_2$ and $O_2$ (Ashraf, 2009). Plants grown under saline conditions tend to have greater activity of SOD (Van Camp et al., 1996; Ashraf, 2009). ROS-induced oxidative stress is also greatly relieved by the activity of APX. APX detoxifies the lethal $H_2O_2$ mainly in chloroplast and cytosol (Asada, 1992; Ashraf, 2009). $H_2O_2$ causes oxidation of ascorbate, resulting in the generation of monodehydroascorbate radical (MDA) and APX acts as biological catalyst in this reaction. Plants are reported to exhibit higher activity of APX under various environmental constraints, for example, water stress, temperature stress, and salinity stress (Noctor and Foyer, 1998; Ashraf, 2009). CAT is another important enzymatic component of the antioxidant system. Inhibitory effects of ROS-induced oxidative damage are significantly mitigated by CAT, which is mainly generated in glyoxisomes and peroxisomes. CAT initiates the conversion of $H_2O_2$ to water and oxygen (Ashraf et al., 2012). Apart from the positive role of enzymatic antioxidants in detoxifying the lethal ROS, nonenzymatic antioxidant enzymes also significantly neutralize the toxic effects of ROS (Ashraf, 2009). Tocopherols, glutathione, flavonoids, ascorbic acid, and carotenoids are the compounds known for nonenzymatic detoxification of ROS (Schafer et al., 2002; Tausz and Grill, 2000; Ashraf, 2009). Ascorbic acid also known as vitamin C is a water-soluble nonenzymatic antioxidant that exists in various compartments of the cell. Plants have a lower endogenous content of ascorbic acid under normal conditions, but a significant increase in this biomolecule is recorded in plants under environmental constraints (Arrigoni and de Tullio, 2000; Ashraf, 2009). Ascorbic acid scavenges ROS by donating electrons in enzymatic and nonenzymatic biochemical reactions. Superoxide, hydrogen peroxide, and singlet oxygen are scavenged by ascorbic acid. During reactions catalyzed by APX, ascorbic acid is oxidized to neutralize $H_2O_2$ that results in the maintenance of cell membrane integrity (Thomas et al., 1992). Tocopherols are also known for their nonenzymatic detoxification of ROS (Holländer-Czytko et al., 2005; Ashraf, 2009). Apart from the existence in plant organs, tocopherols are also an integral part of the lipid bilayer in plasma membranes, and a decline in the levels of tocopherols brings about disintegration of membrane functioning. Plants are reported to possess four isoforms of tocopherols ($\alpha$-, $\beta$-, $\gamma$-, $\delta$-). Among these isomers of tocopherols, vitamin E ($\alpha$-tocopherol) possesses the highest antioxidant potential (Kamal-Eldin and Appelqvist, 1996). Under saline conditions, tocopherols indirectly maintain photosynthetic efficiency of plants by protecting thylakoid membranes. Singlet oxygen (lethal ROS) is readily scavenged by tocopherols (Foyer, 1992; Ashraf, 2009). One molecule of tocopherol is reported to scavenge 120 molecules of singlet oxygen (Munné-Bosch, 2005; Ashraf, 2009). Concentration of different nonenzymatic antioxidants and activities of various enzymatic antioxidants is increased in plants under saline stress. However, this salt-induced variation in the antioxidant system is not the same at different growth stages. For example, exposure of two genetically diverse spring wheat cultivars (salt-tolerant S-24 and salt-sensitive MH-97) to salinity stress in nutrient solution resulted in a consistent rise in the activities of SOD, POD, CAT, and APX in salt-tolerant cv. S-24 at different growth stages (vegetative, boot, and reproductive). The increase or decrease pattern was not consistent in salt-sensitive cv. MH-97 at different growth stages. Overall, the maximal activities enzymatic antioxidants were recorded in wheat plants at the vegetative growth stages. In addition, higher levels of ascorbic acid were found in cv. S-24 at the reproductive stage (Ashraf et al., 2012). Similar salt-induced perturbation in nonenzymatic and enzymatic antioxidants in wheat cultivars (salt-tolerant cv. Kharchia 65 and salt-sensitive cv. HD 2687) was also variable at different growth stages. Salinity stress resulted in significant decline in endogenous levels of ascorbic acid and increase in the activities of different antioxidant enzymes (SOD, GR, and APX) at different growth stages (30 and 40 days after sowing). Higher concentrations of nonenzymatic antioxidants and activities of enzymatic antioxidants were recorded in salt-tolerant cv. Kharchia 65 than that of salt-sensitive cv. HD 2687 (Sairam and Srivastava, 2002). Likewise, Sairam et al. (2002) reported salt-induced increase in the activity of antioxidant enzymes (CAT, GR, and SOD) in salt-tolerant cv. Kharchia 65 and salt-sensitive cv. KRL- 19 at different growth stages. Salt-tolerant cv. Kharchia 65 possessed greater activity of antioxidant enzymes as compared to salt-sensitive cv. KRL 19. Likewise, when salt-tolerant Kharchia 65 and KRL 19 and salt-sensitive HD 2009 and HD 2687 wheat cultivars were exposed to salinity, a significant decline in ascorbic acid and rise in the activities of GR, APX, and SOD were recorded at two different growth stages. Salt-tolerant cv. Kharchia 65 exhibited higher endogenous levels of ascorbic acid and activities of antioxidant enzymes, while lower levels of ascorbic acid and activities of antioxidant enzymes were recorded in salt-sensitive cv. HD 2687 (Sairam et al., 2005). Pea plants grown in saline hydroponic culture also exhibited increase in the activity of MDHAR, SOD, and APX at the early growth stage (Hernández et al., 1999).

Available literature indicates that both the enzymatic and nonenzymatic antioxidants play vital roles in the detoxification of lethal ROS. However, salt-induced modulation in the antioxidant system of plants is variable at different growth stages. Crop plants engineered for enhanced synthesis of SOD exhibited tolerance to salinity, for example, rice (Tanaka et al., 1999), Chinese cabbage (Tseng et al., 2007), tobacco (Badawi et al., 2004), and *Arabidopsis* (Wang et al., 2004). Most of the crop plants are tested for salinity tolerance at early growth stages under controlled environmental conditions. Therefore, there is need to study alteration in antioxidant response at different growth stages so as to characterize the crop plants for salt tolerance.

## 5.5 ENGINEERING PLANTS FOR OVERPRODUCTION OF ORGANIC OSMOLYTES

Plants subjected to salinity stress exhibit significantly higher cellular levels of organic osmolytes. Plants utilize these organic osmolytes in maintaining cell turgor by osmotic adjustment under saline regimes. Among various organic osmolytes, genes involved in the synthesis of GB and proline have been introduced in a wide range of crop plants by genetic engineering for the induction of salinity stress (Ashraf and Foolad, 2007). Plants could be categorized as salt tolerant or sensitive on the basis of their potential to produce organic osmolytes. However, cellular levels of these organic compounds vary among plant species. Some plants have the ability to synthesize a large quantity of organic osmolytes, while there are some plants in which minimal or no production of these compounds is found (Ashraf and Akram, 2009; Ashraf and Foolad, 2007). For example, natural accumulation of GB has been reported in a number of plant species, for example, barley (*H. vulgare*), sorghum (*S. bicolor*), wheat (*T. aestivum*), sugar beet (*B. vulgaris*), and spinach (*S. oleracea*) (Weimberg et al., 1984; Fallon and Phillips, 1989; Rhodes and Hanson, 1993; Yang et al., 2004). In contrast, there are a number of plant species that lack the potential to synthesize GB naturally under normal or environmental stresses, for example, rice (*O. sativa*), tobacco (*N. tabacum*), arabidopsis (*A. thaliana*), and mustard (*Brassica* spp.) (Ashraf and Foolad, 2007). Therefore, genetically modified crop plants with gene and/or genes catalyzing the rate limiting steps in the biosynthesis of organic osmolytes exhibit a great degree of tolerance to different abiotic stresses (Ashraf and Akram, 2009). For instance, the gene for $\Delta$1-pyrroline-5-carboxylate synthetase involved in the biosynthesis of proline has been overexpressed in a number of crop plants, for example, tobacco (Hong et al., 2000), wheat (Sawahel and Hassan, 2002), potato (Hmida-Sayari et al., 2005), and rice (Su and Wu, 2004). These genetically modified plants accumulated elevated levels of proline as compared to the wild relatives and exhibited greater plant biomass under saline conditions (Ashraf and Akram, 2009). Genes responsible for

the synthesis of enzymes such as choline oxidase and choline dehydrogenase governing the biosynthesis of GB have been introduced in different plant species, for example, *Arabidopsis* (Huang et al., 2000), *B. juncea* (Prasad et al., 2000), cabbage (Bhattacharya et al., 2004), and rice (Sakamoto et al., 1998). Salt tolerance of these genetically modified plants was positively correlated with the enhanced potential to synthesis GB as compared to their wild relatives (Ashraf and Akram, 2009).

## 5.6 CONCLUSIONS AND FUTURE PERSPECTIVES

Plants suffer severe nutritional deficiency under elevated levels of NaCl in the growth medium. This imbalance could be attributed to the $Na^+$-induced inhibited uptake of mineral nutrients, particularly $K^+$. Crop plants do not have a great degree of genetic variation in the gene pool that is considered as the main hindrance for plant breeders. Moreover, wild relatives of crop plants are an important source of salinity tolerance genes, but these genes could not be transferred easily to crop plants largely due to the reproductive barrier. This makes the conventional approach time-consuming and labor intensive. In addition, the transfer of undesired genes along with salt-tolerant genes is often a problem associated with the conventional approach. Thus, conventional breeding has not been proven to be a successful approach for generating salt-tolerant plants. However, as an alternative approach, genetic engineering is being utilized for introducing salinity tolerance–related genes in crop plants. Salinity tolerance in plants is a multigenic trait that could not be easily improved by genetic engineering. In addition, there are several hindrances in terms of different legal, social, and political barriers that restrain the release of genetically modified plants for field trials. Overall, most of the salinity-tolerant plants produced through genetic engineering have not successfully showed tolerance at field level. Under the circumstances, the use of various physiological and biochemical approaches to overcome salt intolerance could be an easy, low-cost, efficient, and thus attractive technique. A number of researchers have effectively used various physiological and biochemical approaches including exogenous use of different chemicals to induce salt tolerance in different crops. A plethora of literature shows that increases in total carbohydrates (sugars), proline,

GB, and polyamines are directly linked with the salt tolerance in different crop species. Of them, GB and proline have been extensively used by the researchers for the induction of tolerance in plants against abiotic stresses. There is a need to study alteration in antioxidant response at different growth stages so as to characterize the crop plants for salt tolerance. Furthermore, metabolomics and metabolomic studies could unravel the importance and role as well as the use of physiochemical factors for the inductance of salt tolerance in plants.

## REFERENCES

Abbas, W., M. Ashraf, and N.A. Akram. 2010. Alleviation of salt-induced adverse effects in eggplant (*Solanum melongena* L.) by glycinebetaine and sugarbeet extracts. *Scientia Horticulturae* 125(3):188–195.

Abdeshahian, M., M. Nabipour, and M. Meskarbashee. 2010. Chlorophyll fluorescence as criterion for the diagnosis salt stress in wheat (*Triticum aestivum*) plants. *International Journal of Chemical and Biological Engineering* 4:184–186.

Abdullah, Z. and R. Ahmad. 1990. Effect of pre- and post-kinetin treatments on salt tolerance of different potato cultivars growing on saline soils. *Journal of Agronomy and Crop Science* 165(2–3):94–102.

Abideen, Z., H.W. Koyro, B. Huchzermeyer, M.Z. Ahmed, B. Gul, and M.A. Khan. 2014. Moderate salinity stimulates growth and photosynthesis of Phragmiteskarka by water relations and tissue specific ion regulation. *Environmental and Experimental Botany* 105:70–76.

Abul-Naas, A.A. and M.S. Omran. 1974. Salt tolerance of seventeen cotton cultivars during germination and early seedling development. *Zeitschrift Fur Acker Und Pflanzenbau* 140:229–236.

Afzal, I., S.M.A. Basra, A. Hameed, and M. Farooq. 2006. Physiological enhancements for alleviation of salt stress in wheat. *Pakistan Journal of Botany* 8:1649–1659.

Afzal, I., S.M. Basra, and A. Iqbal. 2005. The effects of seed soaking with plant growth regulators on seedling vigor of wheat under salinity stress. *Journal of Stress Physiology & Biochemistry* 1(1):6–14.

Agami, R.A. 2013. Alleviating the adverse effects of NaCl stress in maize seedlings by pretreating seeds with salicylic acid and 24-epibrassinolide. *South African Journal of Botany* 88:171–177.

Agami, R.A. 2014. Applications of ascorbic acid or proline increase resistance to salt stress in barley seedlings. *Biologia Plantarum* 58(2):341–347.

Agarwal, S. and R. Shaheen. 2007. Stimulation of antioxidant system and lipid peroxidation by abiotic stress in leaves of *Momordica charantia*. *Brazilian Journal of Plant Physiology* 19:149–161.

Agastian, P., S.J. Kingsley, and M. Vivekanandan. 2000. Effect of salinity on photosynthesis and biochemical characteristics in mulberry genotypes. *Photosynthetica* 38:287–290.

Alamgir, A.N.M. and M.Y. Ali. 1999. Effect of salinity on leaf pigments, sugar and protein concentrations and chloroplast ATPase activity of rice (*Oryza sativa* L.). *Photosynthetica* 28:145–149.

Alamgir, A.N.M., M. Musa, and M.Y. Ali. 2007. Some aspects of mechanisms of NaCl stress tolerance in the seedlings of four rice genotypes. *Bangladesh Journal of Botany* 36:181–184.

Aldesuquy, H., Z. Baka, B. Mickky. 2014. Kinetin and spermine mediated induction of salt tolerance in wheat plants: Leaf area, photosynthesis and chloroplast ultrastructure of flag leaf at ear emergence. *Egyptian Journal of Basic and Applied Sciences* 1(2):77–87. doi: http://dx.doi.org/10.1016/j.ejbas.2014.03.002.

Aldesuquy, H.S., A.T. Mankarios, and H.A. Awad. 1998. Effects of some antitranspirants on growth, metabolism and productivity of saline-treated wheat plants: Induction of stomatal closure, inhibition of transpiration and improvement of leaf turgidity. *Acta Botanica Hungarica* 41:1–10.

Aldesuquy, H.S. 2000. Effect of indol-3-yl acetic acid on photosynthetic characteristics of wheat flag leaf during grain filling. *Photosynthetica* 38(1):135–141.

Alexandratos, N. and J. Bruinsma. 2012. World agriculture towards 2030/2050: The 2012 revision. ESA Working paper. Rome, Food and Agriculture Organization, Rome, Italy.

Ali, Q. and M. Ashraf. 2008. Modulation of growth, photosynthetic capacity and water relations in salt stressed wheat plants by exogenously applied 24-epibrassinolide. *Plant Growth Regulation* 56(2):107–116.

Alia, P., P. Saradhi, and P. Mohanty. 1993. Proline in relation to free radical production in seedlings of *Brassica juncea* raised under sodium chloride stress. *Plant and Soil* 155/156:497–500.

Ali, A.A. and F. Alqurainy. 2006. Activities of antioxidants in plants under environmental stress. In: Motohashi, N. (ed.). *The Lutein-Prevention and Treatment for Diseases*. Kerala, India: Transworld Research Network. pp. 187–256.

Álvarez, S., M.J. Gómez-Bellot, M. Castillo, S. Bañón, and M.J. Sánchez-Blanco. 2012. Osmotic and saline effect on growth, water relations, and ion uptake and translocation in *Phlomis purpurea* plants. *Environmental and Experimental Botany* 78:138–145.

Amini, F. and A.A. Ehsanpour. 2005. Soluble proteins, proline, carbohydrates and Na$^+$/Cl$^-$ changes in two tomato (*Lycopersicon esculentum* Mill.) cultivars under in vitro salt stress. *American Journal of Biochemistry and Biotechnology* 1:212–216.

Amirjani, M.R. 2010a. Effect of NaCl on some physiological parameters of rice. *European Journal of Biological Sciences* 3(1):6–16.

Amirjani, M.R. 2010b. Salinity and photochemical efficiency of wheat. *International Journal of Botany* 6(3):273–279.

Anuradha, S. and S.S.R. Rao. 2001. Effect of brassinosteroids on salinity stress induced inhibition of seed germination and seedling growth of rice (*Oryza sativa* L.). *Plant Growth Regulation* 33(2):151–153.

Apse, M.P., G.S. Aharon, W.A. Snedded, and E. Blumwald. 1999. Salt tolerance conferred by overexpression of a vacuolar Na$^+$/H$^+$ antiporter in *Arabidopsis*. *Science* 285:1256–1258.

Arrigoni, O. and M.C. de Tullio. 2000. The role of ascorbic acid in cell metabolism: Between gene-directed functions and unpredictable chemical reactions. *Journal of Plant Physiology* 157:481–488.

Asada, K. 1992. Ascorbate peroxidase: A hydrogen peroxide-scavenging enzyme in plants. *Physiologia Plantarum* 85:235–241.

Asada, K. and M. Takahashi. 1987. Production and scavenging of active oxygen in photosynthesis. In: Kyle, D.J. et al. (eds.). *Photoinhibition*. Amsterdam, the Netherlands: Elsevier. pp. 227–287.

Ashraf, M. 2001. Relationships between growth and gas exchange characteristics in some salt-tolerant amphidiploid *Brassica* species in relation to their diploid parents. *Environmental and Experimental Botany* 45:155–163.

Ashraf, M. 2003. Relationships between leaf gas exchange characteristics and growth of differently adapted populations of Blue panicgrass (*Panicum antidotale* Retz.) under salinity or waterlogging. *Plant science* 165(1):69–75.

Ashraf, M. 2004. Some important physiological selection criteria for salt tolerance in plants. *Flora* 199:361–376.

Ashraf, M. 2009. Biotechnological approach of improving plant salt tolerance using antioxidants as markers. *Biotechnology Advances* 27:84–93.

Ashraf, M. and J.W. O'Leary. 1996a. Responses of some newly developed salt-tolerant genotypes of spring wheat to salt stress. 1. Yield components and ion distribution. *Journal of Agronomy and Crop Science* 176:91–101.

Ashraf, M. and J.W. O'Leary. 1996b. Responses of some newly evolved salt-tolerant genotypes of spring wheat to salt stress. 2. Water relations and gas exchange. *Acta Botanica Neerlandica* 45:29–39.

Ashraf, M. and M.R. Foolad. 2007. Roles of glycine betaine and proline in improving plant abiotic stress resistance. *Environmental and Experimental Botany* 59:206–216.

Ashraf, M. and N.A. Akram. 2009. Improving salinity tolerance of plants through conventional breeding and genetic engineering: An analytical comparison. *Biotechnology Advances* 27:744–752.

Ashraf, M. and N. Parveen. 2002. Photosynthetic parameters at the vegetative stage and during grain development of two hexaploid cultivars differing in salt tolerance. *Biologia Plantarum* 45:401–407.

Ashraf, M. and M. Shahbaz. 2003. Assessment of genotypic variation in salt tolerance of early CIMMYT hexaploid wheat germplasm using photosynthetic capacity and water relations as selection criteria. *Photosynthetica* 41:273–280.

Ashraf, M.A., M. Ashraf, and M. Shahbaz. 2012. Growth stage-based modulation in antioxidant defense system and proline accumulation in two hexaploid wheat (*Triticum aestivum* L.) cultivars differing in salinity tolerance. *Flora* 207:388–397.

Ashraf, M., H.R. Athar, P.J.C. Harris, and T.R. Kwon. 2008. Some prospective strategies for improving crop salt tolerance. *Advances in Agronomy* 97:45–110.

Ashraf, M.A., M.S.A. Ahmad, M. Ashraf, F. Al-Qurainy, and F.M.Y. Ashraf. 2011. Alleviation of waterlogging stress in upland cotton (*Gossypium hirsutum* L.) by exogenous application of potassium in soil and as a foliar spray. *Crop and Pasture Science* 62:25–38.

Athar, H., A. Khan, and M. Ashraf. 2008. Exogenously applied ascorbic acid alleviates salt-induced oxidative stress in wheat. *Environmental and Experimental Botany* 63(1):224–231.

Ayers, A.D., J.W. Brown, and L. Wadleigh. 1952. Salt tolerance of barley and wheat in soil plots receiving several salinization regimes. *Agronomy Journal* 44:307–310.

Badawi, G.H., Y. Yamauchi, E. Shimada, R. Sasaki, N. Kawano, K. Tanaka, and K. Tanaka. 2004. Enhanced tolerance to salt stress and water deficit by overexpressing superoxide dismutase in tobacco (*Nicotiana tabacum*) chloroplasts. *Plant Science* 166:919–928.

Badia, D. and A. Meiri. 1994. Tolerance of two tomato cultivars (*Lycopersicum esculentum* Mill) to soil salinity during emergence phase. *Agrimedia* 124:301–310.

Baker, N.R. 2008. Chlorophyll fluorescence: A probe of photosynthesis in vivo. *Annual Review of Plant Biology* 59:89–113.

Baker, N.R. and E. Rosenqvist. 2004. Applications of chlorophyll fluorescence can improve crop production strategies: An examination of future possibilities. *Journal of Experimental Botany* 55:1607–1621.

Bano, A. and N. Aziz. 2003. Salt and drought stress in wheat and the role of ABA. *Pakistan Journal of Botany* 35:871–883.

Banu, M.N.A., M.A. Hoque, M. Watanabe-Sugimoto, K. Matsuoka, Y. Nakamura, Y. Shimoishi, and Y. Murata. 2009. Proline and glycinebetaine induce antioxidant defense gene expression and suppress cell death in cultured tobacco cells under salt stress. *Journal of Plant Physiology* 166(2):146–156.

Bao-Yan, A.N., L. Yan, L. Jia-Rui, Q. Wei-Hua, Z. Xian-Sheng, and Z. Xin-Qi. 2008. Expression of a vacuolar $Na^+/H^+$ antiporter gene of alfalfa enhances salinity tolerance in transgenic *Arabidopsis*. *Acta Agronomica Sinica* 34:557–564.

Bartels, D. and R. Sunkar. 2005. Drought and salt tolerance in plants. *Critical Reviews in Plant Sciences* 24:23–58.

Bastam, N., B. Baninasab, and C. Ghobadi. 2013. Improving salt tolerance by exogenous application of salicylic acid in seedlings of pistachio. *Plant Growth Regulation* 69:275–284.

Beale, S. and P. Castelfranco. 1974. The biosynthesis of delta-aminolevulinic acid in higher plants. II. Formation of 14 C-delta-aminolevulinic acid from labelled precursors in greening plant tissues. *Plant Physiology* 53:296–297.

Benzarti, M., K.B. Rejeb, D. Messedi, A.B. Mna, K. Hessini, M. Ksontini, and Debez, A. 2014. Effect of high salinity on Atriplexportulacoides: Growth, leaf water relations and solute accumulation in relation with osmotic adjustment. *South African Journal of Botany* 95:70–77.

Bernstein, L. and H.E. Hayward. 1958. Physiology of salt tolerance. *Annual Review of Plant Physiology* 51:875–878.

Bhattacharya, R.C., M. Maheswari, V. Dineshkumar, P.B. Kirti, S.R. Bhat, and V.L. Chopra. 2004. Transformation of *Brassica oleracea* var. *capitata* with bacterial betA gene enhances tolerance to salt stress. *Scientia Horticulturae* 100:215–227.

Blumwald, E. 2000. Salt transport and salt resistance in plants and other organisms. *Current Opinion in Cell Biology* 12:431–434.

Bohnert, H.J. and R.G. Jensen. 1996. Strategies for engineering water stress tolerance in plants. *Trends in Biotechnology* 14:89–97.

Boriboonkaset, T., C. Theerawitaya, N. Yamada, A. Pichakum, K. Supaibulwatana, S. Cha-Um, T. Takabe, and C. Kirdmanee. 2013. Expression levels of some starch metabolism related genes in flag leaf of two contrasting rice genotypes exposed to salt stress. *Protoplasma* 250:1157–1167.

Botia, P., J.M. Navarro, A. Cerda, and V. Martinez. 2005. Yield and fruit quality of two melon cultivars irrigated with saline water at different stage of development. *European Journal of Agronomy* 23:243–253.

Burman, U., B.K. Garg, and S. Kathju. 2003. Water relations, photosynthesis and nitrogen metabolism of Indian mustard (*Brassica juncea* Czern & Coss) grown under salt and water stress. *Journal of Plant Biology* 30:55–60.

Bybordi, A. 2012. Effect of ascorbic acid and silicium on photosynthesis, antioxidant enzyme activity, and fatty acid contents in canola exposure to salt stress. *Journal of Integrative Agriculture* 11(10):1610–1620.

Carassay, L.R., D.A. Bustos, A.D. Golberg, and E. Taleisnik. 2012. Tipburn in salt-affected lettuce (*Lactuca sativa* L.) plants results from local oxidative stress. *Journal of Plant Physiology* 169(3):285–293.

Carden, D.E., D.J. Walker, T.J. Flowers, and A.J. Miller. 2003. Single-cell measurements of the contributions of cytosolic $Na^+$ and $K^+$ to salt tolerance. *Plant Physiology* 131:676–683.

Carpici, E.B., N. Celik, G. Bayram, and B.B. Asik. 2010. The effects of salt stress on the growth, biochemical parameter and mineral element content of some maize (*Zea mays* L.) cultivars. *African Journal of Biotechnology* 9:6937–6942.

Carter, C.T., C.M. Grieve, and J.A. Poss. 2005. Salinity effects on emergence, survival, and ion accumulation of *Limonium perezii*. *Journal of Plant Nutrition* 28:1243–1257.

Castillo, J.M., A.E. Rubio-Casal, S. Redondo, A.A. Álvarez-López, T. Luque, C. Luque, F.J. Nieva, E.M. Castellanos, and M.E. Figueroa. 2005. Short-term responses to salinity of an invasive cordgrass. *Biological Invasions* 7:29–35.

Cha-Um, S. and C. Kirdmanee. 2009. Effect of salt stress on proline accumulation, photosynthetic ability and growth characters in two maize cultivars. *Pakistan Journal of Botany* 41:87–98.

Chakrabarti, N. and S. Mukherji. 2003. Alleviation of NaCl stress by pretreatment with phytohormones in *Vigna radiata*. *Biologia Plantarum* 46(4):589–594.

Chandra Babu, R., P. Srinivasan, N. Natarajaratnam, and S. Rangasamy. 1985. Relationship between leaf photosynthetic rate and yield in blackgram (*Vigna mungo* L. Hepper) genotypes. *Photosynthetica* 19:159–163.

Chaparzadeh, N., R.A. Khavari-Nejad, F. Navari-Izzo, and A. Izzo. 2003. Water relations and ionic balance *Calendula officinalis* L. under salinity conditions. *Agrochimica* 47:69–79.

Chaudhuri, K. and M.A. Choudhuri. 1997. Effects of short term NaCl stress on water relations and gas exchange of two jute species. *Biologia Plantarum* 40:373–380.

Chen, C.H., M. Chen, C.C. Lin, and C.H. Kao. 2001. Regulation of proline accumulation in detached rice leaves exposed to excess copper. *Plant Science* 201:283–290.

Chookhampaeng, S. 2011. The effect of salt stress on growth, chlorophyll content proline content and antioxidative enzymes of pepper (*Capsicum annuum* L.) seedling. *European Journal of Scientific Research* 49:103–109.

Chrispeels, M.J. and Sadava, D.E. 2003. *Plants, Genes and Crop Biotechnology*, 2nd edn. Boston, MA: Jones & Bartlett.

Cornish, K.J., W.E. Radin, L. Turcotte, Z. Lu, and E. Zeiger. 1991. Enhanced photosynthesis and stomatal conductance of Pima cotton (*Gossypium barbadense* L.) bred for increased yield. *Plant Physiology* 97:484–489.

Crosbie, T.M. and R.B. Pearce. 1982. Effects of recurrent phenotypic selection for high and low photosynthesis on agronomic traits in two maize populations. *Crop Science* 22:809–813.

Curtis, P.S. and A. Läuchli. 1986. The role of leaf area development and photosynthetic capacity in determining growth of kenaf under moderate salt stress. *Australian Journal of Plant Physiology* 13:553–565.

da Silva, E.N., R.V. Ribeiro, S.L. Ferreira-Silva, R.A. Viégas, and J.A.G. Silveira. 2011. Salt stress induced damages on the photosynthesis of physic nut young plants. *Scientia Agricola* (Piracicaba, Braz.) 68:62–68.

Dawood, M.G. and M.E. El-Awadi. 2015. Alleviation of salinity stress on *Vicia faba* L. plants via seed priming with melatonin. *Acta Biológica Colombiana* 20(2):223–235.

de Azevedo Neto, A.D.D., J.T. Prisco, J. Enéas-Filho, C.F.D. Lacerda, J.V. Silva, P.H.A.D. Costa, and E. Gomes-Filho. 2004. Effects of salt stress on plant growth, stomatal response and solute accumulation of different maize genotypes. *Brazilian Journal of Plant Physiology* 16(1):31–38.

de Bruxelles, G.L., W.J. Peacock, E.S. Dennies, and R. Dolferus. 1996. Abscisic acid induces the alcohol dehydrogenase gene in *Arabidopsis. Plant Physiology* 111:381–391.

De Lacerda, C.F., J., Cambraia, M.A. Oliva, H.A. Ruiz, and J.T. Prisco. 2003. Solute accumulation and distribution during shoot and leaf development in two sorghum genotypes under salt stress. *Environmental and Experimental Botany* 49:107–120.

Del Amor, F.M., V. Martinez, and A. Cerda. 2001. Salt tolerance of tomato plants as affected by stage of plant development. *Hortscience* 36:1260–1263.

Denby, K. and C. Gehring. 2005. Engineering drought and salinity tolerance in plants: Lessons from genome-wide expression profiling in *Arabidopsis. Trends in Biotechnology* 11:547–552.

Desikan, R., S.A.H. Mackerness, J.T. Hancock, and S.J. Neill. 2001. Regulation of the *Arabidopsis* transcriptome by oxidative stress. *Plant Physiology* 127:159–172.

Dhanapackiam, S. and M.H. Ilyas. 2010. Effect of NaCl salinity on growth, nodulation and total nitrogen in *Sesbania grandiflora. Indian Journal of Science and Technology* 3:87–89.

Dionisio-Sese, M.L. and S. Tobita. 2000. Effects of salinity on sodium content and photosynthetic responses of rice seedlings differing in salt tolerance. *Journal of Plant Physiology* 157:54–58.

Doganlar, Z.B., K. Demir, H. Basak, and I. Gul. 2010. Effects of salt stress on pigment and total soluble protein contents of three different tomato cultivars. *African Journal of Agricultural Research* 5:2056–2065.

Dong, C.-J., X-L. Wang, and Q.-M. Shang. 2011. Salicylic acid regulates sugar metabolism that confers tolerance to salinity stress in cucumber seedlings. *Scientia Horticulturae* 129:629–636.

Duan, J., J. Li, S. Guo, and Y. Kang. 2008. Exogenous spermidine affects polyamine metabolism in salinity-stressed *Cucumis sativus* roots and enhances short-term salinity tolerance. *Journal of Plant Physiology* 165:1620–1635.

Efimova, M.V., A.L. Savchuk, J.A.K. Hasan, R.P. Litvinovskaya, V.A. Khripach, V.P. Kholodova, and Vl.V. Kuznetsov. 2014. Physiological mechanisms of enhancing salt tolerance of oilseed rape plants with brassinosteroids. *Russian Journal of Plant Physiology* 61:733–743.

Egbichi, I., M. Keyster, and N. Ludidi. 2014. Effect of exogenous application of nitric oxide on salt stress responses of soybean. *South African Journal of Botany* 90:131–136.

El-Feky, S.S. and S.A. Abo-Hamad. 2014. Effect of exogenous application of brassinolide on growth and metabolic activity of wheat seedlings under normal and salt stress conditions. *Annual Research & Review in Biology* 4:3687–3698.

Fallon, K.M. and R. Phillips. 1989. Responses to water stress in adapted carrot cell suspension cultures. *Journal of Experimental Botany* 40:681–687.

Fedina, I.S. and T.D. Tsonev. 1997. Effect of pretreatment with methyl jasmonate on the response of *Pisum sativum* to salt stress. *Journal of Plant Physiology* 151:735–740.

Fidalgo, F., A. Santos, I. Santos, and R. Salema. 2004. Effects of long-term salt stress on antioxidant defence systems, leaf water relations and chloroplast ultrastructure of potato plants. *Annals of Applied Biology* 145:185–192.

Flowers, T.J., P.F. Troke, and A.R. Yeo. 1977. The mechanism of salt tolerance in halophytes. *Annual Review of Plant Biology* 28:89–121.

Foyer, M.J. 1992.The antioxidant effects of thylakoid vitamin E (α-tocopherol). *Plant Cell and Environment* 15:381–382.

Freitas, J.B.S., R.M. Chagas, I.M.R. Almeida, F.R. Cavalcanti, and J.A.G. Silveira. 2001. Expression of physiological traits related to salt tolerance in two contrasting cowpea cultivars. *Documentos Embrapa MeioNorte* 56:115–118.

Gadallah, M.A.A. 1999. Effects of proline and glycine-betaine on *Vicia faba* responses to salt stress. *Biologia Plantarum* 42:249–257.

Garg, B.K., U. Burman, and S. Kathju. 2006. Influence of thiourea on photosynthesis, nitrogen metabolism and yield of clusterbean (*Cyamopsis tetragonoloba* (L.) Taub.) under rainfed conditions of Indian arid zone. *Plant Growth Regulation* 48:237–245.

Genard, H., J. Le Saos, J. Hillard, A. Tremolieres, and J. Boucaud. 1991. Effect of salinity on lipid composition, glycine betaine content and photosynthetic activity in chloroplasts of *Suaeda maritime. Plant Physiology and Biochemistry* 29:421–427.

Gibberd, M.R., N.C. Turner, and R. Storey. 2002. Influence of saline irrigation on growth, ion accumulation and partitioning, and leaf gas exchange of carrot (*Daucus carota* L.). *Annals of Botany* 90:715–724.

Gill, P.K., A.D. Sharma, P. Singh, and S.S. Bhullar. 2002. Osmotic stress-induced changes in germination, growth and soluble sugar content of *Sorghum bicolor* (L.) Moench seeds. *Bulgarian Journal of Plant Physiology* 28(3–4):12–25.

Gomez, J.M., A. Jimenez, E. Olmos, and F. Sevilla. 2004. Location and effects of long-term NaCl stress on superoxide dismutase and ascorbate peroxidase isoenzymes of pea (*Pisum sativum* cv. Puget) chloroplasts. *Journal of Experimental Botany* 55:119–130.

Gopa, R. and B.K. Dube. 2003. Influence of variable potassium on barley metabolism. *Annals of Agricultural Research* 24:73–77.

Grattan, S.R. and C.M. Grieve. 1999. Mineral nutrient acquisition and response by plants grown in saline environments. In: Pessarakli, M. (ed.). *Handbook of Plant and Crop Stress.* New York: Marcel Dekker.

Gu, D., X. Liu, M. Wang, J. Zheng, W. Hou, and G. Wang. 2008. Overexpression of ZmOPR1 in *Arabidopsis* enhanced the tolerance to osmotic and salt stress during seed germination. *Plant Science* 174:124–130.

Gulzar, S., M.A. Khan, and I.A. Ungar. 2003. Effects of salinity on growth, ionic content, and plant–water status of *Aeluropus lagopoides. Communications in Soil Science and Plant Analysis* 34:1657–1668.

Gupta, K.J., M. Stoimenova, and W.M. Kaiser. 2005. In higher plants, only root mitochondria, but not leaf mitochondria reduce nitrite to NO, in vitro and in situ. *Journal of Experimental Botany* 56:2601–2609.

Gupta, S., M.K. Chattopadhyay, P. Chatterjee, B. Ghosh, and D.N. SenGupta. 1998. Expression of abscisic acid-responsive element binding protein in salt tolerant indica rice (*Oryza sativa* L. cv. Pokkali). *Plant Molecular Biology* 137:629–637.

Gómez-Cadenas, A., F.R. Tadeo, and E. Primo-Millo, and M. Talon. 1998. Involvement of abscisic acid and ethylene in the response of citrus seedlings to salt shock. *Physiologia Plantarum* 103:475–484.

Gómez-Cadenas, A., V. Arbona, J. Jacas, E. Primo-Millo, and M. Talon. 2002. Abscisic acid reduces leaf abscission and increases salt tolerance in citrus plants. *Journal of Plant Growth Regulation* 21:234–240.

Habib, N., M. Ashraf, Q. Ali, and R. Perveen. 2012. Response of salt stressed okra (*Abelmoschus esculentus* Moench) plants to foliar-applied glycine betaine and glycine betaine containing sugarbeet extract. *South African Journal of Botany* 83:151–158.

HalaEzzat Mohamed, A. and G. Saber Mohamed Ismail. 2014. Tomato fruit quality as influenced by salinity and nitric oxide. *Turkish Journal of Botany* 38:122–129.

Hawkins, H.J. and O.A.M. Lewis. 1993. Combination effect of NaCl salinity, nitrogen form and calcium concentration on the growth and ionic content and gaseous properties of *Triticum aestivum* L. cv. Gamtoos. *New Phytologist* 124:161–170.

Heidari, M. 2009. Antioxidant activity and osmolyte concentration of sorghum (*Sorghum bicolor*) and wheat (*Triticum aestivum*) genotypes under salinity stress. *Asian Journal of Plant Science* 8:240–244.

Heidari, M. and S. Sarani. 2012. Growth, biochemical components and ion content of chamomile (*Matricariachamomilla* L.) under salinity stress and iron deficiency. *Journal of the Saudi Society of Agricultural Sciences* 11(1):37–42.

Heidari-Sharifabad, H.H. and H. Mirzaie-Nodoushan. 2006. Salinity-induced growth and some metabolic changes in three *Salsola* species. *Journal of Arid Environments* 67:715–720.

Hernández, J.A., A. Campillo, A. Jiménez, J.J. Alarcón, and F. Sevilla. 1999. Response of antioxidant systems and leaf water relations to NaCl stress in pea plants. *New Phytologist* 141:241–251.

Hernández, J.A., E. Olmos, F.J. Corpas, F. Sevilla, and L.A. del Río. 1995. Salt-induced oxidative stress in chloroplast of pea plants. *Plant Science* 105:151–167.

Hernández, J.A. and M.S. Almansa. 2002. Short-term effects of salt stress on antioxidant systems and leaf water relations of pea leaves. *Physiologia Plantarum* 115:251–257.

Hester, M.W., I.A. Mendelssohn, and K.L. McKee. 2001. Species and population variation to salinity stress in *Panicum hemitomon, Spartina patens,* and *Spartina alterniflora*: morphological and physiological constraints. *Environmental and Experimental Botany* 46:277–297.

Heuer, B. and Z. Plaut. 1989. Photosynthetic and osmotic adjustment of two sugarbeet cultivars grown under saline conditions. *Journal of Experimental Botany* 40:437–440.

Hideg, E. 1997. Hideg, free radical production in photosynthesis under stress conditions. In: Pessarakli, M. (ed.). *Photosynthesis,* 2nd edn. New York: CRC Press. pp. 911–930.

Hilda, P., R. Graciela, A. Sergio, M. Otto, R. Ingrid, P.C. Hugo, T. Edith, M.D. Estela, and A. Guillermina. 2003. Salt tolerant tomato plants showincreased levels of jasmonic acid. *Plant Growth Regulation* 41:149–158.

Hmida-Sayari, A., R. Gargouri-Bouzid, A. Bidani, L. Jaoua, A. Savoure, and S. Jaoua. 2005. Overexpression of Δ1-pyrroline-5-carboxylate synthetase increases proline production and confers salt tolerance in transgenic potato plants. *Plant Science* 169:746–752.

Ho, L.-W., T.T. Yang, S.S. Shieh, G.E. Edwards, and H.E. Yen. 2010. Reduced expression of a vesicle trafficking-related ATPase SKD1 decreases salt tolerance in *Arabidopsis. Functional Plant Biology* 37:962–973.

Holländer-Czytko, H., J. Grabowski, I. Sandorf, K. Weckermann, and E.W. Weiler. 2005. Tocopherol content and activities of tyrosine aminotransferase and cystine lyase in *Arabidopsis* under stress conditions. *Journal of Plant Physiology* 62:767–770.

Hong Z., K. Lakkineni, Z. Zhang, and D.P.S. Verma. 2000. Removal of feedback inhibition of 1-pyrroline-5-carboxylate synthetase results in increased proline accumulation and protection of plants from osmotic stress. *Plant Physiology* 122:1129–1136.

Huang, J., R. Hirji, L. Adam, K. Rozwadowski, J. Hammerlindl, W. Keller, and G. Selvaraj. 2000. Genetic engineering of glycinebetaine production toward enhancing stress tolerance in plants: Metabolic limitations. *Plant Physiology* 122:747–756.

Husain, S., R. Munns, and A.G. Condon. 2003. Effect of sodium exclusion trait on chlorophyll retention and growth of durum wheat in saline soil. *Australian Journal of Agricultural Research* 54:589–597.

Iqbal, M. and M. Ashraf. 2005a. Presowing seed treatment with cytokinins and its effect on growth, photosynthetic rate, ionic levels and yield of two wheat cultivars differing in salt tolerance. *Journal of Integrative Plant Biology* 47:1315–1325.

Iqbal, M. and M. Ashraf. 2005b. Changes in growth, photosynthetic capacity and ionic relations in spring wheat (*Triticum aestivum* L.) due to pre-sowing seed treatment with polyamines. *Plant Growth Regulation* 46:19–30.

Iqbal, M. and M. Ashraf. 2013a. Salt tolerance and regulation of gas exchange and hormonal homeostasis by auxin-priming in wheat. *Pesquisa Agropecuía Brasileira, Brasília* 48:1210–1219.

Iqbal, M. and M. Ashraf. 2013b. Alleviation of salinity-induced perturbations in ionic and hormonal concentrations in spring wheat through seed preconditioning in synthetic auxins. *Acta Physiologiae Plantarum* 35:1093–1112.

Iqbal, M. and M. Ashraf. 2013c. Gibberellic acid mediated induction of salt tolerance in wheat plants: Growth, ionic partitioning, photosynthesis, yield and hormonal homeostasis. *Environmental and Experimental Botany* 86:76–85.

Iqbal, M., M. Ashraf, and A. Jamil. 2006a. Seed enhancement with cytokinins: Changes in growth and grain yield in salt stressed wheat plants. *Plant Growth Regulation* 50:29–39.

Iqbal, M., M. Ashraf, S. Rehman, and E.S. Rha. 2006b. Does polyamine seed pretreatment modulate growth and levels of some plant growth regulators in hexaploid wheat (*Triticum aestivum* L.) plants under salt stress? *Botanical Studies* 47:239–250.

ISAAA. 2006. International Service for the Acquisition of Agri-Biotech Application. Global Status of Commercialization Biotech/GM Crops.

Jacoby, B. 1999. Mechanisms involved in salt tolerance of plants. In: Pessarakli, M. (ed.). *Handbook of Plant and Crop Stress,* 2nd edn. New York: Marcel Dekker Inc., pp. 97–123.

Jayakannan, M., J. Bose, O. Babourina, Z. Rengel, and S. Shabala. 2013. Salicylic acid improves salinity tolerance in *Arabidopsis* by restoring membrane potential and preventing salt-induced $K^+$ loss via a GORK channel. *Journal of Experimental Botany* 64(8):2255–2268. doi:10.1093/jxb/ert085.

Kaiser, W.M. 1979. Reversible inhibition of the Calvin cycle and activation of oxidative pentose phosphate cycle in isolated intact chloroplasts by hydrogen peroxide. *Planta* 145: 377–382.

Kamal-Eldin, A. and L. Appelqvist. 1996. The chemistry and antioxidant properties of tocopherols and tocotrienols. *Lipids* 31:671–701.

Karimi, G., M. Ghorbanli, H. Heidari, R.A. Khavarinejad, and M.H. Assareh. 2005. The effects of NaCl on growth, water relations, osmolytes and ion content in *Kochia prostrata*. *Biologia Plantarum* 49:301–304.

Karpinski, S., H. Gabryś, A. Mateo, B. Karpinska, and P.M. Mullineaux. 2003. Light perception in plant disease defence signalling. *Current Opinion in Plant Biology* 6:390–396.

Kauser, R., H.R. Athar, and M. Ashraf. 2006. Chlorophyll fluorescence. A potential indicator for rapid assessment of water stress tolerance in canola (*Brassica napus* L.). *Pakistan Journal of Botany* 38:1501–1509.

Kavi-Kishor, P.B., S. Sangam, R.N. Amrutha, P. Sri Laxmi, K.R. Naidu, S.S. Rao, S. Rao, K.J. Reddy, P. Theriappan, and N. Sreeniv. 2005. Regulation of proline biosynthesis, degradation, uptake and transport in higher plants: Its implications in plant growth and abiotic stress tolerance. *Current Science* 88:424–438.

Keutgen, A. and E. Pawelizik. 2008. Quality and nutritional value of strawberry fruit under long term salt stress. *Food Chemistry* 107:1413–1420.

Khalid, K.A. and W. Cai. 2011. The effects of mannitol and salinity stresses on growth and biochemical accumulations in lemon balm. *Acta Ecologica Sinica* 31(2):112–120.

Khan, A.A., T. Mcneilly, and F.M. Azhar. 2001. Stress tolerance in crop plants. *International Journal of Agriculture and Biology* 3:250–255.

Khan, M.A., B. Gul, and D.J. Weber. 2000. Germination responses of *Salicornia rubra* to temperature and salinity. *Journal of Arid Environments* 45:207–214.

Khan, M.A., L.A. Ungar, and A.M. Showalter. 1999. Effects of salinity on growth, ion content, and osmotic relations in *Halopyrum mucronatum* (L.) Stapf. *Journal of Plant Nutrition* 22:191–204.

Khan, M.A., M.U. Shirazi, M.A. Khan, S.M. Mujtaba, E. Islam, S. Mumtaz, A. Shereen, R.U. Ansari, and M.Y. Ashraf. 2009. Role of proline, K/Na ratio and chlorophyll content in salt tolerance of wheat (*Triticum aestivum* L.). *Pakistan Journal of Botany* 41:633–638.

Khan, M.H. and S.K. Panda. 2008. Alterations in root lipid peroxidation and antioxidative responses in two rice cultivars under NaCl-salinity stress. *Acta Physiologiae Plantarum* 30:89–91.

Khan, M.N., M.H. Siddiqui, F. Mohammad, M.M.A. Khan, and M. Naeem. 2007. Salinity induced changes in growth, enzyme activities, photosynthesis, proline accumulation and yield in linseed genotypes. *World Journal of Agricultural Sciences* 3:685–695.

Khatun, S., C.A. Rizzo, and T.J. Flowers. 1995. Genotypic variation in the effect of salinity on fertility in rice. *Plant and Soil* 173:239–250.

Khavarinejad, R.A. and Y. Mostofi. 1998. Effects of NaCl on photosynthetic pigments, saccharides, and chloroplast ultrastructure in leaves of tomato cultivars. *Photosynthetica* 35:151–154.

Korkmaz, A., R. Şirikçi, F. Kocaçınar, Ö. Değer, and A.R. Demirkırıan. 2012. Alleviation of salt-induced adverse effects in pepper seedlings by seed application of glycinebetaine. *Scientia Horticulturae* 148:197–205.

Krishnan, R.R. and B.D.R. Kumari. 2008. Effect of n-triacontanol on the growth of salt stressed soybean plants. *Journal of Bioscience* 19:53–62.

Krishnaraj, S., B.T. Mawson, E.C. Yeung, and T.A. Thorpe. 1993. Utilization of induction and quenching kinetics of chlorophyll *a* fluorescence for in vivo salinity screening studies in wheat (*Triticum aestivum* vars. Kharchia-65 and Fielder). *Canadian Journal of Botany* 71:87–92.

Kurban, H., H. Saneoka, H.K. Nehira, R. Adilla, G.S. Premachandra, and K. Fujita. 1999. Effect of salinity on growth, photosynthesis and mineral composition in leguminous plant *Alhagi pseudalhagi*. *Soil Science and Plant Nutrition* 45:851–862.

Lakshmi, A., S. Ramanjulu, K. Veeranjaneyulu, and C. Sudhakar. 1996. Effect of NaCl on photosynthesis parameters in two cultivars of mulberry. *Photosynthetica* 32:285–289.

Laloi, C., K. Apel, and A. Danon. 2004. Reactive oxygen signalling: The latest news. *Current Opinion in Plant Biology* 7:323–328.

Lamb, C. and R.A. Dixon. 1997. The oxidative burst in plant disease resistance. *Annual Review of Plant Biology* 48:251–275.

Lin, Y., Z. Liu, Q. Shi, X. Wang, M. Wei, and F. Yang. 2012. Exogenous nitric oxide (NO) increased antioxidant capacity of cucumber hypocotyl and radicle under salt stress. *Scientia Horticulturae* 142:118–127.

Lloyd, J., J.P. Syvertsen, and P.E. Kriedemann. 1987. Salinity effects on leaf water relations and gas exchange of 'Valentia' orange, *Citrus sinensis* (L.) Osbeck, on rootstocks with different salt exclusion characteristics. *Australian Journal of Agricultural Research* 40:359–369.

Lobato, A.K.S., B.G. Santos Filho, R.C.L. Costa, M.C. Gonçalves-Vidigal, E.C. Moraes, C.F. Oliveira Neto, V.L.F. Rodrigues, F.J.R. Cruz, A.S. Ferreira, J.D. Pita, and A.G.T. Barreto. 2009. Morphological, physiological and biochemical responses during germination of the cowpea (*Vigna unguiculata* Cv. Pitiuba) seeds under salt stress. *World Journal of Agricultural Sciences* 5:590–596.

Locy, R.D., C.C. Chang, B.L. Nielsen, and N.K. Singh. 1996. Photosynthesis in salt-adapted heterotrophic tobacco cells and regenerated plants. *Plant Physiology* 110:321–328.

Loreto, F., M. Centritto, and K. Chartzoulakis. 2003. Photosynthetic limitations in olive cultivars with different sensitivity to salt stress. *Plant, Cell and Environment* 26:595–601.

Lu, C., N. Qiu, Q. Lu, B. Wang, and T. Kuang. 2002. Does salt stress lead to increased susceptibility of photosystem II to photo inhibition and changes in photosynthetic pigment composition in halophyte *Suaeda salsa* grown outdoors. *Plant Science* 163:1063–1068.

Lutts, S.J., M. Kinet, and J. Bouharmont. 1995. Changes in plant response to NaCl during development of rice (*Oryza sativa* L.) varieties differing in salinity resistance. *Journal of Experimental Botany* 46:1843–1852.

Läuchli, A. and Epstein. 1990. Plant responses to saline and sodic conditions. In: Tanji, K.K. (ed.). *Agricultural Salinity Assessment and Management.* ASCE manuals and reports on engineering practice No, 71. New York: ASCE. pp. 113–137.

Läuchli, A. and S.R. Grattan. 2007. Plant growth and development under salinity stress. In: Jenks, M.A., P.M. Hasegawa, and J.S. Mohan. (eds.). *Advances in Molecular Breeding towards Drought and Salt Tolerant Crops.* Berlin, Germany: Springer. pp. 1–32.

Maas, E.V., G.J. Hoffman, G.D. Chaba, J.A. Poss, and M.C. Shannon. 1983. Salt sensitivity of corn at various growth stages. *Irrigation Science* 4:45–57.

Maas, E.V., J.A. Poss, and G.J. Hoffman. 1986. Salinity sensitivity of sorghum at three growth stages. *Irrigation Science* 7:1–11.

Maas, E.V. and J.A. Poss. 1989a. Salt sensitivity of wheat at various growth stages. *Irrigation Science* 10:29–40.

Maas, E.V. and J.A. Poss. 1989b. Salt sensitivity of cowpea at various growth stages. *Irrigation Science* 10:313–320.

Maas, E.V. and S.R. Grattan. 1999. Crop yields as affected by salinity. In: Pessarakli, M. (ed.). *Handbook of Plant and Crop Stress.* New York: Marcel Dekker. pp. 55–108.

Madan, S., H.S. Nainawatee, R.K. Jain, R.K., and J.B. Chowdhury. 1995. Proline and proline metabolizing enzymes in *in-vitro* selected NaCl-tolerant *Brassica juncea* L. under salt stress. *Annals of Botany* 76:51–57.

Mandhania, S., S. Madan, and V. Sawhney. 2006. Antioxidant defense mechanism under salt stress in wheat seedlings. *Biologia Plantarum* 227:227–231.

Mansour, M.M.F., K.H.A. Salama, and M.M. Al Mutawa. 2003. Transport proteins and salt tolerance in plants. *Plant Science* 164:891–900.

Mansour, M.M.F. 2000. Nitrogen containing compounds and adaptation of plants to salinity stress. *Biologia Plantarum* 43:491–500.

Marschner, H. 1995. *Mineral Nutrition of Higher Plants,* 2nd edn. New York: Academic Press. p. 889.

Mateo, A., P. Muhlenbock, C. Rusterucci, C.C. Chang, Z. Miszalski, B. Karpinska, J.E. Parker, P.M. Mullineaux, and S. Karpinski. 2004. Lesion simulating disease 1 is required for acclimation to conditions that promote excess excitation energy. *Plant Physiology* 136:2818–2830.

Mauromicale, G. and P. Licandro. 2002. Salinity and temperature effects on germination, emergence and seedling growth of globe artichoke. *Agronomie* 22:443–450.

McCue, K.F. and A.D. Hanson. 1990. Salt inducible betaine aldehyde dehydrogenase from sugar beet: cDNA cloning and expression. *Trends in Biotechnology* 8:358–362.

McNeil, S.D., M.L. Nuccio, D. Rhodes, Y. Shachar-Hill, and A.D. Hanson. 2000. Radiotracer and computer modeling evidence that phospho-base methylation is the main route of choline synthesis in tobacco. *Plant Physiology* 123:371–380.

Millar, A.A., M.E. Duysen, and G.E. Wilkinson. 1968. Internal water balance of barley under soil moisture stress. *Plant Physiology* 43:968–972.

Mittler, R. 2002. Oxidative stress, antioxidants and stress tolerance. *Trends in Plant Science* 7:405–410.

Mittova, V., M. Tal, M. Volokita, and M. Guy. 2003. Up-regulation of the leaf mitochondrial and peroxisomal antioxidative systems in response to salt-induced oxidative stress in the wild salt tolerant tomato species. *Plant, Cell and Environment* 26:845–856.

Mohanty, A., H. Kathuria, A. Ferjani, A. Sakamoto, P. Mohanty, N. Murata, and A.K. Tyagi. 2002. Transgenics of an elite indica rice variety *Pusa basmati* 1 harboring the codA gene are highly tolerant to salt stress. *Theoretical and Applied Genetics* 106:51–57.

Moradi, F. and A.M. Ismail. 2007. Responses of photosynthesis, chlorophyll fluorescence and ROS-scavenging systems to salt stress during seedling and reproductive stages in rice. *Annals of Botany* 99:1161–1173.

Munns, R., H. Greenway, R. Delane, and J. Gibbs. 1982. Ion concentration and carbohydrate status of the elongating leaf tissue of *Hordeum vulgare* growing at high external NaCl. II. Cause of the growth reduction. *Journal of Experimental Botany* 33:574–583.

Munns, R., R.A. Hare, R.A. James, and G.J. Rebetzke. 2000. Genetic variation for salt tolerance of durum wheat. *Australian Journal of Agricultural Research* 51:69–74.

Munns, R. 2002. Comparative physiology of salt and water stress. *Plant, Cell and Environment* 25:239–250.

Munns, R. 2005. Genes and salt tolerance: Bringing them together. *New Phytologist* 167:645–663.

Munns, R. 2009. Strategies for crop improvement in saline soils. In: Ashraf, M., M. Ozturk, and H.R. Athar. (eds.). *Springer Salinity and Water Stress: Improving Crop Efficiency. Tasks for Vegetation Science*, Vol. 44. Amsterdam, the Netherlands: Springer. pp. 99–110.

Munns, R. 2011. Plant adaptations to salt and water stress: Differences and commonalities. *Advances in Botanical Research* 57:1–32.

Munns, R. and M. Tester. 2008. Mechanisms of salinity tolerance. *Annual Review of Plant Biology* 59:651–681.

Munné-Bosch, S. 2005. The role of α-tocopherol in plant stress tolerance. *Journal of Plant Physiology* 162:743–748.

Murata, N., S. Takahashi, Y. Nishiyama, and S.I. Allakhverdiev. 2007. Photoinhibition of photosystem 2 under environmental stress. *Biochimica et Biophysica Acta (BBA)* 1767:414–421.

Mühling, K.M. and A. Läuchli. 2002. Effect of salt stress on growth and cation compartmentation in leaves of two plant species differing in salt tolerance. *Journal of Plant Physiology* 159:137–146.

Nawaz, K., K. Hussain, A. Majeed, F. Khan, S. Afghan, and K. Ali. 2010. Fatality of salt stress to plants: Morphological, physiological and biochemical aspects. *African Journal of Biotechnology* 9:5475–5480.

Nawaz, K. and M. Ashraf. 2007. Improvement in salt tolerance of maize by exogenous application of glycinebetaine: Growth and water relations. *Pakistan Journal of Botany* 39:1647–1653.

Nedjimi, B. 2011. Is salinity tolerance related to osmolytes accumulation in *Lygeum spartum* L. seedlings?. *Journal of the Saudi Society of Agricultural Sciences* 10(2):81–87.

Nemati, I., F. Moradi, S. Gholizadeh, M.A. Esmaeili, and M.R. Bihamta. 2011. The effect of salinity stress on ions and soluble sugars distribution in leaves, leaf sheaths and roots of rice (*Oryza sativa* L.) seedlings. *Plant Soil and Environment* 57(1):26–33.

Netondo, G.W., J.C. Onyango, and E. Beck. 2004. Sorghum and salinity: I. Response of growth, water relations and ion accumulation to NaCl salinity. *Crop Science* 44:797–805.

Niu, X., M.L. Narasimhan, R.A. Salzman, R.A. Bressan, and P.M. Hasegawa. 1993. NaCl regulation of plasma membrane H+-ATPase gene expression in glycophyte and a halophyte. *Plant Physiology* 103:713–718.

Noctor, G. and C. Foyer. 1998. Ascorbate and glutathione: Keeping active oxygen under control. *Annual Review of Plant Biology* 49:249–279.

Nounjan, N., P.T. Nghia, and P. Theerakulpisut. 2012. Exogenous proline and trehalose promote recovery of rice seedlings from salt-stress and differentially modulate antioxidant enzymes and expression of related genes. *Journal of Plant Physiology* 169(6):596–604.

Nuccio, M.L., B.L. Russell, K.D. Nolte, B. Rathinasabapathi, D.A. Gage, and A.D. Hanson. 1998. The endogenous choline supply limits glycine betaine synthesis in transgenic tobacco expressing choline monooxygenase. *Plant Journal* 16:487–496.

Orabi, S.A. and M.T. Abdelhamid. 2014. Protective role of α-tocopherol on two Viciafaba cultivars against seawater-induced lipid peroxidation by enhancing capacity of anti-oxidative system. *Journal of the Saudi Society of Agricultural Sciences*.

Padan, E., M. Venturi, Y. Gerchman, and N. Dover. 2001. Na+/H+ antiporters. *Biochimica et Biophysica Acta (BBA)* 1505:144–157.

Palta, J.P. 1992. Mechanisms for obtaining freezing stress resistance in herbaceous plants, In: *Plant Breeding in the 1990s: Proceedings of a Symposium Held at North Carolina State University*, Raleigh, NC, March 1991. pp. 219–250.

Parida, A.K. and A.B. Das. 2005. Salt tolerance and salinity effects on plants: A Review. *Ecotoxicology and Environmental Safety* 60:324–349.

Pastori, G.M. and C.H. Foyer. 2002. Common components, networks, and pathways of cross-tolerance to stress. The central role of "redox" and abscisic acid-mediated controls. *Plant Physiology* 129:7460–7468.

Pearson, G.A. and A.D. Ayers. 1996. Relative salt tolerance of rice during germination and early seedling development. *Soil Science* 102:151–156.

Pedranzani, H., G. Racagni, S. Alemano, O. Miersch, I. Ramírez, H. Peña-Cortés, E. Taleisnik, E. Machado-Domenech, and G. Abdala. 2003. Salt tolerant tomato plants show increased levels of jasmonic acid. *Plant Growth Regulation* 41:149–158.

Percival, G.C. 2004. Evaluation of physiological tests as predictors of young tree establishment and growth. *Journal of Arboriculture* 30:80–92.

Perveen, S., M. Shahbaz, and M. Ashraf. 2012. Changes in mineral composition, uptake and use efficiency of salt stressed wheat (*Triticum aestivum* L.) plants raised from seed treated with triacontanol. *Pakistan Journal of Botany* 44:27–35.

Pettigrew, W.T. and W.R. Meredith. 1994. Leaf gas exchange parameters vary among cotton genotypes. *Crop Science* 34:700–705.

Piqueras, A., J.M. Hernández, E. Olmos, E. Hellín, and F. Sevilla. 1996. Changes in antioxidant enzymes and organic solutes associated with adaptation of citrus cells to salt stress. *Plant Cell, Tissue and Organ Culture* 45:53–60.

Polle, A. 2001. Dissecting the superoxide dismutase–ascorbate–glutathione pathway by metabolic modeling: Computer analysis as a step towards flux analysis. *Plant Physiology* 126:445–462.

Popova, L.P., Z.G. Stoinova, and L.T. Maslenkova. 1995. Involvement of abscisic acid in photosynthetic process in *Hordeum vulgare* L. during salinity stress. *Journal of Plant Growth Regulation* 14:211–218.

Prasad, K.V.S.K., P. Sharmila, P.A. Kumar, and P. Pardha Saradhi. 2000. Transformation of *Brassica juncea* (L.) Czern with bacterial coda gene enhances its tolerance to salt stress. *Molecular Breeding* 6:489–499.

Rady, M.M., B. Varma, and S.M. Howladar. 2013. Common bean (*Phaseolus vulgaris* L.) seedlings overcome NaCl stress as a result of presoaking in Moringaoleifera leaf extract. *Scientia Horticulturae* 162:63–70.

Rajasekaran, L.R., D. Aspinall, G.P. Jones, and L.G. Paleg. 2001. Stress metabolism. IX. Effect of salt stress on trigonelline accumulation in tomato. *Canadian Journal of Plant Science* 81:487–498.

Rajesh, A., R. Arumugam, and V. Venkatesalu. 1998. Growth and photosynthetic characteristics of *Ceriops roxburghiana* under NaCl stress. *Photosynthetica* 35:285–287.

Ranganayakulu, G.S., G. Veeranagamallaiah, and C. Sudhakar. 2013. Effect of salt stress on osmolyte accumulation in two groundnut cultivars (*Arachishypogaea* L.) with contrasting salt tolerance. *African Journal of Plant Science* 7(12):586–592.

Ratner, A. and B. Jacoby. 1976. Effect of KC, its counter anion, and pH on sodium efflux from barley root tips. *Journal of Experimental Botany* 27:843–852.

Rawson, H.M., R.A. Richards, and R. Munns. 1988. An examination of selection criteria for salt tolerance in wheat, barley and triticale genotypes. *Australian Journal of Agricultural Research* 39:759–772.

Rhodes, D. and A.D. Hanson. 1993. Quaternary ammonium and tertiary sulfonium compounds in higher plants. *Annual Review of Plant Physiology and Plant Molecular Biology* 44:357–384.

Robinson, S.P., W.J.S. Downton, and J.A. Milliouse. 1983. Photosynthesis and ion content of leases and isolated chloroplasts of salt-stressed spinach. *Plant Physiology* 73:238–242.

Robinson, S.P. and G.P. Jones. 1986. Accumulation of glycine betaine in chloroplasts provides osmotic adjustment during salt stress. *Australian Journal of Plant Physiology* 13:659–668.

Rogers, M.E. and C.L. Noble. 1992. Variation in growth and ion accumulation between two selected populations of *Trifolium repens* L. differing in salt tolerance. *Plant and Soil* 146:173–178.

Romero-Aranda, R., T. Soria, and J. Cuartero. 2001. Tomato plant water uptake and plant water relationships under saline growth conditions. *Plant Science* 160:265–272.

Sairam, R.K., G.C. Srivastava, S. Agarwal, and R.C. Meena. 2005. Differences in antioxidant activity in response to salinity stress in tolerant and susceptible wheat genotypes. *Biologia Plantarum* 49:85–91.

Sairam, R.K., V.K. Rao, and G.C. Srivastava. 2002. Differential response of wheat genotypes to long term salinity stress in relation to oxidative stress, antioxidant activity and osmolyte concentration. *Plant Science* 163:1037–1046.

Sairam, R.K. and G.C. Srivastava. 2002. Changes in antioxidant activity in subcellular fractions of tolerant and susceptible wheat genotypes in response to long term salt stress. *Plant Science* 162:897–904.

Sakamoto, A., Alia, and N. Murata. 1998. Metabolic engineering of rice leading to biosynthesis of glycinebetaine and tolerance to salt and cold. *Plant Molecular Biology* 38:1011–1019.

Salama, S., S. Trivedi, M. Busheva, A.A. Arafa, G. Garab, and L. Erdei. 1994. Effects of NaCl salinity on growth, cation accumulation, chloroplast structure and function in wheat cultivars differing in salt tolerance. *Journal of Plant Physiology* 144:241–247.

Santos, C. and G. Caldeira. 1999. Comparative responses of *Helianthus annuus* plants and calli exposed to NaCl: I. Growth rate and osmotic regulation in intact plants and calli. *Journal of Plant Physiology* 155:769–777.

Santos, C.L.V., A. Campos, H. Azevedo and G. Caldeira. 2001. in situ and in vitro senescence induced by KCl stress: Nutritional imbalance, lipid peroxidation and antioxidant metabolism. *Journal of Experimental Botany* 52:351–360.

Santos, C.V. 2004. Regulation of chlorophyll biosynthesis and degradation by salt stress in sunflower leaves. *Scientia Horticulturae* 103:93–99.

Sawahel, W.A. and A.H. Hassan. 2002. Generation of transgenic wheat plants producing high levels of the osmoprotectant proline. *Biotechnology Letters* 24:721–725.

Schafer, R.Q., H.P. Wang, E.E. Kelley, K.L. Cueno, S.M. Martin, and G.R. Buettner. 2002. Comparing carotene, vitamin E and nitric oxide as membrane antioxidants. *Biological Chemistry* 383:671–681.

Seemann, J.R. and C. Chritchly. 1985. Effects of salt stress on the growth ion content, stomatal behavior and photosynthetic capacity of a salt sensitive species *Phaseolus vulgaris* L. *Planta* 164:151–162.

Semida, W.M., R.S. Taha, M.T. Abdelhamid, and M.M. Rady. 2014. Foliar-applied α-tocopherol enhances salt-tolerance in *Vicia faba* L. plants grown under saline conditions. *South African Journal of Botany* 95:24–31.

Sestak, Z. and P. Stiffel. 1997. Leaf age related differences in chlorophyll fluorescence. *Photosynthetica* 33:347–369.

Shabala, S.N., S.I. Shabala, A.I. Martynenko, O. Babourina, and I.A. Newman. 1998. Salinity effect on bioelectric activity, growth, $Na^+$ accumulation and chlorophyll fluorescence of maize leaves: A comparative survey and prospects for screening. *Australian Journal of Plant Physiology* 25:609–616.

Shah, S.S. 2011. Comparative effects of 4-Cl-IAA and kinetin on photosynthesis, nitrogen metabolism and yield of black cumin (*Nigella sativa* L.). *Acta Botanica Croatica* 70(1):91–97.

Shahbaz, M., Z. Mushtaq, F. Andaz, and A. Masood. 2013. Does proline application ameliorate adverse effects of salt stress on growth, ions and photosynthetic ability of eggplant (*Solanum melongena* L.)?. *Scientia Horticulturae* 164:507–511.

Shahbaza, M., N. Noreena, and S. Perveen. 2013. Triacontanol modulates photosynthesis and osmoprotectants in canola (*Brassica napus* L.) under saline stress. *Journal of Plant Interactions* 8:350–359.

Sharma, L., E. Ching, S. Saini, R. Bhardwaj, and P.K. Pati. 2013. Exogenous application of brassinosteroid offers tolerance to salinity by altering stress responses in rice variety *Pusa basmati-1*. *Plant Physiology and Biochemistry* 69:17–26.

Sharma, P.K. and D.O. Hall. 1991. Interaction of salt stress and photoinhibition on photosynthesis in barley and sorghum. *Journal of Plant Physiology* 138:614–619.

Shen, B., R.B. Jensen, and H.J. Bohnert. 1997. Mannitol protects against oxidation by hydroxyl radicals. *Plant Physiology* 115:527–532.

Shi, H., B.H. Lee, S.J. Wu, and J.K. Zhu. 2003. Overexpression of a plasma membrane $Na^+/H^+$ antiporter gene improves salt tolerance in *Arabidopsis thaliana*. *Nature Biotechnology* 21:81–85.

Shu, S., L.Y. Yuan, S.R. Guo, J. Sun, and Y.H. Yuan. 2013. Effects of exogenous spermine on chlorophyll fluorescence, antioxidant system and ultrastructure of chloroplasts in *Cucumis sativus* L. under salt stress. *Plant Physiology and Biochemistry* 63:209–216.

Siddiqi, E.H. and M. Ashraf. 2008. Can leaf water relation parameters be used as selection criteria for salt tolerance in safflower (*Carthamus tinctorius* L.). *Pakistan Journal of Botany* 40:221–228.

Singh, P., N. Singh, K.D. Sharma, and M.S. Kuhad. 2010. Plant water relations and osmotic adjustment in *Brassica* species under salinity stress. *Journal of American Science* 6:1–4.

Skirycz, A., H. Claeys, S. De Bodt, A. Oikawa, S. Shinoda, M. Andriankaja, K. Maleux, N.B. Eloy, F. Coppens, S.-D. Yoo, K. Saito, and D. Inzé. 2011. Pause-and-stop: The effects of osmotic stress on cell proliferation during early leaf development in *Arabidopsis* and a role for ethylene signaling in cell cycle arrest. *Plant Cell* 23:1876–1888.

Smillie, R.M. and R. Nott. 1982. Salt tolerance in crops plants monitored by chlorophyll fluorescence in vivo. *Plant Physiology* 70:1049–1054.

Sobhanian, H., R. Razavizadeh, Y. Nanjo, A.A. Ehsanpour, F.R. Jazii, and N. Motamed. 2010. Proteome analysis of soybean leaves, hypocotyls and roots under salt stress. *Proteome Sciences* 8:19–25.

Spollen, W.G., R.E. Sharp, I.N. Saab, and Y. Wu. 1993. Regulation of cell expansion in roots and shoots at low water potentials. In: Smith, J.A.C. and H. Griffiths (eds.). *Water Deficits: Plant Responses from Cell to Community*. Oxford, U.K.: BIOS Scientific Publishers. pp. 37–52.

Stepien, P. and Klobus, G. 2006. Water relations and photosynthesis in *Cucumis sativus* L. leaves under salt stress. *Biologia Plantarum* 50:610–616.

Su, J. and R. Wu. 2004. Stress-inducible synthesis of proline in transgenic rice confers faster growth under stress conditions than that with constitutive synthesis. *Plant Science* 166:941–948.

Summart, J., P. Thanonkeo, S. Panichajakul, P. Prathepha, and M.T. McManus. 2010. Effect of salt stress on growth, inorganic ion and proline accumulation in Thai aromatic rice, Khao Dawk Mali 105, callus culture. *African Journal of Biotechnology* 9:145–152.

Taiz, L. and E. Zeiger. 2010. *Plant Physiology*, 5th edn. Sunderland, U.K.: Sinauer Associates.

Tajbakhsh, M., M. Zhou, Z. Chen, and N.J. Mendham. 2006. Physiological and cytological response of salt-tolerant and non-tolerant barley to salinity during germination and early growth. *Australian Journal of Experimental Agriculture* 46:555–562.

Tanaka, Y., T. Hibino, Y. Hayashi, A. Tanaka, S. Kishitani, T. Takabe, S. Yokota, and T. Takabe. 1999. Salt tolerance of transgenic rice overexpressing yeast mitochondrial Mn-SOD in chloroplasts. *Plant Science* 148:131–138.

Tausz, M. and D. Grill. 2000. The role of glutathione in stress adaptation of plants. *Phyton* 40:111–118.

Temple, M.D., G.G. Perrone, and L.W. Dawes. 2005. Complex cellular responses to reactive oxygen species. *Trends in Cell Biology* 15:319–326.

Tezara, W., V. Mitchell, S.P. Driscoll, and D.W. Lawlor. 2002. Effects of water deficit and its interaction with $CO_2$ supply on the biochemistry and physiology of photosynthesis in sunflower. *Journal of Experimental Botany* 53:1781–1791.

Thiel, G., J. Lynch, and A. Läuchli. 1988. Short term effects of salinity on the turgor and elongation of growing barley leaves. *Journal of Plant Physiology* 132:38–44.

Thomas, J.C., E. McElwain, and H.J. Bohnert. 1992. Convergent induction of osmotic stress-responses. ABA and cytokinin and the effect of NaCl. *Plant Physiology* 100:416–423.

Tseng, M.J., C.W. Liu, and J.C Yiu. 2007. Enhanced tolerance to sulfur dioxide and salt stress of transgenic Chinese cabbage plants expressing both superoxide dismutase and catalase in chloroplasts. *Plant Physiology and Biochemistry* 45:1–12.

Tunçtürk, M., R. Tunçtürk, B. Yildirim, and V. Çiftçi. 2011. Effect of salinity stress on plant fresh weight and nutrient composition of some canola (*Brassica napus* L.) cultivars. *African Journal of Biotechnology* 10:1827–1832.

Turan, M.A., A.H.A. Elkarim, A. Taban, and S. Taban. 2010. Effect of salt stress on growth and ion distribution and accumulation in shoot and root of maize plant. *African Journal of Agricultural Research* 5(7):584–588.

Turan, M.A., A.H.A. Elkarim, N. Taban, and S. Taban. 2009. Effect of salt stress on growth, stomatal resistance, proline and chlorophyll concentrations on maize plant. *African Journal of Agricultural Research* 4:893–897.

U.S. Salinity Laboratory Staff. 1954. Saturated soil paste. Diagnosis and improvement of saline and alkali soils. Agr. Handbook 60, Washington, D.C: USDA.

Vaidyanathan, R., S. Kuruvilla, and G. Thomas. 1999. Characterization and expression pattern of an abscisic acid and osmotic stress responsive gene from rice. *Plant Science* 140:21–30.

Van Camp, W., K. Capiau, M. Van Montagu, D. Inze, and L. Slooten. 1996. Enhancement of oxidative stress tolerance in transgenic tobacco plants over-producing Fe-superoxide dismutase in chloroplasts. *Plant Physiology* 112:1703–1714.

Varshney, K.A., L.P. Gangwar, and N. Goel. 1988. Choline and betaine accumulation in *Trifolium alexandrinum* L. during salt stress. *Egyptian Journal of Botany* 31:81–86.

Verma, D., S.L. Singla-Pareek, D. Rajagopal, M.K. Reddy, and S.K. Sopory. 2007. Functional validation of a novel isoform of $Na^+/H^+$ antiporter from *Pennisetum glaucum* for enhancing salinity tolerance in rice. *Journal of Biosciences* 32:621–628.

Walia, H., C. Wilson, P. Condamine, X. Liu, A.M. Ismail, L. Zeng, S.I. Wanamaker, J. Mandal, J. Xu, X. Cui, and T.J. Close. 2005. Comparative transcriptional profiling of two contrasting rice genotypes under salinity stress during the vegetative growth stage. *Plant Physiology* 139:822–835.

Wang, Y., Y. Ying, J. Chen, and X. Wang. 2004. Transgenic *Arabidopsis* overexpressing Mn-SOD enhanced salt-tolerance. *Plant Science* 167:671–677.

Wang, Y. and N. Nil. 2000. Changes in chlorophyll, ribulose bisphosphate carboxylase-oxygenase, glycine betaine content, photosynthesis and transpiration in *Amaranthus tricolor* leaves during salt stress. *Journal of Horticultural Science and Biotechnology* 75:623–627.

Weinberg, R., H.R. Lerner, and A. Poljakoff-Mayber. 1984. Changes in growth of water soluble concentrations in *Sorghum bicolor* stressed with sodium and potassium salt. *Physiologia Plantarum* 62:472–480.

Wenxue, W., P.E. Bilsborrow, P. Hooley, D.A. Fincham, E. Lombi, and B.P. Forster. 2003. Salinity induced difference in growth, ion distribution and partitioning in barley between the cultivar Maythorpe and its derived mutant Golden Promise. *Plant Soil* 250:183–191.

Wilson, C., S.M. Lesch, and C.M. Grieve. 2000. Growth stage modulates salinity tolerance of New Zealand spinach (*Tetragonia tetragonioides*, Pall) and Red Orach (*Atriplex hortensis* L.) *Annals of Botany* 85:501–509.

Win, K.T., A.Z. Oo, T. Hirasawa, T. Ookawa, and H. Yutaka. 2011. Genetic analysis of Myanmar *Vigna* species in responses to salt stress at the seedling stage. *African Journal of Biotechnology* 10:1615–1624.

Winicov, I. and J.D. Button. 1991. Accumulation of photosynthesis gene transcripts in response to sodium chloride by salt-tolerant alfalfa cells. *Planta* 183:478–483.

Wu, C.A., G.D. Yang, Q.W. Meng, and C.C. Zheng. 2004. The cotton *GhNHX1* gene encoding a novel putative tonoplast $Na^+/K^+$ antiporter plays an important role in salt stress. *Plant and Cell Physiology* 45:600–607.

Wyn Jones, R.G., J. Gorham, and E. McDonnell. 1984. Organic and inorganic solute contents as selection criteria for salt tolerance in the Triticeae. In: Staples, R. and Toennissen, G.H. (eds.). *Salinity Tolerance in Plants: Strategies for Crop Improvement.* New York: Wiley & Sons. pp. 189–203.

Xue, Z.Y., D.Y. Zhi, G.P. Xue, H. Zhang, Y.X. Zhao, and G.M. Xia. 2004. Enhanced salt tolerance of transgenic wheat (*Triticum aestivum* L.) expressing a vacuolar $Na^+/H^+$ antiporter gene with improved grain yields in saline soils in the field and a reduced level of leaf $Na^+$. *Plant Science* 167:849–859.

Yadav, S., M. Irfan, A. Ahmad, and S. Hayat. 2011. Causes of salinity and plant manifestations to salt stress. *Journal of Environmental Biology* 32:667–685.

Yang, W.J., P.I. Rich, J.D. Axtell, K.V. Wood, C.C. Bonham, G. Ejeta, M.V. Mickelbart, and D. Rhodes. 2004. Genotypic variation for glycinebetaine in sorghum. *Crop Science* 43:162–169.

Yang, Y.W., R.J. Newton, and F.R. Miller. 1990. Salinity tolerance in sorghum. I. Whole plant response to sodium chloride in *S. bicolor* and *S. halepense*. *Crop Science* 30:775–781.

Yu-Mei, W., M. Yu-Ling, and N. Naosuke. 2004. Changes in glycine betaine and related enzyme contents in *Amaranthus tricolor* under salt stress. *Journal of Plant Physiology and Molecular Biology* 30: 496–502.

Zeng, L., M.C. Shannon, and S.M. Lesch. 2001. Timing of salinity stress affects rice growth and yield components. *Agricultural Water Management* 48: 191–206.

Zhang, G.H., Q. Su, L.J. An, and S. Wu. 2008. Characterization and expression of a vacuolar $Na^+/H^+$ antiporter gene from the monocot halophyte *Aeluropuslittoralis*. *Plant Physiology and Biochemistry* 46:117–126.

Zheng, C., D. Jiang, T. Dai, Q. Jing, and W. Cao. 2010. Effects nitroprusside, a nitric oxide donor, on carbon and nitrogen metabolism and the activity of the antioxidation system in wheat seedlings under salt stress. *Acta Ecologica Sinica* 30:1174–1183.

Zheng, Y., Z. Wang, X. Sun, A. Jia, G. Jiang, and Z. Li. 2008. Higher salinity tolerance cultivars of winter wheat relieved senescence at reproductive stage. *Environmental and Experimental Botany* 62:129–138.

# CHAPTER SIX

# Plant Cell and Organellar Proteomics and Salinity Tolerance in Plants

*Suping Zhou and Theodore W. Thannhauser*

## CONTENTS

*Abstract.* Soil salinity is among the major environmental problems causing significant reduction in agricultural production on arid and semiarid lands. Plants growing under saline conditions are exposed to excessive concentrations of toxic ions (e.g., $Na^+$ and $Cl^-$) and imbalance of essential ions such as $K^+$, $Ca^{2+}$, and $Mg^{2+}$ as well as micronutrients. Alteration in the proteomes associated with various functional subcellular organelles is the base for the development of salinity tolerance mechanisms. Proteomics, combining high performance mass spectrometry and bioinformatics approaches, is used to identify proteomes within different subcellular compartments. Quantitative proteomics provides insights into protein abundances for different biological processes. Based on the changes in relative protein abundance, testable hypotheses concerning biological processes regulated by those proteins and their importance in salt tolerance are developed. This chapter reviews recent advances in subcellular proteomics analysis and their application in the identification of proteins for salt-tolerance and the development of the association network between protein expression and functional cellular pathways affecting salt tolerance properties.

*Keywords:* Salt stress, quantitative proteomics, organelle proteomics, cell wall proteomics

## 6.1 INTRODUCTION

Increasing soil salinity characterized by excessive levels of $Na^+$, $K^+$, $Ca^{2+}$, $Mg^{2+}$, and $Cl^-$ ions is a growing problem in agricultural production in arid and semiarid regions of the world (Gelbard 1985).

In recent years, the regions at the risk of soil salinization are expanding to areas that used to receive sufficient rainfall but now are experiencing more frequent and prolonged periods of drought throughout the plant's growth season. Due to the high evapotranspiration rate, water-soluble salts in deep groundwater are transported and accumulate in upper soil layer where plant roots are located (Rahimian and Poormohammadi 2012). Together with the overuse of fertilizers, these phenomena have led to changing levels of soil salinization in many large agricultural countries in the world (Rengasamy 2008; Zhou and Li 2013).

Plants vary greatly in the irresponses to soil salinity. Salt tolerances range widely from the sensitive glycophytic to the very tolerant halophytic species. Most of the nonhalophyte plants can adapt to low or moderate salinities, but their growth is severely limited above 200 mM NaCl (Hasegawa et al. 2000). Nonetheless, there is also genetic variation in salt tolerance within species. Landraces, wild germplasms, and varieties with distinctive salt tolerance traits have been identified in rice (*Oryza sativa*), maize (*Zea mays* subsp. *mays*) (Gao and Lin 2013; Shelden and Roessner 2013), tomatoes (*Solanum lycopersicum*, *S. pimpinellifolium* and *S. chilense*) (Nveawiah-Yoho et al. 2013; Sun et al. 2010; Zhou et al. 2009a, 2011a), sorghum (*Sorghum bicolor*) (Khalil 2013; Reddy et al. 2010), and many other plant species.

Plant responses to salinity involve very intricate processes, from gene and protein expression in various cellular pathways to alteration in basic biosynthetic functions, which ultimately leads to the development of tolerant physiological properties and phenotypic traits (reviewed by Kumar et al. 2013; Nveawiah-Yoho et al. 2013). Proteins are functionally active molecules that can alter the chemical environment and are often the major players in biological processes. Thus, their abundance influences greatly the rate at which cellular processes adapt to a changing environment (Muntel et al. 2014). Proteomic studies of the response to salt stress have shown that changes in proteome composition in terms of alterations in the relative abundance of proteins, protein post-translational modifications, and protein–protein interactions have a crucial impact on the modification of plant cell structure and metabolism altering the salt tolerance properties in wheat (*Triticum durum* and *T. aestivum*), barley (*Hordeum vulgare*), maize (*Z. mays*) (reviewed by Zhang et al. 2012) and tomatoes (Figure 6.1) (Buying et al. 2007; Nveawiah-Yoho et al. 2013; Zhou et al. 2009a, 2011a).

Proteome analysis has become an indispensable source of information about protein expression, the existence of splice variants, and erroneous or incomplete prediction of gene structures in databases (Baginsky and Gruissem 2007). Typically plant genomes contain a much large number of proteins then are identifiable using existing mass spectrometry (MS) platforms due to a range of technical problems, including the large dynamic range of protein abundances (Corthals et al. 2000) and the phenomenon of peptide under sampling (Wang et al. 2010). *Arabidopsis thaliana*, which has the smallest genome of all flowering plants, contains 27,000 genes and 35,000 proteins they encode. Therefore, proteomics at the whole tissue level is limited by several factors such as protein abundance, size, hydrophobicity, and other electrophoretic properties. Fortunately, many cellular reactions are constrained within specific subcellular compartments, such as plasma membranes (PM), vacuolar membranes, Golgi membranes, mitochondria (MT), chloroplasts (CP), and nuclei. The complexity of these subcellular proteomes are much reduced than those of whole tissue. Furthermore, the proteomes of organelles comprise a focused set of proteins that fulfills discrete but varied cellular functions. Therefore, the analyses of cell organelle proteomes provide additional and important information about protein localization and pathway compartmentalization (Mo et al. 2003; Pandey et al. 2006; Taylor et al. 2003). In essence, subcellular proteomics divides the whole plant proteome into discrete fractions, each containing a reduced protein complement for a particular subcellular compartment. The identification of these proteins using MS analysis and database searching is greatly simplified compared to the case when whole cell extracts are used. In this chapter, we will discuss the impact of soil salinity on the proteomes of several important subcellular organelles.

## 6.2 PROTEOMICS TECHNOLOGY FOR THE IDENTIFICATION AND QUANTIFICATION OF SUBCELLULAR PROTEOMES

The central platform for subcellular proteomics is high performance mass spectrometry (MS), subcellular organelle isolation, protein extraction, separation and enrichment, genomic databases, and bioinformatics (Brewis and Brennan 2010; Millar and Taylor 2014). There are a range of workflows available for protein (or peptide) separation prior to tandem MS and subsequent bioinformatics analysis to achieve protein identifications. Two-dimensional

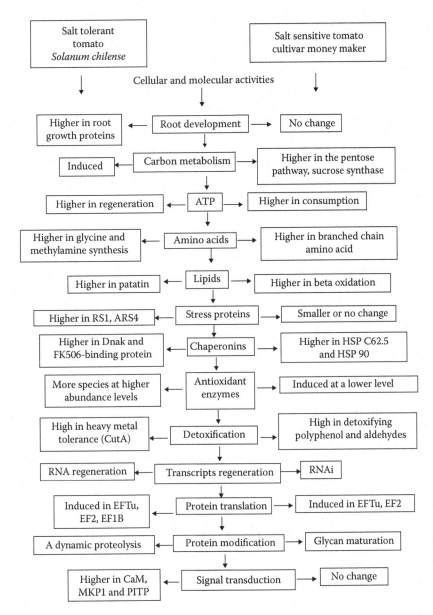

**Figure 6.1.** Effect of salt stress on cellular functional pathways in tomato varieties with contrasting tolerance properties. (Modified from Nveawiah-Yoho P. et al., *J. Am. Soc. Hort. Sci.*, 138, 382, 2013.)

(2D) electrophoresis and subsequent MS, liquid chromatography-MS (LC-MS), and GeLC-MS are used for the identification of individual protein. The most widely used quantitative proteomics techniques include difference gel electrophoresis, isobaric tags for relative and absolute quantification, or stable isotope labeling by amino acids (a review by Brewis and Brennan 2010).

Peltier et al. (2000, 2002) have developed a procedure for the isolation of intact chloroplasts followed by separation into soluble and peripheral proteins in the thylakoids. Proteomes in different subcompartments of thylakoids are characterized using a MS-based approach (Tomizioli et al. 2014). For the isolation of MT, there are many versions of commercial kits, or they can be prepared by releasing of the organelles (using either mechanical disruption or cell wall enzyme digestion procedures) and then followed by purification based on phase partition using discontinuous Percoll gradients (Albertsson et al. 1982; Hrubec et al. 1985; Rhee et al. 2013). Similar procedures can be used for the isolation of plant nuclei (Henfrey and Slater 1988), and the method has been optimized for purification of the respective organelles from leaves (Sikorskaite et al. 2013). A method for the isolation of intact vacuole described by Robert et al. (2007) uses osmotic and thermal disruption of mesophyll-derived *Arabidopsis* protoplasts, followed by a density gradient fractionation of the cellular content. The purified product is basically free of significant contamination from other endomembrane compartments. Several methods have been developed for the isolation of plant cell membrane proteins, with the focus to overcome hydrophobic properties of those proteins and the selection of buffers that are compatible with proteomics analysis (Ephritikhine et al. 2004; Komatsu et al. 2007; Lee et al. 2011). Cytosolic protein can be extracted following a procedure described by Gonzalo et al. (2014). Recently there is an increasing interest in cell wall protein (CWP) metabolism. The isolation and fractionation as well as proteomics analysis are described in several publications (Albenne et al. 2013; Catalá et al. 2011; Jamet et al. 2006, 2008; Ruiz-May and Rose 2013).

Proteomes are highly complex, and identifying and quantifying low-abundance proteins is a significant issue. Several methods have been developed to improve the identification of low-abundance proteins, including using differential detergent solubilization to separate hydrophobic and hydrophilic parts of a subcellular proteomes (Anatrace 2014; Le Maire et al. 2000; VanAken et al. 1986), depletion of high abundance proteins (Bandow 2010; Millioni et al. 2011), and sample enrichment for phosphoproteomics (Fíla and Honys 2012; Hsu et al. 2009; Thingholm and Jensen 2009; Walther and Rakwal 2009).

Quantitative proteomics requires performing a well-designed experiment, which has sufficient number of biological replicates (3–4 to ensure a >0.85 analysis power), following a stringent protein extraction procedure, performing a high-quality MS analysis to produce reliable proteome-wide data, and using the robust statistical analysis procedures (Gatto et al. 2014; Okekeogbu et al. 2014; Zhou et al. 2013). Gatto et al. (2014) have developed a freely available software that implement innovative state-of-the-art analysis pipelines from experimental designs, MS platforms, and quantitation techniques. These analysis strategies can be used for the identification of dynamic changes in sub-cellular localization by comparing and contrasting data describing different biological conditions.

## 6.3 SUBCELLULAR PROTEOMICS AND SALT TOLERANCE IN PLANTS

### 6.3.1 Nuclear Proteomes and Salt Tolerance

The nucleus is the subcellular organelle that contains nearly all the genetic information governing all the major cellular activities in plant. Nuclear proteins (NPs) are predicted to comprise about 10%–20% of total cellular proteins, in accord with our knowledge of the number and variety of important cellular functions that occur in this compartment (Moriguchi et al. 2005; Narula et al. 2013). NPs are involved in transcriptional regulation of nuclear genes, and thus they control expression of both cellular proteins and the NPs themselves. The nuclear proteome is comprised of the high-abundance proteins (such as histones and high mobility group proteins) and lower-abundance proteins (such as transcriptional factors, transcription regulators, and chromatin binding and remodeling proteins) (Lelong et al. 2012; Li et al. 2008). It is dominated by proteins involved in signaling and gene regulation (Pandey et al. 2006). About one-fourth of NPs is involved in DNA replication and transcription (Pandey et al. 2006).

With respect to salt tolerance, one area of particular interest in nuclear proteome are those proteins involved in the regulation of genome stability, including gene transcription factors (TFs) and proteins with upstream regulatory functions in controlling the expression of stress-related genes. Salt impacts genome stability by inducing mutations that could lead to modulation in salt tolerance. High Cl⁻ concentrations induce an increase in the level of DNA strand breaks and homologous recombination rates in *A. thaliana* plants. Transgenerational changes with higher expression of AtRad51 recombinase and reduced AtKu70 (a protein [gene] associated with the repair of double-strand breaks in the DNA) were found to be associated with elevated tolerance to salt (Boyko et al. 2010).

TFs are activator or repressor proteins that bind to promoter regions of stress-regulated genes and thus control the expression of those genes. The comprehensive identification of a TF "proteome" will remain challenging because of the large number of low-abundance and posttranslationally modified proteins involved (Jiang et al. 2009). Specialized purification and analysis methods have been developed over the last decades, which facilitate the study of the TF proteome. TF proteins are selectively purified by affinity techniques using the DNA response element as the basis for highly specific binding. One such affinity method called "trapping" enables purification of TF in a highly purified state. The electrophoretic mobility shift assay (EMSA) is the most important assessment method of TFs because it provides both measures of the affinity and amount of the TF present. Southwestern blotting and DNA–protein cross-linking allow in vitro estimates of DNA-binding-protein mass, while chromatin immunoprecipitation allows confirmation of promoter binding in vivo. Two-dimensional gel electrophoresis (2-DE) methods, and 3-DE methods that combines EMSA with 2-DE, allow further resolution of TFs. The synergy of highly selective purification and analytical strategies has led to an explosion of knowledge about the TF proteome and the proteomes of other DNA- and RNA-binding proteins (Jiang et al. 2009). Ding et al. (2013), using an enrichment method based on an affinity reagent composed of a synthetic DNA containing a concatenated tandem array of consensus TFREs (catTFRE) for the majority of TF families, identified as many as 400 TFs from a single cell line and a total of 878 TFs from 11 cell types, covering more than 50% of the gene products that code for the DNA-binding TFs in the human genome (Ding et al. 2013). This technique can also be adopted to identify the salt-induced TF proteomes in plants.

Regulation of TFs plays a key role in salt tolerance. Evidences from studies in recent years have found that overexpression of a several TFs can enhance salt tolerance in transgenic plants, for example, the overexpression of transcriptional factor LcMYB1 mediated signaling and enhanced the salt tolerance of sheep grass (*Leymus chinensis*) (Cheng et al. 2013). LcMYB1 acts in concert with DREB, a TF and a dehydration-responsive element binding protein. It can interact, specifically, with the DRE/C-repeat (DRE/CRT) cis-acting element contained in the promoter region of many stress-inducible genes and can therefore control the expression of many stress-inducible genes in plants, increasing tolerance to drought, low temperature, and high salt (Wang et al. 2006). Overexpression of WRKY TF, a key component in the abscisic acid (ABA)

signaling pathway, enhanced salt tolerance in *Chrysanthemum* (Liu et al. 2014). Aside from the positive regulation, there are negative TF regulators as well. Rice *OsAP23* encodes for AP2/ERF TFs; it was proposed to have a negative effect on salt tolerance because transgenic plants overexpressing the gene were more susceptible to salt stress (Zhuang et al. 2013). A quantitative proteomics study of tomatoes with contrasting salt tolerance properties found that the effect of salt treatments on TF protein expression is directly associated with certain genotypic traits (Nveawiah-Yoho et al. 2013). TFs are the master regulatory elements that orchestrate the gene expression network for the expression of tolerance/sensitivity traits. A growing body of experimental evidences shows that the expression of these TFs are regulated by the epigenetic mechanisms.

Epigenetics is the study of changes in gene expression or cellular phenotype that do not change the DNA sequence. MS-based proteomics is used to identify epigenetic mechanisms and various approaches such as genome-wide profiling of DNA-binding proteins, histone posttranslational modifications or nucleosome positions, chromatin-associated proteins, and multisubunit complexes (Bartke et al. 2013; Han and Garcia 2013). Salinity stress was shown to affect the methylation status of four TFs (one MYB, one b-ZIP, and two AP2/DREB family members), DNA methylation, and/or histone modification, thus controlling the activation/repression of those TFs. These, in turn, regulate the expression of salt stress tolerance genes, ultimately resulting in altered salt tolerance (Song et al. 2012). A study on salt-treated *A. thaliana* indicates that chromatin modification-mediated epigenetics plays important roles in establishment of transgenerational stress memory (Farahani 2013). In summary, both proteins involved in epigenetics mechanisms and TFs are major components of nuclear proteomes associated with salt tolerance.

## 6.3.2 Chloroplast Proteomes and Salt Tolerance

Chloroplasts are plant cell organelles of cyanobacterial origin. Their main role is photosynthesis and amino acid biosynthesis (Kleffmann et al. 2004). Using tandem MS, 690 different proteins were identified from purified *A. thaliana* chloroplasts. A proteomics analysis of high-purity envelope membrane extracts from *A. thaliana* chloroplasts led to the identification of more than 100 proteins, which were found to be involved in ion and metabolite transport, components of the protein import machinery, and chloroplast lipid metabolism (Ferro et al. 2003). Several databases have been developed that contain information of chloroplast proteins, for example, the Plant Protein Database or the SubCellular Proteomic Database. Furthermore, AT_CHLORO is a comprehensive chloroplast proteome database with curated subplastidial localization in *A. thaliana* (Ferro et al. 2010).

Chloroplasts are thought to be more sensitive to salinity than other organelles. Both the light-dependent reactions in the two reaction centers (photosystem I [PSI] and II [PSII]) of the thylakoid membrane and carbon fixation in the stroma are vulnerable to excessive $Na^+$ concentrations, and the resultant ion imbalance. Using bioinformatics modeling, ribulose-1,5-bisphosphate carboxylase/oxygenase (RUBISCO) and superoxide dismutase were suggested to be among the major targets of salt stress (Chen et al. 2009). One of the most serious injuries of excessive $Na^+$ level is the disruption of the electron transport chain in PSII. The resultant photoinhibition exaggerated oxidative damage to the membrane system. Some proteins of the electron transport chain members (ferredoxin-NADP (+) reductase and quinone oxidoreductase-like protein) are induced by salt stress (Zhou et al. 2009a). Proteomic analysis of a halophytic mangrove species (*Kandelia candel*) indicates that light-dependent reactions are maintained by upregulating the levels of component proteins on the electron transport chain at both moderate and high salinity, and the Calvin cycle remains functional at moderate salinity but declining at high salinity (Kumar et al. 2013; Wang et al. 2014). For the dark reaction, RUBISCO activase was repressed in the sensitive tomato cultivar "Money Maker" by 200 mM NaCl treatment (Zhou et al. 2009a), but it was found to be induced in the tolerant "Roma" variety along with RUBSICO large subunit and several other enzymes involved in carbohydrate metabolism reactions coupled with generation of energy molecules (Manaa et al. 2013).

## 6.3.3 Mitochondrial Proteomes and Salt Tolerance

The MT proteome is partially comprised of proteins encoded by the nuclear genome that are translated in the cytosol and then imported into the organelle, and those encoded by MT subgenome and translated intramurally. The MT contains a complete suite of proteins for DNA replication, gene transcription, transcript processing, and protein translation. Proteomics studies have shown that salt stress affects protein components of the mitochondrial translation machinery such as elongation factor Tu, chaperonins CPN60, and other ribosomal proteins (Nveawiah-Yoho et al. 2013; Wojtyla et al. 2013).

Because MT are the major source for the regeneration of reactive oxygen species (ROS), the moderation in MT biogenesis is considered to be an important mechanism for maintaining normal metabolic status, especially under adverse conditions. Protein translation within the organelle seems to play a key role in the monitoring of mitochondrial homeostasis and might have a role in establishing mitochondrial abundance within a cell (Battersby and Richter 2013). The downstream effect of an impaired mitochondrial translation machinery can affect the function of the organelles (Richter et al. 2013) as several proteins involved in the respiratory chain reactions and the biogenesis of MT are synthesized in the organelle (Okimoto et al. 1992). A majority number of the mitochondrial proteins are encoded by nuclear genes. These proteins are translated in the endoplasmic reticulum (ER) as precursors and then enter the MT through the translocase of the outer membrane (TOM) and the sorting and assembly machinery complex in the outer membrane. They are sorted in the intermembrane space and then imported into matrix with the assistance of the translocases of the inner membrane (TIM 22, Tim50, Tim23, and Tim17) and the associated presequence translocase-associated motor.

Recent studies show that the correlation between NP translation and importation into MT are very important for maintaining functional MT, especially under stress conditions. One of the most detrimental secondary effects induced by salt stress is the "oxidative burst," and the ROS are mostly generated in MT during the oxidative respiration process. A recent study in mammalian cells demonstrated that the protein import translocase Tim17A was actively degraded to reduce the amount of protein imported into the MT when protein translation in the cytosol was attenuated in stressed cells.

It was suggested that this is a major mechanism to control the oxidative stress within the MT to protect its integrity and function (Rainbolt et al. 2013). A similar mechanism may be employed by plants to control the oxidative stress injury from salt and other stresses.

A study on salt-treated lupine embryos found that Tim17/22 and mitochondrial-processing peptidase are also affected by the treatment (Wojtyla et al. 2013). In salt-treated tomato, subunits of the TOM complex were affected by salt stress (Nveawiah-Yoho et al. 2013). Bioinformatics analysis indicates that Pam18/Tim14 and Pam16/Tim16 homologs might play roles in senescence and abiotic stress responses in *A. thaliana* (Chen et al. 2013). Taken together, these results indicate that salt affects the import of proteins into the MT, and the altered MT proteomes have important function in developing tolerance mechanisms.

In the MT, one of the unavoidable by-products of aerobic respiration is superoxide formation (Puntarulo et al. 1988). Through a variety of reactions, superoxide leads to the formation of hydrogen peroxide, hydroxyl radicals, and other ROS molecules. MT function depends on the balance of ROS and antioxidants. However, under salt stress, influx of sodium through non-selective cation channels (NCCs) will activate the NADPH-oxidase-dependent $O_2^-$ generation (Monetti et al. 2014). The elevated ROS levels and ionic disturbance lead to an increase in mitochondrial membrane permeability and efflux of cytochrome c into the cytosol that results in the execution of programmed cell death (PCD) (De Pinto et al. 2012; Monetti et al. 2014). PCD involving mitochondrial proteins was also confirmed in a proteomic analysis of plant PCD (Swidzinski et al. 2004). In salt-treated rice "WYJ 8th" root tip cells, changes in mitochondrial proteins result in release of cytochrome c and the destruction of electron transport, resulting in an oxidative burst. They are therefore considered as candidates for PCD regulation in salt-treated root tip of rice (Chen et al. 2009).

### 6.3.4 Endomembrane Proteomes and Salt Stress

The endomembrane system is composed of the different membranes that are suspended within a eukaryotic cell (Smith 1997). Under this category are nuclear envelope, the ER, the Golgi apparatus, lysosomes, vacuoles, vesicles, and the cell membrane (PM), which surrounds the cells to separate the interior of the cell from the outside environment (Singleton 1999).

Most of the ion channels and ion transporters are PM proteins. These together with proteins involved in signal perception and transduction are the most well studied with respect to salt tolerance (Blumwald 2000; Kumar et al. 2013). The PM $Na^+/H^+$ antiporter (including the salt overly sensitive [SOS] proteins and homologs) regulates sodium efflux in roots and the long-distance transport of sodium from roots to shoots (Wu et al. 1996). Therefore, expression of these protein affects photosynthesis activity and thus the overall health of plants is under the stress condition. Tonoplast $Na^+/H^+$ antiporters and endosomal $Na^+/H^+$ antiporter function to deposit the toxic ions into the vacuole (Bassil et al. 2012). The $Ca^{2+}/H^+$ antiporter (CAX) is a pump that helps to maintain intracellular $Ca^{2+}$ ion homeostasis, and the $Na^+/K^+$ symporter in the PM assists active $Na^+$ extrusion out of the cytosol (Popova et al. 2006; Waters et al. 2013). Several channel proteins, including the NCCs that have been proposed as an entry gate of $Na^+$ into the plant cell, control the amount of root $Na^+$ uptake (Demidchik and Maathuis 2007). Other channel proteins affecting salt tolerance include the anion channels and $H^+/Cl^-$ antiporter (Nakamura et al. 1997) and inward-rectifying $K^+$ channels (KIRC), which mediate the influx of $K^+$ through the PM and selectively accumulate $K^+$ over $Na^+$ upon PM hyperpolarization (Golldack et al. 2003; Kumar et al. 2013; Muller-Rober et al. 1995).

Protein synthesis takes place on the ER system in cytosol. Among the components of the protein translation machinery, the abundance of ribosomes and translation activators responsible for fidelity of protein translation have been correlated with salt tolerance in tomato (Nveawiah-Yoho et al. 2013). Proteins for the modification and folding of nascent proteins (chaperones) and complexes of intracellular vesicles for targeting proteins and other macromolecules via either the secretory or the endocytic systems into subcellular spaces are also sensitive to salt stress (Kumar et al. 2013; Nveawiah-Yoho et al. 2013).

Soluble NSF attachment protein receptor (SNARE) family is one of the key regulators of endomembrane trafficking and executes appropriate membrane fusion between the transport vesicles and the membrane of the target organelle. SNARE proteins have R-SNAREs and Q-SNAREs based on their structural features. Q-SNAREs include syntaxin and SNAP-25. The *trans*-Golgi network (TGN)-localized Qa-SNAREs (belonging to syntaxin) and vesicle SNARES (v-SNAREs) play important roles in abiotic stress tolerance in *A. thaliana* (Leshem et al. 2006; Uemura et al. 2012). A recent study on *A. thaliana* cell suspension culture under high salt conditions (Hamaji et al. 2009) found that knockout of the vacuolar Qa-SNARE (a soluble, N-ethylmaleimide-sensitive attachment protein receptor) protein VAM3/SYP22 caused an increase in salt tolerance with $Na^+$ accumulating more in roots and less in the shoots of the mutant plants. The critical role of intracellular protein trafficking for salt tolerance was also demonstrated in a study of the halophytic ice plant (*Mesembryanthemum crystallinum* L.). Ion transporters were moved to the functional sites on PM with the assistance of the intracellular membrane traffic system immediately upon exposure to salt treatment, thus controlling $Na^+$ flux into cells (Jou et al. 2014). Recent studies show that the dynamics of vesicle trafficking is an important mechanism to avoid injuries from salt ion toxicity and oxidative cellular stress.

Under NaCl-dominated salt stress, the key to plant survival is maintaining a low cytosolic $Na^+$ level or $Na^+/K^+$ ratio. The vacuole has two important physiological roles in a plant's tolerance to salinity. One is the compartmentalization of $Na^+$ into the tonoplast to reduce ion content and the resultant toxicity in the cytoplasm space. A second function is to store and break down the material and recycle denatured proteins and other macromolecules. Proteomics studies in this field focus on identifying functional proteins that control vacuolar $Na^+$ sequestration (the hydrophobic membrane proteins) and various proteolytic enzymes in the soluble proteomes within vacuole lumen.

The vacuole proteome is comprised of the membrane and soluble fractions. A proteomics study of pure vacuole isolated from *A. thaliana* cell culture found 416 proteins identified from the membrane fraction, 195 were considered integral membrane proteins based on the presence of one or more predicted transmembrane domains, and 110 transporters and related proteins were also identified (91 putative transporters and 19 proteins related to the V-ATPase pump) (Jaquinod et al. 2007). Sequestration of sodium into vacuole is attributed to members of the family of $Na^+/H^+$ exchangers and is driven by the acidic pH gradient inside the vacuole generated by the vacuolar $H^+$-ATPase (V-ATPase). The *A. thaliana* SOS protein is regulated through V-ATPase activity that affects ion transport during salt stress and thus promotes salt tolerance (Batelli et al. 2007). This same protein was induced by salt treatment (200 mM NaCl) in tomato leaves (Zhou et al. 2009a). A targeted quantitative proteomics experiment conducted on *M. crystallinum* plants revealed that the vacuole contains glycolytic enzymes such as aldolase and enolase, along with subunits of the vacuolar $H^+$-ATPase V-ATPase. Presumably, the role of the glycolytic proteins in the tonoplast is presumably to help channel ATP to the V-ATPase to enhance the pump activity (Barkla et al. 2009). One mechanism shared in both tolerant glycophytes and halophytes is the efficient internal sequestration of $Na^+$ in the vacuoles (Shabala 2013). A functional genomics study suggested that each V-$H^+$-ATPase subunit (VHA-B, -E, -G) plays a role in enhancing salt tolerance in plants (He et al. 2014).

The soluble vacuolar proteome fraction contains a variety of asparagine-, cysteine-, and serine-type proteases including the protein of responsive to dehydration 21 (RD21), aleurain, and carboxypeptidase Y (Ahmed et al. 2000; Carter et al. 2004; Rojo et al. 2001; Yamada et al. 2001). RD19 and RD21 may alter their expression levels in response to changes in the osmotic potential of plant cells (Koizumi et al. 1993). Using the yeast, two-hybrid approach, *in vitro* and *in vivo* immunoprecipitation, cell-free protein degradation, and *in vitro* ubiquitination assays, RD21 was identified as a substrate protein of AtAIRP3/LOG2, which plays dual functions in ABA-mediated drought stress responses and in an amino acid export pathway in *Arabidopsis* (Kim and Kim 2013). Aleurain and aleurain-like proteins are annotated as response to salt stress and hyperosmotic response proteases in the TAIR database (The Arabidopsis Information Resource 2013). These results demonstrate that the soluble vacuolar proteome also has an important role in salt tolerance or in relieving salt-induced hyperosmotic stress.

### 6.3.5 Cytosolic Proteomes and Salt Tolerance

A cytosolic proteome is comprised of water-soluble proteins that do not bind to cell membranes or the cytoskeleton but dissolved in the aqueous phase of the cytoplasm. Protein complexes such as ubiquitin-proteasomes and carboxysomes belong to this category, and these soluble proteins are synthesized on free ribosomes in the cytosol, whereas secretory/integral membrane proteins are synthesized on ER-bound ribosomes (Hummel et al. 2012; Ito et al. 2011).

Under normal conditions, the cytosol has a high concentration of $K^+$ ions and a low concentration of $Na^+$ ion (Lang 2007). One of the detrimental effects of salinity is the high cytosolic content of $Na^+$ and $Cl^-$, which are both toxic to the cell, and another is the negative osmotic pressure resulting from water loss. These properties and the abundance of cytosolic proteins are not well studied due to the technical difficulty of isolating pure cytosolic proteins devoid of membrane protein contaminants and proteins leaching from various cellular compartments during the isolation process. As new methods for the isolation of this group of proteins and MS-based protein identification are developed, the role of the cytosolic proteome in salinity tolerance will be better understood (Estavillo et al. 2011, 2014; Ito et al. 2014).

### 6.3.6 Cell Wall Proteomics and Salt Stress

Plant cell walls are highly dynamic composite structures composed of polysaccharides, proteins, and phenolic and lipid compounds in some cell or tissue types, which encapsulate all plant cells. These complex matrices reside in the aqueous environment of the apoplast, defined as the area outside the PM. Thus, the "wall" can be envisioned as both a structure and a compartment. The large numbers of CWPs include structural proteins that are immobilized within the wall matrix and soluble proteins that are mobile to varying degrees within the apoplast (Rose and Lee 2010). The term CWP is used here to describe all proteins secreted from the protoplast.

Although CWPs typically comprise only 5%–10% of the cell wall dry mass (Cassab and Varner 1998), they have many essential functions, including defining and modifying cell wall structure, apoplastic metabolism, regulating cell growth and division, signal transduction, and responses to environmental stimuli (Huckelhoven 2007; Roberts 2001; Rose and Lee 2010; Zhu et al. 2006). The characterization of the cell wall proteome therefore has great potential importance for many aspects of basic plant biology, as well as practical applications that involve cell wall biology, including bioenergy, pathogen resistance, and various quality traits of agriculturally important crops. Unfortunately, unlike intracellular organelles, the cell wall is not bounded by a single distinct membrane, thus maximizing the numbers of proteins identified in the cell wall proteome, while minimizing contamination from the intracellular proteins has been a major technical hurdle. While highly effective protocols have been developed to isolate purified fractions of some intracellular organelles (Rose et al. 2004), tissue disruptions that compromise the integrity of the PM inevitably leads to contamination of the wall fraction with intracellular proteins (Rose and Lee 2010; Rose et al. 2004; Thannhauser et al. 2013). To reduce this problem, alternative procedures have been developed that minimize disruption of the PM, such as using cell cultures and the direct analysis of apoplastic fluids isolated by centrifugal or pressure dehydration (Konozy et al. 2012; Lee et al. 2004). While the former represents a highly artificial system, the latter approaches are somewhat laborious and only capture nonbound proteins that are mobile within the apoplast.

Another strategy that has been used to isolate highly enriched samples of CWPs involves lectin affinity chromatography to isolate and fractionate N-glycoproteins. N-glycosylation is a common posttranslational modification of plant proteins that traverse the secretory pathway, many of which will ultimately reside in the cell wall. Minic et al. (2007) used a top-down approach, coupling tissue homogenization with N-glycoprotein isolation using affinity chromatography with the lectin concanavalin A (Con A) and 2-DE-based fractionation, prior to LC-MS, to identify 102 presumably glycosylated CWPs from *A. thaliana* (Minic et al. 2007). While 2-DE has been successfully applied to plant proteomic research (Macarisin et al.

2009; Zhou et al. 2009a,b, 2011a,b), it has several well-known limitations including an incompatibility with extremely hydrophobic proteins or those of extreme pI and molecular weight. Catala et al. (2011) utilized an improved strategy for the identification of CWPs, coupling Con A enrichment for glycoproteins with 2D LC (Catalá et al. 2011). Using this approach, nearly 80% of the 133 proteins identified were predicted to be associated with the cell wall.

CWPs play essential roles in determining the size and shape of cells, and remodeling during cell growth, and thus the plasticity as well as elasticity in morphogenesis of the whole plants. In higher plants, the cell wall proteome is comprised of three groups of structural proteins: glycine-rich proteins (GRPs) that contain more than 60% glycine (Mousavi and Hotta 2005), extensins, and proline-rich proteins. Except for the GRPs, all the CPWs are glycosylated (O-glycosylation) proteins (Cassab and Varner 1998). Enzymes of xyloglucan endotransglucosylase-hydrolases, expansins, and polygalacturonases affect cell wall loosing thus affecting the rate and direction of root growth under salt stress (Irshad et al. 2008). Cell wall proteomes also contain stress receptors, such as wall-associated kinases (WAKs), proline-rich extensin-like receptor kinases, lectin receptor kinases, and leucine-rich repeats receptor-like kinases, different receptor-like kinases in signal transduction pathways in salt stress, expansins that act on cellulose and xyloglucans, and a number of enzymes belonging to the esterase family such as pectin methylesterases (EC 3.1.1.11) and acetylesterases (EC 3.1.1.6), which catalyze esterification and de-esterification reactions. The WAKs play roles in stress responses including salinity and cell expansion during plant development (Kohorn and Kohorn 2012; see a review by Zagorchev et al. 2014).

## 6.4 SUMMARY AND FUTURE PERSPECTIVE

In this chapter, we have discussed the importance of subcellular proteomes and their relationship with salt tolerance in plants. Some of the proteins have been identified using subcellular proteomic approaches, while others have been localized to a particular subcellular organelle based on homology to proteins known to be components of particular organelles. A large number

of these proteins are thought to be involved in tolerance to environmental stress, including salt stress (Ghosh and Xu 2014; Hakeem et al. 2012; Zagorchev et al. 2014).

Plants are considered to have about 40 different cell types (Martin et al. 2001). Plant tissues including roots, stems, and leaves are made of layers of cells with differentiated functions. Therefore, the same subcellular organelles from different cells should contain variations in their protein complements, including the types and relative abundance level of component proteins. The high complexity of most subcellular proteomes is a major factor affecting the number of proteins that can be identified and quantified from individual samples because only a subset of all peptides in a sample can be assayed in a single LC-MS analysis (Ahrens et al. 2010; Bantscheff et al. 2012). The limitation of analytical performance of the mass spectrometers used for proteomics applications restricted the level of detail available to facilitate our understanding of the kinetics associated with particular biological processes (Pereira et al. 2013). Nevertheless, the technology and instrumentation are developing quickly, and newer, more sensitive and faster scanning instruments are constantly being developed. The coupling of these with new analytical strategies such as data-independent analysis techniques such as MS[E] and SWATH holds great promise for the comprehensive cataloging of subcellular proteomes.

Advances in protein isolation procedures also provide reasons to be optimistic. Techniques such as laser-capture microdissection (LCM) have been developed (Isenberg et al. 1976) for the investigation of cells from histological tissue sections (Emmert-Buck et al. 1996). Clearly, proteomics analysis in combination with LCM provides the possibility to identify protein expression at the level of cell type. This approach can minimize the dilution effect and noise caused by the cellular complexity associated with tissues and organs (Kerk et al. 2003). Single-cell-type proteomic analyses have been reported for guard cells (Zhao et al. 2008), trichomes (Kryvych et al. 2011; Wienkoop and Weckwerth 2006; Zhao et al. 2008) and pollen (Holmes-Davis et al. 2005; Sarhadi et al. 2012), tomato cuticles (Matas et al. 2011; Petit et al. 2013), soybean (*Glycine max*) root hairs (Wan et al. 2005), and tobacco (*Nicotiana tabacum*) trichomes (Amme et al. 2005). Application of single-cell-type proteomics will generate more coherent data to build protein expression maps involving cell-type-specific localization of proteins (Petricka et al. 2012). The application of subcellular proteomics techniques of single-cell-type samples will significantly reduce the complexity of proteomes within each biological sample. Together with the use of various subfractionation methods, such as UPLC separation (Yang et al. 2011), gel-eluted liquid fraction entrapment electrophoresis (GeLC-MS/MS) (Piersma et al. 2013), in combination with high-performance MS instrumentation, it is expected to increase the number of the proteins identified and form the bases for developing a systematic protein expression map for salt stress in plants.

## ACKNOWLEDGMENTS

We apologize to all colleagues whose original works could not be cited due to space constraints.

## REFERENCES

Ahrens, C.H., E. Brunner, E. Qeli, K. Basler, and R. Aebersold. 2010. Generating and navigating proteome maps using mass spectrometry. *Nat. Rev. Mol. Cell Biol.* 11:789–801.

Albenne, C., H. Canut, and E. Jamet. 2013. Plant cell wall proteomics: The leadership of *Arabidopsis thaliana*. *Front. Plant Sci.* 4:111.

Albertsson, P.A., B. Andersson, C. Larsson, and H.E. Akerlund. 1982. Phase partition—A method for purification and analysis of cell organelles and membrane vesicles. *Methods Biochem. Anal.* 28:115–150.

Ahmed, S.U., E. Rojo, V. Kovaleva, S. Venkataraman, K.E. Dombrowski, K. Matsuoka, and N.V. Raikhel. 2000. The plant vacuolar sorting receptor AtELP is involved in transport of NH2-terminal propeptide-containing vacuolar proteins in *Arabidopsis thaliana*. *J. Cell Biol.* 149:1335–1344.

Amme, S., T. Rutten, M. Melzer, G. Sonsmann, J.P.C. Vissers, B. Schlesier, and H.P. Mock. 2005. A proteome approach defines protective functions of tobacco leaf trichomes. *Proteomics* 5:2508–2518.

Anatrace. 2014. Detergents and their uses in membrane protein science. http://wolfson.huji.ac.il/purification/PDF/detergents/ANATRACE_DetergentsUse.pdf. Accessed on May 30, 2014.

Baginsky, S. and W. Gruissem. 2007. Current status of *Arabidopsis thaliana* proteomics. In Šamaj, J. and J.J. Thelen (ed.), *Plant Proteomics*. Springer-Verlag, Berlin, Germany, pp. 105–115.

Bandow, J.E. 2010. Comparison of protein enrichment strategies for proteome analysis of plasma. *Proteomics* 10:1416–1425.

Bantscheff, M., S. Lemeer, M.M. Savitski, and B. Kuster. 2012. Quantitative mass spectrometry in proteomics: Critical review update from 2007 to the present. *Anal. Bioanal. Chem.* 404:939–965.

Barkla, B.J., R. Vera-Estrella, M. Hernández-Coronado, and O. Pantoja. 2009. Quantitative proteomics of the tonoplast reveals a role for glycolytic enzymes in salt tolerance. *Plant Cell* 21:4044–4058.

Bartke, T., J. Borgel, and P.A. DiMaggio. 2013. Proteomics in epigenetics: New perspectives for cancer research. *Brief Funct. Genomics* 12:205–218.

Bassil, E., A. Coku, and E. Blumwald. 2012. Cellular ion homeostasis: Emerging roles of intracellular NHX $Na^+/H^+$ antiporters in plant growth and development. *J. Exp. Bot.* 63:5727–5740.

Batelli, G., P.E. Verslues, F. Agius, Q. Qiu, H. Fujii, S. Pan, K.S. Schumaker, S. Grillo, and J.K. Zhu. 2007. SOS2 promotes salt tolerance in part by interacting with the vacuolar $H^+$-ATPase and upregulating its transport activity. *Mol. Cell Biol.* 27:7781–7790.

Battersby, B.J. and U. Richter. 2013. Why translation counts for mitochondria—Retrograde signaling links mitochondrial protein synthesis to mitochondrial biogenesis and cell proliferation. *J. Cell Sci.* 126(Pt 19):4331–4338.

Blumwald, E. 2000. Sodium transport and salt tolerance in plants. *Curr. Opin. Cell Biol.* 2:43–434.

Boyko, A., A. Golubov, A. Bilichak, and I. Kovalchuk. 2010. Chlorine ions but not sodium ions alter genome stability of *Arabidopsis thaliana*. *Plant Cell Physiol.* 51:1066–1078.

Brewis, I. and P. Brennan. 2010. Proteomics technologies for the global identification and quantification of proteins. *Adv. Protein Chem. Struct. Biol.* 80:1–44.

Carter, C., S. Pan, J. Zouhar, E.L. Avila, T. Girke, and N.V. Raikhel. 2004. The vegetative vacuole proteome of *Arabidopsis thaliana* reveals predicted and unexpected proteins. *Plant Cell* 16:3285–3303.

Cassab, G.I. and J.E. Varner. 1998. Cell wall proteins. *Annu. Rev. Plant Physiol. Plant Mol. Biol.* 39:321–353.

Catalá, C., K.J. Howe, S. Hucko, J.K.C. Rose, and T.W. Thannhauser. 2011. Towards characterization of the glycoproteome of tomato (*Solanum lycopersicum*) fruit using Concanavalin A lectin affinity chromatography and LC-MALDI-MS/MS analysis. *Proteomics* 11:1530–1544.

Chen, X., B. Ghazanfar, A.R. Khan, S. Hayat, and Z. Cheng. 2013. Comparative analysis of putative orthologues of mitochondrial import motor subunit: Pam18 and Pam16 in plants. *PLoS ONE* 8(10): e78400.

Chen, X., Y. Wang, J. Li, A. Jiang, Y. Cheng, and W. Zhang. 2009. Mitochondrial proteome during salt stress-induced programmed cell death in rice. *Plant Physiol. Biochem.* 47:407–415.

Cheng, L., X. Li, X. Huang, T. Ma, Y. Liang, X. Ma, X. Peng et al. 2013. Overexpression of sheepgrass R1-MYB transcription factor LcMYB1 confers salt tolerance in transgenic *Arabidopsis*. *Plant Physiol. Biochem.* 70:252–260.

Corthals, G.L., V.C. Wasinger, D.F. Hochstrasser, and J.C. Sanchez. 2000. The dynamic range of protein expression: A challenge for proteomic research. *Electrophoresis* 21:1104–1115.

Demidchik, V. and F.J. Maathuis. 2007. Physiological roles of nonselective cation channels in plants: From salt stress to signaling and development. *New Phytol.* 175:387–404.

De Pinto, M.C., V. Locato, and L. de Gara. 2012. Redox regulation in plant programmed cell death. *Plant Cell Environ.* 35:234–244.

Ding, C., D.W. Chan, W. Liu, M. Liu, D. Li, L. Song, C. Li et al. 2013. Proteome-wide profiling of activated transcription factors with a concatenated tandem array of transcription factor response elements. *Proc. Natl. Acad. Sci. U.S.A.* 110:6771–6776.

Ephritikhine, G., M. Ferro, and N. Rolland. 2004. Plant membrane proteomics. *Plant Physiol. Biochem.* 42:943–962.

Estavillo, G.M., P.A. Crisp, W. Pornsiriwong, M. Wirtz, D. Collinge, C. Carrie, E. Giraud et al. 2011. Evidence for a SAL1-PAP chloroplast retrograde pathway that functions in drought and high light signaling in *Arabidopsis*. *Plant Cell* 23:3992–4012.

Estavillo, G.M., Y. Verhertbruggen, H.V. Scheller, B.J. Pogson, J.L. Heazlewood, and J. Ito. 2014. Isolation of the plant cytosolic fraction for proteomic analysis. *Methods Mol. Biol.* 1072:453–467.

Emmert-Buck, M.R., R.F. Bonner, P.D. Smith, R.F. Chuaqui, Z. Zhuang, S.R. Goldstein, R.A. Weiss, and L.A. Liotta. 1996. Laser capture microdissection. *Science* 274:998–1001.

Farahani, B.N.M.A. 2013. Epigenetics of stress adaptation in *Arabidopsis*: The case of histone modification. Graduate thesis. http://ir.lib.uwo.ca/cgi/viewcontent.cgi?article=2512&context=etd. (Accessed May 13, 2015.)

Ferro, M., S. Brugière, D. Salvi, D. Seigneurin-Berny, M. Court, L. Moyet, C. Ramus et al. 2010. AT_CHLORO, a comprehensive chloroplast proteome database with subplastidial localization and curated information on envelope proteins. *Mol. Cell Proteomics* 9:1063–1084.

Ferro, M., D. Salvi, S. Brugiere, S. Miras, S. Kowalski, M. Louwagie, J. Garin, J. Joyard, and N. Rolland. 2003. Proteomics of the chloroplast envelope membranes from *Arabidopsis thaliana*. *Mol Cell Proteomics* 2:325–345.

Fíla, J. and D. Honys. 2012. Enrichment techniques employed in phosphoproteomics. *Amino Acids* 43:1025–1047.

Gao, J.P. and H.X. Lin. 2013. QTL analysis and map-based cloning of salt tolerance gene in rice. *Methods Mol. Biol.* 956:69–82.

Gatto, L., L.M. Breckels, T. Burger, D.J. Nightingale, A.J. Groen, C. Campbell, C.M. Mulvey, A. Christoforou, M. Ferro, and K.S. Lilley. 2014. A foundation for reliable spatial proteomics data analysis. *Mol. Cell Proteomics* pii: mcp.M113.036350.

Gelbard, DE. 1985. Managing salinity lessons from the past. *J. Soil Water Conserv.* 40:329–331.

Ghosh, D. and J. Xu. 2014. Abiotic stress responses in plant roots: A proteomics perspective. *Front. Plant Sci.* 5:6.

Golldack, D., F. Quigley, C.B. Michalowski, U.R. Kamasani, and H.J. Bohnert. 2003. Salinity stress tolerant and sensitive rice (*Oryza sativa* L.) regulate AKT1-typepotassium channel transcripts differently. *Plant Mol. Biol.* 51:71–81.

Gonzalo, M., G.M. Estavillo, Y. Verhertbruggen, H.V. Scheller, B.J. Pogson, J.L. Heazlewood, and J. Ito. 2014. Isolation of the plant cytosolic fraction for proteomic analysis. *Methods Protoc.* 1072:453–467.

Hakeem, K.R., R. Chandna, P. Ahmad, M. Iqbal, and M. Ozturk. 2012. Relevance of proteomic investigations in plant abiotic stress physiology. *OMICS* 16:621–35.

Hamaji, K., M. Nagira, K. Yoshida, M. Ohnishi, Y. Oda, T. Uemura, T. Goh et al. 2009. Dynamic aspects of ion accumulation by vesicle traffic under salt stress in *Arabidopsis*. *Plant Cell Physiol.* 50:2023–2033.

Han, Y. and B.A. Garcia. 2013. Combining genomic and proteomic approaches for epigenetics research. *Epigenomics* 5:439–452.

Hasegawa, P.M., R.A. Bressan, J.K. Zhu, and H.J. Bohnert. 2000. Plant cellular and molecular responses to salinity. *Annu. Rev. Plant Physiol. Plant Mol. Biol.* 51:463–499.

He, X., X. Huang, Y. Shen, and Z. Huang. 2014. Wheat V-H(+)-ATPase subunit genes significantly affect salt tolerance in *Arabidopsis thaliana*. *PLoS ONE* 9(1):e86982.

Henfrey, R.D. and R.J. Slater. 1988. Isolation of plant nuclei. *Methods Mol. Biol.* 4:447–452.

Holmes-Davis, R., C.K. Tanaka, W.H. Vensel, W.J. Hurkman, and S. McCormick. 2005. Proteome mapping of mature pollen of *Arabidopsis thaliana*. *Proteomics* 5:4864–4884.

Hrubec, T.C., J.M. Robinson, and R.P. Donaldson. 1985. Isolation of mitochondria from soybean leaves on discontinuous percoll gradients. *Plant Physiol.* 77:1010–1012.

Hsu, J.L., L.Y. Wang, S.Y. Wang, C.H. Lin, K.C. Ho, F.K. Shi, and I.F. Chang. 2009. Functional phosphoproteomic profiling of phosphorylation sites in membrane fractions of salt-stressed *Arabidopsis thaliana*. *Proteome Sci.* 7:42.

Huckelhoven, R. 2007. Cell wall associated mechanisms of disease resistance and susceptibility. *Annu. Rev. Phytopathol.* 45:101–127.

Hummel, M., J.H. Cordewener, J.C. de Groot, S. Smeekens, A.H. America, and J. Hanson. 2012. Dynamic protein composition of *Arabidopsis thaliana* cytosolic ribosomes in response to sucrose feeding as revealed by label free MSE proteomics. *Proteomics* 12:1024–1038.

Irshad, M., H. Canut, G. Borderies, R. Pont-Lezica, and E. Jamet. 2008. A new picture of cell wall protein dynamics in elongating cells of *Arabidopsis thaliana*: Confirmed actors and newcomers. *BMC Plant Biol.* 8:94.

Isenberg, G., W. Bielser, W. Meier-Ruge, W. Meier-Ruge, and E. Remy. 1976. Cell surgery by laser micro-dissection: A preparative method. *J. Microsc.* 107:19–24.

Ito, J., T.S. Batth, C.J. Petzold, A.M. Redding-Johanson, A. Mukhopadhyay, R. Verboom, E.H. Meyer, A.H. Millar, and J.L. Heazlewood. 2011. Analysis of the arabidopsis cytosolic proteome highlights subcellular partitioning of central plant metabolism. *J. Proteome Res.* 10:1571–1582.

Ito, J., H.T. Parsons, and J.L. Heazlewood. 2014. The *Arabidopsis* cytosolic proteome: The metabolic heart of the cell. *Front. Plant Sci.* 5:21.

Jamet, E., C. Albenne, G. Boudart, M. Irshad, H. Canut, and R. Pont-Lezica. 2008. Recent advances in plant cell wall proteomics. *Proteomics* 8: 893–908.

Jamet, E., H. Canut, G. Boudart, and R.F. Pont-Lezica. 2006. Cell wall proteins: A new insight through proteomics. *Trends Plant Sci.* 11:33–39.

Jaquinod, M., F. Villiers, S. Kieffer-Jaquinod, V. Hugouvieux, C. Bruley, J. Garin, and J. Bourguignon. 2007. A proteomics dissection of *Arabidopsis thaliana* vacuoles isolated from cell culture. *Mol. Cell Proteomics* 6:394–412.

Jiang, D., H.W. Jarrett, and W.E. Haskins. 2009. Methods for proteomic analysis of transcription factors. *J. Chromatogr. A* 1216:6881–6889.

Jou, Y., C.P. Chiang, and H.E. Yen. 2013. Changes in cellular distribution regulate SKD1 ATPase activity in response to a sudden increase in environmental salinity in halophyte ice plant. *Plant Signal Behav.* 8(12). pii: e27433.

Kerk, N.M., T. Ceserani, S.L. Tausta, I.M. Sussex, and T.M. Nelson. 2003. Laser capture microdissection of cells from plant tissues. *Plant Physiol.* 132:27–35.

Khalil, R.M.A. 2013. Molecular and biochemical markers associated with salt tolerance in some sorghum genotypes. *World Appl. Sci. J.* 22:459–469.

Kim, J.H. and W.T. Kim. 2013. The *Arabidopsis* RING E3 ubiquitin ligase AtAIRP3/LOG2 participates in positive regulation of high-salt and drought stress responses. *Plant Physiol.* 162:1733–1749.

Kleffmann, T., D. Russenberger, A. von Zychlinski, W. Christopher, K. Sjölander, W. Gruissem, and S. Baginsky. 2004. The *Arabidopsis thaliana* chloroplast proteome reveals pathway abundance and novel protein functions. *Curr. Biol.* 14:354–362.

Kohorn, B.D. and S.L. Kohorn. 2012.The cell wall-associated kinases, WAKs, as pectin receptors. *Front. Plant Sci.* 3:88.

Koizumi, M., K. Yamaguchi-Shinozaki, H. Tsuji, and K. Shinozak. 1993. Structure and expression of two genes that encode distinct drought-inducible cysteine proteinases in *Arabidopsis thaliana*. *Gene* 129:175–182.

Komatsu, S., H. Konishi, and M. Hashimoto. 2007. The proteomics of plant cell membranes. *J. Exp. Bot.* 58:103–112.

Konozy, E.H.E., H. Rogniaux, M. Causse, and M. Faurobert. 2012. Proteomic analysis of tomato (*Solanum lycopersicum*) secretome. *J. Plant Res.* 126:251–266.

Kryvych, S., S. Kleessen, B. Ebert, B. Kersten, and J. Fisahn. 2011. Proteomics—The key to understanding systems biology of *Arabidopsis* trichomes. *Phytochemistry* 72:1061–1070.

Kumar, K., M. Kumar, S.R. Kim, H. Ryu, and Y.G. Cho. 2013. Insights into genomics of salt stress response in rice. *Rice* 6:27.

Lang, F. 2007. Mechanisms and significance of cell volume regulation. *J. Am. Coll. Nutr.* 26 (5 Suppl.): 613S–623S.

Lee, J., H. Lee, J. Kim, S. Lee, D.H. Kim, S. Kim, and I. Hwang. 2011. Both the hydrophobicity and a positively charged region flanking the C-terminal region of the transmembrane domain of signal-anchored proteins play critical roles in determining their targeting specificity to the endoplasmic reticulum or endosymbiotic organelles in *Arabidopsis* cells. *Plant Cell* 23:1588–1607.

Lee, S., R.S. Saravanan, C.M.B. Damasceno, H. Yamane, B. Kim, and J.K.C. Rose. 2004. Digging deeper into the plant cell wall proteome. *Plant Physiol. Biochem.* 42:979–988.

Lelong, C., M. Chevallet, H. Diemer, S. Luche, A. Van Dorsselaer, and T. Rabilloud. 2012. Improved proteomic analysis of nuclear proteins, as exemplified by the comparison of two myeloid cell lines nuclear proteomes. *J. Proteomics* 77:577–602.

Le Maire, M., P. Champeil, and J.V. Moller. 2000. Interaction of membrane proteins and lipids with solubilizing detergents. *Biochim. Biophys. Acta* 1508:86–111.

Leshem, Y., N. Melamed-Book, O. Cagnac, G. Ronen, Y. Nishri, M. Solomon, G. Cohen, and A. Levine. 2006. Suppression of *Arabidopsis* vesicle-SNARE expression inhibited fusion of $H_2O_2$-containing vesicles with tonoplast and increased salt tolerance. *Proc. Natl. Acad. Sci. U.S.A.* 103:18008–180013.

Li, G., B.R. Nallamilli, F. Tan, and Z. Peng. 2008. Removal of high-abundance proteins for nuclear subproteome studies in rice (*Oryza sativa*) endosperm. *Electrophoresis* 29:604–617.

Liu, Q.L., K.D. Xu, Y.Z. Pan, B.B. Jiang, G.L. Liu, Y. Jia, and H.Q. Zhang. 2014. Functional Analysis of a novel chrysanthemum WRKY transcription factor gene involved in salt tolerance. *Nat. Methods* 10:570–576.

Macarisin, D., M. Wisiewski, C. Bassett, and T.W. Thannhauser. 2009. Proteomic analysis of B-aminobutyric acid priming and ABA-induction of drought resistance in crabapple (*Malus pumila*): Effect on general metabolism, the phenylpropanoid pathway and cell wall enzymes. *Plant Cell Environ.* 32:1612–1631.

Manaa, A., H. Mimouni, S. Wasti, E. Gharbi, S. Aschi-Smiti, M. Faurobert, and H.B. Ahmed. 2013. Comparative proteomic analysis of tomato (*Solanum lycopersicum*) leaves under salinity stress. *Plant Omics* 6:268–277.

Martin, C., K. Bhatt, and K. Baumann. 2001. Shaping in plant cells. *Curr. Opin. Plant Biol.* 4:540–549.

Matas, A.J., T.H. Yeats, G.J. Buda, Y. Zheng, S. Chatterjee, T. Tohge, L. Ponnala et al. 2011. Tissue- and cell-type specific transcriptome profiling of expanding tomato fruit provides insights into metabolic and regulatory specialization and cuticle formation. *Plant Cell* 23:3893–3910.

Millar, A.H. and N. Taylor. 2014. Subcellular proteomics-where cell biology meets protein chemistry. *Front Plant Sci.* 5:55.

Millioni, R., S. Tolin, L. Puricelli, S. Sbrignadello, G.P. Fadini, P. Tessari, and G. Arrigoni. 2011. High abundance proteins depletion vs low abundance proteins enrichment: Comparison of methods to reduce the plasma proteome complexity. *PLoS ONE* 6(5):e19603.

Minic, Z., E. Jamet, L. Negroni, P. Arsene Der Garabedian, M. Zivy, and L. Jouanin. 2007. A sub-proteome of *Arabidopsis thaliana* mature stems trapped on Concanavalin A is enriched in cell wall glycoside hydrolases. *J. Exp. Bot.* 58: 2503–2512.

Mo, B., Y.C. Tse, and L. Jiang. 2003. Organelle identification and proteomics in plant cells. *Trends Biotechnol.* 21:331–332.

Monetti, E., T. Kadono, D. Tran, E. Azzarello, D. Arbelet-Bonnin, B. Biligui, J. Briand, T. Kawano, S. Mancuso, and F. Bouteau. March 2014. Deciphering early events involved in hyperosmotic stress-induced programmed cell death in tobacco BY-2 cells. *J. Exp. Bot.* 65(5):1361–1375.

Moriguchi, K., T. Suzuki, Y. Ito, Y. Yamazaki, Y. Niwa, and N. Kurata. 2005. Functional isolation of novel nuclear proteins showing a variety of subnuclear localizations. *Plant Cell* 17: 389–403.

Mousavi, A. and Y. Hotta. 2005. Glycine-rich proteins a class of novel proteins. *Appl. Biochem. Biotechnol.* 120:169–174.

Muller-Rober, B., J. Ellenberg, N. Provart, L. Willmitzer, H. Busch, D. Becker, P. Dietrich, S. Hoth, and R. Hedrich. 1995. Cloning and electrophysiological analysis of KST1, an inward rectifying $K^+$ channel expressed in potato guard cells. *EMBO J.* 14:2409–2416.

Muntel, J., V. Fromion, A. Goelzer, S. Maaβ, U. Mäder, K. Büttner, M. Hecker, and D. Becher. 2014. Comprehensive absolute quantification of the cytosolic proteome of *Bacillus subtilis* by data independent, parallel fragmentation in liquid chromatography/mass spectrometry (LC/MS(E). *Mol. Cell Proteomics* 13:1008–1019.

Nakamura, T., S. Yokota, Y. Muramoto, K. Tsutsui, Y. Oguri, K. Fukui, and T. Takabe. 1997. Expression of a betaine aldehyde dehydrogenase gene in rice, a glycinebetaine nonaccumulator, and possible localization of its protein in peroxisomes. *Plant J.* 11:1115–1120.

Narula, K., A. Datta, N. Chakraborty, and S. Chakraborty. 2013. Comparative analyses of nuclear proteome: Extending its function. *Front. Plant Sci.* 4:100.

Nveawiah-Yoho, P., J. Zhou, M. Palmer, R. Sauve, S. Zhou, K.J. Howe, T. Fish, and T.W. Thannhauser. 2013. Identification of proteins for salt tolerance using a comparative proteomics analysis of tomato accessions with contrasting salt tolerance. *J. Am. Soc. Hort. Sci.* 138:382–394.

Okekeogbu, I., Z. Ye, S.R. Sangireddy, H. Li, S. Bhatti, D. Hui, S. Zhou, K.J. Howe, T. Fish, Y. Yang, and T.W. Thannhauser. 2014. Effect of aluminum treatment on proteomes of radicles of seeds derived from Al-treated tomato plants. *Proteomes* 2:169–190.

Okimoto, R., J.L. Macfarlane, D.O. Clary, and D.R. Wolstenholm. 1992. The mitochondrial genomes of two nematodes, *Caenorhabditis elegans* and *Ascaris suum. Genetics* 130:471–498.

Ouyang, B., T. Yang, H. Li, L. Zhang, Y. Zhang, J. Zhang, Z. Fei, and Z. Ye. 2007. Identification of early salt stress response genes in tomato root by suppression subtractive hybridization and microarray analysis. *J. Exp. Bot.* 8:507–520.

Pandey, A., M.K. Choudhary, D. Bhushan, A. Chattopadhyay, S. Chakraborty, A. Datta, and N. Chakraborty. 2006. The nuclear proteome of chickpea (*Cicer arietinum* L.) reveals predicted and unexpected proteins. *J. Proteome Res.* 5:3301–3311.

Peltier, J.B., O. Emanuelsson, D.E. Kalume, J. Ytterberg, G. Friso, A. Rudella, D.A. Liberles, L. Söderberg, P. Roepstorff, G. von Heijne, and K.J. van Wijk. 2002. Central functions of the lumenal and peripheral thylakoid proteome of *Arabidopsis* determined by experimentation and genome-wide prediction. *Plant Cell* 14:211–236.

Peltier, J.B., G. Friso, D.E. Kalume, P. Roepstorff, F. Nilsson, I. Adamska, and K.J. van Wijk. 2000. Proteomics of the chloroplast: Systematic identification and targeting analysis of lumenal and peripheral thylakoid proteins. *Plant Cell* 12:319–341.

Pereira, F., X. Niu, and A.J. deMello. 2013. A nano LC-MALDI mass spectrometry droplet interface for the analysis of complex protein samples. *PLoS ONE* 8(5):e63087.

Petit, J., C. Bres, D. Just, V. Garcia, J.P Mauxion, D. Marion, B. Bakan, J. Joubès, F. Domergue, and C. Rothan. 2013. Analyses of tomato fruit brightness mutants uncover both cutin-deficient and cutin-abundant mutants and a new hypomorphic allele of GDSL lipase. *Plant Physiol.* 164:888–906.

Petricka, J.J., C.M. Winter, and P.N. Benfey. 2012. Control of *Arabidopsis* root development. *Annu. Rev. Plant Biol.* 63:583–590.

Piersma, S.R., M.O. Warmoes, M. de Wit, I. de Reus, J.C. Knol, and C.R. Jiménez. 2013. Whole gel processing procedure for GeLC-MS/MS based proteomics. *Proteome Sci.* 11(1):17.

Popova, L.G., A.G. Kornilova, G.A. Shumkova, I.M. Andreev, and Y.V. Balnokin. 2006. Na⁺-transporting ATPase in the plasma membrane of halotolerant microalga *Dunaliella maritima* operates as a $Na^+$ uniporter. *Russ. J. Plant Physiol.* 53:474–480.

Puntarulo, S., R.A. Sanchez, and A. Boveris. 1988. Hydrogen peroxide metabolism in soybean embryonic axes at the onset of germination. *Plant Physiol.* 86:626–630.

Rainbolt, T.K., N. Atanassova, J.C. Genereux, and R.L. Wiseman. 2013. Stress-regulated translational attenuation adapts mitochondrial protein import through Tim17A degradation. *Cell Metab.* 18: 908–919.

Rahimian, M.H. and S. Poormohammadi. 2012. Assessing the impact of climate change on evapotranspiration and soil salinization. In W. Leal Filho (ed.), *Climate Change and the Sustainable Use of Water Resources, Climate Change Management*. Springer-Verlag, Berlin, Germany, pp. 69–76.

Reddy, B.V.S., A.A. Kumar, P.S. Reddy, M. Ibrahim, B. Ramaiah, A.J. Dakheel, S. Ramesh, and L. Krishnamurthy. 2010. Cultivar options for salinity tolerance in sorghum. *SAT eJournal* 8:1–5. ejournal. icrisat.org.

Rengasamy, P. 2008. Salinity in the landscape: A growing problem in Australia. *Geotimes* Mar. 2008. http://www.geotimes.org/mar08/article.html?id=feature_salinity.html. Accessed on April 20, 2014.

Rhee, H.W., P. Zou, N.D. Udeshi, J.D. Martell, V.K. Mootha, S.A. Carr, and A.Y. Ting. 2013. Proteomic mapping of mitochondria in living cells via spatially restricted enzymatic tagging. *Science* 339: 1328–1331.

Richter, U., T. Lahtinen, P. Marttinen, M. Myöhänen, D. Greco, G. Cannino, H.T. Jacobs, N. Lietzén, T.A. Nyman, and B.J. Battersby. 2013. A mitochondrial ribosomal and RNA decay pathway blocks cell proliferation. *Curr. Biol.* 23:535–541.

Robert, S., J. Zouhar, C. Carter, and N. Raikhel. 2007. Isolation of intact vacuoles from *Arabidopsis* rosette leaf-derived protoplasts. *Nat. Protoc.* 2:259–262.

Roberts, K. 2001. How the cell wall acquired a cellular context. *Plant Physiol.* 125:127–130.

Rojo, E., C.S. Gillmor, V. Kovaleva, C.R. Somerville, and N.V. Raikhel. 2001. VACUOLELESS1 is an essential gene required for vacuole formation and morphogenesis in *Arabidopsis*. *Dev. Cell* 1:303–310.

Rose, J.K.C., S. Bashir, J.J. Giovannoni, M.M. Jahn, and R.S. Saravanan. 2004. Tackling the plant proteome: Practical approaches, hurdles and experimental tools. *Plant J.* 39:715–733.

Rose, J.K.C. and S.J. Lee. 2010. Straying off the highway: Trafficking of secreted plant proteins and complexity in the plant cell wall proteome. *Plant Physiol.* 153:433–436.

Ruiz-May, E. and J.K.C. Rose. 2013. Progress toward the tomato fruit cell wall proteome. *Front. Plant Sci.* 4:159.

Sarhadi, E., M.M. Bazargani, A.G. Sajise, S. Abdolahi, N.A. Vispo, M. Arceta, G.M. Nejad, R.K. Singh, and G.H. Salekdeh. 2012. Proteomic analysis of rice anthers under salt stress. *Plant Physiol. Biochem.* 58: 280–287.

Shabala, S. 2013. Learning from halophytes: Physiological basis and strategies to improve abiotic stress tolerance in crops. *Ann. Bot.* 112:1209–1221.

Shelden, M.C. and U. Roessner. 2013. Advances in functional genomics for investigating salinity stress tolerance mechanisms in cereals. *Front. Plant Sci.* 4:123.

Sikorskaite, S., M.L. Rajamäki, D. Baniulis, V. Stanys, and J.P. Valkonen. 2013. Protocol: Optimised methodology for isolation of nuclei from leaves of species in the Solanaceae and Rosaceae families. *Plant Methods* 9:31.

Singleton, P. 1999. *Bacteria in Biology, Biotechnology and Medicine* (5th edn.). New York: Wiley.

Smith, A.L. 1997. *Oxford Dictionary of Biochemistry and Molecular Biology.* Oxford (Oxfordshire), U.K.: Oxford University Press. p. 206.

Song, Y., D. Ji, S. Li, P. Wang, Q. Li, and F. Xiang. 2012. The dynamic changes of DNA methylation and histone modifications of salt responsive transcription factor genes in soybean. *PLoS ONE* 7(7): e41274.

Sun, W., X. Xu, H. Zhu, A. Liu, L. Liu, J. Li, and X. Hua. 2010. Comparative transcriptomic profiling of a salt-tolerant wild tomato species and a salt-sensitive tomato cultivar. *Plant Cell Physiol.* 51:997–1006.

Swidzinski, J.A., C.J. Leaver, and L.J. Sweetlove. 2004. A proteomic analysis of plant programmed cell death. *Phytochemistry* 65:1829–1838.

Taylor, S.W., E. Fahy, and S.S. Ghosh. 2003. Global organellar proteomics. *Trends Biotechnol.* 21:82–88.

Thannhauser, T.W., M. Shen, R. Sherwood, K.J. Howe, T. Fish, Y. Yang, W. Chen, and S. Zhang. 2013. A workflow for large-scale empirical identification of cell wall N-linked glycoproteins of tomato (*Solanum lycopersicum*) fruit by tandem mass spectrometry. *Electrophoresis* 34(16):2417–2431.

The Arabidopsis Information Resource. 2013. Locus: AT5G60360. https://www.arabidopsis.org/servlets/TairObject?name=AT5G60360&type=locus. (Accessed April 6, 2015.)

Thingholm, T.E. and O.N. Jensen. 2009. Enrichment and characterization of phosphopeptides by immobilized metal affinity chromatography (IMAC) and mass spectrometry. *Methods Mol. Biol.* 527:47–56.

Tomizioli, M., C. Lazar, S. Brugiere, T. Burger, D. Salvi, L. Gatto, L. Moyet et al. 2014. Deciphering thylakoid sub-compartments using a mass spectrometry-based approach. *Mol. Cell. Proteomics.* 13(8):2147–2167. doi:10.1074/mcp.M114.040923.

Uemura, T., T. Ueda, and A. Nakano. 2012. The physiological role of SYP4 in the salinity and osmotic stress tolerances. *Plant Signal Behav.* 7:1118–1120.

VanAken, T., S. Foxall-VanAken, S. Castleman, and S. Ferguson-Miller. 1986. Alkyl glycoside detergents: Synthesis and applications to the study of membrane proteins. *Methods Enzymol.* 125:27–35.

Wan, J., M. Torres, A. Ganapathy, J. Thelen, B.B. DaGue, B. Mooney, D. Xu, and G. Stacey. 2005. Proteomic analysis of soybean root hairs after infection by *Bradyrhizobium japonicum*. *Mol. Plant Microbe Interact.* 18:458–467.

Wang, H., T. Chang-Wong, H.Y. Tang, and D.W. Speicher. 2010. Comparison of extensive protein fractionation and repetitive LC-MS/MS analyses on depth of analysis for complex proteomes. *J. Proteome Res.* 9:1032–1040.

Wang, L., X. Liu, M. Liang, F. Tan, W. Liang, Y. Chen, Y. Lin, L. Huang, J. Xing, and W. Chen. 2014. Proteomic analysis of salt-responsive proteins in the leaves of mangrove *Kandelia candel* during short-term stress. *PLoS ONE* 9(1):e83141.

Wang, P.R., X.J. Deng, X.L. Gao, J. Chen, J. Wan, H. Jiang, and Z.J. Xu. 2006. Progress in the study on DREB transcription factor. *Yi Chuan.* 28:369–374.

Walther, D. and R. Rakwal. 2009. Plant phosphoproteomics: An update. *Proteomics* 9:964–988.

Waters, S., M. Gilliham, and M. Hrmova. 2013. Plant high-affinity potassium (HKT) transporters involved in salinity tolerance: Structural insights to probe differences in ion selectivity. *Int. J. Mol. Sci.* 14:7660–7680.

Wienkoop, S. and W. Weckwerth. 2006. Relative and absolute quantitative shotgun proteomics: Targeting low-abundance proteins in *Arabidopsis thaliana*. *J. Exp. Bot.* 57:1529–1535.

Wojtyla, L., A. Kosmala, and M. Garnczarska. 2013. Lupine embryo axes under salinity stress. II. Mitochondrial proteome response. *Acta Physiol. Plant* 35:2383–2392.

Wu, S.J., L. Ding, and J.K. Zhu. 1996. SOS1, a genetic locus essential for salt tolerance and potassium acquisition. *Plant Cell* 8:617–627.

Yamada, K., R. Matsushima, M. Nishimura, and I. Hara-Nishimura. 2001. A slow maturation of a cysteine protease with a granulin domain in the vacuoles of senescing *Arabidopsis* leaves. *Plant Physiol.* 127:1626–1634.

Yang, Y., X. Qiang, K. Owsiany, S. Zhang, T.W. Thannhauser, and L. Li. 2011. Evaluation of different multidimensional LC-MS/MS pipelines for iTRAQ-based proteomic analysis of potato tubers in response to cold storage. *J. Proteome Res.* 10: 4349–4886.

Zagorchev, L., P. Kamenova, and M. Odjakova. 2014. The role of plant cell wall proteins in response to salt stress. *Sci. World J.* 764089.

Zhang, H., B. Han, T. Wang, S. Chen, H. Li, Y. Zhang, and S. Dai. 2012. Mechanisms of plant salt response: Insights from proteomics. *J. Proteome Res.* 11: 49–67.

Zhao, Z., W. Zhang, B.A. Stanley, and S.M. Assmann. 2008. Functional proteomics of *Arabidopsis thaliana* guard cells uncovers new stomatal signaling pathways. *Plant Cell* 20:3210–3226.

Zhou, G.P. and K. Doctor. 2003. Subcellular location prediction of apoptosis proteins. *Proteins* 50: 44–48.

Zhou, H. and W. Li. 2013. The effects of oasis ecosystem hydrological processes on soil salinization in the lower reaches of the Tarim River, China. *Ecohydrology* 6:1009–1020.

Zhou, S., R. Sauve, T. Fish, and T.W. Thannhauser. 2009a. Salt-induced and salt-suppressed proteins in tomato leaves. *J. Am. Soc. Hort. Sci.* 134:289–294.

Zhou, S., R. Sauve, and T.W. Thannhauser. 2009b. Proteome changes induced by aluminium stress in tomato roots. *J. Exp. Bot.* 60:1849–1857.

Zhou, S., R.J. Sauvé, Z. Liu, S. Reddy, S. Bhatti, S.D. Hucko, T. Fish, and T.W. Thannhauser. 2011a. Identification of salt-induced changes in leaf and root proteomes of the wild tomato, *Solanum chilense*. *J. Am. Soc. Hort. Sci.* 136:288–302.

Zhou, S., R.J. Sauvé, Z. Liu, S. Reddy, S. Bhatti, S.D. Hucko, Y. Yang, T. Fish, and T.W. Thannhauser. 2011b. Heat-induced proteome changes in tomato leaves. *J. Am. Soc. Hort. Sci.* 136:219–226.

Zhou, S., M. Plamer, J. Zhou, S. Bhatti, K. Howe, T. Fish, and T.W. Thannhauser. 2013. Differential root proteome expression in tomato genotypes with contrasting drought tolerance exposed to dehydration. *J. Am. Soc. Hort. Sci.* 138:131–141.

Zhu, J., S. Chen, S. Alvarez, V.S. Asirvatham, D.P. Schachtman, Y. Wu, and R. Sharp. 2006. Cell wall proteome in the maize primary root elongation zone. I. Extraction and identification of water-soluble and lightly ionically bound proteins. *Plant Physiol.* 140:311–325.

Zhuang, J., H.H. Jiang, F. Wang, R.H. Peng, Q.H. Yao, and A.S. Xiong. 2013. A rice OsAP23, functioning as an AP2/ERF transcription factor, reduces salt tolerance in transgenic *Arabidopsis*. *Plant Mol. Biol. Rep.* 31:1336–1345.

# CHAPTER SEVEN

# Heat Shock Protein and Salinity Tolerance in Plants

*Shanlin Yu, Na Chen, Xiaoyuan Chi, Lijuan Pan, Mingna Chen,*
*Tong Wang, Chuantang Wang, Zhen Yang, and Mian Wang*

## CONTENTS

*Abstract.* Heat shock proteins (HSPs) are highly conserved proteins found in plants that consist of stress-inducible and constitutively expressed proteins. These proteins play a key role in promoting correct refolding and preventing aggregation of denatured proteins under stress conditions by functioning as molecular chaperones or proteases. The function of HSPs is usually emphasized under heat stress condition; however, the versatile expression patterns strongly suggest that HSPs may be important for other developmental and stress conditions, especially salinity condition. The expressions of HSPs are regulated by heat shock factors (Hsfs) through their interaction with heat shock elements presented in the HSP promoter region. Plant Hsfs play important roles not only in heat stress but also in other biotic and abiotic stresses.

*Keywords:* Heat shock proteins; Salt tolerance; Expression regulation

## 7.1 INTRODUCTION

Salinity condition is one of the major abiotic stresses in the nature, and many higher plants are seriously damaged by the high salinity in soil that impairs the growth, development, and production of plants (Walia et al. 2006; Wang et al. 2003). High concentration of salts affects plants in two main ways: salts in the soil disturb the capacity of roots to extract water, and salts within the plant itself can be toxic, resulting in an inhibition of many physiological and biochemical processes such as nutrient uptake and assimilation (Hasegawa et al. 2000; Munns 2002; Munns and Tester 2008).

Salt tolerance of plants has been of interest for a long time and has resulted in a considerable body of data from studies using physiological, genetic, and cytogenetic approaches (Ellis et al. 2002). Recent large-scale approaches, including microarrays (Ozturk et al. 2002; Seki et al. 2001) and differential display (Ueda et al. 2002), have been employed to identify genes responding to salinity stress in plants. Also, molecular characterization of salinity stress in plants has indicated the involvement of multiple genes responsive to salinity (Walia et al. 2006). Many of these genes have been identified, functions assigned, and their relation to salt tolerance determined. Plant gene expression is regulated during salt stress by both transcriptional and posttranscriptional mechanisms (Hurkman 1992).

Heat shock proteins (HSPs) are highly conserved proteins found in plants that consist of stress-inducible and constitutively expressed proteins (Parsell and Lindquist 1993). These proteins play a key role in promoting correct refolding and preventing aggregation of denatured proteins under stress conditions by functioning as molecular chaperones or proteases (Hartl and Hayer-Hartl 2002; Li et al. 2012; Mujacic et al. 2004). The function of HSPs is usually emphasized under heat stress condition; however, the versatile expression patterns strongly suggest that HSPs may be important for other developmental and stress conditions, especially salinity condition (Gomathi and Vasantha 2006; Gonzalez and Bradley 1995; Guan et al. 2004; Wang et al. 2004).

## 7.2 FUNCTIONS OF HEAT SHOCK PROTEINS IN SALT TOLERANCE REGULATION OF PLANTS

Abiotic stresses such as drought, salt stress, temperature stress, heavy metal pollution, and oxidative stress affect plant growth and development, which is the main cause of the global crop production (Zou et al. 2007). In the long process of evolution, plants have evolved a spectrum of molecular programs to adapt to environmental stresses. To survive, plants undergo dramatic changes in physiological and molecular mechanisms (Wang et al. 2014). For instance, HSPs are stimulated in response to a wide array of stress conditions and perform a fundamental role in protecting plants against abiotic stresses (Ahuja et al. 2010; Timperio et al. 2008).

HSPs are known to be induced, following exposure to increasing temperature, and were first discovered in *Drosophila* in 1962. These proteins have subsequently been reported in many organisms, including plants (Yang et al. 2014). Plant HSPs can be classified into five major categories based on molecular mass: Hsp100 (ClpB), Hsp90, Hsp70 (DnaK), Hsp60, and small heat shock protein (sHsp) (Jiao et al. 2008; Siddique et al. 2008).

The Hsp100/Clp proteins contain one or two conserved ATPases associated with various cellular activities domains and act as unfoldases or disaggregases. A subset of the Hsp100 family of proteins (including ClpA, ClpX, and HslU) are coaxially stacked with a ring protease such as ClpP or HslV, and their functions are to unfold proteins and deliver them to the protease. Another subset of Hsp100 proteins (ClpB, Hsp104, and Hsp101) function as disaggregases and have the unique ability, together with the Hsp70 system, to recover proteins from both amorphous and amyloid aggregates (Clare and Saibil 2013).

Among all HSPs, Hsp90 and Hsp70 are the most conserved and abundant in cells and have thus been extensively studied (Taipale et al. 2010). The Hsp90 family contains three domains: an N-terminal ATP binding domain (25 kD), a central connecting domain (35 kD), and a C-terminal dimerization domain (12 kD), which form constitutive homodimers (Sun et al. 2014). Hsp70 are ATP-dependent chaperones having a conserved ~44-kD N-terminal ATPase domain (also called nucleotide binding domain), a ~18-kD substrate binding domain (SBD), and a ~10-kD variable C-terminal "lid." The flexible C-terminal lid assists in holding the substrates at SBD. ATPase domain performs a regulatory role and SBD associates to hydrophobic regions exposed in nonnative substrates (Sarkar et al. 2012).

The sHsp family is the most abundant and important family (Borges and Ramos 2005), and sHsp-bound proteins are effective substrates for refolding by other chaperones (Lee and Vierling 2000). All plant sHsp genes are nuclear encoded and can be divided into six different classes based on homology and specific organelle localization of the encoded proteins. Class I and II sHsps are located in the cytoplasm or the nucleus, while the others are located in the endoplasmic reticulum, chloroplasts, mitochondria, and endomembrane (Banzet et al. 1998; Dafny-Yelin et al. 2008). Homologues of the sHsp family have been detected and characterized in maize (Cao et al. 2010), cotton (Maqbool et al. 2010), tomato (Firon et al. 2009), agave (Luján et al. 2009), *Arabidopsis* (Dafny-Yelin et al. 2008), sugarcane (Tiroli et al. 2007), tobacco (Hamilton and Coleman 2001), and carrot (Malik et al. 1999).

Thermal stress is one of the most common factors inducing the expression of HSPs. Under normal environmental conditions, the majority of HSPs, especially sHsp, is not expressed, but the expression was rapidly induced by heat stimulation. The research findings in soybean showed that HSPs could be detected with heat shock treatment by 7–10 min, and the expression of HSPs continuously increased in 2–3 h, then decreased, which was not detected after 6 h (Kimpel et al. 1990).

HSPs are essential for plant survival in adverse environmental conditions, especially supraoptimal temperatures. But thermal stress is not the only factor inducing the expression of HSPs, such as drying/water stress, cold, heavy metal ions, ultraviolet light, and other factors, which all can cause the expression of HSPs (Sun et al. 2002). Recent microarray studies in *Arabidopsis* revealed that a subset of sHsp genes was induced by various stresses such as salt, drought, chilling, oxidative stress, and wounding (Guan et al. 2004; Hamilton and Heckathorn 2001; Liu et al. 2006). Gao et al. (2012) reported that ThHSP18.3 was strongly induced in *Tamarix hispida* by treatment of salinity (NaCl). Sun et al. (2001) observed that *AtHSP17.6A*-over-expressing plants showed increased tolerance to salt stress. Moreover, transgenic *Arabidopsis thaliana* that constitutively expressed RcHSP17.8 exhibited increased tolerance to salt stress (Jiang et al. 2009). Liu et al. (2006) reported that r*Hsp90* mRNA was induced in rice by treatment of salinity. Yang et al. (2014) reported that under NaCl stress, all nine ThsHSPs genes were upregulated. D*cHsp17.7*, a sHsp in carrot (*Daucus carota* L.), performs molecular chaperone activity in salt-stressed transgenic *Escherichia coli* and is involved in tolerance not only to thermal stresses but also to other abiotic stresses, such as salinity (Song and Ahn 2010). Overexpression of alfalfa mitochondrial HSP23 in prokaryotic and eukaryotic model systems confers enhanced tolerance to salinity and arsenic stress (Lee et al. 2012). Montero-Barrientos et al. (2010) inserted an *Hsp70* gene of *Trichoderma harzianum* into *A. thaliana* and effectively improved the transgenic *Arabidopsis* tolerance to high salinity, high temperature, and other abiotic stresses.

Then, how do heat shock proteins play a role in salt tolerance regulation of plants? Under adversity stress, HSPs can maintain the natural structure of protein stabilizing. A direct damage of the stress to the cell is to make the protein denaturation and loss of normal function. Therefore, it is necessary to maintain function of structural proteins; to prevent agglomeration of nonnatural protein, refolding of denatured proteins; to restore its function structure; and to remove potentially harmful denatured protein under adversity stresses. HSP molecular chaperone systems play an important role in the process (Wang et al. 2004). All kinds of HSPs in plant cells cooperate with each other and form an anastomosing network system. The sHsps of plants combine with the partial denatured protein by the form of oligomers under the stress, to prevent irreversible agglomeration (Basha et al. 2004; Friedrich et al. 2004; Wagner et al. 2005), and then restore the folded state under the action of other HSP molecular chaperones (Low et al. 2000; Mogk

et al. 2003). The Hsp100 protein family plays the role of the disintegrating protein aggregates (Lee et al. 2007; Sakamoto 2006). The Hsp100 protein family will disintegrate the aggregation formed by the denatured proteins or misfolded proteins, which followed to be refolded by other HSP molecular chaperones (Mogk et al. 2003). Various kinds of HSPs do their best and mutual cooperation and jointly safeguard the function and structure of the proteins.

## 7.3 FUNCTION MECHANISM OF HSPS IN PLANT SALT STRESS REGULATION

Plants synthesize a wide array of sHsps that are divided into six nuclear multigene families based on their sequence alignments and immunological cross-reactivity (Wang et al. 2004). Each gene family encodes proteins found in a distinct cellular compartment. The high diversification of plant sHsps probably reflects a molecular adaptation to stress conditions that are unique to plants (Waters et al. 1996). Increasing data suggest a strong correlation between sHsp accumulation and plant tolerance to stress (Waters et al. 1996). Maize mitochondrial sHsps were shown to improve mitochondrial electron transport during salt stress, mainly by protection of the NADH: ubiquinone oxidoreductase activity (Scharf et al. 2001). Stress tolerance assays of transgenic rice plants over-expressing O*sHsp17.0* and O*sHsp23.7* demonstrated higher germination ability compared to wild-type (WT) plants when subjected to mannitol and NaCl. And phenotypic analysis showed that transgenic rice lines displayed a higher tolerance to drought and salt stress compared to WT plants (Zou et al. 2012). The *AtHSP17.6A* expression was induced by heat and osmotic stress, as well as during seed development, and overexpression of *AtHSP17.6A* increased salt and drought tolerance in *Arabidopsis* (Sun et al. 2001). Furthermore, heterologous expression of D*cHsp17.7* in *E. coli* exhibited a higher survival rate than control *E. coli* under salt stress and suggests that DcHsp17.7 performs molecular chaperone activity in salt-stressed transgenic *E. coli* (Song and Ahn 2011).

The Hsp60 family is primarily a mitochondrial protein. It varies in size from 58 to 65 kD and is important for the native folding of proteins (Xu et al. 2012). Thus, the chaperone activity appears to play central roles in defense ability and response to stress in addition to normal cells (Vabulas et al. 2001; Ye et al. 2012). The CCTa, from the mangrove plant *Bruguiera sexangula*, a Group II chaperonin (Hsp60; the term chaperonin was first suggested to describe a class of molecular chaperones that are evolutionarily homologous to *E. coli* GroEL), enhances salt and osmotic stress tolerance of *E. coli* transformants (Wang et al. 2004; Yamada et al. 2002).

Hsp70 chaperones, together with their co-chaperones (e.g., DnaJ/Hsp40 and GrpE), make up a set of prominent cellular machines that assist with a wide range of protein-folding processes in almost all cellular compartments (Bechtold et al. 2008). This family protein represents one of the most highly conserved classes of HSPs and has molecular weights in the range of 68–75 kD (Ye et al. 2012). Hsp70 has essential functions in preventing aggregation and in assisting refolding of nonnative proteins under both normal and stress conditions (Pegoraro et al. 2011). Cytosolic Hsp70 plays important roles under abiotic and biotic stresses (Jungkunz et al. 2011). The overexpression of Hsp70 genes correlates positively with the acquisition of thermotolerance and results in enhanced tolerance to salt, water, and high-temperature stress in plants (Ono et al. 2001; Sung and Guy 2003). However, the cellular mechanisms of Hsp70 function under stress conditions are not fully understood (Wang et al. 2004, 2014). In 2014, Wang et al. identified 13 domains and 28 motifs that are potential target sites for Hsp70s in rice. In this study, it was found that nearly 46% of the Hsp70 interactors (197 out of 430) contained protein kinase domains, including PKC, PKA, ASK/M AP3K, MAP2K, MAPK, and CDK. And the expression level of approximately 81% of those protein kinases (159 out of 197) had a strong negative correlation (PCC < −0.90) with that of Hsp70s. This was consistent with previous reports. Hsp70 directly interact with PKC, ASK/MAP3K, and CDK and inhibit the activities of junamino-terminal kinase, ASK/MAP3K, MAPK, and CDK (Gao and Newton 2002; Kumar et al. 2011; Park et al. 2002; Wang et al. 2004). Ding et al. (1998) have shown that overexpression of Hsp70 significantly suppressed the enzymatic activities of PKA and PKC. Therefore, it is likely that Hsp70s indiscriminately downregulate the activity of various protein kinases (Wang et al. 2014).

Hsp90 is an abundant and highly conserved molecular chaperone in eukaryotic cells that is essential for cell viability. Further characterization of Hsp90 expression revealed that although some members of the family are constitutively expressed, others are stress inducible. In plants, Hsp90 is strongly induced by various abiotic stresses, such as salt, drought, low temperature, heat shock, heavy metal ions, and alkaline stress (Liu et al. 2006; Milioni and Hatzopoulos 1997; Song et al. 2009a). These proteins have been isolated and cloned from many plant species. A rice (*Oryza sativa* L) 90 kD HSP (rHsp90) gene plays an important role in salt stress (Liu et al. 2006). It was identified by screening rice root cDNAs that were upregulated under carbonate ($NaHCO_3$) stress. Further studies showed that rHsp90 mRNA accumulated following exposure to several abiotic stresses, including salts (NaCl, $NaHCO_3$, and $Na_2CO_3$), desiccation (using polyethylene glycol [PEG]), high pH (8.0 and 11.0), and high temperature (42°C and 50°C). Yeast (*Saccharomyces cerevisiae*) overexpressing rHsp90 exhibited greater tolerance to NaCl, $Na_2CO_3$, and $NaHCO_3$, and tobacco seedlings overexpressing rHsp90 could tolerate salt concentrations as high as 200 mM NaCl, whereas untransformed control seedlings could not (Liu et al. 2006). In addition, six genes (OsHsp93.04, OsHsp71.10, OsHsp71.18, OsHsp72.57, OsHsp24.15, and OsHsp18.03) were upregulated under salt stress condition (Ye et al. 2012). In halophyte (*Salicornia europaea*), SeHsp90 is essential for tolerance to salt stresses (Xu 2013). Salt resistance analysis inferred that the growth of recombinant strain (pET-SeHSP90) was significantly better than that of control strain. And growth advantage is more significant with salinity increase (Xv 2013). In *Arabidopsis*, AtHsp90s are essential for tolerance to biotic and abiotic stresses (Ye et al. 2012). However, overexpression of three AtHsp90 isoforms (*AtHsp90.2*, *AtHsp90.5*, and *AtHsp90.7*) in *A. thaliana* enhanced plant sensitivity to salt and drought stresses, and the results implied that different cellular compartments of Hsp90s in *Arabidopsis* might be involved in abiotic stresses by different functional mechanisms (Song et al. 2009b).

Hsp100/Clp proteins have been reported in many plant species, such as *Arabidopsis*, soybean, tobacco, rice, maize (*Zea mays*), Lima bean (*Phaseolus lunatus*), and wheat (Agarwal et al. 2001; Keeler et al. 2000). Like many other HSPs/chaperones, Hsp100/Clp family chaperones are often constitutively expressed in plants, but their expression is developmentally regulated and is induced by different environmental assaults, such as heat, cold, dehydration, high salt, or dark-induced etiolation

(Agarwal et al. 2001; Keeler et al. 2000; Queitsch et al. 2000; Wang et al. 2004).

In general, the protective effects of HSPs/chaperones can be attributed to the network of the chaperone machinery, in which many chaperones act in concert. During stress, many enzymes and structural proteins undergo deleterious structural and functional changes. Therefore, maintaining proteins in their functional conformations, preventing aggregation of nonnative proteins, refolding of denatured proteins to regain their functional conformation, and removing of nonfunctional but potentially harmful polypeptides (arising from misfolding, denaturation, or aggregation) are particularly important for cell survival under stress. Thus, the different classes of HSPs/chaperones cooperate in cellular protection and play complementary and sometimes overlapping roles in the protection of proteins from stress.

## 7.4 EXPRESSION REGULATION OF HSPS BY HEAT SHOCK TRANSCRIPTION FACTORS

Transcription of HSPs is regulated by heat shock transcription factors (Hsfs). In *Arabidopsis*, rice, and wheat, more than 20 Hsfs have so far been identified. Characterized by the presence of a highly conserved DNA-binding domain and an oligomerization domain, Hsfs recognize the multiple inverted repeats of the nGAAn sequence, the known heat shock elements present in the promoter regions of HSPs (Xue et al. 2014). Recently, Scharf et al. (2012) published a full review of the structure, function, and evolution of the plant Hsf family.

Plant Hsfs fall into three categories, Hsfs A, Hsfs B, and Hsfs C. Class A Hsfs possess the exclusive C-terminal activation domain with AHA motifs, whereas class B and C Hsfs lack such an activation domain (Xiang et al. 2013).

In some plant species, several salt-responsive Hsfs have been found, but their direct relationship with salt tolerance has not been established. Wheat *TaHsf* genes of A2, A6, A8, B1, C1, and C2 groups in the shoot and A2, A4, A6, and C1 groups in the root were upregulated under salt stress (Xue et al. 2014), with some of these increases in the expression levels in salt-stressed plants being remarkable. An over 10-fold increase in the A2d mRNA level was noted in the salt-stressed shoot, while a 10-fold increase in the A4 group expression levels was seen in the salt-stressed root (Xue

et al. 2014). Likewise, in *Populus trichocarpa*, the transcript level of *poptrHsfA4b* increased to about sixfold higher than that in the control after 1 h NaCl (100 mM) treatment (Zhang et al. 2013).

There is reliable evidence showing Hsfs confer salt tolerance in plants. *HsfA2* was reported to be a key regulator in the induction of the defense system under several stresses and conferred enhanced tolerance to heat, high light, salt, oxidative, osmotic, and anoxia stresses in *Arabidopsis* (Xiang et al. 2013). Overexpressing rice gene *OsHsfA2e* resulted in improved salt stress tolerance in transgenic *Arabidopsis* (Yokotani et al. 2008). A *Hsf* gene expressing preferentially in developing seed tissues of wheat grown under high temperatures was cloned and named as *TaHsfA2d* owing to its high sequence similarity to rice *OsHsfA2d* (Chauhan et al. 2013). Transgenic *Arabidopsis* plants overexpressing *TaHsfA2d* showed considerable tolerance toward high temperature, salinity, and drought (Chauhan et al. 2013).

Although ectopic overexpression of *SlHsfA3* from tomato conferred increased thermotolerance in germination in transgenic *Arabidopsis*, unexpectedly, it led to salt hypersensitivity (Li et al. 2013).

In contrast to the fact that many members of Class A Hsfs have been identified as activators of transcription, the functions of class B and C Hsfs have not been fully recognized (Xiang et al. 2013). The report of Xiang et al. (2013) suggested that OsHsfB2b functions as a negative regulator in response to drought and salt stresses in rice, and its B3 repression domain might be necessary for the repressive activity. In their study, expression of OsHsfB2b, a member of class B Hsfs in rice (*O. sativa*), was strongly induced by heat, salt, abscisic acid, and PEG treatments. Overexpressing OsHsfB2b in rice significantly decreased drought and salt tolerances, whereas tolerances were enhanced in the OsHsfB2b-RNAi transgenic rice (Xiang et al. 2013).

Now that Hsfs regulate HSPs, what is behind Hsfs? The studies of Pérez-Salamó et al. (2014) and Li et al. (2014) may provide some clues. Pérez-Salamó and colleagues clearly showed that the HsfA4A conferred salt tolerance and was regulated by oxidative stress and the mitogen-activated protein kinases MPK3 and MPK6 (Pérez-Salamó et al. 2014). The results demonstrated that estradiol-dependent induction of HsfA4A conferred enhanced tolerance to salt and oxidative agents, while inactivation of HsfA4A resulted in hypersensitivity to salt stress in *A. thaliana* (Pérez-Salamó et al. 2014). The data suggested that HsfA4A was a substrate of the MPK3/MPK6 signaling and that it regulated stress responses in *Arabidopsis* (Pérez-Salamó et al. 2014). In the study of Li et al. (2014), to investigate the mechanism of how glycine betaine (GB) influences the expression of HSPs, WT tobacco seedlings pretreated with exogenous GB and BADH-transgenic tobacco plants that accumulated GB in vivo were studied during NaCl stress (Li et al. 2014). GB was found to enhance the intracellular free calcium ion concentration and upregulate the expression of the *calmodulin* (*CaM*) and *Hsf* genes resulting in potentiated levels of HSPs (Li et al. 2014).

## REFERENCES

Agarwal, M., Katiyar-Agarwal, S., Sahi, C., Gallie, D.R., and Grover, A. 2001. *Arabidopsis thaliana* Hsp100 proteins: Kith and kin. *Cell Stress Chaperones* 6: 219–224.

Ahuja, I., de Vos, R.C., Bones, A.M., and Hall, R.D. 2010. Plant molecular stress responses face climate change. *Trends in Plant Science* 15: 664–674.

Banzet, N., Richaud, C., Deveaux, Y., Kazmaier, M., Gagnon, J., and Triantaphylidès, C. 1998. Accumulation of small heat shock proteins, including mitochondrial HSP22, induced by oxidative stress and adaptive response in tomato cells. *Plant Journal* 13: 519–527.

Basha, E., Lee, G.J., Demeler, B., and Vierling, E. 2004. Chaperone activity of cytosolic small heat shock proteins from wheat. *European Journal of Biochemistry* 271: 1426–1436.

Bechtold, U., Richard, O., Zamboni, A., Gapper, C., Geisler, M., Pogson, B., Karpinski, S., and Mullineaux, P.M. 2008. Impact of chloroplastic and extracellular sourced ROS on high light responsive gene expression in *Arabidopsis*. *Journal of Experimental Botany* 59: 121–133.

Borges, J.C. and Ramos, C.H.I. 2005. Protein folding in the cell. *Protein and Peptide Letters* 12: 256–261.

Cao, Z., Jia, Z., Liu, Y., Wang, M., Zhao, J., Zheng, J., and Wang, G. 2010. Constitutive expression of *ZmsHSP* in *Arabidopsis* enhances their cytokinin sensitivity. *Molecular Biology Reports* 37: 1089–1097.

Chauhan, H., Khurana, N., Agarwal, P., Khurana, J.P., and Khurana, P. 2013. A seed preferential heat shock transcription factor from wheat provides abiotic stress tolerance and yield enhancement in transgenic *Arabidopsis* under heat stress environment. *PLoS One* 8: e79577.

Clare, D.K. and Saibil, H.R. 2013. ATP-driven molecular chaperone machines. *Biopolymers* 99: 846–859.

Dafny-Yelin, M., Tzfira, T., Vainstein, A., and Adam, Z. 2008. Nonredundant functions of sHSP-CIs in acquired thermotolerance and their role in early seed development in *Arabidopsis*. *Plant Molecular Biology* 67: 363–373.

Ding, X.Z., Tsokos, G.C., and Kiang, J.G. 1998. Overexpression of HSP-70 inhibits the phosphorylation of HSF1 by activating protein phosphatase and inhibiting protein kinase C activity. *FASEB Journal* 12: 451–459.

Ellis, R.P., Forster, B.P., Gordon, D.C., Handley, L.L., Keith, R.P., Lawrence, P., Meyer, R. et al. 2002. Phenotype/genotype associations for yield and salt tolerance in a barley mapping population segregating for two dwarfing genes. *Journal of Experimental Botany* 53: 1163–1176.

Firon, N., LaBonte, D., Villordon, A., McGregor, C., Kfir, Y., and Pressman, E. 2009. Botany and physiology: storage root formation and development. In: Loebstein, G. and Thottappilly, G. (eds.). *The Sweetpotato*. pp. 13–26.

Friedrich, K.L., Giese, K.C., Buan, N.R., and Vierling, E. 2004. Interactions between small heat shock protein subunits and substrate in small heat shock protein-substrate complexes. *Journal of Biological Chemistry* 279: 1080–1089.

Gao, C.Q., Jiang, B., Wang, Y.C., Liu, G.F., and Yang, C.P. 2012. Overexpression of a heat shock protein (ThHSP18.3) from *Tamarix hispida* confers stress tolerance to yeast. *Molecular Biology Reports* 39: 4889–4897.

Gao, T. and Newton, A.C. 2002. The turn motif is a phosphorylation switch that regulates the binding of Hsp70 to protein kinase C. *Journal of Biological Chemistry* 277: 31585–31592.

Gomathi, R. and Vasantha, S. 2006. Change in nucleic acid content and expression of salt shock proteins in relation to salt tolerance in sugarcane. *Sugar Tech* 8: 124–127.

Gonzalez, C.R.M. and Bradley, B.P. 1995. Are there salinity stress proteins? *Mar Enivron Res* 35: 79–83.

Guan, J.C., Jinn, T.L., Yeh, C.H., Feng, S.P., Chen, Y.M., and Lin, C.Y. 2004. Characterization of the genomic structures and selective expression profiles of nine class I small heat shock protein genes clustered on two chromosomes in rice (*Oryza sativa* L.). *Plant Molecular Biology* 56: 795–809.

Hamilton, E.W. and Coleman, J.S. 2001. Heat-shock proteins are induced in unstressed leaves of *Nicotiana attenuata* (Solanaceae) when distant leaves are stressed. *American Journal of Botany* 88: 950–955.

Hamilton, E.W. and Heckathorn, S.A. 2001. Mitochondrial adaptations to NaCl. Complex I is protected by anti-oxidants and small heat shock proteins, whereas complex II is protected by proline and betaine. *Plant Physiology* 126: 1266–1274.

Hartl, F.U. and Hayer-Hartl, M. 2002. Molecular chaperones in the cytosol: From nascent chain to folded protein. *Science* 295: 1852–1858.

Hasegawa, P.M., Bressan, R.A., Zhu, J.K., and Bohnert, H.J. 2000. Plant cellular and molecular responses to high salinity. *Annual Review of Plant Physiology and Plant Molecular Biology* 51: 463–499.

Hurkman, W.J. 1992. Effect of salt stress on plant gene expression: A review. *Plant and Soil* 146: 145–151.

Jiang, C., Xu, J., Zhang, H., Zhang, X., Shi, J., Li, M., and Ming, F. 2009. A cytosolic class I small heat shock protein, RcHSP17.8, of *Rosa chinensis* confers resistance to a variety of stresses to *Escherichia coli*, yeast and *Arabidopsis thaliana*. *Plant Cell & Environment* 32: 1046–1059.

Jiao, W., Hong, W., Li, P., Sun, S., Ma, J., Qian, M., Hu, M., and Chang, Z. 2008. The dramatically increased chaperone activity of small heat-shock protein IbpB is retained for an extended period of time after the stress condition is removed. *Biochemical Journal* 410: 63–70.

Jungkunz, I., Link, K., Vogel, F., Voll, L.M., Sonnewald, S., and Sonnewald, U. 2011. AtHsp70-15 deficient *Arabidopsis* plants are characterized by reduced growth, a constitutive cytosolic protein response and enhanced resistance to TuMV. *Plant Journal* 66: 983–995.

Keeler, S.J., Boettger, C.M., Haynes, J.G., Kuches, K.A., Johnson, M.M., Thureen, D.L., Keeler, C.L. Jr, and Kitto, S.L. 2000. Acquired thermotolerance and expression of the HSP100/ClpB genes of lima bean. *Plant Physiology* 123: 1121–1132.

Kimpel, J.A., Nagao, R.T., Goekjian, V., and Key, J.L. 1990. Regulation of the heat shock response in soybean seedlings. *Plant Physiology* 94: 988–995.

Kumar, M., Rawat, P., Khan, S.Z., Dhamija, N., Chaudhary, P., Ravi, D.S., and Mitra, D. 2011. Reciprocal regulation of human immunodeficiency virus-1 gene expression and replication by heat shock proteins 40 and 70. *Journal of Molecular Biology* 410: 944–958.

Lee, G.J. and Vierling, E. 2000. A small heat shock protein cooperates with heat shock protein 70 systems to reactivate a heat-denatured protein. *Plant Physiology* 122: 189–198.

Lee, K.W., Cha, J.Y., Kim, K.H., Kim, Y.G., Lee, B.H., and Lee, S.H. 2012. Overexpression of alfalfa mitochondrial HSP23 in prokaryotic and eukaryotic model systems confers enhanced tolerance to salinity and arsenic stress. *Biotechnology Letters* 34: 167–174.

Lee, U., Rioflorido, I., Hong, S.W., Larkindale, J., Waters, E.R., and Vierling, E. 2007. The *Arabidopsis* ClpB/Hsp 100 family of proteins: Chaperones for stress and chloroplast development. *Plant Journal* 49: 115–127.

Li, J.W., Zhang, H., Hu, J.B., Liu, J.Q., and Liu, K.K. 2012. A heat shock protein gene, *CsHsp45.9*, involved in the response to diverse stresses in cucumber. *Biochemical Genetics* 50: 565–578.

Li, M., Guo, S., Xu, Y., Meng, Q., Li, G., and Yang, X. 2014. Glycine betaine-mediated potentiation of HSP gene expression involves calcium signaling pathways in tobacco exposed to NaCl stress. *Physiologia Plantarum* 150: 63–75.

Li, Z., Zhang, L., Wang, A., Xu, X., and Li, J. 2013. Ectopic overexpression of SlHsfA3, a heat stress transcription factor from tomato, confers increased thermotolerance and salt hypersensitivity in germination in transgenic *Arabidopsis*. *PLoS One* 8: e54880.

Liu, D., Zhang, X., Cheng, Y., Takano, T., and Liu, S. 2006. rHsp90 gene expression in response to several environmental stresses in rice (*Oryza sativa* L.). *Plant Physiology and Biochemistry* 44: 380–386.

Low, D., Brandle, K., Nover, L., and Forreiter, C. 2000. Cytosolic heat stress proteins Hsp17.7 class I and Hsp17.3 class II of tomato act as molecular chaperones *in vivo*. *Planta* 211: 575–582.

Luján, R., Lledías, F., Martínez, L.M., Barreto, R., Cassab, G.I., and Nieto-Sotelo, J. 2009. Small heat-shock proteins and leaf cooling capacity account for the unusual heat tolerance of the central spike leaves in *Agave tequilana* var. Weber. *Plant Cell & Environment* 32: 1791–1803.

Malik, M.K., Slovin, J.P., Hwang, C.H., and Zimmerman, J.L. 1999. Modified expression of a carrot small heat shock protein gene, HSP17.7, results in increased or decreased thermotolerance. *Plant Journal* 20: 89–99.

Maqbool, A., Abbas, W., Rao, A.Q., Irfan, M., Zahur, M., Bakhsh, A., Riazuddin, S., and Husnain, T. 2010. *Gossipium arboreum GHSP26* enhances drought tolerance in *Gossipium hirsutum*. *Biotechnology Progress* 26: 21–25.

Milioni, D. and Hatzopoulos, P. 1997. Genomic organization of HSP90 gene family in *Arabidopsis*. *Plant Molecular Biology* 35: 955–961.

Mogk, A., Deurling, E., Vorderfulbecke, S., Vierling, E., and Bukau, B. 2003. Small heat shock proteins, ClpB and the DnaK system form a functional triade in reversing protein aggregation. *Molecular Microbiology* 50: 585–595.

Montero-Barrientos, M., Hermosa, R., Cardoza, R.E., Gutiérrez, S., Nicolás, C., and Monte, E. 2010. Transgenic expression of the *Trichoderma harzianum hsp70* gene increases *Arabidopsis* resistance to heat and other abiotic stresses. *Journal of Plant Physiology* 167: 659–665.

Mujacic, M., Bader, M.W., and Baneyx, F. 2004. *Escherichia coli* Hsp31 functions as a holding chaperone that cooperates with the DnaK-DnaJ-GrpE system in the management of protein misfolding under severe stress conditions. *Molecular Microbiology* 51: 849–859.

Munns, R. 2002. Comparative physiology of salt and water stress. *Plant, Cell & Environment* 25: 239–250.

Munns, R. and Tester, M. 2008. Mechanisms of salinity tolerance. *Annual Review of Plant Biology* 59: 651–681.

Ono, K., Hibino, T., Kohinata, T., Suzuki, S., Tanaka, Y., Nakamura, T., Takabe, T., and Takabe, T. 2001. Overexpression of DnaK from a halotolerant cyanobacterium *Aphanothece halophytica* enhances the high-temperature tolerance of tobacco during germination and early growth. *Plant Science* 160: 455–461.

Ozturk, Z.N., Talame, V., Deyhoyos, M., Michalowski, C.B., Galbraith, D.W., Gozukirmizi, N., Tuberosa, R., and Bohnert, H.J. 2002. Monitoring large-scale changes in transcript abundance in drought- and salt-stressed barley. *Plant Molecular Biology* 48: 551–573.

Park, H.S., Cho, S.G., Kim, C.K., Hwang, H.S., Noh, K.T., Kim, M.S., Huh, S.H. et al. 2002. Heat shock protein hsp72 is a negative regulator of apoptosis signal-regulating kinase 1. *Molecular and Cellular Biology* 22: 7721–7730.

Parsell, D.A. and Lindquist, S. 1993. The function of heat-shock proteins in stress tolerance: Degradation and reactivation of damaged proteins. *Annual Review of Genetics* 27: 437–496.

Pegoraro, C., Mertz, L.M., Carlos da Maia, L., Rombaldi, C.V., and Costa de Oliveira, A. 2011. Importance of heat shock proteins in maize. *Journal of Crop Science Biotechnology* 14: 85–95.

Pérez-Salamó, I., Papdi, C., Rigó, G., Zsigmond, L., Vilela, B., Lumbreras, V., Nagy, I. et al. 2014. The heat shock factor A4A confers salt tolerance and is regulated by oxidative stress and the mitogen-activated protein kinases MPK3 and MPK6. *Plant Physiology* 165: 319–334.

Queitsch, C., Hong, S.W., Vierling, E., and Lindquist, S. 2000. Heat stress protein 101 plays a crucial role in thermotolerance in *Arabidopsis*. *Plant Cell* 12: 479–492.

Sakamoto, W. 2006. Protein degradation machineries in plastids. *Annual Review of Plant Biology* 57: 599–621.

Sarkar, N.K., Kundnani, P., and Grover, A. 2012. Functional analysis of Hsp70 superfamily proteins of rice (*Oryza sativa*). *Cell Stress and Chaperones* 18: 427–437.

Scharf, K.D., Berberich, T., Ebersberger, I., and Nover, L. 2012. The plant heat stress transcription factor (Hsf) family: Structure, function and evolution. *Biochimica et Biophysica Acta (BBA)—Gene Regulatory Mechanisms* 1819: 104–119.

Scharf, K.D., Siddique, M., and Vierling, E. 2001. The expanding family of *Arabidopsis thaliana* small heat stress proteins and a new family of proteins containing α-crystallin domains (ACD proteins). *Cell Stress Chaperones* 6: 225–237.

Seki, M., Narusaka, M., Abe, H., Kasuga, M., Yamaguchi-Shinozaki, K., Carninci, P., Hayashizaki, Y., and Shinozaki, K. 2001. Monitoring the expression pattern of 1,300 *Arabidopsis* genes under drought and cold stresses by using a full-length cDNA microarray. *Plant Cell* 13: 61–72.

Siddique, M., Gernhard, S., von Koskull-Doring, P., Vierling, E., and Scharf, K.D. 2008. The plant sHSP superfamily: Five new members in *Arabidopsis thaliana* with unexpected properties. *Cell Stress Chaperones* 13: 183–197.

Song, H.M., Fan, P.X., and Li, Y.X. 2009a. Overexpression of organellar and cytosolic AtHSP90 in *Arabidopsis thaliana* impairs plant tolerance to oxidative stress. *Plant Molecular Biology Reporter* 27: 342–349.

Song, H.M., Zhao, R.M., Fan, P.X., Wang, X.C., Chen, X.Y., and Li, Y.X. 2009b. Overexpression of *AtHsp90.2*, *AtHsp90.5* and *AtHsp90.7* in *Arabidopsis thaliana* enhances plant sensitivity to salt and drought stresses. *Planta* 229: 955–964.

Song, N.H. and Ahn, Y.J. 2010. DcHsp17. 7, a small heat shock protein from carrot, is upregulated under cold stress and enhances cold tolerance by functioning as a molecular chaperone. *HortScience* 45: 469–474.

Song, N.H. and Ahn, Y.J. 2011. DcHsp17.7, a small heat shock protein in carrot, is tissue-specifically expressed under salt stress and confers tolerance to salinity. *NewBiotechnology* 28: 698–704.

Sun, W., Bernard, C., van de Cotte, B., Van, M.M., and Verbruggen, N. 2001. At-HSP17.6A, encoding a small heat-shock protein in *Arabidopsis*, can enhance osmotolerance upon overexpression. *Plant Journal* 27: 407–415.

Sun, W., van Montagu, M., and Verbruggen, N. 2002. Small heat shock proteins and stress tolerance in plants. *Biochimica et Biophysica Acta* 1577: 1–9.

Sun, Y., Sheng, Y., Bai, L.X., Zhang, Y.J., Xiao, Y.F., Xiao, L.B., Tan, Y.G., and Shen, Y.M. 2014. Characterizing heat shock protein 90 gene of *Apolygus lucorum* (Meyer-Dür) and its expression in response to different temperature and pesticide stresses. *Cell Stress and Chaperones*. doi: 10.1007//s12192-014–0500-0.

Sung, D.Y. and Guy, C.L. 2003. Physiological and molecular assessment of altered expression of Hsc70-1 in *Arabidopsis*. Evidence for pleiotropic consequences. *Plant Physiology* 132: 979–987.

Taipale, M., Jarosz, D.F., and Lindquist, S. 2010. Hsp90 at the hub of protein homeostasis: Emerging mechanistic insights. *Nature Reviews Molecular Cell Biology* 11: 515–528.

Timperio, A.M., Egidi, M.G., and Zolla, L. 2008. Proteomics applied on plant abiotic stresses: Role of heat shock proteins (HSP). *Journal of Proteomics* 71: 391–411.

Tiroli, A.O. and Ramos, C.H.I. 2007. Chemical and biophysical characterization of small heat shock proteins from sugarcane. Involvement of a specific region located at the N-terminus with substrate specificity. *International Journal of Biochemistry & Cell Biology* 39: 818–831.

Ueda, W.S., Nakamura, T., and Takabe, T. 2002. Analysis of salt-inducible genes in barley roots by differential display. *Journal of Plant Research* 115: 119–130.

Vabulas, R.M., Ahmad-Nejad, P., da Costa, C., Miethke, T., Kirschning, C.J., Haucker, H., and Wagner, H. 2001. Endocytosed HSP60s use toll-like receptor 2 (TLR2) and TLR 4 to activate the toll/interleukin-1 receptor signaling pathway in innate immune cells. *Journal of Biological Chemistry* 276: 31332–31339.

Wagner, D., Schneider-Mergener, J., and Forreiter, C. 2005. Analysis of chaperone function and formation of hetero-oligomeric complexes of Hsp18.1 and Hsp17.7, representing two different cytoplasmic sHsp classes in *Pisum sativum*. *Journal of Plant Growth Regulation* 24: 226–237.

Walia, H., Wilson, C., Wahid, A., Condamine, P., Cui, X.P., and Close, T.J. 2006. Expression analysis of barley (*Hordeum vulgare* L.) during salinity stress. *Functional & Integrative Genomics* 6: 143–156.

Wang, H., Miyazaki, S., Kawai, K., Deyholos, M., Galbraith, D.W., and Bohnert, H.J. 2003. Temporal progression of gene expression responses to salt shock in maize roots. *Plant Molecular Biology* 52: 873–891.

Wang, W.X., Vinocur, B., Shoseyov, O., and Altman, A. 2004. Role of plant heat-shock proteins and molecular chaperones in the abiotic stress response. *Trends in Plant Science* 9: 244–252.

Wang, Y.F., Lin, S.K., Song, Q., Li, K., Tao, H., Huang, J., Chen, X.H., Que, S.F., and He, H.Q. 2014. Genome-wide identification of heat shock proteins (Hsps) and Hsp interactors in rice: Hsp70s as a case study. *BMC Genomics* 15: 344.

Waters, E.R., Lee, G.J., and Vierling, E. 1996. Evolution, structure and function of the small heat shock proteins in plants. *Journal of Experimental Botany* 47: 325–338.

Xiang, J., Ran, J., Zou, J., Zhou, X., Liu, A., Zhang, X., Peng, Y., Tang, N., Luo, G., and Chen, X. 2013. Heat shock factor OsHsfB2b negatively regulates drought and salt tolerance in rice. *Plant Cell Reports* 32: 1795–1806.

Xu, Q.H. and Qin, Y. 2012. Molecular cloning of heat shock protein 60 (PtHSP60) from *Portunus trituberculatus* and its expression response to salinity stress. *Cell Stress and Chaperones* 17: 589–601.

Xue, G.P., Sadat, S., Drenth, J., and McIntyre, C.L. 2014. The heat shock factor family from *Triticum aestivum* in response to heat and other major abiotic stresses and their role in regulation of heat shock protein genes. *Journal of Experimental Botany* 65: 539–557.

Xu, K. 2013. Molecular cloning, Expression and Salt tolerance analysis of *SeHSP90* gene from halophyte *Salicornia europaea*. Anhui Agricultural University, academic dissertation. ID: S10Q71010380.

Yamada, A., Sekiguchi, M., Mimura, T., and Ozeki, Y. 2002. The role of plant CCT a in salt- and osmotic-stress tolerance. *Plant Cell Physiology* 43: 1043–1048.

Yang, G.Y., Wang, Y.C., Zhang, K.M., and Gao, C.Q. 2014. Expression analysis of nine small heat shock protein genes from *Tamarix hispida* in response to different abiotic stresses and abscisic acid treatment. *Molecular Biology Reports* 41: 1279–1289.

Ye, S.F., Yu, S.W., Shu, L.B., Wu, J.H., Wu, A.Z., and Luo, L.J. 2012. Expression profile analysis of 9 heat shock protein genes throughout the life cycle and under abiotic stress in rice. *Chinese Science Bulletin* 57: 336–343.

Yokotani, N., Ichikawa, T., Kondou, Y., Matsui, M., Hirochika, H., Iwabuchi, M., and Oda, K. 2008. Expression of rice heat stress transcription factor OsHsfA2e enhances tolerance to environmental stresses in transgenic *Arabidopsis*. *Planta* 227: 957–967.

Zhang, H., Yang, J., Chen, Y., Mao, X., Wang, Z., and Li, C. 2013. Identification and expression analysis of the heat shock transcription factor (HSF) gene family in *Populus trichocarpa*. *Plant Omics Journal* 6: 415–424.

Zou, J., Chen, X.B., Liu, A.L., Gao, G.F., and Zhu, M.L. 2007. Plant heat shock proteins and crop abiotic stress tolerance improvement. *Plant Physiology Communications* 43: 981–985.

Zou, J., Liu, C., Liu, A., Zou, D., and Chen, X. 2012. Overexpression of OsHsp17.0 and OsHsp23.7 enhances drought and salt tolerance in rice. *Journal of Plant Physiology* 169: 628–635.

CHAPTER EIGHT

# Transcription Factors and Salt Stress Tolerance in Plants

*Shanlin Yu, Na Chen, Xiaoyuan Chi, Lijuan Pan, Mingna Chen,*
*Tong Wang, Chuantang Wang, Zhen Yang, and Mian Wang*

## CONTENTS

*Abstract.* Salt stress is a major environmental factor that adversely affects plant growth, development and crop yields. During the response and adaptation to salt stress, there are many changes in biochemical and physiological processes. Many genes are activated, leading to accumulation of numerous proteins involved in resistance to salt stress. The expression of stress-induced genes is largely regulated by specific transcription factors (TFs). Typically, the TFs are capable of activating or repressing transcription of multiple target genes. So far, various TFs and *cis*-acting elements contained in stress-responsive promoters have been described. These TFs and *cis*-motifs function not only as molecular switches for gene expression but also as terminal points of signal transduction in the signaling processes. In this section, we summarize the research progress of major TFs involved in plant salt stress regulation. We focus on the function and mechanism of the TFs including ERF/AP2, MYB, and NAC.

## 8.1 INTRODUCTION

Salt stress afflicts plant agriculture in many parts of the world, particularly irrigated land (Epstein et al. 1980). Crop growth and productivity are often compromised by salt stress, which results in a range of morphological, physiological, biochemical, and molecular changes in the plant (Bhatnagar-Mathur et al. 2008; Sakamoto et al. 2004). The perception of salt stress operates through various sensors, which initiate a cascade of transcription, consequently leading to the production of protective proteins and metabolites (Bartels and Sunkar 2005). Transcription factors (TFs) are important components for regulating salt-responsive genes. (Nakashima et al.

2009) Large numbers of TFs that include activators, coactivators and suppressors have already been identified. Among them, TFs play critical roles in plant responses to salt stress via transcriptional regulation of the downstream genes responsible for plant tolerance to salt challenges. They constitute a redundant family of transcriptional regulators in plants with 134 members in the *Arabidopsis* genome, 94 for indica rice genome, and 113 in japonica (Kiełbowicz -Matuk 2012). Many TFs are related proteins that share the homologous DNA-binding domain and are classified in families based on their DNA-binding domains, such as the MYB-like proteins (containing helix–turn–helix (HTH) motifs), the MADS domain proteins, the homeobox proteins, the basic region/leucine zipper (bZIP) proteins or the zinc finger proteins (ZFPs) (Fujimoto et al. 2000).

Previous studies have revealed some key components that control and modulate salt stress adaptive pathways that include TFs ranging from bZIP, AP2/ERF, and MYB proteins to general TFs. (Hu et al. 2008; Takasaki et al. 2010). In this review, we focus on recent advances in salt stress-related TFs and TF-based engineering of increased salt adaptation.

## 8.2 AP2/ERF FAMILY TFs AND SALT STRESS

The APETALA2/ethylene-responsive element binding factor (AP2/ERF) family is a large group of plant-specific TFs containing AP2/ERF-type DNA-binding domains, and family members are encoded by 145 loci in *Arabidopsis* (Sakuma et al. 2002) and 170 loci in rice (Muhammad et al. 2012). This domain was first found in the *Arabidopsis* homeotic gene APETALA 2 (Jofuku et al. 1994), and a similar domain was found in tobacco (*Nicotiana tabacum*), ethylene-responsive element binding proteins (Ohme-Takagi and Shinshi 1995). These domains, which consist of approximately 60 residues, are closely related (Weigel 1995). AP2/ERF family proteins are plant-specific TFs, and the lowest plant in which an AP2/ERF family protein has been found is the green alga *Chlamydomonas reinhardtii* (Shigyo et al. 2006). Sequences that are homologous to the AP2/ERF domain have been found in bacterial and viral endonucleases, and thus, it has been hypothesized that the AP2/ERF domain was transferred from a cyanobacterium by endosymbiosis or from a bacterium or a virus by other lateral gene transfer events (Magnani et al. 2004).

Sakuma et al. (2002) classified AP2/ERF TFs into five subfamilies: AP2 (APETALA2), RAV (related to ABI3/VP1), dehydration-responsive element binding protein (DREB), ERF (ethylene-responsive factor), and others, according to the number and similarity of the DNA-binding domains. Members of the AP2 subfamily contain double AP2/ERF domain, and members of the RAV subfamily contain an AP2/ERF domain and an additional B3 DNA-binding domain, while the other members contain only a single AP2/ERF domain (Riechmann et al. 2000; Sakuma et al. 2002). The AP2 subfamily in seed plants can be further classified into the AP2 group and the ANT group (Shigyo and Ito 2004). The DREB and ERF subfamilies were both divided into six subgroups, A-1 to A-6 and B-1 to B-6, respectively. DREB1/CBFs belong to subgroup A-1, and DREB2s belong to subgroup A-2. Three of four "other" members are similar to AP2 subfamily members and seem to have an atypical second AP2/ERF domain, while the fourth "other" protein (AT4G13040) has a unique type of AP2/ERF domain (Sakuma et al. 2002).

Despite the relatively high sequence conservation of the AP2/ERF domain, the number of DNA elements bound by different AP2/ERF TFs is extremely wide (Masaki et al. 2005; Ohme-Takagi and Shinshi 1995; Welsch et al. 2007). The DREB group recognizes the dehydration-responsive element (DRE) with a core motif of A/GCCGAC (Jiang et al. 1996), while the ERF group typically binds to the cis-acting element AGCCGCC, known as the GCC box (Ohme-Takagi and Shinshi 1995). In the AP2 family, a binding element gCAC(A/G) N(A/T)TcCC(a/g)ANG(c/t) has been reported for the protein AINTEGUMENTA (ANT) (Nole-Wilson and Krizek 2000), and the same sequence was shown to be bound by five other AP2 members (Gong et al. 2007). This DNA element is quite distinct from the consensus sequence CCGA/CC bound by the DREB/ERF group and, surprisingly, is not composed of two similar half sites, since the AP2 proteins contain a double AP2 domain (Krizek 2003). Members of RAV subfamily contain an AP2/ERF domain in C-terminal regions and a B3-domain in N-terminal regions. Pepper CARAV1 could recognize and bind to CAACA and CACCTG motifs and activate the reporter gene in yeast (Sohn et al. 2006).

A variety of AP2/ERF genes were successfully identified and investigated in angiosperms (including both monocotyledonous and dicotyledonous types), gymnosperms, and microorganisms (encompassing cyanobacteria, ciliates, and viruses) (Magnani et al. 2004; Shigyo et al. 2006; Xu et al. 2008b). For example, the genome-wide analysis identified at least 145,149 and 170 AP2/ERF genes in *Arabidopsis*, grape, and rice, respectively (Licausi et al. 2010; Muhammad et al. 2012; Sakuma et al. 2002). In *Arabidopsis*, expression of the AtDREB1/CBF genes is quickly and transiently induced by cold, but not by dehydration and high-salt stresses (Liu et al. 1998; Shinwari et al. 1998). In contrast, the *AtDREB2* genes are induced by dehydration and high-salt stresses, but not by cold stress (Liu et al. 1998; Nakashima et al. 2000). Later, three novel DREB1/CBF-related and six novel DREB2-related genes in the *Arabidopsis* genome were reported, but their expression levels were low under the stress conditions (Sakuma et al. 2002). The AtDREB1A, AtDREB1B, and AtDREB1C proteins are major TFs in cold-inducible gene expression, and AtDREB2A and AtDREB2B proteins are TFs in high-salt- and drought-inducible gene expression in *Arabidopsis*, respectively. However, one of the CBF/DREB1 genes, AtCBF4/DREB1D, was weakly induced by osmotic stress, suggesting the existence of cross talk between the CBF/DREB1 and the DREB2 pathways (Haake et al. 2002).

Eighty-five ERF genes have been identified from tomato. Among them, SlERF68 was upregulated ≥17-fold during salt and oxidation stress; however, SlERF80 was found to upregulate ≥400-fold during salt stress. Out of 10 upregulated genes and three downregulated genes responsive to salt stress, five and two genes, respectively, belonged to DREB family (Sharma et al. 2010). In rice, OsDREB1A was induced within 5 h after salt treatment. OsDREB2A was induced within 24 h after dehydration and salt stress, and it responded faintly to abscisic acid (ABA) and cold stress (Dubouzet et al. 2003). Similarly, PgDREB2A transcripts were also induced by cold, drought, and salt stresses (Huang et al. 2004). A DRE binding TF, AhDREB1 from a halophyte *Atriplex hortensis*, was strongly expressed by 200 mM NaCl in roots (Shen et al. 2003). In hot pepper, CaDREBLP1 was rapidly induced by dehydration, high salinity, and to a lesser extent mechanical wounding but not by cold stress (Hong and Kim 2005). In *Glycine max* (L.) Merr,

the transcriptions of GmDREBa and GmDREBb were induced by salt, drought, and cold stresses in leaves of soybean seedlings. The expression of GmDREBc was not significantly affected in leaves but apparently induced in roots by salt, drought, and ABA treatments (Li et al. 2005). CaDREBLP1 was found to be induced by water-deficit and salt stresses in hot pepper, more like the expression patterns of DREB2-type genes (Hong and Kim 2005). Moss PpDREB1 and *Caragana korshinskii* CkDREB were induced by drought, high-salt, and low-temperature stresses and by ABA treatment (Liu et al. 2007; Wang et al. 2011). Expression of JERF3 in tomato is mainly induced by ethylene (ET), jasmonic acid (JA), cold, salt, or ABA (Wang et al. 2004). Expression of chickpea CAP2 was induced by dehydration and by treatment by NaCl, ABA, and auxin (Shukla et al. 2006). Transcription of the wheat TaERF1 gene was induced not only by drought, salinity, and low-temperature stresses and exogenous ABA, ET and salicylic acid (SA) but also by infection with *Blumeria graminis* f. sp. Tritici (Xu et al. 2007b). Expression of soybean GmERF3 was induced by treatments with high salinity, drought, ABA, SA, JA, ET, and soybean mosaic virus (Zhang et al. 2009). The expression of *Arabidopsis* RAP2.6 is responsive to ABA and different stress conditions such as high salt, osmotic stress, and cold (Zhu et al. 2010). Expression of dwarf apple MbDREB1 was induced by cold, drought, and salt stress and also in response to exogenous ABA (Yang et al. 2011). Transcript level of AtERF71/HRE2 was highly increased by anoxia, NaCl, mannitol, ABA, and MV treatments (Park et al. 2011). Expression of tomato SlERF5 was induced by abiotic stress, such as high salinity, drought, flooding, wounding, and cold temperatures (Pan et al. 2012). In peanut, six ERF family TF genes (AhERF1–6) were cloned, and their expression levels during abiotic stress in peanuts were analyzed (Chen et al. 2012c). The expression of AhERF4 gene increased soon after salinity treatment began and exhibited an increase of nearly 150-fold after the roots were treated for 3 h. The transcripts of AhERF6 accumulated dramatically under salt stress, with the highest abundance at 48 h in the leaves and 12 h in the roots, and the greatest increases were about 50-fold and 18-fold, respectively. Wan et al. (2014) identifies 63 ERF unigenes from peanut. The expression analysis revealed that of the 63 genes, 41 genes were related to high salt.

DREBs and ERFs are two major subfamilies of the AP2/ERF family and play crucial roles in the regulation of abiotic- and biotic-stress responses. DREBs show variation in some conserved motifs and biological functions in divergent species and are dichotomized as DREB1 and DREB2 types, which are involved in separate signal transduction pathways under abiotic stresses (Dubouzet et al. 2003). Expression of DREB1-type genes was specifically induced by low-temperature stress in *Arabidopsis* and rice (Dubouzet et al. 2003; Liu et al. 1998; Lucas et al. 2011; Sakuma et al. 2002). In contrast, DREB2-type genes respond to dehydration and high-salt stresses (Liu et al. 1998; Dubouzet et al. 2003). Many DREB1- or DREB2-type genes inserted into plants by transformation were capable of improving multiple abiotic tolerances in agricultural crops. Overexpression of *AtDREB1A* in rice plants resulted in improved tolerance to drought and salinity (Oh et al. 2005). Expression of a sweet cherry DREB1/CBF ortholog, PaCIG-B, in *Arabidopsis* confers salt and freezing tolerance (Kitashiba et al. 2004). Overexpression of rice *OsDREB1A*, *OsDREB1B*, *OsDREB1D*, and *OsDREB1F* or any one of the three *Arabidopsis* DREB1s increases salt, drought, or low-temperature tolerance in *Arabidopsis* or rice (Dubouzet et al. 2003; Ito et al. 2006; Wang et al. 2008; Zhang et al. 2009b). Constitutive overexpression of *OsDREB1B* in transgenic tobacco shows no negative effect on plant growth. Transgenic tobacco plants show higher biomass accumulation and rapid root elongation under high salt stress (Gutha and Reddy 2008). Overexpression of cotton GhDREB in wheat resulted in improved tolerance to drought, high-salt, and freezing stresses through accumulating higher levels of soluble sugar and chlorophyll in leaves after stress treatments (Gao et al. 2009). Overexpression of *Caragana korshinskii CkDBF* in transgenic tobacco plants resulted in higher tolerance to high salinity and osmotic stresses and induction of a downstream target gene under normal conditions (Wang et al. 2010). Compared with wild-type (WT) plants, transgenic *Arabidopsis* overexpressing dwarf apple MbDREB1 showed increased tolerance to low-temperature, drought, and salt stresses (Yang et al. 2011). Other DREB subgroup genes have also been reported to be stress responsive and/or to confer stress tolerance in transgenic plants. Overexpression of HARDY, a subgroup A-4 AP2/ERF gene from *Arabidopsis*, improves drought and salt tolerance in transgenic *Trifolium alexandrinum* L. (Abogadallah

et al. 2011). Overexpression of PpDBF1 (A-5) gene from the moss *Physcomitrella patens* induces expression of stress-responsive genes and reduces sensitivity to high salinity in transgenic tobacco (Liu et al. 2007). Overexpression of the chickpea *CAP2* gene in transgenic tobacco enhanced growth and tolerance to dehydration and salt stress (Shukla et al. 2006).

Many ERFs are known to be involved in responses to pathogens. However, some genes belonging to the ERF subfamily have also been shown to play a role in abiotic stress acclimation of plants. Overexpression of ERF subfamily genes such as soybean GmERF3, tomato JERF3 and TERF1, tobacco Tsi1 and OPBP1, and *Arabidopsis* AtERF71/HRE2 can confer salt tolerance in transgenic plants (Park et al. 2001, 2011; Guo et al. 2004; Huang et al. 2004; Wang et al. 2004; Zhang et al., 2009). Overexpression of the SlERF5 gene of tomato and the BrERF4 gene of Chinese cabbage enhanced salt and drought tolerance of transgenic plants (Pan et al. 2012; Seo et al. 2010). Overexpression of TaERF1 and LeERF3b improves abiotic stress (cold, drought, and salt) tolerance in transgenic plants (Chen et al. 2008; Xu et al. 2007). Overexpression of AhERF019 enhanced tolerance to drought, heat, and salt stresses in *Arabidopsis* (Wan et al. 2014). Overexpression of RAP2.6 confers hypersensitivity to exogenous ABA and abiotic stresses during seed germination and early seedling growth in *Arabidopsis*. The ABA content in RAP2.6 overexpressor lines decreased after being treated with salt (Zhu et al. 2010).

Some studies showed that AP2 or RAV subfamily might be also involved in stress responses. *Brassica napus* RAV-1-HY15 is induced by cold, NaCl, and PEG treatments (Zhuang et al. 2011). Overexpression of the *CARAV1* gene in transgenic *Arabidopsis* plants induced some pathogenesis-related (PR) genes and enhanced resistance against infection by *Pseudomonas syringae* and osmotic stresses by high salinity and dehydration (Sohn et al. 2006). In addition, Lee et al. (2010a) identified that pepper CaRAV1 protein physically interacted with the oxidoreductase CaOXR1. Overexpression of *CaRAV1* and *CaOXR1/CaRAV1* conferred resistance to the biotrophic oomycete *Hyaloperonospora arabidopsidis* infection and high tolerance to high salinity and osmotic stress in *Arabidopsis*.

Various TFs, including AP2/ERFs, have been reported to engage in ABA-mediated gene expression (Fujita et al. 2011). Recently, DREB1A/CBF3, DREB2A, and DREB2C proteins have been reported to physically interact with ABA-responsive element binding protein (AREB)/ABA-responsive element binding factor (ABF) proteins (Lee et al. 2010b), which supports the view that DREB/CBFs and AREB/ABFs may interact to control ABA-regulated gene expression. DREB genes display responses to exogenous ABA, such as barley HvDRF1 and wheat TaDREB2 and TaAIDFa (Egawa et al. 2006; Xu et al. 2008c; Xue and Loveridge 2004). ABA was shown to be involved in the regulation of DREB activity and increased promoter activity (Kizis and Pages 2002; Xu et al. 2008c). In addition, several DREBs have been reported to be positive and negative mediators of ABA and sugar responses, mainly during germination and the early seedling stage (Fujita et al. 2011). Recently, it was demonstrated that the CE1-like CACCG motif in the ABI5 promoter is required for ABI4 transactivation in Arabidopsis protoplasts (Bossi et al. 2009). Therefore, ABA may play an important role in regulating the transcriptional activity of DREBs. ERF genes integrate different pathogen and disease-related stimulus (i.e., ET, JA, and SA) signal pathways, such as ET and JA pathways (Gutterson and Reuber 2004). In addition, some ERFs were activated by ABA (Lee et al. 2005; Qin et al. 2004; Wang et al. 2004; Zhang et al. 2004, 2010), indicating a cross talk pathway between abiotic and biotic stress responses, although interactions are antagonistic between the ABA, SA, and the JA/ET signaling pathways. Some studies showed that AP2 subfamily was responsive to ABA. ADAP is considered to be a positive regulator of ABA responses, because knockout mutant plants show partial insensitivity to ABA and decreased drought tolerance (Lee et al. 2009). CHOTTO1 is probably involved in the ABA-dependent repression of gibberellic acid (GA) biosynthesis (Yano et al. 2009) as well as sugar and nitrate responses in an ABI4-dependent manner during seed germination in Arabidopsis (Yamagishi et al. 2009). ABI3/VP1, related to RAV subfamily, mediates ABA responses and is involved in regulation of seed maturation and ABA-regulated gene expression in plant seeds (Sakata et al. 2010; Zeng et al. 2003).

Transcriptional and posttranscriptional regulation plays an essential role in the adaptation of cellular functions to the environmental changes (Saibo et al. 2009; Walley and Dehesh 2010). For example, the activity of DREB1s/CBFs is mainly regulated at the transcriptional level. In contrast, the activity of DREB-2-type TFs was reported to be subject to posttranscriptional modification,

phosphorylation responses, or alternative splicing (Mizoi et al. 2012; Xu et al. 2011).

At present, AP2/ERFs-mediated signal transduction mechanism and pathway is not clear. Some AP2/ERFs regulate gene expression indirectly by either interacting physically with other proteins or activating TFs. DBF1 interacted with DBF1-interactor protein 1, a protein that contains a conserved R3H single-strand DNA-binding domain, enhances maize rab17 promoter activity, and controls the levels of target gene expression during stress conditions (Saleh et al. 2006). Interaction of grape VvDREB with ASR transcription factor VvMSA in the nucleus regulating the expression of a glucose transporter (Saumonneau et al. 2008). ORC1 and a basic helix–loop–helix (bHLH) factor cooperatively mediate JA-elicited nicotine biosynthesis in tobacco (De Boer et al. 2011). Schramm et al. (2008) found that overexpression of DREB2A directly regulated the expression of heat-shock TF HsfA3 that activated expression of many heat-inducible genes in the transcriptional cascade.

As a whole, AP2/ERF family TFs, by cross-talking with each other, are likely to regulate the developmental, physiological, and biochemical responses of plants to a variety of environmental stress conditions, including those occurring in combination with other abiotic and biotic stresses. It has been possible to engineer stress tolerance in transgenic plants by manipulating the expression of AP2/ERFs. This opens an excellent opportunity to develop stress-tolerant crops in the future. However, further investigation is required to understand their regulation and functional interactions, together with identification of new factors in more plant species.

## 8.3 MYB TFs AND SALT STRESS

The MYB family of proteins is large, functionally diverse, and represented in all eukaryotes. Most MYB proteins function as TFs with varying numbers of MYB domain repeats conferring their ability to bind DNA. Members of this family function in a variety of plant-specific processes, and they are proved to be key factors in regulatory networks controlling abiotic stresses including salt stress.

MYB proteins are characterized by a highly conserved DNA-binding domain: the MYB domain. This domain generally consists of one to four or more imperfect MYB repeats (R), which can function synergistically or individually in DNA binding and protein–protein interactions, respectively. Each MYB repeat is approximately 52 amino acids (AAs) long and contains three regularly spaced tryptophan (or aliphatic) residues that together form a hydrophobic core (Kanei-Ishii et al. 1990). Each MYB repeat forms three α–helices (Ogata et al. 1996). The second and third helices of each repeat build a HTH structure with three regularly spaced tryptophan (or hydrophobic) residues, forming a hydrophobic core in the 3D HTH structure (Ogata et al. 1996). The third helix of each repeat is the "recognition helix" that makes direct contact with DNA and intercalates in the major groove (Jia et al. 2004). During DNA contact, two MYB repeats are closely packed in the major groove, so that the two recognition helices bind cooperatively to the specific DNA sequence motif.

The MYB family is divided into different types according to the number of MYB repeat(s) (one, two, three, or four): 4R-MYB has four repeats, 3R-MYB (R1R2R3-MYB) has three consecutive repeats, R2R3-MYB has two repeats, and the MYB-related type usually, but not always, has a single MYB repeat (Dubos et al. 2010; Jin and Martin 1999; Rosinski and Atchley 1998). All four classes are found in plants, representing the taxon with the highest diversity of MYB proteins.

Many MYB family proteins are involved in unfavorable environmental response, and some of which are proved to be related with plant salt stress regulation. A genome-wide analysis identified at least 155 and 197 MYB genes in rice and Arabidopsis, respectively (Katiyar et al. 2012). Expression analysis indicated that 14 (9.03%) OsMYB genes were upregulated under salt stress in rice, and 56.85% AtMYB genes were downregulated, and 35.02% AtMYB genes were upregulated in salt stress, respectively (Katiyar et al. 2012). Zhang et al. (2011a) identified 60 MYB genes in wheat. Expression analysis indicated that 16 genes were responded to salt stress, among which, 14 were upregulated and two were downregulated. Liao et al. (2008b) identified 156 GmMYB genes in soyabean (G. max) of which the expression of 43 genes changed on treatment with ABA, salt, drought, and/or cold stress. Chen et al. (2014) identified 30 MYB genes of peanut and found the expression of eight genes was induced during salt stress in peanut roots. Many MYB genes in other plants were also responded to salt stress such as BcMYB1, ScMYBAS1, and MdSIMYB1 (Chen et al. 2005; Prabu and Theertha 2012; Wang et al. 2014a). Most MYB genes responded to salt stress was R2R3 MYB members. MYB3R proteins constitute a rather small subfamily in plants. So far, only a few plant MYB3R genes including OsMYB3R-2 and TaMYB3R1 have been identified to be involved in plant salt stress regulation (Cai et al. 2011; Dai et al. 2007). MYB-related MYB proteins form a relatively larger subfamily in plant; however, little was known about the function of this subfamily in plant salt stress regulation. Arabidopsis MYB-related proteins AtMYB2 (having single MYB repeat) and AtMYBL (having two MYB repeats) are both salt stress inducible in expression. Recently, expression of several MYB-related genes in wheat and peanut was proved to be induced under salt stress (Chen et al. 2014; Zhang et al. 2011b).

Function analysis proved some MYB proteins could influence the salt resistance of transgenic plant. Most of functional MYB proteins involved in plant salt stress are R2R3 type. Arabidopsis R2R3-MYB genes such as AtMYB44/AtMYBR1, AtMYB41, AtMYB15, and AtMYB20 were all responded to salinity stress in mRNA level, and overexpression of these genes could enhance plant resistance to salt stress in Arabidopsis (Cui et al. 2013; Ding et al. 2009; Jung et al. 2008; Lippold et al. 2009). Expression of R2R3 MYB genes TaMYB32, TaMYB33, TaMYB56-B, and TaMYB73 were both upregulated by salt stress in wheat, and the overexpressing Arabidopsis plants were more tolerant to salt stress than wild-type plants (Qin et al. 2012; Zhang et al. 2011a). Overexpression of other R2R3 MYB proteins such as OsMYB2 of rice, MdSIMYB1 of apples, PScMYBAS1 of sugarcane, and PeMYB2 of Phyllostachys edulis could all enhance salt stress tolerance in transgenic plants (Prabu and Prasad 2012; Wang et al. 2014; Xiao et al. 2013; Yang et al. 2012). One 3R MYB OsMYB3R-2 was proved to have functions in rice salt stress regulation (Dai et al. 2007).

Several MYB-related proteins were also involved in plant salt stress regulation. An R-R-type MYB-related protein gene AtMYBL was activated during salt stress in Arabidopsis, and overexpressing this gene can improve seed germination rate under saline stress (Zhang et al. 2011b). AmMYB1 is a single-repeat MYB TF in Avicennia marina, and its transcript level increases after NaCl and ABA treatments in the leaves. AmMYB1 overexpression confers better salt tolerance in transgenic tobacco plants (Ganesan et al. 2012). Sheep grass R1-MYB TF gene LcMYB1 was induced by salt stress, and overexpression of this gene confers salt tolerance in transgenic *Arabidopsis* (Cheng et al. 2013b). Expression of *OsMYB48-1*, a MYB-related protein gene of rice, was slightly induced by high salinity treatment. Overexpression of *OsMYB48-1* in rice significantly improved tolerance to simulated salinity stress (Xiong et al. 2014). In contrast to the R2R3-MYB genes, most MYB-related genes have been studied functionally to be involved in the maintenance of circadian rhythms (Kuno et al. 2003; Wang and Tobin 1998) and control of cellular morphogenesis (Kirik et al. 2004; Schellmann et al. 2002). Members of this type have attracted little attention concerning their roles in plant abiotic stress regulation. As a larger subfamily, more MYB-related members may be proved to participate in salt stress regulation of plants in the future.

Some MYB proteins may negatively regulate salt resistance in plant. For example, the expression of *AtMyb73* was upregulated by salt stress but not by other stresses in *Arabidopsis*; however, *atmyb73* ko mutant plants exhibited higher survival rates compared to WT (Col-0) plants, which indicated that AtMyb73 is a negative regulator in response to salt stress in *Arabidopsis thaliana* (Kim et al. 2013).

Despite the similar AA sequences, the function mechanisms of MYB proteins in plant salt stress regulation are not exactly the same. Some MYB members such as AtMYB44, CmMYB2, and TaMYB33 exercise their functions through ABA-dependent pathway (Jung et al. 2008; Qin et al. 2012; Shan et al. 2012). For example, transgenic *Arabidopsis* overexpressing *AtMYB44* and *AtMYB15* is more sensitive to ABA and has a more rapid ABA-induced stomatal closure response than the WT (Ding et al. 2009; Jung et al. 2008). AtMYB44 and AtMYB20 down-regulate the expression of PP2Cs, the negative regulator of ABA signaling, and enhance salt tolerance in *Arabidopsis* (Cui et al. 2013; Jung et al. 2008). The upregulation of *AtAAO3* along with the downregulation of *AtABF3*, *AtABI1*, and *TaMYB33* in the overexpression *Arabidopsis* indicated that ABA synthesis was elevated while its signaling was restricted, which suggests that TaMYB33 enhances salt tolerance partially through superior ability for osmotic balance reconstruction and reactive oxygen species (ROS) detoxification (Qin et al. 2012). Other R2R3 MYB members such as OsMYB2 of rice and

CmMYB2 of chrysanthemum all play their functions in ABA-dependent signaling pathway during salt stress regulation (Shan et al. 2012; Yang et al. 2012). The MYB-related OsMYB48-1 also exercises its function through ABA-dependent signaling pathway (Xiong et al. 2014). Some MYB proteins play their roles through ABA-independent pathway. They may activate some abiotic stress-related genes downstream of DREB/CBF signaling pathway. The expression of some cold-stress-responsive genes, such as *DREB1A/CBF3* and *COR15a*, was found to be elevated in the TaMYB56-B-overexpressing *Arabidopsis* plants compared to the WT, which indicated that TaMYB56-B may act in ABA-independent pathway in plant stress response (Zhang et al. 2012). Overexpression of *Phyllostachys edulis PeMYB2* in *Arabidopsis* could maintain higher survival rate of transgenic plants during salt stress. Expression analysis of saline stress response marker genes in the transgenic and WT plants under the salt stress condition showed that PeMYB2 regulated the expression of *NXH1*, salt overly sensitive 1 (*SOS1*), *RD29A*, and *COR15A* (Xiao et al. 2013). Several MYB proteins play their functions through both ABA-dependent and ABA-independent pathways. For example, soybean GmMYB76 and GmMYB177 could both promote salt stress tolerance in transgenic *Arabidopsis*. Plants overexpressing these three genes exhibited reduced ABA sensitivity, suggesting that the three GmMYB genes could function as negative regulators of ABA signaling. At the same time, both the two GmMYB genes promoted the expression of *DREB2A*, *RD17*, *P5CS*, *rd29B*, *ERD10*, and *COR78* genes (Liao et al. 2008b). *Arabidopsis* plants overexpressing *TaMYB73* had superior germination ability under NaCl and ABA treatments. The expression of signalling genes including both *AtCBF3* and *AtABF3*, suggesting TaMYB73 is most likely to be a secondary player in the regulation of stress-responsive genes in ABA-dependent and ABA-independent pathways (He et al. 2012). A 3R-MYB OsMYB3R-2 of rice was also indicated to be involved in plant salt tolerance regulation through both ABA-dependent and ABA-independent signaling pathways (Dai et al. 2007).

Other mechanisms also existed because of the multiple roles of MYB family proteins. Analysis in *atmyb73* ko mutant plants determined that the accumulation of (SOS) transcripts, *SOS1* and *SOS3*, was higher in *atmyb73* ko and *atmyb73* eko plants than in WT plants in response to 300mM NaCl treatment, which indicate that AtMyb73 is a negative regulator of SOS induction in response to salt stress in *A. thaliana* (Kim et al. 2013). The overexpression of sheep grass *LcMYB1* enhanced the expression levels of P5CS1 but inhibited other salt stress response gene markers. These findings demonstrate that LcMYB1 influences the intricate salt stress response signaling networks by promoting different pathways than the classical DREB1A-mediated signaling pathway (Cheng et al. 2013b). Overexpressing *AmMYB1* of mangrove tree in tobacco could enhance salt stress tolerance by reducing chlorosis in leaf discs and symptoms of salt stress like wilting and yellowing in whole plants (Ganesan et al. 2012). Zhang et al. (2011b) reported that AtMYBL functions in the leaf senescence process and thus modulates an abiotic stress response in *Arabidopsis*. Lippold et al. (2009) indicated that *AtMyb41* was induced by osmotic stress in an ABA-dependent manner, and it is involved in distinct cellular processes, including control of primary metabolism and negative regulation of short-term transcriptional responses to osmotic stress. Cheng et al. (2013c) reported that *IbMYB1* transgenic potato plants showed high DDPH radical scavenging activity and high PS II photochemical efficiency compared with the control line after treatment of 400 mM NaCl, which indicated enhanced IbMYB1 expression affects secondary metabolism and then leads to improved tolerance ability in transgenic plants. Microarray analysis showed that the overexpression of *JAmyb* stimulates the expression of several defense-associated genes and several TFs involved in the JA-mediated stress response are also regulated by JAmyb, which suggest that JAmyb plays a role in JA-mediated abiotic stress response in rice (Yokotani et al. 2013).

Several studies were also focused on the regulatory factors of MYB proteins during abiotic stress. AtMYB44 is a substrate protein of the rapidly stress-induced mitogen-activated protein kinase MPK3. Recently, AtMYB44 was characterized to be phosphorylation-dependent positive regulator in salt stress signaling. A mechanistic model of MAPK-MYB-mediated enhancement in the antioxidative capacity and stress tolerance is proposed (Persak and Pitzschke 2014). Prabu and Prasad (2012) analyzed the promoter flanking the 50 ScMYBAS1 coding region using the

uidA reporter gene. The results indicated that the promoter could respond to dehydration, salinity, cold, wounding, GA, SA, and methyl jasmonic acid, which suggested ScMYBAS1 may mediate multiple signaling transduction pathways. Nagaoka and Takano (2003) found that salt-tolerance-related protein STO binds to a Myb TF homologue and confers salt tolerance in *Arabidopsis*.

As a whole, MYB proteins play an important role in controlling various plant processes, and new insights have been obtained into the mechanisms that control MYB protein activities and gene expression profiles, and several target genes have been determined. However, because of the complicated functions of MYB proteins, a lot of work is required to fully characterize the roles of all MYB proteins in regulatory networks and inferring functions in more plant species. Once more data are available, it will be interesting to establish how the control of specific target genes relates to the salt stress regulation that MYB factors control.

## 8.4 NAC (NAM-ATAF1, 2-CUC2) TFs INVOLVED IN PLANT SALT STRESS

The major NAC pathway is active in response to abiotic stress that have been identified and well elucidated in *A. thaliana* and rice (Nakashima et al. 2009; Yamaguchi-Shinozaki and Shinozaki 2005, 2006). The NAC TF family is widely distributed in plants, but so far has not been found in other eukaryotes (Olsen et al. 2005).

Large-scale abiotic stress-responsive expression analysis indicated that NAC family proteins may have important functions in plant salt stress acclimation. For example, out of 88 NAC transcription members in genome of *Medicago truncatula*, 36 members were upregulated in roots during salt stress treatment (Li et al. 2009). A microarray analysis of the root transcriptome following NaCl exposure detected that 23 ANAC genes were induced and seven genes were reduced, respectively, by a threshold of twofold (Jiang and Deyholos 2006). In crops, such as rice, Fang et al. (2008) systematically analyzed the NAC family and identified 140 putative ONAC or ONAC-like TFs, among which 19 genes were upregulated by salt stress. Computational prediction assumed that there are at least 205 NAC or NAC-like TFs members in soybean, among which 8 genes were characterized to be induced by high salinity (Mochida et al. 2009; Pinheiro et al. 2009; Tran et al. 2009).

Functional analysis revealed potentials of NAC in improvement of plant salt stress tolerance. NAC TFs enhance stress tolerance in the model plant *Arabidopsis*. The expression of *Arabidopsis AtNAC2* was induced by salt stress, and this induction required ET and auxin signaling pathway. Overexpression of *AtNAC2* could maintain the number of lateral roots in transgenic lines during salt stress, which indicate that AtNAC2/ANAC092 may be a TF incorporating the environmental and endogenous stimuli into the process of plant lateral root development (He et al. 2005). Recently, *ANAC092* gene also demonstrated an intricate overlap of ANAC092-mediated gene regulatory networks during salt-promoted senescence and seed maturation (Balazadeh et al. 2010). Trangenic *Arabidopsis* overexpressing a salt-inducible rice NAC gene, the *ONAC063*, showed enhanced tolerance to high salinity and osmotic pressure by similar mechanisms as ONAC063-upregulated genes were almost similar to those upregulated by ANAC019, ANAC055, or ANAC072 (Yokotani et al. 2009). Overexpression of *TaNAC2* and *TaNAC67* resulted in pronounced enhanced tolerances to salt stress in *Arabidopsis* through enhancing expression of multiple abiotic stress-responsive genes and improving physiological traits, including strengthened cell membrane stability and retention of higher chlorophyll contents (Mao et al. 2012, 2014). In addition, many NAC TFs isolated from other plants were also involved in salt resistance regulation in transgenic lines. These members included *EcNAC1* of finger millet, *GmNAC11* and *GmNAC20* of soybean, and *DgNAC1* of chrysanthemum (Hao et al. 2011; Liu et al. 2011; Ramegowda et al. 2012).

Following the discovery of potential use of NAC TFs to improve stress tolerance in *Arabidopsis*, a number of important successes have reported the application of NAC TFs in genetic engineering of important crops, such as cultivated rice. Transgenic rice overexpressing the stress-inducible *SNAC1* or *SNAC2/OsNAC6* gene displayed salt tolerance in both genes (Hu et al. 2006, 2008; Nakashima et al. 2007). The *SNAC1* transgenic rice was more sensitive to ABA (Hu et al. 2006). More recently, Liu et al. (2014b) proved that overexpression of *SNAC1* improves salt tolerance by enhancing root development and reducing transpiration rate in transgenic cotton. Overexpression of *SNAC2/OsNAC6* could enhance expression of a large number of genes encoding proteins with predicted stress tolerance functions such as detoxification, redox homeostasis, and proteolytic degradation as well (Hu et al. 2008; Nakashima et al. 2007). Overexpression of another rice NAC gene, the *ONAC045*, whose expression is induced by high salinity and ABA treatment, showed significantly enhanced tolerance to salt at the seedling stage. At least, expression levels of two stress-responsive genes, *OsLEA3-1* and *OsPM1*, were upregulated in ONAC045 transgenic lines (Zheng et al. 2009). However, expressions of *OsLEA3-1* and *OsPM1* were not affected in either *SNAC1* or *SNAC2* transgenic rice, suggesting that ONAC45 TF improves stress tolerance of transgenic plants in different pathway than SNAC1 and SNAC2/OsNAC6 (Zheng et al. 2009). Moreover, the potential transcriptional target genes of SNAC1 and SNAC2 are also different. There is no overlapping between the two sets of genes up- or downregulated in the two overexpression plants, respectively (Hu et al. 2008). The core DNA-binding sites for the putative SNAC1 and SNAC2 target genes are the same, but comparison of the flanking sequences of the core DNA-binding sites in the putative SNAC1 and SNAC2 target genes revealed different conserved flanking sequences of the core binding sites of genes targeted by SNAC1 and SNAC2. Together, these results may suggest that different stress-responsive NAC TFs may activate the transcription of a different set of target genes, thus conferring diverse functions that jointly lead to stress tolerance.

As we know, most TFs are localized in cell nucleus to exercise their functions; however, some membrane-bound transcription factors (MTTFs) were identified in recent years (Chen et al. 2008). Six of the NAC factors—NTM1, NTL8, NTL6, NTL9, ANAC013, and ANAC017—had been demonstrated to be membrane bounded (Chen et al. 2008; DeClercq et al. 2013; Kim et al. 2006, 2007, 2008; Ng et al. 2013; Seo and Park 2010; Yoon et al. 2008). NAC-MTTFs appear to be localized to different membranes, including the nuclear/endoplasmic reticulum or the plasma membrane (De Clercq et al. 2013; Ng et al. 2013; Kim et al. 2006, 2007;

Yoon et al. 2008). Among these MTTFs, NTL6, and NTL8 had been found to regulate salt stress signaling impinging on flowering or seed germination pathways (Kim et al. 2007, 2008). Under salt stress, the repression of flowering locus T was attenuated, albeit mildly in an *ntl8* mutant (Kim et al. 2007). In addition, the germination rate under salt stress in seeds overexpressing the truncated form of NTL8 was decreased and that of an *ntl8* mutant was increased, suggesting that NTL8 regulates salt responses in seed germination (Kim et al. 2008). Similarly to NTL8, transformants constitutively expressing an active form of NTL6 exhibited a hypersensitive response to ABA and high salinity in seed germination (Seo et al. 2010).

As a whole, there is strong evidence, even in field trials, that transgenic rice plants harboring NAC genes have enhanced stress tolerance, suggesting that NAC TFs are promising candidate genes for genetic engineering of different crops aimed at improving their productivity under adverse conditions.

## 8.5 INVOLVEMENT OF PLANT WRKY TFs IN SALT STRESS RESPONSES

The WRKY TF family is one of the largest families of plant transcriptional regulators and form integral parts of signaling networks that modulate many processes in plants, but are most notably in coping with diverse biotic and abiotic stresses. Previous studies have demonstrated that the WRKY factors are involved in responses to salt and other abiotic stresses (Chen et al. 2012b; Wei et al. 2008; Zhou et al. 2008). Modification of the expression patterns of WRKY genes and/ or changes in their activity contribute to the elaboration of various signaling pathways and regulatory networks and then finally alter the sensitivity of plant to salt stress as well as other stresses (Chen et al. 2012b; Rushton et al. 2010). In addition, it has been shown that WRKY proteins function via interactions with a diverse array of protein partners and autoregulation or cross-regulation is extensively recorded among WRKY genes, which helps us understand the complex mechanisms of signaling and transcriptional reprogramming controlled by WRKY proteins (Rushton et al. 2010).

The defining feature, or signature, of WRKY TFs is their DNA-binding domain. This is called the WRKY domain after the almost invariant WRKY AA sequence (Rushton et al. 2010). The WRKY domain, which is about 60 residues in length, is defined by the conserved AA sequence WRKYGQK at its N-terminal end, together with a novel zinc finger (ZF)-like motif at the C-terminus (Eulgem et al. 2000). In a few WRKY proteins, the WRKY AA sequences have been replaced by WRRY, WSKY, WKRY, WVKY, or WKKY (Xie et al. 2005). The structure of the ZF is either $Cx_{4-5}Cx_{22-23}HxH$ or $Cx_7Cx_{23}HxC$. Both of these two domains are vital for the high binding affinity of WRKY TFs to the consensus *cis*-acting elements termed the W box (TTGACT/C), although alternative binding sites have been identified (Chen et al. 2012b). Almost all WRKY TFs show binding preference to the cognate *cis*-acting element; however, the binding site preferences are also partly determined by additional adjacent DNA sequences outside of the TTGACY-core motif (Ciolkowski et al. 2008).

Based on the number of WRKY domains and the structure of their ZFs, the WRKY TFs initially were divided into three groups. The first group (I) has two WRKY domains; group II has one WRKY domain containing the same Cys2-His2 ZF motif; and group III has one WRKY domain containing the different Cys2-His/Cys Cys2-His2 ZF motif (Chen et al. 2010). The group II factors were further divided into subgroups a–e based on the primary AA sequence (Rushton et al. 2010). Later, analysis based purely on phylogenetic data showed that the WRKY family in higher plants is more accurately divided into groups I, IIa + IIb, IIc, IId + IIe, and III with the group II genes not being monophyletic (Rushton et al. 2008; Zhang and Wang 2005). Besides both of the highly conserved WRKY domain and the Cys2-His/Cys or Cys2-His2 ZF motif, the WRKY factors still consist of the structures as follows: leucine zippers, putative basic nuclear localization signals (NLS), serine–threonine-rich region, proline-rich region, glutamine-rich region, kinase domains, and TIR-NBS-LRRs (Chen et al. 2012b). Due to owning such characterizations of structures, the WRKY proteins can play their appropriate roles in gene expression regulation.

The WRKY TF family is originally believed to be plant specific, but they are later found in *Giardia lamblia*, a primitive eukaryote; *Dictyostelium* discoideum, a slime mold closely related to the lineage of animals and fungi; and the green alga C. reinhardtii, an early branching of plants, which implies an earlier origin (Pan et al. 2009; Zhang and Wang 2005). The WRKY family, which is one of the ten largest families of TFs, is found throughout the higher plants (Ulker and Somssich 2004). The WRKY gene family consists of almost 200 members in soybean, over 100 members in rice, and 74 representatives in Arabidopsis (Wu et al. 2005, Ulker and Somssich 2004). But the family members decreased to 37 in the moss Physcomitrella patens and only a single WRKY TF in the unicellular green alga C. reinhardtii (Rushton et al. 2010). They appear to have been lost in yeast, prokaryotes, and animal lineages. That indicated a lineage-specific expansion of the WRKY gene family during the course of plant evolution.

In higher plants, genes from all the three major WRKY groups that contain seven subfamilies (groups I, IIa, IIb, IIc, IId, IIe, and III) are found. However, it was found that all the WRKY genes found in the unicellular eukaryote Chlamydomonas, the nonphotosynthetic eukaryotic slime mold D. discoideum, the unicellular protist G. lamblia, lower plant Physcomitrella patens, and Ceratopteris richardii belong to group I (Zhao et al. 2012). The single WRKY present in the green alga C. reinhardtii also belongs to group I (Zhao et al. 2012). Base on these, group I could be the apparent ancestral type of WRKY genes. The elucidation of the evolution and duplicative expansion of the WRKY genes should provide valuable information on their functions.

The WRKY family factors participate in the control of a wide variety of biological processes such as seed development, dormancy and germination, plant development, and senescence, as well as response of biotic and abiotic stress, and some of which are proved to be associated with salt stress regulation in plant. Bo et al. (2007) identified three WRKY family genes (DY523446, DY523772, and DY523811) that were related with salt stress response, and the genes were all upregulated after 0.5 h salt stress treatment in two tomato cultivars and reached the expression peaks at 6 h of treatment. In upland cotton, the GhWRKY41 gene was also up-regulated after salt treatment, but it displayed later response and also peaked to its expression level at 6 h. The expression levels of the gene went down after 12 h of treatment, and the other

two GhWRKY genes homologous to AtWRKY17 and AtWRKY41 also enhanced their expression levels after salt stress (Zhang et al. 2011d). AtWRKY25 and AtWRKY33 in Arabidopsis and two WRKY TFs (POPTR_0008s09140.1, POPTR_0016s14490.1) in Populus were also upregulated after NaCl treatment (Chen et al. 2012d; Li et al. 2009b). Interestingly, under salt stress, OsWRKY45-1 expression was first induced and then suppressed in rice, whereas OsWRKY45-2 expression was first suppressed, then induced, and later suppressed again. The results suggest that the two alleles may be involved in rice responses to salt stress, but they may function differently (Tao et al. 2011). In addition, there were many other WRKY family genes in plants that were induced by salt stress, such as MusaWRKY71 transcripts in banana (Shekhawat et al. 2011); GmWRKY13, GmWRKY21, and GmWRKY54 in soybean (Zhou et al. 2008); DgWRKY1 and DgWRKY3 gene from chrysanthemum (Liu et al. 2013, 2014); TcWRKY53 in tobacco (Wei et al. 2008); AtWRKY2 in Arabidopsis (Jiang and Yu, 2009); and GhWRKY39 and ZmWRKY33 (Li et al. 2013a; Shi et al. 2014). Previous study showed that the isolated GhWRKY genes expressed differentially in different tissues, suggesting their impact on plant development, and under NaCl treatment, expressions of the GhWRKY genes are upregulated in part of the tissues, suggesting their tissue specific under salt stress responses (Zhou et al. 2014). The immediate-early expression behavior of WRKY genes assures the successful transduction of the signals to activate adaptive responses and regulation of stress-related genes and finally result in plant salt stress tolerance.

Recently, it has become clear that WRKY TFs play key roles in both the repression and derepression of responses to salt stress as well as other abiotic stresses (Rushton et al. 2010). It was showed that AtWRKY25 and AtWRKY33 double mutants showed moderately increased NaCl sensitivity, while over-expression of either AtWRKY25 or AtWRKY33 led to increased Arabidopsis NaCl tolerance (Li et al. 2011). The tobacco plants overexpressed with DgWRKY1 or DgWRKY3 increase salt tolerance compared with wild-type tobacco plants (Liu et al. 2013, 2014). The DgWRKY1-or DgWRKY3-overexpressed transgenic lines exhibited less accumulation of hydrogen peroxide ($H_2O_2$) and malondialdehyde under salt stress and less antioxidant enzyme activity, including peroxidase (POD), superoxide dismutase, and catalase, than the WT under both controlled conditions and salt stress. In addition, there was greater upregulation of the ROS-related enzyme genes (NtSOD, NtPOD, and NtCAT) in transgenic lines under normal or salt conditions. These findings suggest that DgWRKY1 and DgWRKY3 play positive regulatory roles in salt stress response (Liu et al. 2013, 2014). Similarly, overexpression of OsWRK45 and OsWRK72 enhanced salt and drought tolerance of Arabidopsis plants (Qiu and Yu 2009; Song et al. 2010). Overexpression of ZmWRKY33 in Arabidopsis also enhanced salt stress tolerance in the transgenic plants (Li et al. 2013). Based on the analyses, many WRKY members positively response to salt stress of plants, but some members are negative regulators during this progress. Zhou et al. (2008) found that plants that have overexpressed GmWRKY54 showed enhanced salt and drought tolerance; however, overexpression of GmWRKY13 led to increased sensitivity to salt and mannitol stresses. In addition, it was suggested that AtWRKY18 and AtWRKY60 have negative effect on salt and osmotic stress responses, but have a positive effect on ABA responses (Chen et al. 2010). OsWRKY13 is shown that it negatively regulates rice tolerance to salt and cold stresses, but positively regulates disease resistance (Qiu et al. 2008). These results also indicate that many WRKY genes could simultaneously play roles in multiple stress-induced signalling pathways.

Further evidences have come from altered plant responses to different abiotic stresses by changing a WRKY gene expression. Notably, ecotopic expression of OsWRKY45 in Arabidopsis resulted in plants with elevated tolerance to salt and drought stress, enhanced resistance to virulent P. syringae, increased PR1 expression, but decreased sensitivity toward ABA signalling (Qiu and Yu 2009). Transgenic Arabidopsis plants overexpressing TaWRKY2 exhibited salt and drought tolerance compared with controls, and overexpression of TaWRKY19 conferred tolerance to salt, drought, and freezing stresses in transgenic plants (NIU et al. 2012). GmWRKY54-transgenic Arabidopsis plants were tolerant to salt and drought, and transgenic plants overexpressing GmWRKY13 showed increased sensitivity to salt and mannitol stress, but decreased sensitivity to ABA, when compared with WT plants (Zhao et al. 2012). Constitutive expression of BcWRKY46 in tobacco under the control of the CaMV35S promoter reduced the susceptibility of transgenic tobacco to freezing, ABA, salt, and dehydration stresses, and the results suggest that BcWRKY46 plays an important role in responding to ABA and abiotic stress (Wang et al. 2012). Transgenic banana plants overexpressing MusaWRKY71 displayed enhanced tolerance toward oxidative and salt stress as indicated by better photosynthesis efficiency (Fv/Fm) and lower membrane damage of the assayed leaves (Shekhawat and Ganapathi 2013). Furthermore, the constitutive overexpression of GhWRKY39 in Nicotiana benthamiana displayed enhanced tolerance to salt and oxidative stress and increased transcription of plants. Moreover, GhWRKY39-overexpressing plants conferred greater resistance to bacterial and fungal pathogen infections, and the expression of several PR genes was significantly increased. The transgenic plants also exhibited less $H_2O_2$ accumulation than WT plants following pathogen infection (Shi et al. 2014). These results suggest that GhWRKY39 may mediate the cross talk between abiotic and biotic stresses. A single WRKY gene often simultaneously plays role in responses of several stress factors, indicating its diverse regulatory function during responding to plant stresses.

Initiating signals of salt stress interaction with gene expression are involved in at least two types of signal transduction: ABA dependent and ABA independent. WRKY TFs are identified as key components in ABA signaling networks (Rushton et al. 2012). In Arabidopsis, WRKY TFs appear to act downstream of at least two ABA receptors: the cytoplasmic PYR/PYL/RCAR protein phosphatase 2C-ABA complex and the chloroplast envelope–located ABAR–ABA complex. In vivo and in vitro promoter-binding studies show that the target genes for WRKY TFs that are involved in ABA signalling include well-known ABA-responsive genes such as ABF2, ABF4, ABI4, ABI5, MYB2, DREB1a, DREB2a, and RAB18. Additional well-characterized stress-inducible genes such as RD29A and COR47 are also found in signalling pathways downstream of WRKY TFs (Rushton et al. 2012). These insights reveal that some WRKY TFs are positive regulators of ABA-mediated stress responses. Conversely, some WRKY TFs are negative regulators of ABA-mediated stress responses. OsWRKY45-1-overexpressing lines showed reduced ABA sensitivity, and OsWRKY45-2-overexpressing lines showed increased ABA sensitivity and reduced salt stress tolerance. The different roles that OsWRKY45-1 and OsWRKY45-2 played in ABA signalling and salt stress adaptation could be due to their transcriptional mediation of different signaling pathways (Tao et al. 2011). However, in response to salt stress, few WRKY TFs have been demonstrated to be ABA independent, such as AtWRKY25; its transcript accumulation does not require ABA (Jiang and Deyholo 2009). These suggest the complex functional mechanism of WRKY factors.

The earlier researches have clearly revealed that WRKY factors form a complex and highly interconnected regulatory network (Eulgem and Somssich 2007). WRKY factor functions via interactions with a diverse array of protein partners, including MAP kinases, MAP kinase kinases, 14-3-3 proteins, calmodulin, histone deacetylases, resistance proteins, and other WRKY TFs. WRKY genes exhibit extensive autoregulation and cross-regulation that facilitates transcriptional reprogramming in a dynamic web with built-in redundancy (Chi et al. 2013; Rushton et al. 2010). However, the regulatory mechanism of WRKY factors in salt stress was less reported by date. It has been reported that overexpression of GmWRKY54 enhances tolerance to salt and drought stress, possibly through the regulation of DREB2A, RD29B, and STZ/Zat10 (Zhou et al. 2008). Previous study appears that AtWRKY60 might be a direct target of AtWRKY40 and AtWRKY18 because induction of AtWRKY60 is almost completely abolished in wrky18 and wrky40 mutants, and both AtWRKY40 and AtWRKY18 proteins recognize a cluster of W box sequences in the AtWRKY60 promoter. The result suggested that an AtWRKY18/AtWRKY40 heterocomplex may regulate the expression of the AtWRKY60 gene and that homo- and heterodimer complexes of these three group IIa WRKY proteins then regulate ABA responses (Chen et al. 2010). WRKY proteins function via protein–protein interaction, and autoregulation or cross-regulation is extensively recorded among WRKY genes, which help us understand the complex mechanisms of signaling and transcriptional reprogramming controlled by WRKY proteins.

As a whole, WRKY genes act as important TFs superfamily, and they involved in the response to salt stress as well as other abiotic stresses. They participate in plant growth and development and material metabolic pathways and have

a complex and important role in regulation. To date, researches on the WRKY TFs have focused on gene cloning and expression and gene functions via gene absence or overexpression. However, their roles in salt stress response as well as other stress responses remain obscure, and the response mechanisms are especially poor known. Considering the size of this gene family, the WRKY factors will keep us both fascinated and busy in the coming years.

## 8.6 OTHER TFs INVOLVED IN PLANT SALT STRESS

Although the vast majority of stress-responsive genes remain to be identified, in the signal transduction networks involved in the conversion of stress signal perception to stress-responsive gene expression, various TFs have been described. These TFs function not only as molecular switches for gene expression, but also as terminal points of signal transduction in the signaling processes (Nakashima et al. 2009; Tran et al. 2007; Yamaguchi-Shinozaki and Shinozaki 2005, 2006). In addition to the salt stress–related TFs mentioned earlier, other TFs such as bZIF, bHLH, and NAC were also proved to play important roles in plant salt stress regulation (Golldack et al. 2011; Rodriguez-Uribe et al. 2011).

### 8.6.1 Involvement of Plant $C_2H_2$-Type Zinc-Finger TFs in Plant Salt Stress Responses

ZFPs form a relatively large family of transcriptional regulators in plants (Takatsuji 1999). Based on the number and the location of characteristic residues, the ZFPs are classified into several types, such as $C_2H_2$, $C_2HC$, $C_2HC_5$, $C_3HC_4$, CCCH, $C_4$, $C_4HC_3$, $C_6$, and $C_8$ (Takatsuji 1999). Plant $Cys_2/His_2$ type of ZFPs consists of one to five ZFs. According to the number, type, and arrangement of ZFs, the $C_2H_2$-type ZFPs can be divided into three classes: A, B, and C. In plants, most $C_2H_2$-type ZFs belong to the class C with a highly conserved QALGGH motif containing a single ZF or several dispersed ZFs, also called Q-type ZFs. Another unique feature of plant $C_2H_2$-ZFPs is the presence of long variable length spacers that separate the neighboring fingers (Takatsuji 1999). Based on the number of AA residues separating the two invariant His of ZFs: three, four, or five residues, the class C of $C_2H_2$-type ZFs can be further grouped into subset $C_1$, $C_2$, and $C_3$, respectively (Englbrecht et al. 2004)

The domain of $C_2H_2$-type ZF, also called the TFIIIA-type ZFPs or the Kruppel-like ZFPs, which has approximately 30 AA residues, constitutes one of the largest TF families in eukaryotes (Pabo et al. 2001). Many plant $C_2H_2$-type ZFPs have three characteristic regions outside the ZF motifs. One is the N-terminal B-box region acting as a NLS. The second is a short, leucine-rich region with a core sequence EXEXXAXCLXXL (L-box) located between the B-box and the first ZF domain (Sakamoto et al. 2000). The third is an EAR-motif, a short hydrophobic part in the C-terminal region of a protein with a core sequence DLNL (DLN-box). The latter two are thought to play roles in the protein–protein interactions or in maintaining the folded structure (Ciftci-Yilmaz et al. 2007)

In fact, many plant $C_2H_2$-type ZFPs have been shown to play an important role in regulating defense response to different stresses, including Arabidopsis ZAT7, ZAT10/STZ, ZAT12, AZF1, AZF2, or AZF3 that possess the EAR domain (Ciftci-Yilmaz et al. 2007; Kim et al. 2004; Sakamoto et al. 2004; Vogel et al. 2005). Some plant $C_2H_2$-type ZFPs act as transcriptional activators under stress conditions, but the identification and localization of the possible activation domains have been illucidated for very few of them (Sakamoto et al. 2000; Wang et al. 2009). For instance, a serine-rich region located between the B-box and the L-box motifs in some stress-related $C_2H_2$-type ZFPs has been suggested to function as an activation domain (Sakamoto et al. 2000). Furthermore, it has been confirmed that the repressive effect of the DLN region of the ZAT10/STZ TF played an important role in transactivation of the transcriptional activator DREB1A (Sakamoto et al. 2004). Another stress-related $C_2H_2$-type ZFP called ZAT2-3 containing the EAR-motif from petunia functions as an active repressor. Moreover, using the mutated ZPT2-3 as an effector caused increased expression of a LUC reported gene (Sugano et al. 2003).

Many studies revealed that most of $C_2H_2$-type ZFPs may regulate responses to multiple abiotic stresses. Altering the expression of the genes encoding $C_2H_2$-type transcription improved stress tolerance in Arabidopsis, rice, tobacco, potato, poplar and petunia (Huang et al. 2007; Kim et al. 2001a,b; Sugano et al. 2003; Gourcilleau et al. 2011; Tian et al. 2010). Several studies have reported that overexpression of some $C_2H_2$-type ZFP genes resulted in both the activation of some stress-related genes and enhanced tolerance to salt stress (Kim et al., 2001b; Sakamoto et al., 2000, 2004; Sugano et al., 2003). In Arabidopsis, involvement of several $C_2H_2$-type ZFPs, AZF1, AZF2, AZF3, AZT7, ZAT12, and ZAT10, in drought and salt stress response, has been well characterized. High-salt treatment resulted in elevated levels of expression of all of these genes (Kodaira et al. 2011). ZAT10 and ZAT12 have broad and similar responses to multiple abiotic stresses in all parts of Arabidopsis. Enhanced expression of ZAT10 and ZAT12 genes in response to many different stress conditions suggests that both proteins are somehow involved in plant responses to all of these stresses (Iida et al. 2000; Chen et al. 2002; Zimmermann et al. 2004; Rizhsky et al. 2004; Ciftci-Yilmaz and Mittler 2008). AZF2 and STZ were strongly induced by high-salt stress, and ABA treatment (Xu et al. 2007). Only AZF2 expression was strongly induced by ABA treatment, where the time course of the induction was similar to that caused by high salinity. In situ localization showed that AZF2 mRNA accumulated in the elongation zone of the roots under the salt stress condition (Sakamoto et al. 2000). Constitutive expression of Zat10 was found to elevate the expression of reactive oxygen-defense transcripts and to enhance the tolerance of plants to salinity and osmotic stress. Surprisingly, knockout and RNAi mutants of Zat10 were also more tolerant to osmotic and salinity stress. Further analysis showed that ZAT10 played a dual role in response to abiotic stresses (Mittler et al. 2006). In addition, expression of the ZAT7 gene seems to be more specific, with an enhanced transcript level in roots only during salt stress (Zimmermann et al. 2004; Ciftci-Yilmaz and Mittler 2008). The expression of a novel potato $C_2H_2$-type ZFP gene, StZFP1, increased salt stress resistance in transgenic plant (Tian et al. 2010). The CgZFP1 gene expression in the chrysanthemum leaf was strongly induced by salinity or drought, but not by ABA. Yeast-one hybrid assay showed that CgZFP1 possesses transcriptional activation ability, heterologous expression of CgZFP1 conferred tolerance of transgenic Arabidopsis plants to both salinity and drought stresses (Gao et al. 2012).

Moreover, the expression analysis showed that rice $ZFP_{182}$ gene was markedly induced in the seedlings by 150 mM NaCl and 0.1 mM ABA treatments. Overexpression of $ZFP_{182}$ in transgenic tobacco and rice both increased plant tolerance to salt stress (Huang et al. 2007).

In *Eucalyptus grandis*, seven $C_2H_2$-type ZF genes (EgrZFP1-7) were cloned. The expression of EgrZFP1-6 was found to be enhanced by 200 mM NaCl, whereas the expression of EgrZFP7 was inhibited (Wang et al. 2014). The overexpression of stress-inducible $C_2H_2$-type TFs activated the expression of many target genes conferring plants improved salt tolerance. Some $C_2H_2$ TFs also mediate cross talk between signaling pathways in salt stress response (Kiełbowicz-Matuk 2012). Many TFs such as NAC, MYB, bZIP, and ZFPs have been well characterized with their roles in the regulation of salt responses. There may be cross talk and an overlap among salt-signal transduction pathways. In *Arabidopsis*, a deletion or a mutation of the EAR motif of Zat7 abolishes salinity tolerance. Moreover, ZAT7 was reported to positively mediate salt stress tolerance through regulation of defense responsive genes such as WRKY70, AOX1, COR78, and NHX (Ciftci-Yilmaz et al. 2007). Overexpression of C2H2-type ZFP252 in rice elevated the expression of stress defense genes of OsDREB1A, OsLea3, OsP5CS, and Os-ProT1 under salt and drought stress and enhanced rice tolerance to salt and drought (Xu et al. 2008a). They found that overexpression of $ZFP_{252}$ in rice increased the amount of free proline and soluble sugar, elevated the expression of stress defense genes, and enhanced rice tolerance to salt and drought stresses. Similarly, rice overexpressing a $C_2H_2$-type ZFP179 elevated the expression of OsDREB2A, OsP5CS, OsProT, and OsLea3 and increased POD activity under salt stress, thereby enhancing tolerance to the salt stress (Sun et al. 2010). Thus, there are no general rules regarding which of the $C_2H_2$-type ZFPs transcription regulators activate or repress particular classes of the salt stress–related genes. Several types of transcription regulators can stimulate or inhibit one group of stress-associated genes, or several TFs can regulate expression of the same gene (Kiełbowicz-Matuk 2012).

All these results indicate that $C_2H_2$-type ZFPs are important factors that link different signal transduction pathways in response to salt stress and participate in several cellular processes. These results also suggest that there are multiple signaling pathways involved in the response to salt stress treatment. Only a few $C_2H_2$-type ZFPs seem to be associated with response to salt stress. In summary, salt stress response networks in plants are very complex and contain unidentified proteins and pathways that require further investigation. Individual $C_2H_2$-type ZFPs TFs containing the EAR-motif seem to be particularly important, but the cooperation between different $C_2H_2$- type ZFPs appears to be a fundamental principle of the integrated $C_2H_2$-type ZFPs machinery (Kiełbowicz-Matuk 2012).

### 8.6.2 bZIP TFs and Their Role in Conferring Salt Stress Tolerance to Plants

bZIPs organize a large family that have been described in *Arabidopsis* (75), rice (89), sorghum (92), soybean (131), and recently maize (125; Wei et al. 2012). All the members of this family contain a bZIP domain. The bZIP family was subdivided according to their sequence similarities and functional features, resulting in 10 groups named A to I, plus S in *Arabidopsis* (Jakoby et al. 2002; Nijhawan et al. 2007; Wei et al. 2012). While many bZIPs can form homodimers, bZIP members classified in different groups can be combined through heterodimerization to form specific bZIP pairs with distinct functionalities.

Previous studies have indicated that bZIP proteins are key regulators involved in salt stress adaption. A lot of bZIP proteins isolated from diverse species including *Arabidopsis*, rice, tomato, soybean, maize, and wheat enhanced the salt tolerance of the transgenic plants (Cheng et al. 2013a; Gao et al. 2011; Hsieh et al. 2010; Kobayashi et al. 2008; Li et al. 2013; Liao et al. 2008a; Liu et al. 2014; Xiang et al. 2008; Yáñez et al. 2009; Ying et al. 2012; Zhang et al. 2011c; Zou et al. 2008).

The function mechanisms of bZIPs have been studied well. Some members of the bZIP-type protein family are AREB and ABF, which act as major TFs in ABA-responsive gene expression under salt stress conditions in *Arabidopsis*. For example, bZIP factors SlAREB1, ScAREB1, SpAREB1, AtABI5, OsABI5, and ABF2-4 in *Arabidopsis* all have a key regulatory role in ABA signaling under salt stress (Kim et al. 2004; Tezuka et al. 2013;

Yáñez et al. 2009; Zou et al. 2008). A key regulator of salt stress adaptation, the group F bZIP TF bZIP24, was identified by differential screening of salt-inducible transcripts in *A. thaliana* and a halophytic *Arabidopsis*-relative model species (Yang et al. 2009). In addition, AtbZIP24 shows salt-inducible subcellular retargeting to the nucleus and formation of homodimers suggesting that molecular dynamics of bZIP factors could mediate new signaling connections within the complex cellular signaling network (Yang et al. 2009). RNAi-mediated repression of the factor conferred increased salt tolerance to *Arabidopsis*. The improved tolerance was mediated by stimulated transcription of a wide range of stress-inducible genes involved in cytoplasmic ion homeostasis, osmotic adjustment, and plant growth and development, which demonstrated a central function of bZIP24 in salt tolerance by regulating multiple mechanisms that are essential for stress adaptation (Yang et al. 2009).

In addition, numerous bZIPs were proved to control signal transduction pathways by molecular reorganization and by posttranslational mechanisms (Jindra et al. 2004; Miller 2009; Schütze et al. 2008). Specific homodimerizations and heterodimerizations within the class of bZIP TFs as well as modular flexibility of the interacting proteins and posttranslational modifications might determine the functional specificity of bZIP factors in cellular transcription networks (Liao et al. 2008c; Liu et al. 2013; Miller 2009). The phosphorylation of bZIP proteins seems also important for their function. For example, potato StABF1 is phosphorylated in response to ABA and salt stress in a calcium-dependent manner, and a potato CDPK isoform (StCDPK2) had been identified to phosphorylate StABF1 in vitro (Muñiz García et al. 2012). The three factors AREB1, AREB2, and ABF3 can form homodimers and heterodimers as well as interact with a SnRK2 protein kinase suggesting ABA-dependent phosphorylation of the proteins (Yoshida et al. 2010).

Similar to NAC TFs, one bZIP TF involved in salt stress was also demonstrated to be membrane-bounded. Most bZIP MTTFs are anchored in the membrane of endoplasmic reticulum. In response to stress, cytosolic components of the TFs are released by proteolysis and move to the nucleus where they promote the upregulation of stress response genes (Liu et al. 2008; Srivastava

et al. 2014). One such stress sensor/transducer is *Arabidopsis* AtbZIP17, which is activated in response to salt stress. Under salt stress conditions, the stress-inducible expression of the activated AtbZIP17 enhanced salt tolerance as demonstrated by chlorophyll bleaching and seedling survival assays (Liu et al. 2008).

### 8.6.3 bHLH TFs and Their Role in Conferring Salt Stress Tolerance to Plants

The bHLH family TFs have been intensively studied in plants and animals (Sonnenfeld et al. 2005; Toledo-Ortiz et al. 2003). With the genome-wide analysis of the bHLH TF family in plants, 162 bHLH genes in *Arabidopsis* and 167 in rice were identified (Li et al. 2006). This family is defined by the bHLH signature domain, which is evolutionarily conserved (Ferre-D'Amare et al. 1994; Murre et al. 1989). Plant bHLH proteins bind to the E-box (CANNTG) motif in gene promoters; a consensus core element called G-box (CACGTG) is the most common form (Li et al. 2006).

Although numerous bHLH members were identified in *Arabidopsis* and rice, only a few of them had been revealed to be involved in plant salt stress regulation. *Chrysanthemum dichrum* CdICE1 and tomato SlICE1a both contain conserved bHLH domain, and overexpression of these two genes could both improve the tolerance of transgenic plants to salinity (Chen et al. 2012a; Feng et al. 2013). Jiang et al. (2009) identified three salt-inducible bHLH proteins: bHLH41, bHLH42/TT8, and bHLH92. Among which, bHLH92 was highly induced at the transcript level by a wide range of abiotic stresses, and the response to NaCl was quantitatively the highest. Overexpression of bHLH92 moderately increased the tolerance to NaCl and osmotic stresses. However, knockout mutants of this gene failed to show significant differences from WT plants in the root elongation assay under NaCl stress, suggesting that bHLH92 function is not required for full tolerance to NaCl and may therefore be redundant with other bHLH proteins. *OrbHLH001* and *OrbHLH2* were cloned from Dongxiang wild rice (Li et al., 2010; Zhou et al., 2009). Overexpression of OrbHLH001 and OrbHLH2 enhances tolerance to salt stress in *Arabidopsis* (Li et al., 2010; Zhou et al., 2009). Examination of the expression of cold-responsive genes in transgenic *Arabidopsis* showed that the function of OrbHLH001 differs from that of ICE1 and is independent of a CBF/DREB1 cold-response pathway (Li et al. 2010). However, overexpression of *OrbHLH2* in *Arabidopsis* improved salt tolerance by enhancing the expression level of DREB1A/CBF3 and its downstream target genes, but the ABA signal pathway was not affected in transgenic *Arabidopsis*, which suggest that OrbHLH2 probably function in salt response through an ABA-independent pathway (Zhou et al. 2009). These results suggest that different homologues of ICE1 may mediate the regulation of salt tolerance in a different signal response process. And what's more, Chen et al. (2013) indicated that overexpression of *OrbHLH001* could also confer salt tolerance in transgenic rice plants. OrbHLH001 protein exercise its function by inducing the expression of *OsAKT1*, another quantitative trait loci controlling K+ uptake into the root and the Na+/K+ ratio in salt stress, to regulate the Na+/K+ ratio in OrbHLH001-overexpressed plants (Chen et al. 2013).

Guan et al. (2013) found a nuclear-localized calcium-binding protein, RSA1 (SHORT ROOT IN SALT MEDIUM 1), which is required for salt tolerance, and identified its interacting partner, RITF1, a bHLH TFs. They show that RSA1 and RITF1 regulate the transcription of several genes involved in the detoxification of ROS generated by salt stress and that they also regulate the SOS1 gene that encodes a plasma membrane Na+/H+ antiporter essential for salt tolerance. This study discovered a novel nuclear calcium-sensing and calcium-signaling pathway that is important for gene regulation and salt stress tolerance.

As more members in the complex systems in stress response are reported, the function of bHLH TFs will be better understood.

### 8.6.4  Other TFs Related to Salt Stress in Plants

Numerous TFs were found to be involved in plant salt tolerance regulation using large-scale screening methods such as microarray hybridization. Except for the TFs referred earlier, a lot of other TFs may be involved in plant salt stress regulation. These TFs usually contain the conserved domain such as C2C2-DOF, GARP, GRAS, MADS, PHD, SBP, C3HC4-type RING finger, HSF, MYC, ZIM,

and LBD (Gruber et al. 2009; Ji et al. 2012; Jiang and Deyholos 2006; Ma et al. 2010; Srivastava et al. 2010; Wei et al. 2014).

Several members of the aforementioned family TFs were indicated to confer plant salt stress tolerance. For example, a poplar GRAS gene, PeSCL7, enhanced tolerance to salt treatments in transgenic *Arabidopsis*. Corrales et al. (2014) reported a group of five tomato DOF (DNA binding with One Finger) genes, SlCDF1-5. SlCDF1-5 genes exhibited distinct diurnal expression patterns and were differentially induced in response to osmotic, salt, heat, and low-temperature stresses. *Arabidopsis* plants overexpressing SlCDF1 or SlCDF3 showed increased salt tolerance. In addition, the expression of various stress-responsive genes, such as COR15, RD29A, and RD10, were differentially activated in the overexpressing lines.

However, most of the TFs screened by microarray lacked function analysis in plant salt stress regulation. Therefore, more and more TFs will be proved to function in salt stress regulation in the future.

## REFERENCES

Abogadallah, G.M., Nada, R.M., Malinowski, R., and Quick, P. 2011. Over-expression of HARDY, an AP2/ERF gene from *Arabidopsis*, improves drought and salt tolerance by reducing transpiration and sodium uptake in transgenic *Trifolium alexandrinum* L. *Planta* 233:1265–1276.

Balazadeh, S., Wu, A., and Mueller-Roeber, B. 2010. Salt-triggered expression of the ANAC092-dependent senescence regulon in *Arabidopsis thaliana*. *Plant Signaling & Behavior* 5:733–735.

Bartels, D. and Sunkar, R. 2005. Drought and salt tolerance in plants. *Critical Reviews in Plant Sciences* 24:23–58.

Bhatnagar-Mathur, P., Vadez, V., and Sharma, K. 2008. Transgenic approaches for abiotic stress tolerance in plants: Retrospect and prospects. *Plant Cell Reports* 27:411–424.

Bo, O.Y., Yang, T., Li, H., Zhang, L., Zhang, Y., Zhang, J., and Ye, Z. 2007. Identification of early salt stress response genes in tomato root by suppression subtractive hybridization and microarray analysis. *Journal of Experimental Botany* 58:507–520.

Bossi, F., Cordoba, E., Dupre, P., Mendoza, M.S., Roman, C.S., and Leon, P. 2009. The *Arabidopsis* ABA-INSENSITIVE (ABI) 4 factor acts as a central transcription activator of the expression of its own gene, and for the induction of ABI5 and SBE2.2 genes during sugar signaling. *Plant Journal* 59:359–374.

Cai, H., Tian, S., Liu, C., and Dong, H. 2011. Identification of a MYB3R gene involved in drought, salt and cold stress in wheat (*Triticum aestivum* L.). *Gene* 485:146–152.

Chen, B.J., Wang, Y., Hu, Y.L., Wu, Q., and Lin, Z.P. 2005. Cloning and characterization of a drought inducible MYB gene from *Boea crassifolia*. *Plant Science* 168:493–500.

Chen, G., Hu, Z., and Grierson, D. 2008. Differential regulation of tomato ethylene responsive factor LeERF3b, a putative repressor, and the activator Pti4 in ripening mutants and in response to environmental stresses. *Journal of Plant Physiology* 165:662–670.

Chen, H., Lai, Z., Shi, J., Xiao, Y., Chen, Z., and Xu, X. 2010. Roles of arabidopsis WRKY18, WRKY40 and WRKY60 transcription factors in plant responses to abscisic acid and abiotic stress. *BMC Plant Biology* 10:281.

Chen, L., Chen, Y., Jiang, J., Chen, S., Chen, F., Guan, Z., and Fang, W. 2012a. The constitutive expression of *Chrysanthemum dichrum* ICE1 in *Chrysanthemum grandiflorum* improves the level of low temperature, salinity and drought tolerance. *Plant Cell Reports* 31:1747–1758.

Chen, L., Song, Y., Li, S., Zhang, L., Zou, C., and Yu, D. 2012b. The role of WRKY transcription factors in plant abiotic stresses. *Biochimica et Biophysica Acta (BBA)-Gene Regulatory Mechanisms* 1819:120–128.

Chen, N., Yang, Q., Pan, L., Chi, X., Chen, M., Hu, D., Yang, Z., Wang, T., Wang, M., and Yu, S. 2014. Identification of 30 MYB transcription factor genes and analysis of their expression during abiotic stress in peanut (*Arachis hypogaea* L.). *Gene* 533:332–345.

Chen, N., Yang, Q., Su, M., Pan, L., Chi, X., Chen, M., He, Y. et al. 2012c. Cloning of six ERF family transcription factor genes from peanut and analysis of their expression during abiotic stress. *Plant Molecular Biology Reporter* 30:1415–1425.

Chen, S., Jiang, J., Li, H., and Liu, G. 2012d. The salt-responsive transcriptome of *Populus simonii* × *Populus nigra* via DGE. *Gene* 504:203–212.

Chen, W., Provart, N.J., Glazebrook, J., Katagiri, F., Chang, H.S., Eulgem, T., Mauch, F. et al. 2002. Expression profile matrix of *Arabidopsis* transcription factor genes suggests their putative functions in response to environmental stresses. *Plant Cell* 14:559–574.

Chen, Y., Li, F., Ma, Y., Chong, K., and Xu, Y. 2013. Overexpression of *OrbHLH001*, a putative helix-loop-helix transcription factor, causes increased expression of AKT1 and maintains ionic balance under salt stress in rice. *Journal of Plant Physiology* 170:93–100.

Chen, Y.N., Slabaugh, E., and Brandizzi, F. 2008. Membrane-tethered transcription factors in *Arabidopsis thaliana*: Novel regulators in stress response and development. *Current Opinion in Plant Biology* 11:695–701.

Cheng, L., Li, S., Hussain, J., Xu, X., Yin, J., Zhang, Y., Chen, X., and Li, L. 2013a. Isolation and functional characterization of a salt responsive transcriptional factor, LrbZIP from lotus root (*Nelumbo nucifera* Gaertn). *Molecular Biology Reports* 40:4033–4045.

Cheng, L., Li, X., Huang, X., Ma, T., Liang, Y., Ma, X., Peng, X. et al. 2013b. Overexpression of sheepgrass R1-MYB transcription factor LcMYB1 confers salt tolerance in transgenic *Arabidopsis*. *Plant Physiology and Biochemistry* 70:252–260.

Cheng, Y.J., Kim, M.D., Deng, X.P., Kwak, S.S., and Chen, W. 2013c. Enhanced salt stress tolerance in transgenic potato plants expressing IbMYB1, a sweet potato transcription factor. *Journal of Microbiology and Biotechnology* 23:1737–1746.

Chi, Y., Yang, Y., Zhou, Y., Zhou, J., Fan, B., Yu, J.Q., and Chen, Z. 2013. Protein–protein interactions in the regulation of WRKY transcription factors. *Molecular Plant* 6:287–300.

Ciftci-Yilmaz, S. and Mittler, R. 2008. The zinc finger network of plants. *Cellular and Molecular Life Sciences* 65:1150–1160.

Ciftci-Yilmaz, S., Morsy, M.R., Song, L., Coutu, A., Krizek, B.A., Lewis, M.W., Warren, D., Cushman, J., Connolly, E.L., and Mittler, R. 2007. The EAR-motif of the Cys2/His2—Type zinc finger protein Zat7 plays a key role in the defense response of *Arabidopsis* to salinity stress. *Journal of Biological Chemistry* 282:9260–9268.

Ciolkowski, I., Wanke, D., Birkenbihl, R., and Somssich, I. 2008. Studies on DNA-binding selectivity of WRKY transcription factors lend structural clues into WRKY domain function. *Plant Molecular Biology* 68:81–92.

Corrales, A.R., Nebauer, S.G., Carrillo, L., Fernández-Nohales, P., Marqués, J., Renau-Morata, B., Granell, A. et al. 2014. Characterization of tomato cycling Dof factors reveals conserved and new functions in the control of flowering time and abiotic stress responses. *Journal of Experimental Botany* 65:995–1012.

Cui, M.H., Yoo, K.S., Hyoung, S., Nguyen, H.T., Kim, Y.Y., Kim, H.J., Ok, S.H., Yoo, S.D., and Shin J.S. 2013. An *Arabidopsis* R2R3-MYB transcription factor, AtMYB20, negatively regulates type2C serine/threonine protein phosphatases to enhance salt tolerance. *FEBS Letters* 587:1773–1778.

Dai, X., Xu, Y., Ma, Q., Xu, W., Wang, T., Xue, Y., and Chong, K. 2007. Overexpression of an R1R2R3 MYB gene OsMYB3R-2, increases tolerance to freezing, drought, salt stress in transgenic *Arabidopsis*. *Plant Physiology* 143:739–1751.

De Boer, K., Tilleman, S., Pauwels, L., Vanden Bossche, R., De Sutter, V., Vanderhaeghen, R., Hilson, P., Hamill, J.D., and Goossens, A. 2011. APETALA2/ETHYLENE RESPONSE FACTOR and basic helix-loop-helix tobacco transcription factors cooperatively mediate jasmonate-elicited nicotine biosynthesis. *Plant Journal* 66:1053–1065.

De Clercq, I., Vermeirssen, V., VanAken, O., Vandepoele, K., Murcha, M.W., Law, S.R., Inzé, A. et al. 2013. The membrane-bound NAC transcription factor ANAC013 functions in mitochondrial retrograde regulation of the oxidative stress response in *Arabidopsis*. *PlantCell* 25:3472–3490.

Ding Z., Li, S., An, X., Liu, X., Qin, H., and Wang, D. 2009. Transgenic expression of MYB15 confers enhanced sensitivity to abscisic acid and improved drought tolerance in *Arabidopsis thaliana*. *Journal of Genetics and Genomics* 36:17–29.

Dubos, C., Stracke, R., Grotewold, E., Weisshaar, B., Martin, C., and Lepiniec, L. 2010. MYB transcription factors in *Arabidopsis*. *Trends in Plant Science* 15:573–581.

Dubouzet, J.G., Sakuma, Y., Ito, Y., Kasuga, M., Dubouzet, E.G., Miura, S., Seki, M., Shinozaki, K., and Yamaguchi-Shinozaki, K. 2003. OsDREB genes in rice, *Oryza sativa* L., encode transcription activators that function in drought-, high-salt- and cold-responsive gene expression. *Plant Journal* 33:751–763.

Egawa, C., Kobayashi, F., Ishibashi, M., Nakamura, T., Nakamura, C., and Takumi, S. 2006. Differential regulation of transcript accumulation and alternative splicing of a DREB2 homolog under abiotic stress conditions in common wheat. *Genes & Genetic Systems* 81:77–91.

Englbrecht, C.C., Schoof, H., and Böhm, S. 2004. Conservation, diversification and expansion of $C_2H_2$ zinc finger proteins in the *Arabidopsis thaliana* genome. *BMC Genomics* 5:39.

Epstein, E., Norlyn, J.D., Rush, D.W., Kingsbury, R.W., Kelly, D.B., Cunningham, G.A., and Wrona, A.F. 1980. Saline culture of crops: A genetic approach. *Science* 210:399–404.

Eulgem, T., Rushton, P.J., Robatzek, S., and Somssich, I.E. 2000. The WRKY superfamily of plant transcription factors. *Trends in Plant Science* 5:199–206.

Eulgem T. and Somssich, I.E. 2007. Networks of WRKY transcription factors in defense signaling. *Current Opinion in Plant Biology* 10:366–371.

Fang, Y., You, J., Xie, K., Xie, W., and Xiong, L. 2008. Systematic sequence analysis and identification of tissue-specific or stress-responsive genes of NAC transcription factor family in rice. *Molecular Genetics and Genomics* 280:535–546.

Feng, H.L., Ma, N.N., Meng, X., Zhang, S., Wang, J.R., Chai, S., and Meng, Q.W. 2013. An ovel tomato MYC-type ICE1-like transcription factor, SlICE1a, confers cold, osmotic and salt tolerance in transgenic tobacco. *Plant Physiology and Biochemistry* 73:309–320.

Ferre-D'Amare, A.R., Pognonec, P., Roeder, R.G., and Burley, S.K. 1994. Structure and function of the b/HLH/Z domain of USF. *EMBO Journal* 13:180–189.

Fujimoto, S.Y., Ohta, M., Usui, A., Shinshi, H., and Ohme-Takagi, M. 2000. *Arabidopsis* ethylene-responsive element binding factors act as transcriptional activators or repressors of GCC box-mediated gene expression. *Plant Cell* 12:393–404.

Fujita, Y., Fujita, M., Shinozaki, K., and Yamaguchi-Shinozaki, K. 2011. ABA-mediated transcriptional regulation in response to osmotic stress in plants. *Journal of Plant Research* 124:509–525.

Ganesan, G., Sankararamasubramanian, H.M., Harikrishnan, M., Ganpudi, A., and Parida, A. 2012. A MYB transcription factor from the grey mangrove is induced by stress and confers NaCl tolerance in tobacco. *Journal of Experimental Botany* 63:4549–4561.

Gao, H., Song, A., Zhu, X., Chen, F., Jiang, J., Chen, Y., Sun, Y., Shan, H., Gu, C., Li, P., and Chen, S. 2012. The heterologous expression in *Arabidopsis* of a chrysanthemum Cys2/His2 zinc finger protein gene confers salinity and drought tolerance. *Planta* 235:979–993.

Gao, S.Q., Chen, M., Xia, L.Q., Xiu, H.J., Xu, Z.S., Li, L.C., Zhao, C.P., Cheng, X.G., and Ma, Y.Z. 2009. A cotton (*Gossypium hirsutum*) DRE-binding transcription factor gene, GhDREB, confers enhanced tolerance to drought, high salt, and freezing stresses in transgenic wheat. *Plant Cell Reports* 28:301–311.

Gao, S.Q., Chen, M., Xu, Z.S., Zhao, C.P., Li, L., Xu, H.J., Tang, Y.M., Zhao, X., and Ma, Y.Z. 2011. The soy bean GmbZIP1 transcription factor enhances multiple abiotic stress tolerances in transgenic plants. *Plant Molecular Biology* 75:537–553.

Golldack, D., Lüking, I., and Yang, O. 2011. Plant tolerance to drought and salinity: Stress regulating transcription factors and their functional significance in the cellular transcriptional network. *Plant Cell Reports* 30:1383–1391.

Gong, W., He, K., Covington, M., Dinesh-Kumarb, S.P., Snyder, M., Harmerc, S.L., Zhua, Y.X., and Deng, X.W. 2007. The development of protein microarrays and their applications in DNA-protein and protein-protein interaction analyses of *Arabidopsis* transcription factors. *Molecular Plant* 1:27–41.

Gourcilleau, D., Lenne, C., Armenise, C., Moulia, B., Julien, J.L., Bronner, G., and Leblanc-Fournier, N. 2011. Phylogenetic study of plant Q-type C2H2 zinc finger proteins and expression analysis of poplar genes in response to osmotic, cold and mechanical stresses. *DNA Research* 18:77–92.

Gruber, V., Blanchet, S., Diet, A., Zahaf, O., Boualem, A., Kakar, K., Alunni, B., Udvardi, M., Frugier, F., and Crespi, M. 2009. Identification of transcription factors involved in root apex responses to salt stress in *Medicago truncatula*. *Molecular Genetics and Genomics* 281:55–66.

Guan, Q., Wu, J., Yue, X., Zhang, Y., and Zhu, J. 2013. A nuclear calcium-sensing pathway is critical for gene regulation and salt stress tolerance in *Arabidopsis*. *PLoS Genetics* 9:e1003755.

Guo, Z.J., Chen, X.J., Wu, X.L., Ling, J.Q., and Xu, P. 2004. Overexpression of the AP2/EREBP transcription factor OPBP1 enhances disease resistance and salt tolerance in tobacco. *Plant Molecular Biology* 55:607–618.

Gutha, L.R. and Reddy A.R. 2008. Rice DREB1B promoter shows distinct stress-specific responses, and the overexpression of cDNA in tobacco confers improved abiotic and biotic stress tolerance. *Plant Molecular Biology* 68:533–555.

Gutterson, N. and Reuber, T.L. 2004. Regulation of disease resistance pathways by AP2/ERF transcription factors. *Current Opinion in Plant Biology* 7:465–471.

Haake, V., Cook, D., Riechmann, J.L., Pineda, O., Thomashow, M.F., and Zhang, J.Z. 2002. Transcription factor CBF4 is a regulator of drought adaptation in Arabidopsis. *Plant Physiology* 130:639–648.

Hao, Y.J., Wei, W., Song, Q.X., Chen, H.W., Zhang, Y.Q., Wang, F., Zou, H.F. et al. 2011. Soybean NAC transcription factors promote abiotic stress tolerance and lateral root formation in transgenic plants. *Plant Journal* 68:302–313.

He, X.J., Mu, R.L., Cao, W.H., Zhang, Z.G., Zhang, J.S., and Chen, S.Y. 2005. AtNAC2, a transcription factor downstream of ethylene and auxin signaling pathways, is involved in salt stress response and lateral root development. *Plant Journal* 44:903–916.

He, Y., Li, W., Lv, J., Jia, Y., Wang, M., and Xia, G. 2012. Ectopic expression of a wheat MYB transcription factor gene, *TaMYB73*, improves salinity stress tolerance in *Arabidopsis thaliana*. *Journal of Experimental Botany* 63:1511–1522.

Hong, J.P. and Kim, W.T. 2005. Isolation and functional characterization of the *Ca-DREBLP1* gene encoding a dehydration-responsive element binding-factor-like protein 1 in hot pepper (*Capsicum annuum* L. cv Pukang). *Planta* 220:875–888.

Hsieh, T.H., Li, C.W., Su, R.C., Cheng, C.P., Sanjaya, Tsai, Y.C., and Chan, M.T. 2010. A tomato bZIP transcription factor, SlAREB, is involved in water deficit and salt stress response. *Planta* 231:1459–1473.

Hu, H., Dai, M., Yao, J., Xiao, B., Li, X., Zhang, Q., and Xiong, L. 2006. Overexpressing a NAM, ATAF, and CUC (NAC) transcription factor enhances drought resistance and salt tolerance in rice. *Proceedings of the National Academy of Sciences of the United States of America* 103:12987–12992.

Hu, H., You, J., Fang, Y., Zhu, X., Qi, Z., and Xiong, L. 2008. Characterization of transcription factor gene SNAC2 conferring cold and salt tolerance in rice. *Plant Molecular Biology* 67:169–181.

Huang, J., Yang, X., Wang, M.M., Tang, H.J., Ding, L.Y., Shen, Y., and Zhang, H.S. 2007. A novel rice C2H2-type zinc finger protein lacking DLN-box/EAR-motif plays a role in salt tolerance. *Biochimica et Biophysica Acta* 69:220–227.

Huang, Z., Zhang, Z., Zhang, X., Zhang, H., Huang, D., and Huang R. 2004. Tomato TERF1 modulates ethylene response and enhances osmotic stress tolerance by activating expression of downstream genes. *FEBS Letters* 573:110–116.

Iida, A., Kazuoka, T., Torikai, S., Kikuchi, H., and Oeda, K. 2000. A zinc finger protein RHl41 mediates the light acclimatization response in *Arabidopsis*. *Plant Journal* 24:191–203.

Ito, Y., Katsura, K., Maruyama, K., Taji, T., Kobayashi, M., Seki, M., Shinozaki, K., and Yamaguchi-Shinozaki K. 2006. Functional analysis of rice DREB1/CBF-type transcription factors involved in cold-responsive gene expression in transgenic rice. *Plant Cell Physiology* 47:141–153.

Jakoby, M., Weisshaar, B., Dröge-Laser, W., Vicente-Carbajosa, J., Tiedemann, J., Kroj, T., and Parcy, F. 2002. bZIP transcription factors in *Arabidopsis*. *Trends in Plant Science* 7:106–111.

Ji, X., Wang, Y., and Liu, G. 2012. Expression analysis of MYC genes from *Tamarix hispida* in response to different abiotic stresses. *International Journal of Molecular Sciences* 13:1300–1313.

Jia, L., Clegg, M.T., and Jiang, T. 2004. Evolutionary dynamics of the DNA-binding domains in putative R2R3-MYB genes identified from rice subspecies indica and japonica genomes. *Plant Physiology* 134:575–585.

Jiang, C., Lu, B., and Singh, J. 1996. Requirement of a CCGAC cis-acting element for cold induction of the BN115 gene from winter *Brassica napus*. *Plant Molecular Biology* 30:679–684.

Jiang, W.B. and Yu, D.Q. 2009. *Arabidopsis* WRKY2 transcription factor may be involved in osmotic stress response. *Acta Botanica Yunnanica* 31: 427–432.

Jiang, Y. and Deyholos, M.K. 2006. Comprehensive transcriptional profiling of NaCl-stressed *Arabidopsis* roots reveals novel classes of responsive genes. *BMC Plant Biology* 6:25.

Jiang, Y., Yang, B., and Deyholos, M.K. 2009. Functional characterization of the *Arabidopsis* bHLH92 transcription factor in abiotic stress. *Molecular Genetics and Genomics* 282:503–516.

Jiang, Y.Q. and Deyholo, M.K. 2009. Functional characterization of *Arabidopsis* NaCl-inducible WRKY25 and WRKY33 transcription factors in abiotic stresses. *Plant Molecular Biology* 69: 91–105.

Jin, H. and Martin, C., 1999. Multifunctionality and diversity within the plant MYB-gene family. *Plant Molecular Biology* 41:577–585.

Jindra, M., Gaziova, I., Uhlirova, M., Okabe, M., Hiromi, Y., and Hirose, S. 2004. Coactivator MBF1 preserves the redox-dependent AP-1 activity during oxidative stress in *Drosophila*. *EMBO Journal* 23:3538–3547.

Jofuku, K.D., Boer, B., Montagu, M.V., and Okamuro, J.K. 1994. Control of *Arabidopsis* flower and seed development by the homeotic gene APETALA2. *Plant Cell* 6:1211–1225.

Jung, C., Seo, J.S., Han, S.W., Koo, Y.J., Kim, C.H., Song, S.I., Nahm, B.H., Choi, Y.D., and Cheong, J. 2008. Overexpression of AtMYB44 enhances stomatal closure to confer abiotic stress tolerance in transgenic *Arabidopsis*. *Plant Physiology* 146:623–635.

Kanei-Ishii, C., Sarai, A., Sawazaki, T., Nakagoshi, H., He, D.N., Ogata, K., Nishimura, Y., and Ishii, S. 1990. The tryptophan cluster: A hypothetical structure of the DNA-binding domain of the myb protooncogene product. *Journal of Biological Chemistry* 265:19990–19995.

Katiyar, A., Smita, S., Lenka, S.K., Rajwanshi, R., Chinnusamy, V., and Bansal, K.C. 2012. Genome-wide classification and expression analysis of MYB transcription factor families in rice and *Arabidopsis*. *BMC Genomics* 13:544.

Kiełbowicz-Matuk, A. 2012. Involvement of plant $C_2H_2$-type zinc finger transcription factors in stress responses. *Plant Science* 185–186:78–85.

Kim, J.C., Jeong, J.C., Park, H.C., Yoo, J.H., Koo, Y.D., Yoon, H.W., Koo, S.C., Lee, S.H., Bahk, J.D., and Cho, M.J. 2001a. Cold accumulation of SCOF-1 transcripts is associated with transcriptional activation and mRNA stability. *Molecules and Cells* 12: 204–208.

Kim, J.C., Lee, S.H., Cheong, Y.H., Yoo, C.M., Lee, S.I., Chun, H.J., Yun, D.J. et al. 2001b. A novel cold-inducible zinc finger protein from soybean, SCOF-1 enhances cold tolerance in transgenic plants. *Plant Journal* 25:247–259.

Kim, J.H., Nguyen, N.H., Jeong, C.Y., Nguyen, N.T., Hong, S.W., and Lee, H. 2013. Loss of the R2R3 MYB, At Myb73, causes hyper-induction of the SOS1 and SOS3 genes in response to high salinity in *Arabidopsis*. *Journal of Plant Physiology* 170:1461–1465.

Kim, S., Kang, J.Y., Cho, D.I., Park, J.H., and Kim, S.Y. 2004. ABF2, an ABRE-binding bZIP factor, is an essential component of glucose signaling and its overexpression affects multiple stress tolerance. *Plant Journal* 40:75–87.

Kim, S.G., Kim, S.Y., and Park, C.M. 2007. A membrane-associated NAC transcription factor regulates salt- responsive flowering via flowering locus T in *Arabidopsis*. *Planta* 226:647–654.

Kim, S.G., Lee, A.K., Yoon, H.K., and Park, C.M. 2008. A membrane-bound NAC transcription factor NTL8 regulates gibberellic acid-mediated salt signaling in *Arabidopsis* seed germination. *Plant Journal* 55:77–88.

Kim, Y.S., Kim, S.G., Park, J.E., Park, H.Y., Lim, M.H., Chua, N.H., and Park, C.M. 2006. A membrane-bound NAC transcription factor regulates cell division in *Arabidopsis*. *Plant Cell* 18:3132–3144.

Kirik, V., Simon, M., Huelskamp, M., and Schiefelbein, J. 2004. The ENHANCER OF TRY AND CPC1 (ETC1) gene acts redundantly with TRIPTYCHON and CAPRICE in trichome and root hair cell patterning in *Arabidopsis*. *Developmental Biology* 268:506–513.

Kitashiba, H., Ishizaka, T., Isuzugawa K., Nishimura, K.., and Suzuki, T. 2004. Expression of a sweet cherry DREB1/CBF ortholog in *Arabidopsis* confers salt and freezing tolerance. *Journal of Plant Physiology* 161:1171–1176.

Kizis, D. and Pages, M. 2002. Maize DRE-binding proteins DBF1 and DBF2 are involved in rab17 regulation through the drought responsive element in an ABA-dependent pathway. *Plant Journal* 30:679–689.

Kobayashi, F., Maeta, E., Terashima, A., and Takumi, S. 2008. Positive role of a wheat HvABI5 ortholog in abiotic stress response of seedlings. *Physiologia Plantarum* 134:74–86.

Kodaira, K.S., Qin, F., Tran, L.S., Maruyama, K., Kidokoro, S., Fujita, Y., Shinozaki, K., and Yamaguchi- Shinozaki, K. 2011. *Arabidopsis* Cys2/His2 zinc-finger proteins AZF1 and AZF2 negatively regulate abscisic acid-repressive and auxin-inducible genes under abiotic stress conditions. *Plant Physiology* 157:742–756.

Krizek, B.A. 2003. AINTEGUMENTA utilizes a mode of DNA recognition distinct from that used by proteins containing a single AP2 domain. *Nucleic Acids Research* 31:1859–1868.

Kuno, N., Moller, S.G., Shinomura, T., Xu, X.M., Chua, N.H., and Furuya, M. 2003. The novel MYB protein EARLY-PHYTOCHROME-RESPONSIVE1 is a component of a slave circadian oscillator in *Arabidopsis*. *Plant Cell* 15:2476–2488

Lee, J.H., Kim, D.M., Lee, J.H., Kim, J., Bang, J.W., Kim, W.T., and Pai, H.S. 2005. Functional characterization of NtCEF1, an AP2/EREBP-type transcriptional activator highly expressed in tobacco callus. *Planta* 222:211–224.

Lee, S.C., Choi, D.S., Hwang, I.S., and Hwang, B.K. 2010a. The pepper oxidoreductase CaOXR1 interacts with the transcription factor CaRAV1 and is required for salt and osmotic stress tolerance. *Plant Molecular Biology* 73:409–424.

Lee, S.J., Cho, D.I., Kang, J.Y., and Kim, S.Y. 2009. An ARIA-interacting AP2 domain protein is a novel component of ABA signaling. *Molecules and Cells* 27:409–416.

Lee, S.J., Kang, J.Y., Park, H.J., Kim, M.D., Bae, M.S., Choi, H.I., and Kim, S.Y. 2010b. DREB2C interacts with ABF2, a bZIP protein regulating abscisic acid-responsive gene expression, and its overexpression affects abscisic acid sensitivity. *Plant Physiology* 153:716–727.

Li, D., Su, Z., Dong, J., and Wang, T. 2009a. An expression database for roots of the model legume *Medicago truncatula* under salt stress. *BMC Genomics* 10:517.

Li, F., Guo, S., Zhao, Y., Chen, D., Chong, K., and Xu, Y. 2010. Overexpression of a homopeptide repeat-containing bHLH protein gene (OrbHLH001) from Dongxiang wild rice confers freezing and salt tolerance in transgenic *Arabidopsis*. *Plant Cell Reports* 29:977–986.

Li, H., Gao, Y., Xu, H., Dai, Y., Deng, D., and Chen, J. 2013a. ZmWRKY33, a WRKY maize transcription factor conferring enhanced salt stress tolerances in *Arabidopsis*. *Plant Growth Regulation* 70:207–216.

Li, S.J., Fu, Q.T., Chen, L.G., Huang, W.D., and Yu, D.Q. 2011. *Arabidopsis thaliana* WRKY25, WRKY26, and WRKY33 coordinate induction of plant thermotolerance. *Planta* 233:1237–1252.

Li, S.J., Fu, Q.T., Huang, W.D., and Yu, D.Q. 2009b. Functional analysis of an *Arabidopsis* transcription factor WRKY25 in heat stress. *Plant Cell Report* 28:683–693.

Li, X., Duan, X., Jiang, H., Sun, Y., Tang, Y., Yuan, Z., Guo, J. et al. 2006. Genome-wide analysis of basic/helix-loop-helix transcription factor family in rice and *Arabidopsis*. *Plant Physiology* 141:1167–1184.

Li, X.P., Tian, A.G., Luo, G.Z., Gong, Z.Z., Zhang, J.S., and Chen, S.Y. 2005. Soybean DRE-binding transcription factors that are responsive to abiotic stresses. *Theoretical and Applied Genetics* 110: 1355–1362.

Li, Y., Sun, Y., Yang, Q., Fang, F., Kang, J., and Zhang, T. 2013b. Isolation and characterization of a gene from *Medicago sativa* L., encoding abZIP transcription factor. *Molecular Biology Reports* 40:1227–1239.

Liao, Y., Zhang, J.S., Chen, S.Y., and Zhang, W.K. 2008a. Role of soybean GmbZIP132 under abscisic acid and salt stresses. *Journal of Integrative Plant Biology* 50:221–230.

Liao, Y., Zou, H.F., Wang, H.W., Zhang, W.K., Ma, B., Zhang, J.S., and Chen, S.Y. 2008b. Soybean GmMYB76, GmMYB92, and GmMYB177 genes confer stress tolerance in transgenic *Arabidopsis* plants. *Cell Research* 18:1047–1060.

Liao, Y., Zou, H.F., Wei, W., Hao, Y.J., Tian, A.G., Huang, J., Liu, Y.F., Zhang, J.S., and Chen, S.Y. 2008c. Soybean GmbZIP44, GmbZIP62 and GmbZIP78 genes function as negative regulator of ABA signaling and confer salt and freezing tolerance in transgenic *Arabidopsis*. *Planta* 228:225–240.

Licausi, F., Giorgi, F.M., Zenoni, S., Osti, F., Pezzotti, M., and Perata, P. 2010. Genomic and transcriptomic analysis of the AP2/ERF superfamily in *Vitis vinifera*. *BMC Genomics* 11:719–734.

Lippold, F., Sanchez, D.H., Musialak, M., Schlereth, A., Scheible, W., Hincha, D.K., and Udvardi, M.K. 2009. AtMyb41 regulates transcriptional and metabolic responses to osmotic stress in *Arabidopsis*. *Plant Physiology* 149:1761–1772.

Liu, C., Mao, B., Ou, S., Wang, W., Liu, L., Wu, Y., Chu, C., and Wang, X. 2014a. OsbZIP71, a bZIP transcription factor, confers salinity and drought tolerance in rice. *Plant Molecular Biology* 84:19–36.

Liu, G., Li, X., Jin, S., Liu, X., Zhu, L., Nie, Y., and Zhang, X. 2014b. Over expression of rice NAC gene SNAC1 improves drought and salt tolerance by enhancing root development and reducing transpiration rate in transgenic cotton. *PLoS One* 9:e86895.

Liu, J.X., Srivastava, R., and Howell, S.H. 2008. Stress-induced expression of an activated form of AtbZIP17 provides protection from salt stress in *Arabidopsis*. *Plant Cell Environment* 31:1735–1743.

Liu, N., Zhong, N.Q., Wang, G.L., Li, L.J., Liu, X.L., He, Y.K., and Xia, G.X. 2007. Cloning and functional characterization of PpDBF1 gene encoding a DRE-binding transcription factor from *Physcomitrella patens*. *Planta* 226:827–838.

Liu, Q., Kasuga, M., Sakuma, Y., Abe, H., Miura, S., Yamaguchi-Shinozaki, K., and Shinozaki, K. 1998. Two transcription factors, DREB1 and DREB2, with an EREBP/AP2 DNA binding domain, separate two cellular signal transduction pathways in drought- and low temperature-responsive gene expression, respectively, in *Arabidopsis*. *Plant Cell* 10: 1391–1406.

Liu, Q.L., Xu, K.D., Pan, Y.Z., Jiang, B.B., Liu, G.L., Jia, Y., and Zhang, H.Q. 2014c. Functional analysis of a novel chrysanthemum WRKY transcription factor gene involved in salt tolerance. *Plant Molecular Biology Reporter* 32:282–289.

Liu, Q.L., Xu, K.D., Zhao, L.J., Pan, Y.Z., Jiang, B.B., Zhang, H.Q., and Liu, G.L. 2011. Overexpression of a novel chrysanthemum NAC transcription factor gene enhances salt tolerance in tobacco. *Biotechnology Letters* 33:2073–2082.

Liu, Q.L., Zhong, M., Li, S., Pan, Y.Z., Jiang, B.B., Jia, Y., and Zhang, H.Q. 2013. Overexpression of a chrysanthemum transcription factor gene, DgWRKY3, in tobacco enhances tolerance to salt stress. *Plant Physiology and Biochemistry* 69:27–33.

Lucas, S., Durmaz, E., Akpınar, B.A., and Budak, H. 2011. The drought response displayed by a DRE-binding protein from *Triticum dicoccoides*. *Plant Physiology and Biochemistry* 49:346–351.

Ma, H.S., Liang, D., Shuai, P., Xia, X.L., and Yin, W.L. 2010. The salt- and drought-inducible poplar GRAS protein SCL7 confers salt and drought tolerance in *Arabidopsis thaliana*. *Journal of Experimental Botany* 61:4011–4019.

Magnani, E., Sjölander, K., and Hake, S. 2004. From endonucleases to transcription factors: Evolution of the AP2 DNA binding domain in plants. *Plant Cell* 16:2265–2277.

Mao, X., Chen, S., Li, A., Zhai, C., and Jing, R. 2014. Novel NAC transcription factor TaNAC67 confers enhanced multi-abiotic stress tolerances in *Arabidopsis*. *PLoS One* 9:e84359.

Mao, X., Zhang, H., Qian, X., Li, A., Zhao, G., and Jing, R. 2012. TaNAC2, a NAC-type wheat transcription factor conferring enhanced multiple abiotic stress tolerances in *Arabidopsis*. *Journal of Experimental Botany* 63:2933–2946.

Masaki, T., Mitsui, N., Tsukagoshi, H., Nishii, T., Morikami, A., and Nakamura, K. 2005. Activator of Spomin::LUC1/WRINKLED1 of *Arabidopsis thaliana* transactivates sugar-inducible promoters. *Plant and Cell Physiology* 46:547–556.

Miller, M. 2009. The importance of being flexible: The case of basic region leucine zipper transcriptional regulators. *Current Protein & Peptide Science* 10:244–269.

Mittler, R., Kim, Y., Song, L., Coutu, J., Coutu, A., Ciftci-Yilmaz, S., Lee, H., Stevenson, B., and Zhu, JK. 2006. Gain- and loss-of-function mutations in Zat10 enhance the tolerance of plants to abiotic stress. *FEBS Letters* 580:6537–6542.

Mizoi, J.Y., Shinozaki, K., and Yamaguchi-Shinozaki, K. 2012. AP2/ERF family transcription factors in plant abiotic stress responses. *Biochimica et Biophysica Acta* 1819:86–96.

Mochida, K., Yoshida, T., Sakurai, T., Yamaguchi-Shinozaki, K., Shinozaki, K., and Tran, L.S. 2009. In silico analysis of transcription factor repertoire and prediction of stress responsive transcription factors in soybean. *DNA Research* 16:353–369.

Muhammad, R., He, G., Yang, G., Javeed H., and Yan, X. 2012. AP2/ERF transcription factor in rice: Genome-wide canvas and syntenic relationships between monocots and eudicots. *Evolutionary Bioinformatics* 8:321–355.

Muñiz García, M.N., Giammaria, V., Grandellis, C., Téllez-Iñón, M.T., Ulloa, R.M., and Capiati, D.A. 2012. Characterization of StABF1, a stress-responsive bZIP transcription factor from *Solanum tuberosum* L. that is phosphorylated by StCDPK2 in vitro. *Planta* 235:761–778.

Murre, C., McCaw, P.S., and Baltimore, D. 1989. A new DNA binding and dimerization motif in immunoglobulin enhancer binding, daughterless, MyoD, and myc proteins. *Cell* 56:777–783.

Nagaoka, S., and Takano, T. 2003. Salt tolerance-related protein STO binds to a Myb transcription factor homologue and confers salt tolerance in *Arabidopsis*. *Journal of Experimental Botany* 54: 2231–2237.

Nakashima, K., Ito, Y., and Yamaguchi-Shinozaki, K. 2009. Transcriptional regulatory networks in response to abiotic stresses in *Arabidopsis* and grasses. *Plant Physiology* 149:88–95.

Nakashima, K., Shinwari, Z.K., Sakuma, Y., Seki, M., Miura, S., Shinozaki, K., and Yamaguchi-Shinozaki, K. 2000. Organization and expression of two *Arabidopsis* DREB2 genes encoding DRE-binding proteins involved in dehydration- and high-salinity-responsive gene expression. *Plant Molecular Biology* 42:657–665.

Nakashima, K., Tran, L.S., Van Nguyen, D., Fujita, M., Maruyama, K., Todaka, D., Ito, Y., Hayashi, N., Shinozaki, K., and Yamaguchi-Shinozaki, K. 2007. Functional analysis of a NAC-type transcription factor OsNAC6 involved in abiotic and biotic stress-responsive gene expression in rice. *Plant Journal* 51:617–630.

Ng, S., Ivanova, A., Duncan, O., Law, S.R., VanAken, O., DeClercq, I., Wang, Y. et al. 2013. Membrane-bound NAC transcription factor, ANAC017, mediates mitochondrial retrograde signaling in *Arabidopsis*. *Plant Cell* 25:3450–3471.

Nijhawan, A., Jain, M., Tyagi, A.K., and Khurana, J.P. 2007. Genomic survey and gene expression analysis of the basic leucine zipper transcription factor family in rice. *Plant Physiology* 146:333–350.

Niu, C.F., Wei, W., Zhou, Q.Y., Tian, A.G., Hao, Y.J., Zhang, W.K., and Chen, S.Y. 2012. Wheat WRKY genes *TaWRKY2* and *TaWRKY19* regulate abiotic stress tolerance in transgenic *Arabidopsis* plants. *Plant, Cell and Environment* 35:1156–1170.

Nole-Wilson, S. and Krizek, B.A. 2000. DNA binding properties of the *Arabidopsis* floral development protein AINTEGUMENTA. *Nucleic Acids Research* 21:4076–4082.

Ogata, K., Kanei-Ishii, C., Sasaki, M., Hatanaka, H., Nagadoi, A., Enari, M., Nakamura, H., Nishimura, Y., Ishii, S., and Sarai, A. 1996. The cavity in the hydrophobic core of Myb DNA binding domain is reserved for DNA recognition and trans-activation. *Nature Structural Biology* 3:178–187.

Oh, S.J., Song, S.I., Kim, Y.S., Jang, H.J., Kim, S.Y., Kim, M.J., Kim, Y.K., Nahm, B.H., and Kim, J.K. 2005. *Arabidopsis* CBF3/DREB1A and ABF3 in transgenic rice increased tolerance to abiotic stress without stunting growth. *Plant Physiology* 138:341–351.

Ohme-Takagi, M. and Shinshi, H. 1995. Ethylene-inducible DNA binding proteins that interact with an ethylene-responsive element. *Plant Cell* 7:173–182.

Olsen, A.N., Ernst, H.A., Leggio, L.L., and Skriver, K. 2005. NAC transcription factors: Structurally distinct, functionally diverse. *Trends in Plant Science* 10:79–87.

Pabo, C.O., Peisach, E., and Grant, R.A. 2001. Design and selection of novel Cys2His2 zinc finger proteins. *Annual Review of Biochemistry* 70:313–340.

Pan, Y., Seymour, G.B., Lu, C., Hu, Z., Chen, X., and Chen, G. 2012. An ethylene response factor (ERF5) promoting adaptation to drought and salt tolerance in tomato. *Plant Cell Reports* 31:349–360.

Pan, Y.J., Cho, C.C., and Kao, Y.Y. 2009. A novel WRKY-like protein involved in transcriptional activation of cyst wall protein genes in *Giardia lamblia*. *Journal of Biological Chemistry* 284:17975–17988.

Park, H.Y., Seok, H.Y., Woo, D.H., Lee, S.Y., Tarte V.N., Lee, E.H., Lee, C.H., and Moon, Y.H. 2011. *AtERF71/HRE2* transcription factor mediates osmotic stress response as well as hypoxia response in *Arabidopsis*. *Biochemical and Biophysical Research Communications* 414:135–141.

Park, J.M., Park, C.J., Lee, S.B., Ham, B.K., Shin, R., and Paek, K.H. 2001. Over-expression of the tobacco *Tsi1* gene encoding an EREBP/AP2–Type transcription factor enhances resistance against pathogen attack and osmotic stress in tobacco. *Plant Cell* 13:1035–1046.

Persak, H. and Pitzschke, A. 2014. Dominant repression by *Arabidopsis* transcription factor MYB44 causes oxidative damage and hypersensitivity to abiotic stress. *International Journal of Molecular Science* 15:2517–2537.

Pinheiro, G.L., Marques, C.S., Costa, M.D., Reis, P.A., Alves, M.S., Carvalho, C.M., Fietto, L.G., and Fontes, E.P. 2009. Complete inventory of soybean NAC transcription factors: Sequence conservation and expression analysis uncover their distinct roles in stress response. *Gene* 444:10–23.

Prabu, G. and Prasad, D.T. 2012. Functional characterization of sugarcane MYB transcription factor gene promoter (PScMYBAS1) in response to abiotic stresses and hormones. *Plant Cell Reports* 31:661–669.

Qin, J., Zhao, J., Zuo, K., Cao, Y., Ling, H., Sun, X., and Tang, K. 2004. Isolation and characterization of an ERF-like gene from *Gossypium barbadense*. *Plant Science* 167:1383–1389.

Qin, Y., Wang, M., Tian, Y., He, W., Han, L., and Xia, G. 2012. Over-expression of TaMYB33 encoding a novel wheat MYB transcription factor increases salt and drought tolerance in *Arabidopsis*. *Molecular Biology Reports* 39:7183–7192.

Qiu, D., Xiao, J., Xie, W., Liu, H., Li, X., Xiong, L., and Wang, S. 2008. Rice gene network inferred from expression profiling of plants overexpressing *OsWRKY13*, a positive regulator of disease resistance. *Molecular Plant* 1:538–551.

Qiu, Y. and Yu, D. 2009. Over-expression of the stress-induced *OsWRKY45* enhances disease resistance and drought tolerance in *Arabidopsis*. *Environmental and Experimental Botany* 65:35–47.

Ramegowda, V., Senthil-Kumar, M., Nataraja, K.N., Reddy, M.K., Mysore, K.S., and Udayakumar, M. 2012. Expression of a finger millet transcription factor, EcNAC1, in tobacco confers abiotic stress-tolerance. *PLoS One* 7:e40397.

Riechmann, J.L., Heard, J., Martin, G., Reuber, L., Jiang, C.Z., Keddie, J., Adam, L. et al. 2000. *Arabidopsis* transcription factors: Genome-wide comparative analysis among eukaryotes. *Science* 290:2105–2110.

Rizhsky, L., Davletova, S., Liang, H., and Mittler, R. 2004. The zinc finger protein Zat12 is required for cytosolic ascorbate peroxidase 1 expression during oxidative stress in *Arabidopsis*. *Journal of Biological Chemistry* 279:11736–11743.

Rodriguez-Uribe, L., Higbie, SM., Stewart, J.M., Wilkins, T., Lindemann, W., Sengupta-Gopalan, C., and Zhang, J. 2011. Identification of salt responsive genes using comparative microarray analysis in upland cotton (*Gossypium hirsutum* L.). *Plant Science* 180:461–469.

Rosinski, J.A. and Atchley, W.R. 1998. Molecular evolution of the Myb family of transcription factors: Evidence for polyphyletic origin. *Journal of Molecular Evolution* 46:74–83.

Rushton, D.L., Tripathi, P., Rabara, R.C., Lin, J., Ringler, P., Boken, A.K., and Rushton, P.J. 2012. WRKY transcription factors: Key components in abscisic acid signalling. *Plant Biotechnology Journal* 10:2–11.

Rushton, P.J., Bokowiec, M.T., Han, S., Zhang, H., Brannock, J.F., Chen, X., and Timko, M.P. 2008. Tobacco transcription factors: Novel insights into transcriptional regulation in the Solanaceae. *Plant Physiology* 147:280–295.

Rushton, P.J., Somssich, I.E., Ringler, P., and Shen, Q.J. 2010. WRKY transcription factors. *Trends in Plant Science* 15:247–258.

Saibo, N.J., Lourenc, O.T., and Oliveira, M.M. 2009. Transcription factors and regulation of photosynthetic and related metabolism under environmental stresses. *Annals of Botany* 103:609–623.

Sakamoto, H., Araki, T., Meshi, T., and Iwabuchi, M. 2000. Expression of a subset of the *Arabidopsis* Cys2/His2-type zinc-finger protein gene family under water stress. *Gene* 248:23–32.

Sakamoto, H., Maruyama, K., Sakuma, Y., Meshi, T., Iwabuchi, M., Shinozaki, K., and Yamaguchi-Shinozaki, K. 2004. *Arabidopsis* Cys2/His2-type zinc-finger proteins functions transcription repressors under drought, cold, and high-salinity stress conditions. *Plant Physiology* 136:2734–2746.

Sakata, Y., Nakamura, I., Taji, T., Tanaka, S., and Quatrano, R.S. 2010. Regulation of the ABA-responsive Em promoter by ABI3 in the moss *Physcomitrella patens*: Role of the ABA response element and the RY element. *Plant Signaling & Behavior* 5:1061–1066.

Sakuma, Y., Liu, Q., Dubouzet, J.G., Abe, H., Shinozaki, K., and Yamaguchi-Shinozaki, K. 2002. DNA-binding specificity of the ERF/AP2 domain of *Arabidopsis* DREBs, transcription factors involved in dehydration- and cold-inducible gene expression. *Biochemical and Biophysical Research Communications* 290:998–1009.

Saleh, A., Lumbreras, V., Lopez, C., Puigjaner, E., Kizis, D., and Pagès, M. 2006. Maize DBF1-interactor protein 1 containing an R3H domain is a potential regulator of DBF1 activity in stress responses. *Plant Journal* 46:747–757.

Saumonneau, A., Agasse, A., Bidoven, M.T., Lallemand, M., Cantereau, A., Medici, A., Laloi, M., and Atanassova, R. 2008. Interaction of grape ASR proteins with a DREB transcription factor in the nucleus. *FEBS Letters* 582:3281–3287.

Schellmann, S., Schnittger, A., Kirik, V., Wada, T., Okada, K., Beermann, A., Thumfahrt, J., Jürgens, G., and Hülskamp, M. 2002. TRIPTYCHON and CAPRICE mediate lateral inhibition during trichome and root hair patterning in *Arabidopsis*. *EMBO Journal* 21:5036–5046.

Schramm, F., Arkindale, L., Kiehlmann, E., Ganguli, A., Englich, G., Vierling, E., and vonKoskull-Döring, P. 2008. A cascade of transcription factor DREB2A and heat stress transcription factor HsfA3 regulates the heat stress response of *Arabidopsis*. *Plant Journal* 53:264–274.

Schütze, K., Harter, K., and Chaban, C. 2008. Post-translational regulation of plant bZIP factors. *Trends in Plant Science* 13:247–255.

Seo, P.J. and Park, C.M. 2010. A membrane-bound NAC transcription factor as an integrator of biotic and abiotic stress signals. *Plant Signaling & Behavior* 5:481–483.

Seo, Y.J., Park, J.B., Cho, Y.J., Jung, C., Seo, H.S., Park, S.K., Nahm, B.H., and Song, J.T. 2010. Overexpression of the ethylene-responsive factor gene BrERF4 from *Brassica rapa* increases tolerance to salt and drought in *Arabidopsis* plants. *Molecules and Cells* 30:271–277.

Shan, H., Chen, S., Jiang, J., Chen, F., Chen, Y., Gu, C., Li, P. et al. 2012. Heterologous expression of the chrysanthemum R2R3-MYB transcription factor CmMYB2 enhances drought and salinity tolerance, increases hypersensitivity to ABA and delays flowering in *Arabidopsis thaliana*. *Molecular Biotechnology* 51:160–173.

Sharma, M.K., Kumar, R., Solanke, A.U., Sharma, R., Tyagi, A.K., and Sharma, A.K. 2010. Identification, phylogeny, and transcript profiling of ERF family genes during development and abiotic stress treatments in tomato. *Molecular Genetics and Genomics* 284:455–475.

Shekhawat, U.K.S and Ganapathi, T.R. 2013. *MusaWRKY71* over expression in banana plants leads to altered abiotic and biotic stress responses. *PloS One* 8:e75506.

Shekhawat, U.K.S., Ganapathi, T.R., and Srinivas, L. 2011. Cloning and characterization of a novel stress- responsive WRKY transcription factor gene (*MusaWRKY71*) from *Musa spp.* cv. Karibale Monthan (ABB group) using transformed banana cells. *Molecular Biology Reports* 38:4023–4035.

Shen, Y.G., Zhang, W.K., Yan, D.Q., Du, B.X., Zhang, J.S., Liu, Q., and Chen, S.Y. 2003. Characterization of a DRE-binding transcription factor from a halophyte *Atriplex hortensis*. *Theoretical and Applied Genetics* 107:155–161.

Shi, W., Liu, D., Hao, L., Wu, C.A., Guo, X., and Li, H. 2014. GhWRKY39, a member of the WRKY transcription factor family in cotton, has a positive role in disease resistance and salt stress tolerance. *Plant Cell, Tissue and Organ Culture* 118:17–23.

Shigyo, M., Hasebe, M., and Ito, M. 2006. Molecular evolution of the AP2 subfamily. *Gene* 366:256–265.

Shigyo, M. and Ito, M. 2004. Analysis of gymnosperm two-AP2-domain-containing genes. *Development and Genes Evolution* 214:105–114.

Shinwari, Z.K., Nakashima, K., Miura, S., Kasuga, M., Seki, M., Yamaguchi-Shinozaki, K., and Shinozaki, K. 1998. An *Arabidopsis* gene family encoding DRE/CRT binding proteins involved in low-temperature -responsive gene expression. *Biochemical and Biophysical Research Communications* 250:161–170.

Shukla, R.K., Raha, S., Tripathi, V., and Chattopadhyay, D. 2006. Expression of *CAP2*, an APETALA2-family transcription factor from chickpea, enhances growth and tolerance to dehydration and salt stress in transgenic tobacco. *Plant Physiology* 142:113–123.

Sohn, K.H., Lee, S.C., Jung, H.W., Hong, J.K., and Hwang, B.K. 2006. Expression and functional roles of the pepper pathogen-induced transcription factor RAV1 in bacterial disease resistance, and drought and salt stress tolerance. *Plant Molecular Biology* 61:897–915.

Song, Y., Chen, L., Zhang, L., and Yu, D. 2010. Overexpression of *OsWRKY72* gene interferes in the ABA signal and auxin transport pathway of *Arabidopsis*. *Journal of Biosciences* 35:459–471.

Sonnenfeld, M.J., Delvecchio, C., and Sun, X. 2005. Analysis of the transcriptional activation domain of the *Drosophila tango* bHLH-PAS transcription factor. *Development Genes and Evolution* 215:221–229.

Srivastava, A.K., Ramaswamy, N.K., Suprasanna, P., and D'Souza, S.F. 2010. Genome- wide analysis of thiourea-modulated salinity stress-responsive transcripts in seeds of *Brassica juncea*: Identification of signalling and effector components of stress tolerance. *Annals of Botany* 106:663–674.

Srivastava, R., Deng, Y., and Howell, S.H. 2014. Stress sensing in plants by an ER stress sensor/transducer, bZIP28. *Frontiers in Plant Science* 5:59.

Sugano, S., Kaminaka, H., Rybka, Z., Catala, R., Salinas, J., Matsui, K., Ohme- Takagi, M., and Takatsuji, H., 2003. Stress-responsive zinc finger gene ZPT2-3 plays a role in drought tolerance in petunia. *Plant Journal* 36:830–841.

Sun, S.J., Guo, S.Q., Yang, X., Bao, Y.M., Tang, H.J., Sun, H., Huang, J., and Zhang, H.S. 2010. Functional analysis of a novel Cys2/His2-type zinc finger protein involved in salt tolerance in rice. *Journal of Experimental Botany* 61:2807–2818.

Takatsuji, H. 1999. Zinc-finger proteins: The classical zinc finger emerges in contemporary plant science. *Plant Molecular Biology* 39:1073–1078.

Takasaki, H., Maruyama, K., Kidokoro, S., Ito, Y., Fujita, Y., Shinozaki, K., Yamaguchi- Shinozaki, K., and Nakashima, K. 2010. The abiotic stress-responsive NAC-type transcription factor OsNAC5 regulates stress- inducible genes and stress tolerance in rice. *Molecular Genetics and Genomics* 284:173–183.

Tao, Z., Kou, Y., Liu, H., Li, X., Xiao, J., and Wang, S. 2011. OsWRKY45 alleles play different roles in abscisic acid signalling and salt stress tolerance but similar roles in drought and cold tolerance in rice. *Journal of Experimental Botany* 62:4863–4874.

Tezuka, K., Taji, T., Hayashi, T., and Sakata, Y. 2013. A novel abi5 allele reveals the importance of the conserved Ala in the C3 domain for regulation of downstream genes and salt tolerance during germination in *Arabidopsis*. *Plant Signaling & Behavior* 8:e23455.

Tian, Z.D., Zhang, Y., Liu, J., and Xie, C.H. 2010. Novel potato C2H2-type zinc finger protein gene, StZFP1, which responds to biotic and abiotic stress, plays a role in salt tolerance. *Plant Biology* 12:689–697.

Toledo-Ortiz, G., Huq, E., and Quail, P.H. 2003. The *Arabidopsis* basic/helix–loop–helix transcription factor family. *Plant Cell* 15:1749–1770.

Tran, L.S., Quach, T.N., Guttikonda, S.K., Aldrich, D.L., Kumar, R., Neelakandan, A., Valliyodan, B., and Nguyen, H.T. 2009. Molecular characterization of stress-inducible GmNAC genes in soybean. *Molecular Genetics and Genomics* 281:647–664.

Tran, L-S.P., Nakashima, K., Shinozaki, K., and Yamaguchi-Shinozaki, K. 2007. Plant gene networks in osmotic stress response: From genes to regulatory networks. *Methods in Enzymology* 428:109–128.

Ulker, B. and Somssich, I.E. 2004. WRKY transcription factors: From DNA binding towards biological function. *Current Opinion in Plant Biology* 7:491–498.

Vogel, J.T., Zarka, D.G., Van Buskirk, H.A., Fowler, S.G., and Thomashow, M.F. 2005. Roles of the CBF2 and ZAT12 transcription factors in configuring the low temperature transcriptome of *Arabidopsis*. *Plant Journal* 41:195–211.

Walley, J.W. and Dehesh, K. 2010. Molecular mechanisms regulating rapid stress signaling networks in *Arabidopsis*. *Journal of Integrative Plant Biology* 52:354–359.

Wan, L.Y., Wu, Y.S., Huang, J.Q., Dai, X.F., Lei, Y., Yan, L.Y., Jiang, H.F., Zhang, J.C., Varshney, R.K., and Liao, B.S. 2014. Identification of ERF genes in peanuts and functional analysis of AhERF008 and AhERF019 in abiotic stress response. *Functional & Integrative Genomics* DOI: 10.1007/s10142–014-0381–4.

Wang, Y., Dou, D., Wang, X., Li., Sheng, Y., Hua, C., Cheng, B., Chen, X., Zheng, X., and Wang, Y. 2009. The PsCZF1 gene encoding a C2H2 zinc finger protein is required for growth, development and pathogenesis in *Phytophthora sojae*. *Microbial Pathogenesis* 47:78–86.

Wang, F., Hou, X., Tang, J., Wang, Z., Wang, S., Jiang, F., and Li, Y. 2012. A novel cold-inducible gene from Pak-choi (*Brassica campestris* ssp. chinensis), BcWRKY46, enhances the cold, salt and dehydration stress tolerance in transgenic tobacco. *Molecular Biology Reports* 39:4553–4564.

Wang, H., Huang, Z., Chen, Q., Zhang, Z., Zhang, H., Wu, Y., Huang, D., and Huang, R. 2004. Ectopic over expression of tomato JERF3 in tobacco activates downstream gene expression and enhances salt tolerance. *Plant Molecular Biology* 55:183–192.

Wang, Q., Guan, Y., Wu, Y., Chen, H., Chen, F., and Chu, C. 2008. Over expression of a rice OsDREB1F gene increases salt, drought, and low temperature tolerance in both *Arabidopsis* and rice. *Plant Molecular Biology* 67:589–602.

Wang, R.K., Cao, Z.H., and Hao, Y.J. 2014a. Over expression of a R2R3 MYB gene MdSIMYB1 increases tolerance to multiple stresses in transgenic tobacco an dapples. *Physiologia Plantarum* 150:76–87.

Wang, S., Wei, X.L., Cheng, L.J., and Tong, Z.K. 2014b. Identification of a C2H2-type zinc finger gene family from *Eucalyptus grandis* and its response to various abiotic stresses. *Biologia Plantarum* 58:385–390.

Wang, X., Chen, X., Liu, Y., Gao H., Wang Z., and Sun, G. 2011. CkDREB gene in *Caragana korshinskii* is involved in the regulation of stress response to multiple abiotic stresses as an AP2/EREBP transcription factor. *Molecular Biology Reports* 38:2801–2811.

Wang, X., Dong, J., Liu, Y., and Gao, H. 2010. A novel dehydration-responsive element-binding protein from *Caragana korshinskiiis* involved in the response to multiple abiotic stresses and enhances stress tolerance in transgenic tobacco. *Plant Molecular Biology Reporter* 28:664–675.

Wang, Z.Y. and Tobin, E.M. 1998. Constitutive expression of the CIRCADIAN CLOCK ASSOCIATED 1 (CCA1) gene disrupts circadian rhythms and suppresses its own expression. *Cell* 93:1207–1217.

Wei, B., Zhang, R.Z., Guo, J.J., Liu, D.M., Li, A.L., Fan, R.C., Mao, L., and Zhang, X.Q. 2014. Genome-wide analysis of the MADS-box gene family in *Brachypodium distachyon*. *PLoS One* 9:e84781.

Wei, K., Chen, J., Wang, Y., Chen, Y., Chen, S., Lin, Y., Pan, S., Zhong, X., and Xie, D. 2012. Genome-wide analysis of bZIP- encoding genes in maize. *DNA Research* 19:463–476.

Wei, W., Zhang, Y., Han, L., Guan, Z., and Chai, T. 2008. A novel WRKY transcriptional factor from *Thlaspi caerulescens* negatively regulates the osmotic stress tolerance of transgenic tobacco. *Plant Cell Report* 27:795–803.

Weigel, D. 1995. The APETALA2 domain is related to a novel type of DNA binding domain. *Plant Cell* 7:388–389.

Welsch, R., Maass, D., Voegel, T., Della Penna, D., and Beyer, P. 2007. Transcription factor RAP2.2 and its interacting partner SINAT2: Stable elements in the carotene genesis of *Arabidopsis* leaves. *Plant Physiology* 145:1073–1085.

Wu, K.L., Guo, Z.J., Wang, H.H., and Li, J. 2005. The WRKY family of transcription factors in rice and *Arabidopsis* and their origins. *DNA Research* 12:9–26.

Xiang, Y., Tang, N., Du, H., Ye, H., and Xiong, L. 2008. Characterization of OsbZIP23 as a key player of the basic leucine zipper transcription factor family for conferring abscisic acid sensitivity and salinity and drought tolerance in rice. *Plant Physiology* 148:1938–1952.

Xiao, D.C., Zhang, Z.J., Xu, Y.W., Yang, L., Zhang, F.X., and Wang, C.L. 2013. Cloning and functional analysis of *Phyllostachys edulis* MYB transcription factor PeMYB2. *YiChuan* 35:1217–1225.

Xie, Z., Zhang, Z.L., Zou, X., Huang, J., Ruas, P., Thompson, D., and Shen, Q.J. 2005. Annotations and functional analyses of the rice WRKY gene superfamily reveal positive and negative regulators of abscisic acid signaling in aleurone cells. *Plant Physiology* 137:176–189.

Xiong, H., Li, J., Liu, P., Duan, J., Zhao, Y., Guo, X., Li, Y., Zhang, H., Ali, J., and Li, Z. 2014. Over expression of OsMYB48-1, a novel MYB-related transcription factor, enhances drought and salinity tolerance in rice. *Plos One* 9:e92913.

Xu, D.Q., Huang, J., Guo, S.Q., Yang, X., Bao, Y.M., Tang, H.J., and Zhang, H.S. 2008a. Over expression of a TFIIIA-type zinc finger protein gene ZFP252 enhances drought and salt tolerance in rice (*Oryza sativa* L.). *FEBS Letters* 582:1037–1043.

Xu, S., Wang, X.., and Chen, J. 2007a. Zinc finger protein 1 (ThZF1) from salt cress (*Thellungiella halophila*) is a Cys2/His2-type transcription factor involved in drought and salt stress. *Plant Cell Reports* 26:497–506.

Xu, Z.S., Chen, M., Li, L.C., and Ma, Y.Z. 2008b. Functions of the ERF transcription factor family in plants. *Botany* 86:969–977.

Xu, Z.S., Ni, Z.Y., Liu, L., Nie, L.N., Li, L.C., Chen, M., and Ma, Y.Z. 2008c. Characterization of the *TaAIDFa* gene encoding a CRT/DRE-binding factor responsive to drought, high-salt, and cold stress in wheat. *Molecular Genetics and Genomics* 280:497–508.

Xu, Z.S., Xia, L.Q., Chen, M., Cheng, X.G., Zhang, R.Y., Li, L.C., Zhao, Y.X. et al. 2007b. Isolation and molecular characterization of the *Triticum aestivum* L. ethylene-responsive factor 1 (*TaERF1*) that increases multiple stress tolerance. *Plant Molecular Biology* 65:719–732.

Xu, Z.S., Xia, L.Q., Chen, M., Cheng, X.G., Zhang, R.Y., Xu, Z.S., Chen, M., Li, L.C., and Ma, Y.Z. 2011. Functions and application of the AP2/ERF transcription factor family in crop improvement. *Journal of Integrative Plant Biology* 53:570–585.

Xue, G.P. and Loveridge, C.W. 2004. HvDRF1 is involved in abscisic acid-mediated gene regulation in barley and produces two forms of AP2 transcriptional activators, interacting preferably with a CT-rich element. *Plant Journal* 37:326–339.

Yamagishi, K., Tatematsu, K., Yano, R., Preston, J., Kitamura, S., Taka-hashi, H., McCourt, P., Kamiya, Y., and Nambara, E. 2009. CHOTTO1, a double AP2 domain protein of *Arabidopsis thaliana*, regulates germination and seedling growth under excess supply of glucose and nitrate. *Plant Cell Physiology* 50:330–340.

Yamaguchi-Shinozaki, K. and Shinozaki, K. 2005. Organization of cis-acting regulatory elements in osmotic- and cold-stress-responsive promoters. *Trends in Plant Science* 10:88–94.

Yamaguchi-Shinozaki, K. and Shinozaki, K. 2006. Transcriptional regulatory networks in cellular responses and tolerance to dehydration and cold stresses. *Annual Review of Plant Biology* 57:781–803.

Yáñez, M., Cáceres, S., Orellana, S., Bastías, A., Verdugo, I., Ruiz-Lara, S., and Casaretto, J.A. 2009. An abiotic stress-responsive bZIP transcription factor from wild and cultivated tomatoes regulates stress-related genes. *Plant Cell Reports* 28:1497–1507.

Yang, A., Dai, X., and Zhang, W. 2012. A R2R3-type MYB gene, OsMYB2, is involved in salt, cold, and dehydration tolerance in rice. *Journal of Experimental Botany* 63:2541–2556.

Yang, O., Popova, O.V., Süthoff, U., Lüking, I., Dietz, K.J., and Golldack, D. 2009. The *Arabidopsis* basic leucine zipper transcription factor AtbZIP24 regulates complex transcriptional networks involved in abiotic stress resistance. *Gene* 436:45–55.

Yang, W., Liu, X.D., Chi, X.J., Wu, C.A., Li, Y.Z., Song, L.L., Liu, X.M. et al. 2011. Dwarf apple MbDREB1 enhances plant tolerance to low temperature, drought, and salt stress via both ABA-dependent and ABA-independent pathways. *Planta* 233:219–229.

Yano, R., Kanno, Y., Jikumaru, Y., Nakabayashi, K., Kamiya, Y., and Nam-bara, E. 2009. CHOTTO1, a putative double APETALA2 repeat transcription factor, is involved in abscisic acid-mediated repression of gibberellin biosynthesis during seed germination in *Arabidopsis*. *Plant Physiology* 151:641–654.

Ying, S., Zhang, D.F., Fu, J., Shi, Y.S., Song, Y.C., Wang, T.Y., and Li, Y. 2012. Cloning and characterization of a maize bZIP transcription factor, ZmbZIP72, confers drought and salt tolerance in transgenic *Arabidopsis*. *Planta* 235:253–266.

Yokotani, N., Ichikawa, T., Kondou, Y., Matsui, M., Hirochika, H., Iwabuchi, M., and Oda, K. 2009. Tolerance to various environmental stresses conferred by the salt-responsive rice gene ONAC063 in transgenic *Arabidopsis*. *Planta* 229:1065–1075.

Yokotani, N., Ichikawa, T., Kondou, Y., Iwabuchi, M., Matsui, M., Hirochika, H., and Oda, K. 2013. Role of the rice transcription factor JA my bin abiotic stress response. *Journal of Plant Research* 126:131–139.

Yoon, H.K., Kim, S.G., Kim, S.Y., and Park, C.M. 2008. Regulation of leaf senescence by NTL9-mediated osmotic stress signaling in *Arabidopsis*. *Molecules and Cells* 25:438–445.

Yoshida, T., Fujita, Y., Sayama, H., Kidokoro, S., Maruyama, K., Mizoi, J., Shinozaki, K., and Yamaguchi-Shinozaki, K. 2010. AREB1, AREB2, and ABF3 are master transcription factors that cooperatively regulate ABRE-dependent ABA signaling involved in drought stress tolerance and require ABA for full activation. *Plant Journal* 61:672–685.

Zeng, Y., Raimondi, N., and Kermode, A.R. 2003. Role of an ABI3 homologue in dormancy maintenance of yellow-cedar seeds and in the activation of storage protein and Em gene promoters. *Plant Molecular Biology* 51:39–49.

Zhang, G., Chen, M., Chen, X., Xu, Z., Li, L., Guo, J., and Ma, Y. 2010. Isolation and characterization of a novel EAR-motif-containing gene GmERF4 from soybean (*Glycine max* L.). *Molecular Biology Reports* 37:809–818.

Zhang, G.Y., Chen, M., Li, L.C., Xu, Z.S., Chen, X.P., Guo, J.M., and Ma, Y.Z. 2009. Overexpression of the soybean GmERF3 gene, an AP2/ERF type transcription factor for increased tolerances to salt, drought, and diseases in transgenic tobacco. *Journal of Experimental Botany* 60:3781–3796.

Zhang, H., Huang, Z., Xie, B., Chen, Q., Tian, X., Zhang, X., Zhang, H., Lu, X., Huang, D., and Huang, R. 2004. The ethylene, jasmonate, abscisic acid and NaCl-responsive tomato transcription factor JERF1 modulates expression of GCC box-containing genes and salt tolerance in tobacco. *Planta* 220:262–270.

Zhang, L., Zhao, G., Jia, J., Liu, X., and Kong, X. 2011a. Molecular characterization of 60 isolated wheat MYB genes and analysis of their expression during abiotic stress. *Journal of Experimental Botany* 63:203–214.

Zhang, L., Zhao, G., Xia, C., Jia, J., Liu, X., and Kong, X. 2012. Overexpression of a wheat MYB transcription factor gene, TaMYB56-B, enhances tolerances to freezing and salt stresses in transgenic Arabidopsis. *Gene* 505:100–107.

Zhang, X., Ju, H., Chung, M., Huang, P., Ahn, S., and Kim, C.S. 2011b. The R-R-Type MYB-Like transcription factor, AtMYBL, is Involved in promoting leaf senescence and modulates an abiotic stress response in *Arabidopsis*. *Plant Cell Physiology* 52:138–148.

Zhang, X., Wang, L., Meng, H., Wen, H., Fan, Y., and Zhao, J. 2011c. Maize ABP9 enhances tolerance to multiple stresses in transgenic *Arabidopsis* by modulating ABA signaling and cellular levels of reactive oxygen species. *Plant Molecular Biology* 75:365–378.

Zhang, X., Zhen, J., Li, Z., Kang, D., Yang, Y., Kong, J., and Hua, J. 2011d. Expression profile of early responsive genes under salt stress in upland cotton (*Gossypium hirsutum* L.). *Plant Molecular Biology Reporter* 29:626–637.

Zhang, Y. and Wang, L. 2005. The WRKY transcription factor superfamily: Its origin in eukaryotes and expansion in plants. *BMC Evolutionary Biology* 5:1.

Zhang, Y., Chen, C., Jin, X.F., Xiong, A.S., Peng, R.H., Hong, Y.H., Yao, Q.H., and Chen, J.M. 2009b. Expression of a rice DREB1 gene, OsDREB1D, enhances cold and high-salt tolerance in transgenic *Arabidopsis*. *BMB Reports* 42:486–492.

Zhao, M.Y., Zhang, Z.B., Chen, S.Y., Zhang, J.S., and Shao, H.B. 2012. WRKY transcription factor superfamily: structure, origin and functions. *African Journal of Biotechnology* 11:8051–8059.

Zheng, X., Chen, B., Lu, G., and Han, B. 2009. Overexpression of a NAC transcription factor enhances rice drought and salt tolerance. *Biochemical and Biophysical Research Communications* 379:985–989.

Zhou, J., Li, F., Wang, J.L., Ma, Y., Chong, K., and Xu, Y.Y. 2009. Basic helix–loop–helix transcription factor from wild rice (OrbHLH2) improves tolerance to salt- and osmotic stress in *Arabidopsis*. *Journal of Plant Physiology* 166:1296–1306.

Zhou, L., Wang, N.N., Kong, L., Gong, S.Y., Li, Y., and Li, X.B. 2014. Molecular characterization of 26 cotton WRKY genes that are expressed differentially in tissues and are induced in seedlings under high salinity and osmotic stress. *Plant Cell, Tissue and Organ Culture* 119:141–156.

Zhou, Q.Y., Tian, A.G., Zou, H.F. Xie, Z.M., Lei, G., Huang, J., Wang, C.M., and Chen, S.Y. 2008. Soybean WRKY-type transcription factor genes, GmWRKY13, GmWRKY21, and GmWRKY54, confer differential tolerance to abiotic stresses in transgenic *Arabidopsis* plants. *Plant Biotechnology Journal* 6:486–503.

Zhu, Q., Zhang, J.T., Gao, X.S., Tong J.H., Xiao L.T., Li W.B., and Zhang, H.X. 2010. The *Arabidopsis* AP2/ERF transcription factor RAP2.6 participates in ABA, salt and osmotic stress responses. *Gene* 457:1–12.

Zhuang, J., Sun, C.C., Zhou, X.R., Xiong, A.S., and Zhang, J. 2011. Isolation and characterization of an AP2/ERF-RAV transcription factor BnaRAV-1-HY15 in *Brassica napus* L. HuYou15. *Molecular Biology Reports* 38:3921–3928.

Zimmermann, P., Hirsch-Hoffmann, M., Hennig, L., and Gruissem, W. 2004. Genevestigator. *Arabidopsis* microarray database and analysis toolbox. *Plant Physiology* 136:2621–2632.

Zou, M., Guan, Y., Ren, H., Zhang, F., and Chen, F. 2008. A bZIP transcription factor, OsABI5, is involved in rice fertility and stress tolerance. *Plant Molecular Biology* 66:675–683.

# CHAPTER NINE

# Glyoxalase System and Salinity Stress in Plants

*Preeti Singh and Neeti Dhaka*

## CONTENTS

**Abstract.** The crop productivity and quality of plants is adversely affected by salinity stress. Salt stress imposes ionic imbalance and hyperosmotic stress in plants, which ultimately elicit secondary stresses such as oxidative damage. Accumulation of a cytotoxic compound methylglyoxal (MG) is a general stress response in plants during salinity stress. MG is a dicarbonyl compound and causes degradation of proteins by formation of advanced glycation end products. Therefore, in order to survive under stressful conditions, plants must upregulate MG detoxification process to avoid cellular damage. Glyoxalase system plays a crucial role in abiotic stress tolerance by maintaining an appropriate level of MG by regulating GSH-based reactive oxygen species detoxification. Recent studies have shown that glyoxalase systems are important for stress tolerance in plants and modulation of the detoxification pathway making the plants more tolerant to various abiotic stresses. In this chapter, we summarize the current knowledge on the glyoxalase enzymes and their proposed role during salinity stress tolerance.

*Keywords:* glyoxalase enzymes, methylglyoxal, reactive oxygen species, salt stress

## 9.1 INTRODUCTION

### 9.1.1 Salt Stress and Its Impact on Plants

Food insecurity has increased in recent times leading to competing claims for land, water, labor, energy, and capital and increased pressure to improve crop production per unit of land (Godfray et al., 2010; Varshney et al., 2011). Plants are sessile organisms that continuously encounter adverse environmental conditions during growth in their natural environment. Among such environmental stresses, drought and salinity are the greatest environmental constraints worldwide for agriculture that decrease crop yields by 50% worldwide and annual losses estimated at billions of dollars (Wang et al., 2003; Mittler, 2006). Soil salinity covers an estimated 45 million hectares of irrigated land worldwide (FAO, 2004). The scenario is even more serious due to climate changes associated with global warming and irrigation practices (Munns and Tester, 2008). Global temperatures are likely to increase by 2.5°C–4.3°C by the end of the century (IPCC, 2007), with significant effects on food production and malnutrition (Varshney et al., 2011). Soils containing sufficient salt in the root zone to impose the growth of crop plants are defined as saline soils. According to the USDA Salinity Laboratory saline soils are those having electrical conductivity (EC) of 4 dS m$^{-1}$ or more. EC is the electrical conductivity of the "saturated paste extract" that is of the solution extracted from a soil sample after being mixed with sufficient water to produce a saturated paste. Soil salinity has been divided in four different categories based on the saline irrigation water. EC of 0.25 dS m$^{-1}$ is defined as low salinity, medium salinity (0.25–0.75 dS m$^{-1}$), high salinity (0.75–2.25 dS m$^{-1}$), and very high salinity with an EC exceeding 2.25 dS m$^{-1}$ (U.S. Salinity Laboratory Staff, 1954).

Salinity stress causes changes in various physiological and metabolic processes and ultimately inhibits crop production depending on the severity and duration of the stress (Gupta and Huang, 2014). The primary consequences of salt stress are the reduction in the production of new leaves and inhibition of shoot growth. The initial response of salt stress occurs in the form of osmotic stress because of high salt accumulation in the soil and plants. Osmotic stress causes various changes in the physiological processes, like stomatal closure, membrane disruption, and imbalance between reactive oxygen species (ROS) production and antioxidant enzymes (Rahnama et al., 2010).

The second phase of salinity stress is the "ionic phase" that causes accumulation of ions in plant tissues like Na$^+$, Ca$^{2+}$ and Mg$^{2+}$, Cl$^-$, SO4$^{2-}$, and HCO$_3^-$. Among them, Na$^+$ and Cl$^-$ ions are considered as the most important ions (Hasegawa et al., 2000) as entry of both ions into the plant cells leads to ionic imbalance. High Na$^+$ concentration inhibits uptake of K$^+$ ions and thus perturbs the Na$^+$/K$^+$ ratio. K$^+$ ion is an essential element for growth and development that results into lower productivity and may even lead to death (James et al., 2011). Another consequence of salinity stress is the production of ROS, such as singlet oxygen, superoxide, hydroxyl radical, and hydrogen peroxide (Ahmad and Prasad, 2012). Salinity-induced ROS formation can lead to oxidative damages in various cellular components such as proteins, lipids, and DNA, interrupting vital cellular functions of plants.

### 9.1.2 Mechanism of Salinity Tolerance

To combat these stresses, plants have evolved a series of complex responsive mechanisms, to perceive and respond to external stimuli via multiple signaling pathways. Since abiotic stress tolerance is multigenic and quantitative in nature (Collins et al., 2008), a major challenge exists to understand the mechanisms by which environmental signals are perceived by plants and adaptive responses are activated. This is of critical importance for devising rational breeding and transgenic strategies for developing stress tolerance crops.

Plants develop various physiological and biochemical mechanisms in order to survive in soils with high salt concentration. These strategies include (1) ion homeostasis and compartmentalization, (2) control of ion uptake by roots and transport into leaves, (3) synthesis of osmolytes and compatible solutes, (4) change in photosynthetic pathway, (5) alteration in membrane structure, (6) induction of antioxidative enzymes and antioxidant compounds, and (7) induction of plant hormones.

#### 9.1.2.1 Ion Homeostasis and Compartmentalization

Maintenance of ion homeostasis and compartmentalization is crucial for plant growth during salt stress (Gupta and Huang, 2014). Glycophytic plants restrict uptake of salt and adjust their osmotic potential by synthesizing compatible solutes, while halophytes compartmentalize the accumulated salts in cell vacuoles and thus maintain high cytosolic $K^+/Na^+$ ratio in their cells. Two types of antiporters are involved in $Na^+$ efflux and compartmentalization. The first is a plasma membrane $Na^+/H^+$ antiporter also known as salt overly sensitive 1 (SOS1), and the second one is tonoplast $Na^+/H^+$ antiporter (NHX1) (Zhu, 2003; Rodríguez-Rosales et al., 2009). The $Na^+$ ions present in the cytosol are transported to the vacuoles by $Na^+/H^+$ antiporter (NHX1). AtNHX1 from *Arabidopsis* was the first reported $Na^+/H^+$ antiporter and suggested to compartmentalize $Na^+$ ions in vacuoles (Rodríguez-Rosales et al., 2009). To date, a large number of $Na^+/H^+$ antiporter homologues have been isolated from many plant sources, that is, *Atriplex gmelini* (Hamada et al., 2001), *Suaeda salsa* (Li et al., 2007), *Halostachys caspica* (Guan et al., 2011), *Salicornia brachiata* (Jha et al., 2011), *Karelinia caspica* (Liu et al., 2012), *Leptochloa fusca* (Rauf et al., 2014), and *Kosteletzkya virginica* (Wang et al., 2014). Transgenic plants overexpressing $Na^+/H^+$ antiporter genes from different plant sources were able to grow well and set seeds under higher salt conditions (Apse et al., 1999; Ohta et al., 2002; Yin et al., 2004; Luming et al., 2006). Two types of $H^+$ pumps are present in the vacuolar membrane: vacuolar-type $H^+$-ATPase (V-ATPase) and the vacuolar pyrophosphatase (V-PPase) (Dietz et al., 2001; Wang et al., 2001).

The role of SOS stress signaling pathway in ion homeostasis and salt tolerance has been documented in the literature (Sanders, 2000). The SOS pathway consists of three major proteins, SOS1, SOS2, and SOS3. SOS1 acts as a plasma membrane $Na^+/H^+$ antiporter and regulates $Na^+$ efflux and helps in long-distance transport of $Na^+$ from root to shoot. Overexpression of this protein confers salt tolerance in plants (Shi et al., 2002). SOS2 gene is a serine/threonine kinase and gets activated by $Ca^+$ signals (Liu et al., 2000). The third protein SOS3 is a myristoylated $Ca^+$-binding protein and plays essential role in salt tolerance (Ishitani et al., 2000).

### 9.1.2.2 Synthesis of Compatible Osmolytes

Compatible osmolytes are chemically small molecules, water soluble, and nontoxic to cells. These are synthesized intracellular in response to salt stress and help in stabilizing proteins and protecting cellular structures by balancing the osmotic pressure of the cell (Yancey et al., 1982). Osmoprotectants like proline (Szabados and Savoure, 2010; Nounjan et al.,

2012), glycine betaine (GB) (Ashraf and Foolad, 2007; Chen and Murata, 2008), sugar (Siringam et al., 2011; Redillas et al., 2012), and polyols (Palma et al., 2013; Saxena et al., 2013) are common candidates for ionic adjustments in diverse organisms. These solutes act as low-molecular-weight chaperones because of their hydrophilic nature and also help in ROS scavenging (Gupta and Haung, 2014). It has been reported in the literature that proline and GB are the best known compatible solutes and get accumulated in response to drought and salinity stress (Munns, 2002; Szabados and Savoure, 2010).

Proline is the most widely distributed osmolyte accumulated in plants. In plant systems, pyrroline-5-carboxylase synthetase catalyses the production of proline from glutamate. The suggested role of proline under stress conditions involves maintenance of osmoticum and stabilization of enzymes and membranes, and it also acts as a reservoir of energy and nitrogen for utilization during salinity stress conditions (Bandurska, 1993; Gzik, 1996). GB is an amphoteric quaternary ammonium compound ubiquitously present in prokaryotic and eukaryotic cells and is electrically neutral over a wide range of pH. It is extremely soluble in water despite a nonpolar moiety consists of 3-methyl groups. GB not only acts as osmoregulator but also interacts both with hydrophobic and hydrophilic domains of the enzymes and protein complexes, thereby stabilizing their structure and protecting the photosynthetic apparatus from stress damages (Cha and Kirdmanee, 2010). It has also been suggested that GB plays a role as a scavenger of ROS generated during various stresses (Ashraf and Foolad, 2007). Metabolic engineering of GB biosynthesis in plants led to improved salt, drought, and extreme temperature stress tolerance in non-GB-accumulator plants (Chen and Murata, 2008).

### 9.1.2.3 Modulation of Antioxidant Machinery

A secondary effect of salt stress in plant systems is the generation of ROS. Salt stress indirectly imposes a water-deficit condition on plants and causes osmotic effects on a wide variety of metabolic activities. This water deficit leads to the formation of ROS such as superoxide ($O_2^{\cdot-}$), hydrogen peroxide ($H_2O_2$), hydroxyl radical ($^{\cdot}OH$), and singlet oxygen ($^1O_2$) (Halliwell and Gutteridge, 1986). The alleviation of this oxidative damage is imperative for the survival of plants under salt stress. Plants possess both enzymatic and nonenzymatic mechanisms for scavenging of ROS. Nonenzymatic antioxidants involve ascorbic acid (vitamin C), glutathione (GSH), alpha-tocopherols (vitamin E), carotenoids, and flavonoids (Mittler et al., 2004). The enzymatic mechanisms are designated to minimize the concentration of $O_2^-$ and $H_2O_2$. Plants employ a diverse array of enzymes such as superoxide dismutase (SOD) (EC 1.15.1.1), catalase (CAT) (EC 1.11.1.6), ascorbate peroxidase (APX) (EC 1.11.1.11), glutathione S-transferase (EC 2.5.1.18), and glutathione peroxidase (GPX) (EC 1.11.1.9) to scavenge ROS. Superoxide radical is regularly synthesized in the chloroplast and mitochondria and scavenged by SOD results in the production of $H_2O_2$. CAT, APX, and a variety of peroxidases catalyze the breakdown of $H_2O_2$ to water and oxygen (Chang et al., 1984; Garratt et al., 2002). There have been many reports on overexpression of antioxidant enzymes in transgenic plants for salt stress tolerance. The upregulation of SOD is implicated in combating oxidative stress caused by biotic and abiotic stress and has a critical role in the survival of plants under stressed environments. Significant increase in SOD activity under salt stress has been observed in various plants, namely, mulberry (Harinasut et al., 2003), *Cicer arietinum* (Kukreja et al., 2005), and *Lycopersicon esculentum* (Gapinska et al., 2008). Overexpression of a Mn-SOD in transgenic *Arabidopsis* plants also showed increased salt tolerance (Wang et al., 2004). It has also been noted that overexpression of APX in *Nicotiana tabacum* chloroplasts enhanced plant tolerance to salt and water deficit (Badawi et al., 2004). Transgenic *Arabidopsis* plants overexpressing *OsAPXa* or *OsAPXb* gene exhibited increased salt tolerance. It was found that the overproduction of *OsAPXb* enhanced and maintained APX activity to a much higher extent than *OsAPXa* in transgenic plants under different NaCl concentrations (Lu et al., 2007).

### 9.1.3 MG: A Toxic Marker under Stress in Plants

The glyoxalase system and ROS detoxification systems play an integral part in stress tolerance because excessive accumulation of methylglyoxal (MG) and production of ROS is a common factor in both the abiotic and biotic stresses in plants (Hossain et al., 2011). Unchecked ROS and MG accumulation is highly toxic to plants as these can react with proteins, lipids, and nucleic acids leading to irreparable metabolic catastrophe. Plants have an array of enzymatic and nonenzymatic scavenging pathways or detoxification systems that cooperate to counter the deleterious effects of ROS and MG. In plants, MG is detoxified mainly via the glyoxalase system. Besides detoxification of MG, the glyoxalase system also plays a role in oxidative stress tolerance by recycling reduced GSH that would be trapped nonenzymatically by MG to form hemithioacetal, thereby maintaining GSH homeostasis. The results of numerous recent studies have shown that the alleviation of oxidative damage and increased resistance to abiotic stresses are often correlated with the more efficient glyoxalase systems.

MG is a highly reactive and mutagenic $\alpha\beta$-dicarbonyl aldehyde compound involved in protein modifications through formation of advanced glycation end products and inactivation of antioxidant defense pathways (Wu and Juurlink, 2002; Hoque et al., 2010). It can react with the side chains of amino acids, arginine, lysine, and cysteine and with the base guanine, adenine, and cytosine (Hossain et al., 2011). It also causes increased sister chromatid exchange, endoreduplication, DNA strand breaks, and point mutations (Chaplen, 1998). It can be synthesized enzymatically or nonenzymatically by the elimination of phosphate from triose phosphate intermediate of glycolysis, glyceraldehydes-3-phosphate, and dihydroxyacetone phosphate (Richard, 1993).

Several detoxification pathways are known in prokaryotes and eukaryotes, which prevent and control the accumulation of MG at the level of detoxification. It is mainly catabolized by two major enzymatic pathways, the glyoxalase pathway (Thornalley, 1990) and an alternate pathway involving aldose/aldehyde reductases (Vander Jagt and Hunsaker, 2003). GSH-dependent glyoxalase system is considered to be the major pathway involved in MG catabolism in all organisms. Aldose reductase converts MG to acetol using NADPH and NADH as a cofactor (Hossain et al., 2011).

#### 9.1.3.1 MG Induction under Abiotic and Biotic Stresses

MG is endogenously produced in all biological systems, including higher plants. Under stress and disease conditions, MG accumulation is found to be increased in animals, mammals, yeast, and bacterial and plant systems (Thornally, 1990; Wu and Juurlink, 2002; Yadav et al., 2005). Induction of MG level in response to various abiotic stresses in different crop species like rice, *Pennisetum*, tobacco, and *Brassica* seedlings has been reported for the first time by Yadav et al. (2005). MG is also shown to be induced by hormones (2, 4-D, abscisic acid [ABA]) and white light within 24 h of stress in pumpkin seedlings. Induction of MG to biotic stresses has also been reported in the literature. MG level was found to be upregulated by 2.5-fold in maize genotype (G4666) susceptible to fungus infected with *Aspergillus flavus* (Chen et al., 2004). Approximately, twofold increases were observed in MG content in tobacco BY-2 cells against 200 mM NaCl stress (Banu et al., 2010). MG level was also found to be increased in tobacco plants exposed to salt and drought stress (Kumar et al., 2013; Ghosh et al., 2014). The rapid increase in the level of MG in plants under biotic and abiotic stresses suggests that it is a general stress response and MG might act as a signal for plants to respond to stress.

### 9.1.4 Glyoxalase System

For almost a hundred years, the glyoxalase system has been known to exist in animals. This pathway is ubiquitously present from prokaryotes to eukaryotes and is vital for many biological functions (Thornalley, 1990). This system was independently discovered as a catabolic pathway involved in the conversion of glucose to D-lactate, but in plant systems, it was reported in the late twentieth century (Kaur et al., 2014). It comprises two enzymes, glyoxalase I (Gly I) (EC 4.4.1.5, lactogluthione lyase) and glyoxalase II (Gly II) (EC 3.1.2.6, hydroxyacylglutathione hydrolase), and a catalytic amount of reduced GSH as a cofactor. Gly I is the first enzyme of the pathway that catalyses the formation of S-lactoylglutathione (SLG) from hemithioacetal, which is the product of a non-enzymatic reaction between GSH and MG. Gly II catalyses the hydrolysis of SLG to D-lactate with the regeneration of GSH that is used again in the glyoxalase catalyzed reaction. Glyoxalases are present as multigene families in plants. In *Arabidopsis*, 11 homologues of Gly I and 5 isoforms of Gly II are present, while in rice, 11 *Gly* I genes and 3 *Gly* II have been reported (Kaur et al., 2014). The presence of different isoforms of glyoxalases suggests an important role of glyoxalase system in plants.

#### 9.1.4.1 Glyoxalase Enzymes

##### 9.1.4.1.1 GLYOXALASE I

Glyoxalase I is a member of the vicinal oxygen chelate superfamily that uses a divalent metal ion to perform catalysis. The catalytic reaction involves base-catalyzed shielded-proton transfer from C2 to C2 of the hemithioacetal to form an ene diol intermediate and rapid ketonization to the thioester product. Gly I is present in the cytosol of cells and, its activity prevents the accumulation of reactive $\alpha$-oxoaldehydes and, thereby, suppresses $\alpha$-oxoaldehyde mediated glycation defense. Gly I activity has been detected in almost all organisms studied so far including all human tissues, rat, mouse, sheep, and monkey (Larsen et al., 1985; Thornalley, 1993); however, it is absent in some organisms like the protozoans *Entamoeba histolytica*, *Giardia lamblia*, and *Trypanosoma brucei* (Sousa Silva et al., 2012). Among plants, the glyoxalase I activity was first reported from Douglas fir needles (Smits and Johnson, 1981). Deswal and Sopory (1998) have showed the Gly I activity in various parts of the plant including roots, shoots, seeds, cotyledons, and different floral organs. Gly I is preferentially localized in cytosol; however, peroxisomal localization has also been reported in *Arabidopsis* (Quan et al., 2010).

Structurally the human, bacterial, and plant Gly I enzymes are dimeric in nature. Both, hetero- and homodimers have been reported from animals, while in microorganisms, Gly I exist in monomeric form (Thornalley, 1990). The active site is situated in a barrel that is only formed at the dimer interface and consists of $Zn^{2+}$ or $Ni^{+2}$ metal center and the binding site for prosthetic group and hemithioacetal substrate (Cameron et al., 1997). The side chains involved are Glu33 and Glu99 from one subunit and His126 and Glu172 from the other. In mammals and higher eukaryotes, Gly I functions as a dimeric metalloenzyme, using $Zn^{2+}$ as a cofactor, while $Ni^{2+}$ is needed for its activation in *Escherichia coli* (Clugston et al., 1998).

Gly I has been shown to play a role in animal tumor growth and is involved in the regulation of the cell cycle in microorganisms (Rhee et al., 1986). Inhibition of Gly I with specific inhibitors leads to the accumulation of $\alpha$-oxoaldehydes to cytotoxic levels. In higher plants (*Pisum sativum*), Gly I activity was found to be correlated with the mitotic index, and in *Datura* callus culture, its activity increased with increase in DNA synthesis (Ramaswamy et al., 1984).

Previous reports revealed that *Gly* I gene of plants could enhance salt stress tolerance. Initially, Espartero et al. (1995) showed two- to threefold increase in Gly I mRNA and polypeptide levels in various tissues of tomato plants treated with NaCl. *Gly I* gene expression from *Brassica* was also shown to be upregulated by salinity, mannitol, and heavy metal tolerance (Veena et al., 1999). Induction of the Gly I activity was also observed in the pumpkin seedlings subjected to white light, salinity, MG, and heavy metal stresses (Hossain et al., 2009). In wheat seedlings, the expression of Gly I transcript was induced under NaCl stress (100 mM), while it was repressed at higher NaCl concentrations that may be because of posttranscriptional or translational regulation of Gly I in salt responses (Lin et al., 2010). Microarray analysis of salt-tolerant and salt-sensitive cultivars of tomato has revealed the salt-induced upregulation of *Gly I* gene in the tolerant cultivar (Sun et al., 2010).

### 9.1.4.1.2 GLYOXALASE II

Glyoxalase II (E.C.1.2.6, hydroxyacylglutathione hydrolase) is a metalloenzyme that is structurally related to other metallohydrolases such as metallo-β-lactamases and aryl sulfatase involved in cellular chemical detoxification. Gly II catalyses the hydrolysis of SLG to D-lactic acid and regenerates the reduced GSH consumed in the Gly I catalyzed reaction (Uotila, 1989). Gly II has an imidazole group that is significant for its catalytic activity. A nucleophilic residue is responsible for the attack on the carbonyl carbon of the thiolester substrate forming an acylated enzyme with concomitant release of GSH (Ball and Vander Jagt, 1981). The amino acid sequence deduced from the *Arabidopsis thaliana* cDNA showed 54% identity with that of the human enzyme having eight histidine residues conserved (Ridderstrom and Mannervik, 1997). Characterization of the recombinant enzyme revealed that the plant and human enzyme are not only structurally but also functionally similar.

Like Gly I, Gly II has been described in various organisms, from prokaryotes over plants to mammals (Ridderstrom et al., 1996; Maiti et al., 1997; Cho et al., 1998; Trincao et al., 2006; Yadav et al., 2006; Campos-Bermudez et al., 2007). Gly II enzyme is present both in cytoplasm and mitochondria in animals as well as in plants (Maiti et al., 1997). Multiple forms of mitochondrial Gly II are located in the matrix and in the intermembrane space (Talesa et al., 1989). In plants, Gly II has been purified from

a variety of sources, and detailed characterization has been done in *A. thaliana* (Ridderstrom and Manneervik, 1997), *Zea mays* (Norton et al., 1989), and *Brassica juncea* (Saxena et al., 2005). Five genes have been identified encoding nine Gly II proteins in *A. thaliana* (Maiti et al., 1997). The in silico analysis of the promoter region of *Osgly II* has indicated the presence of specific *cis*-regulatory sequence such as heat shock element and low-temperature-responsive element, further supporting the multiple stress inducibility of this gene (Yadav et al., 2006).

Unlike in animals, the Gly II from plants has an acidic pI and is monomeric. Three isoforms of glyoxalase with an acidic pI (4.7, 4.8, and 5.0) have been reported from *Aloe vera* (Norton et al., 1989). Gly II purified from cytosol and mitochondria of spinach leaves has been shown to exist in multiple forms having pI 5.3, 5.8, and 6.2 (cytosol) and pI 4.8 (mitochondria). The active site contains a binuclear Zn-binding site and a water molecule in the form of a hydroxide ion, which might act as the nucleophile catalyzing the hydrolysis (Cameron et al., 1999), whereby the Tyr 175 residue contributes to the binding of the substrate (Ridderstrom et al., 2000).

The upregulation of *Gly II* gene has been shown in B. *juncea* under salinity and heavy metal stress by Saxena et al. (2005). About 2.5-fold increase in the transcript level of *Gly II* was observed after 24 h treatment, which further increased to 5.0-fold after 48 h of 800 mM NaCl treatment. Transcript analysis in rice showed that *Osgly II* gene expression is stimulated within 15 min in response to salinity, desiccation, and temperature stresses as well as treatment with ABA or salicylic acid (Yadav et al., 2006).

### 9.1.4.1.3 GLYOXALASE III

Another enzyme, glycoxalase III (Gly III) (EC 4.2.1.130), converts MG directly into D-lactate without requiring GSH or any other cofactor (Misra et al., 1995). In E. *coli*, its expression increases during stationary growth phase of bacteria and is regulated by rpoS (RNA polymerase sigma factor). This enzyme exhibits a higher activity than Gly I and Gly II and represents the main system for MG detoxification in bacterial cells (Benov et al., 2004). It also acts as a heat-inducible molecular chaperone (heat shock protein-31, Hsp31) and helps in preventing aggregation of newly unfolded proteins (Subedi et al., 2011).

Crystal structure analysis revealed three potential active sites: a hydrophobic substrate-binding site for the chaperone activity, a potential protease-like catalytic site, and a 2-histidine-1-carboxylate motif, which resembles with the metal-binding site from GLO1 that could be the binding site for MG (Zhao et al., 2003).

In human cells, Gly III (DJ-1, PARK7) is found to be a molecular chaperone associated with Parkinson's disease (Tao and Tong, 2003). Gly III activity has also been reported in mouse and *Caenorhabditis elegans* (Lee et al., 2012). Recently, the presence of Gly III (DJ-1) in *Arabidopsis* characterized as a family consisting of six members has been reported, which is capable of converting MG into D-lactate directly. Overexpression of one of the members has been shown to confer MG resistance to bacterial cells lacking Gly I and Gly III (Kwon et al., 2013). The role of Gly III gene in providing salt stress tolerance in plants has not been explored so far.

## 9.1.5 Glutathione: A Cofactor

GSH is a multifunctional metabolite present in microorganisms, animals, and plants. It is synthesized by two ATP-dependent reactions. $\gamma$-glutamylcysteine synthetase ($\gamma$-ECS) is the first enzyme of the pathway, which catalyzes the formation of $\gamma$-glutamyl cysteine from L-glutamate and L-cysteine. In the second step, glutathione synthetase (GSH-S) catalyzes the formation of GSH by addition of glycine to $\gamma$-glutamyl cysteine. It is an important antioxidant and is a cofactor of enzyme GPX that catalyzes the detoxification of $H_2O_2$ (Hunaiti and Soud, 2000). In plants, the GPX expression is induced under stress conditions. APX and CAT are predominant enzymes involved in the detoxification of $H_2O_2$. However, GPX is independently involved in detoxification of organic peroxidases. During stress, oxidation of GSH occurs that exerts feedback inhibition on $\gamma$-ECS and results in enhanced GSH biosynthesis. Cellular GSH/GSSG ratio is maintained by dimeric flavor protein glutathione reductase (GR), which uses NADPH cofactor to reduce GSSG to GSH (Halliwell and Foyer, 1978). GSH has been shown to alter the transcripts of enzymes involved in detoxification mechanism, for example, Cu–Zn dismutases, GR, and 2-cys-peroxiredoxins (Wingsle and Karpinski, 1996; Baier and Dietz, 1997).

### 9.1.5.1 Relationship between Glyoxalase System and ROS Detoxification System in Modulating Salt Stress Tolerance

Most of the abiotic stresses (drought, salt, heat, and cold) disrupt the metabolic balance of cells and trigger the production of ROS, which ultimately results in oxidative stress. During stress, an increase in ROS and MG impairs the redox balance of the cell. Plants have an array of enzymatic and nonenzymatic detoxification systems, which function as an extremely efficient cooperative system to counter the deleterious effects of ROS and MG. These antioxidants include the enzymes, SOD, CAT, APX, monodehydroascorbate reductase (MDHAR), dehydroascorbate reductase (DHAR), GR, and water-soluble compounds, such as AsA and GSH. The ascorbate–glutathione (AsA-GSH) cycle involves APX, MDHAR, DHAR, GR, AsA, and GSH in a series of cyclic reactions to detoxify $H_2O_2$ and regenerate AsA and GSH.

A close relationship between MG and ROS metabolism was first described by Yadav et al. (2005) in a transgenic system. Overexpression of the glyoxalase pathway genes (Gly I and Gly II) in transgenic tobacco and rice plants has been found to lower MG levels and maintain higher levels of the reduced GSH, thereby decreasing ROS production under salinity stress conditions, which leads to improved stress tolerance. The transgenic tobacco plants overexpressing glyoxalase pathway enzymes resist an increase in the level of MG that increased to over 70% in wild-type plants under salinity stress. These plants showed enhanced basal activity of various GSH-related antioxidative enzymes that increased further upon salinity stress. These plants suffered minimal salinity stress-induced oxidative damage measured in terms of the lipid peroxidation. The reduced GSH content was high in these transgenic plants and they also maintained a higher reduced to oxidized glutathione (GSH/GSSG) ratio under salinity conditions. Manipulation of GSH ratio by exogenous application of GSSG retarded the growth of non-transgenic plants, whereas transgenic plants sustained their growth. These results suggest that the overexpression of glyoxalases strengthens the antioxidant response along with keeping MG levels in check, which confers tolerance under salt stress.

### 9.1.6 Glyoxalase Pathway Engineering in Transgenic Plants for Salt Stress Tolerance

A successful attempt of engineering abiotic stress tolerance in plants using the *Gly I* was made by Veena et al. (1999) (Table 9.1). Transgenic tobacco and rice plants overexpressing *Gly I* gene from B. juncea showed improved salt and heavy metal tolerance (Verma et al., 2005). Bhomkar et al. (2008) have reported that overexpression of *Gly I* gene driven by the novel constitutive Cestrum viral promoter conferred enhanced tolerance to MG and salt stress in the leguminous crop plant *Vigna mungo*. Overexpression of the same *Gly I* gene under inducible promoter Rd29A in *Arabidopsis* imparted salinity tolerance to transgenic plants (Roy et al., 2008). *Gly I* gene isolated from other sources has also been shown to provide salinity tolerance to transgenic tobacco plants (Lin et al., 2010; Wu et al., 2012).

Similarly, the efficacy of the *Gly II* gene in imparting abiotic stress tolerance has also been investigated. Transgenic plants overexpressing rice *Gly II* gene showed tolerance to high NaCl concentrations (Singla-Pareek et al., 2003, 2008; Saxena et al., 2011; Wani and Gosal, 2011).

Overexpression of both *Gly I* and *Gly II* genes together has also been shown to confer enhanced salt tolerance in tobacco as compared to the single gene (Singla-Pareek et al., 2003). Overexpression of *Gly I* and *Gly II* genes together showed enhanced salt stress tolerance by decreasing oxidative stress in transformed tomato plants (Alvarez Viveros et al., 2013).

## 9.2 CONCLUSIONS AND FUTURE PROSPECTS

Due to the rising problem of salinity in modern agriculture, genetic engineering toward developing salt tolerant crops is challenging and gaining the primary importance for future crop improvement programs. This chapter has briefly summarized the role of the glyoxalase system in MG detoxification that has been targeted for manipulation in the endeavor to develop crop plants with improved salt, heavy metal, and MG resistance. The ubiquitous and highly conserved nature of the glyoxalase system suggests its fundamental importance in biological

TABLE 9.1

*Glyoxalase genes that have been overexpressed to improve abiotic stress tolerance in crops.*

| Transgene | Gene isolated from | Transgenic crop | Promoters used | Transgenic plant performance | References |
|---|---|---|---|---|---|
| Gly I | Brassica | Tobacco | CaMV 35S | Improved salinity stress tolerance | Veena et al. (1999) |
| Gly II | Rice | Tobacco | 35S | Confers salinity tolerance | Singla-Pareek et al. (2003) |
| Gly I + Gly II | Brassica + rice | Tobacco | 35S | Salt and heavy metal stress | Singla-Pareek et al. (2003) |
| Gly I | Brassica | Rice | | Salt tolerance | Verma et al. (2005) |
| Gly I + Gly II | Brassica + rice | Tobacco | 35S | Enhanced salinity and MG stress tolerance | Yadav et al. (2005) |
| Gly I + Gly II | Brassica + rice | Tobacco | 35S | Enhanced heavy metal tolerance | Singla-Pareek et al. (2006) |
| Gly I | Brassica | Vigna mungo | CmYLCV | Salinity stress tolerance | Bhomkar et al. (2008) |
| Gly I | Brassica | Arabidopsis | Rd29A | Improved salinity stress tolerance | Roy et al. (2008) |
| Gly II | Rice | Rice | 35S | MG and salt stress tolerance | Singla-Pareek et al. (2008) |
| Gly I | Wheat | Tobacco | 35S | Confers salt and heavy metal tolerance | Lin et al. (2010) |
| Gly II | Rice | Brassica | CaMV35S | Salt stress tolerance | Saxena et al. (2011) |
| Gly I | Beta vulgaris | Tobacco | 35S | Improved salinity and mannitol stress tolerance | Wu et al. (2012) |
| Gly I + Gly II | Brassica + Pennisetum | Tomato | 35S | Salt tolerance | Alvarez Viveros et al. (2013) |

systems. However, the complex nature of the glyoxalase system, the existence of several other detoxification mechanisms for MG, and the existence of MG synthesis in eukaryotes and prokaryotes suggest that there are other, perhaps more, fundamental functions for the glyoxalase system. Detailed studies in this area will provide further insights to unanswered questions in the near future.

## REFERENCES

Ahmad, P. and Prasad, M.N.V. (2012) *Abiotic Stress Responses in Plants: Metabolism, Productivity and Sustainability*, Springer, New York.

Alvarez Viveros, M.F., Inostroza-Blancheteau, C., Timmermann, T., Gonzalez, M., and Arce-Johnson, P. (2013) Overexpression of GlyI and GlyII genes in transgenic tomato (*Solanum lycopersicum* Mill.) plants confers salt tolerance by decreasing oxidative stress. *Molecular Biology Reports* 40: 3281–3290.

Apse, M.P., Aharon, G.S., Snedden, W.A., and Blumwald, E. (1999) Salt tolerance conferred by overexpression of a vacuolar Na⁺/H⁺ antiporter in *Arabidopsis*. *Science* 285: 1256–1258.

Ashraf, M. and Foolad, M.R. (2007) Roles of glycine betaine and proline in improving plant abiotic stress resistance. *Environmental and Experimental Botany* 59: 206–216.

Badawi, G.H., Yamauchi, Y., Shimada, E., Sasaki, R. et al. (2004) Enhanced tolerance to salt stress and water deficit by over-expressing superoxide dismutase in tobacco (*Nicotiana tabacum*) chloroplasts. *Plant Science* 166: 919–928.

Baier, M. and Dietz, K.J. (1997) The plant 2-cys peroxiredoxin BAS1 is a nuclear-encoded chloroplast protein: Its expressional regulation, phylogenetic origin, and implications for its specific physiological function in plants. *The Plant Journal* 12: 179–190.

Ball, J.C. and Vander Jagt, D.L. (1981) S-2-hydroxylacylglutathione hydrolase (glyoxalase II): Active-site mapping of a nonserine thiolesterase. *Biochemistry* 20: 899–905.

Bandurska, H. (1993) In vivo and in vitro effect of proline on nitrate reductase activity under osmotic stress in barley. *Acta Physiologiae Plantarum* 15: 83–88.

Banu, M.N.A., Hoque M.A., Watamable-Sugimoto, M., Islam, M.A., Uraji, M., Matsuoka, M., Nakamura, Y., and Murata, Y. (2010) Proline and glycine betaine ameliorated NaCl stress via scavenging of hydrogen peroxide and methylglyoxal but not superoxide or nitric oxide in tobacco cultured cells. *Bioscience, Biotechnology and Biochemistry* 74: 2043–2049.

Benov, L., Sequeira, F., and Beema, A.F. (2004) Role of rpoS in the regulation of glyoxalase III in *Escherichia coli*. *Acta Biochimica Polonica* 51: 857–860.

Bhomkar, P., Upadhyay, C.P., Saxena, M., Muthusamy, A., Prakash, N.S., Poggin, K., Hohn, T., and Sarin, N.B. (2008) Salt stress alleviation in transgenic *Vigna mungo* L. Hepper (blackgram) by overexpression of the glyoxalase I gene using a novel Cestrum yellow leaf curling virus (CmYLCV) promoter. *Molecular Breeding* 22: 169–181.

Cameron, A.D., Olin, B., Ridderstrom, M., Mannervik, B., and Jones, T.A. (1997) Crystal structure of human glyoxalase I—Evidence for gene duplication and 3D domain swapping. *EMBO Journal* 16: 3386–3395.

Cameron, A.D., Ridderstrom, M., Olin, B., and Mannervik, B. (1999) Crystal structure of human glyoxalase II and its complex with a glutathione thiol ester substrate analogue. *Structure* 7: 1067–1078.

Campos-Bermudez, V.A., Leite, N.R., Krog, R., Costa-Filho, A.J. et al. (2007) Biochemical and structural characterization of *Salmonella typhimurium* glyoxalase II: New insights into metal ion selectivity. *Biochemistry* 46: 11069–11079.

Cha, U.S. and Kirdmanee, C. (2010) Effect of glycine-betaine on proline, water use, and photosynthetic efficiencies and growth of rice seedlings under salt stress. *Turkish Journal of Agriculture and Forestry* 34: 517–527.

Chang, H., Siegel, B.Z., and Siegel, S.M. (1984) Salinity induced changes in isoperoxidase in taro, *Colocasia esculenta*. *Phytochemistry* 23: 233–235.

Chaplen, F.W. (1998) Incidence and potential implications of the toxic metabolite methylglyoxal in cell culture: A review. *Cytotechnology* 26: 173–183.

Chen, T.H.H. and Murata, N. (2008) Glycine betaine: An effective protectant against abiotic stress in plants. *Trends in Plant Science* 13: 499–505.

Chen, Z.Y., Brown, R.L., Damann, K.E., and Cleveland, T.E. (2004) Identification of a maize kernel stress-related protein and its effect on aflatoxin accumulation. *Phytopathology* 94: 938–945.

Cho, M.Y., Bae, C.D., Park, J.B., and Lee, T.H. (1998) Purification and cloning of glyoxalase II from rat liver. *Experimental Biology and Medicine* 30: 53–57.

Clugston, S.L., Barnard, J.F.J., Kinach, R., Miedema, D. et al. (1998) Overproduction and characterization of a dimeric non-zinc glyoxalase I from *Escherichia coli*: Evidence for optimal activation by nickel ions. *Biochemistry* 37: 8754–8763.

Collins, N.C., Tardieu, F., and Tuberosa, R. (2008) Quantitative trait loci and crop performance under abiotic stress: Where do we stand? *Plant Physiology* 147: 469–486.

Deswal, R. and Sopory, S.K. (1998) Biochemical and immunochemical characterization of *Brassica juncea* glyoxalase I. *Phytochemistry* 49: 2245–2253.

Dietz, K.J., Tavakoli, N., Kluge, C., Mimura, T. et al. (2001) Significance of the V-type ATPase for the adaptation to stressful growth conditions and its regulation on the molecular and biochemical level. *Journal of Experimental Botany* 52: 1969–1980.

Espartero, J., Sanchez-Aguayo, I., and Pardo, J.M. (1995) Molecular characterization of glyoxalase-I from a higher plant: Upregulation by stress. *Plant Molecular Biology* 29: 1223–1233.

FAO (Food, Agriculture Organization of the United Nations). (2004) *FAO Production Yearbook*, FAO, Rome, Italy.

Gapinska, M., Sklodowska, M., and Gabara, B. (2008) Effect of short- and long-term salinity on the activities of antioxidative enzymes and lipid peroxidation in tomato roots. *Acta Physiologiae Plantarum* 30: 11–18.

Garratt, L.C., Janagoundar, B.S., Lowe, K., Anthony, P., Power, J.B., and Davey, M.R. (2002) Salinity tolerance and antioxidant status in cotton cultures. *Free Radical Biology and Medicine* 33: 502–511.

Ghosh, A., Pareek, A., Sopory, S.K., and Singla-Pareek, S.L. (2014) A glutathione responsive rice glyoxalase II, OsGLYII-2, functions in salinity adaptation by maintaining better photosynthesis efficiency and anti-oxidant pool. *The Plant Journal* 80: 93–105.

Godfray, H.C., Beddington, J.R., Crute, I.R., Haddad, L. et al. (2010) Food security: The challenge of feeding 9 billion people. *Science* 327: 812–818.

Guan, B., Hu, Y., Zeng, Y., Wang, Y., and Zhang, F. (2011) Molecular characterization and functional analysis of a vacuolar Na$^+$/H$^+$ antiporter gene (HcNHX1) from *Halostachys caspica*. *Molecular Biology Reports* 38: 1889–1899.

Gupta, B. and Haung, B. (2014) Mechanism of salinity tolerance in plants: Physiological, biochemical, and molecular characterization. *International Journal of Genomics*, 2014, Article ID 701596.

Gzik, A. (1996) Accumulation of proline and pattern of α-amino acids in sugar beet plants in response to osmotic, water and salt stress. *Environmental and Experimental Botany* 36: 29–38.

Halliwell, B. and Foyer, C.H. (1978) Properties and physiological function of a glutathione reductase purified from spinach leaves by affinity chromatography. *Planta* 139: 9–17.

Halliwell, B. and Gutteridge, J.M.C. (1986) Oxygen free radicals and iron in relation to biology and medicine: Some problems and concepts. *Archives of Biochemistry and Biophysics* 246: 501–514.

Hamada, A., Shono, M., Xia, T., Ohta, M., Hayashi, Y., Tanaka, A., and Hayakawa, T. (2001) Isolation and characterization of a Na$^+$/H$^+$ antiporter gene from the halophyte *Atriplex gmelini*. *Plant Molecular Biology* 46: 35–42.

Harinasut, P., Poonsopa, D., Roengmongkol, K., and Charoensataporn, R. (2003) Salinity effects on antioxidant enzymes in mulberry cultivar. *Science Asia* 29: 109–113.

Hasegawa, P.M., Bressan, R.A., Zhu, J.K., and Bohnert, H.J. (2000) Plant cellular and molecular responses to high salinity. *Annual Review of Plant Biology* 51: 463–499.

Hoque, M.A., Uraji, M., Banu, M.N., Mori, I.C., Nakamura, Y., and Murata, Y. (2010) The effects of methylglyoxal on glutathione S-transferase from *Nicotiana tabacum*. *Bioscience, Biotechnology, and Biochemistry* 74: 2124–2126.

Hossain, M.A., Hossain, M.Z., and Fujita, M. (2009). Stress-induced changes of methylglyoxal level and glyoxalase I activity in pumpkin seedlings and cDNA cloning of glyoxalase I gene. *Australian Journal of Crop Science* 3:53–64.

Hossain, M.A., da Silva, J.A.T., and Fujita, M. (2011) Glyoxalase system and reactive oxygen species detoxification system in plant abiotic stress response and tolerance: An intimate relationship, in *Abiotic Stress in Plants-Mechanisms and Adaptations*, A.K. Shanker and B. Venkateswarlu, Eds., pp. 235–266, INTECH-Open Access Publisher, Rijeka, Croatia.

Hunaiti, A.A. and Soud, M. (2000) Effect of lead concentration on the level of glutathione, glutathione S-transferase, reductase and peroxidase in human blood. *Science of the Total Environment* 248: 45–50.

IPCC (2007) Climate change. The physical science basis. Contribution of Working Groups I, II and III to the Fourth Assessment Report of the Intergovernmental Panel on Climate Change, Cambridge University Press, Cambridge, U.K.

Ishitani, M., Liu, J., Halfter, U., Kim, C.S., Shi, W., and Zhu, J.K. (2000) SOS3 function in plant salt tolerance requires N-myristoylation and calcium binding. *The Plant Cell* 12: 1667–1677.

James, R.A., Blake, C., Byrt, C.S., and Munns, R. (2011) Major genes for Na$^+$ exclusion, Nax1 and Nax2 (wheat HKT1;4 and HKT1; 5), decrease Na$^+$ accumulation in bread wheat leaves under saline and waterlogged conditions. *Journal of Experimental Botany* 62: 2939–2947.

Jha, A., Joshi, M., Yadav, N.S., Agarwal, P.K., and Jha, B. (2011) Cloning and characterization of the *Salicornia brachiata* Na$^+$/H$^+$ antiporter gene SbNHX1 and its expression by abiotic stress. *Molecular Biology Reports* 38: 1965–1973.

Kaur, C., Singla-Pareek, S.L., and Sopory, S.K. (2014) Glyoxalase and methylglyoxal as biomarkers for plant stress tolerance. *Critical Reviews in Plant Sciences* 33: 429–456.

Kukreja, S., Nandval, A.S., Kumar, N., Sharma, S.K., Sharma, S.K., Unvi, V., and Sharma, P.K. (2005) Plant water status, H$_2$O$_2$ scavenging enzymes, ethylene evolution and membrane integrity of *Cicer arietinum* roots as affected by salinity. *Biologia Plantarum* 49: 305–308.

Kumar, D., Singh, P., Yusuf, M.A., Upadhyaya, C.P., Roy, S.D., Hohn, T., and Sarin, N.B. (2013) The *Xerophyta viscose* aldose reductase (*ALDRXV4*) confers enhanced drought and salinity tolerance to transgenic tobacco plants by scavenging methylglyoxal and reducing the membrane damage. *Molecular Biotechnology* 54: 292–303.

Kwon, K., Choi, D., Hyun, J.K., Jung, H.S., Baek, K., and Park, C. (2013) Novel glyoxalases from *Arabidopsis thaliana*. *FEBS Journal* 280: 3328–3339.

Larsen, K., Aronsson, A.C., Marmstal, E., and Mannervik, B. (1985) Immunological comparison of glyoxalase I from yeast and mammals with quantitative determination of the enzyme in human tissues by radioimmunoassay. *Comparative Biochemistry and Physiology B* 82: 625–638.

Lee, J.Y., Song, J., Kwon, K., Jang, S., Kim, C., Baek, K., Kim, J., and Park, C. (2012) Human DJ-1 and its homologs are novel glyoxalases. *Human Molecular Genetics* 21: 3215–3225.

Li, J., Jiang, G., Huang, P., Ma, J., and Zhang, F. (2007) Overexpression of the Na$^+$/H$^+$ antiporter gene from *Suaeda salsa* confers cold and salt tolerance to transgenic *Arabidopsis thaliana*. *Plant Cell Tissue and Organ Culture* 90: 41–48.

Lin, F., Xu, J., Shi, J., Li, H., and Li, B. (2010) Molecular cloning and characterization of a novel glyoxalase I gene *TaGly I* in wheat (*Triticum aestivum* L.). *Molecular Biology Reports* 37: 729–735.

Liu, J., Ishitani, M., Halfter, U., Kim, C.S., and Zhu, J.K. (2000) The *Arabidopsis thaliana SOS2* gene encodes a protein kinase that is required for salt tolerance. *Proceedings of the National Academy of Sciences of the USA* 97: 3730–3734.

Liu, L., Zeng, Y., Pan, X., and Zhang, F. (2012) Isolation, molecular characterization, and functional analysis of the vacuolar Na$^+$/H$^+$ antiporter genes from the halophyte *Karelinia caspica*. *Molecular Biology Reports* 39: 1–10.

Lu, Z.Q., Liu, D., and Liu, S.K. (2007) Two rice cytosolic ascorbate peroxidases differentially improve salt tolerance in transgenic *Arabidopsis*. *Plant Cell Reports* 26: 1909–1917.

Luming, T., Conglin, H., Rong, Y., Ruifang, L. et al. (2006) Overexpression of AtNHX1 confers salt-tolerance of transgenic tall fescue. *African Journal of Biotechnology* 5: 1041–1044.

Maiti, M.K., Krishnasamy, S., Owen, H.A., and Makaroff, C.A. (1997) Molecular characterization of glyoxalase II from *Arabidopsis thaliana*. *Plant Molecular Biology* 35: 471–481.

Misra, K., Banerjee, A.B., Ray, S., and Ray, M. (1995) Glyoxalase III from *Escherichia coli*: A single novel enzyme for the conversion of methylglyoxal into D-lactate without reduced glutathione. *Biochemical Journal* 305: 999–1003.

Mittler, R. (2006) Abiotic stress, the field environment and stress combination. *Trends in Plant Science* 11: 15–19.

Mittler, R., Vanderauwera, S., Gollery, M., and Breusegem, F.V. (2004) Reactive oxygen gene network of plants. *Trends in Plant Science* 9: 490–498.

Munns, R. (2002) Comparative physiology of salt and water stress. *Plant, Cell and Environment*, 25: 239–250.

Munns, R. and Tester, M. (2008) Mechanisms of salinity tolerance. *Annual Review of Plant Biology* 59: 651–681.

Norton, S.J., Principato, G.B., Talesa, V., Lupattelli, M., and Rosi, G. (1989) Glyoxalase II from *Zea mays*: Properties and inhibition study of the enzyme purified by use of new affinity ligand. *Enzyme* 42: 189–196.

Nounjan, N., Nghia, P.T., and Theerakulpisut, P. (2012) Exogenous proline and trehalose promote recovery of rice seedlings from salt-stress and differentially modulate antioxidant enzymes and expression of related genes. *Journal of Plant Physiology* 169: 596–604.

Ohta, M., Hayashi, Y., Nakashima, A., Hamada, A., Tanaka, A., Nakamura, T., and Hayakawa, T. (2002) Introduction of a Na$^+$/H$^+$ antiporter gene from *Atriplex gmelini* confers salt tolerance in rice. *FEBS Letters* 532: 279–282.

Palma, F., Tejera, N., and Lluch, C. (2013) Nodule carbohydrate metabolism and polyols involvement in the response of *Medicago sativa* to salt stress. *Environmental and Experimental Botany* 85: 43–49.

Quan, S., Switzenberg, R., Reumann, S., and Hu, J. (2010) In vivo subcellular targeting analysis validates a novel peroxisome targeting signal type 2 and the peroxisomal localization of two proteins with putative functions in defense in *Arabidopsis*. *Plant Signaling and Behavior* 5: 151–153.

Rahnama, A., James, R.A., Poustini, K., and Munns, R. (2010) Stomatal conductance as a screen for osmotic stress tolerance in durum wheat growing in saline soil. *Functional Plant Biology* 37: 255–263.

Ramaswamy, O., Pal, S., Mukherjee, S.G., and Sopory, S.K. (1984) Correlation of glyoxalase I activity with cell proliferation in *Datura callus* culture. *Plant Cell Reports* 3: 121–124.

Rauf, M., Shahzad, K., Ali, R., Ahmad, M. et al. (2014) Cloning and characterization of Na+/H+ antiporter (*LfNHX1*) gene from a halophyte grass *Leptochloa fusca* for drought and salt tolerance. *Molecular Biology Reports* 41: 1669–1682.

Redillas, M.C., Park, S.H., Lee, J.W., Kim, Y.S. et al. (2012) Accumulation of trehalose increases soluble sugar contents in rice plants conferring tolerance to drought and salt stress. *Plant Biotechnology Reports* 6: 89–96.

Rhee, H.I., Murata, K., and Kimura, A. (1986) Purification and characterization of glyoxalase I from *Pseudomonas putida*. *Biochemical and Biophysical Research Communications* 147: 931–938.

Richard, J.P. (1993) Mechanism for the formation of methylglyoxal from triose phosphates. *Biochemical Society Symposia* 21: 549–553.

Ridderstrom, M. and Mannervik, B. (1997) Molecular cloning and characterization of the thiolesterase glyoxalase II from *Arabidopsis thaliana*. *Biochemical Journal* 322: 449–454.

Ridderstrom, M., Saccucci, F., Hellman, U., Bergman, T., Principato, G., and Mannervik, B. (1996) Molecular cloning, heterologous expression, and characterization of human glyoxalase II. *Journal of Biological Chemistry* 271: 319–323.

Ridderstrom, M., Jemth, P., Cameron, A.D., and Mannervik, B. (2000) The active-site residue Tyr-175 in glyoxalase II contributes to binding of glutathione derivatives. *Biochimica et Biophysica ACTA-Protein Structure and Molecular Enzymology* 1481: 344–348.

Rodríguez-Rosales, M.P., Gálvez, F.J., Huertas, R., Aranda, M.N., Baghour, M., Cagnac, O., and Venema, K. (2009) Plant NHX cation/proton antiporters. *Plant Signaling & Behavior* 4: 265–276.

Roy, S.D., Saxena, M., Bhomkar, P.S., Pooggin, M., Hohn, T., and Bhalla-Sarin, N. (2008) Generation of marker free salt tolerant transgenic plants of *Arabidopsis thaliana* using the *gly I* gene and *cre* gene under inducible promoters. *Plant Cell Tissue and Organ Culture* 95: 1–11.

Sanders, D. (2000) Plant biology: The salty tale of *Arabidopsis*. *Current Biology* 10:486–488.

Saxena, M., Bisht, R., Roy, S.D., Sopory, S.K., and Bhalla-Sarin, N. (2005) Cloning and characterization of a mitochondrial glyoxalase II from *Brassica juncea* that is upregulated by NaCl, Zn and ABA. *Biochemical and Biophysical Research Communications* 336: 813–819.

Saxena, M., Roy, S.B., Singla-Pareek, S.L., Sopory, S.K., and Bhalla-Sarin, N. (2011) Overexpression of the *glyoxalase II* gene leads to enhanced salinity tolerance in *Brassica juncea*. *Open Plant Science Journal* 5: 23–28.

Saxena, S.C., Kaur, H., Verma, P., Petla, B.P. et al. (2013) Osmoprotectants: Potential for crop improvement under adverse conditions, in *Plant Acclimation to Environmental Stress*, N. Tuteja and S.S. Gill, Eds., pp. 197–232, Springer, New York.

Shi, H., Quintero, F.J., Pardo, J.M., and Zhu, J.K. (2002) The putative plasma membrane Na$^+$/H$^+$ antiporter SOS1 controls long-distance Na+ transport in plants. *The Plant Cell* 14: 465–477.

Singla-Pareek, S.L., Reddy, M.K., and Sopory, S.K. (2003) Genetic engineering of the glyoxalase pathway in tobacco leads to enhanced salinity tolerance. *Proceedings of the National Academy of Sciences of USA* 100: 14672–14677.

Singla-Pareek, S.L., Yadav, S.K., Pareek, A., Reddy, M.K., and Sopory, S.K. (2006) Transgenic tobacco overexpressing glyoxalase pathway enzymes grow and set viable seeds in zinc spiked soils. *Plant Physiology* 140: 613–623.

Singla-Pareek, S.L., Yadav, S.K., Pareek, A., Reddy, M.K., and Sopory, S.K. (2008) Enhancing salt tolerance in a crop plant by overexpression of glyoxalase II. *Transgenic Research* 17: 171–180.

Siringam, K., Juntawong, N., Chaum, S., and Kirdmanee, C. (2011) Salt stress induced ion accumulation, ion homeostasis, membrane injury and sugar contents in salt-sensitive rice (*Oryza sativa* L. spp. indica) roots under iso-osmotic conditions. *African Journal of Biotechnology* 10: 1340–1346.

Smits, M.M. and Johnson, M.A. (1981) Methylglyoxal: Enzyme distributions relative to its presence in Douglas-fir needles and absence in Douglas-fir needle callus. *Archives of Biochemistry and Biophysics* 208: 431–439.

Sousa Silva, M., Ferreira, A.E., Gomes, R., Tomas, A.M., Ponces Freire, A., and Cordeiro, C. (2012) The glyoxalase pathway in protozoan parasites. *International Journal of Medical Microbiology* 302: 225–229.

Subedi, K.P., Choi, D., Kim, I., Min, B., and Park, C. (2011) Hsp31 of *Escherichia coli* K-12 is glyoxalase III. *Molecular Microbiology* 81: 926–936.

Sun, W., Xu, X., Zhu, H., Liu, A., Liu, L., Li, J., and Hua, X. (2010) Comparative transcriptomic profiling of a salt-tolerant wild tomato species and a salt-sensitive tomato cultivar. *Plant and Cell Physiology* 51: 997–1006.

Szabados, A. and Savoure, A. (2010) Proline: A multifunctional amino acid. *Trends in Plant Science* 15: 89–97.

Talesa, V., Uotila, L., Koivusalo, M., Principato, G., Giovannini, E., and Rosi, G. (1989) Isolation of glyoxalase II from two different compartments of rat liver mitochondria. Kinetic and immunochemical characterization of the enzymes. *Biochimica et Biophysica Acta* 993: 7–11.

Tao, X. and Tong, L. (2003) Crystal structure of human DJ-1, a protein associated with early onset Parkinson's disease. *Journal of Biological Chemistry* 278: 31372–31379.

Thornalley, P.J. (1990) The glyoxalase system: New developments towards functional characterization of a metabolic pathway fundamental to biological life. *Biochemical Journal* 269: 1–11.

Thornalley, P.J. (1993) The glyoxalase system in health and disease. *Molecular Aspects of Medicine* 14: 287–371.

Trincao, J., Sousa, S.M., Barata, L., Bonifacio, C. et al. (2006) Purification, crystallization and preliminary X-ray diffraction analysis of the glyoxalase II from *Leishmania infantum*. *Acta Crystallographica. Section F, Structural Biology Communications* 62: 805–807.

Uotila, L. (1989) Glutathione thiol esterases, in *Glutathione: Chemical, Biochemical and Medical Aspects, Coenzymes and Cofactors*, Volume III, Part A, D. Dolphin, R. Poulson, and O. Avramovic, Eds., pp. 767–804, Wiley-Interscience, New York.

U.S. Salinity Laboratory Staff. (1954) Reclamation and improvement of saline and sodic soils, in *USDA Handbook 60*, Riverside, CA.

Vander Jagt, D.L. and Hunsaker, L.A. (2003) Methylglyoxal metabolism and diabetic complications: Roles of aldose reductase, glyoxalase-I, betaine aldehyde dehydrogenase and 2-oxoaldehyde dehydrogenase. *Chemico-Biological Interactions* 143–144: 341–351.

Varshney, R.K., Bansal, K.C., Aggarwal, P.K., Datta, S.K., and Craufurd, P.Q. (2011) Agricultural biotechnology for crop improvement in a variable climate: Hope or hype? *Trends in Plant Science* 16: 363–371.

Veena , Reddy, V.S., and Sopory, S.K. (1999) Glyoxalase I from *Brassica juncea*: Molecular cloning, regulation, and its overexpression confer tolerance in transgenic tobacco under stress. *The Plant Journal* 17: 385–395.

Verma, M., Verma, D., Jain, R.K., Sopory, S.K., and Wu, R. (2005) Overexpression of *glyoxalase I* gene confers salinity tolerance in transgenic japonica and indica rice plants. *Rice Genetics News Letter* 22: 58–62.

Wang, B., Luttge, U., and Ratajczak, R. (2001) Effects of salt treatment and osmotic stress on V-ATPase and V-PPase in leaves of the halophyte *Suaeda salsa*. *Journal of Experimental Botany* 52: 2355–2365.

Wang, H., Tang, X., Shao, C., Shao, H., and Wang, H. (2014) Molecular cloning and bioinformatics analysis of a new plasma membrane $Na^+/H^+$ antiporter gene from the halophyte *Kosteletzkya virginica*. *The Scientific World Journal*, Article ID 141675.

Wang, W., Vinocur, B., and Altman, A. (2003) Plant responses to drought, salinity and extreme temperatures: Towards genetic engineering for stress tolerance. *Planta* 218: 1–14.

Wang, Y., Ying, Y., Chen, J., and Wang, X.C. (2004) Transgenic *Arabidopsis* overexpressing Mn-SOD enhanced salt-tolerance. *Plant Science* 167: 671–677.

Wani, S.H. and Gosal, S.S. (2011) Introduction of *OsglyII* gene into *Oryza sativa* for increasing salinity tolerance. *Biologia Plantarum* 55: 536–540.

Wingsle, G. and Karpinski, S. (1996) Differential redox regulation by glutathione of glutathione reductase and CuZn-superoxide dismutase gene expression in *Pinus sylvestris* L. needles. *Planta* 198: 151–157.

Wu, C., Ma, C., Pan, Y., Gong, S., Zhao, C., Chen, S., and Li, H. (2012) Sugar beet M14 *glyoxalase I* gene can enhance plant tolerance to abiotic stresses. *Journal of Plant Research* 126: 415–425.

Wu, L. and Juurlink, B.H. (2002) Increased methylglyoxal and oxidative stress in hypertensive rat vascular smooth muscle cells. *Hypertension* 39: 809–814.

Yadav, S.K., Singla-Pareek, S.L., Reddy, M.K., and Sopory, S.K. (2005) Transgenic tobacco plants overexpressing glyoxalase enzymes resist an increase in methylglyoxal and maintain higher reduced glutathione levels under salinity stress. *FEBS Letters* 579: 6265–6271.

Yadav, S.K., Singla-Pareek, S.L., Kumar, M., Pareek, A. et al. (2006) Characterization and functional validation of glyoxalase II from rice. *Protein Expression and Purification* 51: 126–132.

Yancey, P.H., Clark, M.E., Hand, S.C., Bowlus, R.D., and Somero, G.N. (1982) Living with water stress: Evolution of osmolyte systems. *Science* 217: 1214–1222.

Yin, X.Y., Yang, A.F., Zhang, K.W., and Zhang, J.R. (2004) Production and analysis of transgenic maize with improved salt tolerance by the introduction of AtNHX1 gene. *Acta Botanica Sinica* 46: 854–861.

Zhao, Y., Liu, D., Kaluarachchi, W.D., Bellamy, H.D., White, M.A., and Fox, R.O. (2003) The crystal structure of *Escherichia coli* heat shock protein YedU reveals three potential catalytic active sites. *Protein Science* 12: 2303–2311.

Zhu, J.K. (2003) Regulation of ion homeostasis under salt stress. *Current Opinion in Plant Biology* 6: 441–445.

CHAPTER TEN

# ROS Production, Scavenging, and Signaling under Salinity Stress

*Yun Fan, Jayakumar Bose, Meixue Zhou, and Sergey Shabala*

## CONTENTS

*Abstract.* Reactive oxygen species (ROS) are often considered as a common denominator for most known abiotic and biotic stresses affecting plant performance under adverse environmental conditions. This list includes salinity—one of the major abiotic stresses that result in the multibillion-dollar penalties to agricultural crop production around the globe. While the causal relationship between stress-induced ROS production and plant adaptive responses to salinity is beyond any doubt, specific details of ROS production and signaling are still far from being fully understood. This chapter summarizes recent advances in the field and describes mechanisms and signal transduction pathways linking ROS production and signaling with plant adaptation to salinity stress.

*Keywords:* ROS, salinity, antioxidants, scavenging, signalling, halophytes, signature

## 10.1 INTRODUCTION

The global annual losses in agricultural production from salt-affected land are in excess of U.S.$12 billion and rising (Qadir et al. 2008, Shabala 2013). Most crops are not able to handle large amounts of salt without a negative impact on their growth and metabolism, so understanding the physiological mechanisms by which plants deal with salinity is essential to minimize these detrimental effects and reduce the earlier penalties (Shabala and Munns 2012). While traditionally salinity stress was associated with two major physiological constraints, namely, osmotic stress and specific ion toxicity (Munns and Tester 2008), research over the last decade has suggested that this list should be complemented by at least one more component: detrimental effects of oxidative stress resulting from increased reactive oxygen species (ROS) production under saline conditions. Importantly, such ROS production occurs not only in leaf but also in root tissues (Luna et al. 2000, Mittler 2002, Miller et al. 2008) and has rapid and direct impact on cell metabolism. For example, *Arabidopsis* root exposure to 100 mM NaCl resulted in a rapid 2.5- to 3-fold increase in hydroxyl radical generation (Demidchik et al. 2010), regulating ion channels' activity and having a major impact of cytosolic ion homeostasis and cell fate. Salinity-induced programmed cell death has been reported in plant roots within 1 h after stress onset (reviewed in Shabala 2009) and was causally related to increased ROS production in root tissues (reviewed in Shabala and Pottosin 2014). This chapter summarizes the recent advances in this area and describes mechanisms and signal transduction pathways linking ROS production and signaling with plant adaptation to salinity stress.

## 10.2 ROS PRODUCTION IN SUBCELLULAR COMPARTMENTS UNDER SALINITY STRESS

ROS are partially reduced or activated derivatives of oxygen ($O_2$). Among major ROS species are singlet oxygen ($^1O_2$), superoxide anion ($O_2^{\cdot-}$), hydrogen peroxide ($H_2O_2$), and hydroxyl radical ($OH\cdot$). ROS are continuously produced in different cellular compartments as by-products of aerobic metabolism such as respiration and photosynthesis. Among them, $H_2O_2$ is the most stable ROS,

with a half-life of an order of minutes (Pitzschke et al. 2006). Because of this stability and its small size, $H_2O_2$ is widely used by plant cells as the second messenger in stress-signaling and developmental pathways. Under normal growth conditions, ROS are produced at low level and increase dramatically when plants are stressed. For example $H_2O_2$ in chloroplasts increased from 0.5 μM to 5–15 μM in plants exposed to stresses (Polle 2001). Due to the earlier role in cell signaling, the equilibrium between ROS production and removal by antioxidative defense components is strictly controlled. However, during environmental stresses like salinity, this homeostasis is often disrupted, with accelerated ROS production rate. This results in an oxidative stress imposed on plants (Zhu 2001). Several cellular compartments contribute to the elevated ROS production under salinity stress (Table 10.1); this includes chloroplasts, peroxisomes, and mitochondria (Mittler et al. 2004).

### 10.2.1 Chloroplasts

The photosystem I (PSI) and photosystem II (PSII) reaction centers in chloroplast thylakoids are major sites of ROS production (Pfannschmidt 2003, Gill and Tuteja 2010). Oxygen generated during photosynthesis can accept electrons and lead to the formation of the superoxide $O_2^{\cdot-}$. Normally, the electron flow from photosystem goes to $NADP^+$ and reduces it to NADPH. After that, the electron flow comes into Calvin cycle and reduces $CO_2$ (Gill and Tuteja 2010). Under salinity stress, plants decrease stomatal conductance to cut down the water loss. This results in limited $CO_2$ availability for carbon fixation by the Calvin cycle. As a result, light capture exceeds the demand for photosynthesis, leading to overloading of the electron transport chain (ETC) and elevated generation of superoxide ions $O_2^{\cdot-}$ in PSI by the Mehler reaction at the antenna pigments (Asada 2006, Ozgur et al. 2013). In addition, singlet oxygen $^1O_2$ is produced at PSII by excited triplet-state chlorophyll at the P680 reaction center, when the ETC is overreduced under high light intensities (Laloi et al. 2004, Asada 2006).

### 10.2.2 Mitochondria

Mitochondria function as energy factories, with their respiratory ETC components

## TABLE 10.1
*Reactive oxygen species production and scavenging under salinity stress.*

| Primary subcompartments | Reactive oxygen species | Production | Scavenging |
|---|---|---|---|
| *Chloroplast* (main ROS production site in leaves under light; results from limited $CO_2$ availability and excessive light captures during stress) | Singlet oxygen ($^1O_2$); half-life: 1 μs | PSII overreduction under high light intensities | α-tocopherols, carotenoids |
| | Superoxide anion ($O_2^-$); half-life: 1 μs | PSI Mehler reaction | SOD (FeSOD, Cu/ZnSOD) |
| | Hydrogen peroxide ($H_2O_2$); half-life: 1 ms | Dismutation of $O_2^-$ by SOD | APX, GPX, PrX, AA, GSH |
| | Hydroxyl radical (OH·); half-life: 1 ns | Haber–Weiss reaction ($H_2O_2$ reaction with $O_2^-$), Fenton reaction ($H_2O_2$ reaction with $Fe^{2+}$) | Flavonoids |
| *Mitochondria* (owing to electron leakage from mETC; respiration rate is accelerated under salt stress for energetically supporting photosynthesis, ion exclusion, and tissue tolerance) | $O_2^-$ | Complex I and III on mitochondria ETC | SOD (MnSOD) |
| | $H_2O_2$ | Dismutation of $O_2^-$ by SOD | APX, CAT, GPX, PrX, AA, GSH |
| | OH· | Haber–Weiss reaction, Fenton reaction | Flavonoids |
| *Peroxisomes* (contribute a large portion of ROS generation; originates from increased photorespiration under salt stress conditions) | $O_2^-$ | Cytochrome b on peroxisome ETC, glyoxisomal photorespiration, xanthine oxidase | SOD (MnSOD) |
| | $H_2O_2$ | Glycolate oxidase reaction in photorespiration, the fatty acid β-oxidation, and the enzymatic reaction of flavin oxidases | CAT, APX, AA, GSH |
| | OH· | Haber–Weiss reaction, Fenton reaction | Flavonoids |
| *Apoplast, plasma membrane* (increased ROS levels in the apoplast could be used for signaling purposes) | $O_2^-$ | NADPH oxidase in membrane | SOD |
| | $H_2O_2$ | Dismutation of $O_2^-$ by SOD, peroxidase | APX, CAT, AA, GSH |
| | OH· | Haber–Weiss reaction, Fenton reaction | Flavonoids |

SOURCE: Based on Karuppanandian, T. et al., *Aust. J. Crop Sci.*, 5, 709, 2011; Ozgur, R. et al., *Funct. Plant Biol.*, 40, 832, 2013.

being another main source of ROS production (Rhoads et al. 2006). Due to decreased photosynthesis during salinity stress, mitochondrial respiration rate increases to compensate reduced ATP synthesis in chloroplast (Atkin and Macherel 2009). Advanced mitochondrial respiration could contribute to ROS overproduction by bringing electrons from cytochrome ETC to $O_2$ (Norman et al. 2004). Complexes I and III in the mitochondrial ETC (mtETC) are major ROS generation sites in mitochondria (Moller 2001). Under salt stress, overreduction of the ubiquinone (UQ) pool from complexes I and III enables electron leakage from mtETC to $O_2$ leading to increased $O_2^{·-}$ production (Noctor et al. 2007, Miller et al. 2010). Superoxide $O_2^{·-}$ is moderately reactive and could be further reduced to $H_2O_2$ by superoxide dismutases (SODs) (Moller 2001,

Quan et al. 2008). $H_2O_2$ can react with reduced $Fe^{2+}$ to produce highly toxic hydroxyl radicals OH·, which could penetrate membranes and leave the mitochondrion (Rhoads et al. 2006). In nongreen tissues like roots, mitochondria are a major source of ROS production, while in leaves, chloroplast and peroxisome contribution is higher, especially under light conditions (Foyer and Noctor 2003, Rhoads et al. 2006).

### 10.2.3 Peroxisomes

Peroxisomes are small organelles bounded by a single bilayer lipid membrane. They are highly metabolically plastic and involved in numerous types of oxidative metabolism, in a tissue-specific manner. Peroxisomes are also major sites of intracellular ROS production. Peroxisomes generate $H_2O_2$ at high rates in various metabolic processes including glycolate oxidase reaction in photorespiration, the fatty acid $\beta$-oxidation, and the enzymatic reaction of flavin oxidases (del Rio et al. 2002, Foyer and Noctor 2003). Under abiotic stress like salinity, reduced stomatal conductance and water availability decrease the ratio of $CO_2$ to $O_2$ in mesophyll cells. This results in photorespiration and increased glycolate accumulation in chloroplasts. Glycolate oxidase reaction in peroxisomes contributes the majority of $H_2O_2$ production during photorespiration (Karpinski et al. 2003). There are two major sites for of $O_2^{\cdot-}$ generation in peroxisomes: one is an organelle matrix, where xanthine oxidase catalyzes the oxidation of xanthine and hypoxanthine (Corpas et al. 2001); another one is peroxisome membranes, where $O_2^{\cdot-}$ is produced by the peroxisome ETC composed of a flavoprotein NADH and cytochrome b (delRio et al. 2002). Increased ROS production and inhibited antioxidant mechanisms were observed in peroxisomes under salinity stress (Mittova et al. 2003, Nyathi and Baker 2006).

### 10.2.4 Apoplast and ER

The apoplast is also an important $H_2O_2$ production site in response to salinity (Hernandez et al. 2001). There are many ROS-forming enzymes present in an apoplastic space such as NADPH oxidases, cell wall–associated oxidase peroxidases, germin-like oxalate oxidases, and amine oxidases (Mittler 2002, Hu et al. 2003). *AtRbohD*

and *AtRbohF* encoding two NADPH oxidases were shown to be involved in apoplastic ROS production (Torres and Dangl 2005). Other additional sources of ROS production in the cell may include detoxifying reactions catalyzed by cytochromes in cytoplasm and endoplasmic reticulum (Urban et al. 1997).

## 10.3 ROS SCAVENGING UNDER SALINITY STRESS

### 10.3.1 Enzymatic and Nonenzymatic Detoxification

ROS detoxification in plants involves two major avenues: enzymatic and nonenzymatic scavenging mechanisms. Major ROS-scavenging enzymes are SOD, ascorbate peroxidase (APX), catalase (CAT), glutathione peroxidase (GPX), and peroxidase (POD). There are four enzymatic pathways: (1) water–water cycle ($O_2^{\cdot-}$—$H_2O_2$—$H_2O$), mainly functioned by SOD and APX; (2) ascorbate–GSH cycle ($H_2O_2$—$H_2O$), which is composed of APX, monodehydroascorbate reductase (MDAR), dehydroascorbate reductase (DHAR), and glutathione reductase (GR); (3) GPX cycle ($H_2O_2$—$H_2O$), involving GPX and GR; and (4) CAT ($H_2O_2$—$H_2O$), which mainly exist in peroxisomes (Willekens et al. 1997). SOD (especially isoforms Cu/ZnSOD and FeSOD) first dismutates superoxide to $H_2O_2$ (Asada 1999). Then, APX, CAT, GPX, and POD detoxify $H_2O_2$ to $H_2O$. MDAR, DHAR, and GR are responsible for the reduction of monodehydroascorbate, dehydroascorbate, and GSH (GSSG), which are produced in the ascorbate–GSH cycle and GPX cycle (Apel and Hirt 2004).

However, some highly toxic ROS such as $^1O_2$ and OH· cannot be scavenged by enzymatic means and require involvement of nonenzymatic pathways. Nonenzymatic antioxidants include low-molecular-weight antioxidants such as ascorbic acid (AsA), GSH, and tocopherols, as well as some compatible solutes such as proline and glycine-betaine. Accumulation of these nonenzymatic scavengers during stresses is a common phenomenon and captures more attention now. AsA serves as an electron donor to various important reactions and was shown to play an important role in protecting photosynthesis in *Arabidopsis* during salt stress (Huang et al. 2005). GSH plays a role in AsA regeneration, and the ratio of GSH to its oxidized form GSSG is important for maintaining redox equilibrium during $H_2O_2$ degradation (Hong-Bo et al. 2008). Both AsA and GSH are abundant in all cell compartments (Noctor and Foyer 1998), and better ascorbate–GSH cycle operation was shown in the halophyte (Shalata et al. 2001). α-Tocopherol (vitamin E) detoxifies $^1O_2$ and lipid peroxyl radicals, thus preventing lipid peroxidation under abiotic stress (Szarka et al. 2012). Apart from its function in osmotic adjustment during stress, proline also has roles in quenching ROS ($^1O_2$, $H_2O_2$, OH·), stabilizing ROS-scavenging enzymes, and protecting plant tissues against damages caused by ROS (Matysik et al. 2002). Glycine betaine accumulates predominantly in chloroplast and protects photosynthetic apparatus during oxidative stress (Ashraf and Foolad 2007). Additional nonenzymatic antioxidants like soluble sugars, trehalose, carotenoids, polyamines, polyols, and polyphenols also contribute to ROS regulation, and most of them were observed to accumulate more in halophytic plants (see review by Bose et al. 2014).

## 10.3.2 ROS Detoxification in Intracellular Compartments under Stress

### 10.3.2.1 Chloroplast

In chloroplasts, membrane-attached Cu/ZnSOD and FeSOD convert superoxide into $H_2O_2$, which can be further converted to $H_2O$ by a membrane-bound thylakoid ascorbate peroxidase (tylAPX) (Table 10.1). Application of $H_2O_2$ was shown to advance the oxidation of quinone A (QA), which is the primary plastoquinone (PQ) electron acceptor. This could lead to an increase in the photosynthetic electron transport flow and the decrease in the generation of $^1O_2$ during stress. Therefore, the water–water cycle also has a positive role in reducing $^1O_2$ photoproduction (Asada 2006, Moller et al. 2007). ROS that escape this cycle in thylakoids would undergo the stromal ascorbate–GSH cycle and detoxification by SOD, GPX, peroxiredoxin in the stroma (Mittler et al. 2004). In halophytes, FeSOD content correlates with salt tolerance (Jithesh et al. 2006b), and Cu/ZnSOD also plays important roles in salt tolerance (Prashanth et al. 2008). In addition, glycine betaine stabilizes the structure and function of the oxygen-evolving complex

of PSII and protects photosynthetic apparatus under high salt stress (Papageorgiou and Murata 1995). Proline could maintain a low ratio of NADPH to NADP⁺, decrease $^1O_2$ production from PSI, and lessen the damage of $^1O_2$ and OH· on PSII under stress (Szabados and Savoure 2010). α-Tocopherol, a predominant form in green tissues, is present in chloroplast stroma and thylakoid membranes and plays an important role in decreasing ROS generation (mainly $^1O_2$) in chloroplasts during stress (Szarka et al. 2012). It was shown that controlling of ROS production and scavenging in chloroplasts is essential for salt tolerance (Hernandez et al. 2001, Tseng et al. 2007). Therefore, to control ROS production in leaves during salt stress, the plant should employ mechanisms to maintain photosynthesis under limited $CO_2$ supply or apply effective electron sinks to avoid ROS overproduction (Bose et al. 2014).

### 10.3.2.2 Mitochondria

SOD (MnSOD) and other components of the ascorbate–GSH cycle (Table 10.1) also exist in mitochondria (Mittler et al. 2004). Moreover, mitochondrial alternative oxidase (AOX) functions on preventing oxidative damage in this organelle. Changes of ROS levels during stress disturb complex I in mitochondria, thus triggering a signaling pathway where AOX and MnSOD are key enzymes (Foyer and Noctor 2005b). Higher mitochondrial SOD was observed in salt-tolerant plants under salinity stress (Mittova et al. 2003). Since mitochondrial respiration is another major source of ROS generation during salt stress, plants employ AOX to maintain the reduction state of UQ, prevent its overreduction, and lower ROS production. AOX removes electrons from the UQ pool, which could be used to reduce $O_2$ to $O_2^{·-}$, and then uses the electrons to reduce $O_2$ to $H_2O$ (Mittler 2002). Furthermore, proline could stabilize mitochondrial respiration through protecting complex II of the mETC during stress (Szabados and Savoure 2010).

### 10.3.2.3 Peroxisomes, Cytosol, and Apoplast

CATs are localized mainly in peroxisomes and are the major ROS-scavenging enzyme responsible for $H_2O_2$ during photorespiration (Mittler et al. 2004). Under stress, ROS originating from photorespiration, fatty acid oxidation, or other reactions in peroxisomes could be detoxified by SOD (MnSOD), CAT, and AsA–GSH cycle (Moller 2001, Mittler et al. 2004, Miller et al. 2010). Cu/ZnSOD and APX are also present in the cytosol, but here they are encoded by a different set of genes (Mittova et al. 2003). In addition, ROS-scavenging pathways in the apoplast and vacuole may also exist, although the details are scarce. Changes in the apoplastic ROS-scavenging activities such as SOD positively correlated with plant salt tolerance (Hernandez et al. 2001). Apoplastic polyamines were shown to potentiate ROS-induced $K^+$ efflux and correlated with plant salt tolerance (Velarde-Buendia et al. 2012). However, accumulation of $H_2O_2$ in apoplast seems to be involved in plant's acclimation responses to salt stress (Hernandez et al. 2001).

Although ROS production and scavenging pathways are mostly separated among different compartments (Table 10.1), $H_2O_2$ can diffuse through membranes mediated by aquaporins, and some antioxidants like AsA and GSH can also be transported between them (Henzler and Steudle 2000, Mittler et al. 2004). ROS production in chloroplasts during stress induced cytosolic but not chloroplastic ROS-scavenging mechanisms (Karpinski et al. 1997). Therefore, ROS produced at a specific cellular site can affect other cellular compartments and alter their gene expression patterns during stress. The cytosol and peroxisomes may act as a *buffer zone* to regulate ROS levels, which reach cellular compartments under normal or stress conditions (Mittler 2002).

### 10.3.2.4 Redundancy and Avoidance

As mentioned earlier, ROS scavenging in one compartment could affect others. Moreover, there is a redundancy among different ROS-scavenging systems (Mittler 2002). For instance, plants with inhibited APX induced SOD, CAT, and GR, while plants with suppressed CAT induced APX, GPX, and AOX to compensate for the loss of CAT (Rizhsky et al. 2002). Also, while plants employ different mechanisms to detoxify over-produced ROS during stress, the same result may be achieved by avoiding ROS overproduction. This may include avoiding excessive light due to changes in leaf angle/positioning, utilizing some physiological adaptations such as a switch between different modes of carbon assimilation

(C4 and CAM metabolism), and applying some molecular mechanisms to rearrange the photosynthetic apparatus (Mittler 2002).

## 10.4 ROS SIGNALING ASSOCIATED WITH SALT STRESS

### 10.4.1 Salinity-Induced ROS Signatures

ROS are continuously produced in aerobic metabolic processes (Rhoads et al. 2006) and, apart from having toxic effects, can also act as signaling molecules. Therefore, in addition to having efficient ROS-scavenging mechanisms to detoxify excess ROS under stress, plant cells need to be able to sense ROS concentration changes and distinguish them from metabolic disturbances, to activate stress responses (Mittler et al. 2004). Among several ROS, $H_2O_2$ is the most stable one, having half-life about 1 ms (as compared with half-lives of $^1O_2$, $O_2^{\cdot-}$, and $OH\cdot$ being in μs or ns range) (Moller et al. 2007). Also, only $H_2O_2$ can diffuse and cross to adjacent subcellular compartments or cells. This transport is mediated by the special class of aquaporins called peroxiporins (Henzler and Steudle 2000, Bienert et al. 2006). Numerous evidences suggested that $H_2O_2$ plays important roles in plants under severe environmental conditions including biotic and abiotic stresses (Dat et al. 2000). Thus, due to its relatively longer life and high permeability across membranes, $H_2O_2$ is now widely accepted as the second messenger for stress-induced signaling events (Neill et al. 2002, Quan et al. 2008).

It has been shown that $H_2O_2$ plays dual functions in plant cells: acting as a signal molecule in stress acclimation process when at low concentration and triggering programmed cell death at high concentration (Dat et al. 2000, Mittler et al. 2004). $H_2O_2$ plays signaling roles in various adaptive processes including abscisic acid (ABA)-induced stomatal opening and closure (Neill et al. 2002), plant senescence (Bhattacharjee 2005), regulation of cell expansion (Suzuki et al. 1999), photorespiration and photosynthesis (Noctor and Foyer 1998), programmed cell death (Mittler 2002), and cell cycle (Mittler et al. 2004) and also in plant growth and development (Foreman et al. 2003). Moreover, it was indicated that $H_2O_2$ could activate mitogen-activated protein kinase (MAPK) protein (Suzuki et al. 1999), modulate activities of other components in signaling networks such as transcription factors (TFs), protein kinases, and protein phosphatases (Cheng and Song 2006). Changes of $H_2O_2$-responsive transcripts in *Arabidopsis* roots and shoots were detected under salinity and osmotic stress (Miller et al. 2010). Several $H_2O_2$-responsive regulators were found, and some transcripts exhibited salt-stress specificity such as CREK1 or WRKY70. $H_2O_2$-induced TF expression mainly appeared in salt-stressed roots while accumulated predominantly in shoots during osmotic stress. At the same time, elevated expression of TFs or protein kinases occurred earlier in osmotic stress than salinity stress. This could be explained by the slower ionic stress in salt-stressed plants (Munns and Tester 2008). In addition to being an important signal molecule, $H_2O_2$ also activates or interacts with other signal molecules such as $Ca^{2+}$, salicylic acid (SA), ABA, jasmonic acid (JA), nitric oxide (NO), and ethylene (Quan et al. 2008). For example $H_2O_2$ participates in long-distance signal transduction induced by various abiotic stimuli (Miller et al. 2009), which needs interaction with SA (Wrzaczek et al. 2013). It was proposed that halophytes had enough SOD in stock to induce $H_2O_2$ rapidly, which could operate stress response signaling similar to $Ca^{2+}$ "signatures." Once the signaling has been activated, $H_2O_2$ "signatures" may be abolished by APX and CAT, which is similar as the $Ca^{2+}$ efflux systems such as $Ca^{2+}$-ATPase to restore the cytosolic $H_2O_2$ or $Ca^{2+}$ levels (Bose et al. 2014).

However, apart from $H_2O_2$ "signature" in signaling network under stress, other ROS have been largely ignored. As mentioned earlier, other ROS such as $^1O_2$, $O_2^{\cdot-}$, and OH· have much shorter half-life and are too reactive to act as specific signals or carry signal long distance, while $H_2O_2$ is perfectly able to act as messenger. $H_2O_2$ produced by NADPH oxidase (respiratory burst oxidase homologues [RBOH]) proteins in extracellular spaces could form an "ROS wave" (Mittler et al. 2011). Nevertheless, superoxide ($O_2^{\cdot-}$) generated by RBOH proteins can also trigger specific signaling events that are different from those activated by $H_2O_2$ (Suzuki et al. 2011, Wrzaczek et al. 2013). Besides, a singlet oxygen ($^1O_2$)-specific signaling pathway was reported previously (op den Camp et al. 2003). This possesses an interesting question: do these "ROS signatures" have the specificity for signaling roles? Under natural stress conditions, it is hard to test the biological activity of each ROS separately since they could be produced simultaneously. Neither $H_2O_2$ nor other ROS is suitable for selective activation of a specific reaction or a certain organelle. It was proposed that oxidized peptides coming from oxidative damaged proteins could be specific secondary ROS messengers (Moller and Sweetlove 2010). Different ROS possess distinct reactivities, which result in different protein modifications (Moller et al. 2007) and, hence, may allow ROS-species-specific or even a subcompartment-specific gene regulation (Moller and Sweetlove 2010). This may explain the different signaling pathways induced by different ROS species.

## 10.4.2 Cross Talk between ROS and Other Signaling Systems

While ROS signaling is complex and not fully understood, it is a well-established fact that ROS can interact with other second messengers such as $Ca^{2+}$ to induce downstream signaling events such as binding to calmodulin (CaM), phospholipid signaling, and activation of MAPKs (Mittler et al. 2004). A general signal transduction pathway starts with signaling perception followed by second messengers like ROS and $Ca^{2+}$. After that, $Ca^{2+}$ sensor activates a phosphorylation cascade, which will further target transcription factors controlling stress-responsive gene expression (Figure 10.1). Thus, changes of stress-induced gene expression lead to plant adaptation locally as individual cells or systemically as a whole organism, which needs some hormones involved in the systemic acquired resistance (SAR) (Foyer and Noctor 2005a, Huang et al. 2012).

Plasma membranes play important roles in stress signal perception and transmission by initiating signaling transduction pathways either directly or indirectly (Huang et al. 2012). Several mechanisms were proposed to mediate ROS perception including oxidative modification of proteins at amino acids like cysteine or methionine (Wang et al. 2012), lipid or fatty acid oxidation (Lopez-Perez et al. 2009), and monitoring redox status (Cuypers et al. 2011). Proteins with oxidative damage could serve as signals before degradation and then are rapidly targeted by proteasomes (Liu et al. 2012). ROS can oxidize redox-sensitive proteins through some redox-sensitive molecules like GSH or thioredoxins in some ways; these redox-sensitive proteins then function through downstream signaling events (Foyer and Noctor 2005a). ROS signaling integrates with many other signaling events including $Ca^{2+}$ signaling, redox responses, and protein kinase. ROS accumulation

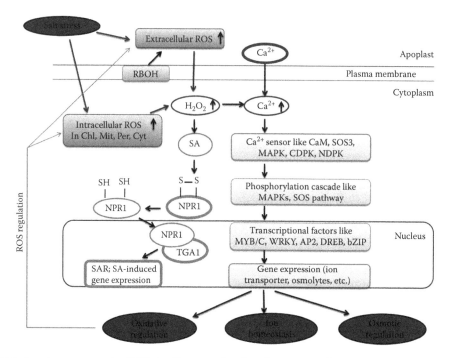

**Figure 10.1.** Reactive oxygen species (ROS) signaling network under salinity stress. Salt stress–induced ROS overproduction (extra- and intracellular) acts as the second messenger and activates $Ca^{2+}$ signaling and phosphorylation cascade, which will further activate transcriptional factors and induce gene expression. Oxidative regulation gene expression in turn controls ROS balance, which forms an ROS loop. Moreover, $H_2O_2$ participates in salicylic acid–induced systemic acquired resistance. (Based on Foyer and Noctor *Plant Cell Environ.* 28, 2005a; *Plant Cell* 17, 1866, 2005b and Wrzaczek, M. et al., Curr. Opin. *Plant Biol.*, 16, 575, 2013.)

was found to occur before these signaling events sometimes, while it was considered as results of them in other instances (Mittler et al. 2011).

### 10.4.2.1 Ca²⁺ Signaling

$Ca^{2+}$ homeostasis plays crucial roles in adaptation mechanisms of plants, and it is closely related to ROS. Apart from ROS produced in subcellular organelles, extracellular ROS production derives for communication from at least two sources in apoplast such as cell wall peroxidases and NADPH oxidase, which is known as "respiratory burst oxidase homologues" (RBOH) in plants (Kwak et al. 2003, Pitzschke et al. 2006). During stress, $O_2^{\cdot-}$ produced by RBOH is dismutated to $H_2O_2$, which serves as messenger operating signaling networks (Figure 10.1). On one hand, $H_2O_2$ could cross the membrane through aquaporins. On the other hand, $H_2O_2$ could activate $K^+$ efflux and $Ca^{2+}$ influx into cytoplasm as the second messenger (Ordonez et al. 2014, Richards et al. 2014). N-terminal (intracellular) of RBOH has several phosphorylation sites together with

two $Ca^{2+}$-binding EF-hand domains. Both $H_2O_2$ and $Ca^{2+}$ could activate or inactivate proteins or channels and trigger downstream signaling networks (Shabala and Pottosin 2014). $H_2O_2$ production requires $Ca^{2+}$ influx, which could activate membrane-anchored NADPH oxidase followed by $K^+$ and anion efflux (Wrzaczek et al. 2013). $H_2O_2$ activated $Ca^{2+}$ influx leading to stomata closure in *Arabidopsis* (Pei et al. 2000). Moreover, $Ca^{2+}$ inhibitors delayed the endogenous $H_2O_2$ accumulation (Lecourieux et al. 2002), suggesting a positive redox feedback loop modulating channel activities. Since plant NADPH oxidase is $Ca^{2+}$ regulated and calcium channels are redox regulated directly or indirectly, the perturbation of either signature could alter both (Sagi and Fluhr 2001, Lecourieux et al. 2002). Therefore, it is hard to tell whether ROS or $Ca^{2+}$ occurs first during stress, and more likely, they act together in the regulatory loops (Miller et al. 2009, Wrzaczek et al. 2013).

RBOH is composed of six conserved transmembrane-spanning domain, intracellular N-terminal domain, and extracellular C-terminal domain, which contains cytosolic FAD- and NADPH-binding regions (Baxter et al. 2014). N-terminal domain consists of two EF-hand motifs for $Ca^{2+}$-binding activation (Ogasawara et al. 2008) and also some phosphorylation sites that are crucial for RBOH (B–F) activities. RBOHD could be activated by CPK5 phosphorylation (Dubiella et al. 2013) and phosphatidic acid (PA) binding, which is produced by phospholipase $D\alpha1$ ($PLD\alpha1$) (Zhang et al. 2009). RBOHF can be activated by OPEN STOMATA 1 (OST1) kinase (Sirichandra et al. 2009) and CBL1/9–CIPK26 complex phosphorylation (Drerup et al. 2013). Consequently, NADPH oxidase activity can be regulated by $Ca^{2+}$ binding and protein phosphorylation, while they could also be induced by ROS. In addition, a serine/threonine protein kinase (OXI1) was shown to play important role in ROS sensing and MAPK3/6 activation by $Ca^{2+}$. OXI1 could be activated by phosphoinositide-dependent kinase 1 through phospholipase C/D–phosphatidic acid (PLC/D-PA) pathway and further activate MAPK3/6 to trigger defense responses during stress (Rentel et al. 2004). It was shown that PLD could also be activated by osmotic stress through a G protein (Frank et al. 2000).

The list of known $Ca^{2+}$ sensors is long and includes, among others, CaM, calcium-dependent protein kinases (CDPKs), MAPKs, nucleotide diphosphate kinase 2 (AtNDPK2), and SOS3 protein. $Ca^{2+}$/CaM has been shown to increase $H_2O_2$ production through activating NADPH oxidase (Harding et al. 1997) and can also downregulate $H_2O_2$ levels by activating catalase (Yang and Poovaiah 2002). CDPKs are serine/threonine protein kinases with a C-terminal $Ca^{2+}$-binding domain. Similar to MAPK activation, CDPK could be activated by various abiotic stresses (Hwang et al. 2000). SOS3 family has been shown to participate in ion homeostasis. In *Arabidopsis*, AtMPK3 and AtMPK6 could be activated by $H_2O_2$ through the activity of *Arabidopsis* NPK1-related protein kinase 1, which is an MAPKKK (Kovtun et al. 2000), and AtNDPK2 is also induced by $H_2O_2$ (Moon et al. 2003). Since AtNDPK2 specifically interacts with AtMPK3/6, and its overexpression leads to increased antioxidant gene expression, it was proposed that AtNDPK2 regulates cellular redox state possibly through activating MAPK cascade (Laloi et al. 2004). The *Arabidopsis* genome contains about 60 MAPKKKs, 10 MAPKKs, and 20 MAPKs, and this may be a good explanation for the cross talk during abiotic stress signaling (Huang et al. 2012). MAPKs cascade could be activated and involved in a broad range of stresses (Chinnusamy

et al. 2004). The effects of ROS on MAPK cascade lead to activation of transcriptional factors.

ROS induce the activation of various transcriptional factors including ethylene response factor (ERF), heat shock factors, basic region leucine zipper (bZIP), zinc-finger protein Zat, nonexpressor of pathogenesis-related genes 1 (NPR1), TGACG-sequence-specific binding protein (TGA), and dehydration-responsive element-binding protein (DREB), Myb families. Knockout analysis indicated that Zat12 is required for Apx1 expression (Rizhsky et al. 2004). MPK3/6 and one of its targets ERF6 are also involved in the regulation of ROS (Meng et al. 2013). Redox-sensitive transcription factor NPR1 could be induced by SA and participate in SA-mediated SAR, which needs cooperation with $H_2O_2$ (Foyer and Noctor 2009). Most of these transcriptional factors are redox regulated, and this suggests that ROS-dependent transcriptional regulation does not exclusively rely on TFs (Wrzaczek et al. 2013).

### 10.4.2.2 SOS Pathway

The SOS pathway is considered to be essential for salt stress signaling and adaptation (Zhu 2002). SOS3 is a $Ca^{2+}$-binding protein with an N-myristoylation motif and three $Ca^{2+}$-binding EF hands. SOS3 responds to salt within seconds and senses specific salt stress–induced $Ca^{2+}$ signal (Ishitani et al. 2000). SOS2 is a Ser/Thr kinase with an N-terminal catalytic kinase domain and a C-terminal regulatory domain (Liu et al. 2000). In the regulatory region of SOS2, there are two motifs: one (positive) is FISL motif for SOS3 binding; another (negative) one is protein phosphatase interaction (PPI) motif for ABA-insensitive 2 (ABI2) binding (Bertorello and Zhu 2009). ABI2 functions as deactivating SOS2 or SOS1 in this pathway (Halfter et al. 2000). SOS1 is a plasma membrane $Na^+/H^+$ antiporter, which is involved in $Na^+$ removal from cell (Shi et al. 2000, Shabala and Mackay 2011). During salt stress, SOS3 is activated by $Ca^{2+}$ and then forms SOS3–SOS2 complex on FISL domain, which will further phosphorylate and activate SOS1 $Na^+/H^+$ antiporter regulating ion homeostasis (Batelli et al. 2007).

It was shown that SOS2 can interact with nucleoside diphosphate kinase (NDPK2) on FISL domain, which could be induced by oxidative stress and participate in $H_2O_2$-induced activation of MPK3 and MPK6 (Moon et al. 2003, Valderrama et al. 2007,

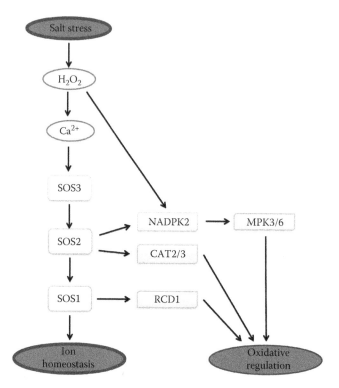

**Figure 10.2.** A cross talk between reactive oxygen species and salt overly sensitive (SOS) signaling pathways. SOS2 interacts with CAT2/3 and NADPK2, while SOS2 interacts with RCD1 leading to oxidative regulation. (Based on Ozgur, R. et al., *Funct. Plant Biol.*, 40, 832, 2013.)

Ozgur et al. 2013). SOS2 can also interact with CAT2/3 in a stress-dependent manner (Valderrama et al. 2007). Moreover, the SOS2 interaction with ABI2 on PPI motif indicates a connection between ABA signaling and SOS pathway (Ohta et al. 2003). Interactions among SOS2, CAT2/3, and ABI2 imply a possible protein complex including three of them (Ozgur et al. 2013). In addition, C-terminal of SOS1 can interact with radical-induced cell death (RCD1), which is also a regulator of oxidative stress response (Katiyar-Agarwal et al. 2006). RCD1 could localize in nuclei and regulate some common stress-induced TFs. It can also move to cytoplasm and interact with C-terminal tail of SOS1, which is located in plasma membranes (Katiyar-Agarwal et al. 2006). All of these interactions provide evidences for a connection between ROS signaling and SOS pathway (Figure 10.2).

### 10.4.2.3 Hormone Regulation

#### 10.4.2.3.1 ROS CROSS TALK WITH SA LEADING TO SAR

Increasing evidence suggests that plant hormones are localized downstream of ROS signal and that $H_2O_2$ induces accumulation of some plant hormones (Kiffin et al. 2006, Huang et al. 2012). The best case for cross talk of SA and ROS regulation is ROS-induced interaction of transcriptional factors NPR1 and TGA, while they also serve as SA receptor (Wu et al. 2012). Under uninduced conditions, NPR1 is localized in the cytosol in oxidized form as a complex composed by several disulfide-bonded intermolecular oligomers (Figure 10.1). Once the cellular redox state is changed under stress conditions, the disulfide bonds of NPR1 are reduced by thioredoxin, and the newly formed monomeric NPR1 move into the nucleus (Mou et al. 2003). The monomeric NPR1 further interacts with another transcription factor TGA, which is responsible for transcriptional control of genes involved in SAR. Similar to NPR1, TGA is in oxidized state with a disulfide bridge between two conserved Cys residues in normal conditions (Despres et al. 2003). During SAR, $H_2O_2$-induced SA accumulation results in TGA reduction and interaction with NPR1 in nucleus (Figure 10.1). This interaction enables TGA bind to SA regulatory sequence or promoter region of some pathogenesis-related gene, which will lead to stress-induced gene activation and SAR (Despres et al. 2003).

Osmotic stress induced by salt or drought is regulated by at least two pathways: one is ABA-dependent pathway, which involves transcriptional factors like AREB/ABF, bZIP, and MYB/MYC; another one is ABA-independent pathways involving TFs such as DREB2 and CBF4 (Huang et al. 2012). It was shown that ROS are crucial signals in the stomatal closure signaling network, which is induced by ABA through $Ca^{2+}$-permeable channels in plasma membrane (Pei et al. 2000). ABA-induced stomata closure partially depends on NADPH oxidase activity (Pei et al. 2000). Open stomata 1 (ost1) protein kinase mutant disturbed in ABA-induced ROS production can still regulate stomata closure in response to $H_2O_2$ (Mustilli et al. 2002). Phosphatidylinositol 3-phosphate (PI3P) could also regulate stomatal closure through ABA-induced ROS production (Park et al. 2003). In addition, ABA was shown to not only advance antioxidant gene expression such as CAT and APX (Zhang et al. 2006) but also increase $H_2O_2$ levels (Zhang et al. 2001). Increased $H_2O_2$ level during ABA treatment indicated that $H_2O_2$ could be a downstream component in ABA signaling network (Guan et al. 2000). ABA induces $H_2O_2$ production in stomatal guard cells, which will activate $Ca^{2+}$ influx and regulate stomatal closure (Guan et al. 2000), whereas $H_2O_2$ may regulate ABA signaling through compromising the negative effect of PP2C in ABA pathway (Vranova et al. 2000).

JA, SA, and ethylene are all crucial regulators in defense-response signaling network and SAR induction (Pieterse and Van Loon 2004). NaCl and SA treatments similarly induced ROS accumulation in *Arabidopsis* roots (Boursiac et al. 2008). Methyl jasmonate can elevate $H_2O_2$ production in guard cells resulting in stomata closure (Suhita et al. 2004). Gibberellin signaling was also found to interact with the ROS network and regulate ROS levels via stimulating the destruction of the nuclear growth-repressing DELLA proteins, which control antioxidant enzyme transcription (Achard et al. 2008). Overexpression of osmotic- and oxidative-stress-responsive ERF exhibited decreased ROS accumulation (Wu et al. 2008). Ethylene is also linked with the ROS signaling network. Exogenous application of $H_2O_2$ promotes ethylene production in a concentration-dependent manner (Ievinsh and Tillberg 1995). Furthermore, $H_2O_2$ production is required for ethylene-dependent salt stress–induced abscission, while indole-3-acetic acid (IAA) compromises abscission by inhibiting $H_2O_2$ generation (Sakamoto et al. 2008).

## 10.5 ROS SIGNALING AND TOLERANCE IN HALOPHYTES

### 10.5.1 ROS Signaling in Halophytes

Salt-induced ROS production is a double-edged sword known to provoke signaling as well as hypersensitive response in plants. In order to use ROS (especially $H_2O_2$) as a signaling molecule, precise temporal and spatial regulation of ROS production is needed. The superior ability of halophytes to use $H_2O_2$ as the signaling molecule in comparison with glycophytes has been demonstrated in a few studies. For example, salt-induced accumulation of $H_2O_2$ reached maximum within 4 h of salt stress and rapidly declined afterward in the leaves of *Cakile maritima* (a halophyte), while in a glycophyte (*Arabidopsis thaliana*), $H_2O_2$ continued to accumulate even after 72 h of salt stress (Ellouzi et al. 2011). This suggests that halophyte species indeed use $H_2O_2$ as a signal and have efficient antioxidant mechanism to decrease $H_2O_2$ concentration after the signaling is completed. In a more recent study, enhanced generation of chloroplastic $H_2O_2$ was shown to be the key contributor of salt stress preparedness in *Thellungiella salsuginea* in comparison with *A. thaliana* (Wiciarz et al. 2014).

The interaction of $H_2O_2$ with transition metals, in particular with iron, will result in the formation of highly reactive OH· radical through the Fenton reaction (Guern et al. 1989). This reaction is undesirable for the $H_2O_2$-mediated signal propagation because the OH· radical has been shown to damage cell structures; it also directly activates a range of $Na^+$-, $K^+$-, and $Ca^{2+}$-permeable cation channels (Demidchik et al. 2010, Zepeda-Jazo et al. 2011) altering cytosolic $K^+/Na^+$ ratio and triggering programmed cell death (Shabala 2009, Demidchik et al. 2010). Hence, decreasing the concentration of free atoms of transition metals in cells is an important component of $H_2O_2$-mediated signaling during salt stress. Plants in general use metal-binding proteins (e.g., ferritin) to decrease the availability of free iron. Involvement of ferritin in $H_2O_2$-mediated signal propagation is inferred based on the fact that iron-induced increase in the abundance of ferritin mRNA (Fer1) is completely abolished when antioxidants (GSH or N-acetylcysteine) were applied; at the same time, $H_2O_2$ application enhanced Fer1 transcript accumulation in derooted maize plantlets (Lobréaux et al. 1995, Savino et al. 1997). Salt stress increased the ferritin deposits in all the cellular organelles of halophytic *Mesembryanthemum crystallinum*, where $H_2O_2$ is known to accumulate (Paramonova et al. 2004). Likewise, accumulation of triplicated transferrin-like protein has also been noted in a halophytic green micro-alga (*Dunaliella salina*) that grows only in sea salt fields (Liska et al. 2004). Interestingly, transient increase in the transcripts of ferritin gene (Fer1) in the leaves of a mangrove *Avicennia marina* has been observed up to 12 h of salt stress but declined afterward (Mehta et al. 2005, Jithesh et al. 2006a) suggesting that halophytes may use ferritin to prevent OH• radical formation at the early phase of $H_2O_2$-mediated signaling. To validate this hypothesis, accumulation pattern of ferritin during salt stress must be compared between glycophytes and halophytes.

Once $H_2O_2$-mediated stress signaling is completed, signal termination may be achieved by bringing $H_2O_2$ concentration back to the control level. In this step, enzymatic detoxification of ROS might play a key role. This notion has to be validated through comparison studies involving halophytes and glycophytes at different time points and at various plant tissues.

## 10.5.2 ROS Tolerance in Halophytes

The threshold salt concentrations required to inflict the oxidative damage (lipid peroxidation—TBARS content) seem to be higher for the halophytes than the glycophytes, because the first sign of lipid peroxidation is noticed in halophytes only when the salt concentration was in excess of 150 mM (Ozgur et al. 2013), but this salt concentration is lethal to majority of the glycophytes. This difference is possible only if halophytes have mechanisms to prevent the buildup of toxic ROS either by decreasing ROS production or by upregulating enzymatic and nonenzymatic means of ROS detoxification. Indeed, halophytes possess such mechanisms. Halophytes synthesize and/or modify proteins (e.g., 33 KDa Mn-stabilizing

proteins, CP24 protein) and lipids (e.g., increase in the sulfolipid content, modification in fatty acid profile of sulfoquinovosyldiacylglycerol) that can protect or stabilize the photosystems I and II thereby potentially decreasing the production of $O_2{}^{.-}$ and $^1O_2$ from these photosystems (Ramani et al. 2004, Wang et al. 2008, Peng et al. 2009, Sengupta and Majumder 2009, 2010). However, such synthesis and/or modification of proteins and lipids was not observed in related glycophytes (reviewed in Bose et al. 2014). Interestingly, halophytes also have a capacity to switch between different modes of carbon fixation, especially to C4 and crassulacean acid metabolism (CAM) following salt stress in an effort to decrease the ROS production (Bose et al. 2014). It was shown that both C3 (e.g., *Portulacaria afra* or *M. crystallinum*) and C4 (e.g., *Portulaca oleracea*) halophyte species switched to CAM during salt stress (Kennedy 1977, Ting and Hanscom 1977, Cushman et al. 1990, Grieve and Suarez 1997, Lin 2009). Likewise, *Atriplex lentiformis* (a halophyte) showed a shift from the C3 to C4 mode of carbon fixation in response to salinity stress (Meinzer and Zhu 1999).

Halophytes also use alternative electron sinks such as plastid terminal oxidase and AOX to prevent ROS generation from $O_2$ (Bose et al. 2014). For instance, halophyte *T. salsuginea* diverted up to 30% of total PSII electron flow via PQ to $O_2$ producing water instead of $O_2{}^{.-}$ (Stepien and Johnson 2009). Similarly, 3.1- and 1.8-fold increase in an AOX expression was observed in halophytic *Populus euphratica* following 200 mM salt shock and gradual salt adoption, respectively (Ottow et al. 2005). Such expression will remove electrons from the UQ pool of mitochondria and transfer them to oxygen to form $H_2O$ (Millar et al. 2011).

## 10.6 CONCLUSION

As commented in our recent review (Bose et al. 2014), "the idea of improving salinity stress tolerance by increasing antioxidant activity has long been, and still remains, attractive to plant breeders." Unfortunately, while the causal relationship between oxidative and salinity stress is beyond any doubt, the dual role of ROS and the complexity of ROS signaling severely handicap the progress in the field. The concept "the higher antioxidant activity the better to plant" is clearly not working, and high tissue specificity and time dependence of $H_2O_2$ production led to a proposal that $H_2O_2$ "signatures" may operate in plant signaling networks (Bose et al. 2014), in addition to well-known cytosolic calcium "signatures" (Dodd et al. 2010). Apart from that, various ROS species may directly regulate a wide array of $Na^+$-, $K^+$-, and $Ca^{2+}$-permeable ion channels (Demidchik et al. 2002, 2003, 2007, 2010, Zepeda-Jazo et al. 2011, Richards et al. 2014), impacting cytosolic ionic homeostasis and cell metabolism and ultimately determining cell's fate (Shabala 2009, 2013). Understanding the subtle details of this regulation and finding a right balance between ROS production and scavenging will remain a great challenge for the next generation of plant biologists and breeders.

## REFERENCES

Achard, P., J. P. Renou, R. Berthome, N. P. Harberd and P. Genschik. 2008. Plant dellas restrain growth and promote survival of adversity by reducing the levels of reactive oxygen species. Current Biology 18: 656–660.

Apel, K. and H. Hirt. 2004. Reactive oxygen species: Metabolism, oxidative stress, and signal transduction. Annual Review of Plant Biology 55: 373–399.

Asada, K. 1999. The water-water cycle in chloroplasts: Scavenging of active oxygens and dissipation of excess photons. Annual Review of Plant Physiology and Plant Molecular Biology 50: 601–639.

Asada, K. 2006. Production and scavenging of reactive oxygen species in chloroplasts and their functions. Plant Physiology 141: 391–396.

Ashraf, M. and M. R. Foolad. 2007. Roles of glycine betaine and proline in improving plant abiotic stress resistance. Environmental and Experimental Botany 59: 206–216.

Atkin, O. K. and D. Macherel. 2009. The crucial role of plant mitochondria in orchestrating drought tolerance. Annals of Botany 103: 581–597.

Batelli, G., P. E. Verslues, F. Agius, Q. Qiu, H. Fujii, S. Pan, K. S. Schumaker, S. Grillo and J. K. Zhu. 2007. SOS2 promotes salt tolerance in part by interacting with the vacuolar H+-Atpase and upregulating its transport activity. Molecular and Cellular Biology 27: 7781–7790.

Baxter, A., R. Mittler and N. Suzuki. 2014. ROS as key players in plant stress signalling. Journal of Experimental Botany 65: 1229–1240.

Bertorello, A. M. and J. K. Zhu. 2009. Sik1/SOS2 networks: Decoding sodium signals via calcium-responsive protein kinase pathways. Pflugers Archiv-European Journal of Physiology 458: 613–619.

Bhattacharjee, S. 2005. Reactive oxygen species and oxidative burst: Roles in stress, senescence and signal transduction in plants. Current Science 89: 1113–1121.

Bienert, G. P., J. K. Schjoerring and T. P. Jahn. 2006. Membrane transport of hydrogen peroxide. Biochimica Et Biophysica Acta-Biomembranes 1758: 994–1003.

Bose, J., A. Rodrigo-Moreno and S. Shabala. 2014. ROS homeostasis in halophytes in the context of salinity stress tolerance. Journal of Experimental Botany 65: 1241–1257.

Boursiac, Y., J. Boudet, O. Postaire, D. T. Luu, C. Tournaire-Roux and C. Maurel. 2008. Stimulus-induced downregulation of root water transport involves reactive oxygen species-activated cell signalling and plasma membrane intrinsic protein internalization. Plant Journal 56: 207–218.

Cheng, Y. L. and C. P. Song. 2006. Hydrogen peroxide homeostasis and signaling in plant cells. Science in China Series C: Life Sciences 49: 1–11.

Chinnusamy, V., K. Schumaker and J. K. Zhu. 2004. Molecular genetic perspectives on cross-talk and specificity in abiotic stress signalling in plants. Journal of Experimental Botany 55: 225–236.

Corpas, F. J., J. B. Barroso and L. A. del Rio. 2001. Peroxisomes as a source of reactive oxygen species and nitric oxide signal molecules in plant cells. Trends in Plant Science 6: 145–150.

Cushman, J. C., C. B. Michalowski and H. J. Bohnert. 1990. Developmental control of crassulacean acid metabolism inducibility by salt stress in the common ice plant. Plant Physiology 94: 1137–1142.

Cuypers, A., K. Smeets, J. Ruytinx, K. Opdenakker, E. Keunen, T. Remans, N. Horemans, N. Vanhoudt, S. Van Sanden, F. Van Belleghem, Y. Guisez, J. Colpaert and J. Vangronsveld. 2011. The cellular redox state as a modulator in cadmium and copper responses in Arabidopsis thaliana seedlings. Journal of Plant Physiology 168: 309–316.

Dat, J., S. Vandenabeele, E. Vranova, M. Van Montagu, D. Inze and F. Van Breusegem. 2000. Dual action of the active oxygen species during plant stress responses. Cellular and Molecular Life Sciences 57: 779–795.

del Rio, L. A., F. J. Corpas, L. M. Sandalio, J. M. Palma, M. Gomez and J. B. Barroso. 2002. Reactive oxygen species, antioxidant systems and nitric oxide in peroxisomes. Journal of Experimental Botany 53: 1255–1272.

Demidchik, V., H. C. Bowen, F. J. M. Maathuis, S. N. Shabala, M. A. Tester, P. J. White and J. M. Davies. 2002. Arabidopsis thaliana root non-selective cation channels mediate calcium uptake and are involved in growth. Plant Journal 32: 799–808.

Demidchik, V., T. A. Cuin, D. Svistunenko, S. J. Smith, A. J. Miller, S. Shabala, A. Sokolik and V. Yurin. 2010. *Arabidopsis* root K$^+$-efflux conductance activated by hydroxyl radicals: Single-channel properties, genetic basis and involvement in stress-induced cell death. *Journal of Cell Science* 123: 1468–1479.

Demidchik, V., S. N. Shabala, K. B. Coutts, M. A. Tester and J. M. Davies. 2003. Free oxygen radicals regulate plasma membrane Ca$^{2+}$ and K$^+$-permeable channels in plant root cells. *Journal of Cell Science* 116: 81–88.

Demidchik, V., S. N. Shabala and J. M. Davies. 2007. Spatial variation in H$_2$O$_2$ response of *Arabidopsis thaliana* root epidermal Ca$^{2+}$ flux and plasma membrane Ca$^{2+}$ channels. *Plant Journal* 49: 377–386.

Despres, C., C. Chubak, A. Rochon, R. Clark, T. Bethune, D. Desveaux and P. R. Fobert. 2003. The *Arabidopsis* NPR1 disease resistance protein is a novel cofactor that confers redox regulation of DNA binding activity to the basic domain/leucine zipper transcription factor TGA1. *Plant Cell* 15: 2181–2191.

Dodd, A. N., J. Kudla and D. Sanders. 2010. The language of calcium signaling. *Annual Review of Plant Biology* 61: 593–620.

Drerup, M. M., K. Schlucking, K. Hashimoto, P. Manishankar, L. Steinhorst, K. Kuchitsu and J. Kudla. 2013. The calcineurin B-like calcium sensors CBL1 and CBL9 together with their interacting protein kinase CIPK26 regulate the arabidopsis NADPH oxidase RBOHF. *Molecular Plant* 6: 559–569.

Dubiella, U., H. Seybold, G. Durian, E. Komander, R. Lassig, C. P. Witte, W. X. Schulze and T. Romeis. 2013. Calcium-dependent protein kinase/NADPH oxidase activation circuit is required for rapid defense signal propagation. *Proceedings of the National Academy of Sciences of the United States of America* 110: 8744–8749.

Ellouzi, H., K. Ben Hamed, J. Cela, S. Munne-Bosch and C. Abdelly. 2011. Early effects of salt stress on the physiological and oxidative status of *Cakile maritima* (halophyte) and *Arabidopsis thaliana* (glycophyte). *Physiologia Plantarum* 142: 128–143.

Foreman, J., V. Demidchik, J. H. F. Bothwell, P. Mylona, H. Miedema, M. A. Torres, P. Linstead, S. Costa, C. Brownlee, J. D. G. Jones, J. M. Davies and L. Dolan. 2003. Reactive oxygen species produced by NADPH oxidase regulate plant cell growth. *Nature* 422: 442–446.

Foyer, C. H. and G. Noctor. 2003. Redox sensing and signalling associated with reactive oxygen in chloroplasts, peroxisomes and mitochondria. *Physiologia Plantarum* 119: 355–364.

Foyer, C. H. and G. Noctor. 2005a. Oxidant and antioxidant signalling in plants: A re-evaluation of the concept of oxidative stress in a physiological context. *Plant Cell and Environment* 28: 1056–1071.

Foyer, C. H. and G. Noctor. 2005b. Redox homeostasis and antioxidant signaling: A metabolic interface between stress perception and physiological responses. *Plant Cell* 17: 1866–1875.

Foyer, C. H. and G. Noctor. 2009. Redox regulation in photosynthetic organisms: Signaling, acclimation, and practical implications. *Antioxidants & Redox Signaling* 11: 861–905.

Frank, W., T. Munnik, K. Kerkmann, F. Salamini and D. Bartels. 2000. Water deficit triggers phospholipase d activity in the resurrection plant *Craterostigma plantagineum*. *Plant Cell* 12: 111–123.

Gill, S. S. and N. Tuteja. 2010. Reactive oxygen species and antioxidant machinery in abiotic stress tolerance in crop plants. *Plant Physiology and Biochemistry* 48: 909–930.

Grieve, C. and D. Suarez. 1997. Purslane (*Portulaca oleracea* L.): A halophytic crop for drainage water reuse systems. *Plant and Soil* 192: 277–283.

Guan, L. Q. M., J. Zhao and J. G. Scandalios. 2000. Cis-elements and trans-factors that regulate expression of the maize cat1 antioxidant gene in response to aba and osmotic stress: H$_2$O$_2$ is the likely intermediary signaling molecule for the response. *Plant Journal* 22: 87–95.

Guern, J., Y. Mathieu, A. Kurkdjian, P. Manigault, J. Manigault, B. Gillet, J. C. Beloeil and J. Y. Lallemand. 1989. Regulation of vacuolar pH of plant-cells.2. A P-31 NMR-study of the modifications of vacuolar pH in isolated vacuoles induced by proton pumping and cation H$^+$ exchanges. *Plant Physiology* 89: 27–36.

Halfter, U., M. Ishitani and J. K. Zhu. 2000. The *Arabidopsis* SOS2 protein kinase physically interacts with and is activated by the calcium-binding protein SOS3. *Proceedings of the National Academy of Sciences of the United States of America* 97: 3735–3740.

Harding, S. A., S. H. Oh and D. M. Roberts. 1997. Transgenic tobacco expressing a foreign calmodulin gene shows an enhanced production of active oxygen species. *Embo Journal* 16: 1137–1144.

Henzler, T. and E. Steudle. 2000. Transport and metabolic degradation of hydrogen peroxide in *Chara corallina*: Model calculations and measurements with the pressure probe suggest transport of H$_2$O$_2$ across water channels. *Journal of Experimental Botany* 51: 2053–2066.

Hernandez, J. A., M. A. Ferrer, A. Jimenez, A. R. Barcelo and F. Sevilla. 2001. Antioxidant systems and $O_2^-/H_2O_2$ production in the apoplast of pea leaves. Its relation with salt-induced necrotic lesions in minor veins. *Plant Physiology* 127: 817–831.

Hong-Bo, S., C. Li-Ye, S. Ming-An, C. A. Jaleel and M. Hong-Mei. 2008. Higher plant antioxidants and redox signaling under environmental stresses. *Comptes Rendus Biologies* 331: 433–441.

Hu, X., D. L. Bidney, N. Yalpani, J. P. Duvick, O. Crasta, O. Folkerts and G. H. Lu. 2003. Overexpression of a gene encoding hydrogen peroxide-generating oxalate oxidase evokes defense responses in sunflower. *Plant Physiology* 133: 170–181.

Huang, C. H., W. L. He, J. K. Guo, X. X. Chang, P. X. Su and L. X. Zhang. 2005. Increased sensitivity to salt stress in an ascorbate-deficient arabidopsis mutant. *Journal of Experimental Botany* 56: 3041–3049.

Huang, G. T., S. L. Ma, L. P. Bai, L. Zhang, H. Ma, P. Jia, J. Liu, M. Zhong and Z. F. Guo. 2012. Signal transduction during cold, salt, and drought stresses in plants. *Molecular Biology Reports* 39: 969–987.

Hwang, I., H. Sze and J. F. Harper. 2000. A calcium-dependent protein kinase can inhibit a calmodulin-stimulated $Ca^{2+}$ pump (ACA2) located in the endoplasmic reticulum of *Arabidopsis*. *Proceedings of the National Academy of Sciences of the United States of America* 97: 6224–6229.

Ievinsh, G. and E. Tillberg. 1995. Stress-induced ethylene biosynthesis in pine needles: A search for the putative 1-aminocyclopropane-1-carboxylic acid-independent pathway. *Journal of Plant Physiology* 145: 308–314.

Ishitani, M., J. P. Liu, U. Halfter, C. S. Kim, W. M. Shi and J. K. Zhu. 2000. SOS3 function in plant salt tolerance requires N-myristoylation and calcium binding. *Plant Cell* 12: 1667–1677.

Jithesh, M., S. Prashanth, K. Sivaprakash and A. Parida. 2006a. Monitoring expression profiles of antioxidant genes to salinity, iron, oxidative, light and hyperosmotic stresses in the highly salt tolerant grey mangrove, *Avicennia marina* (Forsk.) vierh. By MRNA analysis. *Plant Cell Reports* 25: 865–876.

Jithesh, M. N., S. R. Prashanth, K. R. Sivaprakash and A. K. Parida. 2006b. Antioxidative response mechanisms in halophytes: Their role in stress defence. *Journal of Genetics* 85: 237–254.

Karpinski, S., C. Escobar, B. Karpinska, G. Creissen and P. M. Mullineaux. 1997. Photosynthetic electron transport regulates the expression of cytosolic ascorbate peroxidase genes in *Arabidopsis* during excess light stress. *Plant Cell* 9: 627–640.

Karpinski, S., H. Gabrys, A. Mateo, B. Karpinska and P. M. Mullineaux. 2003. Light perception in plant disease defence signalling. *Current Opinion in Plant Biology* 6: 390–396.

Karuppanapandian, T., J. C. Moon, C. Kim, K. Manoharan and W. Kim. 2011. Reactive oxygen species in plants: Their generation, signal transduction, and scavenging mechanisms. *Australian Journal of Crop Science* 5: 709–725.

Katiyar-Agarwal, S., J. Zhu, K. Kim, M. Agarwal, X. Fu, A. Huang and J. K. Zhu. 2006. The plasma membrane $Na^+/H^+$ antiporter SOS1 interacts with RCD1 and functions in oxidative stress tolerance in *Arabidopsis*. *Proceedings of the National Academy of Sciences of the United States of America* 103: 18816–18821.

Kennedy, R. A. 1977. The effects of NaCl-, polyethyleneglycol-, and naturally-induced water stress on photosynthetic products, photosynthetic rates, and $CO_2$ compensation points in $C_4$ plants. *Zeitschrift für Pflanzenphysiologie* 83: 11–24.

Kiffin, R., U. Bandyopadhyay and A. M. Cuervo. 2006. Oxidative stress and autophagy. *Antioxidants & Redox Signaling* 8: 152–162.

Kovtun, Y., W. L. Chiu, G. Tena and J. Sheen. 2000. Functional analysis of oxidative stress-activated mitogen-activated protein kinase cascade in plants. *Proceedings of the National Academy of Sciences of the United States of America* 97: 2940–2945.

Kwak, J. M., I. C. Mori, Z. M. Pei, N. Leonhardt, M. A. Torres, J. L. Dangl, R. E. Bloom, S. Bodde, J. D. G. Jones and J. I. Schroeder. 2003. NADPH oxidase AtrbohD and Atrbohf genes function in ROS-dependent ABA signaling in *Arabidopsis*. *Embo Journal* 22: 2623–2633.

Laloi, C., K. Apel and A. Danon. 2004. Reactive oxygen signalling: The latest news. *Current Opinion in Plant Biology* 7: 323–328.

Lecourieux, D., C. Mazars, N. Pauly, R. Ranjeva and A. Pugin. 2002. Analysis and effects of cytosolic free calcium increases in response to elicitors in *Nicotiana plumbaginifolia* cells. *Plant Cell* 14: 2627–2641.

Lin, C. C. 2009. The importance of environmental factors in the induction of crassulacean acid metabolism (CAM) expression in facultative CAM species. *Journal of Undergraduate Life Sciences* 3: 64–66.

Liska, A. J., A. Shevchenko, U. Pick and A. Katz. 2004. Enhanced photosynthesis and redox energy production contribute to salinity tolerance in *Dunaliella* as revealed by homology-based proteomics. *Plant Physiology* 136: 2806–2817.

Liu, Y. M., J. S. Burgos, Y. Deng, R. Srivastava, S. H. Howell and D. C. Bassham. 2012. Degradation of the endoplasmic reticulum by autophagy during endoplasmic reticulum stress in *Arabidopsis*. *Plant Cell* 24: 4635–4651.

Liu, J. P., M. Ishitani, U. Halfter, C. S. Kim and J. K. Zhu. 2000. The *Arabidopsis thaliana* SOS2 gene encodes a protein kinase that is required for salt tolerance. *Proceedings of the National Academy of Sciences of the United States of America* 97: 3730–3734.

Lobréaux, S., S. Thoiron and J. F. Briat. 1995. Induction of ferritin synthesis in maize leaves by an iron-mediated oxidative stress. *The Plant Journal* 8: 443–449.

Lopez-Perez, L., M. D. Martinez-Ballesta, C. Maurel and M. Carvajal. 2009. Changes in plasma membrane lipids, aquaporins and proton pump of broccoli roots, as an adaptation mechanism to salinity. *Phytochemistry* 70: 492–500.

Luna, C., L. G. Seffino, C. Arias and E. Taleisnik. 2000. Oxidative stress indicators as selection tools for salt tolerance in *Chloris gayana*. *Plant Breeding* 119: 341–345.

Matysik, J., Alia, B. Bhalu and P. Mohanty. 2002. Molecular mechanisms of quenching of reactive oxygen species by proline under stress in plants. *Current Science* 82: 525–532.

Mehta, P. A., K. Sivaprakash, M. Parani, G. Venkataraman and A. K. Parida. 2005. Generation and analysis of expressed sequence tags from the salt-tolerant mangrove species *Avicennia marina* (forsk) vierh. *Theoretical and Applied Genetics* 110: 416–424.

Meinzer, F. C. and J. Zhu. 1999. Efficiency of $C_4$ photosynthesis in *Atriplex lentiformis* under salinity stress. *Functional Plant Biology* 26: 79–86.

Meng, X. Z., J. Xu, Y. X. He, K. Y. Yang, B. Mordorski, Y. D. Liu and S. Q. Zhang. 2013. Phosphorylation of an ERF transcription factor by *Arabidopsis* MPK3/MPK6 regulates plant defense gene induction and fungal resistance. *Plant Cell* 25: 1126–1142.

Millar, A. H., J. Whelan, K. L. Soole and D. A. Day. 2011. Organization and regulation of mitochondrial respiration in plants. *Annual Review of Plant Biology* 62: 79–104.

Miller, G., K. Schlauch, R. Tam, D. Cortes, M. A. Torres, V. Shulaev, J. L. Dangl and R. Mittler. 2009. The plant NADPH oxidase RBOHD mediates rapid systemic signaling in response to diverse stimuli. *Science Signaling* 2: ra45.

Miller, G., V. Shulaev and R. Mittler. 2008. Reactive oxygen signaling and abiotic stress. *Physiologia Plantarum* 133: 481–489.

Miller, G., N. Suzuki, S. Ciftci-Yilmaz and R. Mittler. 2010. Reactive oxygen species homeostasis and signalling during drought and salinity stresses. *Plant Cell and Environment* 33: 453–467.

Mittler, R. 2002. Oxidative stress, antioxidants and stress tolerance. *Trends in Plant Science* 7: 405–410.

Mittler, R., S. Vanderauwera, M. Gollery and F. Van Breusegem. 2004. Reactive oxygen gene network of plants. *Trends in Plant Science* 9: 490–498.

Mittler, R., S. Vanderauwera, N. Suzuki, G. Miller, V. B. Tognetti, K. Vandepoele, M. Gollery, V. Shulaev and F. Van Breusegem. 2011. ROS signaling: The new wave? *Trends in Plant Science* 16: 300–309.

Mittova, V., M. Tal, M. Volokita and M. Guy. 2003. Up-regulation of the leaf mitochondrial and peroxisomal antioxidative systems in response to salt-induced oxidative stress in the wild salt-tolerant tomato species *Lycopersicon pennellii*. *Plant Cell and Environment* 26: 845–856.

Moller, I. M. 2001. Plant mitochondria and oxidative stress: Electron transport, NADPH turnover, and metabolism of reactive oxygen species. *Annual Review of Plant Physiology and Plant Molecular Biology* 52: 561–591.

Moller, I. M., P. E. Jensen and A. Hansson. 2007. Oxidative modifications to cellular components in plants *Annual Review of Plant Biology* 58: 459–481.

Moller, I. M. and L. J. Sweetlove. 2010. Ros signalling: Specificity is required. *Trends in Plant Science* 15: 370–374.

Moon, H., B. Lee, G. Choi, S. Shin, D. T. Prasad, O. Lee, S. S. Kwak, D. H. Kim, J. Nam, J. Bahk, J. C. Hong, S. Y. Lee, M. J. Cho, C. O. Lim and D. J. Yun. 2003. NDP kinase 2 interacts with two oxidative stress-activated mapks to regulate cellular redox state and enhances multiple stress tolerance in transgenic plants. *Proceedings of the National Academy of Sciences of the United States of America* 100: 358–363.

Mou, Z., W. H. Fan and X. N. Dong. 2003. Inducers of plant systemic acquired resistance regulate npr1 function through redox changes. *Cell* 113: 935–944.

Munns, R. and M. Tester. 2008. Mechanisms of salinity tolerance. *Annual Review of Plant Biology* 59: 651–681.

Mustilli, A. C., S. Merlot, A. Vavasseur, F. Fenzi and J. Giraudat. 2002. *Arabidopsis* OST1 protein kinase mediates the regulation of stomatal aperture by abscisic acid and acts upstream of reactive oxygen species production. *Plant Cell* 14: 3089–3099.

Neill, S., R. Desikan and J. Hancock. 2002. Hydrogen peroxide signalling. *Current Opinion in Plant Biology* 5: 388–395.

Noctor, G., R. De Paepe and C. H. Foyer. 2007. Mitochondrial redox biology and homeostasis in plants. *Trends in Plant Science* 12: 125–134.

Noctor, G. and C. H. Foyer. 1998. Ascorbate and gluta-thione: Keeping active oxygen under control. *Annual Review of Plant Physiology and Plant Molecular Biology* 49: 249–279.

Norman, C., K. A. Howell, A. H. Millar, J. M. Whelan and D. A. Day. 2004. Salicylic acid is an uncoupler and inhibitor of mitochondrial electron transport. *Plant Physiology* 134: 492–501.

Nyathi, Y. and A. Baker. 2006. Plant peroxisomes as a source of signalling molecules. *Biochimica Et Biophysica Acta-Molecular Cell Research* 1763: 1478–1495.

Ogasawara, Y., H. Kaya, G. Hiraoka, F. Yumoto, S. Kimura, Y. Kadota, H. Hishinuma, E. Senzaki, S. Yamagoe, K. Nagata, M. Nara, K. Suzuki, M. Tanokura and K. Kuchitsu. 2008. Synergistic activation of the *Arabidopsis* NADPH oxidase atrbohd by $Ca^{2+}$ and phosphoryla-tion. *Journal of Biological Chemistry* 283: 8885–8892.

Ohta, M., Y. Guo, U. Halfter and J. K. Zhu. 2003. A novel domain in the protein kinase SOS2 mediates interaction with the protein phosphatase 2C AB12. *Proceedings of the National Academy of Sciences of the United States of America* 100: 11771–11776.

op den Camp, R. G. L., D. Przybyla, C. Ochsenbein, C. Laloi, C. H. Kim, A. Danon, D. Wagner, E. Hideg, C. Gobel, I. Feussner, M. Nater and K. Apel. 2003. Rapid induction of distinct stress responses after the release of singlet oxygen in *Arabidopsis*. *Plant Cell* 15: 2320–2332.

Ordonez, N. M., C. Marondedze, L. Thomas, S. Pasqualini, L. Shabala, S. Shabala and C. Gehring. 2014. Cyclic mononucleotides modulate potassium and calcium flux responses to $H_2O_2$ in *Arabidopsis* roots. *Febs Letters* 588: 1008–1015.

Ottow, E. A., M. Brinker, T. Teichmann, E. Fritz, W. Kaiser, M. Brosché, J. Kangasjärvi, X. Jiang and A. Polle. 2005. *Populus euphratica* displays apoplas-tic sodium accumulation, osmotic adjustment by decreases in calcium and soluble carbohydrates, and develops leaf succulence under salt stress. *Plant Physiology* 139: 1762–1772.

Ozgur, R., B. Uzilday, A. H. Sekmen and I. Turkan. 2013. Reactive oxygen species regulation and anti-oxidant defence in halophytes. *Functional Plant Biology* 40: 832–847.

Papageorgiou, G. C. and N. Murata. 1995. The unusu-ally strong stabilizing effects of glycine betaine on the structure and function of the oxygen-evolving photosystem-ii complex. *Photosynthesis Research* 44: 243–252.

Paramonova, N. V., N. I. Shevyakova and V. V. Kuznetsov. 2004. Ultrastructure of chloroplasts and their storage inclusions in the primary leaves of *Mesembryanthemum crystallinum* affected by putres-cine and NaCl. *Russian Journal of Plant Physiology* 51: 86–96.

Park, K. Y., J. Y. Jung, J. Park, J. U. Hwang, Y. W. Kim, I. Hwang and Y. Lee. 2003. A role for phosphati-dylinositol 3-phosphate in abscisic acid-induced reactive oxygen species generation in guard cells. *Plant Physiology* 132: 92–98.

Pei, Z. M., Y. Murata, G. Benning, S. Thomine, B. Klusener, G. J. Allen, E. Grill and J. I. Schroeder. 2000. Calcium channels activated by hydrogen peroxide mediate abscisic acid signalling in guard cells. *Nature* 406: 731–734.

Peng, Z., M. Wang, F. Li, H. Lv, C. Li and G. Xia. 2009. A proteomic study of the response to salinity and drought stress in an introgression strain of bread wheat. *Molecular and Cellular Proteomics* 8: 2676–2686.

Pfannschmidt, T. 2003. Chloroplast redox signals: How photosynthesis controls its own genes. *Trends in Plant Science* 8: 33–41.

Pieterse, C. M. and L. Van Loon. 2004. NPR1: The spider in the web of induced resistance signaling pathways. *Current Opinion in Plant Biology* 7: 456–464.

Pitzschke, A., C. Forzani and H. Hirt. 2006. Reactive oxygen species signaling in plants. *Antioxidants & Redox Signaling* 8: 1757–1764.

Polle, A. 2001. Dissecting the superoxide dismutase-ascorbate-glutathione-pathway in chloroplasts by metabolic modeling. Computer simulations as a step towards flux analysis. *Plant Physiology* 126: 445–462.

Prashanth, S. R., V. Sadhasivam and A. Parida. 2008. Over expression of cytosolic copper/zinc superox-ide dismutase from a mangrove plant *Avicennia marina* in indica rice var pusa basmati-1 confers abiotic stress tolerance. *Transgenic Research* 17: 281–291.

Qadir, M., A. Tubeileh, J. Akhtar, A. Larbi, P. S. Minhas and M. A. Khan. 2008. Productivity enhancement of salt-affected environments through crop diver-sification. *Land Degradation & Development* 19: 429–453.

Quan, L. J., B. Zhang, W. W. Shi and H. Y. Li. 2008. Hydrogen peroxide in plants: A versatile molecule of the reactive oxygen species network. *Journal of Integrative Plant Biology* 50: 2–18.

Ramani, B., H. Zorn and J. Papenbrock. 2004. Quantification and fatty acid profiles of sulfoli-pids in two halophytes and a glycophyte grown under different salt concentrations. *Zeitschrift Fur Naturforschung C: A Journal of Biosciences* 59: 835–842.

Rentel, M. C., D. Lecourieux, F. Ouaked, S. L. Usher, L. Petersen, H. Okamoto, H. Knight, S. C. Peck, C. S. Grierson, H. Hirt and M. R. Knight. 2004. Oxi1 kinase is necessary for oxidative burst-mediated signalling in *Arabidopsis*. *Nature* 427: 858–861.

Rhoads, D. M., A. L. Umbach, C. C. Subbaiah and J. N. Siedow. 2006. Mitochondrial reactive oxygen species. Contribution to oxidative stress and interorganellar signaling. *Plant Physiology* 141: 357–366.

Richards, S. L., A. Laohavisit, J. C. Mortimer, L. Shabala, S. M. Swarbreck, S. Shabala and J. M. Davies. 2014. Annexin 1 regulates the $H_2O_2$-induced calcium signature in *Arabidopsis thaliana* roots. *Plant Journal* 77: 136–145.

Rizhsky, L., S. Davletova, H. J. Liang and R. Mittler. 2004. The zinc finger protein Zat12 is required for cytosolic ascorbate peroxidase 1 expression during oxidative stress in *Arabidopsis*. *Journal of Biological Chemistry* 279: 11736–11743.

Rizhsky, L., E. Hallak-Herr, F. Van Breusegem, S. Rachmilevitch, J. E. Barr, S. Rodermel, D. Inze and R. Mittler. 2002. Double antisense plants lacking ascorbate peroxidase and catalase are less sensitive to oxidative stress than single antisense plants lacking ascorbate peroxidase or catalase. *Plant Journal* 32: 329–342.

Sagi, M. and R. Fluhr. 2001. Superoxide production by plant homologues of the gp91(phox) NADPH oxidase. Modulation of activity by calcium and by tobacco mosaic virus infection. *Plant Physiology* 126: 1281–1290.

Sakamoto, M., I. Munemura, R. Tomita and K. Kobayashi. 2008. Involvement of hydrogen peroxide in leaf abscission signaling, revealed by analysis with an in vitro abscission system in capsicum plants. *Plant Journal* 56: 13–27.

Savino, G., J.-F. Briat and S. Lobréaux. 1997. Inhibition of the iron-induced zmfer1 maize ferritin gene expression by antioxidants and serine/threonine phosphatase inhibitors. *Journal of Biological Chemistry* 272: 33319–33326.

Sengupta, S. and A. L. Majumder. 2009. Insight into the salt tolerance factors of a wild halophytic rice, *Porteresia coarctata*: A physiological and proteomic approach. *Planta* 229: 911–929.

Sengupta, S. and A. L. Majumder. 2010. *Porteresia coarctata* (Roxb.) Tateoka, a wild rice: A potential model for studying salt-stress biology in rice. *Plant, Cell and Environment* 33: 526–542.

Shabala, S. 2009. Salinity and programmed cell death: Unravelling mechanisms for ion specific signalling. *Journal of Experimental Botany* 60: 709–711.

Shabala, S. 2013. Learning from halophytes: Physiological basis and strategies to improve abiotic stress tolerance in crops. *Annals of Botany* 112: 1209–1221.

Shabala, S. and A. Mackay. 2011. Ion transport in halophytes. *Adv Bot Res* 57: 151–199.

Shabala, S. and R. Munns. 2012. Salinity stress: Physiological constraints and adaptive mechanisms. In: S. Shabala (ed.). *Plant Stress Physiology*. CAB International, Oxford, U.K. pp. 59–93.

Shabala, S. and I. Pottosin. 2014. Regulation of potassium transport in plants under hostile conditions: Implications for abiotic and biotic stress tolerance. *Physiologia Plantarum* 151: 257–279.

Shalata, A., V. Mittova, M. Volokita, M. Guy and M. Tal. 2001. Response of the cultivated tomato and its wild salt-tolerant relative *Lycopersicon pennellii* to salt-dependent oxidative stress: The root antioxidative system. *Physiologia Plantarum* 112: 487–494.

Shi, H. Z., M. Ishitani, C. S. Kim and J. K. Zhu. 2000. The *Arabidopsis thaliana* salt tolerance gene sos1 encodes a putative $Na^+/H^+$ antiporter. *Proceedings of the National Academy of Sciences of the United States of America* 97: 6896–6901.

Sirichandra, C., D. Gu, H. C. Hu, M. Davanture, S. Lee, M. Djaoui, B. Valot, M. Zivy, J. Leung, S. Merlot and J. M. Kwak. 2009. Phosphorylation of the *Arabidopsis* AtrbohF NADPH oxidase by OST1 protein kinase. *Febs Letters* 583: 2982–2986.

Stepien, P. and G. N. Johnson. 2009. Contrasting responses of photosynthesis to salt stress in the glycophyte *Arabidopsis* and the halophyte *Thellungiella*: Role of the plastid terminal oxidase as an alternative electron sink. *Plant Physiology* 149: 1154–1165.

Suhita, D., A. S. Raghavendra, J. M. Kwak and A. Vavasseur. 2004. Cytoplasmic alkalization precedes reactive oxygen species production during methyl jasmonate- and abscisic acid-induced stomatal closure. *Plant Physiology* 134: 1536–1545.

Suzuki, N., G. Miller, J. Morales, V. Shulaev, M. A. Torres and R. Mittler. 2011. Respiratory burst oxidases: The engines of ROS signaling. *Current Opinion in Plant Biology* 14: 691–699.

Suzuki, K., A. Yano and H. Shinshi. 1999. Slow and prolonged activation of the p47 protein kinase during hypersensitive cell death in a culture of tobacco cells. *Plant Physiology* 119: 1465–1472.

Szabados, L. and A. Savoure. 2010. Proline: A multifunctional amino acid. *Trends in Plant Science* 15: 89–97.

Szarka, A., B. Tomasskovics and G. Banhegyi. 2012. The ascorbate-glutathione-alpha-tocopherol triad in abiotic stress response. *International Journal of Molecular Sciences* 13: 4458–4483.

Ting, I. P. and Z. Hanscom. 1977. Induction of acid metabolism in *Portulacaria afra*. *Plant Physiology* 59: 511–514.

Torres, M. A. and J. L. Dangl. 2005. Functions of the respiratory burst oxidase in biotic interactions, abiotic stress and development. *Current Opinion in Plant Biology* 8: 397–403.

Tseng, M. J., C. W. Liu and J. C. Yiu. 2007. Enhanced tolerance to sulfur dioxide and salt stress of transgenic Chinese cabbage plants expressing both superoxide dismutase and catalase in chloroplasts. *Plant Physiology and Biochemistry* 45: 822–833.

Urban, P., C. Mignotte, M. Kazmaier, F. Delorme and D. Pompon. 1997. Cloning, yeast expression, and characterization of the coupling of two distantly related *Arabidopsis thaliana* NADPH-cytochrome p450 reductases with p450 cyp73a5. *Journal of Biological Chemistry* 272: 19176–19186.

Valderrama, R., F. J. Corpas, A. Carreras, A. Fernandez-Ocana, M. Chaki, F. Luque, M. V. Gomez-Rodriguez, P. Colmenero-Varea, L. A. del Rio and J. B. Barroso. 2007. Nitrosative stress in plants. *Febs Letters* 581: 453–461.

Velarde-Buendia, A. M., S. Shabala, M. Cvikrova, O. Dobrovinskaya and I. Pottosin. 2012. Salt-sensitive and salt-tolerant barley varieties differ in the extent of potentiation of the ROS-induced $K^+$ efflux by polyamines. *Plant Physiology and Biochemistry* 61: 18–23.

Vranova, E., C. Langebartels, M. Van Montagu, D. Inze and W. Van Camp. 2000. Oxidative stress, heat shock and drought differentially affect expression of a tobacco protein phosphatase 2c(1). *Journal of Experimental Botany* 51: 1763–1764.

Wang, M.-C., Z.-Y. Peng, C.-L. Li, F. Li, C. Liu and G.-M. Xia. 2008. Proteomic analysis on a high salt tolerance introgression strain of *Triticum aestivum/Thinopyrum ponticum*. *Proteomics* 8: 1470–1489.

Wang, Y., J. Yang and J. Yi. 2012. Redox sensing by proteins: Oxidative modifications on cysteines and the consequent events. *Antioxidants & Redox Signaling* 16: 649–657.

Wiciarz, M., B. Gubernator, J. Kruk and E. Niewiadomska. 2014. Enhanced chloroplastic generation of $H_2O_2$ in stress-resistant *Thellungiella salsuginea* in comparison to *Arabidopsis thaliana*. *Physiologia Plantarum* 153(3): 467–476. DOI: 10.1111/ppl.12248.

Willekens, H., S. Chamnongpol, M. Davey, M. Schraudner, C. Langebartels, M. VanMontagu, D. Inze and W. VanCamp. 1997. Catalase is a sink for $H_2O_2$ and is indispensable for stress defence in C-3 plants. *Embo Journal* 16: 4806–4816.

Wrzaczek, M., M. Brosche and J. Kangasjarvi. 2013. ROS signaling loops—Production, perception, regulation. *Current Opinion in Plant Biology* 16: 575–582.

Wu, Y., D. Zhang, J. Y. Chu, P. Boyle, Y. Wang, I. D. Brindle, V. De Luca and C. Despres. 2012. The *Arabidopsis* NPR1 protein is a receptor for the plant defense hormone salicylic acid. *Cell Reports* 1: 639–647.

Wu, L. J., Z. J. Zhang, H. W. Zhang, X. C. Wang and R. F. Huang. 2008. Transcriptional modulation of ethylene response factor protein JERF3 in the oxidative stress response enhances tolerance of tobacco seedlings to salt, drought, and freezing. *Plant Physiology* 148: 1953–1963.

Yang, T. and B. W. Poovaiah. 2002. Hydrogen peroxide homeostasis: Activation of plant catalase by calcium/calmodulin. *Proceedings of the National Academy of Sciences of the United States of America* 99: 4097–4102.

Zepeda-Jazo, I., A. M. Velarde-Buendía, R. Enríquez-Figueroa, J. Bose, S. Shabala, J. Muñiz-Murguía and I. Pottosin. 2011. Polyamines interact with hydroxyl radicals in activating $ca^{2+}$ and $K^+$ transport across the root epidermal plasma membranes. *Plant Physiology* 157: 2167–2180.

Zhang, Y. D., Z. Y. Wang, L. D. Zhang, Y. F. Cao, D. F. Huang and K. X. Tang. 2006. Molecular cloning and stress-dependent regulation of potassium channel gene in Chinese cabbage (*Brassica rapa* ssp. Pekinensis). *Journal of Plant Physiology* 163: 968–978.

Zhang, X., L. Zhang, F. C. Dong, J. F. Gao, D. W. Galbraith and C. P. Song. 2001. Hydrogen peroxide is involved in abscisic acid-induced stomatal closure in *Vicia faba*. *Plant Physiology* 126: 1438–1448.

Zhang, Y. Y., H. Y. Zhu, Q. Zhang, M. Y. Li, M. Yan, R. Wang, L. L. Wang, R. Welti, W. H. Zhang and X. M. Wang. 2009. Phospholipase dalpha 1 and phosphatidic acid regulate NADPH oxidase activity and production of reactive oxygen species in ABA-mediated stomatal closure in *Arabidopsis*. *Plant Cell* 21: 2357–2377.

Zhu, J. K. 2001. Plant salt tolerance. *Trends in Plant Science* 6: 66–71.

Zhu, J. K. 2002. Salt and drought stress signal transduction in plants. *Annual Review of Plant Biology* 53: 247–273.

CHAPTER ELEVEN

# Hydrogen Peroxide Mediated Salt Stress Tolerance in Plants

## SIGNALLING ROLES AND POSSIBLE MECHANISMS

*Mohammad Anwar Hossain, Soumen Bhattacharjee,*
*Ananya Chakrabarty, David J. Burritt, and Masayuki Fujita*

CONTENTS

*Abstract.* Soil salinity is one of the major abiotic stresses that threaten sustainable crop production worldwide. In addition to ionic and osmotic stress, salt stress triggers the accumulation of reactive oxygen species (ROS) that can cause oxidative damage to macromolecules, leading to growth retardation or even plant death. Among the various ROS, $H_2O_2$ has received the most attention from the scientific community in the last decade. $H_2O_2$, a central modulator of stress signal transduction pathways, activates multiple defense responses that reinforce resistance to various abiotic and biotic stresses in plants. Numerous recent studies on plants have confirmed that pre-treatment or priming of seeds or seedlings with appropriate levels of $H_2O_2$ can modulate abiotic stress tolerance by regulating multiple stress responsive

pathways. Although much is known about $H_2O_2$ mediated salt stress tolerance, little is known about how $H_2O_2$ is sensed by plants and how $H_2O_2$ induces an inductive pulse that helps to protect plants from salinity stress. In-depth knowledge of the signal transduction pathways associated with $H_2O_2$ accumulation is essential if the processes of oxidative signalling and redox regulation of gene expression are to be fully understood in plants under salt stress. In this chapter we summarize the current understanding of $H_2O_2$-mediated enhanced salt stress tolerance. We also aim to highlight the signalling roles of $H_2O_2$ in salt stress adaptation.

*Keywords:* Salt stress, oxidative stress, hydrogen peroxide, priming, stress tolerance

## 11.1 INTRODUCTION

Salinity is one of the most brutal environmental factors limiting the productivity of crop plant worldwide. Above 20% of the agricultural lands in the world are affected by high salinity, and this percentage is expected to be further increased in the near future (Munns 2005). Hence, there is an urgent need to develop improved varieties that are more resilient to salt stress. Salinity affects plant growth and development by imposing ionic and osmotic stress on plants, causing ion toxicity, affecting the activity of major cytosolic enzymes, and leading to an abrupt increase in reactive oxygen species (ROS) and finally oxidative damage to plant macromolecules (Chen et al. 2007; Pandolfi et al. 2012; Hossain and Fujita 2013). To survive under salt stress conditions, plants have evolved intricate adaptive or tolerance mechanisms including the synthesis of osmolytes, proper maintenance of ionic balance, ROS detoxification, and other responses (Wu et al. 2013). Intensive research over the last decade has gradually unraveled the mechanisms that underlie how plants react to soil salinity, but many aspects are still a matter of intensive research. Among the various ROS, hydrogen peroxide ($H_2O_2$) has received the most attention from the scientific community in the last decade. Hydrogen peroxide, a central modulator of stress signal transduction pathways, activates multiple defense responses that reinforce resistance to various environmental stresses in plants (Petrov and Van Breusegem 2012). It has been reported that balanced $H_2O_2$ metabolism is central in controlling NaCl-induced oxidative stress tolerance (Talukdar 2013). Additionally, plenty of recent studies in plants have demonstrated that the appropriate level of $H_2O_2$ pretreatment can enhance abiotic stress tolerance including salt stress through the modulation of multiple different plant physiological processes (Fedina et al. 2009; Gondim et al. 2010; Ma et al. 2012; Hossain and Fujita 2013; Ashfaque et al. 2014). In this chapter, we summarize our current understanding of the possible mechanisms associated with $H_2O_2$-induced enhanced salinity tolerance. We also aim to highlight the signaling roles of $H_2O_2$ in shaping the outcome of salt stress adaptation reactions.

## 11.2 GENERATION OF HYDROGEN PEROXIDE IN PLANT CELLS UNDER SALT STRESS

Hydrogen peroxide is generated by two-electron reduction of $O_2$, catalyzed by certain oxidases or indirectly via reduction or dismutation of superoxide ($O2^{\bullet-}$) that is formed by oxidases, peroxidases, or photosynthetic and respiratory electron transport chains (ETCs) (Mittler et al. 2004; Sagi and Fluhr 2006). These reactions generate $H_2O_2$ at several subcellular compartments of the cell (Figure 11.1), and the impact of $H_2O_2$ will be strongly influenced by the extent to which the potent antioxidative system allows its accumulation. Salt stress can result in greater production of ROS that can exceed the capacity of the plant's antioxidant defense mechanisms, and an imbalance in intracellular ROS content is established that results in oxidative stress (Hossain et al. 2011, 2013a,b; Sharma et al. 2012; Hossain and Fujita 2013; Bose et al. 2013). While an increase in ROS levels induces a metabolic response in plants in order to eliminate them, this metabolic response is highly dependent on the plant species, plant-growing conditions, and the type and duration of the stress.

Chloroplasts, where photosynthesis takes place, are the prime sources of ROS in plant cells. In chloroplasts, various forms of ROS ($^1O_2$, $O_2^{\bullet-}$ and $H_2O_2$) are generated from several locations, such as the ETC, photosystem I (PS I), and photosystem II (PS II). $O_2^{\bullet-}$, which is produced mainly by electron leakage from Fe-S centers of PS I or reduced ferredoxin (Fd) to $O_2$ (Mehler reaction), is then converted to $H_2O_2$ by superoxide dismutase (SOD) (Gechev et al. 2006). $O_2^{\bullet-}$ can also be produced by leaking of electrons to molecular oxygen from the ETCs of PS I and II (Sgherri et al. 1996). Singlet oxygen ($^1O_2$) can also be produced by excited-chlorophyll formation in the PS II reaction center and in the antennae system (Asada 2006). In response to salt stress, stomatal conductance in plants decreases to avoid excessive water loss that leads to a decrease in the internal $CO_2$ concentration (Ci) and slows down the reduction of $CO_2$ by the Calvin cycle and induces photorespiration (Abogadallah 2010). The limitation of $CO_2$ fixation and the induction of the oxygenase activity of RuBisCO enhances photorespiration resulting majority of $H_2O_2$ production (Szarka et al. 2012).

As mentioned earlier, mitochondrial respiration is another major source of salt-induced ROS

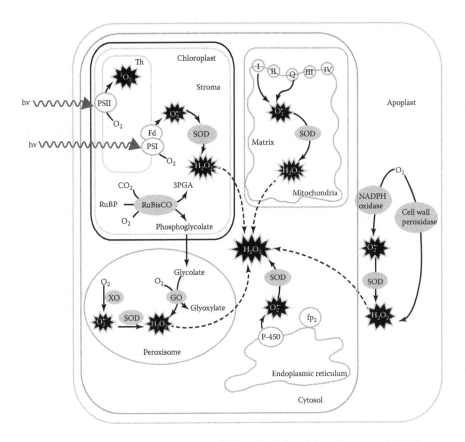

**Figure 11.1.** Major sites and sources of $H_2O_2$ production of plant cells. (Adapted from Hossain et al. 2011.)

production. The mitochondrial ETC consists of several dehydrogenase complexes that reduce a common pool of ubiquinone (Q). ROS production is likely to occur mostly in complex I and the Q zone (Blokhina and Fagerstedt 2006) where $O_2^{\bullet-}$ is generated and rapidly catalyzed into $H_2O_2$. Several enzymes present in the mitochondrial matrix can also produce ROS, either directly or by feeding electrons to ETC (Sharma et al. 2012). Increased mitochondrial ROS formation due to ETC perturbations was observed in plants exposed to salt stress (Hernandez et al. 1993; Mittova et al. 2003). Although, mitochondrial ROS production is much lower compared to chloroplast ROS production, mitochondrial ROS are important regulators of a number of cellular processes, including stress adaptation and programmed cell death (Robson and Vanlerberghe 2002). Furthermore, mitochondria may play a major role in interorganelle cross talk under environmental/oxidative stress by signaling with chloroplasts (Millar et al. 2001) and by inducing altered nuclear gene

expression through mitochondria-to-nucleus signaling (Rhoads et al. 2006).

The peroxisomes are important sites for the production of large quantities of $H_2O_2$. There are two sites of $O_2^{\bullet-}$ production in peroxisomes (del Rio et al. 2002), one of them is the organelle matrix and the other is peroxisome membrane. In this organelle, $H_2O_2$ can be formed directly from $O_2$ by photorespiratory glycolate oxidase or by other enzyme systems such as xanthine oxidase coupled to SOD (Mhamdi et al. 2010). Apart from the normal routes discussed earlier, ROS are also generated and accumulated via plasma membrane–localized NADPH oxidase, cell wall peroxidase, germin-like oxalate oxidase, and amine oxidase (Ma et al. 2013). NADPH oxidases and cell wall peroxidases are the prime sources of $O_2^{\bullet-}$ and $H_2O_2$ and have been found to be stimulated during salt stress, contributing significantly to ROS generation (Sagi and Fluhr 2006; Abogadallah 2010). A redox imbalance associated with environmental stresses such as salinity, extremes of temperature increased rate of overall metabolism that eventually upregulates $H_2O_2$ production in plant cell (Bhattacharjee 2012b, 2013). ROS are normally scavenged immediately at their sites of production by locally present antioxidants. However, when the local antioxidant capacity cannot cope with ROS production, $H_2O_2$ can leak in the cytosol and diffuse to other compartments.

## 11.3 ENZYMATIC AND NONENZYMATIC CONTROL OF HYDROGEN PEROXIDE IN PLANT CELLS

In most cases, $H_2O_2$ formed after reduction of superoxide radicals catalyzed by different isoforms of SOD present in different subcellular organelles. The concentration of $H_2O_2$ must be tightly controlled in plants cells to reduce the chance of oxidative damage and also to induce signal transduction. In parallel, a vast network of antioxidant defense system (Figure 11.2) is constantly on the alert for raising $H_2O_2$ concentrations and provides effective scavenging for it (Gachev et al. 2006; Miller et al. 2010; Petrov and Van Breusegem 2012). In plants, the main enzymatic $H_2O_2$ scavengers are catalase (CAT), ascorbate peroxidase (APX), various types of peroxiredoxins (PRXs), glutathione/thioredoxin peroxidases (GPXs), glutathione S-transferases (GSTs), and secretory peroxidase (Mittler et al. 2004; Mahamdi et al. 2010; Miller et al. 2010; Petrov and

Van Breusegem 2012). In all cases, these enzymes are encoded by multiple genes. CATs are most notably distinguished from the other enzymes in not requiring a reductant as they catalyze a dismutation reaction. CATs are highly expressed enzymes, particularly in certain plant cell types, and are thus an integral part of the plant antioxidative system. Together with APX, CATs are also distinguished from many other peroxide-metabolizing enzymes by their high specificity for $H_2O_2$ but weak activity against organic peroxides. CATs have a very fast turnover rate, but a much lower affinity for $H_2O_2$ than APX and PRX. APX is the main enzyme responsible for $H_2O_2$ removal in the chloroplast, peroxisomes, and mitochondria. APX uses ascorbate as its specific electron donor to reduce $H_2O_2$ to water (Asada 1992). GST and GPX catalyzes the GSH-dependent reduction of $H_2O_2$ and organic peroxides, including lipid peroxides to $H_2O$ or alcohols. The ability of plant GPX to scavenge $H_2O_2$ is decreased, largely due to its Cys residue without selenium. Additionally, nonenzymatic antioxidant content compounds like glutathione, ascorbic acid, tocopherol, and flavonoids also have role on $H_2O_2$ detoxification. Ascorbate present in the chloroplast, cytosol and vacuole, and apoplastic spaces in leaf cells has the fundamental role in the removal of $H_2O_2$. Reduced glutathione can react directly with ROS to detoxify them or can scavenge peroxide as a cofactor of GPX. The ascorbate-glutathione cycle is the most important $H_2O_2$-detoxifying system

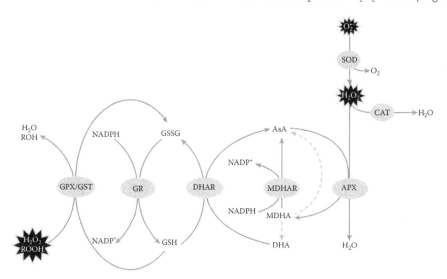

Figure 11.2. Reactive oxygen species ($H_2O_2$) detoxification systems in plants. Abbreviations are defined in the text. (Adapted from Hossain and Fujita 2013.)

in plant cells. The delicate balance between ROS production and scavenging that allows this duality in function to exist in plants is thought to be orchestrated by a large network of genes that tightly regulates ROS production and scavenging (Mittler et al. 2004; Miller et al. 2008; Petrov and Van Breusegem 2012).

## 11.4 HYDROGEN PEROXIDE AND SALINITY STRESS TOLERANCE

There are several lines of evidence suggesting strongly that $H_2O_2$ initiates a signal transduction process for acquisition of tolerance to abiotic and biotic stresses or the induction of cross stress tolerance phenomena. Exogenous application of $H_2O_2$ can induce tolerance to salinity, drought, chilling, high temperatures, and heavy metal stresses all of which cause elevated $H_2O_2$ production. Our recent study showed that cold shock (5 h) induced intracellular increases in $H_2O_2$ level modulating salt- and drought stress–induced oxidative protection of mustard seedlings (Hossain et al. 2013a). Plenty of recent studies in plants have demonstrated that pretreatment with exogenous $H_2O_2$ induces salt stress tolerance. Uchida et al. (2002) have studied the effects of $H_2O_2$ and nitric oxide (NO) oxidative stress tolerance induced by salt and heat stress. Their results showed that pretreating of rice seedlings with low levels (<10 µM) of $H_2O_2$ or NO permitted the survival of more green leaf tissue, and of higher quantum yield for PS II, than in nontreated controls, under salt and heat stresses. It was also shown that the pretreatment induces not only active oxygen scavenging enzymes activities but also expression of transcripts for stress-related genes encoding sucrose phosphate synthase, $\Delta'$-pyrroline-5-carboxylate synthase, and small heat shock protein 26. These results suggest that $H_2O_2$ and NO can increase both salt and heat tolerance in rice seedlings by acting as signal molecules for the response.

Azevedo-Neto et al. (2005) reported that addition of $H_2O_2$ to the nutrient solution induces salt tolerance by enhanced activities of antioxidants and reduced peroxidation of membrane lipids in the leaves and roots of maize plants as an acclimation response. Wahid et al. (2007) reported that exogenous $H_2O_2$ improves salinity tolerance in *Triticum aestivum*. Seeds were soaked in $H_2O_2$ (1–120 µM, 8 h) and were subsequently grown in a saline condition (150 mM NaCl). Levels of $H_2O_2$ in

seedlings arising from $H_2O_2$-treated seeds grown under salinity were markedly lower than salinized controls, suggesting the operation of antioxidant system in them. These seedlings also exhibited better photosynthetic capacity. Moreover, the $H_2O_2$ treatment improved leaf water relations and maintained turgor and improved the $K^+/Na^+$ ratio. Exogenous $H_2O_2$ treatment also enhanced membrane properties, with greatly reduced relative membrane permeability and less ion leakage. Surprisingly, the expression of two heat-stable (stress) proteins (32 and 52 kDa) was observed in $H_2O_2$ pretreated seedlings. Fedina et al. (2009) reported that pretreated *Hordeum vulgare* seedlings with $H_2O_2$ (1 and 5 mM) followed by exposure to 150 mM NaCl for 4 and 7 days showed higher rate of $CO_2$ fixation with lower MDA and $H_2O_2$ contents in comparison with the seedlings subjected NaCl stress only. In addition, $Cl^-$ content in the leaves of NaCl-treated plants is considerably less than in pretreated plants. These results clearly indicated that $H_2O_2$ metabolism is involved as a signal in the processes of salt tolerance.

Gondim et al. (2010) evaluate the effects of $H_2O_2$ on germination and acclimation of a triple maize hybrid (BRS3003) plants subject to salt stress through three consecutive experiments. In the first experiment, $H_2O_2$ accelerated the germination percentage of seeds at 100 mM, but not at 500 mM. In the second experiment, the pretreatment of seeds was observed to induce a pronounced increase in APX and CAT enzyme activity after 30 h of soaking in $H_2O_2$. It was also observed that GPX activity was smaller in the seeds soaked in $H_2O_2$ for 12, 24, 30, 36, and 42 h, in relation to those soaked in distilled water. The SOD activity was not affected by the pretreatment of seeds, except for the 24 h treatment. Only one CAT isoform was detected. In the third experiment, seeds were pretreated with 36 h soaking in 100 mM $H_2O_2$ solution or in distilled water and later cultivated in Hoagland's nutrient solution or nutrient solution with 80 mM NaCl. The results showed the pretreatment of seeds with $H_2O_2$-induced acclimation of the plants to salinity. It decreased the deleterious effects of salt stress on the growth of maize. In addition, the differences in antioxidative enzyme activities may explain the increased tolerance to salt stress of plants originated from $H_2O_2$-pretreated seeds.

Li et al. (2011) reported that exogenously applied $H_2O_2$ (0.05 µM) decreased the MDA content, enhanced the GSH content, and increased the activities of SOD, peroxidase (POD), CAT, and APX in salt-stressed wheat seedlings. A similar response in *Suaeda fruticosa* (a halophyte) was also found, indicating that subcellular defense mechanisms are enhanced by exogenous application of $H_2O_2$ (Hameed et al. 2012). Increases in the activities of SOD and CAT following the exogenous application of $H_2O_2$ (0.5 mM) was also observed in oat plants under salt stress (Xu et al. 2008).

Gondim et al. (2012) evaluated the effects of a $H_2O_2$ leaf spraying pretreatment on plant growth and investigated the antioxidant mechanisms involved in the response of these plants to salt stress. They found that salinity reduced maize seedling growth when compared to controls and that $H_2O_2$ foliar spraying was effective in minimizing this effect. Analysis of the antioxidative enzymes CAT, guaiacol peroxidase (GPOX), APX, and SOD revealed that $H_2O_2$ spraying increased the activities of these enzymes. CAT was the most responsive of enzyme to $H_2O_2$, with higher activity earlier (48 h) after treatment, while GPX and APX were responded much later (240 h after treatment). Increased CAT activity appeared to be linked to gene expression regulation. Lower MDA levels were detected in plants with higher CAT activity, which may result from the protective function of this enzyme. Overall, we can conclude that pretreatment with $H_2O_2$ by leaf spraying reduces the deleterious effects of salinity on seedling growth and lipid peroxidation. These responses could be attributed to the ability of $H_2O_2$ to induce antioxidant defenses, especially CAT activity. Yadav et al. (2011) observed that seeds of *Capsicum annuum* primed with $H_2O_2$ (1.5 mM) showed enhance tolerance to salt stress (200 mM NaCl, 10 days). The plants grown from primed seeds flowered earlier and also produced more number of fruits.

Later on, Gondim et al. (2013) also studied the influence of exogenous $H_2O_2$ on AsA and GSH metabolism, plant growth, relative water content (RWC), relative chlorophyll content, and gas exchange in maize plants under NaCl stress. Photosynthesis and transpiration, stomatal conductance, and intercellular $CO_2$ concentration were strongly decreased by salt stress; however, the decrease was lower in plants sprayed with $H_2O_2$. The improved gas exchange in $H_2O_2$-sprayed stressed plants correlated positively with higher RWC and relative chlorophyll content and lower leaf $H_2O_2$ accumulation under NaCl stress conditions. Ascorbate and glutathione did not play any obvious effects as nonenzymatic antioxidants in the ROS scavenging. They concluded that salt tolerance induced by $H_2O_2$ leaf pretreatment is attributed to a reduction in the $H_2O_2$ content and maintenance of RWC and chlorophyll in maize leaves. These characteristics allow maize plants to maintain high rates of photosynthesis under salt stress and improve the growth.

Recently, Ashfaque et al. (2014) conducted an experiment to determine the role of $H_2O_2$ in the alleviation of salt stress in wheat (*T. aestivum* L.). Treatment of plants with $H_2O_2$ significantly influenced the parameters both under nonsaline and salt stress. The application of both 50 and 100 μM $H_2O_2$ reduced the severity of salt stress through the reduction in $Na^+$ and $Cl^-$ content and the increase in proline content and N assimilation. This resulted in increased water relations, photosynthetic pigments, and growth under salt stress. However, maximum alleviation of salt stress was noted with 100 μM $H_2O_2$ and 50 μM $H_2O_2$ proved less effective. Under nonsaline condition, also application of $H_2O_2$ increased all the studied parameters. The treatment of 100 μM $H_2O_2$ maximally benefitted the wheat plants under nonsaline condition and alleviated the effects of salt stress. The treatment of $H_2O_2$ increased proline content that might help increase photosynthetic pigments and growth under salt stress. From the aforementioned discussions, it can be concluded that sufficient accumulation of intracellular $H_2O_2$ or exogenous application modulated salinity-induced oxidative stress tolerance.

Very recently, Sathivaraj et al. (2014) have found that *Panax ginseng* seedlings acclimated with 100 μM $H_2O_2$ for 2 days showed enhanced salinity tolerance by increasing the activities of APX, CAT, and guaiacol peroxidase. The oxidative parameter such as MDA level and endogenous $H_2O_2$ and $O_2^{\bullet-}$ was lower in the $H_2O_2$- acclimated salt-stressed seedlings. Seedling dry weight and chlorophyll and carotenoid contents were also favorably modulated in $H_2O_2$ acclimating salt-stressed seedlings. The aforementioned example clearly demonstrated that $H_2O_2$ priming induces salt stress tolerance in plants by modulating different plant physiological and metabolic processes like photosynthesis, proline accumulation, and ROS detoxification that ultimately lead to better growth and development. Importantly, ROS metabolism plays a pivotal role in changing oxygen relative impact on cells and allowing cell resistance mechanisms leading to cross-tolerance.

## 11.5 HYDROGEN PEROXIDE: AN IMPORTANT COMPONENT OF REDOX-SIGNALING NETWORK IN PLANT CELL

Rapid development of functional genomics and proteomics along with availability of complete genome sequences of *Arabidopsis thaliana* and *Oryza sativa* has revolutionized our understanding of ROS biology and $H_2O_2$ signaling in plants. The production of ROS, particularly $H_2O_2$ in plant cell, functions as double-edged sword; where the basal level of $H_2O_2$ is important to maintain housekeeping metabolic activities (Halliwell 2006; Bhattacharjee 2012b), while the overproduction of $H_2O_2$ could always lead to oxidative damage that may eventually cause termination of the functional life of a plant (Apel and Hirt 2004). Among the spectrum of ROS, $H_2O_2$ receives most attention of the scientific community in the last decade for its chemistry and regulatory activities. The long half-life of $H_2O_2$ (1 ms) and its ability of migration through cell membrane facilitate its signaling function (Bhattacharjee 2012b). $H_2O_2$ is found to be involved in a number of signaling cascades in plants under adverse environmental conditions (Neill et al. 2002; Bhattacharjee 2012b), triggering host-specific responses like triggering MAP kinase activity and stomatal closure, initiating programmed cell death (Desikan et al. 1996), cellular elongation (Foreman et al. 2003), and differentiation (Tsukagoshi et al. 2010) as well as responses to broad spectrum of environmental stimuli (Dat et al. 2000).

Oxidative stress, a condition in which redox homeostasis of the cell is disturbed in the direction of prooxidants due to accumulation of ROS, is a challenge faced by all aerobic organisms under unfavorable environmental cues, including salinity stress. Although, overaccumulation of ROS was originally considered detrimental to cells, it is now widely recognized that redox regulation involving ROS like $H_2O_2$ is a key component modulating cellular activities (Neill et al. 2002; Bhattacharjee 2012b; Petrov and Van Breusegen 2012). The primary molecular events associated with signal transduction leading to changes in gene expression under environmental odds are extremely complicated and largely unknown. It is however becoming gradually evident that gene expression associated with stress acclimation is sensitive to redox state of the cell. Out of many components that contribute to the redox balance of the cell, two factors have been shown to be crucial in mediating stress responses. Generation of $H_2O_2$ and thiol/disulfide exchange reactions involving glutathione pool are now regarded as central components of signal transduction associated with environmental stresses.

A diverse range of environmental stimuli including soil salinity leads to transient rise in $H_2O_2$ levels in plant cells. In such situations, $H_2O_2$ may be utilized by the cell as signal that perceives that environmental threats and subsequently relay it to the downstream effectors. Unlike $Ca^{2+}$ signaling, which is predominantly regulated by storage and release of $Ca^{2+}$, $H_2O_2$ signaling is controlled by its production and scavenging. Subcellular organelles exhibiting redox activities or intense rate of electron flow, such as chloroplast, mitochondria, and microbodies, happen to be the major sources of $H_2O_2$ in plant cell (Figure 11.3). A redox imbalance, associated with environmental stresses such as salinity and extremes of temperature, increased rate of overall metabolism, which eventually upregulates $H_2O_2$ production in plant cell (Bhattacharjee 2012a, 2013). The physiological effect of $H_2O_2$ basically depends on its titer, site of production, developmental stages, exposure to stress, etc. At high concentration, a manifestation of deteriorative events associated with senescence is triggered, whereas at low concentration, a signaling process is evoked (Galvez-Valdiveso and Mullineaux 2010, Miller et al 2010, Bhattacharjee 2012b). $H_2O_2$ mediated cell death is essential for some developmental processes and environmental responses including hypersensitive reaction, leaf senescence, allelopathic interactions, and aleurone cell death (Bais et al. 2003).

$H_2O_2$ generated in chloroplasts, particularly under environmental stresses such as salinity, heat, and chilling stress, may give rise to retrograde signals (Figure 11.3). The signaling event mediated by $H_2O_2$ is not well clarified and might cause passive diffusion of $H_2O_2$ that even can interact or

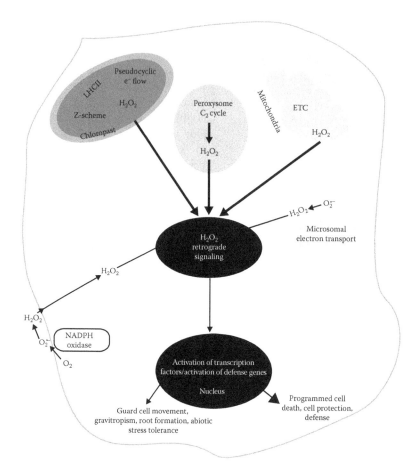

**Figure 11.3.** Generation of $H_2O_2$ in plant cell showing the organeller contribution and subsequent retrograde signaling in the regulation of gene expression in plant cell.

cross talk with other pathways like ABA, SA, and $C_2H_4$ (Galvez-Valdivieso and Mullineaux 2010). The spilling of $H_2O_2$ from chloroplasts is not only supported by known out of cytosolic APX1, which causes development of hypersensitivity of photosynthetic apparatus to excess photochemical stress (Davletova et al. 2005), but also demonstrated by in vitro experiment (Mubaraksina et al. 2010). Aquaporins of the chloroplast membrane are particularly involved in the passage of $H_2O_2$ for evoking retrograde signaling under stress (Petrov and Van Breusegen 2012). Although the regulation of aquaporins in controlling the efflux of $H_2O_2$ is poorly understood, its indirect role in retrograde signaling influencing nuclear gene expression is becoming gradually evident (Petrov and Van Breusegen 2012). More probably, $H_2O_2$ seems to be sensed by components of specific redox-sensitive proteins, which mediate the signals to the nucleus, ultimately influencing gene expression.

## 11.6 REDOX-SENSITIVE PROTEINS AS $H_2O_2$ SENSORS FOR DECIPHERING $H_2O_2$ SIGNALING IN PLANT CELL

In plants, generation of ROS occurs under diverse range of conditions, and it appears that ROS accumulation in specific tissues and appropriate quantities is of benefit to plants and can mediate cross-tolerance toward other stresses. ROS, specifically $H_2O_2$, is found to be involved in plant defense response, affecting both gene expression and activities of proteins such as MAP kinase, which in turn functions as regulators of transcription (Desikan et al. 1999, 2000). In spite of the fact that ROS and cellular redox state are known to control expression of plant genes, the signaling pathway(s) involving transcription factors or promoter elements specific for the redox regulation are still to be identified. There are, however, several candidates for promoter elements as well

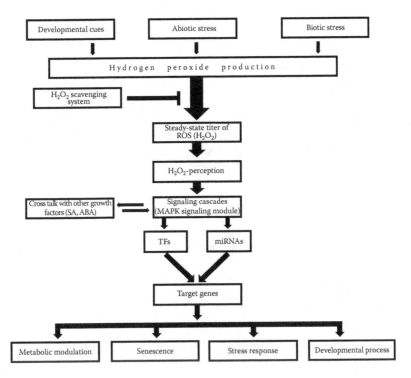

Figure 11.4. A schematic representation of major signaling components in hydrogen peroxide signaling network and their possible outcome in plant cell.

as DNA-binding factors that act as redox response elements (Figure 11.4).

Redox-sensitive proteins have been suggested as possible downstream mediators of such $H_2O_2$ signaling. In fact, plant cell can sense and transduce $H_2O_2$ signaling into appropriate cellular responses through the involvement of redox-sensitive proteins (Figure 11.4). Redox-sensitive proteins mainly operate through reversible oxidation/reduction events, thereby switching on or off, depending on cellular redox state. Plant cells can sense, transduce, and translate the ROS signals into appropriate cellular responses through the involvement of redox-sensitive proteins. Redox-sensitive proteins mainly operate through reversible oxidation/reduction thereby switching "on" and "off" depending on the cellular redox state.

ROS can oxidize the redox-sensitive proteins directly (Mylona and Polidoros 2010) or indirectly via some other ubiquitous redox-sensitive molecules like glutathione or thioredoxins, which control the cellular redox state (Neill et al. 2002; Mylona and Polidoros 2010). Therefore, redox-sensitive proteins are susceptible to oxidation and reduction that depends on the titer of ROS and or

redox state of the cell. Redox-sensitive proteins further execute their function via downstream signaling components like kinases, phosphatases, and transcription factors (Figure 11.5). In some cases, ROS directly oxidize the target proteins, particularly peroxyredoxins and thioredoxins and subsequently the transcription factors (Mylona and Polidoros 2010; Nishiyama et al. 2001). In fact, most of the redox regulation of gene expression is mediated by a family of protein disulfide oxidoreductases, namely, thioredoxins, PRXs, glutaredoxins, and protein disulfide isomerases (Rouhier et al. 2009). Thioredoxins are small (approximately 12 kDa) protein with S=S reducing activity. They oxidized directly by ROS or indirectly by PRXs (thioredoxin peroxidase). Thioredoxins may be reduced by thioredoxin reductase, by NADPH-dependent enzymes. There is ample evidence to suggest that thioredoxins and other similar proteins are enzymatic mediators of the regulatory effects of ROS at transcriptional levels (Mahalingam and Fedoroff 2003). It is found that UV irradiation promotes translocation of thioredoxin to the mammalian nucleus where it activates stress-related transcription factors like NF–kB and AP–1 by enhancing DNA binding (Mahalingam and Fedoroff 2003).

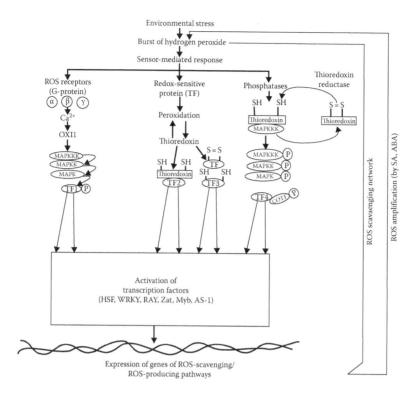

**Figure 11.5.** Schematic diagram showing possible mechanism of action of reactive oxygen species $H_2O_2$ in sensor mediated response leading to gene expression in plants.

Thioredoxin can bind directly with NF–kB p[50] and interacts indirectly with AP–1 through redox factor 1 (Mylona and Polidoros 2010). On the other hand, plants possess chloroplast, mitochondria, and cytosol redox-regulating system for controlling expression of genes (Mahalingam and Fedoroff 2003; Rouhier et al. 2009). Fd, being the component of PS I of z-scheme of electron flow, gets photoreduced during photosynthesis. The reducing power of Fd is then subsequently transformed to thioredoxin by Fd-Trx reductase, which then interacts with target enzymes. There are distinct classes of "thioredoxin target" including the "transcription factors." One of the best studied redox-regulatory plant proteins is a class of RNA-binding proteins that control translation or stability of chloroplastic mRNA under the exposure of photochemical energy by the way of Fd–thioredoxin system (Mahalingam and Fedoroff 2003).

Redox-sensitive transcription factors that orchestrate downstream cascades seem to be the most promising candidate as "$H_2O_2$ sensor" in plant cell (Miller et al. 2008). Heat shock factors are shown to be responsive to oxidative stress and $H_2O_2$ detection (Miller et al. 2010). Other

transcription factors that are associated with $H_2O_2$ signaling include WRKY, MFB, ZAT, and bZIP (Figure 11.3) (Petrov and Van Breusegen 2012).

$H_2O_2$ perception and sensing might also involve oxidation of methionine to methionine sulfoxide (MetSo) and initiation of oxidative signals that change protein phosphorylation (Hardin et al. 2009). Exogenously applied $H_2O_2$ might cause serine phosphorylation in vivo, whereas overexpression of cytoplasmic MetSo repair enzyme exhibits opposite effect and enhancement of phosphorylation. A redox methionine-dependent switch would represent a fast and efficient way to relay oxidative signal mediated by $H_2O_2$. However, identification and characterization of proteins with methionine residues ultimately help us to unfold the redox-regulatory metabolisms.

One of the examples of induction of defense genes, controlled by redox balance of the cell, is GSTs, which catalyze the conjugation of GSH to a variety of hydrophobic electrophilic compounds that otherwise attack important cellular macromolecules. Compounds with bound GSH undergo cellular detoxification pathway (Daniel 1993). The signal by which the expression of GST gene is regulated is believed to be a prooxidant state of the cells, probably resulting from a reduced GSH content (Daniel 1993). The promoter element responsible for the induction of the Ya subunit in mouse GST by electrophilic compounds consists of two adjacent AP1-like sites (Friling et al. 1992). The consensus sequence of this site is AGACA (A/T) (A/T) GC and is called an antioxidant-responsive element or electrophile-responsive element. Two adjacent AP1-like sites are also present in the *Arabidopsis* GST 6 gene and constitute a promoter element (Chen et al. 1996). This promoter element is at least in part required for GST 6 inductions by $H_2O_2$, SA, and auxins (Chen and Singh 1999). A single antioxidant-responsive element has recently been identified in the promoter of a maize CAT gene (CAT 1) and was found to bind molecular factors from senescing scutella that accumulate CAT 1 transcripts, probably as a result of oxidative stress (Polidoros and Scandalios 1999).

The G box (CACGTG) is a ubiquitous *cis*-element present in many plant genes and is thought to mediate response to diverse environmental stimuli, including primarily redox changes (Dröge-Laser et al. 1997; Menkens et al. 1995). Together with the H box (CCTACC), the G box functions in the activation of phenylpropanoid biosynthetic genes. Transcription of at least two of these genes that encode phenylalanine ammonia-lyase and chalone synthase is under redox control (mainly controlled by GSH) (Wingate et al. 1988).

## 11.7 $H_2O_2$ VIS-À-VIS $CA^{2+}$ AS SECOND MESSENGER IN REDOX-REGULATORY SIGNALING PATHWAY IN PLANTS

Like $Ca^{2+}$, $H_2O_2$ possesses several features typical to second messenger in plant cell. The most important of which is the tight control of $H_2O_2$ titer through a redox network that comprises $H_2O_2$ producer, scavenger, sensor, and subset of redox-sensitive genes. Apart from it, its small size and mobility between cellular compartments make it a more convenient second messenger for evoking redox-regulatory pathway and physiological response. Its ability to cross talk with other signaling compounds further make it a versatile signaling molecule (Figure 11.4). A positive or negative feedback loop with other growth factors exists that ultimately control different physiological outcomes including the synthesis of $H_2O_2$ itself

(dePinto et al. 2006). A positive feedback provides a way to amplify the initial signal, while the negative feedback ensures the blockage in order to prevent excessive oxidative damage (Figure 11.4).

Regulation of $Ca^{2+}$ homeostasis is one of the main targets of $H_2O_2$ signaling (Petrov and Van Breusegen 2012). Numerous works support the view that $H_2O_2$ and $Ca^{2+}$ homeostasis in plants cell are interdependent, with some cases showing regulation of $Ca^{2+}$ fluxes by $H_2O_2$, while in others, $Ca^{2+}$ affects $H_2O_2$ metabolism. Movement of the stomatal guard cell is probably one of the widely studied processes to corroborate $H_2O_2$-mediated $Ca^{2+}$ signaling. In fact, ABA induces the production of $H_2O_2$ in guard cells that results in increase in cytosolic $Ca^{2+}$ ($[Ca^{2+}]_{cyt}$ in intact guard cell (Pei et al 2000). A biphasic increase of cytosolic $Ca^{2+}$ is also evident in *Arabidopsis* seedlings treated with $H_2O_2$ (Rentel and Knight 2004). The short-time requirement of $Ca^{2+}$ peak under the inductive pulse is considered as one of the earliest responses of $H_2O_2$ signaling. Wu et al. (2010) showed that spermidine oxidase–derived $H_2O_2$ regulates pollen membrane hyperpolarization-activated $Ca^{2+}$ channels in order to induce pollen tube growth.

On the contrary, cytoplasmic $Ca^{2+}$ is in turn able to trigger changes in $H_2O_2$ level. It is also evident that $H_2O_2$ synthesis requires a continuous cytoplasmic influx, which activates NADPH oxidases located at plasma membrane (Lamb and Dixon, 1997). Experiments with *Arabidopsis* infected with *Pseudomonas syringae* demonstrated that NADPH oxidase inhibitor does not change $Ca^{2+}$ homeostasis, but the application of lanthanum chloride (a calcium channel blocker) suppresses $H_2O_2$ accumulation and subsequent hypersensitive response (Grant et al. 2000), strongly supporting the view that $Ca^{2+}$ works upstream of $H_2O_2$ in signaling cascades (Lachaud et al. 2011).

$Ca^{2+}$ homeostasis also regulate antioxidative defense in plant. A rise in concentration of intracellular $Ca^{2+}$ causes efficient detoxification of $H_2O_2$ involving detoxification enzymes including $Ca^{2+}$-sensitive CAT3. Application of $Ca^{2+}$ channel blocker ($LaCl_3$), $Ca^{2+}$ chelator (EGTA), and calmodulin inhibitor (trifluoperazine) to germinating *Amaranthus* seeds causes significant reduction of $H_2O_2$-scavenging enzymes (Bhattacharjee 2008), strongly supporting $Ca^{2+}$ regulation of $H_2O_2$ titer in plant tissue. Ca-CaM signaling pathway also regulates a number of different target proteins in signaling cascades including MAP kinase (MAP kinase 8). MAP kinase pathway in turn negatively regulates $H_2O_2$ synthesis by upregulating the expression of RbohD.

## 11.8 CONCLUSION AND FUTURE PERSPECTIVE

Hydrogen peroxide production is inevitable during plant growth, development, and resistance responses. As an element of oxidative stress, $H_2O_2$ has a deleterious impact on cell components when accumulated in excess, and simultaneously, it acts as an inducer of protective mechanisms, especially at the early stages of plant stress response. A plant operates mechanisms particularly under environmental stress to utilize ROS, especially $H_2O_2$, for signaling purposes that confer acclamatory stress tolerance including salt stress through the modulation of osmotic adjustment, ROS detoxification, photosynthetic C fixation, and hormonal regulation. Since excessive accumulation of $H_2O_2$ leads to deteriorative events leading to cell death, regulatory conditions operate both at metabolic and genomic levels to control the titer of $H_2O_2$ required for operating signaling cascades. Evidence to date suggests a central role of $H_2O_2$ in intracellular and systemic signaling processes giving rise to successful acclimation and tolerance to biotic and abiotic stresses. It is likely that redox poise of the cell may control cellular events through interactions with $H_2O_2$ and redox-sensitive molecules such as GSH, AsA, thioredoxins, and PRXs. Several other signaling molecules, for example, SA, ABA, jasmonate, NO, and $Ca^{2+}$, were proposed to interact with $H_2O_2$ in regulation of plant defense responses at both the gene and protein levels (Petrov and Van Breusegem 2012). Although comprehensive work has been done in the last decade to understand $H_2O_2$ physiology and components associated with signaling network, our knowledge is far from complete. The complete unfolding of $H_2O_2$ physiology, particularly $H_2O_2$ sensing, identification of detailed components of $H_2O_2$ signaling network, and $H_2O_2$ cross talk with other growth factors, is absolutely necessary and is of immense significance in context to the performance of the crop under salt stress. The use of transgenic plants that are impaired in $H_2O_2$ generation or increased $H_2O_2$ production and the isolation of $H_2O_2$-signaling mutants will be invaluable in elucidating further biological roles of $H_2O_2$ in specific cells and in response to salt or other abiotic stresses. Postgenomic developments in transcriptomics and proteomics will facilitate further insights into cellular responses to $H_2O_2$ in the near future. Future work will no doubt reveal novel signaling roles for $H_2O_2$ and its interaction with ROS-scavenging system including AsA-GSH cycle. Understanding the subtle and sensitive mechanism of plants to fine tune $H_2O_2$ titer and their associated signaling cascades holds the key to improving future agriculture.

## ACKNOWLEDGMENT

The authors wish to acknowledge the University Grants Commission, New Delhi, India, for financial assistance in the form of the Major Research Project (F.No 41–429/2012[SR], dated 16.07.2012).

## REFERENCES

Abogadallah, G.M. 2010. Antioxidative defense under salt stress. *Plant Signaling and Behavior* 5:369–374.

Apel, K. and H. Hirt. 2004. Reactive oxygen species: Metabolism, oxidative stress, and signal transduction. *Annual Review of Plant Biology* 55:373–399.

Asada, K. 1992. Ascorbate peroxidase—A hydrogen peroxide-scavenging enzyme in plants. *Physiologia Plantarum* 85:235–241.

Asada, K. 2006. Production and scavenging of reactive oxygen species in chloroplast and their functions. *Plant Physiology* 141:391–396.

Ashfaque, F., M.I.R. Khan, and N.A. Khan. 2014. Exogenously applied $H_2O_2$ promotes proline accumulation, water relations, photosynthetic efficiency and growth of wheat (*Triticum aestivum* L.) under salt stress. *Annual Research and Review in Biology* 4:105–120.

Azevedo-Neto, A.D., J.T. Prisco, J. Eneas-Filho, J.V.R. Medeiros, and E. Gomes-Filho. 2005. Hydrogen peroxide pre-treatment induces salt stress acclimation in maize plants. *Journal of Plant Physiology* 162:1114–1122.

Bais, H.P, R. Vepachedu, S. Gilroy, R.M. Callaway, and J.M. Vivanco. 2003. Allelopathy and exotic plant invasion: From molecules and genes to species interactions. *Science* 301:1377–1380.

Bhattacharjee, S. 2008. Calcium-dependent signaling pathway in the heat-induced oxidative injury in *Amaranthus lividus* L. *Biologia Plantarum* 52(01):137–140.

Bhattacharjee, S. 2012a. An inductive pulse of hydrogen peroxide pretreatment restores redox-homeostasis and mitigates oxidative membrane damage under extremes of temperature in two rice cultivars (*Oryza sativa* L., cultivars Ratna and SR 26B). *Plant Growth Regulation* 68:395–410.

Bhattacharjee, S. 2012b. The language of reactive oxygen species signalling in plants. *Journal of Botany* 2:01–22.

Bhattacharjee, S. 2013. Heat and chilling induced disruption of redox homeostasis and its regulation by hydrogen peroxide in rice (*Oryza sativa* L., Cultivar Ratna). *Physiology and Molecular Biology of Plants* 19:199–207.

Blokhina, O. and K. Fagerstedt. 2006. Oxidative stress and antioxidant defenses in plants. In *Oxidative Stress, Disease and Cancer*, ed. K.K. Singh, pp. 151–199. London, U.K.: Imperial College Press.

Bose, J., A. Rodrigo-Moreno, and S. Shabala. 2013. ROS homeostasis in halophytes in the context of salinity stress tolerance. *Journal of Experimental Botany*. 65: 1241–1257.

Chen, W., G. Chao, and K.B. Singh. 1996. The promoter of a $H_2O_2$-inducible, *Arabidopsis* glutathione S-transferase gene contains closely linked OBF- and OBP1-binding sites. *The Plant Journal* 10:955–963.

Chen, W. and K.B. Singh. 1999. The auxin, hydrogen peroxide and salicylic acid induced expression of the *Arabidopsis* GST 6 promoter is mediated in part by an OCS element. *The Plant Journal* 19:667–674.

Chen, Z., I.I. Pottosin, T.A. Cuin et al. 2007. Root plasma membrane transporters controlling $K^+/Na^+$ homeostasis in salt stressed barley. *Plant Physiology* 145:1714–1725.

Daniel, V. 1993. Glutathione S-transferases: Gene structure and regulation of expression. *Critical Review of Biochemistry and Molecular Biology* 28:173–207.

Dat, J., S. Vandenabeele, E. Vranová, M. Van Montagu, D. Inzé, and F. Van Breusegem. 2000. Dual action of the active oxygen species during plant stress responses. *Cellular and Molecular Life Sciences* 57:779–795.

Davletova, S., L. Rizhsky, H. Liang et al. 2005. Cytosolic ascorbate peroxidase 1 is a central component of the reactive oxygen gene network of *Arabidopsis*. *The Plant Cell* 17:268–281.

Desikan, R., A. Clarke, and S.J. Neill. 1999. $H_2O_2$ activates a MAP kinase-like enzyme in *Arabidopsis thaliana* suspension cultures. *Journal of Experimental Botany* 50:1863–1866.

Desikan, R., J.T. Hancock, M.J. Coffey, and S.J. Neill. 1996. Generation of active oxygen in elicited cells of Arabidopsis thaliana is mediated by a NADPH oxidase-like enzyme. *Federation of European Biochemical Societies Letters* 382:213–217.

Desikan, R., S.J. Neill, and J.T. Hancock. 2000. Hydrogen peroxide-induced gene expression in *Arabidopsis thaliana*. *Free Radical Biology and Medicine* 28:773–778.

Dröge-Laser, W., A. Kaiser, W.P. Lindsay et al. 1997. Rapid stimulation of a soybean protein-serine kinase that phosphorylates a novel bZip DNA-binding protein, G/HBF-1, during the induction of early transcription-dependent defenses. *EMBO Journal* 16:726–734.

Fedina, I.S., D. Nedeva, and N. Çiçek. 2009. Pretreatment with $H_2O_2$ induces salt tolerance in barley seedlings. *Biologia Plantarum* 53:321–324.

Foreman, J., V. Demidchik, and J.H. Bothwell. 2003. Reactive oxygen species produced by NADPH oxidase regulate plant cell growth. *Nature* 422:442–446.

Friling, R.S., S. Bergelson, and V. Daniel. 1992. Two adjacent AP-1-like binding sites from the electrophile-responsive element of the murine glutathione S-transferase Ya subunit gene. *Proceedings of the National Academy of Sciences USA* 89:668–673.

Galvez-Valdivieso, G. and P.M. Mullineaux. 2010. The role of reactive oxygen species in signalling from chloroplasts to the nucleus. *Physiologia Plantarum* 138:430–439.

Gechev, T.S., F. Van Breusegem, J.M. Stone, L. Denev, and C. Laloi. 2006. Reactive oxygen species as signals that modulate plant stress responses and programmed cell death. *BioEssays* 28:1091–1101.

Gondim, F.A., E. Gomez-Filho, J.H. Costa, N.L. Mendes Alencar, and J.T. Prisco. 2012. Catalase plays a key role in salt stress acclimation induced by hydrogen peroxide pretreatment in maize. *Journal of Plant Physiology and Biochemistry* 56:6271

Gondim, F.A., E. Gomes-Filho, C.F. Lacerda, J.T. Prisco, A.D.A. Neto, and E.C. Marques. 2010. Pretreatment with $H_2O_2$ in maize seeds: Effects on germination and seedling acclimation to salt stress. *Brazilian Journal of Plant Physiology* 22:103–112.

Gondim, F.A., M. Rafael de Souza, E. Gomes-Filho, and J.T. Prisco. 2013. Enhanced salt tolerance in maize plants induced by $H_2O_2$ leaf spraying is associated with improved gas exchange rather than with non-enzymatic antioxidant system. *Theoretical and Experimental Plant Physiology* 25:251–260.

Grant, M., I. Brown, S. Adams, M. Knight, A. Ainslie, and J. Mansfield. 2000. The RPM1 plant disease resistance gene facilitates a rapid and sustained increase in cytosolic calcium that is necessary for the oxidative burst and hypersensitive cell death. *The Plant Journal* 23:441–450.

Halliwell, B. 2006. Reactive species and antioxidants. Redox biology is a fundamental theme of aerobic life. *Plant Physiology* 141:312–322.

Hardin, S.C., C.T. Larue, M.H. Oh, V. Jain, and S.C. Huber. 2009. Coupling oxidative signals to protein phosphorylation via methionine oxidation in *Arabidopsis*. *Biochemical Journal* 422:305–312.

Hernandez, J.A., F.J. Corpas, M. Gomez, L.A. Delrio, and F. Sevilla. 1993. Salt-induced oxidative stress mediated by activated oxygen species in pea leaf mitochondria. *Physiologia Plantarum* 89:103–110.

Hossain, M.A. and M. Fujita. 2013. Hydrogen peroxide priming stimulates drought tolerance in mustard (*Brassica juncea* L.). *Plant Gene and Trait* 4:109–123.

Hossain, M.A., M. Hasanuzzaman, and M. Fujita. 2011. Coordinate induction of antioxidant defense and glyoxalase system by exogenous proline and glycinebetaine is correlated with salt tolerance in mung bean. *Frontiers of Agriculture in China* 5:1–14.

Hossain, M.A., M.G. Mostofa, and M. Fujita. 2013a. Cross protection by cold-shock to salinity and drought stress-induced oxidative stress in mustard (*Brassica campestris* L.) seedlings. *Molecular Plant Breeding* 4:50–70.

Hossain, M.A., M.G. Mostofa, and M. Fujita. 2013b. Heat-shock positively modulates oxidative protection of salt and drought-stressed mustard (*Brassica campestris* L.) seedling. *Journal of Plant Science Molecular Breeding* 2:1–14.

Lachaud, C., D. Da Silva, N. Amelot et al. 2011. Dihydrosphingosine-induced programmed cell death in tobacco BY-2 cells is independent of $H_2O_2$ production. *Molecular Plant* 4:310–318.

Lamb, C. and R.A. Dixon. 1997. The oxidative burst in plant disease resistance. *Annual Review of Plant Physiology and Plant Molecular Biology* 48:251–275.

Li, J.T., Z.B. Qiu, X.W. Zhang, and L.S. Wang. 2011. Exogenous hydrogen peroxide can enhance tolerance of wheat seedlings to salt stress. *Acta Physiologiae Plantarum* 33:835–842.

Ma, L., H. Zhang, L. Sun, Y. Jiao, G. Zhang, C. Miao, and F. Hao. 2012. NADPH oxidase AtrbohD and AtrbohF function in ROS-dependent regulation of $Na^+/K^+$ homeostasis in Arabidopsis under salt stress. *Journal of Experimental Botany* 63:305–317.

Ma, N.L., Z. Rahmat, and S. Lam. 2013. A review of the "omics" approach to biomarkers of oxidative stress in *Oryza sativa*. *International Journal of Molecular Sciences* 14:7515–7541.

Mahalingam, R. and N. Fedoroff. 2003. Stress response, cell death and signaling. *Physiologia Plantarum* 119:56–68.

Menkens, A.E., U. Schindler, and A.R. Cashnore. 1995. The G-Box: A ubiquitous regulatory DNA element in plants bound by GBF family of bZIP proteins. *Trends in Biochemical Sciences* 20:506–511.

Mhamdi, A., G. Queval, S. Chaouch, S. Vanderauwera, F. Van Breusegem, and G. Noctor. 2010. Catalase function in plants: A focus on *Arabidopsis* mutants as stress-mimic models. *Journal of Experimental Botany* 61:4197–4220.

Millar, H., M.J. Considine, D.A Day, and J. Whelan. 2001. Unraveling the role of mitochondria during oxidative stress in plants. *International Union of Biochemistry and Molecular Biology Life* 51:201–205.

Miller, G., V. Shulaev, and R. Mittler. 2008. Reactive oxygen signaling and abiotic stress. *Physiologia Plantarum* 133:481–489.

Miller, G., N. Sujuki, S. Ciftci-Yilmaz, and R. Mittler. 2010. Reactive oxygen species homeostasis and signalling during drought and salinity. *Plant Cell and Environment* 33:453–467.

Mittler, R., S. Vanderauwera, M. Gollery, and F.V. Breusegem. 2004. Reactive oxygen gene network of plants. *Trends in Plant Science* 9:490–498.

Mittova, V., M. Tal, M. Volokita, and M. Guy. 2003. Up-regulation of the leaf mitochondrial and peroxisomal antioxidative systems in response to salt-induced oxidative stress in the wild salt-tolerant tomato species *Lycopersicon pennellii*. *Plant Cell and Environment* 26:845–856.

Mubarakshina, M.M., B.N. Ivanov, I.A. Naydov, W. Hillier, M.R. Badger, and A. Krieger-Liszkay. 2010. Production and diffusion of chloroplastic $H_2O_2$ and its implication to signalling. *Journal of Experimental Botany* 61:3577–3587.

Munns, R. 2005. Genes and salt tolerance: Bringing them together. *New Phytologist* 16:645–663.

Mylona, P.V. and A.N. Polidoros. 2010. ROS regulation and antioxidant genes. In *Reactive Oxygen Species and Antioxidants in Higher Plants*, ed. S.D. Gupta, pp. 1–30. Boca Raton, FL: Science Pub., CRC Press.

Neill, S., R. Desikan, and J. Hancock. 2002. Hydrogen peroxide signalling. *Current Opinion in Plant Biology* 5:388–395.

Nishiyama, A., H. Masutani, H. Nakamura, Y. Nishinaka, and J. Yodi. 2001. Redox regulation by thioredoxin and thioredoxin binding proteins. *IUBMB Life* 52:29–33.

Pandolfi, C., S. Mancuso, and S. Shabala. 2012. Physiology of acclimation to salinity stress in pea (*Pisum sativum*). *Environmental and Experimental Botany* 84:44–51.

Pei, Z.M., Y. Murata, G. Benning et al. 2000. Calcium channels activated by hydrogen peroxide mediate abscisic acid signalling in guard cells. *Nature* 406:731–734.

Petrov, V.D. and F. Van Breusegem. 2012. Hydrogen peroxide-a central hub for information flow in plant cells. *AoB Plants*, pls014.

de Pinto, M.C., A. Paradiso, P. Leonetti, and L. De Gara. 2006. Hydrogen peroxide, nitric oxide and cytosolic ascorbate peroxidase at the crossroad between defence and cell death. *The Plant Journal* 48:784–795.

Polidoros, A.N. and J.G. Scandalios. 1999. Role of hydrogen peroxide and different classes of antioxidants in the regulation of catalase and glutathione S-transferase gene expression in maize (*Zea mays* L.). *Physiologia Plantarum* 106:112–117.

Rentel, M.C. and M.R. Knight. 2004. Oxidative stress-induced calcium signalling in *Arabidopsis*. *Plant Physiology* 135:1471–1479.

Rhoads, D.M., A.L. Umbach, C.C Subbaiah, and J.N. Siedow. 2006. Mitochondrial reactive oxygen species. Contribution to oxidative stress and interorganellar signaling. *Plant Physiology* 141:357–366.

del Rio, L.A., F.J. Corpas, L.M. Sandalio, J.M. Palma, M. Gomez, and J.B. Barroso. 2002. Reactive oxygen species, antioxidant systems and nitric oxide in peroxisomes. *Journal of Experimental Botany* 53:1255–1272.

Robson, C.A. and G.C. Vanlerberghe. 2002. Transgenic plant cells lacking mitochondrial alternative oxidase have increased susceptibility to mitochondria-dependent and independent pathway of programmed cell death. *Plant Physiology* 129:1908–1920.

Rouhier, N., S.D. Lamaire, and J.P. Jaqcquot. 2009. The role of GSH in photosynthetic organisms: The emerging function of glutaredoxins and glutathionylation. *Annual Review of Plant Biology* 59:143–166.

Sagi, M. and R. Fluhr. 2006. Production of reactive oxygen species by plant NADPH oxidases. *Plant Physiology* 141:336–340.

Sathiyaraj, G., S. Srinivasan, Y.J. Kim et al. 2014. Acclimation of hydrogen peroxide enhances salt tolerance by activating defense-related proteins in *Panax ginseng* C.A. Meyer. *Molecular Biology Reports* 41:3761–3771.

Sgherri, C.L.M., C. Pinzino, and F. Navari-Izzo. 1996. Sunflower seedlings subjected to increasing stress by water deficit: Changes in $O_2^{\cdot-}$ production related to the composition of thylakoid membranes. *Physiologia Plantarum* 96:446–452.

Sharma, P., A.B. Jha, R.S. Dubey, and M. Pessarakli. 2012. Reactive oxygen species, oxidative damage, and antioxidative defense mechanisms in plants under stressful conditions. *Journal of Botany* Article ID 217037, 26 p.

Szarka, A., B. Tomasskovies, and G. Banhegyi. 2012. The ascorbate-glutathione-$\alpha$-tocopherol triad in abiotic stress response. *International Journal of Molecular Science* 13:4458–4483.

Talukdar, D. 2013. Balanced hydrogen peroxide metabolism is central in controlling NACL-induced oxidative stress in medicinal legume, fenugreek (*Trigonella foenum-graecum* L.). *Biochemistry and Molecular Biology* 1(2):34–43.

Tsukagoshi, H., W. Busch, and P.N. Benfey. 2010. Transcriptional regulation of ROS controls transition from proliferation to differentiation in the root. *Cell* 143:606–616.

Uchida, A., A.T. Jagendorf, T. Hibino, and T. Takabe. 2002. Effects of hydrogen peroxide and nitric oxide on both salt and heat stress tolerance in rice. *Plant Science* 163:515–523.

Wahid, A., M. Perveen, S. Gelani, and S.M.A. Basra. 2007. Pretreatment of seed with $H_2O_2$ improves salt tolerance of wheat seedlings by alleviation of oxidative damage and expression of stress proteins. *Journal of Plant Physiology* 164:283–294.

Wu, D., S. Cai, M. Chen, L. Ye, Z. Chen, H. Zhang, F. Wu, and G. Zhang. 2013. Tissue metabolic responses to salt stress in wild and cultivated barley. *PLoS ONE* 8(1):e55431.

Wu, J., Z. Shang, X. Jiang et al. 2010. Spermidine oxidase-derived $H_2O_2$ regulates pollen plasma membrane hyperpolarization-activated $Ca^{2+}$-permeable channels and pollen tube growth. *The Plant Journal* 63:1042–1053.

CHAPTER TWELVE

# Plant Oxidative Stress and the Role of Ascorbate–Glutathione Cycle in Salt Stress Tolerance

*Hai Fan and Bao-Shan Wang*

## CONTENTS

*Abstract.* Soil salinity is one of the most emergent environmental problems in the world, which limits crop yield and productivity seriously. Today, about 20% of the world's cultivated land and nearly half of all irrigated lands are affected by salinity (Rhoades and Loveday, 1990). Salt in the soil may lower water potential in the soil and decrease water absorption ability of the plant; salt accumulation in the plant shows lower growth rate and photosynthetic rate, together with damaging syndromes. There are different hypotheses about salt injury and salinity tolerance mechanisms of plants, yet reactive oxygen species (ROS) are regarded as the main source of oxidative damage to cells under salinity stress as well as other environmental stresses such as drought, high temperature, chilling, and flooding. To scavenge the ROS, plants require the activation of complex metabolic activities including antioxidative pathways, of which ascorbate (AsA)–glutathione (GSH) cycle plays a very important role, both enzymatically and nonenzymatically. AsA–GSH cycle was first put forward by Noctor and Foyer (1998), which operates in the cytosol, mitochondria, plastids, and peroxisomes (Noctor and Foyer, 1998). Since GSH, AsA, and NADPH are present in high concentrations in plant cells, it is assumed that the AsA–GSH cycle plays a key role in ROS detoxification. In this chapter, the production of ROS, the family of AsA–GSH cycle, and the role of AsA–GSH cycle in plant salt tolerance are discussed, respectively.

*Keywords:* Antioxidant, Antioxidant enzyme, Ascorbate–glutathione cycle, Reactive oxygen species, Salt stress

## 12.1 ROS

### 12.1.1 ROS and Their Generation in Chloroplasts

Under normal conditions, one oxygen molecule is reduced to two molecules of water by accepting four electrons and four protons at the same time. However, in plant cells, there remain many electron transmitters that can transmit one, two, or three electrons at one time; hence, molecular $O_2$ is partially reduced and gives birth to many oxygen intermediates. The typical oxygen intermediates are singlet oxygen ($^1O_2$), superoxide anion radical ($O_2^{\cdot-}$), hydrogen peroxide ($H_2O_2$), and hydroxyl radical ($\cdot OH$). In contrast to atmospheric oxygen, the oxygen intermediates are very active and are capable of unrestricted oxidation or reduction of various cell metabolites, thus the intermediates are called reactive oxygen species (ROS), active oxygen species (Noctor and Foyer, 1998), or reactive oxygen intermediates (Mittler, 2002).

ROS are natural and unavoidable by-products of aerobic metabolism. There are many potential sources of ROS in the normal metabolism of plant cells, especially in the chloroplasts and mitochondria, in the processes of photosynthesis and respiration, where the oxygen concentration is relatively high. In chloroplasts, photosystem I (PSI) and PSII are the main sites for the generation of superoxide radical and singlet oxygen. In mitochondria, about 0.1%–2% of electrons leak from the electron chain through complexes I and III to give out superoxide radicals. Other sources of ROS are related to the pathways enhanced during abiotic stresses, such as glycolate oxidase in peroxisomes during photorespiration. Apart from these, new sources of ROS have been identified in plants in recent years, which include NADPH oxidases, amine oxidases, and cell-wall-bound peroxidases. These are tightly regulated and participate in the production of ROS during processes such as programmed cell death and pathogen defense (Hammond-Kosack and Jones, 1996; Dat et al., 2000; Grant and Loake, 2000).

Under usual conditions in nature, oxygen is a gas composed of diatomic molecules $O_2$, dioxygen. Triplet is the ground state of $O_2$ ($^3O_2$) since the molecule has two electrons with parallel spins in two antibonding molecular orbits. The reaction of $O_2$ with cell components has constraints. Singlet oxygen, $^1O_2$, is formed as the result of the spin flip of one of unpaired electrons. The transformation of $^1O_2$ to $^3O_2$ is relatively slow; its lifetime in the cell was estimated to be approximately 3 μs (Hatz et al., 2007), and its diffusion distance in chloroplast thylakoids is estimated to be 5.5 nm (Krasnovsky, 1998).

The chloroplast is a prevailing source of $^1O_2$ in living organisms. The main route of $^1O_2$ generation in thylakoids is the transfer of energy from the chlorophyll in triplet state to molecular $O_2$.

The main place of the chlorophyll triplet state formation is PSII in thylakoids; $^1O_2$ is usually formed when the charge recombination in $P680^+Pheo^-$ is increased; this phenomenon often happens when the forward electron transport is limited, such as in the cases of high irradiation and low assimilation rate, when the PQ pool becomes overreduced. This leads to the full reduction of QA and results in a low yield of charge separation due to the electrostatic effect of $QA^-$ on the $P680^+Pheo^-$ radical pair. This is known as closed PSII. However, in this status, still around 15% of charge separation occurs, leading to the formation of the chlorophyll triplet state $^3P680^*$. Triplet chlorophyll passes its energy to $O_2$ and thus gives birth to $^1O_2$.

$$^3O_2 + {}^3P680^* \rightarrow {}^1O_2 + {}^1P680$$

$^1O_2$ readily reacts with lipids, proteins, and pigments and is rapidly quenched by water, which makes its diffusion distance from the site of production shortest among all ROS (Asada, 2006).

Superoxide anion radical ($O_2^{\bullet-}$) is produced if one additional electron is transferred to the antibonding orbital of $O_2$. $O_2^{\bullet-}$ is rather stable even in water; the half-life of $O_2^{\bullet-}$ was found to be close to 15 s at pH 11 (Fujiwara et al., 2006). In the electron chain, superoxide can be generated on the acceptor side of PSI, where the thylakoid-bound (4Fe-4S) clusters X on psaA and psaB or A/B on psaC of the PSI complex is the most likely electron donor to $O_2$. Apart from this, the end electron transmitter, reduced Fd, is a possible source of $O_2^{\bullet-}$, though its rate is very low (redFd + $O_2 \rightarrow$ oxiFd + $O_2^{\bullet-}$, kobs: 0.081 $s^{-1}$ in air-saturated conditions) (Hosein and Palmer, 1983). Fd is reduced by the (4Fe-4S) cluster A/B in psaC of the PSI complex. Normally the photoreduced Fd donates electrons to $NADP^+$ catalyzed with the Fd-$NADP^+$ reductase (FNR) and thus leads to the formation of the assimilation power NADPH. Furthermore, photoreduced Fd reduces the monodehydroascorbate (MDHA) radical at a rapid rate to regenerate AsA from MDHA produced in the APX reaction (discussed later). Besides, reduced Fd passes its electron to plastoquinones for the Fd-dependent cyclic electron flow around PSI. Furthermore, reduced Fd is the electron source of nitrite reductase, thioredoxin reductase, and glutamate synthase. However, the photoproduction of $O_2^{\bullet-}$ increases severalfold when no electron acceptor other than $O_2$ is available (Furbank and Badger, 1983), such as in the case of high irritation and other environmental stresses. When the photon-utilizing capacity is high and $NADP^+$ is available for chloroplasts, the reduced Fd has little chance to interact with $O_2$.

PSII can also generate $O_2^{\bullet-}$. However, the rate is very slow, and the oxygen reduction in this PS can reach only about 1–1.5 µmol $O_2$ mg $Chl^{-1}$ $h^{-1}$ at physiological conditions (Khorobrykh et al., 2002). The donor of electron to $O_2$ in PSII is $Pheo^-$. However, under normal functional conditions, electron transfer from $Pheo^-$ to QA is much faster (300–500 ps; Dekker and Grondelle, 2000), which prevents the electron transfer from $Pheo^-$ to $O_2$. Only if QA is fully reduced under strong stress conditions, this process may occur.

Hydrogen peroxide ($H_2O_2$) is the most stable ROS in aqueous solutions. The dismutation of $O_2^{\bullet-}$ is the main reaction of $H_2O_2$ production in the cell, though glycolate and other substances such as glucose can be directly reduced to $H_2O_2$ by two electron oxidase. In chloroplast stroma, $O_2^{\bullet-}$ is mainly dismutated with the catalysis of superoxide dismutase (SOD) at a rate of $2 \times 10^9$ $M^{-1}$ $s^{-1}$. On the other hand, the production of $H_2O_2$ through the reduction of $O_2^{\bullet-}$ by ascorbic acid or by reduced GSH is also possible, yet, their rates are considerably less than that for SOD-catalyzed dismutation. In chloroplast thylakoid membrane, Mubarakshina et al. (2006) found that $H_2O_2$ was produced with significant rate, and the production increases with an increase in light intensity. And plastoquinol ($PQH_2$) was supposed to reduce $O_2^{\bullet-}$ to give out $H_2O_2$ in the following reaction (Mubarakshina and Ivanov, 2010):

$$PQH_2 + O_2^{\bullet-} \rightarrow PQ^{\bullet-} + H_2O_2$$

$H_2O_2$ can also be produced at the donor and acceptor sides of PSII. At the acceptor side, $H_2O_2$ can be formed outside thylakoids by the dismutation of $O_2^{\bullet-}$ with nonheme iron of PSII. At the donor side, $H_2O_2$ can be formed as an intermediate during water oxidizing cycle operation if this cycle is seriously disrupted (Ivanov, 2012).

Hydroxyl radical ($\cdot$OH) is the most destructive ROS. It is produced in cells by the Fenton and Haber–Weiss reactions. $H_2O_2$ molecule decomposition catalyzed by $Fe^{2+}$ is named as the Fenton reaction, where the reductant of $H_2O_2$ is ferrous iron and the product is ferric iron. Apart from

$Fe^{2+}$, $Cu^+$ and $Mn^{2+}$ can also serve as a Fenton reagent, though their catalytic power is different:

$$Fe^{2+} + H_2O_2 \rightarrow Fe^{3+} + OH^- + {}^{\cdot}OH$$

The Haber–Weiss reaction is the generation of ${}^{\cdot}OH$ from $H_2O_2$ and $O_2^{\cdot-}$. This reaction comprises two steps:

Step one

$$Fe^{3+} + O_2^{\cdot-} \rightarrow Fe^{2+} + O_2$$

Step two, the Fenton reaction

$$Fe^{2+} + H_2O_2 \rightarrow Fe^{3+} + OH^- + {}^{\cdot}OH$$

The net reaction

$$O_2^{\cdot-} + H_2O_2 \rightarrow {}^{\cdot}OH + OH^- + O_2$$

${}^{\cdot}OH$ is among the most reactive species known to chemistry, able to react indiscriminately to cause lipid peroxidation, the denaturation of proteins, and the mutation of DNA (Bowler et al., 1992). In other words, it is able to oxidize almost all cell molecules readily right in the place where it is generated and causes damage. One should be aware that the generation of ${}^{\cdot}OH$ depends on the location of $H_2O_2$ production, as well as the presence of both $Fe^{2+}$ and reductants. In chloroplast stroma, there are pools of iron deposited in a chelated inactive form with chelators such as ferritin. The concentration of $Fe^{2+}$ can be increased when the accumulation of iron either exceeds the chelating ability of chloroplasts or the iron is released from its complex with chelators (Thomas et al., 1985). This phenomenon will increase the production of ${}^{\cdot}OH$ and oxidize the chloroplast and membrane, which often happens under stress conditions.

## 12.1.2 ROS in Mitochondria, Peroxisomes, and Apoplasts

Apart from chloroplasts, ROS are also generated in the mitochondria, the peroxisomes, and the apoplasts, as well as in the nucleus, the plasma membrane, and the endoplasmic reticulum (Overmyer et al., 2003; Ashtamker et al., 2007; Foyer and Noctor, 2009; Jaspers and Kangasjärvi, 2010; Mazars et al., 2010). In mitochondria, where the mitochondrial electron transport chain continuously reduces $O_2$ to water, a four-electron reduction, primary ROS, is generated in the form of $O_2^{\cdot-}$, as a result of monoelectronic reduction of $O_2$. $O_2^{\cdot-}$ is then transformed into more stable $H_2O_2$ in mitochondria through the activity of matrix Mn-SOD, as well as Cu-Zn-SOD in the intermembrane space (Weisiger and Fridovich, 1973a,b). If $H_2O_2$ cannot be timely scavenged by mitochondrial antioxidant systems, $H_2O_2$ may generate the ${}^{\cdot}OH$ through Fenton reaction. ${}^{\cdot}OH$ is highly reactive and generally is believed to act essentially as a damaging molecule. For this reason, mitochondria are believed to have developed efficient $H_2O_2$ removal systems, as well as metal-chelating mechanisms, to prevent the formation of this radical. In the electron transfer chain, complexes III and I are major sites of ROS formation. Complex III receives electrons from reduced coenzyme Q ($UQH_2$) and donates them to cytochrome c. In this process, $UQ^{\cdot-}$ is generated as an intermediate. Because the $UQ^{\cdot-}/UQ$ pair is highly reducing, $O_2^{\cdot-}$ may be formed by electron donation from it. On the other hand, UQ cycling may also be responsible for $O_2^{\cdot-}$ formation within complex I. It was found that myxothiazol, which keeps the ubiquinone pool highly reduced, enhances ROS formation by complex I more significantly than the complex I inhibitor rotenone (Lambert and Brand, 2004).

Peroxisomes are probably the major sites of intracellular $H_2O_2$ production. They can oxidize many substances and give birth to $H_2O_2$. The main metabolic processes responsible for the $H_2O_2$ generation includes the glycolate oxidase reaction, the fatty acid β-oxidation, the enzymatic reaction of flavin oxidases, and the disproportionation of $O_2^{\cdot-}$ radicals. In photorespiration, for example, glycolate oxidase, which converts glycolate to glyxylate, generates $H_2O_2$ by directly oxidizing glycolate with $O_2$. Peroxisomes also produce $O_2^{\cdot-}$ as a consequence of their normal metabolism, which has been identified using biochemical and electron spin resonance spectroscopy methods.

The generation of ROS in apoplasts is mainly related to plasma membrane NADPH oxidases (Suzuki et al., 2011) and transmembrane flavoproteins that oxidize cytoplasmic NADPH, translocate electrons across plasma membrane, and reduce extracellular oxygen to yield $O_2^{\cdot-}$ in the cell wall. The latter is unable to passively cross the lipid bilayer and remain in the apoplast, where it is rapidly converted into $H_2O_2$, either spontaneously or in a reaction catalyzed by SOD.

## 12.2 DAMAGES OF ROS

The production and the removal of ROS must be strictly controlled in order to avoid oxidative stress. However, the equilibrium between production and scavenging of ROS is often perturbed by a number of stressful conditions such as drought, salinity, high irradiation, heavy metals, pathogens, and gaseous pollution. Enhanced level of ROS can cause damage to biomolecules such as lipids, proteins, and DNA, and thus alter intrinsic membrane properties like fluidity, ion transport, loss of enzyme activity, protein cross-linking, inhibition of protein synthesis, DNA damage, and the disruption of normal metabolism of plants.

### 12.2.1 Damage to DNA

ROS are the major source of DNA damage, including nuclear, mitochondrial, and chloroplastic DNA. The mitochondrial and chloroplastic DNA are more susceptible to oxidative damage than nuclear DNA because of the lack of protective protein, histones, and close locations to the ROS-producing systems (Richter, 1992). Oxidative attack to DNA bases generally involves ·OH addition to double bonds, while sugar damage mainly results from hydrogen abstraction from deoxyribose. ·OH is considered as the main ROS injuring DNA, which reacts with all purine and pyrimidine bases and the deoxyribose backbone (Halliwell and Gutteridge, 1999). By interacting with bases, ·OH generates various products such as 8-oxo-7,8 dehydro-2'-deoxyguanosine, hydroxymethyl urea, urea, thymine glycol, thymine and adenine ring-opened, and saturated products (Tsuboi et al., 1998). $^1O_2$ reacts only with guanine, whereas $H_2O_2$ and $O_2^{·-}$ do not react with bases at all (Halliwell and Aruoma, 1991).

Furthermore, oxidative attack on DNA results in oxidation of deoxyribose, breakage of strands, removal of nucleotides, modification of organic bases of the nucleotides, and DNA–protein cross-links. In addition, DNA degradation has also been observed in plants exposed to various environmental stresses such as salinity.

### 12.2.2 Damage to Lipid

·OH can trigger lipid peroxidation if its level is out of control. Lipid peroxidation refers to the oxidative degradation of lipids. It is a process whereby ·OH robs electrons from the lipids in cell membranes. It most often attacks polyunsaturated fatty acids. Two common sites of ROS attack on the phospholipid molecules are the double bond between two carbon atoms and the ester linkage between glycerol and the fatty acid. This process proceeds by a free radical chain reaction mechanism. A single ·OH can result in peroxidation of many polyunsaturated fatty acids, and the reaction stops only when two radicals react and produce a nonradical species or the radical is scavenged by antioxidants or enzymes. The end products of lipid peroxidation are reactive aldehydes, such as malondialdehyde (MDA) and 4-hydroxynonenal. MDA not only is an indicator of membrane lipid peroxidation but also reacts with deoxyadenosine and deoxyguanosine in DNA, forming DNA adducts to them (Marnett, 1999).

### 12.2.3 Damage to Protein

The attack of ROS on proteins may cause protein modification in a variety of ways such as nitrosylation, carbonylation, disulphide bond formation, and glutathionylation. Proteins can also be modified by conjugation with breakdown products of membrane lipid peroxidation (Yamauchi et al., 2008). Under the attack of ROS, site-specific amino acid modification, fragmentation of the peptide chain, aggregation of cross-linked reaction products, and altered electric charge also occur. The amino acids in a peptide differ in their susceptibility to attack by ROS. Thiol groups and sulfur-containing amino acids are especially susceptible to ROS. ROS can rob a hydrogen atom from cysteine residues to form a thiyl radical, which can cross-link to another thiyl radical to form a disulfide bridge. Several metals, including Cd, Pb, and Hg, have been shown to deplete protein-bound thiol groups. Oxygen also can be added to a methionine to form a methionine sulfoxide derivative. Tyrosine is readily cross-linked to form bityrosine products in the presence of ROS (Sharma et al., 2012). All these kinds of modification of proteins have been reported in plants under various stresses such as drought and salinity (Sharma and Dubey, 2005; Tanou et al., 2009).

### 12.2.4 Damage to Chloroplast

As discussed earlier, the destructive action of ROS to chloroplasts is targeted on proteins, DNA, and lipids, which is subjected to alteration even due to small changes in their structure. ·OH is considered as the main ROS injuring DNA. Since the chloroplast genome contains the genes coding some components of photosynthetic electron transfer chain, the breakdown of the operation of such genes can affect the normal electron transfer and, in turn, increase the production of even more $O_2{}^{\cdot-}$. In the thylakoid membrane, ·OH initiates lipid peroxidation that leads to the disturbance of the membrane structure and its function. In the stroma, the target of $O_2{}^{\cdot-}$ is heme-containing enzymes, such as peroxidases (Asada, 1994). The $H_2O_2$ in chloroplasts inhibits photosynthesis due to its oxidation of thiol groups of enzymes involved in carbon fixation Calvin cycle. $^1O_2$ generated in PSII mainly interacts with D1 protein of PSII, though PSII activity can also be destroyed by ROS produced in PSI (Krieger-Liszkay et al., 2011).

## 12.3 SCAVENGING OF ROS

Under normal conditions, ROS are generated at a low level, and there is a balance between their production and elimination. However, the balance between production and elimination of ROS may be perturbed by a number of adverse environmental factors, giving rise to rapid increases in intracellular ROS levels. Higher levels of ROS will attack DNA, lipids, and proteins and thus influence the normal metabolism such as photosynthesis and respiration and in the end lead to plant death. Traditionally, the antioxidative defense system to ROS is classified into two categories: enzymatic system and nonenzymatic systems. Antioxidative enzymes are either those that directly catalyze the processing of ROS, such as SOD, catalase (CAT), and ascorbate peroxidase (APX) or those involved in the regeneration of antioxidants such as monodehydroascorbate reductase (MDHAR or MDAR), dehydroascorbate reductase (DHAR), and glutathione reductase (GR). Many of these enzymes' functions rely on the electrons supplied by reductants called oxidants, the nonenzymatic substances. These substances include AsA, GSH, tocopherols, and carotenoids, among which two low-molecular-weight antioxidants, AsA and GSH, are of the most importance. They play multiple roles in defense reactions, and their cycle endows the plant with enough antioxidative power. They are major assimilate sinks, present in many tissues at millimolar concentrations (Noctor and Foyer, 1998). Furthermore, their level is found to be upregulated in many stress conditions. It is believed that their cycle plays a very important role in plants' adaptation to environmental stress.

### 12.3.1 Antioxidants

Some low-molecular-weight compounds of itself can quench ROS, for example, AsA, GSH, tocopherol, carotenoids, and phenolic compounds. Of these substances, AsA and GSH are of great importance. They can not only scavenge ROS by themselves but also work as cofactors with oxidative enzymes. GSH even assists in maintaining the AsA level of the cell, and the AsA–GSH cycle works to keep the oxireductive status of the cell. Mutants with decreased nonenzymatic antioxidants have been shown to be very sensitive to stress (Gao and Zhang, 2008; Semchuk et al., 2009).

#### 12.3.1.1 Ascorbate

Ascorbate (AsA), also known as vitamin C, is a very important antioxidant in plant and animal cells with a high concentration at millimolar concentrations. It is synthesized in the mitochondria but transported into any place of the cell since it is water soluble. Under normal conditions, AsA is mostly localized in the cytoplasm, chloroplast, and apoplast. In chloroplast, AsA can readily react with $O_2{}^{\cdot-}$, ·OH, and $H_2O_2$. And it is active in protecting and regenerating oxidized tocopherols and carotenes in the thylakoids. Besides, AsA is a cofactor of violaxanthin deepoxidase, thus AsA is very important in the dissipation of excess excitation energy under high light (Smirnoff, 2000), as high light stress is often exaggerated by other stresses that limit a plant's photosynthetic power. Apart from the roles as being antioxidants, AsA plays a key role in the removal of $H_2O_2$ through the AsA–GSH cycle, in which two molecules of AsA are utilized by APX to reduce $H_2O_2$ to water with concomitant generation of MDHA. MDHA is a radical with a short lifetime and can spontaneously dismutate into dehydroascorbate (DHA), and AsA or is reduced to AsA by NADP(H)-dependent enzyme MDHAR. DHA is rapidly reduced to AsA by the enzyme DHAR using reducing equivalents from GSH.

## 12.3.1.2 Tocopherol

Tocopherol, another name of vitamin E, is synthesized in the plastid envelopes and stored mainly in the membranes of chloroplasts (including thylakoid membranes), where it executes the antioxidant function. Tocopherol can react with almost all ROS including $O_2^{\cdot-}$, ·OH, lipid peroxy radicals, $H_2O_2$, and $^1O_2$ (Polle and Rennenberg, 1994; Wang and Jiao, 2000). Tocopherols are known to protect lipids and other membrane components by reacting with ROS in chloroplasts, thus protecting the structure and function of PSII. Tocopherols prevent the chain propagation step in lipid peroxidation by reducing fatty acyl peroxy radicals, which makes it an effective free radical trap. The oxidized tocopherol can be reduced by AsA, GSH, or coenzyme Q. Accumulation of tocopherol has been shown to induce tolerance to chilling, water deficit, and salinity in some plants, while mutants with a reduced tocopherol level showed higher concentration of protein carbonyl groups and GSSG compared to the wild type, indicating the development of oxidative stress (Yamaguchi-Shinozaki K. and Shinozaki, 1994; Munné-Bosch et al., 1999; Bafeel and Ibrahim, 2008).

## 12.3.1.3 Carotenoids

Carotenoids include carotene and xanthophyll; they can efficiently quench the dangerous triplet state of chlorophylls ($^3Chl^*$) to prevent the formation of $^1O_2$ to protect the photosynthetic apparatus. This mostly occurs in the antenna system rather than in the reaction center. β-Carotene can also quench $^1O_2$ directly (Foote and Denny, 1968). It was reported that an *Arabidopsis* mutant lacking zeaxanthin and lutein showed $^1O_2$ accumulation in thylakoids (Alboresi et al., 2011).

## 12.3.1.4 Glutathione

Glutathione (GSH), with three oligopeptides (γ-glutamyl-cysteinyl-glycine), is present in all cell compartments such as cytosol, chloroplasts, mitochondria, endoplasmic reticulum, vacuoles, and apoplasts. It is one of the most important antioxidants in plants. The balance between reductive form GSH and oxidative form GSSG determines the redox status of a cell. Due to its reducing power, GSH plays a crucial role in a lot of biological processes, including cell growth, cell division,

regulation of sulfate transport, signal transduction, conjugation of metabolites, enzymatic regulation, synthesis of proteins and nucleic acids, synthesis of phytochelatins, detoxification of xenobiotics, and the expression of the stress-responsive genes (Foyer et al., 1997; Sharma et al., 2012). GSH can directly quench $O_2^{\cdot-}$, ·OH, and $H_2O_2$. But of more importance, it can participate in the regeneration of AsA via AsA–GSH cycle. The regeneration of GSH from GSSG is through the reduction by GR, which is of vital importance for the maintenance of cell redox status. Altered ratios of GSH/GSSG has been observed in plants under various stresses like drought, salinity, chilling, and metal toxicity (Hefny and Abdel-Kader, 2009).

## 12.3.2 Antioxidant Enzymes

The enzymatic components of the antioxidative defense system comprise of several antioxidant enzymes such as SOD, CAT, and guaiacol peroxidase (GPX), together with the enzymes associated with AsA–GSH cycle: APX, MDHAR, DHAR, and GR. These enzymes cooperate to alleviate the damage by oxidative stress. Table 12.1 shows the antioxidant enzymes that play an important role in scavenging ROS.

## 12.3.2.1 Superoxide Dismutase

SODs are enzymes that catalyze the dismutation of $O_2^{\cdot-}$ into $O_2$ and $H_2O_2$:

$$2O_2^{\cdot-} + 2H^+ \rightarrow 2H_2O_2 + O_2$$

There are three major families of SOD, depending on the metal cofactor they are binding: Cu, Zn type (which binds both copper and zinc), Fe and Mn types (which bind either iron or manganese), and Ni type, which binds nickel. Cu, Zn-SOD is present in three isoforms, which are found in the cytosol, chloroplast, and peroxisome as well as mitochondria. Fe-SOD is found in the chloroplast, while Mn-SOD is localized in mitochondrion. Ni-SOD is found only in prokaryotes.

## 12.3.2.2 Catalase

CAT is a heme-containing enzyme that catalyzes the conversion of two molecules of $H_2O_2$ into water and oxygen. It has high specificity to $H_2O_2$, but its affinity to $H_2O_2$ is lower than APX, and its

**TABLE 12.1**
*Reactive oxygen species–scavenging and reactive oxygen species–detoxifying enzymes.*

| Enzyme | EC Number | Reaction Catalyzed |
|---|---|---|
| Superoxide dismutase | 1.15.1.1 | $O_2^{\bullet-} + O_2^{\bullet-} + 2H^+ \Leftrightarrow 2H_2O_2 + O_2$ |
| Catalase | 1.11.1.6 | $2H_2O_2 \Leftrightarrow O_2 + 2H_2O$ |
| Glutathione peroxidase | 1.11.1.12 | $2GSH + PUFA\text{-}OOH \Leftrightarrow GSSG + PUFA + 2H_2O$ |
| Glutathione S-transferases | 2.5.1.18 | $RX + GSH \Leftrightarrow HX + R\text{-}S\text{-}GSH^a$ |
| Phospholipid-hydroperoxide glutathione peroxidase | 1.11.1.9 | $2GSH + PUFA\text{-}OOH\ (H_2O_2) \Leftrightarrow GSSG + 2H_2O^b$ |
| Ascorbate peroxidase | 1.11.1.11 | $AA + H_2O_2 \Leftrightarrow DHA + 2H_2O$ |
| Guaiacol-type peroxidase | 1.11.1.7 | $Donor + H_2O_2 \Leftrightarrow oxidized\ donor + 2H_2O^c$ |
| Monodehydroascorbate reductase | 1.6.5.4 | $NADH + 2MDHA \Leftrightarrow NAD^+ + 2AA$ |
| Dehydroascorbate reductase | 1.8.5.1 | $2GSH + DHA \Leftrightarrow GSSG + AA$ |
| Glutathione reductase | 1.6.4.2 | $NADPH + GSSG \Leftrightarrow NADP^+ + 2GSH$ |

SOURCE: Blokhina, O. et al., *Ann. Bot.*, 91, 179, 2003.

[a] R may be aliphatic, aromatic, or heterocyclic group; X may be a sulfate, nitrite, or halide group.

[b] Reaction with $H_2O_2$ is slow.

[c] AA acts as an electron donor.

catalytic activity is weak against organic peroxides. As previously discussed, peroxisomes are major sites of $H_2O_2$ production. CAT scavenges $H_2O_2$ generated in this organelle during photorespiratory oxidation and β-oxidation of fatty acids. CAT is believed not to be present in the chloroplasts:

$$2H_2O_2 \rightarrow O_2 + 2H_2O$$

### 12.3.2.3 Guaiacol Peroxidase

GPX is also a heme-containing protein. It preferably oxidizes aromatic electron donors such as guaiacol and pyrogallol at the expense of $H_2O_2$. GPXs are localized in vacuoles, the cell wall, and the cytosol. GPX is associated with many processes such as lignification of cell wall, degradation of IAA, biosynthesis of ethylene, and wound healing. Under various abiotic and biotic stresses, the activity of GPX is ready to increase (Sharma et al., 2012):

$$Donor + H_2O_2 \rightarrow oxidized\ donor + 2H_2O$$

### 12.3.2.4 Ascorbate Peroxidase

APXs are enzymes that detoxify peroxides such as hydrogen peroxide using AsA as a substrate. The reaction they catalyze is the transfer of electrons from AsA to a peroxide, producing MDHA and

water as products. Five chemically and enzymatically distinct isoenzymes of APX have been found at different compartments of higher plants. They are cytosolic, stromal, thylakoidal, mitochondrial, and peroxisomal isoforms (Nakano and Asada, 1987; Sharma and Dubey, 2004). APX contained in chloroplast has two isoforms, thylakoid-bound (t-APX) and soluble stromal (s-APX) ones (Miyake and Asada, 1992). Together with the APX from cytosol, both APXs are highly specific to AsA as the electron donor, while the APXs from the mitochondrion and peroxisome are not so specific to AsA. The higher affinity for $H_2O_2$ in APXs than CAT makes the former efficient scavengers of $H_2O_2$ under environment stress (Wang et al., 1999):

$$AA + H_2O_2 \rightarrow MDHA + 2H_2O$$

## 12.4 ASCORBATE–GLUTATHIONE CYCLE

### 12.4.1 Oxidation of AsA

As briefly discussed previously, under stress even normal conditions, light-driven electrons from PSII that pass through PSI may finally reduce dioxygen and give out $O_2^{\bullet-}$ in the thylakoid membrane; $O_2^{\bullet-}$ is rapidly converted to $H_2O_2$ by the action of SOD or AsA directly. However, dismutation of $O_2^{\bullet-}$ simply serves to convert one destructive ROS to another. $H_2O_2$ is a strong oxidant that rapidly oxidizes thiol groups; it should be quenched immediately.

Though CAT has extremely high catalytic rates on $H_2O_2$, its substrate affinity is low, and it is not found in chloroplasts. The enzyme responsible for $H_2O_2$ reduction is APX. In chloroplasts, t-APX is bound to the thylakoid membrane and is very efficient in reducing $H_2O_2$ to $H_2O$. In this process, t-APX uses two molecules of AsA as reductant, giving birth to two molecules of MDHA.

MDHA is generated not only by the APX reaction, but also by the following reactions: (1) In chloroplasts, the interaction of AsA with $O_2^{\cdot-}$, $\cdot OH$, and thiol radicals when they escape from the scavenging system, and also with organic radicals including tocopherol radical (Noctor and Foyer, 1998). (2) In the thylakoid lumen, accompanying the de-epoxidation of violaxanthin to zeaxanthin via antheraxanthin with violaxanthin de-epoxidase for the downregulation of the quantum yield of PSII under excess photon conditions, AsA is oxidized to MDHA (Neubauer and Yamamoto, 1994; Horton et al., 1996). (3) By the donation of electrons from AsA to PSII when the oxidizing side of PSII is inactivated. And (4) AsA donates an electron to PSI when PSII is inactivated and thus produces MDHA (Noctor and Foyer, 1998). In the cases when the MDHA is produced in the lumen, where it cannot be reduced by appropriate reductants, MDHA is spontaneously disproportionated to DHA and AsA. DHA can easily penetrate through the thylakoid membranes to the stroma where it can be finally reduced, whereas anionic MDHA cannot penetrate through the membrane (Mano et al., 1997).

## 12.4.2 Regeneration of AsA from MDHA and DHA

### 12.4.2.1 Reduction of MDHA by Reduced Ferredoxin

In the thylakoid membrane, before it spontaneously disproportionates to DHA, MDHA can be directly reduced by reduced ferredoxin (RedFd) at a rate of $10^7 \ M^{-1} \ s^{-1}$. Although RedFd is also the electron source for $NADPH^+$, the RedFd is preferably used to reduce MDHA rather than $NADPH^+$, and the reduction rate of MDHA is 34-fold higher than that of $NADP^+$ (Miyake and Asada, 1994), which suggests that MDHA is mainly photoreduced by Fd near the thylakoid membrane:

$$MDHA + RedFd \rightarrow AsA + Fd$$

### 12.4.2.2 Reduction of MDHA with MDHA Reductase

MDHA reductase (MDHAR or MDAR) is a FAD-containing enzyme, and it is not only distributed in chloroplasts but also in cytoplasm, mitochondria, and peroxisomes (Leonardis et al., 1995; Jim´enez et al., 1997). MDAR shows a high specificity to MDHA. The enzyme prefers NADH ($K_m$; 5 μM) rather than NADPH ($K_m$; 22–200 μM) as the electron donor (Sano et al., 1995), even the chloroplastic isoform:

$$2MDHA + NAD(P)H \rightarrow 2AsA + NAD(P)^+$$

Since RedFd more effectively reduces MDA than $NADP^+$, MDA reductase cannot participate in the reduction of MDHA in the thylakoidal scavenging system. Therefore, MDA reductase would function at a site where NAD(P)H is available while RedFd is not. Since the diffusion of Fd from the PSI complex is likely very limited, the localization of MDHA reductase is likely in the vicinity of thylakoid membranes (Asada, 1999).

### 12.4.2.3 Spontaneous Disproportion of MDA

If MDHA generated in the lumen or other compartments of the cell cannot be reduced immediately, it will be disproportionated into DHA and AsA; such reaction happens in the lumen where no RedFd or APX is present. The DHA produced can penetrate through the thylakoid membrane to the stroma to be reduced. This reaction also happens in the stroma:

$$MDHA + MDHA \rightarrow AsA + DHA$$

### 12.4.2.4 Reduction of DHA with DHA Reductase

The DHA disproportionated from MDHA in the stroma as well as the DHA diffused from the lumen must be reduced to AsA, which is indispensable to keep AsA in the reduced state for the scavenging of $H_2O_2$. GSH can nonenzymatically reduce DHA to AsA, yet its rate is too slow to account for the observed reduction of DHA in chloroplasts. In chloroplast, GSH-dependent DHA reductase (DHAR) is the most likely enzyme for the reduction of DHA:

$$DHA + 2GSH \rightarrow AsA + GSSG$$

DHAR utilizes GSH as the electron donor. Its thiol group participates in the reaction. Apart from DHAR, thioredoxin reductase (May et al., 1997) and protein disulfide isomerase (Miyake and Asada, 1992) also show DHA-reducing activity.

### 12.4.2.5 Regeneration of GSH

In the reduction of DHA to AsA, GSH is depleted either nonenzymatically or associated with DHAR, leading to the generation of GSSG. The enzyme responsible for the regeneration of GSH from GSSG is GR, which is contained at about 1.4 μM in the stroma (Mullineaux and Creissen, 1997). Transgenic plants overexpressing GR in chloroplasts show tolerance to oxidative stress (Sofo et al., 2010). These results indicate the importance of the DHA–GSH system to protect against photodamage by active oxygens. Some tripeptide homologues of GSH, in which the carboxyl terminal gly is replaced by other amino acids, such as γ-glu-cys-β-ala (homoglutathione) and γ-glu-cys-ser (hydroxymethylglutathione), have similar roles to GSH. Their oxidized forms can also be reduced by GR (Noctor and Foyer, 1998):

$$GSSG + NADPH \rightarrow NADP^+ + 2GSH$$

$O_2^{\cdot-}$ is mainly generated in the PSI complex. The thylakoidal scavenging system is composed of Cu-Zn-SOD attached on the thylakoids (in several plants, Fe-SOD), thylakoid-bound APX (t-APX), and ferredoxin (Fd). Fd reduces MDA directly to AsA. The stromal scavenging system is composed of Cu-Zn-SOD localized in the stroma, stromal APX (sAPX), MDHAR, DHAR, and GR. NAD(P)H is responsible for the reduction of either MDHA or DHA, and it is regenerated via ferredoxin-NADP⁺ oxidoreductase (FNR). MDA is generated also in the lumen in the reaction of violaxanthin de-epoxidase and is rapidly disproportionated to AsA and DHA. DHA in the lumen penetrates through the thylakoid membranes and is reduced to AsA by the stromal scavenging system (Figures 12.1 and 12.2).

### 12.5  ASA–GSH CYCLE UNDER SALT STRESS

Soil salinity is one of the most harmful environmental problems and a major constraint for crop production. Currently, about 20% of the world's cultivated land is affected by salinity, which

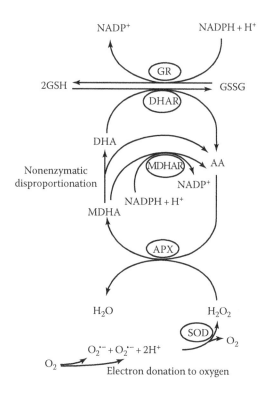

**Figure 12.1.** Ascorbate–glutathione cycle. (From Noctor G. and Foyer C.H., *Annu. Rev. Plant Biol.*, 49, 249, 1998.)

results in the loss of 50% of agricultural yield (Zhu, 2001). Generally salt stress comprises two effects on plants: osmotic stress and ionic stress, which act in collusion to impair plant growth, disturb ion homeostasis, reduce photosynthesis, and affect nitrogen fixation, as well as other key physiological processes (Zhu, 2001; Munns, 2002). Many studies have shown that ROS may participate in the mediation of toxic effects of NaCl on some plants. Evidences prove that membranes are the primary sites of salt injury to plants (Candan and Tarhan, 2003) because ROS can react with unsaturated fatty acids to cause peroxidation of essential membrane lipids in plasmalemma or intracellular organelles. Peroxidation of plasmalemma leads to the leakage of cellular contents. Intracellular membrane damage can affect respiratory activity in mitochondria and carbon-fixing ability in chloroplasts. It is suggested in plants under optimal growth conditions that the antioxidant defense system is efficient in coping with ROS, whereas in plants exposed to salinity or other stressful conditions, the antioxidant system capacity may be overwhelmed by ROS production. For plants to cope with salinity effectively, keeping

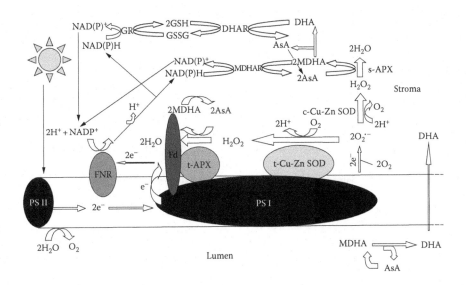

**Figure 12.2.** The generation of $H_2O_2$ and its scavenging by ascorbate–glutathione cycle. (From Sofo A. et al., Regulation of the ascorbate–glutathione cycle in plants under drought stress, in: Anjum, N.A. et al. (eds.), *Ascorbate-Glutathione Pathway and Stress Tolerance in Plants*, Springer, New York, 2010, pp. 137–189.)

the steady-state concentrations of ROS by antioxidant system is critical to prevent oxidative damage in plant cells while simultaneously allowing ROS to perform useful functions as signal molecules.

AsA plays a central role in scavenging ROS, as described previously, which detoxifies $H_2O_2$, ·OH, and $^1O_2$. AsA is also a cofactor of violaxanthin deepoxidase, an enzyme that converts violaxanthin to zeaxanthin under excess light, which is involved in the nonphotochemical quenching of excess excited energy in PSII (Demmig-Adams and Adams, 1990; Eskling et al., 1997). Therefore, AsA plays crucial roles in both scavenging ROS produced in photosynthesis and dissipating excess photons (Demmig-Adams and Adams, 1992; Niyogi, 1999). *vtc-1* and *vtc-2* are mutants of *Arabidopsis thaliana* deficient in AsA synthesis; *vtc-1* mutant produces only 30% of AsA as compared to wild-type levels. Veljovic-Jovanovic found that the photosynthesis and the oxidative system are not disturbed in the *vtc-1* mutant under normal growth conditions; the *vtc-1* mutant is only smaller than the WT and shows retarded flowering and accelerated senescence (Veljovic-Jovanovic et al., 2001). This result suggests that 30% AsA is sufficient to maintain normal plant physiological activity under normal conditions. However, 200 mM NaCl stress resulted in a more significant decrease in the $CO_2$ assimilatory capacity and PSII activity in the *vtc-1* mutant than in the WT, which is paralleled with dramatic oxidative stress in the

*vtc-1* mutant (Huang et al., 2005). Further investigation clarified that despite an elevated GSH pool in the *vtc-1* mutant, the whole AsA content and the reduced form of AsA decreased sharply under salt stress. Furthermore, the activities of MDHAR and DHAR were lower in the *vtc-1* mutant than in the WT under salt stress. Thus, low intrinsic AsA in the *vtc-1* mutant under salt stress probably induced a dramatic decrease in the reduced form of AsA, which resulted in both enhanced ROS contents and decreased NPQ in the *vtc-1* mutant (Huang et al., 2005). *vtc-2* mutant had only 25% of the wild-type AsA level. It also has less NPQ level. After feeding the leaves with 10 mM extraneous AsA, the NPQ and violaxanthin deepoxidation in isolated thylakoids and detached leaves restored (Müller-Moulé et al., 2002).

As described previously, GSH is the crucial reductant for the regeneration of AsA. It is the major source of nonprotein thiols and an important nonenzymatic antioxidant in both plants and animals. It is also involved in the redox sensing and signaling and regulates the protective reaction of plants against salt stress (Nazar et al., 2011). Under salt stress, the content of GSH may be increased or decreased, depending on the plant species or cultivars. Generally speaking, in salt-sensitive plants or cultivars, or under high salinity, the content is downregulated, yet, in salt-tolerant species or cultivars or under low salinity, it is upregulated (Hoque et al, 2007; Hasanuzzaman et al., 2014).

It is anticipated that supplying plants with GSH will increase their salt tolerance, such as in tomato (Akladious and Abbas, 2013) and rice (Wang et al., 2014). Wang et al. treated two rice cultivars with exogenous AsA and GSH. They found that exogenous AsA and GSH enhanced SOD, APX, and GR activities, increased endogenous AsA and GSH contents, and reduced those of $H_2O_2$ and MDA in the chloroplasts of both cultivars under salt stress (200 mM NaCl), but the effects were significantly more pronounced in the salt-tolerant cultivars. GSH acted more strongly than AsA on the plastidial reactive oxygen scavenging systems.

GSH content in plants is found to be strongly influenced by the supply of sulfur (sulfate, $H_2S$, or cysteine). Ruiz and Blumwald (2002) reported that S assimilation rate and biosynthesis of GSH were greatly increased in Brassica napus exposed to saline conditions. Recently, studies on barley (Astolfi and Zuchi, 2013) and mustard (Fatma et al., 2013) have shown that excess S supply protects plants from adverse effects of salt stress by increasing GSH content. Fatma et al. found that 200 mg S $kg^{-1}$ soil supply has the same effect on mustard photosynthesis, redox state, and antioxidant enzyme (SOD, POD, CAT) activities to 1 mM GSH.

SOD is an essential component in the plant's ROS defense mechanism. In the detoxifying process, SOD is the first defense line, which converts superoxide radicals to $H_2O_2$. Overexpression of SOD was found to protect against drought and salinity stress (Bowler et al., 1992; McKersie et al., 1996; Hasegawa et al., 2000; Zhu, 2002). Nevertheless, there is no conclusive evidence. Pitcher et al. (1991) and Tepperman and Dunsmuir (1990) found that overexpression of petunia Cu, Zn-SOD in the chloroplast of tobacco failed to improve tolerance to paraquat or ozone. On the other hand, there are many reports approving the protective role of SOD as an antioxidant enzyme and those reported its role in cross-tolerance for many stresses (Tanaka, 1994; Camp et al., 1996; Tanaka et al., 1999). For example, Tanaka et al. introduced yeast mitochondrial Mn-SOD gene in rice. They found the total SOD activity in the transformant was about 1.7-fold that of the control under non-stressed conditions. The photosynthetic electron transport rates of control and transgenic rice were similar under nonstressed conditions. However, upon salt stress (100 mM NaCl), the SOD activities decreased in both plants, but the decrease was faster in the control plant (Tanaka et al., 1999). Badawi et al. (2004) overexpressed cytosolic Cu, Zn-SOD cDNA of rice into tobacco chloroplasts and found that the transgenic lines showed enhanced tolerance to the active oxygen-generating paraquat and sodium sulfite. Also, the photosynthesis showed enhanced tolerance to salt, water, and PEG stresses over the wild type.

APX and GR play very important roles in keeping AsA–GSH cycle under salt tolerance. Several studies showed that the activity of APX and GR was increased in plants under stress conditions. For example, increased activity of APX and GR was reported in A. thaliana (Guan et al., 2014), Casuarina (Desingh et al., 2006), pea (Hernandez et al., 2000), citrus (Gueta-Dahan et al., 1997), rice (Lin and Kao, 2000), Cicer arietinum (Kukreja et al., 2005), tomato (Mittova et al., 2004), wheat (Sairam et al., 2005), cotton (Desingh and Kanagaraj, 2007), etc. However, the changes of the antioxidant enzyme activities are related to the plant species or cultivars. After treating with 30, 60, 90, or 120 mmol $L^{-1}$ NaCl for 4 weeks, salt stress markedly enhanced APX, CAT, and GR activities in the shoots of salt-tolerant potato, while the activity decreased in salt-sensitive cultivar. APX and GR activities remained unchanged in the roots of salt-tolerant maize and decreased in the salt-sensitive genotype (Azevedo-Neto et al., 2006). Higher APX and GR activities in a salt-tolerant cultivar were also reported in rice (Kim et al., 2005), bean (Yasar et al., 2008), and wheat (Sairam et al., 2005). APX and GR activities in the chloroplasts of the halophyte Suaeda salsa leaves was distinctly increased under 200 mM NaCl (Pang et al., 2005). The relatively high APX and GR activities under salt stress in salt-tolerant species may be an important adaptation mechanism for plants to salinity. Transgenic plants overexpressing APX genes showed higher salt tolerance in Arabidopsis, cotton, and alfalfa with their increased ROS scavenging ability (Guan et al., 2014; Wu et al., 2014; Zhang et al., 2014). Furthermore, overexpression of enzymes related to the synthesis of GSH also confers plants with high tolerance to salinity stress (Qi et al., 2010; Chen et al., 2012).

DHAR takes part in the reduction of DHA to AsA and thus is very important in the AsA–GSH cycle and plant stress tolerance (Yin et al., 2010; Li et al., 2012; Shin et al., 2013). Chen et al. (2003) isolated DHAR cDNA from wheat and expressed it in tobacco and maize; the increase in DHAR expression increased foliar and kernel ascorbic acid levels two- to fourfold and significantly increased the AsA pool and AsA redox state in both tobacco and maize. In addition, the level of GSH, the reductant used by DHAR, was also increased. Expressing an *Arabidopsis* cytosolic DHAR increased AsA content in tobacco about twofold without affecting the AsA redox state (Eltayeb et al., 2006; Yin et al., 2010), whereas the expression of an *Arabidopsis* cytosolic DHAR in *Arabidopsis* increased foliar AsA content by 2- to 4.25-fold and the redox state by 3- to 16-fold (Wang et al., 2010). A modest increase in AsA content was observed in *Arabidopsis* expressing a rice cytosolic DHAR (Ushimaru et al., 2006). The fact that AsA recycling efficiency can be significantly improved through increases in DHAR expression indicates that the level of DHAR expression is rate limiting in the AsA–GSH cycle (Gallie, 2013); thus, increasing DHAR activity endows a plant with high ROS resistance. Lee et al. (2007) introduced Cu, Zn-SOD, APX, and DHAR genes in tobacco chloroplasts. Simultaneous expression of these antioxidant enzymes confers increased protection against MV-induced damage and salt-induced damage in the transgenic plants.

MDHAR is another enzyme for the regeneration of oxidized AsA (MDHA) to reduced AsA form. Though MDHA is mainly directly reduced by reduced Fd in the vicinity of thylakoid, in the places away from thylakoid membrane, the majority of MDHA is assumed to be reduced by MDHAR (Polle, 2001), and the specific activity of MDHAR was found to be 10 times higher than DHAR in both mitochondria and chloroplasts of some plants (Mittova et al., 2003). Oxidative stresses were reported to upregulate the expression of different MDAR isoforms (Ye and Gressel, 2000; Yoon et al., 2004). Overexpression of *Avicennia marina* MDHAR gene in transgenic tobacco seedlings leads to enhanced survival under salt stress. The transgenic plant shows enhanced activity of MDHAR and APX, whereas the activity of DHAR was reduced under salt stressed and unstressed conditions. The transgenic plant shows an enhanced redox state of AsA and reduced levels of MDA, indicating its enhanced tolerance to salt-induced oxidative stress (Kavitha et al., 2010). Overexpression of MDHAR from mangrove (*Ae*MDHAR) also confers salt tolerance on rice (Sultana et al., 2012). The transgenic plants are endowed with higher AsA content and AsA/DHA ratio together with high chlorophyll content.

## 12.6 CONCLUSIONS

Under salt stress, the light utilization ability of plants may be impaired by decreased $CO_2$ assimilation capacity and limited $CO_2$ supply. The absorbed solar energy becomes redundant and aggravates the production of ROS. Oxidative stress induces secondary injury to the plant that seriously decreases crop yield. As a strategy to alleviate the damage, antioxidant systems are triggered to keep the ROS at relatively low level, reducing the damage to plant but not affecting their role in cell signaling. By far, transgenic approaches are widely tested to introduce anti-ROS genes into plants. Manipulation of SOD, POD, GR, and MDAR genes in plants has shown positive results in plants exposed to salt stress. Increasing the contents of antioxidants such as AsA, GSH, and tocopherol by surface application or genetic modification also reduces salinity damage to the plant. All these findings prove the appropriateness of increasing ROS scavenging ability under salt stress. To breed salt-tolerant plant species, comprehensive measures should be taken to increase plant resistance to salt, of which the importance of AsA–GSH should be considered.

## REFERENCES

Akladious S.A., Abbas S.M. 2013. Alleviation of seawater stress on tomato by foliar application of aspartic acid and glutathione. *J Stress Physiol Biochem* 9:282–298.

Alboresi A., Dall'Osto L., Aprile A. et al. 2011. Reactive oxygen species and transcript analysis upon excess light treatment in wild-type *Arabidopsis thaliana* vs a photosensitive mutant lacking zeaxanthin and lutein. *BMC Plant Biol* 11:62.

Asada K. 1994. Production and action of active oxygen species in photosynthesis tissues. In: *Causes of Photooxidative Stress and Amelioration of Defense Systems in Plants.* C.H. Foyer, P.M. Mullineaux (eds.) CRC Press, Boca Raton, FL, pp. 78–104.

Asada K. 1999. The water–water cycle in chloroplasts: Scavenging of active oxygen and dissipation of excess photons. *Annu Rev Plant Physiol Plant Mol Biol* 50:601–639.

Asada K. 2006. Production and scavenging of reactive oxygen species in chloroplasts and their functions. *Plant Physiol* 141:391–396.

Ashtamker C., Kiss V., Sagi M., Davydov O., Fluhr R. 2007. Diverse subcellular locations of cryptogein-induced reactive oxygen species production in tobacco Bright Yellow-2 cells. *Plant Physiol* 143:1817–1826.

Astolfi S., Zuchi S. 2013. Adequate S supply protects barley plants from adverse effects of salinity stress by increasing thiol contents. *Acta Physiol Plant* 35:175–181.

Azevedo-Neto A.D., Ptisco J.T., Eneas-Filho J., Abreu C.E.B., Gomez-Filho E. 2006. Effect of salt stress on antioxidative enzymes and lipid peroxidation in leaves and roots of salt-tolerant and salt-sensitive maize genotypes. *Environ Exp Bot* 56:87–94.

Badawi G.H., Yamauchi Y., Shimada E. 2004. Enhanced tolerance to salt stress and water deficit by overexpressing superoxide dismutase in tobacco (*Nicotiana tabacum*) chloroplasts. *Plant Sci* 166:919–928.

Bafeel S.O., Ibrahim M.M. 2008. Antioxidants and accumulation of α-tocopherol induce chilling tolerance in *Medicago sativa*. *Int J Agric Biol* 10(6):593–598.

Blokhina O., Virolainen E., Fagerstedt K.V. 2003. Antioxidants, oxidative damage and oxygen deprivation stress: A review. *Ann Bot* 91:179–194.

Bowler C., Montagu M.V., Inzé D. 1992. Superoxide dismutase and stress tolerance. *Annu Rev Plant Physiol Plant Mol Biol* 43:83–116.

Camp W.V., Katelijne C., Marc V.M., Inzé D., Slooten L. 1996. Enhancement of oxidative stress tolerance in transgenic tobacco overproducing Fe-superoxide dismutase in chloroplast. *Plant Physiol* 112:1703–1714.

Candan N., Tarhan L. 2003. The correlation between antioxidant enzyme activities and lipid peroxidation levels in *Mentha pulegium* organs grown in $Ca^{2+}$, $Mg^{2+}$, $Cu^{2+}$, $Zn^{2+}$ and $Mn^{2+}$ stress conditions. *Plant Sci* 163:769–779.

Chen J.H., Jiang H.W., Hsieh E.J. et al. 2012. Drought and salt stress tolerance of an *Arabidopsis* glutathione S-transferase U17 knockout mutant are attributed to the combined effect of glutathione and abscisic acid. *Plant Physiol* 158(1):340–351.

Chen Z., Young T.E., Ling J. et al. 2003. Increasing vitamin C content of plants through enhanced ascorbate recycling. *PNAS* 100:3525–3530.

Dat J., Vandenabeele S., Vranová E. et al. 2000. Dual action of the active oxygen species during plant stress responses. *Cell Mol Life Sci* 57:779–795.

Dekker J.P., van Grondelle R. 2000. Primary charge separation in photosystem II. *Photosynth Res* 63(3):195–208.

Demmig-Adams B., Adams III W.W. 1990. The carotenoids zeaxanthin and high-energy-state quenching of chlorophyll fluorescence. *Photosynth Res* 25:187–198.

Demmig-Adams B., Adams III W.W. 1992. Photoprotection and other responses of plants to high light stress. *Annu Rev Plant Physiol Plant Mol Biol* 43: 599–626.

Desingh R., Jutur P.P., Reddy A.R. 2006. Salinity stress-induced changes in photosynthesis and antioxidative systems in three *Casuarina* species. *J Plant Biol* 33:1–2.

Desingh R., Kanagaraj G. 2007. Influence of salinity stress on photosynthesis and antioxidative systems in two cotton varieties. *Gen Appl Plant Physiol* 33:221–234.

Eltayeb A.E., Kawano N., Badawi G.H., Kaminaka H., Sanekata T., Morishima I., Shibahara T., Inanaga S., Tanaka K. 2006. Enhanced tolerance to ozone and drought stresses in transgenic tobacco overexpressing dehydroascorbate reductase in cytosol. *Physiol Plant* 127:57–65.

Eskling M., Arvidsson P.O., Akerlund H.E. 1997. The xanthophyll cycle, its regulation and components. *Physiol Plant* 100:806–816.

Fatma M., Khan M.I.R., Masood A., Khan N.A. 2013. Coordinate changes in assimilatory sulfate reduction are correlated to salt tolerance: Involvement of phytohormones. *Annu Rev Res Biol* 3:267–295.

Foote C.S., Denny R.W. 1968. Chemistry of singlet oxygen. VII. Quenching by β-carotene. *J Am Chem Soc* 90(22):6233–6235.

Foyer C.H., Lopez-Delgado H., Dat J.F., Scott I.M. 1997. Hydrogen peroxide- and glutathione-associated mechanisms of acclimatory stress tolerance and signalling. *Physiol Plant* 100(2):241–254.

Foyer C.H., Noctor G. 2009. Redox regulation in photosynthetic organisms: Signaling, acclimation, and practical implications. *Antioxid Redox Signal* 11:861–905.

Fujiwara K., Kumata H., Kando N., Sakuma E., Aihara M., Morita Y., Miyakawa T. 2006. Flow injection analysis to measure the production ability of superoxide with chemiluminescence detection in natural waters. *Int J Environ Anal Chem* 86(5):337–346.

Furbank R.T., Badger M.R. 1983. Oxygen exchange associated with electron transport and photophosphorylation in spinach thylakoids. *Biochim Biophys Acta* 723:400–409.

Gallie D.R. 2013. The role of l-ascorbic acid recycling in responding to environmental stress and in promoting plant growth. *J Exp Bot* 64(2):433–443.

Gao Q., Zhang L. 2008. Ultraviolet-B-induced oxidative stress and antioxidant defense system responses in ascorbate-deficient vtc1 mutants of *Arabidopsis thaliana*. *J Plant Physiol* 165(2):138–148.

Grant J.J., Loake G.J. 2000. Role of reactive oxygen intermediates and cognate redox signaling in disease resistance. *Plant Physiol* 124:21–29.

Guan Q., Wang Z., Wang X., Takano T., Liu S. 2014. A peroxisomal APX from *Puccinellia tenuiflora* improves the abiotic stress tolerance of transgenic *Arabidopsis thaliana* through decreasing of $H_2O_2$ accumulation. *J Plant Physiol* 175:183–191. doi:10.1016/j.jplph.2014.10.020.

Gueta-Dahan Y., Yaniv Z., Zilinkas B.A., Ben-Hayyim G. 1997. Salt and oxidative stress: Similar and specific responses and their relation to salt tolerance in citrus. *Planta* 203:460–469.

Halliwell B., Aruoma O.I. 1991. DNA damage by oxygen-derived species. Its mechanism and measurement in mammalian systems. *FEBS Lett* 281(1–2):9–19.

Halliwell B., Gutteridge J.M.C. 1999. *Free Radicals in Biology and Medicine*, 3rd ed. Oxford University Press, Oxford, U.K.

Hammond-Kosack K.E., Jones J.D.G. 1996. Resistance gene-dependent plant defense responses. *Plant Cell* 8:1773–1791.

Hasanuzzaman M., Alam M.M., Rahman A., Hasanuzzaman M., Nahar K., Fujita M. 2014. Exogenous proline and glycine betaine mediated upregulation of antioxidant defense and glyoxalase systems provides better protection against salt-induced oxidative stress in two rice (*Oryza sativa* L.) varieties. *BioMed Res Int* 2014:1–17. Article ID 757219.

Hasegawa P.M., Ray A.B., Kang Z.J., Hans J.B. 2000. Plant cellular and molecular responses to high salinity 51:463–499.

Hatz S., Lambert J.D.C., Ogilby P.R. 2007. Measuring the lifetime of singlet oxygen in a single Cell: Addressing the issue of cell viability. *Photochem Photobiol Sci* 6:1106–1116.

Hefny M., Abdel-Kader D.Z. 2009. Antioxidant-enzyme system as selection criteria for salt tolerance in forage sorghum genotypes (*Sorghum bicolor* L. Moench). In: *Salinity and Water Stress*. M. Ashraf, M. Ozturk, H.R. Athar (eds.) Springer, Dordrecht, the Netherlands, pp. 25–36.

Hernandez J.A., Jimenez A., Mullineaux P.M., Sevilla F. 2000. Tolerance of pea (*Pisum sativum* L.) to long term salt stress is associated with induction of antioxidant defenses. *Plant Cell Environ* 23:853–862.

Hoque M.A., Banu M.N., Okuma E., Amako K., Nakamura Y., Shimoishi Y., Murata Y. 2007. Exogenous proline and glycinebetaine increase NaCl-induced ascorbate-glutathione cycle enzyme activities, and proline improves salt tolerance more than glycinebetaine in tobacco Bright Yellow-2 suspension-cultured cells. *J Plant Physiol* 164(11):1457–1468.

Horton P., Ruban A.V., Walters R.G. 1996. Regulation of light harvesting in green plants. *Annu Rev Plant Physiol Plant Mol Biol* 47:655–684.

Hosein B., Palmer G. 1983. The kinetics and mechanism of reaction of reduced ferredoxin by molecular oxygen and its reduced products. *Biochim Biophys Acta* 723:383–390.

Huang C., He W., Guo J., Chang X., Su P., Zhang L. 2005. Increased sensitivity to salt stress in an ascorbate-deficient *Arabidopsis* mutant. *J Exp Bot* 56(422):3041–3049.

Ivanov B. 2012. Oxygen metabolism in chloroplast. In: *Cell Metabolism—Cell Homeostasis and Stress Response*. P. Bubulya (ed.) InTech, Rijeka.

Jaspers P., Kangasjärvi J. 2010. Reactive oxygen species in abiotic stress signaling. *Physiol Plant* 138:405–413.

Jimenez A., Hernandez J.A., Del Reo L.A., Sevilla F. 1997. Evidence for the presence of the ascorbate-glutathione cycle in mitochondria and peroxisomes of pea leaves. *Plant Physiol* 114:272–284.

Kavitha K., George S., Venkataraman G., Parida A. 2010. A salt-inducible chloroplastic monodehydroascorbate reductase from halophyte *Avicennia marina* confers salt stress tolerance on transgenic plants. *Biochimie* 92:1321–1329.

Khorobrykh S.A., Khorobrykh A.A., Klimov V.V., Ivanov B.N. 2002. Photoconsumption of oxygen in photosystem II preparations under impairment of water-oxidation complex. *Biochemistry (Moscow)* 67(6):683–688.

Kim D.W., Rakwal R., Agrawal G.K., Jung Y.H., Shibato J. 2005. A hydroponic rice seedling culture model system for investigating proteome of salt stress in rice leaf. *Electrophoresis* 26:4521–4539.

Krasnovsky A.A. Jr. 1998. Singlet molecular oxygen in photobiochemical systems: IR phosphorescence studies. *Membr Cell Biol* 12:665–690.

Krieger-Liszkay A., Kós P.B., Hideg E. 2011. Superoxide anion radicals generated by methylviologen in photosystem I damage photosystem II. *Physiol Plant* 142(1):17–25.

Kukreja S., Nandwal A.S., Kumar N., Sharma S.K., Sharma S.K., Unvi V., Sharma P.K. 2005. Plant water status, $H_2O_2$ scavenging enzymes, ethylene evolution and membrane integrity of *Cicer arietinum* roots as affected by salinity. *Biol Plant* 49:305–308.

Lambert A.J., Brand M.D. 2004. Inhibitors of the quinone-binding site allow rapid superoxide production from mitochondrial NADH:ubiquinone oxidoreductase (complex I). *J Biol Chem* 279:39414–39420.

Lee Y.P., Kim S.H., Bang J.W. 2007. Enhanced tolerance to oxidative stress in transgenic tobacco plants expressing three antioxidant enzymes in chloroplasts. *Plant Cell Rep* 26:591–598.

Leonardis S.D., Lorenzo G.D., Borraccino G., Dipierro S. 1995. A specific ascorbate free radical reductase isozyme participates in the regeneration of ascorbate for scavenging toxic oxygen species in potato tuber mitochondria. *Plant Physiol* 109:847–851.

Li F., Wu Q.Y., Duan M. et al. 2012. Transgenic tomato plants overexpressing chloroplastic monodehydroascorbate reductase are resistant to salt- and PEG-induced osmotic stress. *Photosynthetica* 50(1):120–128.

Lin C.C., Kao C.H. 2000. Effect of NaCl stress on $H_2O_2$ metabolism in rice leaves. *Plant Growth Regulation* 30:151–155.

Mano J., Ushimaru T., Asada K. 1997. Ascorbate in thylakoid lumen as an endogenous electron donor to photosystem II: Protection of thylakoids from photoinhibition and regeneration of ascorbate in stroma by dehydroascorbate reductase. *Photosynth Res* 53:197–204.

Marnett L.J. 1999. Lipid peroxidation-DNA damage by malondialdehyde. *Mutat Res* 424(1–2):83–95.

May J.M., Mendiratta S., Hill K.E., Burk R.F. 1997. Reduction of dehydroascorbate to ascorbate by the selenoenzyme thioredoxin reductase. *J Biol Chem* 272:22607–22610.

Mazars C., Thuleau P., Lamotte O., Bourque S. 2010. Cross-talk between ROS and calcium in regulation of nuclear activities. *Mol Plant* 3:706–718.

McKersie D.B., Stephen R.B., Erni H., Oliviier L. 1996. Water-deficit tolerance and field performance of transgenic alfalfa overexpressing superoxide dismutase. *Plant Physiol* 111:1177–1181.

Mittler R. 2002. Oxidative stress, antioxidants and stress tolerance. *Trends Plant Sci* 7(9):405–410.

Mittova V., Guy M., Tal M., Volokita M. 2004. Salinity up-regulates the antioxidative system in root mitochondria and peroxisomes of the wild salt-tolerant tomato species *Lycopersicon pennellii*. *J Exp Bot* 55:1105–1113.

Mittova V., Tal M., Volokita M., Guy M. 2003. Up-regulation of the leaf mitochondrial and peroxisomal antioxidative systems in response to salt-induced oxidative stress in the wild salt-tolerant tomato species *Lycopersicon pennellii*. *Plant Cell Environ* 6:845–856.

Miyake C., Asada K. 1992. Thylakoid-bound ascorbate peroxidase in spinach chloroplasts and photoreduction of its primary oxidation product monodehydroascorbate radicals in thylakoids. *Plant Cell Physiol* 33(5):541–553.

Miyake C., Asada K. 1994. Ferredox independent photoreduction of monodehydroascorbate radical in spinach thylakoids. *Plant Cell Physiol* 34:539–549.

Mubarakshina M., Khorobrykh S., Ivanov B. 2006. Oxygen reduction in chloroplast thylakoids results in production of hydrogen Peroxide inside the membrane. *Biochimica et Biophysica Acta* 1757(11):1496–1503.

Mubarakshina M.M., Ivanov B.N. 2010. The production and scavenging of reactive oxygen species in the plastoquinone pool of chloroplast thylakoid membranes. *Physiol Plant* 140(2):103–110.

Müller-Moulé P., Conklin P.L., Niyogi K.K. 2002. Ascorbate deficiency can limit violaxanthin de-epoxidase activity in vivo. *Plant Physiol* 128:970–977.

Mullineaux P.M., Creissen G.P. 1997.Glutathione reductase: Regulation and role in oxidative stress. In: *Oxidative Stress and the Molecular Biology of Antioxidant Defenses*. J.G. Scandalios (ed.) Cold Spring Harbor Laboratory Press, New York, pp. 667–713.

Munné-Bosch S., Schwarz K., Alegre L. 1999. Enhanced formation of α-tocopherol and highly oxidized abietane diterpenes in water-stressed rosemary plants. *Plant Physiol* 121(3):1047–1052.

Munns R. 2002. Comparative physiology of salt and water stress. *Plant Cell Environ* 25:239–250.

Nakano Y., Asada K. 1987. Purification of ascorbate peroxidase in spinach chloroplasts: Its inactivation in ascorbate-depleted medium and reactivation by monodehydroascorbate radical. *Plant Cell Physiol* 28(1):131–140.

Nazar R., Iqbal N., Masood A., Syeed S., Khan N.A. 2011. Understanding the significance of sulfur in improving salinity tolerance in plants. *Environ Exp Bot* 70:80–87.

Neubauer C., Yamamoto H.Y. 1994. Membrane barriers and Mehler-peroxidase reaction limit the ascorbate available for violaxanthin de-epoxidase activity in intact chloroplasts. *Photosynth Res* 39:139–147.

Niyogi K.K. 1999. Photoprotection revisited: Genetic and molecular approaches. *Annu Rev Plant Physiol Plant Mol Biol* 50:333–359.

Noctor G., Foyer C.H. 1998. Ascorbate and glutathione: Keeping active oxygen under control. *Annu Rev Plant Biol* 49:249–279.

Overmyer K., Brosché M., Kangasjärvi J. 2003. Reactive oxygen species and hormonal control of cell death. *Trends Plant Sci* 8:335–342.

Pang C.H., Zhang S.J., Gong Z.Z., Wang B.S. 2005. NaCl treatment markedly enhances $H_2O_2$-scavenging system in leaves of halophyte *Suaeda salsa*. *Physiol Plant* 125:490–499.

Pitcher L.H., Brennan E., Hurley A., Dunsmuir P., Tepperman J.M., Zilinskas B.A. 1991. Overproduction of petunia chloroplastic copper/zinc superoxide dismutase does not confer ozone tolerance in transgenic tobacco. *Plant Physiol* 97:452–455.

Polle A. 2001. Dissecting the superoxide dismutase-ascorbate-glutathione-pathway in chloroplasts by metabolic modeling computer simulations as a step towards flux analysis. *Plant Physiol* 126:445–462.

Polle A.R., Rennenberg H. 1994. Photooxidative stress in trees. In: *Causes of Photooxidative Stress and Amelioration of Defense Systems in Plants*. C.H. Foyer, P.M. Mullineaux (eds.) CRC Press, Boca Raton, FL, pp. 199–209.

Qi Y.C., Liu W.Q., Qiu L.Y. et al. 2010. Overexpression of glutathione S-transferase gene increases salt tolerance of *Arabidopsis*. *Russ J Plant Physiol* 57(2):233–240.

Richter C. 1992. Reactive oxygen and DNA damage in mitochondria. *Mutat Res* 275(3–6):249–255.

Rhoades J.D., Loveday J. 1990. Salinity in irrigated agriculture. In: *American Society of Civil Engineers, Irrigation of Agricultural Crops*. Monograph, Vol. 30. B.A. Steward, D.R. Nielsen (eds.) American Society of Agronomists, Madison, WI, pp. 1089–1142.

Ruiz J.M., Blumwald E. 2002. Salinity-induced glutathione synthesis in *Brassica napus*. *Planta* 214:965–969.

Sairam R.K., Srivastava G.C., Agarwal S., Meena R.C. 2005. Differences in antioxidant activity in response to salinity stress in tolerant and susceptible wheat genotypes. *Biol Plant* 49:85–91.

Sano S., Miyake C., Mikami B., Asada K. 1995. Molecular characterization of monodehydroascorbate radical reductase from cucumber overproduced in *Escherichia coli*. *J Biol Chem* 270:21354–21361.

Semchuk M., Lushchak O.V., Falk J. et al. 2009. Inactivation of genes, encoding tocopherol biosynthetic pathway enzymes, results in oxidative stress in outdoor grown *Arabidopsis thaliana*. *Plant Physiol Biochem* 47(5):384–390.

Sharma P., Dubey R.S. 2004. Ascorbate peroxidase from rice seedlings: Properties of enzyme isoforms, effects of stresses and protective roles of osmolytes. *Plant Sci* 167(3):541–550.

Sharma P., Dubey R.S. 2005. Drought induces oxidative stress and enhances the activities of antioxidant enzymes in growing rice seedlings. *Plant Growth Regulation* 46(3):209–221.

Sharma P., Jha A.B., Dubey R.S., Pessarakli M. 2012. Reactive oxygen species, oxidative damage, and antioxidative defense mechanism in plants under stressful conditions. *J Bot* 2012:Article ID 217037, pp. 1–26.

Shin S.Y., Kim M.H., Kim Y.H. et al. 2013. Co-expression of monodehydroascorbate reductase and dehydroascorbate reductase from *Brassica rapa* effectively confers tolerance to freezing-induced oxidative stress. *Mol Cells* 36(4):304–315.

Smirnoff N. 2000. Ascorbic acid: Metabolism and functions of a multi-facetted molecule. *Curr Opin Plant Biol* 3(3):229–235.

Sultana S., Khew C., Morshed M.M. et al. 2012. Overexpression of monodehydroascorbate reductase from a mangrove plant (AeMDHAR) confers salt tolerance on rice. *J Plant Physiol* 169(3):311–318.

Suzuki N., Miller G., Morales J., Shulaev V., Torres M.A., Mittler R. 2011. Respiratory burst oxidases: The engines of ROS signaling. *Curr Opin Plant Biol* 14:691–699.

Sofo A., Cicco N., Paraggio M. et al. 2010. Regulation of the ascorbate–glutathione cycle in plants under drought stress. In: *Ascorbate-Glutathione Pathway and Stress Tolerance in Plants*. N.A. Anjum, S. Umar, M.-T. Chan (eds.) Springer, New York, pp. 137–189.

Tanaka K. 1994. Tolerance to herbicides and air pollutants. In: *Causes of Photooxidative Stress and Amelioration of Defense Systems in Plants*. C.H Foyer, P.M. Mullineaux (eds.) CRC Press, Boca Raton, FL, pp. 365–378.

Tanaka Y., Hibino T., Hayashi Y. et al. 1999. Salt tolerance of transgenic rice overexpressing yeast mitochondrial Mn-SOD in chloroplasts. *Plant Sci* 148:131–138.

Tanou G., Molassiotis A., Diamantidis G. 2009. Induction of reactive oxygen species and necrotic death-like destruction in strawberry leaves by salinity. *Environ Exp Bot* 65(2–3):270–281.

Tepperman J.M., Dunsmuir P. 1990. Transformed plants with elevated levels of chloroplastic SOD are not more resistant to superoxide toxicity. *Plant Mol Biol* 14:501–511.

Thomas C.E., Morehouse L.A., Aust S.D. 1985. Ferritin and superoxide-dependent lipid peroxidation. *J Biol Chem* 260(6):3275–3280.

Tsuboi H., Kouda H., Takeuchi H. et al. 1998. 8-Hydroxydeoxyguanosine in urine as an index of oxidative damage to DNA in the evaluation of atopic dermatitis. *Br J Dermatol* 138(6):1033–1035.

Ushimaru T., Nakagawa T., Fujioka Y., Daicho K., Naito M., Yamauchi Y., Nonaka H., Amako K., Yamawaki K., Murata N. 2006. Transgenic *Arabidopsis* plants expressing the rice dehydroascorbate reductase gene are resistant to salt stress. *J Plant Physiol* 163:1179–1184.

Veljovic-Jovanovic S.D., Pignocchi C., Noctor G., Foyer C.H. 2001. Low ascorbic acid in the vtc-1 mutant of *Arabidopsis* is associated with decreased growth and intracellular redistribution of the antioxidant system. *Plant Physiol* 127:426–435.

Wang J., Zhang H., Allen R.D. 1999. Overexpression of an *Arabidopsis* peroxisomal ascorbate peroxidase gene in tobacco increases protection against oxidative stress. *Plant Cell Physiol* 40(7):725–732.

Wang R., Liu S., Zhou F. et al. 2014. Exogenous ascorbic acid and glutathione alleviate oxidative stress induced by salt stress in the chloroplasts of *Oryza sativa* L. *Z Naturforsch C J Biosci* 69(5–6):226–236.

Wang S.Y., Jiao H. 2000. Scavenging capacity of berry crops on superoxide radicals, hydrogen peroxide, hydroxyl radicals, and singlet oxygen. *J Agric Food Chem* 48(11):5677–5684.

Wang Z., Xiao Y., Chen W., Tang K., Zhang L. 2010. Increased vitamin C content accompanied by an enhanced recycling pathway confers oxidative stress tolerance in *Arabidopsis*. *J Integr Plant Biol* 52:400–409.

Weisiger R.A., Fridovich I. 1973a. Superoxide dismutase: Organelle specificity. *J Biol Chem* 248:3582–3592.

Weisiger R.A., Fridovich I. 1973b. Mitochondrial superoxide dismutase: Site of synthesis and intramitochondrial localization. *J Biol Chem* 248:4793–4796.

Wu G., Wang G., Ji J. et al. 2014. Cloning of a cytosolic ascorbate peroxidase gene from *Lycium chinense* Mill. and enhanced salt tolerance by overexpressing in tobacco. *Gene* 543(1):85–92.

Yamauchi Y., Furutera A., Seki K. et al. 2008. Malondialdehyde generated from peroxidized linolenic acid causes protein modification in heat-stressed plants. *Plant Physiol Biochem* 46(8–9):786–793.

Yamaguchi-Shinozaki K., Shinozaki K. 1994. A novel cis element in an *Arabidopsis* gene is involved in responsiveness to drought, low-temperature, or high-salt stress. *Plant Cell* 6(2):251–264.

Yasar F., Ellialtioglu S., Yildiz K. 2008. Effect of salt stress on antioxidant defense systems, lipid peroxidation, and chlorophyll content in green bean. *Russ J Plant Physiol* 55:782–786.

Ye B., Gressel J. 2000. Transient, Oxidant-induced antioxidant transcript and enzyme levels correlate with greater oxidant-resistance in paraquat resistant *Conyza bonariensis*. *Planta* 211:50–61.

Yin L., Wang S., Eltayeb A.E., Uddin M.I., Yamamoto Y., Tsuji W., Takeuchi Y., Tanaka K. 2010. Overexpression of dehydroascorbate reductase, but not monodehydroascorbate reductase, confers tolerance to aluminum stress in transgenic tobacco. *Planta* 231:609–621.

Yoon H., Lee H., Lee I., Kim K., Jo J. 2004. Molecular cloning of monodehydroascorbate reductase gene from *Brassica campestris* and analysis of its mRNA level in response to oxidative stress. *Biochem Biphys Acta* 1658:181–186.

Zhang Q., Cui M.A., Xue X. et al. 2014. Overexpression of a cytosolic ascorbate peroxidase gene, OsAPX2, increases salt tolerance in transgenic alfalfa. *J Integr Agricu* 13(11):2500–2507.

Zhu J.K. 2001. Plant salt tolerance. *Trends Plant Sci* 6:66–71.

Zhu J.K. 2002. Salt and drought stress signal transduction in plants. *Annu Rev Plant Biol* 53(2002):247–273.

CHAPTER THIRTEEN

# Salinity Stress Tolerance in Relation to Polyamine Metabolism in Plants

*Malabika Roy Pathak and Shabir Hussain Wani*

## CONTENTS

*Abstract.* Plants meet several environmental stresses during their life cycle, which have negative impacts on their growth and productivity. Salinity is one of the main abiotic stress factors limiting normal plant growth and yield. The inhibition of plant growth by salinity is due to the reduction of water availability and ion accumulation. This leads to morphological, physiological, and metabolic modifications in plants. Several studies at physiological, biochemical, and molecular levels have shown the contribution of naturally occurring plant polyamines (PAs), in conferring salt stress tolerance in plants. PAs are low-molecular-weight, nonprotein polycations, present in all living organisms including plants and have strong binding capacity to negatively charged macromolecules in cellular environment (DNA, RNA, proteins, etc.). Evidences showed that PAs are involved in many physiological processes, such as cell growth and development, and differentiation of plant cells, and respond to stress tolerance under environmental stress situations. In order to become salt tolerant, plants accumulate compatible solutes of which PAs are most important. Commonly known PAs such as putrescine, spermidine, and spermine and their biosynthetic enzymes are greatly involved in conferring salt stress tolerance by acting as cell-signaling molecules. The relationship of plant stress tolerance was noted with the production of PAs as well as stimulation of PA oxidation. Therefore, genetic manipulation of crop plants with genes encoding enzymes of PA biosynthetic pathways may provide better salt stress tolerance to plants. This chapter will discuss the role of PAs in salinity stress tolerance in plants.

*Keywords:* Polyamines, Putrescine, Salinity stress, Spermidine, Spermine, Transgenic plants

## 13.1 INTRODUCTION

The major challenging task of world agriculture is to produce 70% more food for an additional 2.3 billion people by 2050 (FAO 2009). The average yield of major crops have decreased by 70% in the past decade due to adverse effect of abiotic stresses (Gosal et al. 2009). Maintaining crop yields under adverse environmental stresses is probably the major challenge of modern agriculture (Gill and Tuteja 2010). In the recent years, one of the main causes of lower crop productivity is mostly accredited by various abiotic stresses, and it is an important area of concern to switch the increasing food requirements (Shanker and Venkateswarlu 2011). The major abiotic stress factors limiting crop productivity and threatening food security are soil salinity, exposure to high and low temperatures, and drought (Mantri et al. 2012). Among the abiotic stresses, soil salinity is one of the most ruthless environmental factors and a complex phenotypic and physiological phenomenon in plants and imposing ion imbalance or disequilibrium, and hyperionic and hyperosmotic stress. It is disrupting the overall metabolic activities that limit crop productivity worldwide (Munns and Tester 2008; Flowers et al. 2010). Saline soils occupy more than 10% of the land surface and 50% of all irrigated land (Ruan et al. 2010). Soil salinity represents a major limiting factor in decreasing more than 20% worldwide crop production under irrigated land, while more than 40% worldwide irrigated land has already been damaged by salt (Porcel et al. 2012).

Salt stress induces both osmotic stress and ionic stress, which inhibits the plant's normal cell growth and division. To encounter the adverse environment, plants maintain osmotic and ion homeostasis with rapid osmotic and ionic signaling by changing several activities at cellular, physiological, biochemical, and molecular levels (Munns et al. 2012). Salt stress responds by accumulating low-weight amino acids such as proline and arginine, amino acid–derived compounds, and compatible solutes of quaternary amides such as glycinebetaine and several di- and polyamines (PAs) (Sharma and Dietz 2006; Chen and Murata 2008). PAs are assumed to be involved in the adjustment of various plant stress situations by acting as sensor molecular sensor in a coordinated manner (Alcázar et al. 2010). PAs, low-molecular-weight aliphatic amines, widely present in living organisms, are now regarded as a new class of growth substances, which includes spermidine (Spd, a triamine), spermine (Spm, a tetramine), and their obligate precursor putrescine (Put, a diamine), which play a pivotal role in the regulation of plant growth and developmental response as well as physiological responses including abiotic and biotic plant stress replies (Kumar et al. 1997; Bouchereau et al. 1999). The potassium stress-induced Put accumulation

demonstrates the first link between PAs and abiotic stress (Richards and Coleman 1952). Since then, a large number of reports indicate salinity and osmotic stress-induced PA accumulation as well as their physiological and biochemical significance during and after stress (Flores and Galston 1984; Krishnamurty and Bhagwat 1989; Chattopadhyay et al. 1997; Bouchereau et al. 1999). The studies show that the accumulation of PAs is related to its increased biosynthetic enzyme activities (Basu and Ghosh 1991; Besford et al. 1993; Das et al. 1995; Mo and Pua 2002).

The physiological role of PAs in response to salt stress is almost thoroughly understood, and significant progress has been made to establish their protective effect also (Bae et al. 2008). The availability of cDNAs of PA biosynthetic genes and genetic manipulation of PA metabolism have evaluated the physiological roles of PAs in transgenic plants under salinity stress and salinity stress-induced drought stress (Roy and Wu 2001, 2002). The transgenic plants overexpressing different PA biosynthetic genes accumulated different PAs to develop multiple stress tolerance capacity (Capell et al. 2004; Wi et al. 2006; Bassie et al. 2008).

## 13.2 SALINITY STRESS RESPONSE

Salt stress leads to severe inhibition of plant growth and development, membrane damages, ion imbalances due to $Na^+$ and $Cl^-$ accumulation, enhanced lipid peroxidation, and increased production of reactive oxygen species (ROS) like superoxide radicals, hydrogen peroxide, and hydroxyl radicals. Salt stress–induced osmotic and ionic stress is the complex phenomenon that inhibits the plant's normal cell growth, division, and ultimately plant productivity (Khodarahmpour et al. 2012). To encounter the adverse environment, plants maintain osmotic and ion homeostasis with rapid osmotic and ionic signaling. Salinity tolerance in plants is a physiologically multifaceted trait and is attributed to multiple mechanisms. These numerous mechanisms are usually grouped into three major clusters: (1) osmotolerance, (2) sodium exclusion mechanisms, and (3) tissue tolerance mechanisms (Munns and Tester 2008). Despite the significant progress that has been made in elucidating specific details of each of these mechanisms, the relative contribution of the earlier components to overall salinity tolerance attempts by modifying the expression and function of specific genes by molecular means (Plett and Moller 2010). Stress conditions activate both mechanism of acclimation and adaptation by the plant. The survival ability of stress tolerance depends on proper maintenance of structural and functional integrity of the cellular environment (Roychoudhury et al. 2008). Due to the complex nature of salt stress, multiple sensors, rather than a single sensor, are responsible for the perception of stress stimuli to interact in plants. Very often, it was observed that various nitrogen-containing compounds accumulate in the plants in response to environmental stress such as amino acids (arginine, proline), amino acid–derived compounds, amides (glutamine, asparagines), ammonium, quaternary ammonium (glycinebetaine), and PAs (Sharma and Dietz 2006; Chen and Murata 2008; Munns et al. 2012). These osmoprotectants protect plants by osmotic adjustment, detoxification of ROS, and stabilization of the quaternary structure of proteins (Bohnert and Jensen 1996). The osmoprotectants play an important role during salinity stress-induced osmotic stress (Rontein et al. 2002). Among these osmoprotectants, PAs are identified as key protective elements in maintaining cellular components and membrane protein structures by functioning as molecular chaperons (Alcázar et al. 2010; Takahashi and Kakeh 2010; Hussain et al. 2011). Substantial amount of work showed that accumulation of PAs under salt stress situation is a very common phenomenon in response to stress tolerance (Bouchereau et al. 1999; Gill and Tuteja 2010). The genetic manipulation of crop plants with gene-encoding enzymes of PA biosynthetic pathways provides better stress tolerance capacity (Roy and Wu 2002; Capell et al. 2004; Alcázar et al. 2006). Physiological and molecular studies at transcriptional, translational, and transgenic plant levels have shown the pronounced involvement of naturally occurring plant PAs, in controlling, conferring, and modulating abiotic stress tolerance in plants (Alcázar et al. 2010; Gill and Tuteja 2010; Ahmed et al. 2012; Pathak et al. 2014; Shi and Chen 2014). The reviews indicated several abiotic stress-related modulation of PA biosynthetic pathway and their gene expression by studding loss-of-function mutants and transgenic overexpression of PA biosynthetic genes in various plant species. This chapter is concerned about the protective role of PAs, their accumulation, and modulation under salt stress condition in transgenic plant level.

## 13.3 POLYAMINES IN RELATION TO PLANT STRESS

PAs are naturally occurring low-molecular-weight, nonprotein aliphatic compounds, present in all plants, which play an important role in many physiological processes. The common di- and PAs, Put, Spd, and Spm, are good controller of plant cell growth, development, morphogenesis, embryogenesis, organ formation, senescence, and biotic and abiotic stress responses (Kumar et al. 1997; Bouchereau et al. 1999; Alcázar et al. 2010). Except these common di- and PAs, several uncommon diamines (1, 3-diaminopropane, Dap; cadaverine, Cad) and PAs (norspermidine, Nor-Spd; homospermidine, norspermine, Nor-Spm; homospermine, thermospermine, caldopentamine, caldohexamine, homocaldopentamine, and homocaldohexamine) also accumulated in plants and microorganisms under different stresses (Bagga et al. 1991, Phillips and Kuehn 1991; Roy and Ghosh 1996). At physiological pH, mostly, they remain in free or conjugated form at cellular environment and show strong binding capacity to negatively charged DNA, RNA, different protein molecules, and several phenolic compounds and possess the ability to stabilize membrane structures (Liu et al. 2007; Hussain et al. 2011). These ionic interactions are important in regulating the structure and function of biological macromolecules, as well as their synthesis in vivo (Jacob and Stetler 1989). PAs are involved in many processes of plant growth and development, and their levels are critical for a number of developmental processes including cell division, somatic embryogenesis, root growth, flower initiation, and flower and fruit development (Evans and Malmberg 1989; Slocum 1991). PAs play an important role as a modulator of plant regulators under environmental stress situations by protecting plants from salinity and drought-related adverse environmental stress situation (Shi et al. 2007; Do et al. 2014). PAs are well known for their antisenescence and antioxidant properties; those collectively help plants to develop defense response against environmental stresses (Bouchereau et al. 1999), which includes salinity, osmotic stress, drought, metal toxicity, oxidative stress, and temperature (Basu and Ghosh 1991; Roy and Ghosh 1996; Nayyar and Chander 2004; Demetriou et al. 2007; Liu et al. 2007; Cuevas et al. 2008; Groppa and Benavides 2008; Choudhary et al. 2010). Exogenous application of PAs showed enhanced stress tolerance (Chattopadhayay et al. 2002; Yiu et al. 2009; Zhang et al. 2010). Several environmental stress-induced PA accumulation associated with increased arginine decarboxylase (ADC) activities was reviewed by Bouchereau et al. (1999). Evaluation of the complete genome sequence of *Arabidopsis* facilitated the use of global "omic" tactics in the documentation of target genes in PA biosynthesis and signaling pathways (Alcazar et al. 2010).

## 13.4 POLYAMINE BIOSYNTHESIS AND CATABOLISM IN PLANTS

Put, Spd, and Spm are the most abundant Pas, and their biosynthetic pathways are regulated by a limited number of key enzymes. Diamine Put synthesis is the first step of PA biosynthesis. In plants, two different routes of Put biosynthesis are known, one is ornithine decarboxylase (ODC; EC 4.1.1.17) pathway, where ornithine is directly converted to Put; and another one is ADC (EC 4.1.1.19) pathway, which is an indirect pathway of arginine conversion to Put via two intermediates. The first product of arginine via ADC is agmatine, which is converted into Put through an intermediate N-carbamoylputrescine. The two pathways of Put synthesis are tissue specific, under developmental and environmental control. The synthesis of Put from arginine with the involvement of ADC is usually associated with plant responses to abiotic stress (Bouchereau et al. 1999; Capell et al. 2004). The implication of PAs in stress responses has been revealed by transgenic genetic approaches mostly in *Arabidopsis thaliana*, which contains two different genes encoding ADC (*ADC1 and ADC2*), while *ADC2* is mostly related to salt stress responses (Urano et al. 2004; Alcazar et al. 2006; Alet et al. 2011). The two alternate routes of Put synthesis show differential expression.

The diamine, Put, is the obligate precursor of triamine (Spd) and tetramine (Spm). Spd and Spm are synthesized by successive attachment of aminopropyl first to Put and then to Spd. Aminopropyl is formed by decarboxylation of S-adenosylmethionine (SAM) by S-adenosylmethionine decarboxylase (SAMDC, EC 4.1.1.50), which is a rate-limiting enzyme of Spd and Spm biosyntheses. These reactions are catalyzed by aminopropyltransferases, such as Spd synthase (SPDS, EC 2.5.1.16) and Spm synthase (SPMS, EC 2.5.1.22). Genes encoding enzymes for the PA biosynthesis pathway have been cloned and characterized from various plant species (Bell and Malmberg 1990; Kasukabe et al. 2006; Kuznetsov and Shevyakova 2007).

PA catabolism efficiently regulates the level of free PA in cell, which can perform an important physiological role under normal and stress situations. Endogenous and exogenous PAs are catabolized by two oxidative enzymes. Diamine oxidase (DAO; EC 1.4.3.6) and PA oxidase (PAO; EC 1.5.3.3) are responsible for oxidizing the diamine Put and PAs such as Spd and Spm, respectively (Phillips and Kuehn 1991; Tiburcio et al. 2012). DAO converts Put into 1-pyrroline, $NH_3$, and $H_2O_2$, while PAO oxidizes Spd and Spm producing, 1-pyrolline, 1,3-diaminopropane (DAP), and aminopropyl pyrroline together with $H_2O_2$ (Kusano et al. 2007). The genes encoding enzymes DAO and PAO have been isolated, cloned, and characterized. The biosynthesis of uncommon PAs from 1,3-diaminopropane, which is derived by the action of PAO and serves as the substrate for successive, repetitious reactions, adds an aminopropyl group donated from dcSAM to form norspermidine, norspermine, and caldopentamine. It is not known whether these reactions are catalyzed by independent APT enzymes, each specific for a unique substrate or, alternatively, by a single APT enzyme with broad specificity to transfer an aminopropyl group from dcSAM to each PA precursor (Kuehn et al. 1990). The biosynthetic and catabolic pathways of PAs have been shown in Figure 13.1.

## 13.5 ACCUMULATIVE AND PROTECTIVE ROLE OF POLYAMINES UNDER SALINE STRESS

Richards and Coleman (1952) first time reported the accumulation of Put in barley plants when exposed to potassium starvation. Large number of studies reported the accumulation of different PAs under different environmental stress situations. Plants can increase salinity tolerance by modifying the biosynthesis of PAs. Under salt stress, diverse changes of PA levels were reported depending on species, varieties, and plant tissue. Differential response of PAs and diamine Put in relation to salt stress of salt-sensitive and salt-tolerant cultivar of rice (*Oryza sativa*) has been reported (Krishnamurthy and Bhagat 1989; Basu and Ghosh 1991; Krishnamurthay

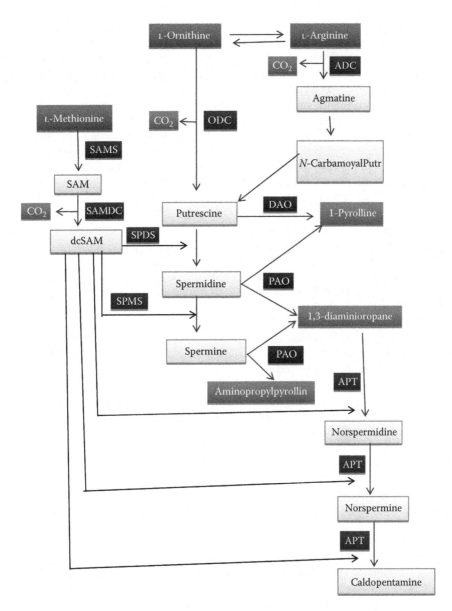

Figure 13.1. Common and uncommon polyamine biosynthetic and main catabolic pathways in plants. *Notes:* ADC, arginine decarboxylase; APT, aminopropyltransferase; dcSAM, decarboxylates S-adenosylmethionine; DAO, diaminoxidase; ODC, ornithine decarboxylic acid; PAO, polyamineoxidase; SAM, S-adenosylmethionine; SAMDC, S-adenosylmethionine decarboxylase; SPDS, spermidine synthase; SPMS, spermine synthase. (Modified from Bouchereau, A. et al., *Plant Sci.*, 140, 103, 1999.)

1991; Quinet et al. 2010). There is considerable genetic variability of salt tolerance power in rice (*O. sativa* L.) in terms of accumulation of various PAs under salinity stress situation. PAs act as modulators of salt stress tolerance and influence the plant performance under saline condition (Roy and Wu 2001). NaCl stress-induced growth inhibition of rice was reduced by exogenous application of Put (Chattopadhayay et al. 2002; Ndayiragiji and Lutts 2006; Shu et al. 2012)

suggesting the protective role of Put during stress. Salt-tolerant rice cultivars drastically accumulated high levels of Spd and Spm resulting in enhanced level of total PA (Prakash and Prathapasenan 1988; Krishnamurthy and Bhagwat 1989). Other studies showed Spd and Spm accumulation during salt and osmotic stress in sorghum (*Sorghum bicolor*), maize (*Zea mays*), and tomato (*Lycopersicon esculentum*) seedling (Erdei et al. 1996) and salt stress response in sunflower

(Mutlu and Bozcuk 2005). The effect of salinity stress on the activity of ADC, as well as its transcript level, was compared in salt-sensitive and salt-tolerant rice cultivars at mRNA level. In salt-tolerant cultivar, ADC activity as well as its transcript level increased during prolonged stress of 12 h at 150 mM NaCl stress. The sensitive rice cultivars showed low abundance of transcript at 200 mM NaCl stress as well as low level of Spd and Spm, growth inhibition, ion imbalance, and oxidative damage (Chattaopadhaya et al. 1997). Similarly, increased level of Spd and Spm with increased tolerance to salt stress was observed in maize and *A. thaliana* grown in constant 200 mM NaCl-stressed condition. Increased level of Spd and Spm as well as increased ADC and SPDS expression level was detected at molecular-level study by reverse transcriptase polymerase chain reaction in *A. thaliana* (Jiménez-Bremont et al. 2007; Tassoni et al. 2008). In most of the studies concerning salt stress, the induced PA response has been assumed mainly on ADC activation and differential response in long-term and short-term exposure of stress situations (Bouchereau et al. 1999).

Flores and Galston (1984) showed the accumulation of Put and Spd in cereal cells and protoplasts by exposure to various osmotic stress similar to salt stress–induced effects. Put accumulation and ADC activity were prevented by ADC inhibitor (α-difluoromethylarginine) but were not affected by ODC inhibitor (α-difluoromethylornithine), which gives clear idea of the activation of ADC pathway in PA synthesis under salt stress-induced osmotic stress and increment of ADC mRNA (Borrell et al. 1996). Aziz et al. (1997) showed responses of osmotic stress in rape leaf disks as well as whole plant level by increasing Put and diaminopropane (Dap) while decreasing Spm. Both saline stress–induced ADC and ODC pathways were observed in *Brassica napus* (Das et al. 1995). Increased level of free Put and ADC activity was reported in apple callus under 200 Mm NaCl stress (Liu et al. 2006). The stress-induced PA pathways differ in monocot and dicot (Aziz et al. 1999). The study of PA level and their enzyme activity during salt stress in PA-deficient mutants of *A. thaliana* (spe1-1, spe2-2) demonstrated decreased PA formation due to lower ADC activity, which ultimately leads to reduced salt tolerance (Kasinathan and Wingler 2004). Salt and osmotic stress–induced PA accumulation and increased ADC activity were

reported in roots and leaves in *Lupinus luteus* seedlings (Legocka and Kluk 2005). The translocation and accumulation of Put and other PAs in organ-specific manner during salt and osmotic stress conditions represented their probable role in adaptation. Transcriptional profile of PA biosynthetic genes and PA metabolism during salinity stress–induced drought acclimation response in *A. thaliana* was studied and their PA metabolic canalization differed (Alcazar et al. 2010). PAs are scavengers of ROS to regulate cell membrane stability under salinity stress in rice seedling by acting as antioxidant (Ghosh et al. 2012).

## 13.6 TRANSGENIC STUDIES

Advancement of molecular biology with emphasis on cloning of genes related to PA biosynthesis pathway as well as plant genetic engineering techniques made it possible to evaluate the regulatory functions of PAs in adverse environmental stress situations through transgenic plant study (Alcázar et al. 2010; Gill and Tituja 2010; Ahmed et al. 2012). Study at physiological, molecular, and mutant levels in different plants indicated that ADC genes are operative to the overproduction of PAs for enhanced stress tolerance (Kumar et al. 1997; Urano et al. 2004; Alcazar et al. 2010). Manipulation of PA levels in several plants leads to improved salt tolerance as observed in transgenic rice by the overexpression of *ADC* gene (Roy and Wu 2001; Capell et al. 2004) and several other plants (Raman and Rajam 2007). The implication of PAs in stress responses has been revealed by genetic approaches mostly in *A. thaliana*, which contains two different genes encoding ADC (*ADC1* and *ADC2*). Differential expression of two genes encoding ADC (*ADC1 and ADC2*) was observed in *A. thaliana* in response to environmental stresses (Alet et al. 2011). In the last two decades, several studies were carried out to manipulate PA levels by transferring different PA biosynthetic genes under the control of constitutive and inducible promoters. Different levels of mRNA accumulation of *ADC* were observed in different transgenic plants such as tobacco (Burtin and Michael 1997; Masgrau et al. 1997), rice (Noury et al. 2000; Roy and Wu 2001; Capell et al. 2004), eggplant (Raman and Rajam 2007), and *A. thaliana* (Alet et al. 2011). Transgenic tobacco plants containing oat *ADC* gene under the control of tetracycline-inducible promoter were reported with altered phenotypes (Masgrau et al. 1997). Transgenic rice plants with ABA-induced *Tritordeum* SAMDC produced three- to fourfold increase in Spd and Spm than unstressed control-type rice plants under salt stress giving protection to transgenic plants after salt stress in recovery study (Roy and Wu 2002). Transgenic rice plants under drought stress-inducible promoter action accumulated different levels of higher PAs, Spd and Spm and showed tolerance under salinity stress in transgenic plant levels, which indicates salt stress induced drought stress. Constitutive overexpression of human SAMDC in tobacco showed increment in Put and Spd levels that lead to a major tolerance to salt and osmotic stress (Waie and Rajam 2003). Similarly, tobacco plants transformed with carnation SAMDC showed rise in Put and Spd levels and tolerance to high salinity together with other abiotic stresses (Wi et al. 2006). Transgenic rice and tomato plants containing SAMDC gene of hybrid wheat showed accumulation of higher PAs under saline stress conditions in comparison to nontransgenic plants (Roy and Wu 2002; Cheng et al. 2009). Overexpression of apple SPDS gene in pear (*Pyrus communis*) showed altered PA level as well as increased tolerance under multiple abiotic stress situation including salinity stress (Wen et al. 2008). Overexpression of the melon SPDS gene in *A. thaliana* induced the expression of several stress-regulated genes and increased level of stress tolerance under multiple environmental stresses (Kasukabe et al. 2004). These genes included those for stress-responsive transcription factors such as DREB and stress-protective proteins like rd29A. Knowledge of the molecular mechanisms governing plant responses to abiotic stresses has increased considerably during the last decade, and it is clear that protective mechanisms and metabolic networks connected with several upstream and downstream processes involved in stress tolerance are interconnected where PAs play a central role to enhance tolerance to adverse abiotic stress situations (Alcazar et al. 2010; Ahmed et al. 2012).

## 13.7 CONCLUSIONS

The commonly known PAs tend to accumulate together accompanied by an increase in the activities of their biosynthetic enzymes under a range of environmental stresses with specific discussion during salt stress. This highlights how PAs in higher

plants act during environmental stress and how transgenic strategies have improved our understanding of the molecular mechanisms at play.

## REFERENCES

Ahmad P, Kumar A, Gupta A, Hu X, Hakeem KR, Azooz MM, Sharma S. 2012. Polyamines: Role in plants under abiotic stress. In: Ashraf M, Öztürk M, Ahmad MSA, Aksoy AM (eds.), *Crop Production for Agricultural Improvement*, Springer Science+Business Media B.V., Amsterdam, the Netherlands, pp. 491–512.

Alcázar R, Altabella T, Marco F, Bortolotti C, Reymond M, Koncz C, Carrasco P, Tiburcio AF. 2010. Polyamines: Molecules with regulatory functions in plant abiotic stress tolerance. *Planta* 231:1237–1249.

Alcázar R, Marco F, Cuevas JC, Patron M, Ferrando A, Carrasco, P, Tiburcio AF, Altabella T, 2006. Involvement of polyamines in plant response to abiotic stress. *Biotechnol Lett.* 28:1867–1876.

Alet AI, Sanchez DH, Cuevas JC, Del Valle S, Altabella T, Tiburcio AF, Marco F, Ferrando A et al. 2011. Putrescine accumulation in *Arabidopsis thaliana* transgenic lines enhances tolerance to dehydration and freezing stress. *Plant Signal Behav.* 6:278–286.

Aziz A, Martin-Tanguy J, Larher F. 1997. Plasticity of polyamine metabolism associated with high osmotic stress in rape leaf discs and with ethylene treatment. *Plant Growth Regul.* 21:153–163.

Aziz A, Martin-Tanguy J, Larher F. 1999. Salt stress-induced proline accumulation and changes in tyramine and polyamine levels are linked to ionic adjustment in tomato leaf discs. *Plant Sci.* 145:83–91.

Bae H, Kim SH, Moon SK, Richard CS, David L, Mary DS, Natarajan S, Bailey BA. 2008. The drought response of *Theobroma cacao* (cacao) and the regulation of genes involved in polyamine biosynthesis by drought and other stresses. *Plant Physiol Biochem.* 46:174–188.

Bagga S, Dharma A, Phillips GC, Kuehn GD. 1991. Evidence for the occurrence of polyamine oxidase in the dicotyledonous plant *Medicago sativa* L. (alfalfa). *Plant Cell Rep.* 10:550–554.

Bassie L, Zhu C, Romagosa I, Christou P, Capell T. 2008. Transgenic wheat plants expressing an oat arginine decarboxylase cDNA exhibit increases in polyamine content in vegetative tissue and seeds. *Mol Breed.* 22:39–50.

Basu R, Ghosh B. 1991. Polyamines in various rice (*Oryza sativa*) genotypes with respect to sodium chloride salinity. *Physiol Plant.* 82:575–581.

Bell E, Malmberg RL. 1990. Analysis of a cDNA encoding arginine decarboxylase from oat reveals similarity to the *Escherichia coli* arginine decarboxylase and evidence of protein processing. *Mol Gen Genet.* 224:431–436.

Besford RT, Richardson CM, Capell T, Tiburcio AF. 1993. Effects of polyamine on stabilization of molecular complexes in thylakoid membranes of osmotically stresses oat leaves. *Planta* 189:201–206.

Bohnert HJ, Jensen RG. 1996. Strategies for engineering water-stress tolerance in plants. *Trends Biotechnol.* 14:89–97.

Borrell A, Bestford T, Altabella T, Masgrau C, Tiburcio AF. 1996. Regulation of arginine decarboxylase by spermine in osmotically-stressed oat leaves. *Physiol Plant.* 98:105–110.

Bouchereau A, Aziz A, Larher F, Martin-Tanguy J. 1999. Polyamines and environmental challenges: Recent development. *Plant Sci.* 140:103–125.

Burtin D, Michael AJ. 1997. Overexpression of arginine decarboxylase in transgenic plants. *Biochem J.* 325:331–337.

Capell T, Bassie L, Christou P. 2004. Modulation of the polyamine biosynthetic pathway in transgenic rice confers tolerance to drought stress. *Proc Natl Acad Sci USA.* 101:9909–9914.

Chattopadhyay MK, Gupta S, Sengupta DN, Ghosh B. 1997. Expression of arginine decarboxylase in seedlings of indica rice (*Oryza sativa* L.) cultivars as affected by salinity stress. *Plant Mol Biol.* 34:477–483.

Chattopadhayay MK, Tiwari BS, Chattopadhyay G, Bose A, Sengupta DN, Ghosh B. 2002. Protective role of exogenous polyamines on salinity-stressed rice (*Oryza sativa*) plants. *Physiol Plant.* 116:192–199.

Chen TH, Murata N. 2008. Glycinebetaine: An effective protectants against abiotic stress in plants. *Trends Plant Sci.* 13:499–505.

Cheng L, Zou Y, Ding S, Zhang J, Yu X, Cao J, Lu G. 2009. Polyamine accumulation in transgenic tomato enhances the tolerance to high temperature stress. *J Integr Plant Biol.* 51:489–499.

Choudhary SP, Bhardwaj R, Gupta BD, Dutt P, Gupta RK, Kanwar M, Dutt P. 2010. Changes induced by $Cu^{2+}$ and $Cr^{6+}$ metal stress in polyamines, auxins, abscisic acid titers and antioxidative enzymes activities of radish seedlings. *Braz J Plant Physiol.* 22:263–270.

Cuevas JC, Lo'pez-Cobollo R, Alca'zar R, Zarza X, Koncz C, Altabella T, Salinas J, Tiburcio AF, Ferrando A. 2008. Putrescine is involved in *Arabidopsis* freezing tolerance and cold acclimation by regulating abscisic acid levels in response to low temperature. *Plant Physiol.* 148:1094–1105.

Das S, Bose A, Ghosh B. 1995. Effect of salt stress on polyamine metabolism in *Brassica campestris*. *Phytochem.* 39:283–285.

Demetriou G, Neonaki C, Navakoudis E, Kotzabasis K. 2007. Salt stress impact on the molecular structure and function of the photosynthetic apparatus—The protective role of polyamines. *Biochim Biophys Acta.* 1767:272–280.

Do TP, Drechsel O, Heyer AG, Hincha DK, Zuther E. 2014. Changes in free polyamine levels, expression of polyamine biosynthesis genes, and performance of rice cultivars under salt stress: A comparison with responses to drought. *Front Plant Sci.* 5:1–16.

Erdei L, Szegletes Z, Barabas K, Pestenacz A. 1996. Responses in polyamine titer under osmotic and salt stress in sorghum and maize seedlings. *J Plant Physiol.* 147:599–603.

Evans PT, Malmberg RL. 1989. Do polyamines have role in plant development? *Annu Rev Plant Physiol.* 40:235–269.

FAO.2009. *High Level Expert Forum: How to Feed the World in 2050.* Economic and Social Development Department, Food and Agricultural Organization of the United Nations, Rome, Italy.

Flores H, Galston AW. 1984. Osmotic stress-induced polyamine accumulation in cereal leaves. I. Physiological parameters of the response. *Plant Physiol.* 75:102–109.

Flowers T, Galal H, Bromham L. 2010. Evolution of halophytes: Multiple origins of salt tolerance in land plants. *Funct Plant Biol.* 37:604–612.

Ghosh N, Das SP, Mandal C, Gupta S, Das K, Dey N et al. (2012) Variations of antioxidative responses in two rice cultivars with polyamine treatment under salinity stress. *Physiol Mol Biol Plants.* 18:301–313.

Gill SS, Tuteja N. 2010. Polyamines and abiotic stress tolerance in plants. *Plant Signal Behav.* 5:26–33.

Gosal SS, Wani HS, Kang MS. 2009. Biotechnology and drought tolerance. *J Crop Improv.* 23:19–54.

Groppa MD, Benavides MP. 2008. Polyamines and abiotic stress: Recent advances. *Amino Acids.* 34:35–45.

Hussain SS, Ali M, Ahmad M, Siddique KHM. 2009. Polyamine: Natural and engineered abiotic and biotic stress tolerance in plants. *Biotechnol Adv.* 29:300–311.

Jacob ST, Stetler DA. 1989. Polyamines and RNA synthesis. In: Bachrach U, Heimer QM (eds.), *The Physiology of Polyamines in Plants*, CRC Press, Boca Raton, FL, pp. 133–140.

Jiménez-Bremont JF, Ruiz OA, Rodríguez-Kessler M. 2007. Modulation of spermidine and spermine levels in maize seedlings subjected to long-term salt stress. *Plant Physiol Biochem.* 45:812–821.

Kasinathan V, Wingler A. 2004. Effect of reduced arginine decarboxylase activity on salt tolerance and on polyamine formation during salt stress in *Arabidopsis thaliana*. *Physiol Plant.* 121:101–107.

Kasukabe Y, He L, Nada K, Misawa S, Ihara I, Tachibana S. 2004. Overexpression of spermidine synthase enhances tolerance to multiple environmental stresses and up-regulates the expression of various stress regulated genes in transgenic *Arabidopsis thaliana*. *Plant Cell Physiol.* 45:712–722.

Kasukabe Y, Lixiong H, Yuriko W, Motoyasu O, Shimada T, Tachibana S. 2006. Improvement of environmental stress tolerance of sweet potato by introduction of genes for spermidine synthase. *Plant Biotechnol.* 23:75–83.

Khodarahmpour Z, Mansour I, Mohammad M. 2012. Effects of NaCl salinity on maize (*Zea mays* L.) at germination and early seedling stage. *Afr J Biotechnol.* 11(2):298–304.

Krishnamurthy R. 1991. Amelioration of salinity effect in salt tolerant rice (*Oryza sativa* L.) by foliar application of putrescine. *Plant Cell Physiol.* 32:699–703.

Krishnamurthy R, Bhagwat, KA. 1989. Polyamines as modulators of salt tolerance in rice cultivars. *Plant Physiol.* 91:500–504.

Kuehn GD, Rodriguez-Garay B, Bagga S, Phillips CG. 1990. Novel occurrence of uncommon polyamines in higher plants. *Plant Physiol.* 94:855–857.

Kumar A, Altabella T, Taylor MR, Tiburcio AF. 1997. Recent advances in polyamine research. *Trends Plant Sci.* 2:124–130.

Kusano T, Berberich T, Tateda C, Takahashi Y. 2008. Polyamines: Essential factors for growth and survival. *Planta.* 228:367–381.

Kuznetsov V, Shevyakova NI. 2007. Polyamines and stress tolerance of plants. *Plant Stress.* 5:55–71.

Legocka J, Kluk A. 2005. Effect of salt and osmotic stress on changes in polyamine content and arginine decarboxylase activity in *Lupinus luteus* seedlings. *J Plant Physiol.* 162:662–668.

Liu JH, Nada K, Honda C, Kitashiba H, Wen XP, Pang XM, Moriguchi T. 2006. Polyamine biosynthesis of apple callus under salt stress: Importance of the arginine decarboxylase pathway in stress response. *J Exp Bot.* 57:2589–2599.

Liu JH, Kitashiba H, Wang J, Ban Y, Moriguchi T. 2007. Polyamines and their ability to provide environmental stress tolerance to plants. *Plant Biotechnol.* 24:117–126.

Mantri N, Patade V, Penna S, Ford R, Pang E. 2012. Abiotic stress responses in plants: Present and future. In: Ahmad P, Prasad MNV (Eds.), *Abiotic Stress Responses in Plants: Metabolism, Productivity and Sustainability*, Springer, New York, pp. 1–19.

Masgrau C, Altabella T, Farrás R, Flores D, Thompson AJ, Besford RT, Tiburcio AF. 1997. Inducible over-expression of oat arginine decarboxylase in trans-genic tobacco plants. *Plant J.* 11:465–473.

Mo H, Pua EC. 2002. Up-regulation of arginine decarboxylase gene expression and accumulation of polyamines in mustard (*Brassica juncea*) in response to stress. *Physiol Plant.* 114:439–449.

Munns R, Tester M: Mechanisms of salinity tolerance. *Annu Rev Plant Biol.* 59:651–681.

Munns R, James RA, Xu B, Athman A, Conn SJ, Jordans C, Byrt CS. et al. 2012. Wheat grain yield on saline soils is improved by an ancestral $Na^+$ transporter gene. *Nat Biotechnol.* 30(4):360–364.

Mutlu F, Bozcuk S. 2005. Effects of salinity on the contents of polyamines and some other compounds in sunflower plants differing in salt tolerance. *Russian J Plant Physiol.* 52:29–34.

Nayyar H, Chander S. 2004. Protective effects of polyamines against oxidative stress induced by water and cold stress in chickpea. *J Agron Crop Sci.* 190:355–365.

Ndayiragiji A, Lutts S. 2006. Exogenous putrescine reduces sodium and chloride accumulation in NaCl-treated calli of the salt-sensitive rice cultivar I Kong Pao. *Plant Growth Regul.* 48:51–63.

Noury M, Bassie L, Lepin O, Kurek I, Christou P, Capell T. 2000. A transgenic rice cell lineage expressing the oat arginine decarboxylase (adc) cDNA constitutively accumulates putrescine in callus and seeds but not in vegetative tissues. *Plant Mol Biol.* 43:537–544.

Pathak RM, da Silva JAT, Wani SH. 2014. Polyamines in response to abiotic stress tolerance through transgenic approaches. *GM Crops Food.* 5(2):1–10.

Phillips GC, Kuehn GD. 1991. Uncommon polyamines in plants and other organisms. In: Slocum RD, Flores HE (Eds.), *Biochemistry and Physiology of the Polyamines in Plants*, CRC Press, Boca Raton, FL, pp. 121–136.

Plett DC, Moller IS. 2010. $Na^+$ transport in glycophytic plants: What we know and would like to know. *Plant Cell Environ.* 33(4):612–626.

Porcel R, Aroca R, Ruiz-Lozano JM. 2012. Salinity stress alleviation using arbuscular mycorrhizal fungi. A review. *Agron Sust Dev.* 32:181–200.

Prakash L, Prathapasenan G. 1988. Effect of NaCl salinity and putrescine on shoot growth, tissue ion concentration and yield of rice (*Oryza sativa* L. GR3). *J Agric Crop Sci.* 160:325–334.

Quinet M, Ndayiragije A, Lefèvre I, Lambillotte B, Dupont-Gillain CC, Lutts S. 2010. Putrescine differently influences the effect of salt stress on polyamine metabolism and ethylene synthesis in rice cultivars differing in salt resistance. *J Exp Bot.* 61: 2719–2733.

Raman VP, Rajam MV. 2007. Polyamine accumulation in transgenic eggplant enhances tolerance to multiple abiotic stresses and fungal resistance *Plant Biotechnol.* 24:273–282.

Richard FJ, Coleman RG. 1952. Occurrence of putrescine in potassium-deficient barley. *Nature.* 170:460.

Rontein D, Basset G, Hanson AD. 2002. Metabolic engineering of osmoprotectant accumulation in plants. *Metab Eng.* 4:49–56.

Roy M, Ghosh B. 1996 Polyamines, both common and uncommon, under heat stress in rice (*Oryza sativa*) callus. *Physiol Planta.* 98:196–200.

Roy M, Wu R. 2001 Arginine decarboxylase transgene expression and analysis of environmental stress tolerance in transgenic rice. *Plant Sci.* 160:869–875.

Roy M, Wu R. 2002. Overexpression of S-adenosylmethionine decarboxylase gene in rice increases polyamine level and enhances sodium chloride stress tolerance. *Plant Sci.* 163:987–992.

Roychoudhury A, Basu S, Sarkar SN, Sengupta DN. 2008. Comparative physiological and molecular responses of a common aromatic indica rice cultivar to high salinity with non-aromatic indica rice cultivars. *Plant Cell Rep.* 27:1395–1410.

Ruan CJ, DaSilva JAT, Mopper S, Qin P, Lutts S. 2010. Halophyte improvement for salinized world. *Crit Rev Plant Sci.* 29:329–359.

Shanker AK, Venkateswarlu B. 2011. *Abiotic Stress in Plants: Mechanisms and Adaptations*, In Tech Publisher, Rijeka, Croatia, p. 428.

Sharma SS, Dietz KJ. 2006. The significance of amino acid-derived molecules in plant responses and adaptation to heavy metal stress. *J Exp Bot.* 57:711–726.

Shi K, Huang YY, Xia XJ, Zhan YL, Zhou YH, Yu1 JQ. 2008. Protective role of putrescine against salt stress is partially related to the improvement of water relation and nutritional imbalance in cucumber. *J Plant Nutr.* 31:1820–1831.

Shi H, Chen Z. 2014. Improvement of plant abiotic stress tolerance through modulation of polyamine pathway. *J Integr Plant Biol.* 56(2):114–121.

Shu S, Guo SR, Sun J, Yuan LY. 2012. Effects of salt stress on the structure and function of the photosynthetic apparatus in *Cucumis sativus* and its protection by exogenous putrescine. *Physiol Plant.* 146:285–296.

Slocum RD. 1991. Polyamine biosynthesis in plants. In: Slocum RD, Flores HE (eds.), *Biochemistry and Physiology of Polyamines in Plants*, CRC Press, Boca Raton, FL, pp. 22–40.

Takahashi T, Kakeh J. 2010. Polyamines: Ubiquitous polycations with unique roles in growth and stress responses. *Ann Bot.* 105:1–6.

Tassoni A, Franceschetti M, Bagni N. 2008. Polyamines and salt stress response and tolerance in *Arabidopsis thaliana* flowers. *Plant Physiol Biochem.* 46:607–613.

Tiburcio AF, Wollenweber B, Zilberstein A, Koncz C. 2012. Abiotic stress tolerance. *Plant Sci.* 182:1–2.

Urano K, Yoshiba Y, Nanjo T, Ito Y, Seki M, Yamaguchi-Shinozaki K, Shinozaki K. 2004. *Arabidopsis* stress-inducible gene for arginine decarboxylase AtADC2 is required for accumulation of putrescine in salt tolerance. *Biochem Biophys Res Commun.* 313:369–375.

Waie B, Rajam MV. 2003. Effect of increased polyamine biosynthesis on stress responses in transgenic tobacco by introduction of human S-adenosylmethionine gene. *Plant Sci.* 164:727–734.

Wen XP, Pang XM, Matsuda N, Kita M, Inoue H, Hao YJ, Honda C, Moriguchi T. 2008. Over-expression of the apple spermidine synthase gene in pear confers multiple abiotic stress tolerance by altering polyamine titers. *Transgen Res.* 17:251–263.

Wi SJ, Kim WT, Park KY. 2006. Overexpression of carnation S-adenosylmethionine decarboxylase gene generates a broad-spectrum tolerance to abiotic stresses in transgenic tobacco plants. *Plant Cell Rep.* 25:1111–1121.

Yiu JC, Juang LD, Fang DYT, Liu CW, Wu SJ. 2009. Exogenous putrescine reduces flooding-induced oxidative damage by increasing the antioxidant properties of welsh onion. *Sci Horticult.* 120:306–314.

Zhang CM, Zou ZR, Huang Z, Zhang ZX. 2010. Effects of exogenous spermidine on photosynthesis of tomato seedlings under drought stress. *Agric Res Arid Areas.* 3:182–187.

## CHAPTER FOURTEEN

# Metabolomics and Salt Stress Tolerance in Plants

*Anita Mann, Sujit Kumar Bishi, Mahesh Kumar Mahatma, and Ashwani Kumar*

## CONTENTS

*Abstract.* Salinity is one of the most serious factors limiting the productivity of agricultural crops, with adverse effects on germination, plant vigor, and crop yield. During the onset and development of salt stress within a plant, all the major processes such as photosynthesis, protein synthesis, and energy and lipid metabolism are affected, thereby, increasing or decreasing the levels of different metabolites involved in various metabolic processes. The universality of stress responses is probably the most salient feature in plants. The network of interactions between different inputs and signaling channels that is formed in a plant-specific way drives metabolic adjustments that include reactions that are common to all or nearly all plant species. The advancement of metabolomics is providing a detailed fingerprint of metabolites upregulated or downregulated in plant cells during adverse environmental conditions. This database may generate information for the dissection of the plant response to salinity and try to find future applications for ameliorating the impact of salinity on plants, improving the performance of species important to human health and agricultural sustainability.

## 14.1 INTRODUCTION

Salt stress is a major abiotic stress factor limiting growth and productivity of plants in many parts of the world due to increasing use of poor-quality water for irrigation and soil salinization. It is estimated that about 6% of the agricultural land in the world is salt affected (Munns and Tester, 2008). Plants are sessile in nature and complete their life cycle at one place and therefore must be able to adapt to various stresses. Plant adaptation or tolerance to salinity stress involves complex physiological traits, metabolic pathways, and molecular networks.

In the past few years, a wealth of information has been generated for many model organisms, including the plants *Arabidopsis thaliana, Oryza sativa, Zea mays,* and *Medicago truncatula* using genomics, but genome sequence information alone is insufficient to reveal the facts concerning gene function, developmental/regulatory biology, and the biochemical kinetics of life. To elucidate the relationship between genome and phenotype in cells and organisms, concerted efforts have been made in the post-genomic era. It is clear that even a complete understanding of the state of the genes, messages, and proteins in a living system does not reveal its phenotype. Phenotypic plasticity enables plants to withstand various stresses for both short- and long-term scales and is governed by genes that determine not only the character of an organism but also the degree of responsiveness of that character to environmental stimuli (Brunetti et al., 2013). This phenotypic plasticity is measured by a plant's ability to change the way it grows (morphological traits) and functions (metabolic traits). Morphological traits are time-consuming and not always accurate at low expression levels, thus metabolic profiling/metabolomics allows to detect potentially important differences.

A comprehensive understanding of how plants respond to salinity stress at different levels and an integrated approach of combining physiological and biochemical techniques are imperative for the development of salt-tolerant varieties of plants in salt-affected areas, because most of the crop species developed so far through conventional selection and breeding techniques are based on differences in agronomic characters. Recent research has identified various adaptive responses to salinity stress at cellular, metabolic, and physiological levels, although mechanisms underlying salinity tolerance are far from being completely understood.

The plant metabolites involved in biotic/abiotic stress responses include polyols (mannitol and sorbitol), dimethylsulfonium (dimethylsulfopropionate, glycine betaine), sugars (sucrose, trehalose, fructan), and amino acids (proline and ecotine) that serve as osmolytes and osmoprotectant to protect plants under extreme salt, drought and desiccation stress (Shulaev et al., 2008). Higher plants have a remarkable ability to synthesize a vast array of metabolites categorized as primary and secondary metabolites. The primary metabolites that include carbohydrates, proteins, nucleic acids, and lipids are implicated in essential life functions such as nutrition, growth, and reproduction, whereas the secondary metabolites are said to influence ecological interactions between the plant and its environment (Croteau et al., 2000). In the past few decades, rapid advances in biochemical and molecular techniques have established a defined role for these secondary products in ecosystem interactions (Bourgaud et al., 2001), including protection against biotic and abiotic stresses, aid in pollination and seed dispersal, and as allelopathic agents (Croteau et al., 2000).

Phenolic compounds and flavonoids are the major secondary products in plants and increase their tolerance to different environmental stresses (Ali and Abbas, 2003). Under salt stress conditions, for example, many halophytes and glycophytes synthesize those compounds that serve protective cellular functions (Pichersky and Gang, 2000). The plant cells producing such products do not benefit, but accumulation of the secondary metabolites increases the overall fitness of plants and thereby increases their adaptive capacity under adverse growing conditions (van der Fits and Memelink, 2000). The secondary metabolites often increase the activities of antioxidant enzymes that in turn scavenge the reactive oxygen species (ROS), thereby controlling the level of lipid peroxidation, which is an indicator of the oxidative damage of cell membranes under stress conditions (Baque et al., 2012). The products of secondary metabolism are the metabolism of chemicals, which are very rarely found in plants and do not have any specific role in plants functioning. Few compounds like nicotine, which are otherwise toxic, are also produced from secondary metabolism which protects plants from herbivores and also from microbes.

In this "omics" era, metabolomics is still at its infancy, and thus an attempt has been made in this chapter to provide a comprehensive study on metabolites regulating plant adaptation and tolerance to salt stress.

## 14.2 TECHNIQUES USED FOR METABOLOMICS

Metabolomics is the functional assessment of endogenous metabolites and attempts to systematically identify and quantify metabolites from a biological sample. A range of analytical technologies are being used in metabolomics, including gas chromatography–mass spectrometry (GC-MS), capillary electrophoresis–mass spectrometry (CE-MS), liquid chromatography–mass spectrometry (LC-MS), liquid chromatography–electrochemistry mass spectrometry (LC-EC-MS), liquid chromatography–Fourier transform mass spectrometry (LC-FT/MS), nuclear magnetic resonance (NMR) spectroscopy, LC-NMR, direct infusion mass spectrometry, and Fourier-transform infrared and Raman spectroscopies. However, NMR and chromatographic techniques coupled with mass spectrometry (GC-MS and LC-MS) are most widely applied. None of the single analytical platforms can cover the whole metabolome due to the complexity and chemical diversity of plant metabolites that include hydrophilic carbohydrates, polar amino acids, hydrophobic lipids, polar and nonpolar phenols, organic acids, vitamins, and thiols. The combined use of these modern analytical techniques has explained the ideal outcomes in metabolomics and is beneficial to increase the coverage of detected metabolites that cannot be achieved by single-analysis techniques. Integrated platforms have been frequently used to provide sensitive and reliable detection of thousands of metabolites in a biofluid sample (Tugizimana et al., 2013). Briefly, the most widely used techniques are depicted in Table 14.1 with their strengths and limitations.

Initially, plants sense the changes in the environmental condition and activate a network of signaling pathways. In later phases, these signal transduction pathways trigger the production of different proteins and compounds that restore or achieve a new state of homeostasis. From the standpoint of metabolomics, at least three different types of compounds are important for these processes: compounds involved in the acclimation process such as antioxidants or osmoprotectants,

by-products of stress that appear in cells because of disruption of normal homeostasis by the alterations in growth conditions, and signal transduction molecules involved in mediating the acclimation response. Plants normally cope with salinity stress in a number of ways. Among these responses, the following determinants are most studied, which may provide a backbone of metabolites conferring resistance/tolerance to salinity.

## 14.3 OSMOLYTES ACCUMULATION AND PLANT PROTECTION

Osmolytes are a group of chemically diverse organic compounds that are uncharged, polar, and soluble in nature and do not interfere with cellular metabolism even at high concentration. They mainly include proline, glycine betaine, sugar, and polyols. Osmolytes are synthesized and accumulated in varying amounts among different plant species. The concentration of compatible solutes within the cell is maintained either by irreversible synthesis of the compounds or by a combination of synthesis and degradation. As their accumulation is proportional to the external osmolarity, the major functions of these osmolytes are to protect the structure and to maintain osmotic balance within the cell via continuous water influx (Hasegawa et al., 2000). The different osmolytes and their roles in salinity tolerance are discussed next.

### 14.3.1 Amino Acids

Amino acids such as cysteine, arginine, and methionine, which constitute about 55% of total free amino acids, decrease when exposed to salinity stress, whereas proline concentration rises in response to salinity stress (El-Shintinawy and El-Shourbagy, 2001). Intracellular proline, which is accumulated during salinity stress, not only provides tolerance toward stress but also serves as an organic nitrogen reserve during stress recovery. Proline is a proteinogenic amino acid with an exceptional conformational rigidity, essential for primary metabolism, which normally accumulates in large quantities in response to drought or salinity stress. Its accumulation normally occurs in the cytosol where it contributes substantially to the cytoplasmic osmotic adjustment. Proline is synthesized either from glutamate or ornithine. In osmotically stressed cells, glutamate functions

TABLE 14.1
*Comparison of various analytical techniques used for metabolite profiling.*

| Technology/Properties | Application | Advantages | Limitation |
|---|---|---|---|
| GC-MS<br>Sample preparation: −<br>Accuracy: <50 ppm<br>Mass range: <350 Da | Analyses of polar or lipophilic compounds (e.g., sugars, organic acids, tocopherols, vitamins) | High resolution, ideal to resolve complex biological samples, simultaneous analysis of different class of metabolites | Unable to analyze thermolabile molecules, derivatization required for nonvolatiles |
| GC X GC-MS<br>Sample preparation: −<br>Accuracy: <50 ppm<br>Mass range: <350 Da | Similar to GC-MS but with better separation of highly complex compounds | Better separation of coeluting compounds due to column of different polarity and increased accuracy owing to GC X GC | Complex, high cost, and cooling gases |
| SPME-GC-MS<br>Sample preparation: −<br>Accuracy: <50 ppm<br>Mass range: <350 Da | Analysis of volatile compounds (e.g., aroma components, repellents) | Improve sensitivity and reproducibility of GC analysis of volatiles, solvent free, easy to automate | Fragile matrix and has limited lifetime, carryover effects due to repeated use of fiber |
| CE-MS<br>Sample preparation: +<br>Accuracy: <50 ppm<br>Mass range: <350 Da | Analyses of polar compounds (e.g., amino acids, CoA derivates, sugars, organic acids, tocopherols, vitamins) | Useful for complex biological samples, small volume<br><br>high resolution | Complex methodology and quantification, buffer incompatibility, difficulty in interfacing |
| LC-MS<br>Sample preparation: −<br>Accuracy: 50–100 ppm<br>Mass range: <1500 Da | Analyses of secondary metabolites (e.g., carotenoids, flavonoids, glucosinolates, vitamins) | High sensitivity and resolution, derivatization not required, enables analysis of thermo labile metabolites | A few restriction on LC eluent, requires desalting, limited structural information, matrix effect, less reproducibility |
| FT-ICR-MS<br>Sample preparation: +++<br>Accuracy: <1 ppm<br>Mass range: <1500 Da | Direct MS analyses of crude sample mixtures without chromatographic Separations, high-throughput screening of large number of samples | High mass resolution, accurate mass measurement that enables the identification of unknown metabolites by mass-to-charge ratio | Formidable cost<br>High magnetic field |
| NMR<br>Sample preparation: +++<br>Mass range: ≲50 kDa | Nondestructive analyses of abundant metabolites in a sample | Highly selective, absolute and relative quantification high, highly reproducible | Relatively lower sensitivity, high cost |
| Direct-injection-MS<br>Sample preparation: ++<br>Accuracy: 50–100 ppm<br>Mass range: <1500 Da | Nonseparative technique giving a fingerprint of the metabolic content in a biological sample | Rapid conformity of sample, minimal sample preparation | Ion suppression/enhancement by matrix effects a common problem in ESI |
| FAIMS-MS<br>Sample preparation: −<br>Accuracy: 50–100 ppm<br>Mass range: <1500 Da | Next-generation technology to MS; enables selection of specific ions, reducing ion suppression and matrix effects; enables the separation of isobaric compounds in combination with selective MS | Quantitative analysis of inorganic and organometallic compounds, differentiates isomers, reduced background, improved signal-to-noise ratios | Difficulty to achieve reproducibility and cross-calibration between different instruments |

SOURCES: Reproduced from Villas-Bôas, S.G. et al., *Mass Spectrom. Rev.*, 24, 616, 2005; Fernie, A.R. and Schauer, N., *Trends Genet.*, 25, 39, 2008.

as the primary precursor. The biosynthetic pathway comprises two major enzymes: pyrroline carboxylic acid synthetase and pyrroline carboxylic acid reductase. Both these regulatory steps are used to overproduce proline in plants (Sairam and Tyagi, 2004). It functions as an $O_2$ quencher, thereby revealing its antioxidant capability. It has been reported that proline improves salt tolerance in *Nicotiana tabacum* by increasing the activity of enzymes involved in antioxidant defense system (Hoque et al., 2008). Deivanai et al. (2011) also demonstrated that rice seedlings from seeds pretreated with 1 mM proline exhibited improvement in growth during salt stress.

### 14.3.2 Glycine Betaine

Glycine betaine is an amphoteric quaternary ammonium compound ubiquitously found in microorganisms, higher plants, and animals and is electrically neutral over a wide range of pH. Glycine betaine is a nontoxic cellular osmolyte that raises the osmolarity of the cell during stress periods; thus it plays an important role in stress mitigation. Glycine betaine also protects the cell by osmotic adjustment, stabilizes proteins, and protects the photosynthetic apparatus from stress damages and reduction of ROS (Ahmad et al., 2013; Saxena et al., 2013). Accumulation of glycine betaine is found in a wide variety of plants belonging to different taxonomical backgrounds. Glycine betaine is synthesized within the cell from either choline or glycine. Synthesis of glycine betaine from choline is a two-step reaction involving two or more enzymes. Under salt stress, the ultrastructure of the seedling shows several damages such as swelling of thylakoids, disintegration of grana and intergranal lamellae, and disruption of mitochondria. However, these damages were largely prevented when seedlings were pretreated with glycine betaine. When glycine betaine is applied as a foliar spray in a plant subjected to stress, it led to pigment stabilization and increase in photosynthetic rate and growth (Ahmad et al., 2013).

### 14.3.3 Polyols

Polyols are compounds with multiple hydroxyl functional groups available for organic reactions. Sugar alcohols are a class of polyols functioning as compatible solutes, low-molecular-weight chaperones, and ROS-scavenging compounds (Saxena et al., 2013). They can be classified into two major types, cyclic (e.g., pinitol) and acyclic (e.g., mannitol). Mannitol synthesis is induced in plants during stressed periods via the action of NADPH-dependent mannose-6-phosphate reductase. These compatible solutes function as a protector or stabilizer of enzymes or membrane structures that are sensitive to dehydration or ionically induced damage.

It was found that the transformation with bacterial *mltd* gene that encodes for mannitol-1-phosphate dehydrogenase in both *Arabidopsis* and tobacco (*N. tabacum*) plants confers salt tolerance, thereby maintaining normal growth and development when subjected to high levels of salt stress (Binzel et al., 1988). Pinitol is accumulated within the plant cell when the plant is subjected to salinity stress. The biosynthetic pathway consists of two major steps: methylation of myo-inositol, which results in the formation of an intermediate compound, and ononitol, which then undergoes epimerization to form pinitol. Inositol methyl transferase enzyme encoded by *imt* gene plays a major role in the synthesis of pinitol. The transformation of *imt* gene in plants shows a result similar to that observed in the case of *mltd* gene. Thus it can be said that pinitol also plays a significant role in stress alleviation. Accumulation of polyols, either straight-chain metabolites such as mannitol and sorbitol or cyclic polyols such as myo-inositol and its methylated derivatives, is correlated with tolerance to drought and/or salinity, based on polyol distribution in many species of plants (Bohnert et al., 1995).

### 14.3.4 Carbohydrates

Accumulations of carbohydrates such as sugars (e.g., glucose, fructose, fructans, and trehalose) and starch occur under salt stress (Parida et al., 2004). These carbohydrates help in stress mitigation through osmoprotection, carbon storage, and scavenging of ROS. Salt stress–induced increase in the level of reducing sugars (sucrose and fructans) within the cell in a number of plants has been reported (Kerepesi and Galiba, 2000). Besides being a carbohydrate reserve, trehalose accumulation protects organisms against several physical and chemical stresses, including salinity stress. They play an osmoprotective role in physiological responses (Ahmad et al., 2013). Sucrose content was found to increase in tomato (*Solanum lycopersicum*) under salinity due to increased activity of sucrose phosphate synthase (Gao et al., 1998). Sugar content, during salinity stress, has been reported to both increase and decrease in various rice genotypes. It has been observed that starch content decreased in rice roots in response to salinity while it remained fairly unchanged in the shoot (Alamgir and Yousuf, 1999).

Raffinose family oligosaccharides (RFOs) have long been suggested to act as antistress agents in plant tissues. They are the most widely distributed nonstructural carbohydrates in the plant kingdom (Peterbauer et al., 2002). They are nonreducing storage carbohydrates that accumulate in large quantities without affecting primary metabolic processes. A strong correlation between the accumulation of RFOs, primarily raffinose, stachyose, and verbascose, and desiccation tolerance has been revealed in seeds (Horbowicz and Obendorf, 1994; Peters et al., 2007). The levels of galactinol and inositol increased by twofold to threefold in the leaves of salt-tolerant genotype of barley during salt stress. The trisaccharide raffinose (galactosyl-sucrose) is produced by the condensation of galactinol ($\alpha$-galactosyl-myo-inositol) and sucrose to produce raffinose and inositol (Tapernoux-Luthi et al., 2004). Higher levels of inositol accumulation reported in plants during salinity stress (Brosche et al., 2005; Gong et al., 2005; Gagneul et al., 2007) are proposed as an effective free radical scavenger (Smirnoff and Cumbes, 1989). Moreover, overexpression of *Arabidopsis galactinol synthase1* or *galactinol synthase2* in *Arabidopsis* accumulated high levels of galactinol and raffinose and showed higher tolerance to drought and salinity stress (Taji et al., 2002; Nishizawa et al., 2008).

## 14.4 POLYAMINES IN SALT TOLERANCE

Polyamines are small, low-molecular-weight, ubiquitous, polycationic aliphatic molecules widely distributed throughout the plant kingdom. Polyamines play a variety of roles in normal growth and development such as regulation of cell proliferation, somatic embryogenesis, differentiation and morphogenesis, dormancy breaking of tubers and seed germination, development of flowers and fruit, and senescence (Gupta et al., 2013). Increases in the level of polyamines are correlated with abiotic stress tolerance in plants including salinity (Groppa and Benavides, 2008). The most common polyamines that are found within the plant system are diamine putrescine, triamine spermidine, and tetra-amine spermine. The polyamine biosynthetic pathway has

been thoroughly investigated in many organisms including plants and has been reviewed in detail (Martin-Tanguy, 2001; Kusano et al., 2007). Diamine putrescine is the smallest polyamine and is synthesized from either ornithine or arginine by the action of enzyme ornithine decarboxylase and arginine decarboxylase, respectively (Alcazar et al., 2006; Hasanuzzaman et al., 2014).

Application of exogenous polyamine has been found to increase the level of endogenous polyamine during stress. Briefly, the application of exogenous polyamines (Put, Spd, and Spm) enhanced salt tolerance in rice, *Atropa belladonna*, and barley; ozone and salt tolerance in tobacco; chilling and salt tolerance in cucumber; and salt and heavy metal (Cu, Fe, and Ni) tolerance in *Brassica napus*. The positive effects of polyamines have been associated with the maintenance of membrane integrity, regulation of gene expression for the synthesis of osmotically active solutes, reduction in ROS production, and controlling accumulation of $Na^+$ and $Cl^-$ ion in different organs. Overproduction of diamine putrescine, triamine spermidine, and tetra-amine spermine in rice, tobacco, and *Arabidopsis* enhances salt tolerance (Roy and Wu, 2002). It has been reported that exogenous application of polyamines could alleviate salt-induced reduction in photosynthetic efficiency, but this effect depends on polyamine concentration and types and level of stress. Li et al. (2013) also reported that the regulation of Calvin cycle, protein folding assembly, and the inhibition of protein proteolysis by triamine spermidine might play important roles in salt tolerance.

## 14.5 ANTIOXIDANTS IN SALT TOLERANCE

The most common plant stressor in saline soils is NaCl. The observed plant responses to saline conditions generally include osmotic imbalance, which in turn leads to changes in ion concentrations, particularly of potassium and Ca. At higher levels, $Na^+$ and $Cl^-$ have direct toxic effects on membrane structure and enzyme systems (Ashraf and Harris, 2004). This ultimately leads to secondary stresses, such as oxidative stress, linked to the production of toxic ROS accompanied by lipid peroxidation (Qureshi et al., 2005).

Active oxygen species such as superoxide ($O_2^{\cdot-}$), hydrogen peroxide ($H_2O_2$), hydroxyl radical ($\cdot OH$), and singlet oxygen ($^1O_2$) are produced during normal aerobic metabolism when electrons from the electron transport chains in mitochondria and chloroplasts are leaked and react with $O_2$ in the absence of other acceptors (Thompson et al., 1987). ROS are extremely reactive in nature because they can interact with a number of other molecules and metabolites such as DNA, pigments, proteins, lipids, and other essential cellular molecules that lead to a series of destructive processes (Sharma et al., 2012). Thus, ROS are considered as cellular indicators of stresses as well as secondary messengers actively involved in the stress response signaling pathways. Plants have the ability to scavenge/detoxify ROS by producing different types of antioxidants. Antioxidants can be generally categorized into two different types, that is, enzymatic and nonenzymatic. Enzymatic antioxidants include superoxide dismutase (SOD), catalase, ascorbate peroxidase (APX), monodehydroascorbate reductase, dehydroascorbate reductase, and glutathione reductase. The commonly known nonenzymatic antioxidants are glutathione (GSH), ascorbate (AsA), carotenoids, and tocopherols (Asada, 1999; Ashraf, 2009).

There is now conclusive evidence that production of activated oxygen species is enhanced in plants in response to different environmental stresses such as salinity, drought, waterlogging, temperature extremes, high light intensity, herbicide treatment, or mineral nutrient deficiency (Wise and Naylor, 1987; Gossett et al., 1994). Plants containing high concentrations of antioxidants show considerable resistance to the oxidative damage caused by the activated oxygen species (Garratt et al., 2002). Comparing the mechanisms of antioxidant production in salt-tolerant and salt-sensitive plants, Dionisiosese and Tobita (1998) reported a decline in SOD activity and an increase in peroxidase activity in the salt-sensitive rice varieties, Hitomebore and IR28, in response to salt stress. These salt-sensitive varieties also showed an increase in lipid peroxidation and electrolyte leakage as well as $Na^+$ accumulation in the leaves under saline conditions. In contrast, two salt-tolerant rice varieties, Pokkali and Bankat, showed differing protective mechanisms against activated oxygen species under salt stress. Cv. Pokkali showed only a slight increase in SOD but a slight decrease in peroxidase activity, and almost unchanged lipid peroxidation, electrolyte leakage, and $Na^+$ accumulation under saline conditions. In contrast, cv. Bankat showed $Na^+$ accumulation in leaves and symptoms of oxidative damage similar to the salt-sensitive cultivars. Pea plants grown under saline (150 mM NaCl) stress showed an enhancement of both APX activity and S-nitrosylated APX, as well as an increase in $H_2O_2$, NO, and S-nitrosothiol content that can justify the induction of APX activity.

The exogenous application of antioxidants also helps to mitigate the adverse effects of salinity stress in various plant species and promote plant recovery from the stress (Agarwal and Shaheen, 2007; Munir and Aftab, 2011), for example, AsA and GSH react with superoxide radical, hydroxyl radical, and hydrogen peroxide, thereby functioning as a free radical scavenger. It can also participate in the regeneration of AsA via the AsA-GSH cycle. When applied exogenously, GSH helps to maintain plasma membrane permeability and cell viability during salinity stress in *Allium cepa* (Aly-Salama and Al-Mutawa, 2009). The application of GSH and AsA was found to be effective in increasing the height of the plant; branch number; fresh and dry weight of and the content of carbohydrates, phenols, and xanthophylls pigment; and mineral ion content when subjected to saline condition (Rawia et al., 2011). Recently, it has been proposed that there are three main traits in plants, which help them in their adaptation to salinity stress, ion exclusion, tissue tolerance, and salinity tolerance (Roy et al., 2014), and antioxidants seem to have some role in tissue and salinity tolerance mechanisms.

## 14.6 ORGANIC ACIDS

Organic acids, especially those that are part of the tricarboxylic acid (TCA) cycle, are up- and downregulated during salinity stress. The TCA cycle connects glycolysis to amino acid biosynthesis, and it plays an important role in the regulation of respiration and energy generation by producing ATP and NADH. Increased levels of organic acids suggest an increase in the flow of carbon from glycolysis through TCA cycle leading to an increased production of NADH, FADH2, and ATP.

The roots of the more tolerant varieties of rice had lower levels of the TCA cycle intermediates and other organic acids (Zuther et al., 2007). Higher levels of organic acids may be correlated with the ability of the plant to improve its growth under salt stress. Lower levels may be correlated with reduced metabolic activity and, therefore, reduced growth of salinity-sensitive plants (Widodo et al., 2009).

## 14.7 PROTEINS IN SALT TOLERANCE

One of the most extensively studied proteins that are accumulated in response to salt adaptation is osmotin, which was first identified in salt-adapted tobacco cells. At the cellular level, it accumulates in the vacuole of salt-stressed cells. Osmotin is regulated in cells at the transcriptional level by abscisic acid (ABA) application (Singh et al., 1989), but posttranscriptional regulation has been shown to control the protein accumulation (LaRosa et al., 1992). Osmotin polypeptide sequence shows features common to maize α-amylase/trypsin inhibitor. The other most important osmoprotectants are proline and glycine betaine (Holmstrom et al., 2000). Proline accumulation also provides an example of osmoprotectant indicating the upregulation of proteins and key enzymes.

It is believed that genes induced by the application of salt stress are usually regulated at the transcriptional level. A few promoters of these genes have been isolated and studied, and cis-acting elements have been functionally identified and characterized (Yamaguchi-Shinozaki and Shinozaki, 2006). Moreover, a good correlation between mRNA and protein levels has been shown for some genes (e.g., DSP 22 from *Craterostigma plantagineum* and HVA1 from barley). However, in other cases, there is a marked delay in protein induction compared with mRNA accumulation (e.g., rice SALT). On the other hand, changes in mRNA levels are not always followed by similar changes in the corresponding protein (e.g., tobacco osmotin) (LaRosa et al., 1992). A salt-stress-associated protein from citrus, as well as an encoding gene, has been isolated. This protein was demonstrated to be a phospholipid hydroperoxide GSH peroxidase, which had not been identified before in plants (Beeor-Tzahar et al., 1995).

Salt stress induces the expression of the vacuolar $H^{(+)}$-ATPase. For salinity stress tolerance in plants, the vacuolar-type $(H)^{(+)}$-ATPase is of prime importance in energizing $Na^+$ sequestration into the central vacuole and is known to respond to salt stress with increased expression and enzyme activity (Golldack and Dietz, 2001). Salt stress has been shown to induce the expression of a protein having a strong homology to APX in radish, while in leaves of *Vigna unguiculata*, cytosolic

APX was slightly reduced and chloroplastic APX unchanged. It has been reported that, in NaCl tolerant pea cultivars, leaf mitochondrial Mn-SOD and chloroplast Cu/Zn-SOD activities increased under salt stress, while the Cu/Zn-SOD activity remained unchanged (Hernandez et al., 1995). In the salt-sensitive cultivar, neither APX nor chloroplastic SOD was increased by salt, while the cytosolic and mitochondrial SOD even decreased. In salt-sensitive and salt-tolerant cultured citrus cells and leaf tissues, it was shown that only cytosolic Cu/Zn-SOD activity was increased by salt, whereas the activity of other isoforms was unchanged (Gueta-Dahan et al., 1997).

Another essential mechanism of salt tolerance involves the ability to reduce the ionic stress on the plant by minimizing the amount of $Na^+$ that accumulates in the cytosol of cells, particularly those in the transpiring leaves. This process, as well as tissue tolerance, involves up- and downregulation of the expression of specific ion channels and transporters, allowing the control of $Na^+$ transport throughout the plant (Munns and Tester, 2008; Rajendran et al., 2009). $Na^+$ exclusion from leaves is associated with salt tolerance in cereal crops including rice, durum wheat, bread wheat, and barley (James et al., 2011). The proteins encoded by the Nax1 and Nax2 genes are shown to increase the retrieval of $Na^+$ from the xylem in roots, thereby reducing shoot $Na^+$ accumulation. In particular, the Nax1 gene confers a reduced rate of transport of $Na^+$ from root to shoot and retention of $Na^+$ in the leaf sheath, thus giving a higher sheath-to-blade $Na^+$ concentration ratio.

Xylem sap proteins may also play a role in abiotic stress responses, as 39 xylem sap proteins were found to be differentially regulated in maize in response to water stress (Alvarez et al., 2008). Many of the upregulated proteins were cell wall metabolism enzymes, which may function by reinforcing the secondary cell walls of xylem vessels during periods of drought. Generally, the proteins are present in xylem sap at very low concentrations (10–300 μg/mL) (Buhtz et al., 2004; Alvarez et al., 2006); nevertheless, hundreds of protein spots can be detected in xylem sap from *B. napus* and *Z. mays* using 2D gel electrophoresis (Kehr et al., 2005; Alvarez et al., 2006). Until recently, little was known about the identity of xylem sap proteins, but advances in genomics, proteomics, and metabolomics are now facilitating their characterization. To date, xylem sap proteomes have been reported for annual plants including B. *napus*, *Brassica oleracea*, *Cucurbita maxima*, *Cucumis sativus*, *Z. mays*, *Lycopersicon esculentum*, *Glycine max*, and *O. sativa* (Rep et al., 2003; Buhtz et al., 2004; Kehr et al., 2005; Alvarez et al., 2006; Djordjevic et al., 2007; Toshihiko et al., 2008).

Research into γ-aminobutyric acid (GABA), a nonprotein amino acid, has focused on its role as a metabolite, mainly in the context of responses to biotic and abiotic stresses. It is mainly metabolized through a short pathway called the GABA shunt, because it bypasses two steps of the TCA cycle (Bouche et al., 2003). Although differences in the subcellular localization of GABA shunt enzymes in different organisms have been reported, such as in yeast, where succinic semialdehyde dehydrogenase (SSADH) is present in the cytosol, the pathway is composed of three enzymes: the cytosolic GAD and the mitochondrial enzymes GABA transaminase and SSADH. If GABA could activate signaling pathways in a broad range of organisms, then GABA receptors should be present/overexpressed in these organisms because the initial event leading to the activation of a cellular signaling pathway is the binding of a ligand, such as a hormone, to a specific receptor.

Other than the proteins discussed, there are a number of proteins that alter their concentration under stress. Examples of such proteins are programmed cell death–related proteins and, more importantly, proteases (Bouche et al., 2003).

## 14.8 CALCIUM SIGNALING

The changes in intercellular calcium concentration ($Ca^{2+}$) in response to abiotic stresses are well established, suggesting that $Ca^{2+}$ serves as a messenger in normal growth and developmental processes of plants (Reddy and Reddy, 2004). $Ca^{2+}$ channels have been detected in the plasma membrane, vacuolar membrane, endoplasmic reticulum (ER), chloroplast and nuclear membranes of plant cells. In response to osmotic, drought, and salt stress, increased intercellular $Ca^{2+}$ concentration mobilized from intercellular stores has been shown by Sanders et al. (2002). Intercellular and extracellular sources of $Ca^{2+}$ and inositol phosphates have a role to play in the response to salinity. In plant and animal cells, phospholipase C–mediated hydrolysis of phosphatidylinositol-4,5-bisphosphate (PIP2) results in the production of inositol-1,4,5-trisphosphate (IP3) and diacylglycerol. There is some

indirect evidence for the involvement of heterotrimeric G-proteins in promoting PIP2 hydrolysis in plants. Phosphatidylinositol-specific phospholipase C (PI-PLC) activity and genes encoding this enzyme have been characterized from plants. Plant PI-PLC hydrolyses phosphatidylinositol-4,5-bisphosphate into IP3 and diacylglycerol with an absolute requirement for $Ca^{2+}$ (1 mM). It was shown that the expression of one of the phospholipase C genes (AtPLC1) is induced by stresses including dehydration, salinity, and low temperature. Three PI-PLC isoforms (StPLC1–3) have been isolated from guard cell–enriched tissues of potato. Reddy (2001) has reviewed the role of Ca in signaling.

## 14.9 SALT OVERLY SENSITIVE METABOLITES

Salt overly sensitive (SOS) proteins are sensors for calcium signals that turn on the machinery for $Na^+$ export and $K^+/Na^+$ discrimination. In the SOS pathway, a myristoylated calcium-binding protein, SOS3, senses cytosolic calcium changes elicited by salt stress. SOS3 physically interacts with and activates the protein kinase, SOS2. The SOS3/SOS2 kinase complex phosphorylates and activates the transport activity of the plasma membrane $Na^+/H^+$ exchanger encoded by the SOS1 gene. Preliminary results suggest that, in addition to its transport function, SOS1 may also have a regulatory role and may even be a novel sensor for sodium. The SOS stress-signaling pathway was identified to be a pivotal regulator of plant ion homeostasis and salt tolerance (Hasegawa et al., 2000; Sanders et al., 2002). This signaling pathway functionally resembles the yeast calcineurin cascade that controls $Na^+$ influx and efflux across the plasma membrane (Bressan et al., 1998). Expression of an activated form of calcineurin in yeast or plants enhances salt tolerance, further implicating the functional similarity between the calcineurin and the SOS pathways (Mendoza et al., 1996; Pardo et al., 1998). The plasma membrane–localized SOS1, $H^+$ pump, is a P-type ATPase and is primarily responsible for pH and membrane potential gradient across this membrane (Morsomme and Boutry, 2000). A vacuolar-type $H^+$-ATPase and a vacuolar pyrophosphatase generate the pH and membrane potential across the tonoplast (Drozdowicz and Rea, 2001; Maeshima, 2001). The activity of these $H^+$ pumps is increased by salt treatment, and induced gene expression may account for some of the upregulation (Hasegawa et al., 2000; Maeshima, 2001). The plasma membrane $H^+$-ATPase was confirmed as a salt tolerance determinant based on the analyses of phenotypes caused by the semi-dominant aha4-1 mutation (Vitart et al., 2001). The upregulation of the SOS pathway proteins under salt stress leads to salt tolerance. Proteomic studies supporting elevation in the expression of SOS proteins were carried out under salinity (Zorb et al., 2004) and ozone stress (Agarwal et al., 2002).

## 14.10 HORMONE-MEDIATED PROTECTION UNDER SALT STRESS

### 14.10.1 Methyl Jasmonate

Methyl jasmonate (MeJA) and its free acid jasmonic acid, collectively referred to as jasmonates, are important cellular regulators involved in diverse developmental processes, such as seed germination, root growth, fertility, fruit ripening, and senescence. In addition, jasmonates activate plant defense mechanisms in response to insect-driven wounding, various pathogens, and environmental stresses, such as drought, low temperature, and salinity (Creelman and Mullet, 1997). The way in which jasmonates regulate these processes has been studied by looking for gene expression patterns in a wide range of jasmonate-responsive physiological states. Genes involved in jasmonate biosynthesis, secondary metabolism, cell-wall formation, and those encoding stress protective and defense proteins are upregulated by MeJA treatment, while genes involved in photosynthesis, such as ribulose bisphosphate carboxylase/oxygenase, chlorophyll a/b-binding protein, and light-harvesting complex II, are downregulated. These genes are still under intensive analysis using functional genomics and bioinformatics approaches to better define MeJA-mediated signaling pathways and cellular responses. MeJA formation could be one of several important control points for jasmonate-regulated plant responses. This hypothesis was tested with transgenic *Arabidopsis* (*A. thaliana*) overexpressing JMT, where various jasmonate-responsive genes were constitutively expressed in the absence of wounding or jasmonate treatment.

### 14.10.2 Abscisic Acid

The acclimation/adaptation processes are mostly mediated by the plant hormone ABA (Xiong et al., 2002). Hormone levels increase under common stress conditions to trigger metabolic and physiological changes. Numerous genes involved in the acclimation/adaptation processes are up- and/or downregulated under stress conditions. Although not all of them are subjected to ABA regulation, the expression of a large number of them is controlled by ABA. For the past several years, researchers have been trying to identify transcription factors that regulate the expression of ABA/stress-responsive genes via the consensus element, which is generally known as "abscisic acid response element" (ABRE). Researchers have focused on the small subfamily of *Arabidopsis* bZIP proteins referred to as ABRE-binding factors (ABFs) (Choi et al., 2002), whose expression is induced by ABA and by various abiotic stresses (i.e., cold, high salt, and drought) (Bae et al., 2003). To investigate their in vivo roles, transgenic *Arabidopsis* plants that constitutively overexpress each of them have been generated. Their phenotypes were then analyzed with special attention to changes in ABA/stress responses. Each ABF displayed similar, but distinct, phenotypes. Data suggest that ABF3 is probably most important for stress tolerance among the four ABFs (ABF1, ABF2, ABF3, and ABF4). ABF3 overexpression affected the expression levels of ABA/stress-regulated genes (Kang et al., 2002).

The term metabolomics has been defined as the identification and quantitation of all low-molecular-weight metabolites in a given organism, at a given developmental stage, and in a given organ, tissue, or cell type (Fiehn, 2002; Arbona et al., 2010). This is a challenging task due to the wide array of molecules with different structures and chemical properties. For example, it is estimated that a single accession of *Arabidopsis* contains more than 500 metabolites, most of them yet uncharacterized. The metabolomics techniques focus on metabolites with similar and specific chemical properties and are globally known as metabolite profiling only covering up a fraction of the metabolome (Arbona et al., 2010). Thus to achieve a comprehensive coverage of a vast range of metabolites present in the plant kingdom, several analytical techniques consisting of a separation technique coupled to a detection device (usually MS) are combined. Various studies using metabolomic tools in plants during salt stress showed that the biochemical changes involve metabolic pathways that fulfill crucial functions in the plant adaptation to salt-stressing conditions. A few such studies have been summarized in Table 14.2.

TABLE 14.2
*Summary of metabolite profiling of various crops under salt stress.*

| Crop | Technique | Major findings | References |
|------|-----------|----------------|------------|
| Alfalfa | GLC HPLC | Salt stress induced a large increase in the amino acid, proline and carbohydrate, and pinitol. Increased accumulation of proline reflected an osmoregulatory mechanism, and pinitol might contribute to the tolerance. | Fougere et al. (1991) |
| *Arabidopsis* | GC-MS LC-MS | The methylation cycle, the phenylpropanoid pathway, and glycinebetaine biosynthesis were synergetically induced as a short-term response against salt stress while coinduction of glycolysis, sucrose metabolism, and coreduction of the methylation cycle as long-term responses to salt stress were observed. | Kim et al. (2007) |
| Barley | GC-MS | In the more tolerant Sahara plants, the levels of the hexose phosphates, TCA cycle intermediates, and raffinose increased in response to salt while remained unchanged in the more sensitive plants. Increased putrescine and GABA levels in sensitive plants may also represent an onset of senescence and cell damage in the leaf tissue. | Widodo et al. (2009) |
| *Thellungiella* (Salt cress) | LC-ESI-QTOF-MS | Higher levels of basal metabolic configuration (proline and secondary compounds) provide tolerance to environmental cues. | Arbona et al. (2010) |
| Tomato | GC-MS | Salt stress suppressed the accumulation of organic acids (except citrate) in columella tissue of fruit during early stages, whereas organic acid accumulation (except that of malate) was enhanced in pericarp of fruit during the ripening stages. | Yin et al. (2010) |
| Tobacco | NMR | Prolonged salinity with high-dose-induced progressive accumulation of osmolytes, such as proline and myo-inositol, and changes in GABA shunt along with the shikimate-mediated secondary metabolisms with enhanced biosynthesis of aromatic amino acids. | Zhang et al. (2011) |
| Maize | NMR | The levels of alanine, glutamate, asparagine, glycine betaine, and sucrose were increased, and levels of malic acid, trans-aconitic acid, and glucose decreased in a dose-dependent manner in shoots. | Gavaghan et al. (2011) |
| Lotus | GC/EI-TOF-MS | The amino acids proline, serine, threonine, glycine, and phenylalanine, the sugars sucrose and fructose, and myo-inositol increased, whereas organic acids such as citric, succinic, fumaric, erythronic, glycolic, and aconitic acid decreased in response to salt stress. | Sanchez et al. (2011) |
| Soybean | GC-MS and LC-FT/MS | Tolerance for salt due to the synthesis of compatible solutes, induction of ROS scavengers, and induction of plant hormones. | Lu et al. (2013) |
| Barley | GC-MS | Polyols (osmotic adjustment) played important roles in developing salt tolerance in roots along with high levels of sugars and energy, while active photosynthesis in leaves were important for barley to develop salt tolerance. | Wu et al. (2013) |

## 14.11 CONCLUDING REMARKS

In the past few decades, plant breeders have successfully improved salinity tolerance of some crops through conventional selection and breeding techniques, with relatively little direct input from physiologists or biochemists. These achievements undoubtedly relied on the phenotypic characteristics of the plants without a full understanding of the underlying biochemical mechanisms. There is a consensus among scientists that selection is more convenient and practicable if the plant species under test possesses distinctive indicators of salt tolerance, whether at the whole plant, tissue, or cellular level. Thus, there is a pressing need to unravel the cellular mechanisms so as to give meaningful advice to plant breeders. Despite a number of studies on salinity tolerance of plants, neither the metabolic sites at which salt stress damages plants nor the adaptive components of salt tolerance are fully understood. As a result, there are no well-defined plant indicators for salinity tolerance that can practically be used by plant breeders in their breeding programs for improvement of salinity tolerance in a number of agricultural crops. It would be much more valuable if biochemical indicators are specified for individual species rather than generalized for all species. Out of the many biochemical indicators listed earlier, antioxidants and organic solutes such as glycine betaine and proline have recently gained ground, because there are numerous reports in the literature that show that plants containing high concentrations of antioxidants or either of the two organic solutes show considerable resistance to salinity as well as other abiotic stresses.

Stress biologists now feel comfortable with the development of excellent tools in the fields of metabolomics, proteomics, and bioinformatics. Earlier, they used enzymatic kinetics (estimation of precursor/s or product/s) to estimate the concentration of functional proteins, immunological reactions, fractionation methods, and radioactive isotopes. Enzyme levels, for example, estimated in a test tube by kinetic reaction, followed by UV/visible spectrometry analysis, may be in the form of one among hundreds of spots visible on a gel. The unseen proteins are now visible. However, most of these techniques used were economical, and the field of proteomics was not well developed. Now, the advancement in tools of metabolomics has changed the whole scenario. Given the high throughput and high sensitivity of mass spectrometry, coupled via advanced software to metabolites databases, metabolomics is gaining overwhelming response. The metabolome map may serve in future as a reference for a specific plant part/organ, under physiological as well as stress conditions (stress metabolome). It may be helpful to have a record of metabolites as environmental biomarkers and can be beneficial in two major areas, in the understanding of plant metabolism and its regulatory factors and in the development of strategies to utilize the generated knowledge/data for improved traits in plants, mainly for crop and medicinal plants via genetic/metabolic engineering after hunting a gene using reverse genetics. The understanding of plant stress physiology and of the factors that influence it is essential in order to correlate them with the changes in metabolome. Further improvement with innovative additions in metabolomics will surely transform most stress biologists into environmental metabolomists!

## ABBREVIATIONS

CE-MS: Capillary electrophoresis–mass spectrometry
Da: Dalton
FAIMS: Field asymmetric waveform ion mobility spectrometry
FT-ICR: Fourier transform ion-cyclotron resonance
GC/EI-TOF-MS: Gas chromatography coupled to electron impact ionization time of flight-mass spectrometry
GC-MS: Gas chromatography–mass spectrometry
LC-ESI-QTOF-MS: Liquid chromatography–electrospray–quadrupole-time of flight–mass spectrometry
LC-MS: Liquid chromatography–mass spectrometry
ppm: parts per million
SPME: Solid phase microextraction

## REFERENCES

Agarwal, S. and R. Shaheen. 2007. Stimulation of antioxidant system and lipid peroxidation by abiotic stresses in leaves of *Momordica charantia*. *Brazilian Journal of Plant Physiology* 19:149–161.

Agrawal, G.K., R. Rakwal, M. Yonekura, A. Kubo, and H. Saji. 2002. Proteome analysis of differentially displayed proteins as a tool for investigating ozone stress in rice (*Oryza sativa* L.) seedlings. *Proteomics* 2:947–959.

Ahmad, R., C.J. Lim, and S.Y. Kwon. 2013. Glycine betaine: A versatile compound with great potential for gene pyramiding to improve crop plant performance against environmental stresses. *Plant Biotechnology Reports* 7:49–57.

Alamgir, A.N.M. and M. Yousuf Ali. 1999. Effect of salinity on leaf pigments, sugar and protein concentrations and chloroplast ATPase activity of rice (*Oryza sativa* L.). *Bangladesh Journal of Botany* 28:145–149.

Alcazar, R., F. Marco, and J.C. Cuevas. 2006. Involvement of polyamines in plant response to abiotic stress. *Biotechnology Letters* 28:1867–1876.

Ali, R.M. and H.M. Abbas. 2003. Response of salt stressed barley seedlings to phenylurea. *Plant, Soil and Environment* 49:158–162.

Alvarez, S., J.Q. Goodger, E.L. Marsh, S. Chen, V.S. Asirvatham, and D.P. Schachtman. 2006. Characterization of the maize xylem sap proteome. *Journal of Proteome Research* 5:963–972.

Alvarez, S., E.L. Marsh, S.G. Schroeder, and D.P. Schachtman. 2008. Metabolomic and proteomic changes in the xylem sap of maize under drought. *Plant, Cell & Environment* 31:325–340.

Aly-Salama, K.H. and M.M. Al-Mutawa. 2009. Glutathione-triggered mitigation in salt-induced alterations in plasmalemma of onion epidermal cells. *International Journal of Agriculture and Biology* 11:639–642.

Arbona, V., R. Argamasilla, and A. Gómez-Cadenas. 2010. Common and divergent physiological, hormonal and metabolic responses of *Arabidopsis thaliana* and *Thellungiella halophila* to water and salt stress. *Journal of Plant Physiology* 167:1342–1350.

Asada, K. 1999. The water-water cycle in chloroplasts: Scavenging of active oxygens and dissipation of excess photons. *Annual Review of Plant Physiology and Plant Molecular Biology* 50:601–639.

Ashraf, M. 2009. Biotechnological approach of improving plant salt tolerance using antioxidants as markers. *Biotechnology Advances* 27:84–93.

Ashraf, M. and P.J.C. Harris. 2004. Potential biochemical indicators of salinity tolerance in plants. *Plant Science* 166:3–16.

Bae, H., E. Herman, and R. Sicher. 2003. Exogenous trehalose promotes non-structural carbohydrate accumulation and induces chemical detoxification and stress response proteins in *Arabidopsis thaliana* grown in liquid culture. *Plant Science* 168:1293–1301.

Baque, M.A., A. Elgirban, E. Lee, and K. Paek. 2012. Sucrose regulated enhanced induction of anthraquinone, phenolics, flavonoids biosynthesis and activities of antioxidant enzymes in adventitious root suspension cultures of *Morinda citrifolia* (L.). *Acta Physiologiae Plantarum* 34:405–415.

Beeor-Tzahar, T., G. Ben-Hayyim, D. Holland, Z. Faltin, and Y. Eshdat. 1995. A stress-associated citrus protein is a distinct plant phospholipid hydroperoxide glutathione peroxidase. *FEBS Letter* 366:151–155.

Binzel, M.L., F.D. Hess, R.A. Bressan, and P.M. Hasegawa. 1988. Intracellular compartmentation of ions in salt adapted tobacco cells. *Plant Physiology* 86:607–614.

Bohnert, H.J., D.E. Nelson, and R.G. Jensen. 1995. Adaptations to environmental stresses. *Plant Cell* 7:1099–1111.

Bouche, N., B. Lacombe, and H. Fromm. 2003. GABA signaling: A conserved and ubiquitous mechanism. *Trends in Cell Biology* 13:607–610.

Bourgaud, F., A. Gravot, S. Milesi, and E. Gontier. 2001. Production of plant secondary metabolites: A historical perspective. *Plant Science* 161:839–851.

Bressan, R.A., P.M. Hasegawa, and J.M. Pardo. 1998. Plants use calcium to resolve salt stress. *Trends in Plant Science* 3:411–412.

Brosché, M., B. Vinocur, E.R. Alatalo, A. Lamminmäki, T. Teichmann, E.A. Ottow, D. Djilianov et al. 2005. Gene expression and metabolite profiling of *Populus euphratica* growing in the Negev Desert. *Genome Biology* 6:R101.

Brunetti, C., R.M. George, M. Tattini, K. Field, and M.P. Davey. 2013. Metabolomics in plant environmental physiology. *Journal of Experimental Botany* 64:4011–4020.

Buhtz, A., A. Kolasa, K. Arlt, C. Walz, and J. Kehr. 2004. Xylem sap protein composition is conserved among different plant species. *Planta* 219:610–618.

Choi, D.W., E.M. Rodriguez, and T.J. Close. 2002. Barley Cbf3 gene identification, expression pattern, and map location. *Plant Physiology* 129:1781–1787.

Creelman, R.A. and J.E. Mullet. 1997. Biosynthesis and action of jasmonates in plants. *Annual Review of Plant Physiology and Plant Molecular Biology* 48:355–381.

Croteau, R., T.M. Kutchan, and N.G. Lewis. 2000. Natural products (secondary metabolites). In *Biochemistry and Molecular Biology of Plants*. Buchanan, B., W. Gruissem, and R. Jones (eds.), pp. 1250–1268. American Society of Plant Biologists, Rockville, MD.

Deivanai, S., R. Xavier, V. Vinod, K. Timalata, and O.F. Lim. 2011. Role of exogenous proline in ameliorating salt stress at early stage in two rice cultivars. *Journal of Stress Physiology & Biochemistry* 7:157–174.

Dionisiosese, M.L. and S. Tobita. 1998. Antioxidant response of rice seedlings to salinity stress. *Plant Science* 135:1–9.

Djordjevic, M.A., M. Oakes, D.X. Li, C.H. Hwang, C.H. Hocart, and P.M. Gresshoff. 2007. The *Glycine max* xylem sap and apoplast proteome. *Journal of Proteome Research* 6:3771–3779.

Drozdowicz, Y.M. and P.A. Rea. 2001. Vacuolar $H^+$ pyrophosphatases: From the evolutionary backwaters into the mainstream. *Trends in Plant Science* 6:206–211.

El-Shintinawy, F. and M.N. El-Shourbagy. 2001. Alleviation of changes in protein metabolism in NaCl-stressed wheat seedlings by thiamine. *Biologia Plantarum* 44:541–545.

Fernie, A.R. and N. Schauer. 2008. Metabolomics-assisted breeding: A viable option for crop improvement?. *Trends in Genetics* 25:39–48.

Fiehn, O. 2002. Metabolomics—The link between genotypes and phenotypes. *Plant Molecular Biology* 48:155–171.

Fougere, F., D.L. Rudulier, and J.G. Streeter. 1991. Effects of salt stress on amino acid, organic acid, and carbohydrate composition of roots, bacteroids, and cytosol of alfalfa (*Medicago sativa* L.). *Plant Physiology* 96:1228–1236.

Gagneul, D., A. Alnouche, C. Duhazé, R. Lugan, F.R. Larher, and A. Bouchereau. 2007. A reassessment of the function of the so-called compatible solutes in the halophytic plumbaginaceae *Limonium latifolium*. *Plant Physiology* 144:1598–1611.

Gao, Z., M. Sagi, and S.H. Lips. 1998. Carbohydrate metabolism in leaves and assimilate partitioning in fruits of tomato (*Lycopersicon esculentum* L.) as affected by salinity. *Plant Science* 135:149–159.

Garratt, L.C., B.S. Janagoudar, K.C. Lowe, P. Anthony, J.B. Power, and M.R. Davey. 2002. Salinity tolerance and antioxidant status in cotton cultures. *Free Radical Biology and Medicine* 33:502–511.

Gavaghan, C.L., J.V. Li, S.T. Hadfield, S. Hole, J.K. Nicholson, I.D. Wilson, P.W. Howe, P.D. Stanley, and E. Holmes. 2011. Application of NMR-based metabolomics to the investigation of salt stress in maize (*Zea mays*). *Phytochemical Analysis* 22:214–224.

Golldack, D. and K.J. Dietz. 2001. Salt-induced expression of the vacuolar $H^+$ ATPase in the common ice plant is developmentally controlled and tissue specific. *Plant Physiology* 125:1643–1654.

Gong, Q., P. Li, S. Ma, S.I. Rupassara, and H. Bohnert. 2005. Salinity stress adaptation competence in the extremophile *Thellungiella halophila* in comparison with its relative *Arabidopsis thaliana*. *Plant Physiology* 44:826–839.

Gossett, D.R., E.P. Millhollon, and M.C. Lucas. 1994. Antioxidant response to NaCl stress in salt-tolerant and salt-sensitive cultivars of cotton. *Crop Science* 34:706–714.

Groppa, M.D. and M.P. Benavides. 2008. Polyamines and abiotic stress: Recent advances. *Amino Acids* 34:35–45.

Gueta-Dahan, Y., Z. Yaniv, B.A. Zilinskas, and G. Ben-Hayyim. 1997. Salt and oxidative stress: Similar and specific responses and their relation to salt tolerance in citrus. *Planta* 203:460–469.

Gupta, K., A. Dey, and B. Gupta. 2013. Plant polyamines in abiotic stress responses. *Acta Physiologiae Plantarum* 35:2015–2036.

Hasegawa, P.M., R.A. Bressan, J.K. Zhu, and H.J. Bohnert. 2000. Plant cellular and molecular responses to high salinity. *Annual Review of Plant Physiology and Plant Molecular Biology* 51:463–499.

Hernández, J.A., E. Olmos, F.J. Corpas, F. Sevilla, and L.A. Del Río. 1995. Salt-induced oxidative stress in chloroplast of pea plants. *Plant Science* 105:151–167.

Holmström, K., S. Somersalo, A. Mandal, T.E. Palva, and B. Welin. 2000. Improved tolerance to salinity and low temperature in transgenic tobacco producing glycine betaine. *Journal of Experimental Biology* 51:177–185.

Hoque, M.A., M.N.A. Banu, Y. Nakamura, Y. Shimoishi, and Y. Murat. 2008. Proline and glycine-betaine enhance antioxidant defence and methylglyoxal detoxification systems and reduce NaCl-induced damage in cultured tobacco cells. *Journal of Plant Physiology* 165:813–824.

Horbowicz, M. and R.L. Obendorf. 1994. Seed desiccation tolerance and storability: Dependence of flatulence-producing oligosaccharides and cyclitols—Review and survey. *Seed Science Research* 4:385–405.

James, R.A., C. Blake, C.S. Byrt, and R. Munns. 2011. Major genes for $Na^+$ exclusion, Nax1 and Nax2 (wheat HKT1;4 and HKT1;5), decrease $Na^+$ accumulation in bread wheat leaves under saline and waterlogged conditions. *Journal of Experimental Botany* 62(8):2939–2947.

Kang, J.Y., H.I. Choi, M.Y. Im, and S.Y. Kim. 2002. *Arabidopsis* basic leucine zipper proteins that mediate stress-responsive abscisic acid signaling. *Plant Cell* 14:343–357.

Kehr, J., A. Buhtz, and P. Giavalisco. 2005. Analysis of xylem sap proteins from *Brassica napus*. *BMC Plant Biology* 5:11.

Kerepesi, I. and G. Galiba. 2000. Osmotic and salt stress-induced alteration in soluble carbohydrate content in wheat seedlings. *Crop Science* 40:482–487.

Kim, J.K., T. Bamba, K. Harada, E. Fukusaki, and A. Kobayashi. 2007. Time-course metabolic profiling in *Arabidopsis thaliana* cell cultures after salt stress treatment. *Journal of Experimental Botany* 58:415–424.

Kusano, T., K. Yamaguchi, T. Berberich, and Y. Takahashi. 2007. The polyamine spermine rescues *Arabidopsis* from salinity and drought stresses. *Plant Signaling and Behavior* 2:251–252.

LaRosa, P.C., Z. Chen, D.E. Nelson, N.K. Singh, P.M. Hasegawa, and R.A. Bressan. 1992. Osmotin gene expression is posttranscriptionally regulated. *Plant Physiology* 100:409–415.

Li, B., L. He, S. Guo, J. Li, Y. Yang, B. Yan, J. Sun, and J. Li. 2013. Proteomics reveal cucumber Spd-responses under normal condition and salt stress. *Plant Physiology and Biochemistry* 67:7–14.

Lu, Y., H. Lam, E. Pi, Q. Zhan, S. Tsai, C. Wang, Y. Kwan, and S. Ngai. 2013. Comparative metabolomics in *Glycine max* and *Glycine soja* under salt stress to reveal the phenotypes of their offspring. *Journal of Agricultural and Food Chemistry* 61:8711–8721.

Maeshima, M. 2001. Tonoplast transporters: Organization and function. *Annual Review of Plant Physiology and Plant Molecular Biology* 52:469–497.

Martin-Tanguy, J. 2001. Metabolism and function of polyamines in plants: Recent development (new approaches). *Plant Growth Regulation* 34:135–148.

Mendoza, I., F.J. Quintero, R.A. Bressan, P.M. Hasegawa, and J.M. Pardo. 1996. Activated calcineurin confers high tolerance to ion stress and alters the budding pattern and cell morphology of yeast cells. *Journal of Biological Chemistry* 271:23061–23067.

Morsomme, P. and M. Boutry. 2000. The plant plasma membrane $H(+)$-ATPase: Structure, function and regulation. *Biochimica et Biophysica Acta* 1465:1–16.

Munir, N. and F. Aftab. 2011. Enhancement of salt tolerance in sugarcane by ascorbic acid pretreatment. *African Journal of Biotechnology* 10:18362–18370.

Munns, R. and M. Tester. 2008. Mechanisms of salinity tolerance. *Annual Review of Plant Biology* 59:651–681.

Nishizawa, A., Y. Yabuta, and S. Shigeoka. 2008. Galactinol and raffinose constitute a novel function to protect plants from oxidative damage. *Plant Physiology* 147:1251–1263.

Pardo, J.M., M.P. Reddy, S. Yang, A. Maggio, G.-H. Huh, T. Matsumoto, M.A. Coca et al. 1998. Stress signaling through $Ca^{2+}$/calmodulin-dependent protein phosphatase calcineurin mediates salt adaptation in plants. *Proceedings of the National Academy of Sciences USA* 95:9681–9686.

Parida, A.K., A.B. Das, and P. Mohanty. 2004. Investigations on the antioxidative defence responses to NaCl stress in a mangrove, *Bruguiera parviflora*: Differential regulations of isoforms of some antioxidative enzymes. *Plant Growth Regulation* 42:213–226.

Peterbauer, T., J. Mucha, L. Mach, and A. Richter. 2002. Chain elongation of raffinose in pea seeds—Isolation, characterization, and molecular cloning of a multifunctional enzyme catalyzing the synthesis of stachyose and verbascose. *Journal of Biological Chemistry* 277:194–200.

Peters, S., S.G. Mundree, J.A. Thomson, J.M. Farrant, and F. Keller. 2007. Protection mechanisms in the resurrection plant *Xerophyta viscosa* (Baker): Both sucrose and raffinose family oligosaccharides (RFOs) accumulate in leaves in response to water deficit. *Journal of Experimental Botany* 58:1947–1956.

Pichersky, E. and D.R. Gang. 2000. Genetics and biochemistry of secondary metabolites in plants: An evolutionary perspective. *Trends in Plant Science* 10:439–445.

Qureshi, M.I., M. Israr, M.Z. Abdin, and M. Iqbal. 2005. Responses of *Artemisia annua* L. to lead and salt-induced oxidative stress. *Environmental and Experimental Botany* 53:185–193.

Rajendran, K., M. Tester, and S.J. Roy. 2009. Quantifying the three main components of salinity tolerance in cereals. *Plant, Cell & Environment* 32(3):237–249.

Rawia, E.A., L.S. Taha, and S.M.M. Ibrahiem. 2011. Alleviation of adverse effects of salinity on growth, and chemical constituents of marigold plants by using glutathione and ascorbate. *Journal of Applied Sciences Research* 7:714–721.

Reddy, A.S.N. 2001. Calcium: Silver bullet in signaling. *Plant Science* 160:381–404.

Reddy, V.S. and A.S. Reddy. 2004. Proteomics of calcium-signaling components in plants. *Phytochemistry* 65:1745–1776.

Rep M., H.L. Dekker, J.H. Vossen, A.D. De Boer, P.M. Houterman, C.G. De Koster, and B.J. Cornelissen. 2003. A tomato xylem sap protein represents a new family of small cysteine-rich proteins with structural similarity to lipid transfer proteins. *European Journal of Biochemistry* 534:82–86.

Roy, M. and R. Wu. 2002. Overexpression of S-adenosylmethionine decarboxylase gene in rice increases polyamine level and enhances sodium chloride-stress tolerance. *Plant Science* 163:987–992.

Roy, S.J., S. Negrao, and M. Tester. 2014. Salt resistant crop plants. *Current Opinion in Biotechnology* 26:115–124.

Sairam, R.K. and A. Tyagi. 2004. Physiology and molecular biology of salinity stress tolerance in plants. *Current Science* 86:407–421.

Sanchez, D.H., F.L. Pieckenstain, F. Escaray, A. Erban, U. Kraemer, M.K. UdvardI, and J. Kopka. 2011. Comparative ionomics and metabolomics in extremophile and glycophytic *Lotus* species under salt stress challenge the metabolic pre-adaptation hypothesis. *Plant, Cell & Environment* 34:605–617.

Sanders, D., J. Pelloux, C. Brownlee, and J.F. Harper. 2002. Calcium at the crossroads of signaling. *Plant Cell* 14:401–417.

Saxena, S.C., H. Kaur, and P. Verma. 2013. Osmoprotectants: Potential for crop improvement under adverse conditions. In Tuteja, N. and S. S. Gill (eds.), *Plant Acclimation to Environmental Stress*, pp. 197–232. Springer, New York.

Sharma, P., A.B. Jha, R.S. Dubey, and M. Pessarakli. 2012. Reactive oxygen species, oxidative damage, and antioxidative defense mechanism in plants under stressful conditions. *Journal of Botany* 2012:1–26.

Shulaev, V., D. Cortes, G. Miller, and R. Mittler. 2008. Metabolomics for plant stress response. *Physiologiae Plantarum* 132:199–208.

Singh, N.K., D.E. Nelson, D. Kuhn, P.M. Hasegawa, and R.A. Bressan. 1989. Molecular cloning of osmotin and regulation of its expression by ABA and adaptation to low water potential. *Plant Physiology* 90:1096–1101.

Smirnoff, N. and Q.J. Cumbes. 1989. Hydroxyl radical scavenging activity of compatible solutes. *Phytochemistry* 28:1057–1060.

Taji, T., C. Ohsumi, S. Iuchi, M. Seki, M. Kasuga, M. Kobayashi, K. Yamaguchi-Shinozaki, and K. Shinozaki. 2002. Important roles of drought- and cold-inducible genes for galactinol synthase in stress tolerance in *Arabidopsis thaliana*. *The Plant Journal* 29:417–426.

Tapernoux-Luthi, E.M., A. Bohm, and F. Keller. 2004. Cloning, functional expression, and characterization of the raffinose oligosaccharide chain elongation enzyme, galactan:galactan galactosyltransferase, from common bugle leaves. *Plant Physiology* 134:1377–1387.

Thompson, J.E., R.L. Ledge, and R.F. Barber. 1987. The role of free radicals in senescence and wounding. *New Phytologist* 105:317–344.

Toshihiko, A., S. Mikao, N. Ryouhei, Y. Tadakatsu, and Y. Shuichi. 2008. Nano scale proteomics revealed the presence of regulatory proteins including three FT-like proteins in phloem and xylem saps from rice. *Plant and Cell Physiology* 49:767–790.

Tugizimana, F., L.A. Piater, and I.A. Dubery. 2013. Plant metabolomics: A new frontier in phytochemical analysis. *South African Journal of Science* 109:1–11.

van der Fits, L. and J. Memelink. 2000. ORCA3, a jasmonate-responsive transcriptional regulator of plant primary and secondary metabolism. *Science* 289:295–297.

Villas-Bôas, S.G., S. Mas, M. Akesson, J. Smedsgaard, and J. Nielsen. 2005. Mass spectrometry in metabolome analysis. *Mass Spectrometry Reviews* 24:616–646.

Vitart, V., I. Baxter, P. Doerner, and J.F. Harper. 2001. Evidence for a role in growth and salt resistance of a plasma membrane $H^+$-ATPase in the root endodermis. *Plant Journal* 27:191–201.

Widodo, J.H.P., J.H. Patterson, E. Newbigin, M. Tester, A. Bacic, and U. Roessner. 2009. Metabolic responses to salt stress of barley (*Hordeum vulgare* L.) cultivars, Sahara and Clipper, which differ in salinity tolerance. *Journal of Experimental Botany* 60:4089–4103.

Wise, R.R. and A.W. Naylor. 1987. Chilling-enhanced photo-oxidation: Evidence for the role of singlet oxygen and endogenous antioxidants. *Plant Physiology* 83:278–282.

Wu, D., S. Cai, M. Chen, L. Ye, Z. Chen, H. Zhang, F. Dai, F. Wu, and G. Zhang. 2013. Tissue metabolic responses to salt stress in wild and cultivated barley. *PLoS One* 8:e55431.

Xiong, L., H. Lee, M. Ishitani, and J.K. Zhu. 2002. Regulation of osmotic stress-responsive gene expression by the *LOS6/ABA1* locus in *Arabidopsis*. *Journal of Biological Chemistry* 277:8588–8596.

Yamaguchi-Shinozaki, K. and Shinozaki, K. 2006. Transcriptional regulatory networks in cellular responses and tolerance to dehydration and cold stresses. *Annual Review of Plant Biology* 57:781–803.

Yin, Y.G., T. Tominaga, Y. Iijima, K. Aoki, D. Shibata, H. Ashihara, S. Nishimura, H. Ezura, and C. Matsukura. 2010. Metabolic alterations in organic acids and gamma-aminobutyric acid in developing tomato (*Solanum lycopersicum* L.) fruits. *Plant Cell Physiology* 51:1300–1314.

Zhang, J., Y. Zhang, Y. Du, S. Chen, and H. Tang. 2011. Dynamic metabonomic responses of tobacco (*Nicotiana tabacum*) plants to salt stress. *Journal of Proteome Research* 10:1904–1914.

Zorb, C., S. Schmitt, A. Neeb, S. Karl, M. Linder, and S. Schubert. 2004. The biochemical reaction of maize (*Zea mays* L.) to salt stress is characterized by a mitigation of symptoms and not by a specific adaptation. *Plant Science* 167:91–100.

Zuther, E., K. Koehl, and J. Kopka. 2007. Comparative metabolome analysis of the salt response in breeding cultivars of rice. In *Advances in Molecular Breeding toward Drought and Salt Tolerance Crops*. Jenks, M.A., P.M. Hasegawa, and S.M. Jain (eds.), pp. 285–315. Springer-Verlag, Berlin, Germany.

# CHAPTER FIFTEEN

# Plant–Microbe Interaction and Salt Stress Tolerance in Plants

*Neveen B. Talaat and Bahaa T. Shawky*

## CONTENTS

*Abstract.* Excessive salt accumulation in soils is a major ecological and agronomical problem, in particular in arid and semiarid areas. While important physiological insights about the mechanisms of salt tolerance in plants have been gained, the transfer of such knowledge into crop improvement has been limited. The identification and exploitation of soil microorganisms (especially rhizosphere bacteria and mycorrhizal fungi) that interact with plants by alleviating stress opens new alternatives for a pyramiding strategy against salinity as well as new approaches to discover new mechanisms involved in stress tolerance. Considering the kingdom of fungi, arbuscular mycorrhizal fungi (AMF) stand out as the most significant and widespread group of plant growth–promoting microorganisms. Ectomycorrhizal fungi (EMF) are also important symbionts of particular relevance for many woody plants. Considering the kingdom of bacteria, a wide range of microorganisms including different species and strains of *Bacillus*, *Burkholderia*, *Pseudomonas*, and the well-known nitrogen-fixing organisms

*Rhizobium, Bradyrhizobium, Azotobacter, Azospirillum,* and *Herbaspirillum* are classically regarded as important plant growth–promoting rhizobacteria (PGPR). Today, it is widely accepted that AMF, EMF, and PGPR promote plant growth and increase tolerance against stress conditions, at least in part, because they facilitate water and nutrient uptake and distribution as well as alter plant hormonal status, and this ability has been attributed to various mechanisms. This chapter addresses the significance of soil biota in alleviation of salinity stress and their beneficial effects on plant growth and productivity. Moreover, it emphasizes new perspectives and challenges in physiological and molecular studies on salt stress alleviation by soil biota.

## 15.1 INTRODUCTION

Salinity is a major factor reducing crop productivity and a major cause of the abandonment of lands and aquifers for agricultural purposes. Saline soils occupy 7% of the earth's land surface, and increased salinization of arable land will result into 50% land loss by the middle of the twenty-first century (Porcel et al. 2012). At present, out of 1.5 billion ha of cultivated land around the world, about 77 million ha (5%) is affected by excess salt content (Sheng et al. 2008). The direct effects of salt on plant growth may involve (a) reduction in the osmotic potential of the soil solution that reduces the amount of water available to the plant causing physiological drought (to counteract this problem, plants must maintain lower internal osmotic potentials in order to prevent water movement from roots into the plant soil) (Jahromi et al. 2008), (b) toxicity of excessive $Na^+$ and $Cl^-$ ions toward the cell (the toxic effects include disruption to the structure of enzymes and other macromolecules, damage to cell organelles and plasma membrane, disruption of photosynthesis, respiration, and protein synthesis) (Sheng et al. 2008; Hajiboland et al. 2010; Talaat and Shawky 2014a), and (c) nutrient imbalance in the plant cells leading to ion deficiencies (Talaat and Shawky 2011, 2013).

Developing salt-tolerant crops has been a much desired scientific goal, but with little success to date, a few major-determinant genetic traits of salt tolerance have been identified (Schubert et al. 2009). An alternative strategy to improve crop salt tolerance may be to introduce salt-tolerant microbes that enhance crop growth. Indeed, several recent studies have demonstrated that local adaptation of plants to their environment is driven by genetic differentiation in closely associated microbes (Rodriguez and Redman 2008). Attention is focused here on symbiotic relationships such as arbuscular mycorrhizal fungi (AMF), whose hyphal networks ramify throughout the soil and within the plant cells; ectomycorrhizal fungi (EMF), which form a fungal layer around the root system and root intercellular spaces; and root-associated plant growth–promoting rhizobacteria (PGPR). They promote salinity tolerance by employing various mechanisms (Figure 15.1), such as inducing the synthesis and accumulation of compatible solutes to avoid cell dehydration and maintain root water uptake (Nautiyal et al. 2013; Talaat and Shawky 2014a); regulating the ion homeostasis to control ion uptake by roots, compartmentation, and transport into shoots (Talaat and Shawky 2013; Kang et al. 2014); regulating the water uptake and distribution to plant tissues by the action of aquaporins (Marulanda et al. 2010); reducing the oxidative damage through improved antioxidant capacity (Gururani et al. 2013; Talaat and Shawky 2014b); maintaining photosynthesis and protein synthesis at values adequate for plant growth (Golpayegani and Tilebeni 2011; Talaat and Shawky 2014a); altering plant hormonal status; improving rhizospheric and soil conditions (Shawky 1990; Nautiyal et al. 2013); and inducing molecular changes (the expression of genes $Na^+/H^+$ antiporters, PIP, ABA [*Lsnced*], late embryogenesis abundant protein [*Lslea*], and $\Delta$1-pyrroline-5-carboxylate synthetase [*LsP5CS*]) (Ouziad et al. 2006; Jahromi et al. 2008). These mechanisms do not work in isolation but rather in an integrated manner to finally affect the major physiological processes affecting plant growth and productivity under salinity stress. Finally, the commercial and agronomic significance of soil biota that can enhance crop salinity tolerance is briefly considered.

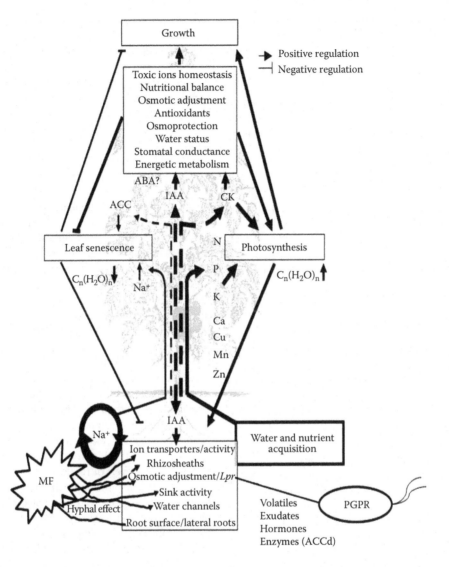

**Figure 15.1.** Salt stress alleviation by mycorrhizal fungi and plant growth-promoting rhizobacteria. Through a hyphal effect or exudation of specific compounds, soil biota alter root physiology by acting on different biochemical and molecular mechanisms that essentially improve water and nutrient acquisition and exclude toxic ions from the rhizosphere. Altered root physiology modifies root–shoot communication and contributes to water, ionic, and hormonal homeostasis in the shoot. This alleviates the salt-induced growth impairment during the osmotic phase of salinity and delays the toxic effect on leaf senescence. In turn, this helps to maintain rhizosphere microbial colonization by maintaining assimilate flow to the roots. Microbes modify the expression and activity of ion and water transporters and increase the effective root surface area *via* fungal hyphae or enhancement of lateral root formation by altering auxin (indole-3-acetic acid) transport and signaling. Some exudates (polysaccharides) sequester toxic ions ($Na^+$), while other metabolites (amino acids, trehalose) and hormones (cytokinins and auxins) can be transported to the shoot to protect growth and photosynthetic machinery, although direct effects on plant metabolism cannot be ruled out. Line thickness indicates the magnitude of the microbial effects, while an arrow or a perpendicular line indicates positive or negative regulation, respectively. (Adapted from Dodd, I.C. and Perez-Alfocea, F., *J. Exp. Bot.*, 1, 2012.)

## 15.2 STRATEGIES USED BY RHIZOSPHERE MICROBES TO CONFER PLANT SALT TOLERANCE

### 15.2.1 Improving Plant Growth and Productivity

Excessive soil salinity affects the establishment, development, and growth of plants, resulting in important losses in productivity (Evelin et al. 2009). However, mycorrhization was found to increase the fitness of the host plant to saline conditions. Improved tolerance of *Acacia nilotica* to salt stress by arbuscular mycorrhiza (AM), *Glomus fasciculatum*, has been shown by Giri et al. (2007), who reported that mycorrhizal seedlings had higher root and shoot dry weight than the nonmycorrhizal ones under salt excess. Improved growth, yield, and quality of fruits of salt-stressed *Cucurbita pepo* plants colonized by *Glomus intraradices* (Colla et al. 2008). Higher plant height, shoot and root dry weight, number of leaves, tillers, and spikes, as well as number and weight of grains were detected in salt-stressed mycorrhizal wheat plants (Talaat and Shawky 2012). Positive influence of AMF on the shoot and root dry weight and leaf area of pepper plants was recorded under saline conditions (Beltrano et al. 2013). Similarly, protective effects of AMF on *Triticum aestivum* plants exposed to salinity were reported by Talaat and Shawky (2014a,b), who illustrated that the beneficial effect of AM symbiosis on plant productivity has been largely attributed to improve water content, enhance plant nutrient acquisition, enhance photosynthetic efficiency and protein synthesis, induce a better osmotic adjustment *via* compatible solutes accumulation, and prevent the oxidative stress by enhancing enzymatic and nonenzymatic antioxidant defense system.

Furthermore, suitable EMF have a high potential to increase plant biomass production on salt-affected soils. Langenfeld-Heyser et al. (2007) reported that EM-forming fungi such as the basidiomycete *Paxillus involutus* have been identified as highly salt tolerant and can attenuate detrimental salt effects in their host plants. Luo et al. (2009) investigated metabolic and transcriptional profiles in EM and non-EM roots of gray poplar (*Populus* × *canescens*) in the presence and absence of osmotic stress imposed by excess salinity. Colonization with the EMF *P. involutus* increased root cell volumes, a response associated with carbohydrate accumulation. The stress-related hormones abscisic acid and salicylic acid were increased, whereas jasmonic acid and auxin were decreased in EM compared with non-EM roots. Auxin-responsive reporter plants showed that auxin decreased in the vascular system.

PGPR have been known to play an essential role in the growth and metabolism of plants (Gururani et al. 2013), and they colonize the rhizosphere and enhance plant growth by increasing plant nutrient uptake *via* multiple mechanisms including biological nitrogen fixation, siderophore production, and phosphate solubilization (Lugtenberg and Kamilova 2009). In addition to causing increases in general plant growth, some PGPR promote root development (Mantelin and Touraine 2004) and alter root architecture by the production of phytohormones such as indole acetic acid (Kloepper et al. 2007), resulting in increased root surface area and numbers of root tips. Such stimulation of roots can aid plant defense against stress conditions. Given that root tips and root surfaces are sites of nutrient uptake, it is likely that one mechanism by which PGPR lead to increased nutrient uptake is *via* stimulation of root development. It has also been suggested that PGPR increase plant uptake of mineral ions *via* stimulation of the proton pump ATPase (Mantelin and Touraine 2004). Barley seedlings inoculated with *Azospirillum brasilense* strain Az39 and subjected to 200 mM NaCl for 18 days were studied by Zawoznik et al. (2011), who reported that the growth rate of uninoculated plants was adversely affected by saline treatment, which was associated with higher putrescine content and lower levels of HvPIP2;1 transcripts in the roots. *Azospirillum* inoculation triggered the transcription of this gene, and inoculated plants may be better prepared to thrive under saline conditions. Moreover, Rojas-Tapias et al. (2012) mentioned that inoculation with *Azotobacter chroococcum* strains C5 and C9 promoted maize growth and alleviated the saline stress, likely through the integration of several mechanisms that improve the plant response; they fixed nitrogen and solubilized phosphate, improved Na$^+$ exclusion and K$^+$ uptake, increased content of polyphenol and chlorophyll, and decreased accumulation of proline.

Additionally, many PGPR can increase plant salt tolerance and can improve plant growth *via* altering plant hormone status by producing auxins and cytokinins or by decreasing plant ethylene levels *via* the bacterial enzyme 1-aminocyclopropane-1-carboxylate (ACC) deaminase (ACCd) that hydrolyzes the immediate ethylene precursor ACC into ammonia and α-ketobutyrate. Various ACCd-containing rhizobacteria were repeatedly shown to promote plant growth, particularly when plants were subjected to environmental stresses likely to stimulate stress-induced ethylene production (Dodd and Perez-Alfocea 2012). In this respect, Cheng et al. (2012) reported that the shoot fresh and dry weights of canola plant treated with salt plus *Pseudomonas putida* UW4 were 1.7-fold and nearly 2-fold, respectively, greater than those of plants treated with salt only. By contrast, plants treated with salt plus the ACCd minus mutant did not show significant differences from the ones treated with salt only, indicating that plant growth enhancement by this bacterium was dependent on having a functional ACCd. Similarly, Ali et al. (2014) studied the ability of ACCd-containing plant growth-promoting bacteria (PGPB) endophytes *Pseudomonas fluorescens* YsS6, *Pseudomonas migulae* 8R6, and their ACCd-deficient mutants on tomato growth in the absence of salt and under two different levels of salt stress (165 and 185 mM), and they showed that plants pretreated with wild-type ACCd-containing endophytic strains were healthier and grew to a much larger size under high salinity stress compared to plants pretreated with the ACCd-deficient mutants or no bacterial treatment. The plants pretreated with ACCd-containing bacterial endophytes exhibit higher fresh and dry biomass, higher chlorophyll contents, and a greater number of flowers and buds than the other treatments. Since the only difference between wild-type and mutant bacterial endophytes was ACCd activity, they concluded that this enzyme is directly responsible for the different behaviors of tomato plants in response to salt stress.

The effect of *Bacillus amyloliquefaciens* NBRISN13 (SN13) inoculation on the growth of the rice plant and expression analysis of related genes under salt stress (200 mM NaCl) was evaluated by Nautiyal et al. (2013). They suggested the significance of SN13 treatment in ameliorating the salt stress possibly through increased colonization and ACCd activity of SN13, increased chlorophyll content, enhanced proline accumulation, and enrichment of osmoprotectant utilizing microbes in the rice rhizosphere along with upregulation or repression of set of stress-responsive genes in leaf blade. Among 14 genes studied by them, 5 (*BADH*, betaine aldehyde dehydrogenase; *EREBP1*, ethylene-responsive element-binding protein; *NADP-ME2*, NADP-malic enzyme; *SERK1*, somatic

embryogenesis receptor-like kinase; and *SOS1*, salt-overly-sensitive1) were upregulated, and 2 (*GIG*, gigantea; *SAPK4*, serine/threonine protein kinase) repressed under salt stress in hydroponic condition. In greenhouse experiment, salt stress resulted in the accumulation of (*MAPK5*, mitogen-activated protein kinase 5) and downregulation of the remaining 13 transcripts. Hence, SN13 was conferred salt tolerance in rice by modulating differential transcription in a set of at least 14 genes.

### 15.2.2 Regulating of Water Uptake and Distribution to Plant Tissues by the Action of Aquaporins

Lowering water potential in salty soils obliges the plants to face the problem of acquiring enough water from the soil (Ouziad et al. 2006); moreover, plant root water uptake capacity (i.e., root hydraulic conductivity) was inhibited, which could be due to changes either in the aquaporin function or in the amount of this protein present in the membrane (Sade et al. 2010). Aquaporins are a group of water-channel proteins that facilitate and regulate the passive movement of water molecules following a water potential gradient through the plasma membrane intrinsic proteins (by PIPs) and the tonoplast intrinsic proteins (by TIPs) (Kruse et al. 2006).

AM symbiosis regulates root hydraulic properties, including root hydraulic conductivity, and these effects have been linked to the regulation of plant aquaporins (Ruiz-Lozano and Aroca 2010). Ouziad et al. (2006) found reduced transcript levels of both a tonoplast and a plasmalemma aquaporin gene in the roots of *Lycopersicon esculentum* colonized by a mixture of *Glomus geosporum* and *G. intraradices*, with this reduction being even greater in plants living under sustained salt stress (0.8% NaCl). Interestingly, AMF colonization resulted in drastic increases of transcript levels of the three aquaporin genes analyzed (*LePIP1*, *LePIP2*, and *LeTIP*) in tomato leaves upon salt stress, while genes encoding two Na$^+$/H$^+$ transporters were unaffected. In the same line, Aroca et al. (2007) reported that mycorrhizal (*G. intraradices* BEG 123) *Phaseolus vulgaris* plants had a greater osmotic root hydraulic conductance under saline (3.1 dS m$^{-1}$) stress than uninoculated plants. Although regulation of root hydraulic properties by AM symbiosis was strongly correlated with the regulation of aquaporin (*PvPIP2*) protein abundance and phosphorylation state, different *PIP* genes were differentially regulated under salinity stress.

Moreover, Jahromi et al. (2008) studied two PIP genes in *Lactuca sativa* plants subjected to increasing salinity levels. The expression of *LsPIP1* and *LsPIP2* genes was downregulated by mycorrhization in the absence of salinity. In contrast, under saline conditions, in mycorrhizal plants, the expression level of the *LsPIP2* gene was almost unaffected, while the expression of the *LsPIP1* gene was upregulated, mainly at 100 mM NaCl. In agreement with such uneven responses, earlier findings obtained by quantitative real-time reverse transcription-polymerase chain reaction (RT-PCR) analysis in *Arabidopsis thaliana* demonstrated that only one of a set of 13 PIP genes was upregulated by cold treatment (most of them were downregulated by this stress), while marked up- or downregulation of PIPs was observed after drought stress and a less severe modulation was detected under high salinity (Jang et al. 2004). It should also be mentioned that overexpression of a plasma membrane aquaporin in transgenic tobacco improved plant vigor under favorable conditions, but not under drought or salt stress (Aharon et al. 2003).

Despite apparent beneficial effects of ectomycorrhizal associations on plant water uptake, the mechanisms involved in regulation of root water flow of EM plants remain unclear. Several studies have demonstrated that EM associations may increase whole root hydraulic conductivity by as much as severalfold (Lee et al. 2010). However, no change (Yi et al. 2008) or even a decrease (Calvo-Polanco et al. 2008) in root hydraulic conductivity by EMF has also been reported. Marjanović et al. (2005) showed increased root transcript levels of three poplar putative water channels belonging to the PIP2 group in the hybrid poplar *Populus tremula* × *tremuloides* inoculated with the EMF *Amanita muscaria* and found about 57% larger water transport capacity in mycorrhized plants than in nonmycorrhized plants. Interestingly, these authors analyzed separately main roots and fine roots and observed different patterns of aquaporin expression between these categories, with distinct responses to mycorrhizal inoculation. Moreover, the effect of EMF *Suillus tomentosus* on water transport properties was studied in jack pine (*Pinus banksiana*) seedlings by Lee et al. (2010), who found that the hydraulic conductivity of root cortical cells and of the whole root system in EM plants was higher by four-fold compared with the non-EM ones. They also showed that hydraulic conductivity of root cortical cells was significantly reduced by 50 mM NaCl,

and, whereas this decline was followed by a quick recovery to the pretreatment level in mycorrhized seedlings, reduction progressed over time in non-mycorrhized seedlings.

Mycorrhizas could affect root water flow through their effects on apoplastic and transcellular pathways. Both EM and AM mycorrhizas produce hyphae that could potentially decrease resistance of roots to water flow. However, the contribution of internal hyphae to apoplastic pathway in mycorrhizal plants remains unclear. The increases in root hydraulic conductivity by mycorrhizas in soybean (Safir et al. 1972) and American elm (Muhsin and Zwiazek 2002) were attributed to apoplastic pathway. On the other hand, hyphal penetration of the root cortex cells in *Populus balsamifera* was shown not to be directly involved in root hydraulic responses (Siemens and Zwiazek 2008). However, there is growing evidence that both AM and EM mycorrhizas may affect the expression of AQPs (aquaporins) (Marjanović et al. 2005). Regulation of AQP activity in mycorrhizal plants *via* methylation, phosphorylation, or changes in cytoplasmic pH has also been suggested (Aroca et al. 2008).

Contribution of rhizospheric bacteria to plant tolerance against salinity has been largely documented, and the mechanisms that may account for these effects were described by Egamberdieva and Islam (2008), Lugtenberg and Kamilova (2009), and Dodd and Perez-Alfocea (2012). However, few studies described the direct relationships between colonization of roots by PGPR and plant water channels. Marulanda et al. (2010) demonstrated that *Zea mays* plants inoculated with the PGPR *Bacillus megaterium* and exposed to salinity (2.59 dS m$^{-1}$) had increased root hydraulic conductance compared with uninoculated plants, which was correlated with increased root expression of two (of the six studied) ZmPIP isoforms. Moreover, plant performance and aquaporin expression of HvPIP2,1 gene in the roots of *Azospirillum*-inoculated barley seedlings growing under saline (200 mM NaCl) and nonsaline conditions were assessed by Zawoznik et al. (2011), and they showed that the growth rate of uninoculated plants was more adversely affected by saline treatment than that of inoculated seedlings; this was associated to higher putrescine contents and lower levels of HvPIP2;1 transcripts in the roots. Of note is that *Azospirillum* inoculation itself triggered the transcription of this PIP aquaporin.

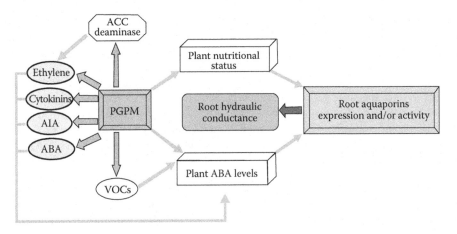

Figure 15.2. Plant growth–promoting microorganisms may affect root hydraulic conductance through a complex network of biochemical interactions. Microbial biosynthesis of plant hormones and microbial-mediated regulation of endogenous plant hormone levels, particularly of abscisic acid, which in turn is involved in aquaporin gene expression, seem to be of capital relevance in these processes. Additionally, changes in the nutritional status of colonized plants and volatile organic compounds released by certain microorganisms may directly or indirectly impact on root aquaporins expression and/or activity and, therefore, on root hydraulic conductance. (Adapted from Groppa, M.D. et al., *Appl. Soil Ecol.*, 61, 247, 2012.)

In fact, little is known of the mechanisms by which soil microbes regulate root expression of specific aquaporin isoforms. Porcel et al. (2012) postulated that ABA could be the signal that differentially regulates the behavior of root hydraulic conductance and aquaporins in AM- and non-AM-inoculated plants under saline conditions. ABA has been shown to regulate a number of genes in plants including AQPs in vesicular-AM associations (Aroca et al. 2008). Furthermore, changes in the nutritional status of colonized plants and volatile organic compounds (VOCs) released by certain microorganisms may directly or indirectly impact on root aquaporins expression and/or activity and, therefore, on root hydraulic conductance (Hamdia et al. 2004; Groppa et al. 2012, Figure 15.2).

### 15.2.3 Inducing the Synthesis and Accumulation of Compatible Solutes

As salts accumulate in soil, the soil water potential becomes more negative, and plants must decrease their water potential to maintain a favorable gradient for water flow from the soil into the roots. The most important mechanism used to achieve such an effect is decreasing the plant osmotic potential by active accumulation of organic ions or solutes, known as osmotic adjustment or osmoregulation (Hoekstra et al. 2001). The inorganic ions that participate in osmotic adjustment are mainly $K^+$

and $Cl^-$, while the most important organic solutes include soluble sugars, proline, glycinebetaine, and polyamines (Flowers and Colmer 2008). Osmoregulation allows cells to maintain turgor and turgor-dependent processes, such as cellular expansion and growth, stomatal opening, and photosynthesis. Osmoprotectors can accumulate to a high level without disturbing intracellular biochemistry, protecting subcellular structures, mitigating oxidative damage caused by free radicals, and maintaining the enzyme activities under saline conditions. It also serves to keep a favorable gradient of water potential, which permits water entrance into the plant (Evelin et al. 2009).

#### 15.2.3.1 Soluble Sugars

Sugars during salt stress can contribute up to 50% of the total osmotic potential and play a leading role in osmoprotection, C storage, and radical scavenging (Cram 1976). Sink organs often accumulate carbohydrates following salinization, as their utilization in growth is constrained. Thus, carbohydrate use, rather than availability, seems to be the first limiting step for salt tolerance (Perez-Alfocea et al. 2010). Several reports have shown a correlation between tolerance to salinity stress and sugar concentration in mycorrhized plants. Sheng et al. (2011) and Talaat and Shawky (2011, 2014a) found sugar accumulation in salt-stressed plants colonized by AMF, and they proposed that the high levels of sugars in AMF-inoculated plants might be the result of an increase in the photosynthetic capacity of the plants and that these sugars contributed to the osmotic adjustment of these plants. However, Beltrano et al. (2013) reported that the total sugar leaf content was higher in mycorrhizal plants, regardless of the stressful situation, and decreased with severe salinity stress. In roots, the total soluble sugar content decreased under high salinity stress conditions (100 and 200 mM NaCl). On the other hand, Sharifi et al. (2007) observed no role of soluble sugars in the responses to salinity of *Glomus etunicatum*–colonized soybean plants. It is not clear whether or not the accumulation of sugars in the roots of AM plants is due to the sink effect of the AMF demanding sugars from shoots (Auge 2000).

### 15.2.3.2 Proline

Proline protects plants against salinity by serving as a nontoxic and protective osmolyte, a sink for energy to regulate redox potentials, a hydroxyl radical scavenger, a solute that protects macromolecules against denaturation, and a means of reducing acidity in the cell (Kavi Kishor et al. 2005). Several studies have reported a higher proline concentration in AM-colonized plants than in non-AM ones at different salinity levels. Sharifi et al. (2007) showed that inoculation of soybean (*Glycine max*) with *G. etunicatum* increased the root and shoot growth of plants grown with 50 and 100 mM NaCl, which was correlated with increased root but decreased shoot proline concentrations. Kumar et al. (2010) reported that better growth of AM-inoculated *Jatropha curcas* compared with noninoculated plants when exposed to salinity (1.7–8.5 dS m$^{-1}$ NaCl for 60 days) may result from increased soluble sugars and proline in the leaves of inoculated plants, allowing maintenance of leaf water status. Similarly, increased proline concentration was positively correlated with mycorrhization of host *T. aestivum* salt-stressed plants (Talaat and Shawky 2011, 2014a). Beltrano et al. (2013) found that the leaves of *Capsicum annuum* mycorrhizal plants accumulated more proline than nonmycorrhizal ones at all salinity levels, while the roots of mycorrhizal plants accumulated more proline, only at higher levels of stress (100 and 200 mM NaCl). They concluded that the proline content increases with increasing salt stress and was significantly higher in leaves than

in roots. Likewise, foliar proline concentrations of mycorrhizal maize were decreased up to fivefold across a range of soil salinities (0.8–2.6 dS m$^{-1}$ NaCl for 55 days; Sheng et al. 2011), while root proline concentrations of mycorrhizal lettuce (*L. sativa*) were decreased in plants grown at 50 mM NaCl (Jahromi et al. 2008). However, mycorrhization did not affect either root or shoot proline concentrations following exposure of *Lotus glaber* to 200 mM NaCl (Sannazzaro et al. 2007).

Enhanced proline synthesis in abiotically stressed plants has been reported in the presence of beneficial bacteria such as *Arthrobacter* and *Bacillus* (Sziderics et al. 2007; Gururani et al. 2013). Whether the increased proline is synthesized *de novo* in the plant or is absorbed from the root zone is not clear, but the introduction of a *proBA* gene derived from *Bacillus subtilis* into *A. thaliana* increased the production of free proline, which was associated with increased salt tolerance in the transgenic plants (Chen et al. 2007). In accordance to Nautiyal et al. (2013) data, higher proline content in plant tissues was further increased by *B. amyloliquefaciens* NBRISN13 inoculation under salinity stress, which concurrently enhances osmoadaptation and facilitates its stress-ameliorating property. However, Hamdia et al. (2004) reported that when salt-stressed maize was inoculated with *Azospirillum*, proline concentration declined significantly. Similarly, Rojas-Tapias et al. (2012) found that inoculation with *A. chroococcum* strains C5 and C9 decreased accumulation of proline in maize plants under saline conditions.

### 15.2.3.3 Glycinebetaine

Glycinebetaine helps in maintaining the osmotic status of the cell and protects against detrimental effects of salinity by acting as an osmoprotectant and plays a role in stabilizing the structures of complex proteins, protecting the transcriptional machinery or maintaining the integrity of membranes (Yang et al. 2008). Accumulation of betaines under salt stress was found to increase when the plant was colonized by AMF (Talaat and Shawky 2014a).

Glycinebetaine is synthesized by only a few PGPR and is actively transported and accumulated as an osmoprotectant. Zhang et al. (2010) reported that the beneficial soil bacterium *B. subtilis* (strain GB03) enhances *Arabidopsis* glycinebetaine synthesis associated with enhanced plant tolerance to osmotic stress.

### 15.2.3.4 *Polyamines*

Polyamines, namely, diamine putrescine (Put), triamine spermidine (Spd), and tetraamine spermine (Spm), are low-molecular-weight, aliphatic polycations found in the cells of all living organisms. Plant polyamines (PAs) can play a role in stress reactions and resistance by its ability to (a) associate with anionic components of the membrane such as phospholipids thereby stabilizing the bilayer surface and retarding membrane deterioration; (b) stabilize the conformation of nucleic acids, resulting in improved translation and protein synthesis; (c) prevent chlorophyll loss and the inhibition of photochemical reactions of photosynthesis; (d) maintain cellular pH and ion homeostasis; (e) improve water and nutrient uptake like phosphorus, nitrogen, and micronutrients; (f) scavenge free radicals; and (g) modify expression of some stress-related genes (Pang et al. 2007; Smith and Read 2008) and thus improve crop productivity. Evidence that soil microbes alter polyamines pool and improve plant growth in salinized crops is less considered. Sannazzaro et al. (2007) showed higher content of total free polyamines in *L. glaber* plants colonized by *G. intraradices* compared to non-AM ones. They suggested that since PAs have been proposed as candidates for the regulation of root development under saline situations, it is possible that AM-colonized plants (which contained higher PA levels and showed improved root growth) were better shaped to cope with salt stress. Moreover, Talaat and Shawky (2013) found that mycorrhization changed the polyamine balance under saline conditions and an increase in putrescine level associated with low contents of spermidine and spermine in salt-sensitive wheat plants was observed, while salt-stressed plants showed a decrease in putrescine and high increase in spermidine and spermine. Moreover, mycorrhizal inoculation significantly reduced the activities of diamine oxidase and polyamine oxidase in salt-stressed wheat plants. It may be inferred that modulation of polyamine pool can be one of the mechanisms used by AMF to improve plant adaptation to saline soils.

### 15.2.4 Regulating the Ionic Homeostasis

Increases in ambient salt concentrations in the rooting medium led to toxic accumulation of $Na^+$ ion in the cytosol, which negatively affects the acquisition and homeostasis of other nutrients and alters ionic relationships (Evelin et al. 2009; Talaat and Shawky 2011, 2013). Microbes can alter root uptake of toxic ions and nutrients by altering host physiology (by regulating ion transporter expression and/or activity) and modifying physical barriers around the roots (more extensive rhizosheaths formed by bacterial exopolysaccharides) or by directly reducing foliar accumulation of toxic ions ($Na^+$, $Cl^-$), while improving the nutritional status of both macro- (N, P, K) and micronutrients (Zn, Fe, Cu, Mn), mostly via unknown mechanisms. Nutrients may also become more accessible to the plant due to microbial-induced changes in rhizosphere pH (organic acid excretion) and/or chelation with organic molecules (siderophores) exuded by microbes (Giri et al. 2007; Sharifi et al. 2007; Dodd and Perez-Alfocea 2012; Talaat and Shawky 2013).

AMF have a positive effect on the composition of mineral nutrients of plants grown under saline conditions by enhancing and/or selective uptake of nutrients (Evelin et al. 2009). Mycorrhizal colonization of host plants prevents $Na^+$ translocation to shoot tissues, while $K^+$ absorption is enhanced under saline conditions; prevention of $Na^+$ accumulation and enhancement of $K^+$ concentration could be a part of the general mechanism of salt stress alleviation in plants by AMF (Sharifi et al. 2007; Talaat and Shawky 2011; Beltrano et al. 2013). Hammer et al. (2011) indicated that AMF *G. intraradices* can selectively take up elements such as $K^+$, $Mg^{2+}$, and $Ca^{2+}$, which act as osmotic equivalents while they avoid the uptake of toxic $Na^+$. A possible buffering effect of AMF on the uptake of $Na^+$ when the content of $Na^+$ is within acceptable limits, which could make it important in the alleviation of salinity stress in the host plants. Moreover, Talaat and Shawky (2013) showed that AMF significantly enhanced wheat productivity under salinity stress by diminishing $Na^+$ uptake and increasing N, P, $K^+$, Fe, Zn, and Cu acquisition. Improved plant nutrition is a quite general beneficial effect and may contribute to the maintenance of homeostasis of toxic ions under saline stress. Additionally, AM-colonized plants grown in salty soils maintained higher $K^+/Na^+$, $Ca^{2+}/Na^+$, $Mg^{2+}/Na^+$, and $Ca^{2+}/Mg^{2+}$ ratios in their tissues than their non-AM counterparts (Giri et al. 2007; Hammer et al. 2011).

Although overexpression of $Na^+/H^+$ and $K^+/H^+$ antiporters improves salt tolerance in plants

(Rodriguez-Rosales et al. 2008), expression analysis of *LeNHX1* (Na$^+$/H$^+$) and *LeNHX2* (K$^+$/H$^+$) antiporter genes under salt (137 mM NaCl) and mycorrhizal colonization with *G. intraradices* and *G. geosporum* in tomato did not reveal changes in gene expression (Ouziad et al. 2006). On the other hand, Porcel et al. (2012) proposed cyclic nucleotide-gated ion channels (CNGCs) as candidate genes for studies of salt stress amelioration in AM-inoculated plants and suggested that CNGCs contribute to sodium reallocation within the plant tissues, assisting the plant in coping with salinity stress.

EMF can also increase plant tolerance to saline conditions by altering plant nutrient acquisition. Calvo-Polanco et al. (2008) found that when American elm (*Ulmus americana*) seedlings were inoculated with *Hebeloma crustuliniforme*, *Laccaria bicolor*, and a mixture of the two fungi, they had severalfold lower leaf concentrations of Na$^+$ compared with the noninoculated plants. All three fungal inocula had similar effects on the responses of elm seedlings to salt treatment.

In the case of PGPR, decreased plant Na$^+$ accumulation could be explained by the excretion of bacterial exopolysaccharides, which bind cations (especially Na$^+$) in roots, thus preventing their transfer to leaves and helping alleviate salt stress in wheat plants (Ashraf et al. 2004). Ryu et al. (2004) reported that among the 600 Arabidopsis genes isolated by transcriptome analysis, transcriptional expression of *High-Affinity K$^+$ Transporter 1* (HKT1), which controls Na$^+$ import in roots, was decreased. HKT1 has been shown to adjust Na$^+$ and K$^+$ levels differentially, depending on the plant tissue. Transcriptional validation revealed that specific bacterial strains of PGPR (bacterial volatile organic compounds; VOCs) downregulated HKT1 expression in roots, but upregulated it in shoot tissues, thereby orchestrating lower Na$^+$ levels in the whole plant. They also suggested that HKT1 functions in shoots to retrieve Na$^+$ from the xylem, thereby facilitating shoot-to-root Na$^+$ recirculation. Overall, plant perception of this bacterial causes a tissue-specific regulation of HKT1 that controls Na$^+$ homeostasis under salt stress. Furthermore, Mayak et al. (2004) found that *Achromobacter piechaudii* increased foliar K and P concentrations by 24% and 62%, respectively, while decreasing Ca and Mg concentrations by 21% and 14%, respectively (averaged over two salt concentrations). They also indicated that although

rhizobacterial inoculation decreased foliar Na concentration (by 24%) when tomato plants were grown with 120 mM NaCl, there was no significant effect at a higher salt concentration (207 mM). It has also been suggested that PGPR increases plant uptake of mineral ions *via* stimulation of the proton pump ATPase (Mantelin and Touraine 2004). Moreover, *Azospirillum* inoculation restricted Na$^+$ uptake and enhanced the uptake of K$^+$ and Ca$^{2+}$ in maize salt-sensitive cultivar (Hamdia et al. 2004). The two PGPR strains, *Pseudomonades* sp. and *Bacillus lentus*, increased P, K, and Ca uptake by basil plant under soil salinity stress compared to the nonsalinity stress control, which means that a PGPR treatment under salinity stress conditions could alleviate the inhibition of plant growth (Golpayegani and Tilebeni 2011). Inoculation with PGPR *Bacillus pumilus* and *Pseudomonas pseudoalcaligenes* in salt-stressed plants could help to alleviate salt stress in the paddy. Inoculated plants showed increased concentrations of N (26%), P (16%), and K (31%) and reduced concentrations of Na (71%) and Ca (36%) as compared to noninoculated plants under saline conditions (Jha and Subramanian 2011). Barley plants inoculated with *A. brasilense* strain Az39 may be better prepared to thrive under saline conditions by altering Na$^+$ and K$^+$ plant acquisition (Zawoznik et al. 2011). Inoculated maize plants with *A. chroococcum* strains C5 and C9 improved Na$^+$ exclusion and K$^+$ uptake, thereby increasing their K$^+$/Na$^+$ ratio under saline conditions (Rojas-Tapias et al. 2012). Sodium contents in salt-treated canola plant tissues (both shoot and root) were increased by about 2.5 to 4-fold. This increase was not altered by the addition of *P. putida* bacteria (either with or without ACCd activity) in roots. On the other hand, bacteria were able to facilitate Na accumulation in shoots and to a greater extent in the presence of *P. putida* (with ACCd activity) (Cheng et al. 2012). Recently, PGPRs like *Burkholderia cepacia* SE4, *Promicromonospora* sp. SE188, and *Acinetobacter calcoaceticus* SE370 have been shown to increase the growth of cucumber plants under saline condition by supporting host plants P and K uptake and by reducing Na acquisition (Kang et al. 2014).

## 15.2.5 Maintenance of Photosynthetic Capacity

During the onset and development of salt stress within a plant, photosynthesis is disrupted, ultimately resulting in the loss of plant productivity (Evelin et al. 2009). The salt-induced limitation of the photosynthetic performance is commonly ascribed to lower net photosynthetic rate and stomatal conductance, reduction in chlorophyll concentration, relative water content, and membrane stability index, as well as impaired photochemical reactions of photosynthesis (Debez et al. 2008; Talaat and Shawky 2014a). However, there have been very few attempts that studied the influence of AM inoculum on photosynthesis under saline conditions. Only a few reports have indicated that AM colonization could enhance relative water content in zucchini leaves (Colla et al. 2008), water potential and photosynthesis of maize plants (Sheng et al. 2008), and chlorophyll concentration in the leaves of several plant species, that is *Sesbania aegyptiaca*, *Sesbania grandiflora*, *L. glaber*, and *C. annuum* (Giri and Mukerji 2004; Sannazzaro et al. 2006; Colla et al. 2008; Beltrano et al. 2013). Hajiboland et al. (2010) showed that mycorrhization improved the net assimilation rates both by elevating stomatal conductance and by protecting photochemical processes of photosystem II (PSII) against salinity. However, the enhancement of the PSII photochemistry by AMF did not occur in tomato plants without salt treatment. Thus, the authors suggested that AM colonization acted only in maintenance of photochemical capacity in stressed leaves and did not increase its potential for energy trapping. Recently, Talaat and Shawky (2014a) demonstrated that AM symbiosis helps wheat plants to alleviate salt stress by enhancing its photosynthetic capacity, thus improving its ability to generate further growth or harvestable biomass. Mycorrhizal colonization enhanced the photosynthetic efficiency under saline conditions by significantly improving chlorophyll content, photochemical reactions of photosynthesis; gas exchange capacity (increased net $CO_2$ assimilation rate and stomatal conductance and decreased intercellular $CO_2$ concentration); carbonic anhydrase activity; acquisition of N, $K^+$, $Na^+$, and $Mg^{2+}$; relative water content; and membrane stability index. Moreover, mycorrhization maintained the photosynthetic apparatus *via* the accumulation of protective osmolytes and/or *via* the upregulation of antioxidant metabolism that suggests interesting biotechnological perspectives for improving agricultural productivity under salinity stress (Talaat and Shawky 2014a,b).

Few studies have investigated the influence of PGPR on photosynthesis. Inoculation with two PGPR strains, *Pseudomonades* sp. and *B. lentus*, into saline soils alleviated the salinity effects on basil plant growth by increasing the chlorophyll content, photosynthetic rate, and stomatal resistance (Golpayegani and Tilebeni 2011). Beneficial effect of PGPR *A. brasilense* and *Pantoea dispersa* inoculation under salinity on pepper (*C. annuum*) growth was related to the alleviation of osmotic stress by maintaining higher stomatal conductance and photosynthetic activities, but without affecting chlorophyll concentration or PSII photochemical efficiency (del Amor and Cuadra-Crespo 2012). However, Gururani et al. (2013) found positive influence of *Bacillus* on the PSII photochemistry of potato plants growing under salt stress condition and suggested that this PGPR was able to confer salt stress tolerance in plants. Moreover, reduction in chlorophyll content due to salt stress and relative recovery in the presence of *B. amyloliquefaciens* NBRISN13 was observed by Nautiyal et al. (2013), who said that the presence of ACCd in bacteria mitigated the effect of salt on chlorophyll. Similar growth-promoting and salt stress tolerance effects were also observed by Kang et al. (2014), who showed that *B. cepacia* SE4, *Promicromonospora* sp. SE188, and *A. calcoaceticus* SE370 inoculation to salt-stressed cucumber plants increased chlorophyll content. Recently, Ali et al. (2014) showed that ACCd-containing PGPB endophytes *P. fluorescens* YsS6 and *P. migulae* 8R6 significantly increased fresh and dry biomass, chlorophyll content, and the number of flowers and buds of tomato plants under salt stress (165 and 185 mM) when compared with plants inoculated with their ACCd mutants.

### 15.2.6 Improving Protein Synthesis

Salinity stress can disrupt the protein synthesis by retarding nitrate uptake, reducing nitrate reductase activity, enhancing the oxygen radicals and $H_2O_2$ production, and increasing $Na^+$ level and/or $Na^+/K^+$ ratio in the cytoplasm (Talaat and Shawky 2012, 2014a). Little information is known about how AMF take up nitrogen and transfer it to their host plants. Govindarajulu et al. (2005) reported that inorganic nitrogen is taken up by the fungal extraradical mycelium (ERM) and assimilated *via* nitrate reductase and the GS -GOGAT (glutamine synthetase; GS–glutamate synthase (glutamate 2-oxoglutarate aminotransferase; GOGAT) cycle. It is then converted into arginine, which is translocated along the coenocytic fungal hyphae from the ERM into the intraradical mycelium (IRM). Arginine is broken down in the IRM, releasing urea and ornithine, which are further broken down by the actions of urease and ornithine aminotransferase. Ammonia released from arginine breakdown passes to the host *via* ammonia channels. Amino acids from ornithine breakdown and/or $NH_4^+$ assimilation in the IRM may be catabolized within the IRM or translocated to the ERM. Application of AMF to salty soils can help in better assimilation of nitrogen in the host plant. Giri and Mukerji (2004) recorded higher accumulation of N in shoots of mycorrhizal *S. grandiflora* and *S. aegyptiaca* than nonmycorrhizal plants. Improved N nutrition may help to reduce the toxic effects of Na ions by reducing its uptake, and this may indirectly help in maintaining the chlorophyll content of the plant. The extraradical mycelia take up inorganic nitrogen from the soil in the form of nitrate and assimilated it *via* nitrate reductase, located in the arbuscule-containing cells (Kaldorf et al. 1998). Recently, Talaat and Shawky (2014a) illustrated that the AM symbiosis raised the grains protein content and ameliorated the deleterious effects of salinity *via* improving membrane stability index, nitrate reductase activity, nitrate content, $K^+$ concentration, and $K^+/Na^+$ ratio as well as *via* preventing the oxidative stress by alleviating the peroxidation of membrane lipids and decreasing the $H_2O_2$ content. They suggested that regulation of protein synthesis can be one of the most important mechanisms used by EM-treated plants to improve plant adaptation to saline soils.

Beneficial effects of PGPR under salinity may be due to alterations in nitrogen metabolism. Hamdia et al. (2004) studied the effect of inoculation with *Azospirillum* spp. on salt-stressed maize plants, and they found a sharp reduction in the activity of nitrate reductase and nitrogenase in shoots and roots by salinity stress. This reduction in nitrate reductase (NR) and nitrogenase (NA) activity was highly significant at all salinity concentrations. *Azospirillum* inoculation stimulated NR and nitrogenase activity in both shoots and roots of maize plants.

### 15.2.7 Source–Sink Relations and Energetic Metabolism

Stimulation of carbohydrate transport and metabolism between source and sink tissues has been proposed as a mechanism to alleviate metabolic feedback inhibitions of photosynthesis, thus avoiding photoinhibition during the osmotic phase of salinity when carbohydrates usually accumulate (Perez-Alfocea et al. 2010). Symbiotic microorganisms can directly modulate source–sink relations by enhancing sink activity *via* increased exchange of carbohydrates and mineral nutrients. Indeed, plant roots become a strong sink for carbohydrates when colonized by AMF, as these fungi can consume up to 20% of the host photosynthate (Heinemeyer et al. 2006). Moreover, about 30% of total respiration by EM-associated *Pinus sylvestris* and *Pinus contorta* seedlings can be attributed to fungal mycelia in the soil (Soderstrom and Read 1987). In mycorrhizal *Plantago lanceolata* plants, the respiratory carbon cost of the mycelium of *Glomus mosseae* ranged between 0.8% and 5% and increased with soil warming, showing a strong dependence on recent photosynthates (Heinemeyer et al. 2006). Therefore, mycorrhizal sink strength influences the whole-plant carbon balance, and fungal colonization can stimulate the rate of photosynthesis sufficiently to compensate for fungal carbon requirements and to eliminate growth reduction (Wright et al. 1998). It has been also proposed that sugar accumulation may also be due to the hydrolysis of starch in inoculated seedlings, as mycelium growth requirements mobilize carbon reserves (Heinemeyer et al. 2006), which could help decrease salinity-induced starch accumulation as a consequence of the inhibition of sink activity in growing tissues (Balibrea et al. 2000). Thus maintenance of an active carbohydrate sink in symbiotic roots, when assimilate transport and use in other sink tissues is impaired, could help maintain the source activity of mature leaves for longer, thereby improving salt tolerance (Perez-Alfocea et al. 2010). Furthermore, Lerat et al. (2003) postulated that in return to enhance mineral nutrition by the AMF, the host plant channels between 4% and 20% of its net photosynthetic C to its mycobionts through the arbuscules that are the sites of exchange for phosphorus, carbon, water, and other nutrients. In support of this idea, the improved salt tolerance of mycorrhizal (*G. mosseae*) maize plants was related to higher accumulation of soluble sugars in the roots as a specific response, independent of plant nutritional (P) status (Feng et al. 2002). This effect could be responsible for improved plant water status, chlorophyll concentration, and photosynthetic capacity by increasing photochemical efficiency (Talaat and Shawky 2014a).

In the case of PGPR, Bertrand et al. (2011) reported that in soybean in symbiosis with

Bradyrhizobium, photosynthates are transported from the leaves to the nodules through the phloem under the form of sucrose. In the nodule, carbon (C) from sucrose is made available to the bacteroides to sustain $N_2$ fixation. In response, the bacteroides release to the plant the ammonia they produce from $N_2$ fixation. In soybean, nitrogen is exported from the nodules to the other plant parts through the xylem under the form of ureides, which are degraded in the leaves to release ammonia and then assimilated by the plant into amino acids. The regulation of this pathway is complex and involves ammonia availability to the plant and feedback inhibition of $N_2$ fixation by high concentration of nitrogenous compounds in the leaves. In fact, they confirmed that a major portion of photosynthates are translocated to the nodules and utilized there in support of $N_2$ fixation, nodule growth and maintenance, and assimilation of ammonia. On the other hand, $N_2$ fixation in nodules is rapidly inhibited under conditions that reduce C availability.

## 15.2.8 Altering Plant Hormonal Status

Following exposure to salinity, the root–shoot ratio is increased (an important adaptive response) due to the rapid inhibition of shoot growth (which limits plant productivity) while root growth is maintained. These changes in plant growth under salinity stress may be regulated *via* changes in phytohormone concentrations (Perez-Alfocea et al. 2010). On the other hand, soil microbes may improve crop salt tolerance by altering hormonal root–shoot signaling that regulates leaf growth and gas exchange (Dodd et al. 2010). Moreover, Dodd and Perez-Alfocea (2012) reported that alteration plant hormonal status to decrease evolution of the growth-retarding and senescence-inducing hormone ethylene (or its precursor 1-aminocyclopropane-1-carboxylic acid) or to maintain source–sink relations, photosynthesis, and biomass production and allocation (by altering indole-3-acetic acid [IAA] and cytokinin biosynthesis) seems to be promising target processes for soil biota-improved crop salt tolerance.

### 15.2.8.1 Auxins

IAA affects plant cell division, extension, and differentiation; stimulates seed and tuber germination; increases the rate of xylem and root development;

controls processes of vegetative growth; initiates lateral and adventitious root formation; mediates responses to light, gravity, and florescence; and affects photosynthesis, pigment formation, biosynthesis of various metabolites, and resistance to stressful conditions. IAA produced by rhizobacteria likely interfere the aforementioned physiological processes of plants by changing the plant auxin pool (Glick 2012). EMF were found to alter the hormonal status of the host plant under saline conditions. Luo et al. (2009) found that ectomycorrhizal (P. involutus) colonization of roots not only significantly increased root salicylic acid (SA) and abscisic acid (ABA) concentrations while decreasing root IAA, jasmonic acid (JA), and JA-isoleucine concentrations under control conditions but also enhanced the salinity-induced accumulation of SA and also prevented or attenuated salinity-induced changes in IAA, JA, and JA-isoleucine. On the other hand, there were no systemic mycorrhizal effects on foliar hormone concentrations (SA, ABA, IAA, and JA-isoleucine), while salinity only increased foliar ABA concentrations and decreased foliar JA-isoleucine concentrations.

Microbial synthesis of the phytohormone auxin has been known for a long time. It is reported that 80% of microorganisms isolated from the rhizosphere of various crops possess the ability to synthesize and release auxins as secondary metabolites (Patten and Glick 1996). It is also suggested that the phytohormone auxin produced by root-colonizing bacteria plays an important role in alleviating salt stress in plants and these organisms should therefore be considered as a seed dressing in field trials to improve growth and yield of crop plants in farms where soil salinity is high (Frankenberger and Arshad 1995). Salt-tolerant IAA-producing bacterial strains *Pseudomonas aurantiaca* TSAU22 and *Pseudomonas extremorientalis* TSAU20 alleviated quite successfully the reductive effect of salt stress on percentage of germination (up to 79%), probably through their ability to produce IAA (Egamberdieva 2009). Moreover, Chaiharn and Lumyong (2011) found that 18.05% from 216 bacterial strains isolated from rice rhizospheric soils in northern Thailand produced IAA and identified the best IAA producer by biochemical testing and 16s rDNA sequence analysis as *Klebsiella* SN 1.1. This strain produced the highest amount of IAA ($291.97 \pm 0.19$ ppm) in culture media supplemented with L-tryptophan. Nautiyal et al. (2013) demonstrated that the plant growth promotion under salinity stress was probably because of auxin-producing properties of B. amyloliquefaciens NBRISN13.

### 15.2.8.2 Abscisic Acid

Salinity increases ABA concentrations in roots, xylem sap, and shoots concomitant with a decrease in transpiration rates (Albacete et al. 2008). Mycorrhization can alter the ABA levels in the host plant. Sannazzaro et al. (2007) observed higher ABA levels in L. glaber plants colonized by G. intraradices than noncolonized ones. They suggested that ABA regulates free spermine pools in the shoots, and the increased spermine content in the mycorrhizal plant may be due to increased ABA content. In contrast, Jahromi et al. (2008) found lower ABA levels in salt-stressed lettuce plants colonized by G. intraradices than in the non-AM plants, which could be attributed to the lower expression of Lsnced gene in these plants. Lsnced encodes for 9-cis-epoxycarotenoid dioxygenase, a key enzyme in the biosynthesis of the stress hormone ABA. Comparing the aforementioned observations, it might be suggested that the effect of AMF species on ABA content varies with the host plants. In contrast, ectomycorrhizal (P. involutus) colonization of the root system of gray poplar (P. × canescens) had no effect on salinity-induced changes in root or shoot ABA concentration (Luo et al. 2009).

In the case of PGPR, immersing cotton (Gossypium hirsutum) seeds in suspensions of P. putida Rs-198 ($10^9$ CFU mL$^{-1}$ for 6 h) prior to planting increased seedling biomass accumulation by 10% and 19% in saline and nonsaline soil, respectively, and prevented any salinity-induced ABA accumulation in cotton seedlings (Yao et al. 2010). Likewise, PGPRs like B. cepacia SE4, Promicromonospora sp. SE188, and A. calcoaceticus SE370 have been shown to decrease ABA level in salt-stressed cucumber plants (Kang et al. 2014). PGPR-treated plants counteracted adverse effects of salt stress by significantly increasing plant growth and producing a low level of ABA compared to control plants.

### 15.2.8.3 Ethylene

Ethylene is an important plant growth regulator that functions in the processes of root initiation, fruit ripening, seed germination, flower wilting, leaf abscission, biosynthesis of other phytohormones, and stress signaling. Plants normally synthesize only small amounts of ethylene, levels that typically confer beneficial effects on plant growth and development (except during fruit ripening when the ethylene concentration is much higher). However, in response to various stresses, there is often a significant rise in endogenous ethylene biosynthesis, called "stress ethylene" (Glick 2012). Many rhizobacteria can alter plant hormone status by decreasing plant ethylene levels via the bacterial enzyme ACCd that hydrolyzes the immediate ethylene precursor ACC into ammonia and α-ketobutyrate. Various ACCd-containing rhizobacteria were repeatedly shown to promote plant growth, particularly when plants were subjected to environmental stresses likely to stimulate stress-induced ethylene production (Dodd and Perez-Alfocea 2012). Since ethylene often acts as a growth inhibitor, it seems likely that decreased ACC levels in planta lowered ethylene production, thereby increasing shoot growth and yield particularly under soil stress conditions (Belimov et al. 2009). These bacteria may enhance the survival of some seedlings, especially during the first few days after the seeds are planted. Egamberdieva and Kucharova (2009) found that only strain Pseudomonas trivialis 3Re27 could utilize ACC as the sole N source and showed high stimulatory effect on the growth of plants in saline soils. Additionally, ethylene-responsive element-binding proteins (EREBP) were reported to have a role in hormone metabolism, ethylene signal transduction, and abiotic stress conditions. Nautiyal et al. (2013) reported 0.37- and 0.80-fold less expression of EREBP in salt-stressed rice plants and 1.5-fold more expression in salt-stressed plants inoculated with B. amyloliquefaciens NBRISN13, which emphasized the role of NBRISN13 in salt stress management.

In accordance to Mayak et al. (2004) data, they showed that the ethylene content in tomato seedlings exposed to high salt was reduced by the application of A. piechaudii, indicating that bacterial ACCd was functional. A. piechaudii, which produces ACC, increased the growth of tomato seedlings by as much as 66% in the presence of high-salt contents. Moreover, the effect of high salt (150 mM NaCl) on the growth of red pepper seedlings in the presence or absence of the plant growth-promoting ACCd-containing halotolerant bacteria was evaluated by Siddikee et al. (2011), who reported that up to 57% of the production of stress ethylene was reduced when inoculated with the bacterium, and the overall biomass of the inoculated plantlets was similar to the no-salt treatment control plant. Similarly, Cheng et al. (2012) indicated that ethylene production rate was significantly enhanced by the addition of salt stress (approximately by 14-fold) but not by P. putida UW4 treatment. The presence of ACCd activity limited this increase in ethylene level and improved the growth of canola plant treated with salt plus P. putida UW4. Ali et al. (2014) showed that ACCd-containing plant PGPB have been reported to facilitate plant growth under salt stress by reducing the ethylene level that is produced as a consequence of saline conditions. Furthermore, rhizobacterial impacts on in planta concentrations of one phytohormone may have feedback effects on the concentration of other hormones. Mayak et al. (2004) reported that decreased ethylene production of plants inoculated with ACCd-containing rhizobacteria may cause feedback reductions of plant ABA levels.

## 15.2.9 Enhancing Antioxidant Defense System

Under salinity, the availability of atmospheric $CO_2$ is restricted because stomatal closure is increased and consumption of NADPH by the Calvin cycle is reduced. When ferrodoxin is overreduced during photosynthetic electron transfer, electrons may be transferred from photosystem I to oxygen to form superoxide radicals ($O_2^{\bullet-}$) by a process known as the Mehler reaction, which initiates chain reactions that produce more harmful oxygen radicals. These include singlet oxygen ($^1O_2$), hydrogen peroxide ($H_2O_2$), and hydroxyl radicals ($OH\bullet$). When these cytotoxic reactive oxygen species (ROS) are produced in excess, they can destroy normal metabolism through oxidative damage of lipids, proteins, and nucleic acids (Miller et al. 2010). Thus, there is a constant need for efficient mechanisms to avoid oxidative damage to cells. Plants have developed a series of enzymatic and nonenzymatic detoxification systems to counteract ROS and protect cells from oxidative damage (Miller et al. 2010; Talaat and Shawky 2014b). The efficient destruction of $O_2^{\bullet-}$ and $H_2O_2$ requires the action of several antioxidant enzymes acting in synchrony. Superoxide is rapidly converted to $H_2O_2$ by the action of superoxide dismutase (SOD; EC 1.15.1.1) (Alscher et al. 2002). However, since $H_2O_2$ is a strong oxidant that rapidly oxidizes thiol groups, it cannot be allowed to accumulate to excess. Catalases (CAT; EC 1.11.1.6) convert

$H_2O_2$ to water and molecular oxygen in peroxisomes. An alternative mode of $H_2O_2$ destruction is via peroxidases (POX; EC 1.11.1.7), which are found throughout the cell and which have a much higher affinity for $H_2O_2$ than catalases (Apel and Hirt 2004). Plants also show high activities for the enzymes of the ascorbate–glutathione cycle in which $H_2O_2$ is scavenged. In the first step of this pathway, ascorbate peroxidase (APX; EC 1.11.1.11), which is the most important peroxidase in $H_2O_2$ detoxification, catalyzes the reduction of $H_2O_2$ to water by ascorbate, and the resultant monodehydroascorbate and dehydroascorbate are reduced back to ascorbate by monodehydroascorbate reductase (MDHAR; EC 1.6.5.4) and by dehydroascorbate reductase (DHAR; EC 1.8.5.1) plus glutathione reductase (GR; EC 1.6.4.2), respectively (Noctor and Foyer 1998).

AM symbiosis helps plants to alleviate salt stress by enhancing antioxidant defense system. Higher activities of CAT and POX were detected in mycorrhizal wheat plants than in nonmycorrhizal ones under saline conditions by Talaat and Shawky (2011). Indeed, AMF increased the activities of antioxidant enzymes and the concentrations of antioxidant molecules under salt stress to protect plant cells from oxidative damage (Wu et al. 2010; Dudhane et al. 2011; Cekic et al. 2012). In support of this idea, Talaat and Shawky (2014b) demonstrated that the activities of antioxidative enzymes (SOD, POX, CAT, and GR) and the concentrations of antioxidant molecules (glutathione and ascorbate) were increased under saline conditions; moreover, these increases were more significant in salt-stressed mycorrhizal plants. They also found that salt stress induced oxidative damage through increased lipid peroxidation, electrolyte leakage, and hydrogen peroxide concentration; however, mycorrhizal colonization altered plant physiology and significantly reduced oxidative damage. Hence, they concluded that elimination of ROS can be one of the mechanisms how AMF improve plant adaptation to saline soils and increase its productivity.

Moreover, the existence of genes encoding different proteins that may be involved in cellular defense against oxidative stress has been described in AMF. This includes genes encoding SOD (Gonzalez-Guerrero et al. 2010); glutaredoxins, which are small proteins with glutathione-dependent disulfide oxidoreductase activity involved in cellular defense against oxidative stress

(Benabdellah et al. 2009); and even a metallo-thionein-encoding gene, which was suggested to play a role in the regulation of the redox status of the ERM of *G. intraradices* (Gonzalez-Guerrero et al. 2007). Some of these genes have been shown to be induced by oxidative stress, but their responses to salinity have never been studied.

PGPR can prevent the deleterious effects of salinity stress by enhancing antioxidant status. The two PGPR strains, *Pseudomonades* sp. and *B. lentus*, alleviated the salinity effects on the antioxidant enzymes APX and GR, along with that on growth of basil plant (Golpayegani and Tilebeni 2011). Induction of antioxidant enzymes (CAT, SOD, and POX) by PGPR *B. pumilus* and *P. pseudoalcaligenes* increases the tolerance of paddy grown under severe salt stress, which could serve as a useful tool to alleviate salt stress in PGPR-inoculated plants (Jha and Subramanian 2011). Induction of antioxidant level, a prerequisite for resistance mechanism, was well demonstrated by the upregulation of catalase (1.6-fold) in salt-stressed rice plants inoculated with *B. amyloliquefaciens* NBRISN13 (SN13) (Nautiyal et al. 2013). Enhanced mRNA expression levels of the various ROS-scavenging enzymes (APX, SOD, CAT, DHAR, and GR) were detected in *Bacillus*-inoculated plants growing under salt stress condition, which contributed to the increased plant tolerance to salt stress (Gururani et al. 2013). However, oxidative stress induced by salinity stress was mitigated by *B. cepacia* SE4, *Promicromonospora* sp. SE188, and *A. calcoaceticus* SE370 in cucumber plants through reduced activities of catalase, peroxidase, polyphenol oxidase, and total polyphenol (Kang et al. 2014).

As the first report, Talaat (2014) studied the contribution of effective microorganisms (EM) (photosynthetic bacteria [*Rhodopseudomonas palustris* and *Rhodobacter sphaeroides*], lactic acid bacteria [*Lactobacillus plantarum*, *L. casei*, and *Streptococcus lactis*], yeast [*Saccharomyces cerevisiae* and *Candida utilis*], actinomycetes [*Streptomyces albus* and *S. griseus*], and fermenting fungi [*Aspergillus oryzae*, *Penicillium* sp., and *Mucor hiemalis*]) to the scavenging capacity of the ascorbate–glutathione cycle in *P. vulgaris* plants grown in salty soils. They found that lipid peroxidation and $H_2O_2$ content were significantly decreased with EM application in both stressed and nonstressed plants. Moreover, activities of APX, GR, MDHAR, and DHAR, contents of ascorbate and glutathione, and ratios of AsA/DHA and GSH/GSSG were significantly enhanced in stressed plants treated with EM. In fact, the EM treatment detoxified the stress generated by salinity and significantly improved plant growth and productivity. Enhancing the $H_2O_2$-scavenging capacity of the ascorbate–glutathione cycle in EM-treated plants may be an efficient mechanism to attenuate the activation of plant defenses.

## 15.3 COMMERCIAL/AGRONOMIC PROSPECTS

Successful commercial application of microbial inoculants to improve crop growth and yield in saline soil implies that the inoculants are also salt tolerant, which highlights the potential of using microorganisms from saline habitats. Certain PGPR, whose ability to colonize the root system is undiminished by salinity, offer considerable potential as inoculants. In this respect, Paul and Nair (2008) reported that the root colonization potential of the salt-tolerant *Pseudomonas* strain was not hampered with higher salinity in soil. As a means of salt tolerance, the strain synthesized the osmolytes in their cytosol. Although some mycorrhizal species (e.g., *Scutellospora calospora*) reach maximum spore germination under high-salt (300 mM NaCl) conditions, spore germination of other AMF was delayed, and the specific rate of hyphal extension was reduced in the presence of NaCl (Jahromi et al. 2008; Talaat and Shawky 2012). The extent to which spore germination was inhibited was not the same for all species investigated but was similar for isolates of the same species and was independent of the salinity of the environment of origin. However, the evidence that propagules from root pieces have a higher capacity to germinate under high salinity than the spores of the same species suggests that specific technologies can be developed to optimize fungal viability under saline conditions in order to optimize colonization of the host plant. One interesting proposition is to pretreat mycorrhizal hyphae with salt prior to inoculation into saline environments, which almost doubled root colonization and stimulated root and shoot growth (10%–15%) of soybean plants exposed to 100 mM NaCl (Sharifi et al. 2007). Thus, parallel programs of independently increasing both microbial and plant salt tolerance may assist the productivity of crops grown in saline environments.

The microbial formulation and application technology are crucial for the development of commercial salt-tolerant bioformulation effective under saline conditions. Microbial mixtures as multitasking inoculants and stress-protecting bioformulations are one alternative to overcome inconsistent in vivo effects. Adesemoye et al. (2009) found that inoculation with mixed strains was more consistent than single-strain inoculations. Another future possibility for efficient inoculation, valid for plants propagated from tissue culture, is to inoculate the salt-tolerating PGPR into the plant cell suspension and regenerate embryos and eventually stress-tolerating plants. These plants will probably be inoculated from their onset as this can be done in a tissue culture laboratory accustomed to sterile and precise work, and this will result in even inoculation of the plants. A potentially promising future application could be the enhancement of salt tolerance of transgenic plants by identification of enzymes and genes involved in the synthesis of novel osmoprotectants found in stress-tolerant microorganisms that can be expected to provide more such opportunities for stress tolerance engineering in agricultural crops (Apse and Blumwald 2002). Further work should be done to highlight the mechanisms of stress tolerance by PGPR in salinized conditions and to uplift the status of stress tolerance mechanisms in plant.

## 15.4 SUSTAINABLE INTENSIFICATION OF AGRICULTURE

The ultimate goal of sustainable agriculture is to develop farming systems that are productive, profitable, energy-conserving, environmentally sound, and conserving of natural resources and that ensure food safety and quality. Microbial fertilizers can increase yield and quality of crops without a large investment of money and labor and can clean the environment and encourage the productive capacity of land. There is a growing interest in the use of soil microbial inoculants as an alternative biological approach to improve soil quality; enhance the growth, yield, and quality of crops; and reduce the inputs of chemical fertilizers and pesticides in agriculture worldwide. The potential of rhizosphere microbes in agriculture and environmental management is significant; they can be used easily and economically to enhance productivity of agricultural systems, especially organic systems and in mitigating environmental pollution.

Indeed, decline in soil fertility, increased soil erosion, and increasing shortage of food are major factors affecting human health in many regions all over the world. Fertilizers are costly and therefore out of reach of most resource poor farmers. There is also increasing evidence that synthetic agrochemicals and fertilizers have caused adverse effects on the environment leading to loss of biodiversity. This observation has promoted the need of introducing methods of farming aimed at reducing health risks including the use of rhizosphere microbes in organic farming. Inoculation of crops with rhizosphere microbes can improve vegetable crop yields and thereby improve food security. Interest in microbial fertilizers has increased, as it would substantially reduce the use of chemical fertilizers and pesticides that often contribute to the pollution of soil–water ecosystems. This environmentally friendly technology claims an enormous amount of tremendous benefits for both soil and plants. Moreover, evolving efficient, low-cost, and easily adaptable methods for salinity stress management is a major challenge. Worldwide, extensive research is being carried out, to develop strategies to cope with salty soils. While most of these technologies are cost-intensive and non-eco-friendly, this chapter indicates that microbial fertilizers can also help crops to cope with saline conditions.

In agreement with Dardanelli et al. (2010), the presence of microorganisms in the soil is critical to the maintenance of soil function, in both natural and managed agricultural soils. The microbes are involved in key processes such as soil structure formation, decomposition of organic matter, toxin removal, and the cycling of elements—carbon, nitrogen, phosphorus, potassium, and sulfur. It is also clear that beneficial microorganisms play key roles in suppressing soilborne plant diseases and in promoting plant growth and changing the vegetation. Efforts should be directed toward maximizing the identified benefits of PGPR or biofertilizers in all developing economies. If the benefits of PGPR in crop production can be maximized, this will certainly help in the fight against hunger.

It is important to recognize that the rhizosphere (the area of the soil adjacent to the root surface) is biologically diverse and that rhizosphere organisms can play a major role in plant resource capture. Most attention has focused on certain bacterial genera that can fix atmospheric nitrogen within a specialized host organ (the legume nodule) and certain fungal genera (mycorrhizae) whose hyphal networks ramify throughout the soil and within the plant and seem particularly important in plant acquisition of relatively immobile nutrients such as phosphorus (P). Rhizosphere bacteria can also play important roles in plant resource capture. In addition, PGPR holds enormous prospects in improved and sustainable plant production, including enhanced plant tolerance to stress, better plant nutrient uptake, and reduced use of chemical inputs. The roles of PGPR in nutrient uptake and stress management are emerging areas in agriculture that is not yet well understood; consequently, the benefits are yet to be maximized anywhere in the world. It is even less explored in many developing economies and may seem entirely new in some regions. Evidently, PGPR will most likely be used in conjunction with "proven technologies" such as mycorrhizae (to augment plant P uptake) and nodulating bacteria (to fix nitrogen). Despite some successes from such co-inoculation (Talaat and Abdallah 2008; Adesemoye et al. 2009), developing compatible microbial mixtures will remain an academic and commercial challenge, because of its usual empirical methodology and the prospect of microbial antagonism. Nevertheless, experiments demonstrating that application of ACCd-containing rhizobacteria to field soils enhanced legume nodulation by indigenous rhizobia (Belimov et al. 2009) suggest that rhizobacteria that mediate plant hormone status may prove a viable technology as independent inocula.

Importantly, regions in developing economies may have to use more of products that are based on local isolates because as emphasized by Adesemoye et al. (2009), no microbial inoculant can be universal for all ecosystems. Rather, biofertilizers' performances may be specific as effectiveness is dependent upon factors like plant type and soil type.

## 15.5 CONCLUSION

Soil salinity disturbs the plant–microbe interaction, which is a critical ecological factor to help further plant growth in degraded ecosystems. The use of specific microbe antagonists that stimulate plant growth and/or promote plant salinity tolerance allows a considerable decrease in the use of agrochemicals that are now being used for plant growth stimulation and prevention of soil toxicity. Development of such a stress-tolerant microbial strain associated with the roots of agronomic crops

can also lead to improved fertility of salt-affected soils. Indeed, the use of beneficial microbes in agricultural production systems started about 60 years ago, and there is now increasing evidence that the use of beneficial microbes can enhance plants' resistance to soil salinity. Most rhizosphere microbes do not have a single mechanism that completely accounts for their beneficial effects on the plant salt tolerance. Some mechanisms by which soil microbial inoculants are able to stimulate plant growth and to prevent damage caused by salinity stress include mobilization of nutrients, production of phytohormones, regulation of water uptake, induction of synthesis and accumulation of compatible solutes, maintenance of photosynthetic capacity, improvement of protein synthesis, enhancement of antioxidant defense system, improvement of rhizospheric and soil conditions, altering the physiological and biochemical properties of the host, defending roots against soilborne pathogens, and inducing molecular changes. However, the interactions among these microbes are still not well understood in field applications under different environments. An understanding of the functions and mechanisms by which these microbes can promote plant growth in salt-stressed environments (e.g., arid region) may provide valuable information on plant–microbe interactions to develop new agricultural technologies that may improve soil ecology and plant development. This chapter already discusses recent developments and advances of the interactions between the plant and the rhizosphere microbes and their mechanism of action under salt-stressed environments. However, there is a need for more studies on plant–microbe interactions and their activities in different regions and ecologies, including stressed environments, for instance, in arid and tropical regions. Availability of more information will enable the development and widespread acceptance of new agricultural technologies, which can improve soil ecology, plant development, and resistance against stress conditions. Moreover, new research will provide farmers with novel control strategies for the development of microbial strains that are more effective and have longer shelf-lives as a "plant growth stimulators" and "biocontrol" to supplement and/or complement in chemical fertilizers and pesticides agriculture and at the same time protect plants against the detrimental effects of salinity and other osmotic stress conditions.

# REFERENCES

Adesemoye, A. O., H. A. Torbert, and J. W. Kloepper. 2009. Plant growth-promoting rhizobacteria allow reduced application rates of chemical fertilizers. *Microb Ecol* 58:921–929.

Aharon, R., Y. Shahak, S. Wininger, R. Bendov, Y. Kapulnik, and G. Galili. 2003. Overexpression of a plasma membrane aquaporin in transgenic tobacco improves plant vigor under favorable growth conditions but not under drought or salt stress. *Plant Cell* 15:439–447.

Albacete, A., M. E. Ghanem, C. Martinez-Andujar, M. Acosta, J. Sanchez-Bravo, V. Martinez, S. Lutts, I. C. Dodd, and F. Perez-Alfocea. 2008. Hormonal changes in relation to biomass partitioning and shoot growth impairment in salinized tomato (*Solanum lycopersicum* L.) plants. *J Exp Bot* 59:4119–4131.

Ali, S., T. C. Charles, and B. R. Glick. 2014. Amelioration of high salinity stress damage by plant growth promoting bacterial endophytes that contain ACC deaminase. *Plant Physiol Biochem* 80:160–167.

Alscher, P. G., N. Erturk, and L. S. Heath. 2002. Role of superoxide dismutases (SODs) in controlling oxidative stress in plant. *J Exp Bot* 53:1331–1341.

Apel, K. and H. Hirt. 2004. Reactive oxygen species: Metabolism, oxidative stress and signal transduction. *Annu Rev Plant Physiol Plant Mol Biol* 55:373–399.

Apse, M. P. and E. Blumwald. 2002. Engineering salt tolerance in plants. *Curr Opin Biotechnol* 3:146–150.

Aroca, R., R. Porcel, and J. M. Ruiz-Lozano. 2007. How does arbuscular mycorrhizal symbiosis regulate root hydraulic properties and plasma membrane aquaporins in *Phaseolus vulgaris* under drought, cold or salinity stresses? *New Phytol* 173:808–816.

Aroca, R., P. Vernieri, and J. M. Ruiz-Lozano. 2008. Mycorrhizal and non-mycorrhizal *Lactuca sativa* plants exhibit contrasting responses to exogenous ABA during drought stress and recovery. *J Exp Bot* 59:2029–2041.

Ashraf, M., S. H. Berge, and O. T. Mahmood. 2004. Inoculating wheat seedling with exopolysaccharide-producing bacteria restricts sodium uptake and stimulates plant growth under salt stress. *Biol Fert Soils* 40:157–162.

Auge, R. M. 2000. Stomatal behaviour of arbuscular mycorrhizal plants. In: Kapulnik, Y. and Douds, D., eds. *Arbuscular mycorrhizas: Physiology and Function*. Kluwer Academic Publishers, Dordrecht, the Netherlands, pp. 201–237.

Balibrea, M. E., J. Dell'Amico, M. C. Boların, and F. Perez-Alfocea. 2000. Carbon partitioning and sucrose metabolism in tomato plants growing under salinity. *Physiol Plant* 110:503–511.

Belimov, A. A., I. C. Dodd, N. Hontzeas, J. C. Theobald, V. I. Safronova, and W. J. Davies. 2009. Rhizosphere bacteria containing 1-aminocyclopropane-1-carboxylate deaminase increase yield of plants grown in drying soil via both local and systemic hormone signaling. *New Phytol* 181:413–423.

Beltrano, J., M. Ruscitti, M. C. Arango, and M. Ronco. 2013. Effects of arbuscular mycorrhiza inoculation on plant growth, biological and physiological parameters and mineral nutrition in pepper grown under different salinity and P levels. *J Soil Sci Plant Nutr* 13:123–141.

Benabdellah, K., M. A. Merlos, C. Azcon-Aguilar, and N. Ferrol. 2009. GintGRX1, the first characterized glomeromycotan glutaredoxin, is a multifunctional enzyme that responds to oxidative stress. *Fungal Genet Biol* 46:94–103.

Bertrand, A., D. Prévost, C. Juge, and F-P. Chalifour. 2011. Impact of elevated $CO_2$ on carbohydrate and ureide concentration in soybean inoculated with different strains of *Bradyrhizobium japonicum*. *Botany* 89:481–490.

Calvo-Polanco, M., J. J. Zwiazek, and M. C. Voicu. 2008. Responses of ectomycorrhizal American elm (*Ulmus americana*) seedlings to salinity and soil compaction. *Plant Soil* 308:189–200.

Cekic, F. O., S. Unyayar, and I. Ortas. 2012. Effects of arbuscular mycorrhizal inoculation on biochemical parameters in *Capsicum annuum* grown under long term salt stress. *Turk J Bot* 36:63–72.

Chaiharn, M. and S. Lumyong. 2011. Screening and optimization of indole-3-acetic acid production and phosphate solubilization from rhizobacteria aimed at improving plant growth. *Curr Microbiol* 62:173–181.

Chen, M., H. Wei, J. Cao, R. Liu, Y. Wang, and C. Zheng. 2007. Expression of *Bacillus subtilis* proAB genes and reduction of feedback inhibition of proline synthesis increases proline production and confers osmotolerance in transgenic *Arabidopsis*. *J Biochem Mol Biol* 40:396–403.

Cheng, Z., O. Z. Woody, B. J. McConkey, and B. R. Glick. 2012. Combined effects of the plant growth-promoting bacterium *Pseudomonas putida* UW4 and salinity stress on the *Brassica napus* proteome. *Appl Soil Ecol* 61:255–263.

Colla, G., Y. Rouphael, M. Cardarelli, M. Tullio, C. M. Rivera, and E. Rea. 2008. Alleviation of salt stress by arbuscular mycorrhizal in zucchini plants grown at low and high phosphorus concentration. *Biol Fert Soils* 44:501–509.

Cram, W. J. 1976. Negative feedback regulation of transport in cells. The maintenance of turgor, volume and nutrient supply. In: Luttge, U. and Pitman, M.G., eds. *Encyclopaedia of Plant Physiology, New Series*, Vol. 2. Springer-Verlag, Berlin, Germany, pp. 283–316.

Dardanelli, M. S., S. M. Carletti, N. S. Paulucci, D. B. Medeot, E. A. Rodriguez Caceres, F. A. Vita, M. Bueno, M. V. Fumero, and M. B. Garcia. 2010. Benefits of plant growth-promoting rhizobacteria and rhizobia in agriculture. In: Maheshwari, D.K., ed. *Plant Growth and Health Promoting Bacteria*. Microbiology Monographs, Vol. 18. Springer, Berlin, Germany, pp. 1–20.

Debez, A., H. W. Koyro, C. Grignon, C. Abdelly, and B. Huchzermeyer. 2008. Relationship between the photosynthetic activity and the performance of Cakile maritima after long-term salt treatment. *Physiol Plant* 133:373–385.

del Amor, F. and P. Cuadra-Crespo. 2012. Plant growth-promoting bacteria as a tool to improve salinity tolerance in sweet pepper. *Funct Plant Biol* 39:82–90.

Dodd, I. C. and F. Perez-Alfocea. 2012. Microbial amelioration of crop salinity stress. *J Exp Bot* 63(9): 3415–3428.

Dodd, I. C., N. Y. Zinovkina, V. I. Safronova, and A. A. Belimov. 2010. Rhizobacterial mediation of plant hormone status. *Ann Appl Biol* 157:361–379.

Dudhane, M. P., M. Y. Borde, and P. K. Jite. 2011. Effect of arbuscular mycorrhizal fungi on growth and antioxidant activity in Gmelina arborea Roxb. under salt stress condition. *Not Sci Biol* 3:71–78.

Egamberdieva, D. 2009. Alleviation of salt stress by plant growth regulators and IAA producing bacteria in wheat. *Acta Physiol Plant* 31:861–864.

Egamberdieva, D. and K. R. Islam. 2008. Salt tolerant rhizobacteria: Plant growth promoting traits and physiological characterization within ecologically stressed environment. In: Ahmad, I., Pichtel, J., and Hayat, S., eds. *Plant-Bacteria Interactions: Strategies and Techniques to Promote Plant Growth*. Wiley-VCH, Weinheim, Germany, pp. 257–281.

Egamberdieva, D. and Z. Kucharova. 2009. Selection for root colonising bacteria stimulating wheat growth in saline soils. *Biol Fert Soils* 45:563–571.

Evelin, H., R. Kapoor, and B. Giri 2009. Arbuscular mycorrhizal fungi in alleviation of salt stress: A review. *Ann Bot* 104:1263–1280.

Feng, G., F. S. Zhang, X. L. Li, C. Y. Tian, C. Tang, and Z. Rengel. 2002. Improved tolerance of maize plants to salt stress by arbuscular mycorrhiza is related to higher accumulation of soluble sugars in roots. *Mycorrhiza* 12:185–190.

Flowers, T. J. and T. D. Colmer. 2008. Salinity tolerance in halophytes. *New Phytol* 179:945–963.

Frankenberger, J. and M. Arshad. 1995. Microbial synthesis of auxins. In: Frankenberger, W.T. and Arshad, M., eds. *Phytohormones in Soils*. Marcel Dekker, New York, pp. 35–71.

Giri, B., R. Kapoor, and K. G. Mukerji. 2007. Improved tolerance of *Acacia nilotica* to salt stress by arbuscular mycorrhiza, *Glomus fasciculatum*, may be partly related to elevated $K^+/Na^+$ ratios in root and shoot tissues. *Microb Ecol* 54:753–760.

Giri, B. and K. Mukerji. 2004. Mycorrhizal inoculant alleviates salt stress in Sesbania aegyptiaca and Sesbania grandiflora under field conditions: Evidence for reduced sodium and improved magnesium uptake. *Mycorrhiza* 14:307–312.

Glick, B. R. 2012. *Plant Growth-Promoting Bacteria: Mechanisms and Applications*. Hindawi Publishing Corporation, Scientifica. 2012:1–15.

Golpayegani, A. and H. G. Tilebeni. 2011. Effect of biological fertilizers on biochemical and physiological parameters of basil (Ociumum basilicm L.) medicine plant. *Amer-Eur J Agric Environ Sci* 11:445–450.

Gonzalez-Guerrero, M., C. Cano, C. Azcon-Aguilar, and N. Ferrol. 2007. GintMT1 encodes a functional metallothionein in Glomus intraradices that responds to oxidative stress. *Mycorrhiza* 17:327–335.

Gonzalez-Guerrero, M., E. Oger, K. Benabdellah, C. Azcon-Aguilar, L. Lanfranco, and N. Ferrol. 2010. Characterization of a CuZn superoxide dismutase gene in the arbuscular mycorrhizal fungus Glomus intraradices. *Curr Genet* 56:265–274.

Govindarajulu, M., P. E. Pfeffer, H. Jin, J. Abubaker, D. D. Douds, J. W. Allen, H. Bucking, P. J. Lammers, and Y. Shachar-Hill. 2005. Nitrogen transfer in the arbuscular mycorrhizal symbiosis. *Letters* 435:819–823.

Gururani, M. A., C. P. Upadhyaya, V. Baskar, J. Venkatesh, A. Nookaraju, and S. W. Park. 2013. Plant growth-promoting rhizobacteria enhance abiotic stress tolerance in Solanum tuberosum through inducing changes in the expression of ROS-scavenging enzymes and improved photosynthetic performance. *J Plant Growth Regul* 32:245–258.

Groppa, M. D., M. P. Benavides, and M. S. Zawoznik. 2012. Root hydraulic conductance, aquaporins and plant growth promoting microorganisms: A revision. *Appl Soil Ecol* 61:247–254.

Hajiboland, R., N. Aliasgharzadeh, S. F. Laiegh, and C. Poschenrieder. 2010. Colonization with arbuscular mycorrhizal fungi improves salinity tolerance of tomato (Solanum lycopersicum L.) plants. *Plant Soil* 331:313–327.

Hamdia, A. B., M. A. Shaddad, and M. M. Doaa. 2004. Mechanisms of salt tolerance and interactive effects of Azospirillum brasilense inoculation on maize cultivars grown under salt stress conditions. *Plant Growth Regul* 44:165–174.

Hammer, E. C., H. Nasr, J. Pallon, P. A. Olsson, and H. Wallander. 2011. Elemental composition of arbuscular mycorrhizal fungi at high salinity. *Mycorrhiza* 21:117–129.

Heinemeyer, A., P. Ineson, N. Ostle, and A. H. Fitter. 2006. Respiration of the external mycelium in the arbuscular mycorrhizal symbiosis shows strong dependence on recent photosynthates and acclimation to temperature. *New Phytol* 171:159–170.

Hoekstra, F. A., E. A. Golovina, and J. Buitink. 2001. Mechanisms of plant desiccation tolerance. *Trends Plant Sci* 6:431–438.

Jahromi, F., R. Aroca, R. Porcel, and J. M. Ruiz-Lozano. 2008. Influence of salinity on the in vitro development of *Glomus intraradices* and on the in vivo physiological and molecular responses of mycorrhizal lettuce plants. *Microb Ecol* 55:45–53.

Jang, J. Y., D. G. Kim, Y. O. Kim, J. S. Kim, and H. Kang. 2004. An expression analysis of a gene family encoding plasma membrane aquaporins in response to abiotic stresses in *Arabidopsis thaliana*. *Plant Mol Biol* 54:713–725.

Jha, Y. and R. B. Subramanian. 2011. Paddy plants inoculated with PGPR show better growth physiology and nutrient content under saline conditions. *Chil J Agric Res* 73:213–219.

Kaldorf, M., E. Schemelzer, and H. Bothe. 1998. Expression of maize and fungal nitrate reductase in arbuscular mycorrhiza. *Mol Plant–Microbe Interact* 11:439–448.

Kang, S., A. Khan, M. Waqas, Y. You, J. Kim, J. Kim, M. Hamayun, and I. Lee. 2014. Plant growth-promoting rhizobacteria reduce adverse effects of salinity and osmotic stress by regulating phytohormones and antioxidants in *Cucumis sativus*. *J Plant Interact* 9:673–682.

Kavi Kishor, P. B., S. Sangam, R. N. Amrutha, P. Sri Laxmi, K. R. Naidu, K. R. Rao, S. Rao, K. J. Reddy, P. Theriappan, and N. Sreenivasulu. 2005. Regulation of proline biosynthesis, degradation, uptake and transport in higher plants: Its implications in plant growth and abiotic stress tolerance. *Curr Sci* 88:424–438.

Kloepper, J. W., A. Gutierrez-Estrada, and J. A. McInroy. 2007. Photoperiod regulates elicitation of growth promotion but not induced resistance by plant growth-promoting rhizobacteria. *Can J Microbiol* 53:159–167.

Kruse, E., N. Uehlein, and R. Kaldenhoff. 2006. The aquaporins. *Genome Biol* 7:206.

Kumar, A., S. Sharma, and S. Mishra. 2010. Influence of arbuscular mycorrhizal (AM) fungi and salinity on seedling growth, solute accumulation, and mycorrhizal dependency of *Jatropha curcas* L. *J Plant Growth Regul* 29:297–306.

Langenfeld-Heyser, R., J. Gao, T. Ducic, P. Tachd, C. Lu, E. Fritz, A. Gafur, and A. Polle. 2007. *Paxillus involutus* mycorrhiza attenuate NaCl-stress responses in the salt-sensitive hybrid poplar *Populus* × *canescens*. *Mycorrhiza* 17:121–131.

Lee, S. H., M. Calvo-Polanco, G. C. Chung, and J. J. Zwiazek. 2010. Role of aquaporins in root water transport of ectomycorrhizal jack pine (*Pinus banksiana*) seedlings exposed to NaCl and fluoride. *Plant Cell Environ* 33:769–780.

Lerat, S., L. Lapointe, S. Gutjahr, Y. Piché, and H. Vierheilig. 2003. Carbon partitioning in a split-root system of arbuscular mycorrhizal plants is fungal and plant species dependent. *New Phytol* 157:589–595.

Lugtenberg, B. and F. Kamilova. 2009. Plant-growth-promoting rhizobacteria. *Annu Rev Microbiol* 63:541–546.

Luo, Z. B., D. Janz, X. Jiang, C. Gobel, H. Wildhagen, Y. Tan, H. Rennenberg, I. Feussner, and A. Polle. 2009. Upgrading root physiology for stress tolerance by ectomycorrhizas: Insights from metabolite and transcriptional profiling into reprogramming for stress anticipation. *Plant Physiol* 151:1902–1917.

Mantelin, S. and B. Touraine. 2004. Plant growth-promoting bacteria and nitrate availability impacts on root development and nitrate uptake. *J Exp Bot* 55:27–34.

Marjanović, Z., U. Nehls, and R. Hampp. 2005. Mycorrhiza formation enhances adaptive response of hybrid poplar to drought. *Ann NY Acad Sci* 1048:496–499.

Marulanda, A., R. Azcon, F. Chaumont, J. Ruiz-Lozano, and R. Aroca. 2010. Regulation of plasma membrane aquaporins by inoculation with a *Bacillus megaterium* strain in maize (*Zea mays* L.) plants under unstressed and salt-stressed conditions. *Planta* 232:533–543.

Mayak, S., T. Tirosh, and B. Glick. 2004. Plant growth-promoting bacteria confer resistance in tomato plants to salt stress. *Plant Physiol Biochem* 42:565–572.

Miller, G., N. Suzuki, S. Ciftci-Yilmaz, and R. Mittler. 2010. Reactive oxygen species homeostasis and signalling during drought and salinity stress. *Plant Cell Environ* 33:453–467.

Muhsin, T. M. and J. J. Zwiazek. 2002. Ectomycorrhizas increase apoplastic water transport and root hydraulic conductivity in *Ulmus americana* seedlings. *New Phytol* 153:153–158.

Nautiyal, C. S., S. Srivastava, P. S. Chauhan, K. Seem, A. Mishra, and S. K. Sopory. 2013. Plant growth-promoting bacteria *Bacillus amyloliquefaciens* NBRISN13 modulates gene expression profile of leaf and rhizosphere community in rice during salt stress. *Plant Physiol Biochem* 66:1–9.

Noctor, G. and C. Foyer. 1998. Ascorbate and glutathione: Keeping active oxygen under control. *Ann Rev Plant Physiol Plant Mol Biol* 49:249–279.

Ouziad, F., P. Wilde, E. Schmelzer, U. Hildebrandt, and H. Bothe. 2006. Analysis of expression of aquaporins and Na$^+$/H$^+$ transporters in tomato colonized by arbuscular mycorrhizal fungi and affected by salt stress. *Environ Exp Bot* 57:177–186.

Pang, X., Z. Zhang, X. Wen, Y. Ban, and T. Moriguchi. 2007. Polyamines, all-purpose players in response to environment stresses in plants. *Plant Stress* 1:173–188.

Paul, D. and S. Nair. 2008. Stress adaptations in a plant growth promoting *Rhizobacterium* (PGPR) with increasing salinity in the coastal agricultural soils. *J Basic Microbiol* 48:1–7.

Patten, C. L. and B. R. Glick. 1996. Bacterial biosynthesis of indole-3- acetic acid. *Can J Microbiol* 42:207–220.

Perez-Alfocea, F., A. Albacete, M. Ghanem, and I. Dodd. 2010. Hormonal regulation of source–sink relations to maintain crop productivity under salinity: A case study of root-to-shoot signalling in tomato. *Funct Plant Biol* 37:592–603.

Porcel, R., R. Aroca, and J. Ruiz-Lozano. 2012. Salinity stress alleviation using arbuscular mycorrhizal fungi: A review. *Agron Sustain Develop* 32:181–200.

Rodriguez, R. and R. Redman. 2008. More than 400 million years of evolution and some plants still can't make it on their own: Plant stress tolerance via fungal symbiosis. *J Exp Bot* 59:1109–1114.

Rodriguez-Rosales, M. P., X. Jiang, F. J. Galvez, M. N. Aranda, B. Cubero, and K. Venema. 2008. Overexpression of the tomato K$^+$/H$^+$ antiporter *LeNHX2* confers salt tolerance by improving potassium compartmentalization. *New Phytol* 179:366–377.

Rojas-Tapias, D., A. Moreno-Galván, S. Pardo-Díaz, M. Obando, D. Rivera, and R. Bonilla. 2012. Effect of inoculation with plant growth-promoting bacteria (PGPB) on amelioration of saline stress in maize (*Zea mays*). *Appl Soil Ecol* 61:264–272.

Ruiz-Lozano, J. M. and R. Aroca. 2010. Modulation of aquaporin genes by the arbuscular mycorrhizal symbiosis in relation to osmotic stress tolerance. In: Sechback, J. and Grube, M, eds. *Symbiosis and Stress*. Springer-Verlag, Berlin, Germany, pp. 359–374.

Ryu, C.-M., M. A. Farag, C-H. Hu, M. S. Reddy, J. W. Kloepper, and P. W. Pare. 2004. Bacterial volatiles induce systemic resistance in *Arabidopsis*. *Plant Physiol* 134:1017–1026.

Sade, N., M. Gebretsadik, R. Seligmann, A. Schwartz, R. Wallach, and M. Moshelion. 2010. The role of tobacco aquaporin1 in improving water use efficiency, hydraulic conductivity, and yield production under salt stress. *Plant Physiol* 152:245–254.

Safir, G. R., J. S. Boyer, and J. W. Gerdemann. 1972. Nutrient status and mycorrhizal enhancement of water transport in soybean. *Plant Physiol* 49:700–703.

Sannazzaro, A. I., M. Echeverria, E. O. Alberto, O. A. Ruiz, and A. B. Menendez. 2007. Modulation of polyamine balance in *Lotus glaber* by salinity and arbuscular mycorrhiza. *Plant Physiol Biochem* 45:39–46.

Sannazzaro, A. I., R. Oscar, A. Edgardo, and M. Ana. 2006. Alleviation of salt stress in *Lotus glaber* by *Glomus intraradices*. *Plant Soil* 285:279–287.

Schubert, S., A. Neubert, A. Schierholt, A. Sumer, and C. Zorb. 2009. Development of salt-resistant maize hybrids: The combination of physiological strategies using conventional breeding methods. *Plant Sci* 177:196–202.

Sharifi, M., M. Ghorbanli, and H. Ebrahimzadeh. 2007. Improved growth of salinity-stressed soybean after inoculation with salt pre-treated mycorrhizal fungi. *J Plant Physiol* 164:1144–1151.

Shawky, B. T. 1990. Effect of azotobacters and azospirilla on germination of seeds of some agricultural crops. *Zentralbl Mikrobiol* 145:209–217.

Sheng, M., M. Tang, H. Chen, B. W. Yang, F. F. Zhang, and Y. H. Huang. 2008. Influence of arbuscular mycorrhizae on photosynthesis and water status of maize plants under salt stress. *Mycorrhiza* 18:287–296.

Sheng, M., M. Tang, F. F. Zhang, and Y. H. Huang. 2011. Influence of arbuscular mycorrhiza on organic solutes in maize leaves under salt stress. *Mycorrhiza* 21:423–430.

Siddikee, M. A., B. R. Glick, P. S. Chauhan, W. J. Yim, and T. Sa. 2011. Enhancement of growth and salt tolerance of red pepper seedlings (*Capsicum annuum* L.) by regulating stress ethylene synthesis with halotolerant bacteria containing 1-aminocyclopropane-1-carboxylic acid deaminase activity. *Plant Physiol Biochem* 49:427–434.

Siemens, J. A. and J. J. Zwiazek. 2008. Root hydraulic properties and growth of balsam poplar (*Populus balsamifera*) mycorrhizal with *Hebeloma crustuliniforme* and *Wilcoxina mikolae* var. mikolae. *Mycorrhiza* 18:393–401.

Smith, S. E. and D. J. Read. 2008. *Mycorrhizal Symbiosis*, 3rd edn. Academic Press, London, U.K.

Soderstrom, B. and D. J. Read. 1987. Respiratory activity of intact and excised ectomycorrhizal mycelial systems growing in unsterilized soil. *Soil Biol Biochem* 19:231–236.

Sziderics, A. H., F. Rasche, F. Trognitz, A. Sessitsch, and E. Wilhelm. 2007. Bacterial endophytes contribute to abiotic stress adaptation in pepper plants (*Capsicum annuum* L.). *Canad J Microbiol* 53:1195–1202.

Talaat, N. B. 2014. Effective microorganisms enhance the scavenging capacity of the ascorbate–glutathione cycle in common bean (*Phaseolus vulgaris* L.) plants grown in salty soils. *Plant Physiol Biochem* 80:136–143.

Talaat, N. B. and A. Abdallah. 2008. Response of faba bean (*Vicia faba* L.) to dual inoculation with *Rhizobium* and VA mycorrhiza under different levels of N and P fertilization. *J Appl Sci Res* 4:1092–1102.

Talaat, N. B. and B. T. Shawky. 2011. Influence of arbuscular mycorrhizae on yield, nutrients, organic solutes, and antioxidant enzymes of two wheat cultivars under salt stress. *J Plant Nutr Soil Sci* 174:283–291.

Talaat, N. B. and B. T. Shawky. 2012. Influence of arbuscular mycorrhizae on root colonization, growth and productivity of two wheat cultivars under salt stress. *Arch Agron Soil Sci* 58:85–100.

Talaat, N. B. and B. T. Shawky. 2013. Modulation of nutrient acquisition and polyamine pool in salt-stressed wheat (*Triticum aestivum* L.) plants inoculated with arbuscular mycorrhizal fungi. *Acta Physiol Plant* 35:2601–2610.

Talaat, N. B. and B. T. Shawky. 2014a. Protective effects of arbuscular mycorrhizal fungi on wheat (*Triticum aestivum* L.) plants exposed to salinity. *Environ Exp Bot* 98:20–31.

Talaat, N. B. and B. T. Shawky. 2014b. Modulation of the ROS-scavenging system in salt-stressed wheat plants inoculated with arbuscular mycorrhizal fungi. *J Plant Nutr Soil Sci* 177:199–207.

Wright, D. P., D. J. Read, and J. D. Scholes. 1998. Mycorrhizal sink strength influences whole plant carbon balance of *Trifolium repens* L. *Plant Cell Environ* 21:881–891.

Wu, Q. S., Y. N. Zou, W. Liu, X. F. Ye, H. F. Zai, and L. J. Zhao. 2010. Alleviation of salt stress in citrus seedlings inoculated with mycorrhiza: Changes in leaf antioxidant defense systems. *Plant Soil Environ* 56:470–475.

Yang, X., Z. Liang, X. Wen, and C. Lu. 2008. Genetic engineering of the biosynthesis of glycinebetaine leads to increased tolerance of photosynthesis to salt stress in transgenic tobacco plants. *Plant Mol Biol* 66:73–86.

Yao, L. X., Z. S. Wu, Y. Y. Zheng, I. Kaleem, and C. Li. 2010. Growth promotion and protection against salt stress by *Pseudomonas putida* Rs-198 on cotton. *Eur J Soil Biol* 46:49–54.

Yi, H., M. Calvo-Polanco, M. MacKinnon, and J. Zwiazek. 2008. Responses of ectomycorrhizal *Populus tremuloides* and *Betula* seedlings to salinity. *Environ Exp Bot* 62:357–363.

Zawoznik, M. S., M. Ameneiros, M. P. Benavides, S. Vazquez, and M. D. Groppa. 2011. Response to saline stress and aquaporin expression in *Azospirillum*-inoculated barley seedlings. *Appl Microbiol Biotechnol* 90:1389–1397.

Zhang, H., C. Murzello, Y. Sun, M. Kim, X. Xie, R. M. Jeter, J. C. Zak, S. E. Dowd, and P. W. Paré. 2010. Choline and osmotic-stress tolerance induced in *Arabidopsis* by the soil microbe *Bacillus subtilis* (GB03). *Mol Plant Microbe Interact* 23:1097–1104.

CHAPTER SIXTEEN

# Molecular Breeding for Salt Stress Tolerance in Plants

*Mirzamofazzal Islam, S.N. Begum, N. Hoque, and M.K. Saha*

CONTENTS

*Abstract.* Soil salinity is a foremost abiotic stress for agriculture, sturdily influencing plant productivity worldwide. The reduction of productivity in saline soils can be overcome by soil reclamation or by improving salt tolerance in target plants. Plant salt tolerance is a complex trait affected by numerous genetic and nongenetic factors, and its improvement via conventional breeding is less noteworthy. Recent advancements in biotechnology have led to the development of more efficient selection paraphernalia to substitute phenotype-based selection. Gene(s) or quantitative trait loci (QTLs) that affect important traits and association with molecular markers have been identified, which could be used as indirect selection criteria to improve breeding effectiveness by means of marker-assisted selection (MAS). While the use of MAS for manipulating simple traits has been rationalized in many plant breeding programs, MAS for improving complex traits seems to be at the beginning stage. In the last decade, a rapid progress has been made toward the development of molecular marker technologies and their application in constructing linkage maps, molecular dissection of the complex agronomical traits, and marker-assisted breeding (MAB). The basis of MAB approach is to transfer a specific allele at the target locus from a donor line or variety to a recipient line/variety while selecting against donor introgressions across the rest of the genomes. Molecular marker and its use permit the genetic dissection of the progeny at each generation and increase the speed of the selection process, thus increasing genetic gain per unit time. The main advantages of MAB are (1) efficient foreground selection for the target locus, (2) efficient background selection for the recurrent parent genome, (3) linkage drag minimization surrounding the locus being introgressed, and (4) rapid breeding of new genotypes with favorable traits. The effectiveness of MAB depends on the availability of closely linked markers and/or flanking markers for the target locus, the size of the population, the number of backcrosses, and the position and number of markers for background selection. Numerous QTLs have been reported for ST in different crop species; however, few commercial cultivars or breeding lines with improved ST have been developed by the use of MAS. Strategies and current status of MAS, MAB, QTL mapping, genetic engineering, and application of RNAi are discussed in this chapter, which are important to learn for the development of salt-tolerant plants.

## 16.1 INTRODUCTION

Salinity is one of the most severe factors restricting the productivity of 45 million hectares (Carillo et al. 2011) of land and agricultural crops grown on it, with unfavorable effects on germination, plant vigor, and crop yield (Munns and Tester 2008). Increasing threat of salinity has become an essential issue linked to the penalty of climate change where uprising $CO_2$ concentration increases temperature, resulting in rising sea-level and enlargement of coastal areas worldwide. High salinity affects plants in more than a few ways: water stress, ion toxicity, nutritional disorders, oxidative stress, alteration of metabolic processes, membrane disorganization, reduction

of cell division and expansion, and genotoxicity (Hasegawa et al. 2000; Munns 2002; Zhu 2007). Together, these effects reduce plant growth, development, and survival. During the onset and development of salt stress within a plant, all the major processes such as photosynthesis, protein synthesis, and energy–lipid metabolism are affected (Parida and Das 2005). During initial exposure to salinity, plants experience water stress, which in turn reduces leaf expansion.

Ionic stress consequences in premature senescence of older leaves and in toxicity symptoms like chlorosis and necrosis in mature leaves due to high $Na^+$, which affects plants by disrupting protein synthesis and interfering with enzyme activity (Hasegawa et al. 2000; Munns 2002; Munns and

Termaat 1986). Many plants have evolved several mechanisms either to exclude salt from their cells or to tolerate its presence within the cells. In this chapter, we mainly discuss soil salinity, its effects on plants, and tolerance mechanisms that permit the plants to withstand stress, with particular emphasis on ion homeostasis, $Na^+$ exclusion, and tissue tolerance, as well as marker-assisted breeding (MAB) and transgenic breeding techniques.

Salt tolerance is a quantitative trait under polygenic control that is influenced by environmental variation. Because of linkage drag, the classical method cannot detect a single gene locus associated with quantitative traits, their locations on chromosomes, or their relationship with other genes. Phenotypic screening is not very reliable and will delay breeding process. Recent advancement of molecular markers through marker-assisted selection (MAS) facilitates to analyze both the simply inherited traits and the quantitative traits and identify the individual genes controlling the traits of interest and genetic modification of these traits and introgression to release crops with improved salt tolerance.

## 16.2 SALINITY AND ITS EFFECT ON PLANT

Accumulation of salts in the surface zone of land is simply called soil salinity, which has a number of causes and varies in different geological and climatic regions where the causes can be natural due to clearing of native vegetation (dryland salinity) or due to irrigation (Munn 2002). Natural salinity is caused by short-term waterlogged conditions after irrigation that produce relatively lower electrical conductivity (EC) 2–4 dS m$^{-1}$, while dryland salinity is caused seasonally when waterlogged conditions with EC 4–8 dS m$^{-1}$ exist (Munns 1993), whereas more than this range of salinity causes high salinity that can be seeping or drying according to season in discharge areas.

Salinity impacts on plants from the seed germination stage to postreproductive stage and causes specific ion toxicity and nutritional disorders (Lauchi and Epstein 1990). Actually, salinity imposes two major stresses on the plant: one is a high osmotic pressure in the soil solution, making it harder for the plant to extract water, and the second is a high NaCl concentration in the soil solution that makes it difficult for the plant to exclude the NaCl while taking up other cations and anions (Carillo et al. 2011). The effects

of these two stresses are seen in sequence. Salinity lowers the water potential of the roots, and this quickly causes reductions in growth.

Accumulation of salts in the shoot and leaf affects stomatal closure, with concomitant increases in leaf temperature and inhibition of shoot elongation (Gupta and Haung 2014). Thus, the primary consequence is the overall reduction in the production of new leaves and a significant reduction in shoot growth, which was termed and described as the "osmotic phase" by Munns and Tester (2008).

Next, a slower effect is the result of salt toxicity in leaves where a salt-sensitive species or genotype differs from a more salt-tolerant one by its inability to prevent salt accumulation in leaves to toxic levels (James et al. 2011), which is due to accumulation of salt over time, especially in the older leaves, causing the premature senescence of those older leaves that is termed the "ionic phase" of salt toxicity, which is due to both the accumulation of salts and the inability of the shoot to tolerate the salt that has accumulated to toxic concentrations (Munns and Tester 2008).

For the last decades, most of the studies indicated that plants were particularly susceptible to salinity during the seedling and early vegetative growth stage as compared to germination, found in rice, wheat, jute, pulses, etc.

Also, screening at germination provides little beginning for assessing crop salt tolerance where most of the germination studies have been conducted in the laboratory using Petri dishes with different extents of saline solution (Rozema and Flowers 2008). Reproductive development is considered less sensitive to salt stress than vegetative growth, although in wheat and rice, salt stress can speed up reproductive growth, hinder spike development, and reduce the yield potential, while in the more salt-sensitive rice, low yield is first and foremost associated with a decline in tillers and by sterile spikelet in some cultivars (Munns 2005). After the salt-sensitive early vegetative growth stage, maximum research recommended that most crops turn into gradually more tolerant as the plants grow older (Lauchli and Epstein 1990).

### 16.2.1 Mechanism of Salinity Tolerance

As salinity has impact on plants in different organs during different stages, so plants adapt many mechanisms to tolerate this stress. The general

TABLE 16.1
*Salinity tolerance mechanism and mode of action in a whole plant.*

| Sl. No. | Mechanism | Mode of action |
|---|---|---|
| 1 | Salt exclusion | Plants do not take up excess salt by selective absorption. |
| 2 | Salt reabsorption | Tolerant varieties absorb excess salt but it is reabsorbed from the xylem and $Na^+$ is not translocated to the shoot. |
| 3 | Root–shoot translocation | Salt tolerance is associated with a high electrolyte content in the roots and a low content in the shoot. |
| 4 | Salt translocation | Tolerant plants have the ability to translocate a lesser proportion of $Na^+$ to the shoot. |
| 5 | Salt compartmentation | Excess salt is transported from younger to older leaves. |
| 6 | Tissue tolerance | Plants absorb salt that properly compartmentalized in vacuoles within leaves in order to lower the harmful effects on plant growth. |

mechanism of salt tolerance in whole plants with mode of action according to the study of Hasegawa et al. (2000), Munns (1993, 2002), and Munns and Tester (2008) is given in Table 16.1

At the cellular level, plants develop various physiological and biochemical mechanisms under salinity stress (Gupta and Haung 2014) including (1) osmotic tolerance (which is regulated by long-distance signals that reduce shoot growth and is triggered before shoot $Na^+$ accumulation), (2) ion exclusion (where $Na^+$ and $Cl^-$ transport processes in roots reduce the accumulation of toxic concentrations of $Na^+$ and $Cl^-$ within leaves), and (3) tissue tolerance (where high salt concentrations are found in leaves but are compartmentalized at the cellular and intracellular levels, especially in the vacuole). Though details of *osmotic tolerance* are still unknown, it must involve rapid, long-distance signaling, perhaps via processes such as reactive oxygen species (ROS) waves (Mittler et al. 2011), $Ca^{2+}$ waves, or even long-distance electrical signaling (Maischak et al. 2010). Differences in osmotic tolerance may be due to differences in this long-distance signaling, or they may involve differences in the initial perception of the salt or differences in the response to the signals, which is still an area of salinity research with many unknowns and where further research is required to obtain a better understanding of osmotic tolerance.

In *ion exclusion*, which is due to the accumulation of $Na^+$ and $Cl^-$ in the leaf blade, plants can reduce toxicity during the ionic phase by reducing accumulation of toxic ions in the leaf blades ($Na^+$ and $Cl^-$ exclusion) and/or by increasing their ability to tolerate the salts that they have failed to exclude from the shoot, such as by compartmentation into

vacuoles (tissue tolerance) where a number of transporters and their controllers are involved at both plasma membrane and tonoplast (Plett and Moller 2010; Tester and Davenport et al. 2003). On the other hand, *tissue tolerance*, involving the removal of $Na^+$ from the cytosol and compartmentalizing it in the vacuole before the ion has a detrimental effect on cellular processes, is also likely to require the synthesis of compatible solutes and higher-level controls to coordinate transport and biochemical processes, thus having a role in both osmoprotection and osmotic adjustment (Flowers and Colmer 2008; Munns and Tester 2008).

## 16.2.2 Genes Involved in Crops for Salt Tolerance

Currently, it is clear that salinity tolerance can be complex and involve multigenic nature (Dewey 1962; Flowers and Colmer 2008). The role of several genes involved in the following categories that were studied from the beginning of molecular breeding for salt tolerance is described in this section.

### 16.2.2.1 Ion Exclusion

The high-affinity potassium transporter (HKT) gene family (Ali et al. 2012; Byrt et al. 2007; Davenport et al. 2007; Hauser and Horie 2010; Horie et al. 2006, 2009; Platten et al. 2006; Xue et al. 2011) and the salt overly sensitive (SOS) pathway (Kudla et al. 2010; Mahajan et al. 2008; Yang et al. 2009) have both been implicated in having an important role in regulating $Na^+$ transport within a plant. Manipulation of the expression of these genes has been frequently reported to alter accumulation of Na$^+$ in the shoot. However, to date, the application of this knowledge into generating successful crop plants in the field has been limited. Between the two families, the HKT1 has perhaps the greatest potential for improving the salinity tolerance of crops, frequently appearing as the most likely candidate for quantitative trait loci (QTLs) when phenotyping for salt tolerance and Na$^+$ exclusion in mutant and mapping populations (Ahmadi et al. 2011; James et al. 2006; Ren et al. 2005). A MAS approach was used successfully to incorporate novel HKT alleles from Triticum monococcum to improve the salinity tolerance of durum wheat (James et al. 2012; Munn et al. 2012; Rus et al. 2006). In contrast, transgenic approaches to improve salinity tolerance using HKT1s have been only moderately successful; an HKT2 has been reported to increase salinity tolerance, although not through Na$^+$ exclusion (Very 2011).

### 16.2.2.2 Shoot Tissue Tolerance

To date, three main mechanisms contributing to shoot tissue tolerance have been targeted: (1) accumulation of Na$^+$ in the vacuole, (2) synthesis of compatible solutes, and (3) production of enzymes catalyzing detoxification of ROS. Genes encoding Na$^+$/H$^+$ antiporters (NHX) and vacuolar H$^+$ pyrophosphatases increase the expression of those proteins (Stuart et al. 2014), which are involved in the synthesis of compatible solutes (such as proline and glycinebetaine), and enzymes responsible for the detoxification of ROS have had differing degrees of success in improving crop salinity tolerance. With a large number of papers reporting success in increasing plant salinity tolerance by improving shoot tissue tolerance, particularly by improving accumulation of Na$^+$ in vacuoles by transgenic approach, it might be persuasive to recommend this is the best mechanism for improving crop performance.

### 16.2.2.3 Osmotic Tolerance

Differences in osmotic tolerance are likely to involve long-distance signaling and control of cell cycle and processes involving perception of signals from the roots in the shoots, which is why genes encoding for osmotic tolerance traits have been considered as potential than those involved in ion exclusion.

The introgression of TmHKT1 (candidate gene for osmotic tolerance) from T. monococcum into the durum wheat, Tamaroi, resulted in a significant improvement in grain yield in salt-stressed, field-grown durum by increasing its ion exclusion, but only in plots with highly saline soils (James et al. 2012; Munn et al. 2012). However, the yield of Tamaroi with TmHKT1 was similar to that observed in the Tamaroi cultivar without the introgressed gene, under low and moderate saline conditions (James et al. 2012; Munn et al. 2012), suggesting that osmotic stress was having a greater effect on the end yield of these plants growing in low to moderate salinity than ionic stress.

### 16.2.2.4 Signaling or Regulatory Pathways

Signaling and regulatory pathways are also involved in salinity tolerance of crops. ROS signaling has recently been shown to be important for regulating plant responses to salinity (Mittler et al. 2011; Suzuki et al. 2011) and is involved in controlling shoot Na$^+$ accumulation by regulating vasculature Na$^+$ concentrations (Suzuki et al. 2011). Ca$^{2+}$ signaling cascade is responsible to control plant growth and development.

In addition, SOS stress signaling pathway plays a role in ion homeostasis where three major genes like SOS1(encode a plasma membrane Na/H antiporter), SOS2 (encode serine or threonine kinase), and SOS3 (myristoylated Ca$^{2+}$-binding protein) are involved, among which SOS1 is essential in regulating Na$^+$ efflux at cellular level and facilitates long-distance transport of Na$^+$ from root to shoot; SOS2 is activated by salt stress Ca$^{2+}$ and SOS3 involves in the SOS stress signaling pathway (Guo et al. 2004; Hasegawa et al. 2000; Ishitani et al. 2000; Liu et al. 2000; Quintero et al. 2002; Sanders 2000; Shi et al. 2000, 2002). Researchers establish that C-terminal regulatory domain of SOS2 protein contains a FISL motif (also known as NAF domain), which is about 21-amino-acid-long sequence, and serves as a site of interaction for Ca$^{2+}$-binding SOS3 protein. This interaction between SOS2 and SOS3 proteins results in the activation of kinase (Guo et al. 2004). The activated kinase then phosphorylates SOS1 protein, thereby increasing its transport activity that was initially identified in yeast (Quintero et al. 2002). The phosphorylated SOS1 results in the increased Na$^+$ efflux, reducing Na$^+$ toxicity (Martimez et al. 2007).

## 16.3 MOLECULAR BREEDING

Molecular plant breeding (MPB) is an evolving science and technology where genetic manipulation is performed at DNA molecule level to get better characters of interest in plants for the development of improved cultivars suited to meet the demands of farmers and consumers. In MPB, DNA markers are used as a substitute for phenotypic selection and to accelerate the releasing process of improved cultivars. It has gradually been evolving from art to science over the last 10,000 years (Xu 2010) preliminary as an ancient art to the present molecular design–based science. MAB, mutation breeding, and genetic engineering are being used as essential tools for MPB, but in near future, DNA chips and other modern technologies including robotics and bioinformatics will play an important role in plant breeding (Xu 2010).

Though MPB found the scientific basis from the Darwin law and Mendelian genetics at the beginning, subsequent advances in our understanding of plant biology, the analysis and induction of genetic variation, cytogenetics, quantitative genetics, molecular biology, biotechnology, and most recently genomics have been successively applied to further increase the scientific base and its application to the plant breeding process (Stoskopf et al. 1993).

In practice, MPB is a three-step process (Stephen and Rita 2008), wherein populations or germplasm collection with useful genetic variation is created or assembled, individuals with superior phenotypes are identified, and improved cultivars are developed from the selected individuals.

Primary goals of plant breeding have typically aimed at improved yields, nutritional quantities, resistance and tolerance to biotic and abiotic stresses as well as other traits of commercial value. MPB also expands genetic diversity from breeding populations, segregating progeny, exotic materials that are not adapted to the target environment, wide interspecific crosses, natural occurring or induced mutations, the introduction of transgenic events, or combinations of these sources (Stephen and Rita 2008). In addition to the increase in favorable gene action due to MPB, it is important to emphasize that QTL studies, when conducted with appropriate scale and precision to identify causal genes responsible for the trait of interest, represent a powerful functional genomics approach. It increases the efficiency of selection and removes the shortfalls of conventional breeding like challenges in measuring phenotypes, identifying individuals with the highest breeding value, environmental problems, extra manpower, etc. (Wenzel 2006).

To meet global needs for sustainable increases in agricultural productivity, MPB will further expand its contribution and impacts. Due to adverse climate change, saline-prone area is increasing day by day, which is why salt-tolerant cultivars are very essential to ensure food security globally. To achieve this aim, the knowledge and expertise of MPB and many logistic and genetic constraints related to stress breeding also need to be resolved.

## 16.4 ADVANTAGES, DISADVANTAGES, AND CHALLENGES OF MOLECULAR BREEDING

Molecular breeding can allow any kinds of trait to be carried out at seedling stage as it deals with genomic DNA and no selfing or testcrossing is needed to detect the traits controlled by recessive alleles, thus saving time and accelerating breeding progress (Guo 2013). Molecular breeding can be not affected by environment as any kinds of experiments are possible to conduct in greenhouses and during off-season, which is very helpful for improvement of some traits like disease or pest resistance and stress tolerance that are expressed only when favorable environmental conditions exist. As one gene may mask the effect of additional genes, in traditional phenotypic selection, it is problematic to distinguish individual genes or loci. But in the case of molecular breeding, for the traits controlled by multiple genes or QTLs, individual genes or QTLs can be identified and selected through MAS at the same time and in the same individuals, and thus molecular breeding is particularly suitable for gene pyramiding (Wang and Chee 2010). Genotypic assays based on molecular markers can be faster, cheaper, and more accurate than conventional phenotypic assays, depending on the traits and conditions, which is why molecular breeding may result in higher effectiveness and higher efficiency in terms of time, resources, and efforts saved. The use of molecular breeding in plants has increased in research for the improvement of crop cultivars particularly since the beginning of the last decade, but there are some constraints to simply inherited traits, such as monogenic or oligogenic resistance to diseases or pests, although quantitative traits were also involved (Collard and Mackill 2008). Based on the investigation of Collard and Mackill (2008) and the viewpoint of plant breeders, there are some reasons for the low impact of molecular breeding, for example not all markers are breeder friendly. This problem may be solved by converting non-breeder-friendly markers to other types of breeder-friendly markers like restriction fragment length polymorphism (RFLP) to sequence-tagged site (STS) and randomly amplified length polymorphism (RAPD) to sequence-characterized amplified region. Not all markers can be applicable across populations due to lack of marker polymorphism or reliable marker–trait association. Multiple mapping populations are helpful in understanding marker allelic diversity and genetic background effects. In addition, QTL positions and effects also need to be validated and reestimated by breeders in their specific germplasm (Heffner et al. 2009).

Besides, false selection may occur due to recombination between the markers and the genes or QTLs of interest. Use of flanking markers or more markers for the target gene/QTL can help in this case. The efficiency of QTL detection is attributed to multiple factors, such as algorithms, mapping methods, number of polymorphic markers, and population type and size (Wang et al. 2012). High-marker-density fine mapping with large populations and well-designed phenotyping across multiple environments may provide more accurate estimates of QTL location and effects. A large number of breeding programs have not been equipped with adequate facilities and conditions for a large-scale adoption of MAB in practice. Molecular breeding requires higher start-up expenses and labor costs than conventional breeding, though molecular breeding cannot replace conventional breeding but is rather a supplementary addition to conventional breeding.

In spite of high-cost, technical, equipmental, and infrastructural requirements in molecular breeding, which are to be considered as major obstacles, developing countries have already used this technology from the time when they felt the demand of food security (Collard and Mackill 2008; Ribaut et al. 2010). But we hope integration of breeding into conventional breeding programs will be an optimistic strategy for crop improvement in the near future by removing all kinds of drawbacks related to theory, technology, and application of such kind of breeding modern programs (Guo 2013).

## 16.5 MOLECULAR BREEDING TOOLS

There are many kinds of tools for molecular breeding like markers and genetic maps. Such tools remove the drawbacks of conventional breeding by measuring the polymorphism at the molecular level based on DNA and their effect on the phenotype.

### 16.5.1 Markers

In molecular biology, markers are genetic traits that are determined by allelic forms and can be used as investigational probes or tags to track an individual, a tissue, a cell, a nucleus, a chromosome, or a gene and transmitted from one generation to another, which ultimately facilitates the study of inheritance and variation. A marker should have codominance, high level of genetic polymorphism, easy detection, features for clear distinct allele, neutral selection, cost-effectiveness, and high duplicability. In a broad sense, markers are classified as classical and DNA markers (Xu 2010) (Figure 16.1).

#### 16.5.1.1 Classical Markers

Classical markers are classified as morphological, protein, and cytological markers. Among them, morphological markers were basis for breeding of plant in the Mendelian era when a particular morphological trait like height, color, and grain size was used as marker to select parental lines in a cross. But segregation of traits in further generations was the starting point of thinking for any genetic analysis of plants. Then important morphological traits were mapped to particular chromosomes using the Mendelian laws of inheritance that represented genetic polymorphisms that were visible as differences in appearance such as distinct differences in response to abiotic and biotic stresses and the presence/absence of specific morphological characteristics. For example there are a total of over 300 morphological markers available for genetic studies in rice (Khush 1987), and more are being created for functional genomics, whereas many morphological marker stocks are also available for tomato, maize, and soybean (Neuffer et al. 1997; Palmer and Shoemaker 1998). Nevertheless, it is complicated to construct a comparatively saturated genetic map because of the limitation in the number of morphological

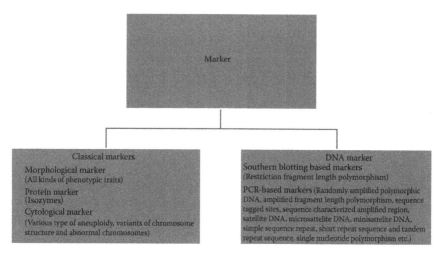

Figure 16.1. Classification of marker used in molecular breeding.

markers with discernible polymorphisms. Due to various deleterious effects on morphological traits by environmental factor, such markers were unable to find acceptability in genetics and plant breeding.

Some plant breeders have also been using cytological markers like number and variants of chromosome structure from different species and particular cytogenetic features (various types of aneuploidy).

The structural features of chromosomes can be shown by chromosome karyotype and bands where the banding patterns are indicated by color, width, order, and position, revealing the difference in distributions of euchromatin and heterochromatin. Such markers have been widely used to identify linkage groups within specific chromosomes and have been extensively applied in physical mapping. But they have constrained applications in genetic diversity analysis and genetic mapping due to limited number and resolution.

To overcome the drawbacks of morphological and cytological markers, plant breeders have used protein or biochemical markers like isozymes that are structural variants of an enzyme and differ from the original enzyme in molecular weight and mobility in an electric field, though they have the same catalytic activity. The difference among isozymes is caused by point mutations resulting from amino acid substitution such that isozymes reflect the products of different alleles rather than different genes, which is why isozymes can be genetically mapped onto chromosomes and then used as genetic markers for mapping important genes.

But the numbers of isozymes are limited and each isozyme can only be identified with a specific stain, so that they cannot be used to construct a complete genetic map (Xu 2010).

### 16.5.1.2 DNA Markers

After the discovery of the double-helix model of DNA, a new revolution started in plant breeding that was actually gene revolution. From 1980 to date, many kinds of DNA markers have been identified to determine a specific character (Xu 2010). DNA markers are defined as a fragment of DNA revealing mutations or variations, which can be used to detect polymorphism between different genotypes or alleles of a gene for a particular sequence of DNA in a population or gene pool.

Actually, a DNA marker is a small region of DNA sequence showing polymorphism (base deletion, insertion, and substitution) between different individuals. Suitable DNA markers should represent genetic polymorphism at the DNA level and should be reproducible across tissues, organs, developmental stages, and environments. They should also be neutral with no effect on the expression of the target trait. Today, molecular markers can be employed not only in MAS, marker-assisted backcrossing (MABC), and marker-assisted gene pyramiding (MAGP) but also in their applications in improved access and utilization of germplasm resources, genetic analysis of breeding populations, parental selection, protection of plant breeder's rights, and comparative mapping (Collard and Mackill 2008).

Simply, DNA markers are categorized on Southern blotting (a kind of nucleic acid hybridization technique) and polymerase chain reaction (PCR). The names of some DNA markers are given in Figure 16.1. RFLP markers were first used by Botstein et al. (1980) in human linkage mapping and subsequently in animal genetics, plant breeding, germplasm characterization, and management. Actually, they are the first generation of DNA markers and a type of Southern blotting–based markers. In plants, as with other living organisms, mutation events like deletion and insertion may occur at restriction sites or between adjacent restriction sites in the genome where gain or loss of restriction sites resulting from mutation at restriction sites within the restriction fragments may cause differences among individuals. RFLP markers are codominant and locus specific, which is why they have high reproducibility and are known as powerful tools for comparative and synteny mapping though they may be problematic as requiring large amounts of high-quality DNA and involving radioactive autoradiography (Konieczny and Ausubel 1993).

In plant molecular biology, among PCR-based marker systems, RAPD marker was the oldest one where the total genomic DNA of a plant is amplified by PCR using a single, short, and random primer that binds to many different loci to amplify random sequences from a complex DNA template that is complementary to it (Williams et al. 1990). The amplified fragments generated by PCR depend on the length and size of both the primer and the target genome where the PCR products up to 3 kb are separated by agarose gel electrophoresis, imaged by ethidium bromide staining and visualized by gel documentation system. Though RAPD has low reproducibility and incapability to detect allelic differences in heterozygotes, they yield high levels of neither polymorphism using a simple and easy technique and requiring radioactive DNA probes nor large amount of pure DNA (Welsh and McClelland 1990).

Amplified fragment length polymorphism (AFLP) markers are another kind of PCR-based marker (Zabeau and Voss 1993) that are visualized by selective PCR amplification of DNA restriction fragments from a total double digest of genomic DNA under stringent conditions like the combination of polymorphism at restriction sites and hybridization of arbitrary primers. Though AFLP primer (17–21 nucleotides in length) is a combination of a synthetic adaptor sequence, a restriction endonuclease recognition sequence, and an arbitrary and nondegenerate *selective* sequence (1–3 nucleotides), the primers are capable of annealing perfectly to their target sequences (the adapter and restriction sites) as well as a small number of nucleotides adjacent to the restriction sites (Vos et al. 1995). AFLPs have relatively more reproducibility and require small quantity of DNA samples where polymorphic information content for biallelic markers is low. Besides this marker system is not cost-efficient, though it may be applicable in biodiversity studies, analysis of germplasm collections, genotyping of individuals, identification of closely linked DNA markers, construction of genetic DNA marker maps, construction of physical maps, gene mapping, and transcript profiling.

## 16.5.2 QTL Approach

Adaptive response of plants to salinity has been unveiled in the past decades. Remarkable advances have been achieved in the field of molecular complexity of the traits responsible for saline susceptibility or tolerance (Blum et al. 1996). Plant breeders usually manipulate the naturally occurring allelic variation for improved crop performance, but adaptive response of genes to salinity was understood through studies of *Arabidopsis* mutants at loci inheritance in the Mendelian fashion (Werner and Finkelstein 1995). The low heritability of salinity, tolerance-related characters, and its quantitative inheritance had hindered the complete understanding of the genetic and physiological bases of yield in crops when exposed to salinity (Blum 1988). Extensive efforts have been made for the manipulation of morphophysiological and biochemical traits of higher heritability than that of yield itself where direct selection for yield in crops gives low response in saline environment (Duvick 2005).

QTLs governing the relevant genetic variation for salinity in cultivated wild germplasm and the dissection of genetic basis of stress tolerance have been possible through the use of molecular tools (Tanksley 1993), which includes cloning, genetic engineering, and MAS of target QTLs for improving yield under saline condition; it is possible through molecular breeding. The rule and function of genes leading the response to salinity have been elucidated by the information generated by sequencing, postgenomic platforms, and bioinformatics (Tubersa et al. 2002). QTLs for salinity tolerance and the number of cloned QTLs relevant for improving tolerance to salinity are very low because identification of major QTLs suitable for cloning faces difficulties. Moreover, phenotypic effects of QTLs and their accurate characterization in the saline environment are unsuccessful in most cases (Salvi and Tuberloya 2005).

### 16.5.2.1 Targeting QTLs for Tolerance to Salinity

A number of QTLs with additive effect for salinity-related traits have been identified. Congenic strains that are different for a small chromosome region flanking the target QTL are required for the characterization and validation of a QTL. Consistency of an effect, its breeding value, and parental alleles present at a QTL region need to be evaluated through testcross that is obtained from the crossing of different tester lines with pairs of near isogenic lines (NILs) (Landi et al. 2007). These lines are needed for positional cloning of QTLs of salinity tolerance through systematic and high-throughput approach (Huang et al. 2006). Effectiveness of such appoaches using QTL discovery and cloning of gene have been prove in maize and tomato (Salvi and Tuberosa 2005; Zamir 2001). Sometimes, thousands of introgressed lines must be produced to understand the complex polygenic traits of salinity (Zhikang et al. 2005). Introgressed lines thought to be containing a significant portion of allelic diversity exist in the primary gene pool of a respective crop that affects some selected complex phenotypes after introgression of a trait.

### 16.5.2.2 Positional Cloning

In NILs, derived from a large segregating population where only the target QTL segregates for the trait, positional cloning is done in such a large population where segregating population allows for the recovery of high number of recombinant in the target region of genome, which gives an ample level of map resolution for cloning of gene. Bulk segregant analysis (Salvi et al. 2002), comparative mapping based on the system of model species, and microarray analysis among crops can help the markers in the target region of a chromosome (Salvi et al. 2005a).

Tightly linked marker becomes anchored to the chromosome after fine mapping, which is extendable if the sequence of the entire genome is available. QTL mapping and cloning can be facilitated by mapping any monomorphic probe onto the map. In the absence of a genome sequence, cloning information is gathered by the screening of a genomic library for genomic sequence or polymorphic genes that cosegregate with QTL. These genomic sequences are then function tested with a number of approaches like association mapping, genetic engineering, or identification knockdown gene. On the other hand, a genome-wide functional screening and validation for almost any crop species could be done by RNA interference (RNAi) and target induced local lesion in genome (TILLING process) (Jenk et al. 2007).

### 16.5.2.3 Association Mapping

QTL detection through association mapping is based on the molecular and phenotypic characterization of unrelated accessions of crops (Flint et al. 2003). Association mapping is a less time-consuming production of large experimental populations for performing QTL analysis and cloning (Buckler and Thornsberry 2002; Morgante and Salamini 2003). The level of linkage disequilibrium (LD) present in a population, the availability and cost of molecular markers, and the presence of population structure among the investigated accessions influenced the applicability of association mapping in crop plants (Flint et al. 2003; Pritchard et al. 2000; Remington et al. 2001). On the other hand, the validation of the role of a candidate gene requires the utilization of panels with much lower LD (<0.01 cM), where a much higher level of genetic resolution is not required. In barley, association mapping has been applied to identify chromosome regions influencing tolerance to salt stress (Jenk et al. 2007; Pakniyat et al. 1997).

### 16.5.2.4 Molecular Maps

Molecular maps determine the order and relative distance of genetic features using molecular markers that are associated with genetic variation or polymorphism due to crossing-over and natural mutation (Xu 2010). Crossing-over is the process by which homologous chromosomes exchange portions of their chromatids during meiosis, resulting in new combinations of genetic information and thus affecting inheritance and increasing genetic diversity. Genes that are present together on the same chromosome tend to be inherited together and are referred to as linked. Genes that are normally linked may be inherited independently during crossing-over. The construction of genetic linkage maps using molecular markers is based on the same principles as those used in the preparation of classical genetic maps.

The sophistication of molecular map construction has developed from the RFLP maps of the 1980s to PCR-based markers of the 1990s to more integrated maps, as a result of the use of different types of molecular markers including genetic markers, over the past decade. Linkage maps have been used in gene mapping for major genes and QTL, MAS, and map-based gene cloning. During the period 1980–1990, molecular maps were developed for many plant species. The first generation of molecular maps has been integrated with conventional genetic maps constructed using morphological and isozyme markers through cytological markers and markers shared by different maps (Xu 2010).

For many crop plants, several molecular maps have been constructed using different populations. These populations are of variable size and structure and maps have been created using different numbers and types of markers. To build an integrated reference or consensus map, the order and genetic distance between specific markers are compared across populations and maps. Stam (1993) developed a computer program, JoinMap, for the construction of genetic linkage maps. Integrated genetic and physical genome maps are extremely valuable for map-based gene isolation and comparative genome analysis and as sources of sequence-ready clones for genome sequencing projects (Xu 2010).

A well-defined correlation between the physical and genetic maps will greatly facilitate molecular breeding efforts through associating candidate genes with important biological or agronomic traits, positional cloning, and comparative analysis across populations and species and whole-genome sequences, which will in turn facilitate the development of various molecular breeding tools. Various methods have been developed for assembling physical maps of complex genomes and integrating them with genetic maps.

To create an integrated genetic and physical map resource for maize, a comprehensive approach was used, which included three core components (Xu 2010). An integrated genetic and physical mapping tool has been developed by the Maize

Mapping Project, Columbia, Missouri, United States (Xu 2010). Contigs that were assembled by fingerprinting and the automated matching of BACs were then anchored onto IBM2 and IBM2 neighbor maps. In the Gramene database, a web-based tool, CMAP, was developed to allow users to view comparisons of genetic and physical maps (Ware et al. 2002). In addition, an integrated bioinformatics tool, the comparative map and trait viewer, was developed to construct consensus maps and compare QTL and functional genomics data across genomes and experiments (Sawkins et al. 2004). All these tools can be used to build integrated maps based on shared markers and a reference map to initiate the process.

## 16.6 PLANT BREEDING FOR SALT TOLERANCE

It is well known that plant species vary in their capacity to tolerate salinity. Among the major crop species, barley, cotton, sugar beet, and canola are the most tolerant; wheat and lucerne are moderately tolerant, while rice and most legume species are sensitive. Besides, salinity tolerance varies within species, especially in outcrossing species. Worldwide vast land is affected by salinity where it becomes difficult to grow salinity-susceptible major food crops like rice, wheat, and most legumes.

When researchers failed to mitigate salinity stress on crop growing by management practices due to several environmental factors, they then looked for some alternatives like breeding to explore salinity tolerance in the plants itself. But what would be the source of salinity-tolerant genetic trait? Then researchers found landraces of crop, which have been cultivated by local farmers in saline-prone areas, worldwide, though those kinds of landraces were not high yielding. Next, the researchers prepared a research design where the breeding approach was to (1) screen collected landraces for salinity tolerance, (2) cross the identified tolerant types with the desired cultivars to introgress salinity-tolerant gene, and (3) select the desired plant types having salinity tolerance as well as other desirable traits from the advancing and segregating generations.

By exploiting the naturally occurring genetic variability that exists within a species, some relatively tolerant cultivars have been developed for crops including rice, wheat, lucerne, white clover, and citrus.

Due to the disturbance in the estimation of phenotypic tolerance under salt stress during seedling, vegetative, and reproductive stages, researchers tried to find alternative technologies that actually reduce the time and labor, provide authentic information in favor of salinity tolerance of plant, and become easy to conduct experiment in any environment.

The development of molecular (DNA) markers did open a new era in plant breeding, as this technology easily identifies salinity-tolerant landraces in a laboratory within few hours. Besides, after crossing with suitable high-yielding genotypes, marker technology also reduced time and labor to select the desired tolerant genotypes as molecular marker identifies the presence of introgressed gene. Furthermore, salinity-tolerant genotype developments become easier due to the utilization of nuclear energy for mutation and genetic engineering. In the following section, we discussed briefly the types of molecular breeding that have been used for several decades worldwide to develop salinity tolerance in plants.

### 16.6.1 MARKER-ASSISTED BREEDING

In MAB, DNA markers are used as a substitute for phenotypic selection and to accelerate the release of improved cultivars (Guo 2013). This technology can provide information on the genotype of a single plant without exposing the plant to stress and is capable of handling large numbers of samples using PCR with ultimate potentials to reduce the time, effort, and expense often associated with physiological screening. In order to use MAS in breeding programs, the markers must be closely linked to the trait and work across different genetic backgrounds. The efficiency of genetic mapping has improved greatly in recent years, with the advent of high-density maps incorporating short sequence repeat (SSR) markers, RFLP markers, AFLP technique, or SNP (Collard et al. 2005).

As described earlier, numerous studies have identified QTLs contributing to salt tolerance at different developmental stages and for different salt tolerance attributes. The challenge, however, has been the utilization of the marker-QTL information in breeding programs for improving crop salt tolerance. In the case of salt tolerance, however, a few studies can be deciphered from the literature where gene or QTL information has been used to develop lines or cultivars

with improved salinity tolerance (ST) through MAB. One example is the recent development of a highly salt-tolerant cultivar of durum wheat by Munns and her research team at CSIRO, Australia (Munns et al. 2012). Durum wheat is generally known for its higher sensitivity to salt stress compared to common bread wheat, and this has been attributed to its inherent lower ability of excluding $Na^+$ from the leaf blade. In an earlier study, Munns et al. (2006) had identified a durum line (known as Line 149) with $Na^+$ exclusion ability similar to that of bread wheat. In a genetic analysis of a cross between Line 149 and a durum wheat accession (Tamaroi) with normal $Na^+$ exclusion ability, two genes, Nax1 and Nax2, were identified for controlling $Na^+$ exclusion (Munns et al. 2003). Nax1 was mapped to the distal region of chromosome 2AL (Lindsay et al. 2004), and Nax2 was mapped to chromosome 5AL (Byrt et al. 2007). In the subsequent studies, these two genes were introduced into various durum wheat lines through MAB. The newly developed durum wheat lines were tested under natural saline fields in northern New South Wales, Australia, and the lines possessing particularly Nax2 were able to produce approximately 25% more yield than the control durum wheat lines under such saline conditions (Munns et al. 2012). The findings of this research are encouraging and provide confidence to the use of MAS for breeding plants for improved salt tolerance. In a similar study of rice, scientists at the International Rice Research Institute (IRRI) identified a single major QTL (*Saltol*) on the short arm of chromosome 1, which explained much of the variation for salt tolerance in a segregating rice population (Bonilla et al. 2002). In the subsequent studies conducted at the Bangladesh Rice Research Institute (BRRI), attempts have been made to transfer *Saltol* to two high-yielding popular commercial rice varieties, BR11 and BR28 (Rahmana et al. 2010).

In this study, various markers closely linked to *Saltol* were identified and used to transfer the QTL to the two commercial varieties via MABC.

In tomato, MAS was employed to develop salt-tolerant breeding lines using an $F_2$ population of a cross between the cultivated tomato (salt sensitive) and a salt-tolerant accession of the tomato wild species *Solanum pimpinellifolium* (Monforte et al. 1997). In this study, a combination of phenotypic selection and MAS was used to develop lines with improved salt tolerance. However, the limited use of markers for improving crop ST is attributable to various reasons, including (1) generally limited efforts that have been made by breeders to develop plants with improved salt tolerance, compared to efforts devoted to other economically important traits such as disease resistance and improved quality and quantity of yield, (2) limited familiarity of many plant breeders with the marker technology, (3) insufficient reliability of the identified QTLs or lack of QTL confirmation, and (4) population specificity of the reported QTLs and their associated markers. However, with the new advancement in marker technology and trait phenotyping and the greater need in crop plants with improved salt tolerance, it is expected that more progress will be made in developing new cultivars with improved salt tolerance in particular via using MAB approaches. Furthermore, once QTLs have been verified for use in MAS, efforts must be made to clone and characterize important QTLs. Cloning of QTLs conferring salt tolerance will not only enhance the functional understanding of the tolerance and the underlying genes and mechanisms (Salvi and Tuberosa 2005) but also provide breeders with precise markers for both breeding purposes and exploitation of allelic variations present in germplasm collections. The latter is particularly crucial for further identification and characterization of desirable genetic resources and enhancement of crop salt tolerance.

### 16.6.1.1 Prerequisites for Success in Marker-Assisted Breeding for Salinity Tolerance

MAB needs more complicated equipment and facilities, along with the origin and status of germplasms like landraces with trait of salinity tolerance. Besides phenotyping is also essential to the success of MAB for salt tolerance. Decision support tools are needed to manage and evaluate crop genetic resources and breeding materials, including genetic diversity and variation analysis and population structure evaluation, and for hybrid crops and use of the genetic diversity to define heterotic groups and predict hybrid performance (Xu 2010). For this, a suitable marker system and reliable markers available are critically important to initiate a MAB program for salinity tolerance in plants. As discussed earlier, suitable markers should have attributes like ease and low cost of use and analysis, small amount of DNA required, codominance, repeatability/reproducibility of results, high levels of polymorphism,

and occurrence and even distribution genome-wide (Guo 2013). In addition, another important desirable attribute for the markers to be used is close association with the genes involved in salinity tolerance. If the markers are located in close proximity to the target gene or present within the gene, the selection of markers will ensure success in the selection of the target gene (Xu 2010). Although they can also be used in plant breeding programs, the number of classical markers possessing these features is very small where DNA markers for polymorphism are available throughout the genome, and their presence or absence is not affected by the environment and usually does not directly affect the phenotype. DNA markers can be detected at any stage of plant growth, but the detection of classical markers is usually limited to certain growth stages (Guo 2013). Therefore, DNA markers are the predominant types of genetic markers for MAB.

Each type of marker has advantages and disadvantages for specific purposes (Xu 2010). Relatively speaking, SSRs have most of the desirable features and thus are the current marker of choice for many crops. SNPs require more detailed knowledge of the specific, single-nucleotide DNA changes responsible for genetic variation among individuals. However, more and more SNPs have become available in many species, and thus they are also considered an important type for MAB for salinity tolerance. For most plant breeding programs, hundreds to thousands of plants/individuals are usually screened for the desired marker patterns by extracting genomic DNA following PCR and gel electrophoresis to identify the corresponding gene or QTL (Guo 2013).

Besides, quick and efficient data processing and management may provide timely and useful reports for breeders. In a MAB program, not only are large numbers of samples handled, but multiple markers for each sample also need to be screened at the same time that the situation requires an efficient and quick system for labeling, storing, retrieving, processing, and analyzing large data sets, and even integrating data sets available from other programs. The development of bioinformatics and statistical software packages provides a useful tool to analyze information for constructing a decision to design a breeding program (Guo 2013). In breeding, the choice of MAB strategy, breeders fluent with MAB, stakeholder involvement, and seed industry for delivery are also needed.

### 16.6.1.2 Salinity-Tolerant Gene Mapping for All Growth Stages

To detect marker–trait associations, it is better to construct a linkage map that provides a framework for choosing markers to use in MAB for salinity tolerance. To use markers and select a salinity trait present in a specific germplasm line, a proper population of segregation for the trait is required to construct a linkage map. Once few markers are found to be associated with the trait in a given population, a dense molecular marker map in a standard reference population will help identify makers that are close to the target gene. If a region is found associated with the tolerance traits of salinity, fine mapping also can be done with additional markers to identify the markers tightly linked to the gene controlling the trait. A favorable genetic map for all growth stages should have an adequate number of evenly spaced polymorphic markers to accurately locate the desired QTLs or genes (Babu et al. 2004).

### 16.6.1.3 Knowledge of Marker–Trait Association

The most important factor for MAB is to gather knowledge of the associations between markers, and salinity tolerance where markers are closely associated with the target traits or tightly linked to the genes can provide sufficient undertaking for the success in practical breeding. The more closely the markers are associated with the traits, the higher the possibility of success and efficiency of use will be that information can be obtained in various ways like gene mapping, QTL analysis, association mapping, classical mutant analysis, linkage or recombination analysis, and bulked segregant analysis (Bernacchi et al. 1998; Guo 2013). Besides, it is also critical to know the linkage situation if either the markers are linked in *cis* or *trans* (coupling or repulsion) with the desired allele of the trait. Even if some markers have been reported to be tightly linked with a QTL, a plant breeder still needs to determine the association of alleles in his own breeding material. This makes QTL information difficult to directly transfer between different materials (Guo 2013).

### 16.6.2 Marker-Assisted Selection

MAS is nothing but a kind of breeding procedure in which DNA marker detection and selection are integrated into a conventional breeding program. Actually, it means traditional breeding with molecular help, which is why it's also called SMART (Selection with Markers and Advanced Reproductive Technology) breeding (Jan 2009). Besides, the selection of individuals with specific alleles for traits controlled by a limited number of loci (up to 6–8) is shortly called MAS (Guo 2013). With respect to important MAS schemes, the following uses of molecular markers in plant breeding can be emphasized.

### 16.6.2.1 Marker-Assisted Evaluation of Breeding Materials

In plant breeding, parent selection is the most important step, which is a combination of phenotypic assessments, pedigree information, breeding records, and chance (Jan 2009). At present, the use of molecular markers enables a marker-assisted genotype evaluation that has the potential to make parental selection more efficient, to expand the gene pool of modern cultivars, and to speed up the development of new varieties (Edwards and Mc Couh 2007; Xu and Crouch 2008).

### 16.6.2.2 Marker-Assisted Gene Introgression and Backcrossing

The process where a gene or a QTL is introgressed (introduced) from an individual to another by crossing between them and then repeatedly backcrossing to recipient parent is called introgression (Jan 2009). If molecular markers can be utilized to identify the presence of the target gene or QTL and to accelerate the return of background genome to recipient type, then this process is called marker-assisted gene introgression, which is very effective for introgressing genes or QTLs from landraces and related wild species, because it reduces both the time needed to produce commercial cultivars and the risk of undesirable linkage drag with unwanted traits of the landrace or wild species (Dwivedi et al. 2007).

On the other hand, marker assisted backcrossing (MABC) is defined as the simplest form of MAS, and at the present, it is the most widely and successfully used method in practical breeding where a limited number of loci for any trait are transferred from one genetic background to another (Guo 2013). MABC aims to transfer one or a few genes or QTLs of interest from one genetic source (serves as the donor parent and may be inferior agronomically or not good enough in comprehensive performance in many cases) into a superior cultivar or elite breeding line (serves as the recurrent parent) to improve the targeted trait. Unlike traditional backcrossing, MABC is based on the alleles of markers associated with or linked to gene(s) or QTL(s) of interest instead of phenotypic performance of target trait (Guo 2013).

For MABC program, therefore, there are two types of selection recognized: foreground selection and background selection (Hospital 2003). In foreground selection, the selection is made only for the marker allele(s) of donor parent at the target locus to maintain the target locus in heterozygous state until the final backcrossing is completed. Then the selected plants are selfed and the progeny plants with homozygous donor parent (DP) allele(s) of selected markers are harvested for further evaluation and release. As described earlier, this is the general procedure of MABC where effectiveness of foreground selection depends on the number of genes/loci involved in the selection, the marker–gene or QTL association or linkage distance, and the undesirable linkage to the target gene or QTL. Recombinant selection is important when the donor is of traditional variety where flanking markers are usually used to select recombinants between the target locus and flanking marker.

In background selection, the selection is made for the marker alleles of recurrent parent in all genomic regions of desirable traits except the target locus or selection against the undesirable genome of the donor parent. The objective is to accelerate the restoration of the RP genome and eliminate undesirable genes introduced from the DP. The progress in recovery of the RP genome depends on the number of markers used in background selection. The more markers evenly located on all the chromosomes are selected for the RP alleles, the faster recovery of the RP genome will be achieved but larger population size and more genotyping will be required as well. In addition, the linkage drag also can be efficiently addressed by background selection using DNA markers, although it is difficult to overcome in a traditional backcrossing program (Guo 2013).

In practice, however, both foreground and background selection is usually conducted in the same program, either simultaneously or successively.

In many cases, they can be performed alternatively even in the same generation. The individuals that have the desired marker alleles for target trait are selected first (foreground selection). Then the selected individuals are screened for other marker alleles again for the RP genome (background selection). The theoretical proportion of the recurrent parent genome after a generation of backcrossing is given by $(2n^{+1} - 1/2n^{+1})$, where n equals the number of backcrosses (Guo 2013).

The efficiency of MABC depends upon several factors, such as the population size for each generation of backcrossing, marker–gene association or the distance of markers from the target locus, number of markers used for target trait and RP background, and undesirable linkage drag. Based on the simulations of 1000 replicates, Hospital (2003) presented the expected results of a typical MABC program, in which heterozygotes were selected at the target locus in each generation and RP alleles were selected for two flanking markers on target chromosome, each located 2 cm apart from the target locus, and for three markers on nontarget chromosomes.

MABC is favorable when phenotyping is difficult and/or expensive or impossible; heritability of the target trait is low; the trait is expressed in late stages of plant development and growth, molecular markers, and MAS in plants; the traits are controlled by genes that require special conditions to express; the traits are controlled by recessive genes; and gene pyramiding is needed for one or more traits (Guo 2013).

To conduct a successful experiment of MABC, researchers must maintain some considerations (Guo 2013; Jan 2009): (1) always producing backcross seeds from some backup plants; (2) roughing off-type plants; (3) avoiding off-type pollen load during dusting; (4) making no mistakes during leaf collection; (5) confirming selection by recollection of leaf samples; (6) uprooting best plants in pots and keeping them inside a crossing house; (7) taking maximum care during DNA dilution, PCR, and gel loading; and (8) being aware of rats, birds, stunting virus, and also cyclones.

### 16.6.2.3 Marker-Assisted Gene Pyramiding

MAGP is an important application of DNA markers in plant breeding that has been proposed and applied already to enhance resistance to disease and insects by selecting for two or more genes at a time, as in rice. Such pyramids have been developed against bacterial blight and blast (Huang et al. 1997; Luo et al. 2012; Singh et al. 2001).

But it is possible to develop salt tolerance in crop variety by pyramiding two or three traits related to it? Actually, pyramiding is the process of combining several genes or QTLs together into a plant variety where using phenotypic selection methods are extremely difficult and sometimes impossible to pyramid the desired traits like salinity and disease resistance. When a variety is protected by one gene with a major effect against salinity stress, it is often not possible to introgress additional resistance genes to the same trait because they show the same phenotype. However, if tolerance genes can be tagged with markers, the number of genes in any plant can be easily determined (Collard and Mackill 2008). Genes can be pyramided through pedigree breeding by crosses involving multiple parental lines containing different favorable alleles or MABC to introgress those alleles into the same genetic background. Pyramiding of multiple gene(s) or QTL(s) may be achieved through different approaches: multiple-parent crossing or complex crossing, backcrossing, and recurrent selection.

A suitable breeding scheme for MAGP depends on the heritability of traits of interest, the number of gene(s) or QTL(s) required for the improvement of traits, the number of parents that contain the required gene(s) or QTL(s), and other factors like marker–gene association, expected duration to complete the plan, and relative cost. Nevertheless, if there are more than four genes or QTLs to be pyramided, complex or multiple crossing and/or recurrent selection may be often preferred (Guo 2013).

### 16.6.2.4 Marker-Assisted Recurrent Selection

Marker-assisted recurrent selection (MARS) is a recurrent selection scheme using molecular markers for the identification and selection of multiple genomic regions involved in the expression of complex traits to assemble the best-performing genotype within a single or across related populations (Ribaut et al. 2010).

It was developed for forward breeding of native genes and QTL for relatively complex traits such as salinity tolerance, disease resistance, and grain

yield (Crosbie et al. 2006; Eathington 2005; Ragot et al. 2000; Ribaut and Betran 1999; Ribaut et al. 2000). Marker-only recurrent selection schemes have been implemented for a variety of traits including grain yield and grain moisture (Eathington 2005) or salinity stress tolerance (Ragot et al. 2000), and multiple traits are being targeted simultaneously. According to the research review of Ragot and Lee (2007), MARS can be applied to (1) rather large sizes of the populations submitted to selection at each cycle, (2) use of flanking versus single markers, (3) selection before flowering, (4) increased number of generations from one to four generations per year, and (5) lower cost of marker data points (Guo 2013).

### 16.6.2.5 Marker-Assisted Hybrid Prediction

Actually, hybrid performance largely depends on the general combining ability (GCA) of the parental lines and the specific combining ability (SCA) between the parents. GCA is defined as an attribute of an inbred line and is measured as the average performance of all hybrids made with that inbred line as a parent. The higher the GCA of an inbred, the higher the average performance of its hybrids where SCA is defined for specific combinations of parents and is measured by the deviation of the hybrid performance from the expected performance as estimated from the GCA of the parents (Guo 2013). As a result, hybrid performance is determined by its parents' GCA and the cross's SCA. Hybrid performance can be measured by heterosis, the performance of a hybrid over their parental lines. For developing new hybrids, it is possible to generate a huge amount of combination after crossing, but due to limited resources, breeders usually are unable to test all combinations in all environments of interest. But it becomes trouble-free to predict hybrid easily using molecular marker technology.

## 16.7 HIGH-THROUGHPUT SELECTION TO IDENTIFY SALT-TOLERANT GENES

Ahead of MAS, there are some methods for high-throughput selection to identify salt-tolerant genes, mutant screening, and misexpression approach, which are described briefly in this section.

### 16.7.1 Mutant Screening Approach

To generate variations, several kinds of chemical and physical mutagens have been usually applied on seeds of corresponding plants. Mutant screens are a great approach for the identification of genes with loss-of-function or change-of-function phenotypes. Theoretically, every gene in the genome can be mutated and the resulting homozygous plants characterized. As such, a species with moderate salt tolerance may be converted into mutants with strong tolerance while exposed to mutagen, due to the chromosomal change, though most mutant screens for salt sensitivity have been performed in *Arabidopsis*, a salt-sensitive species, because of its tractability to molecular genetics approaches. The available genetic and genomics tools in *Arabidopsis* greatly enhance our capability to do map-based cloning. Besides, high-density oligomicroarray chips can even make gene identification possible (Mockler et al. 2005). Salinity-tolerant mutants of rice and wheat would be potentially identified under various salt stress condition at greenhouse or saline-prone fields; however, the cloning of the mutated genes will be challenging because of the lack of genome sequence and genetic markers.

One of the earliest screens for salt tolerance was performed by Saleki and colleagues (Saleki et al. 1993). They screened EMS-mutagenized Columbia seeds for mutants that were tolerant to high salt stress at the germination stage. Three mutants, RS17, RS19, and RS20, were identified and characterized, but none of them showed enhanced salt tolerance in later stages of development as compared to the wild type.

A few similar direct mutant screens performed in *Arabidopsis* later also turned out to have achieved limited success (Quesada et al. 2000; Werner 1995). This is likely because loss-of-function mutations usually cause even greater sensitivity to salinity stress. Thus, salt-tolerant mutants cannot be easily detected in direct screens at the vegetative growth stage.

### 16.7.2 Misexpression Approach

Most of the genes known to be important in salt tolerance display a positive function, that is overexpression of the gene enhances salt tolerance. Several such examples include *AtNHX1*, *SOS1*, *CBF1*, *DREB1*, and *Tsi1*. As such, a high-throughput screen of a population of lines overexpressing thousands of unique genes in model plants such as *Arabidopsis* or rice appears to offer huge potential for the identification of salt-tolerant genes. The selection of candidate genes that are putatively functional in salt tolerance when overexpressed is the first step in this approach if there is a need to prioritize genes for cloning.

A large number of salinity-induced genes have been identified by comparing the gene expression profiles in salt-treated and salt-untreated plants using a number of experiments such as mRNA differential display (Park et al. 2001; Shiozaki et al. 2005), cDNA-AFLP (Chen et al. 2003), and cDNA or oligonucleotide microarray hybridization (Gong et al. 2005; Gu et al. 2004; Kreps et al. 2002; Taji and Seki 2004; Walia et al. 2005).

## 16.8 GENETIC ENGINEERING FOR SALT TOLERANCE

Significant advances in the field of marker technology have been made during the past decade. This technology only reduces the time of conventional breeding, makes an easy selection of donor parents for crossing programs that contain corresponding traits, and finally confirms the presence of the desired trait in recipient genotypes. But this technology cannot be proficient in removing the main drawback of conventional breeding like linkage drag. For example salinity is a multigenic trait where QTLs are associated not only with the aspects of germination, ion transport, and yield but also with unwanted traits (Maris and Eduardo 2002). Besides, environment and genetic backgrounds have a significant influence on the QTL (Asins 2002) that is already identified. This may be an indication that traits are determined by a limited number of sites and/or that genes associated with physiological traits are clustered on chromosomes. However, the fact that a QTL represents many, perhaps hundreds of, genes remains a problem to find out key loci within a QTL.

But improvement of crop plants through genetic engineering (transgenic breeding) is a major approach in the field of molecular breeding (Xu 2010). It can utilize the genes from any organism, by which limitations associated with sexual hybridization can be overcome. On the other hand, transgenic breeding provides a quick approach to pyramid genes of different sources into one genetic background. Transgenic approach has promoted the development of salinity-tolerant cultivars based on specific traits that are controlled by one gene, for example a transcription factor or an important ion channel. Zhang et al. (2001) reported the development of a salinity-tolerant transgenic tomato plant in which overexpression of the vacuolar $Na^+/H^+$ antiporter showed dramatic improvement of vegetative growth and of fruit yield. This antiporter is the only known transporter that would compartmentalize $Na^+$ in the vacuole, where $Na^+$ has little chance of toxic effect on metabolism or is to be transported to younger leaves and fruits.

A number of genes, encoding proteins with known functions in ion transport or in synthesis of compatible organic solutes, as well as genes whose functions are not fully understood, have been used to transform a number of species in efforts to improve salinity tolerance. Despite numerous claims of improved salinity tolerance, poor experimental designs and choices of parameters measured to evaluate tolerance mean that much doubt remains. None of these transgenics have been proven in the field. Since salinity tolerance is a multigenic trait, large improvements based on modification of only one gene could only occur if the gene is a transcription factor and regulates a number of genes that control ion transport or some other processes involved in salinity tolerance.

Conventional MPB represents the principal approach to crop improvement that employs methods such as hybridization, MAS and introgression, as well as induced mutagenesis to randomly modify genomes, which results in creating genetic variation (Xu 2010).

Given this logic, genetic engineering is different from the traditional methods in that any modification can be designed and tailored to achieve the desired effect. This method often fuses promoters and genes to produce expression cassettes that are introduced into plants using bacterial transfer DNAs. Now, it excludes the transfer of neither known allergens nor toxin-encoding genes and analyzes the sequence of insertion sites. The ability to identify rapidly and eliminate plants containing inadvertent fusions or disruptions of genes is not available to traditional plant breeding, where genes can be inactivated through unpredictable transposition of resident mobile elements. The second advantage of transgenic applications is that it generally takes less than a year to transform an existing cultivar with one or several traits. For the same reasons already, genetic engineering has shown improvement of salt tolerance in plants. The methodology of transgenic approach for salt tolerance in plants includes (1) identification and isolation of candidate gene through mRNA extraction, cDNA construction, PCR with gene-specific primers, or high-throughput sequencing; (2) construction of a vector DNA including candidate gene with site-specific promoter region, reporter gene, antibiotic resistance gene, and origin of replication; (3) transformation of candidate gene into recipient plant cell during tissue culture by *Agrobacterium*-mediated, gene gun, or electroporation approach; and (4) screening of transgenic plants through several bioassay techniques to confirm the transgene and its expression at salt stress. Genetic transformation technology enables researchers to achieve gene transfer in precise and predictable manner (Gupta and Huang 2014). Hence, genetic engineering approaches would be useful to manipulate the osmoprotectant biosynthetic pathways for accumulating such molecules that act by scavenging ROS, reducing lipid peroxidation, and maintaining protein structure and functions (Abogadallah 2010; Ashraf 2009; Saxena et al. 2013).

Research on the transformation of plants for improving salinity tolerance is increasing regularly where genes controlling ion transport like regulation of $Na^+$ uptake and compartmentalization are manipulating successfully. The choice of promoters can significantly affect the results from a transgenic manipulation. Consequently, salt stress–tolerant plants could be engineered by successful alteration of the salinity response by engineering novel regulatory targets, appropriate thoughtful of posttranslational modifications that control plant growth performance under salinity stress, overexpression of miRNAs or their targets, maintaining hormone homeostasis let alone pleiotropic property under salinity stress, and applying plant synthetic biology approaches to improve genetic engineering strategies (Cabello et al. 2014; Gupta and Huang 2014; Saxena et al. 2013).

## 16.9 ROLE OF RNA INTERFERENCE TO CONFER SALINITY TOLERANCE

RNAi is important for the genetic improvement of salinity stress tolerance, given the complex mechanism of such kind of tolerance. Plants usually utilize a glut of gene regulation mechanisms in response to several kinds of developmental and environmental cues (Viswanathan et al. 2007). RNAi directly cleavage or indirectly regulate the protein-encoding mRNA, which has a detrimental impact on plant physiology due to salinity stress, thus playing a role in posttranscriptional regulation, translational repression, chromatin remodeling, and DNA methylation. Several stress-responsive small RNAs have been identified in plants, and their roles in oxidative stress tolerance, osmolyte accumulation/osmoprotection, and nutrient starvation response have been established (Achard et al. 2004; Chinnusamy et al. 2005). Under salinity stress, stress-upregulated miRNAs may downregulate their target genes, which are likely negative regulators of stress tolerance, and finally, it may positively regulate stress tolerance (Lippman et al. 2004). Overexpression of RNAi will help to overcome posttranscriptional gene silencing and thus may lead to better expression of engineered trait in transgenic plants. Understanding the roles of RNAi in transcriptome homeostasis, cellular tolerance, and phenological and developmental plasticity of plants under abiotic stress and recovery will help genetic engineering of salinity stress resistance in crop plants (Sunkar et al. 2004; Viswanathan et al. 2007).

## 16.10 SALINITY-TOLERANT CROP PLANTS FROM LAB TO FIELD

The final test for salt tolerance of a crop variety is the evaluation of field performance at replicated sites and seasons (Munns et al. 2006). Identification of yield-related QTLs involves molecular marker analyses for salinity tolerance in populations, and the validation of QTLs is done always in the field of saline-prone area for the expression of the salt-tolerant gene. Populations experience more G × E interaction and population data generate ambiguous result. In many cases, QTLs selected through MAS did not work well in marginal environment (Quarrie et al. 2005; Reyna and Sneller 2001).

As field conditions vary from site to site, not only in soil salinity but also in soil physical and chemical properties, such as sodicity, high pH, and boron, and their interactions, field evaluation must be done over at least 3 years (Munns et al. 2006) because high pH can cause reduced $K^+$ uptake even though it might not affect $Na^+$ uptake (Ahmad 2002; Gregorio et al. 2002) and boron

can affect the subcellular distribution of salt in leaves (Wimmer et al. 2003). Besides, waterlogging worsens the effects of salinity on wheat (Barrett 2003) and may be a major reason why wheat bread for salt tolerance has had little success in farmers' fields in some regions (Hollington et al. 2002). When $O_2$ deficiency occurs in roots in waterlogged soils in species with little aerenchyma or adaptations such as shallow roots, respiration is impaired and may be the cause of the higher salt uptake described by Barrett (2003), especially now that it is known that there are normally high rates of $Na^+$ efflux from the roots to the external solution (Davenport et al. 2005), which is an energy-demanding process. The roots of many wetland species contain a barrier to radial $O_2$ loss in addition to having extensive aerenchyma, and introduction of these traits into wheat may increase its waterlogging tolerance and its ability to provide enough energy to exclude $Na^+$ in waterlogged soils (Colmer et al. 2005). Moreover all breeding lines developed with greater tolerance to salt or any soil constraint should have the ability to yield well under optimal conditions. Efforts by the researcher make it possible to breed more than 30 salt-tolerant crop cultivars, but some are accepted and cultivated by farmers (Flowers 2004; Flowers and Yeo 1995; Gregorio and Cabuslay 2005; Jenks et al. 2007). Participatory plant breeding becomes essential part of variety development and release, which is combined with MABC and MAS and used to introgress stress-tolerant genes in rice (Steele et al. 2006, 2007). Developing salt tolerance requires identification and use of genes from different sources; molecular markers could be used to evaluate the genetic diversity between and within germplasms of crops and uniformity in the released variety (Bajracharya et al. 2006; Mkumbira et al. 2003). Use of molecular markers can save time and cost in molecular breeding of saline-tolerant crop plants. Binadhan-7 is one of the examples of introgression of salinity-tolerant gene in popular mega rice variety where the application of the molecular breeding approach is found successful (Figure 16.2).

Varieties with introgressed genes were evaluated in the farmer's field and selected by the farmers. There are also varieties selected from participatory varietal selection process performing well in the stress environment, which reflexes the efficiency of molecular breeding and consumer need-based selection in stress-tolerant variety development

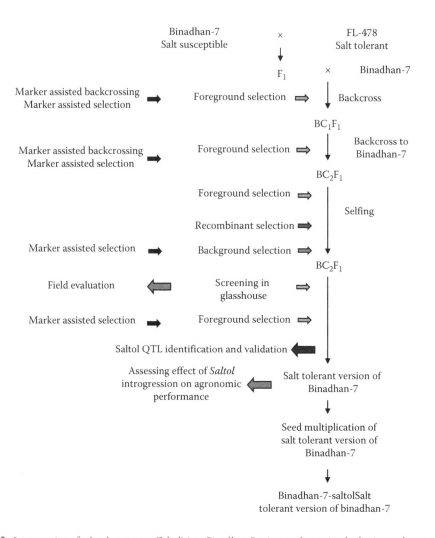

Binadhan-7                    ×          FL-478
Salt susceptible                         Salt tolerant

F₁          ×          Binadhan-7

Marker assisted backcrossing  ➡  Foreground selection ⇨    │ Backcross
Marker assisted selection

BC₁F₁

Marker assisted backcrossing  ➡  Foreground selection ⇨    │ Backcross to
Marker assisted selection                                    Binadhan-7

BC₂F₁

Foreground selection ⇨

Recombinant selection ⇨                   Selfing

Marker assisted selection  ➡  Background selection ⇨

BC₂F₁

Field evaluation  ⬅  Screening in       ⇨
                     glasshouse

Marker assisted selection  ➡  Foreground selection ⇨

Saltol QTL identification and validation ◀

Assessing effect of *Saltol*  ⬅  Salt tolerant version of
introgression on agronomic        Binadhan-7
performance

Seed multiplication of
salt tolerant version of
Binadhan-7

Binadhan-7-saltolSalt
tolerant version of binadhan-7

Figure 16.2. Introgression of salt-tolerant gene (Saltol) into Binadhan-7 using marker-assisted selection and gene mapping.

and improvement (Biggs and Gauchan 2001; Witcombe et al. 2005). Sometimes it is assumed that marker-evaluated selection (MES) is more powerful than QTL analysis when MES is used to evaluate the genotypes chosen from a wide cross, while it can determine the consequences of selection all along the entire genome for every preferred character in a crop (Jenks et al. 2007).

Once the loci contribute to adaptation to a particular ecotype, then crop idiotype could be developed in the second step for breeding of stress-tolerant varieties. Marker-based selection and evaluation proved to be a novel method (Steele et al. 2004) in India and Nepal for selecting a wide range of agronomic character that determined adaptation of a genotype in several and particular ecotypes (Virk et al. 2003; Witcombe

and Virk 2001). Saline environments are greatly diverse and create strong G × E interactions for tolerance. For this reason, an integrated program was developed by Bennett and Khush (2003) to develop varieties with useful alleles that confer specific mechanisms of salt tolerance from multiple donors by gene pyramiding using molecular breeding tools (Jenks et al. 2007).

QTL analysis is an unbiased approach to understand the complex inheritance of quantitative traits into Mendelian-like factors using MAS and the end cloning of the trait of interest. QTL analysis enables us to identify the target chromosome regions and the DNA sequences whose functional polymorphisms in genome control the phenotypic variation (Tankslay 1993). In QTL approach, a mapping population is evaluated with molecular markers like RAPD, FRLPs, SSRs, AFLPs, SNPs, and STS across all individuals for the variation present in population that are reasonable through statistical methods. Identification of major QTL requires the construction of a reference map that is obtained from different crosses and by sharing common polymorphisms in a particular species. The reference map enables us to compare the position of QTLs and cloned genes in mutants used with different objectives (Tuberosa et al. 2002). Naturally occurring QTLs are less drastic than mutant phenotype that is caused by a variant allele in a particular phenotype. In QTL analysis, quantitative and Mendelian genetics do not classify the loci on the basis of phenotypic expression and heritability of the additive effect of the parental alleles segregating in mapping population. Mutant analysis through ecotilling (Comai et al. 2004) allows us to measure the candidate loci expressing phenotypic variability at target loci (Buckllow and Thorsberg 2002).

In a germplasm collection, allelic variation is present for different traits of interest for crop improvement. QTL identification and its cloning are very essential. After the advent of new molecular platforms like positional cloning and association mapping based on LD, these are used to dissect QTL and identify the promising candidate gene sequence for QTL validation. A large number of salinity-related genes, which are valuable sources for the construction of functional maps that can be used to identify the candidate gene for QTLs of the trait of interest, have been identified. Linkage maps with function-specific genes are being utilized for QTL dissection (Jenks et al. 2007).

## 16.11 ACHIEVEMENT AND FUTURE PERSPECTIVE OF MOLECULAR BREEDING

At present, rice productivity in salt-affected areas is very low as around 1 ton ha$^{-1}$. But this potentially increases by at least 3–4 tons ha$^{-1}$ with improved newly developed rice varieties that can withstand soil salinity. Earlier, farmers received good yields from this variety, but nowadays, they are facing serious problems to get a good harvest from these non-salt-tolerant rice varieties during boro season due to increased salinity intrusion in these areas while the trend of salinity intrusion in the coastal belt is still increasing. Scientists have comprehensive rice research and breeding programs to make climate-ready rice varieties that

are more tolerant of submergence, drought, heat, and salinity conditions predicted to increase in frequency and severity with climate change. It is often argued that the future development of the country depends predominantly on the sustainable development of such stress-tolerant rice varieties as well as extension and utilization of such varieties in stress-prone areas.

A new salinity-tolerant paddy species could increase the productivity of coastal small-scale farmers, boost income, and potentially buffer against urban migration. The Strategic Foresight Group writes, The Bangladesh Institute of Nuclear Agriculture (BINA) has created a new type of salinity-tolerant paddy species for coastal districts in the southern region of the country. Salt tolerance capacities of conventional high-yielding rice varieties are below 4 dS m$^{-1}$ (deci-siemens per metre). "Soil salinity in the coastal areas of Bangladesh varies from 2 to 6 dS m$^{-1}$ depending on the distance from the sea.

The development of improved salt-tolerant materials followed a long chain of introduction, selection, and recombination processes, which finally benefited the farmers by increasing their harvest in salt-affected lands. There are many rice varieties that have been released as salt-tolerant varieties in many countries. The Philippines have released many IRRI-developed materials, for example the varieties like IRRI-112 as PSBRc48 (Hagonoy); IRRI 113 as PSBRc50 (Bicol); IRRI 124 as PSBRc84 (Sipocot); IRRI 125 as PSBRc86 (Matnog); IRRI 126 as PSBRc88 (Naga); and IRRI 128 as NSICRc106. In other countries, also many salt-tolerant rice varieties have been released for commercial cultivation like CSR10, CSR13, CSR23, CSR27, CSR30, and CSR36 and Lunishree, Vytilla 1, Vytilla 2, Vytilla 3, Vytilla 4, Panvel 1, Panvel 2, Sumati, Usar dhan (1, 2, and 3) (from India); BRRI dhan 40, BRRI dhan 41, BRRI dhan 47, Binadhan-8, Binadhan-10, BRRI dhan 53, and BRRI dhan 54 (from Bangladesh); and OM2717, OM2517, and OM3242 (from Vietnam).

To improve the productivity and sustainability of agriculture, the development of high-yielding and durable abiotic and biotic stress-tolerant cultivars and climate-resilient crops is essential. Henceforth, understanding the molecular mechanism and dissection of complex quantitative yield and stress tolerance traits is the prime objective in current agricultural biotechnology research. In recent years, tremendous progress has been made in plant genomics and molecular breeding research pertaining to conventional and next-generation whole-genome, transcriptome, and epigenome sequencing efforts; generation of huge genomic, transcriptomic, and epigenomic resources; and development of modern genomics-assisted breeding approaches in diverse crop genotypes with contrasting yield and abiotic stress tolerance traits. Unfortunately, the detailed molecular mechanism and gene regulatory networks controlling such complex quantitative traits are not yet well understood in crop plants. The new high-throughput sequencing technology driving this revolution generates massive amounts of sequence data faster and cheaper than the traditional method. Consequently, it is expected that whole-genome molecular breeding will develop new super crop varieties that contain multiple desirable traits.

## REFERENCES

Abogadallah, G. M. 2010. Antioxidative defense under salt stress. *Plant Signaling and Behavior*, 5(4): 369–374.

Achard, P., Herr, A., Baulcombe, D. C., and Harberd, N. P. 2004. Modulation of floral development by a gibberellin-regulated micro RNA. *Development*, 131: 3357–3365.

Ahmad, M. 2002. Effects of salinity and pH on ion uptake in SARC-1 wheat under hydroponic conditions. In *Prospects for Saline Agriculture*, R. Ahmad and K. A. Malik (eds.). Dordrecht, the Netherlands: Kluwer Academic Publishers.

Ahmadi, N., Negra, S. et al. 2011. Targeted association analysis identified japonica rice varieties achieving $Na^+/K^+$ homeostasis without the allelic make-up of the salt tolerant indica variety NonaBokra. *Theoretical and Applied Genetics*, 123: 881–895.

Ali, Z., Park, H. C. et al. 2012.TsHKT1–2: A HKT1 homolog from the extremophile *Arabidopsis* relative *Thellungiella salsuginea*, shows $K^+$ specificity in the presence of NaCl. *Plant Physiology*, 158: 1463–1474.

Ashraf, M. 2009. Biotechnological approach of improving plant salt tolerance using antioxidants as markers. *Biotechnology Advances*, 27(1): 84–93.

Asins, M. J. 2002. Present and future of quantitative trait locus analysis in plant breeding. *Plant Breeding*, 121: 281–291.

Babu, R., Nair, S. K., Prasanna, B. M., and Gupta, H. S. 2004. Integrating marker-assisted selection in crop breeding-prospects and challenges. *Current Science*, 87: 607–661.

Bajracharya, J., Steele, K. A., Jarvis, D. I., Sthapit, B. R., and Witcombe, J. R. 2006. Rice landrace diversity in Nepal: Variability of agro-morphological traits and SSR markers in landraces from a high-altitude site. *Field Crops Research*, 95: 327–335.

Barrett-Lennard, E. G. 2002. Restoration of saline land through revegetation. *Agricultural Water Management*, 53: 213–226.

Barrett-Lennard, E. G. 2003. The interaction between waterlogging and salinity in higher plants: Causes, consequences and implications. *Plant and Soil*, 253: 35–54.

Bennett, J. and Khush, G. S. 2003. Enhancing salt tolerance in crops through molecular breeding: A new strategy. *Journal of Crop Production*, 7: 11–65.

Bernacchi, D., Beck-Bunn, T. et al. 1998. Advanced backcross QTL analysis of tomato: Evaluation of near-isogenic lines carrying single-donor introgressions for desirable wild QTL-alleles derived from *Lycopersicon hirsutum* and *L. pimpinellifolium*. *Theoretical and Applied Genetics*, 97: 170–180.

Biggs, S. and Gauchan, D. 2001. Resource-poor farmer participation in research: A synthesis of experiences from nine national agricultural research systems. Special series on the Organization and Management of On-Farm Client-Oriented Research (OFCOR). OFCOR-Comparative Study Paper. The Hague, the Netherlands: International Service for National Agricultural Research (ISNAR).

Biggs, S. and Gauchan, D. 2001. Resource-poor farmer participation in research: A synthesis of experiences from nine national agricultural research systems. In Special Series on the Organisation and Management of On-Farm Client-Oriented Research (OFCOR). OFCOR-Comparative Study Paper 3. The Hague, the Netherlands: International Service for National Agricultural Research (ISNAR).

Blum, A. 1988. *Breeding for Stress Environments*. Boca Raton, FL: Taylor & Francis.

Blum, A., Munns, R., Passioura, J. B., and Turner, N. C. 1996. Genetically engineered plants resistant to soil drying and salt stress: How to interpret osmotic relations? *Plant Physiology*, 110: 1051–1053.

Bonilla, P., Dvorak, J., Mackill, D., Deal, K., and Gregorio, G. 2002. RFLP and SSLP mapping of salinity tolerance genes in chromosome 1 of rice (*Oryza sativa* L.), using recombinant inbred lines. *The Philippine Agricultural Scientist*, 85: 68–76.

Botstein, D. R., White, R. L., Skolnick, M., and Davis, R. W. 1980. Construction of a genetic linkage map in man using restriction fragment length polymorphisms. *American Journal of Human Genetics*, 32: 314–331.

Buckler, E. S. I. and Thornsberry, J. M. 2002. Plant molecular diversity and applications to genomics. *Current Opinion in Plant Biology*, 5: 107–111.

Byrt, C. S., Platten, J. D. et al. 2007. HKT1,5-like cation transporters linked to Na⁺ exclusion loci in wheat, Nax2 and Kna1. *Plant Physiology*, 143: 1918–1928.

Cabello, J. V., Lodeyro, A. F., and Zurbriggen, M. D. 2014. Novel perspectives for the engineering of abiotic stress tolerance in plants. *Current Opinion in Biotechnology*, 26: 62–70.

Carillo, P., Grazia, M. A., Pontecorvo, G., Fuggi, A., and Woodrow, P. 2011. Salinity stress and salt tolerance. In *Abiotic Stress in Plants—Mechanisms and Adaptations*, A. Sarker (ed.). InTech, University of Napes, Italy. http://www.intechopen.com/books/abioticstress-in-plants-mechanisms-and-adaptations/salinity-stress-and-salt-tolerance (accessed April 15, 2014).

Chan, S. W., Zilberman, D., Xie, Z., Johansen, L. K., Carrington, J. C., and Jacobsen, S. E. 2004. RNA silencing genes control de novo DNA methylation. *Science*, 303: 1336.

Chen, G. P., Ma, W. S., Huang, Z. J., and Shen, Y. Z. 2003. Isolation and characterization of SIR73 gene fragment in salt resistant mutant of wheat involved in salt stress. *Yi Chuan*, 25(2): 173–176.

Chinnusamy, V., Jagendorf, A., and Zhu, J. K. 2005. Understanding and improving salt tolerance in plants. *Crop Science*, 45: 437–448.

Collard, B. C. Y., Jahufer, M. Z. Z., Brouwer, J. B., and Pang, E. C. K. 2005. An introduction to markers, quantitative trait loci (QTL) mapping and marker-assisted selection for crop improvement: The basic concepts. *Euphytica*, 142: 169–196.

Collard, B. C. Y. and Mackill, D. J. 2008. Marker-assisted selection: An approach for precision plant breeding in the twenty-first century. *Philosophical Transactions on the Royal Society*, 363: 557–572.

Colmer, T. D., Munns, R., and Flowers, T. J. 2005. Improving salt tolerance of wheat and barley: Future prospects. *Australian Journal of Experimental Agriculture*, 45: 1425–1443.

Comai, L., Young, K. et al. 2004. Efficient discovery of DNA polymorphisms in natural populations by Ecotilling. *Plant Journal*, 37: 778–786.

Crosbie, T. M., Eathington, S. R., and Johnson G. R. 2006. Plant breeding: Past, present and future. In *Plant Breeding*, K. R. Lamkey and M. Lee (eds.). Oxford, U.K.: Blackwell Publishing.

Davenport, R., James, R. A., Zakrisson P. A., Tester, M., and Munns, R. 2005. Control of sodium transport in durum wheat. *Plant Physiology*, 137: 807–818.

Davenport, R. J., Munoz, M. A., Jha, D., Essah, P. A., Rus, A. N .A., and Tester, M. 2007. The Na⁺ transporter AtHKT1-1 controls retrieval of Na⁺ from the xylemin *Arabidopsis*. *Plant, Cell and Environment*, 30: 497–507.

Davis, G. L., McMullen, M. D., and Baysdorfer, C. 1999. A maize map standard with sequenced core markers, grass genome reference points and 932 expressed sequence tagged sites (ESTs) in a 1736 locus map. *Genetics*, 152: 1137–1172.

Dewey, D. R. 1962. Breeding crested wheat grass Agropyron desert orum for salt tolerance. *Crop Science*, 2: 403–407.

Duvick, D. N. 2005. Contribution of breeding to yield advances in maize. In *Advances in Agronomy*, D. N. Sparks (ed.). San Diego, CA: Academic Press, pp. 83–145.

Dwivedi, S. L., Crouch, J. H. et al. 2007. The molecularization of public sector crop breeding: Progress, problems and prospects. *Advances in Agronomy*, 95: 163–318.

Eathington, S. R. 2005. Practical applications of molecular technology in the development of commercial maize hybrids. In *Proceedings of the 60th Annual Corn and Sorghum Seed Research Conferences*. Washington, DC: American Seed Trade Association.

Edwards, D., Forster, J. W., Chagné, D., and Batley, J. 2007. What is SNPs? In *Association Mapping in Plants*, N. C. Oraguzie, E. H. A. Rikkerink, S. E. Gardiner, and H. N. D. Silva (eds.). Berlin, Germany: Springer, pp. 41–52.

Edwards, J. D. and Mc Couch, S. R. 2007. Molecular markers for use in plant molecular breeding and germplasm evaluation. In *Marker-Assisted Selection: Current Status and Future Perspectives in Crops, Livestock, Forestry and Fish*, E. P. Guimaraes, J. Ruane, B. D. Scherf, A. Sonnino, and J. D. Dargie (eds.). Rome, Italy: Food and Agriculture Organisation of the United Nations (FAO).

Edwards, K. and Phillips, R. L. 2002. Toward positional cloning of Vgt1, a QTL controlling the transition from the vegetative to the reproductive phase in maize, *Plant Molecular Biology*, 48: 601–613.

Flint, G. S. A., Thornsberry, J. M., and Buckler, E. S. 2003. Structure of linkage disequilibrium in plants. *Annual Review of Plant Biology*, 54: 357–374.

Flowers, T. J. 2004. Improving crop salt tolerance. *Journal of Experimental Botany*, 55: 307–319.

Flowers, T. J. and Colmer, T. D. 2008. Salinity tolerance in halophytes. *New Phytologist*, 179: 945–963.

Flowers, T. J. and Yeo, A. R. 1995. Breeding for salinity resistance in crop plants: where next? *Australian Journal of Plant Physiology*, 22: 875–884.

Freudenreich, C. H., Stavenhagen, J. B., and Zakian, V. A. 1997. Stability of CTG: CAG tri nucleotide repeat in yeast is dependent on its orientation in the genome. *Molecular and Cell Biology*, 4: 2090–2098.

Gong, Q., Li, P. et al. 2005. Salinity stress adaptation competence in the extremophile *Thellungiella halophila* in comparison with its relative *Arabidopsis thaliana*. *Plant Journal*, 44(5): 826–839

Gregorio, G. B. and Cabuslay, G. S. 2005. Breeding for abiotic stress tolerance in rice. In *Abiotic Stresses: Plant Resistance through Breeding and Molecular Approaches*, M. Ashraf and P. J. C. Harris (eds.). New York: Food Products Press/Haworth, pp. 513–544.

Gregorio, G. B., Senadhira, D., Mendoza, R. D., Manigbas, N. L., Roxas, J. P., and Guerta, C. Q. 2002. Progress in breeding for salinity tolerance and associated abiotic stresses in rice. *Field Crops Research*, 76: 91–101.

Gu, R., Fonseca, S. et al. 2004. Transcript identification and profiling during salt stress and recovery of *Populus euphratica*. *Tree Physiology*, 24(3): 265–276.

Guo, L. G. 2013. *Molecular Markers and Marker-Assisted Breeding in Plants*. Intech.

Guo Y., Qiu, Q. S. et al. 2000. Transgenic evaluation of activated mutant alleles of SOS2 reveals a critical requirement for its kinase activity and C-terminal regulatory domain for salt tolerance in *Arabidopsis thaliana*. *Plant Cell*, 16(2): 435–449.

Guo, Y., Qiu, Q. S. et al. 2004. Transgenic evaluation of activated mutant alleles of SOS2 reveals a critical requirement for its kinase activity and C-terminal regulatory domain for salt tolerance in *Arabidopsis thaliana*. *Plant Cell*, 16: 435–449.

Gupta, B. and Huang, B. 2014. Mechanism of salinity tolerance in plants: Physiological, biochemical, and molecular characterization. *International Journal of Genomics*, 2014: 1–18.

Hasegawa, P. M., Bressan, R. A., Zhu, J. K., and Bohnert, H. J. 2000. Plant cellular and molecular responses to high salinity. *Annual Review of Plant Physiology and Plant Molecular Biology*, 51: 463–499.

Hauser, F. and Horie, T. 2010. A conserved primary salt tolerance mechanism mediated by HKT transporters: A mechanism for sodium exclusion and maintenance of high K⁺/Na⁺ ratio in leaves during salinity stress. *Plant Cell Environment*, 33: 552–565.

Heffner, E. L., Sorrells M. E., and Jannink, J. L. 2009. Genomic selection for crop improvement. *Crop Science*, 49: 1–12.

Hollington, P. A., Aktar, J. et al. 2002. Recent advances in the development of salinity and waterlogging tolerant bread wheats. In *Prospects for Saline Agriculture*, R. Ahmad and K. A. Malik (eds.). Dordrecht, the Netherlands: Kluwer Academic Publishers.

Horie, T., Hauser, F., and Schroeder, J. I. 2009. HKT transporter-mediated salinity resistance mechanisms in *Arabidopsis* and monocot crop plants. *Trends Plant Science*, 14: 660–668.

Horie, T., Horie, R., Chan, W. Y., Leung, H. Y., and Schroeder, J. I. 2006. Calcium regulation of sodium hypersensitivities of sos3 and athkt1 mutants. *Plant Cell Physiology*, 47: 622–633.

Hospital, F. 2003. Marker-assisted breeding. In *Plant Molecular Breeding*, H. J. Newbury (ed.). Boca Raton, FL: Blackwell Publishing and CRC Press.

Hospital, F. 2009. Challenges for effective marker-assisted selection in plants. *Genetica*, 136: 303–310.

Huang, N., Angeles, E. R., and Domingo, J. 1997. Pyramiding of bacterial blight resistance genes in rice: Marker-assisted selection using RFLP and PCR. *Theoretical and Applied Genetics*, 95: 313–320.

Huang, S. B., Spielmeyer, W., and Lagudah, E. S. 2006. A sodium transporter (HKT7) is a candidate for Nax1, a gene for salt tolerance in durum wheat. *Plant Physiology*, 142: 1718–1727.

Ishitani, M., Liu, J., Halfter, U., Kim, C. S., Shi, W., and Zhu, J. K. 2000. SOS3 function in plant salt tolerance requires N myristoylation and calcium binding. *Plant Cell*, 12(9): 1667–1677.

James, R. A., Blake, C., Byrt, C. S., and Munns, R. 2011. Major genes for Na⁺ exclusion, Nax1 and Nax2 (wheat HKT1;4 and HKT1;5), decrease Na⁺ accumulation in bread wheat leaves under saline and waterlogged conditions. *Journal of Experimental Botany*, 62(8): 2939–2947.

James, R. A., Blake, C., Zwart, A. B., Hare, R. A., Rathjen, A. J., and Munns, R. 2012. Impact of ancestral wheat sodium exclusion genes Nax1 and Nax2 on grain yield of durum wheat on saline soils. *Functional Plant Biology*, 39: 609–618.

James, R. A., Davenport, R. J., and Munns, R. 2006. Physiological characterization of two genes for Na⁺ exclusion in durum wheat, Nax1 and Nax2. *Plant Physiology*, 142: 1537–1547.

Jan, V. A. 2009. Marker assisted selection: A non-invasive biotechnology alternative to genetic engineering of plant varieties. In *Smart Breeding*. Erwood S. and Truchi N. (eds.) Amsterdam, the Netherlands: Greenpeace International.

Jenks, M. A., Hasegawa, P. M., and Jain, S. M. (ed.). 2007. *Advances in Molecular Breeding toward Drought and Salt Tolerant Crops*. New York: Springer.

Khush, G. S. 1987. List of gene markers maintained in the Rice Genetic Stock Center, IRRI. *Rice Genetics Newsletter*, 4: 56–62.

Konieczny, A. and Ausubel, F. 1993. A procedure for mapping *Arabidopsis* mutations using co-dominant ecotype-specific PCR based markers. *The Plant Journal*, 4: 403–410.

Kreps, J. A., Wu, Y., Chang, H. S., Zhu, T., Wang, X., and Harper, J. F. 2002. Transcriptome changes for *Arabidopsis* in response to salt, osmotic, and cold stress. *Plant Physiology*, 130(4): 2129–2141.

Kudla, J., Batistic, O., and Hashimoto, K. 2010. Calcium signals: The lead currency of plant information processing. *Plant Cell*, 22: 541–563.

Landi, P., Sanguineti, M. C., and Liu, C. 2007. Root-ABA1 QTL affects root lodging, grain yield and other agronomic traits in maize grown under well-watered and water-stressed conditions. *Journal of Experimental Botany*, 58: 319–326.

Lauchli, A. and Epstein, E. 1990. Plant responses to saline and sodic conditions. In *Agricultural Salinity Assessment and Management*, K. K. Tanji (ed.). New York: American Society of Civil Engineers, pp. 113–137.

Lindsay, M. P., Lagudah, E. S., Hare, R. A., and Munns, R. 2004. A locus for sodium exclusion (*Nax1*), a trait for salt tolerance, mapped in durum wheat. *Functional Plant Biology*, 31: 1105–1114.

Lippman, Z., Gendrel, A. V. et al. 2004. Role of transposable elements in heterochromatin and epigenetic control. *Nature*, 430: 471–476.

Liu, J., Ishitani, M., Halfter, U., Kim, C. S., and Zhu, J. K. 2000. The *Arabidopsis thaliana* SOS2 gene encodes a protein kinase that is required for salt tolerance. *Proceedings of the National Academy of Sciences of the United States of America*, 97(7): 3730–3734.

Luo, Y., Sangha, J. S., Wang, S., Yang, J., and Yin, Z. 2012. Marker-assisted breeding of Xa4, Xa21 and Xa27 in the restorer lines of hybrid rice for broad-spectrum and enhanced disease resistance to bacterial blight. *Molecular Breeding*, 30: 1601–1610.

Mahajan, S., Pandey, G. K., and Tuteja, N. 2008. Calcium and salt-stress signaling in plants: shedding light on SOS pathway. *Archives of Biochemistry and Biophysics*, 471: 146–158.

Maischak, H., Zimmermann, M. R., Felle, H. H., Boland, W., and Mithofer, A. 2010. Alamethicin-induced electrical long distance signaling in plants. *Plant Signal Behavior*, 5: 988–990.

Maris, P. A. and Eduardo, B. 2002. Engineering salt tolerance in plants. *Current Opinion in Biotechnology*, 13: 146–150.

Martmez, A. J., Jiang, X. et al. 2007. Conservation of the salt overly sensitive pathway in rice. *Plant Physiology*, 143(2): 1001–1012.

Mittler, R., Vanderauwera, S. et al. 2011. Signaling: the new wave? *Trends Plant Science*, 16: 300–309.

Mittler, R. and Vanderauwera, S. 2012. ROS-mediated vascular homeostatic control of root-to-shoot soil Na delivery in *Arabidopsis*. *EMBO Journal*, 31: 4359–4370.

Mkumbira, J., Chiwona K. L. et al. 2003. Classification of cassava into "bitter" and "cool" in Malawi: From farmers' perception to characterisation by molecular markers. *Euphytica*, 132: 7–22.

Mockler, T. C., Chan, S. et al. 2005. Applications of DNA tiling arrays for whole-genome analysis. *Genomics*, 85(1): 1–15.

Monforte, A., Asins, M. J., and Carbonell, E. A. 1997. Salt tolerance in *Lycopersicon* species. VI. Genotype by salinity interaction in quantitative trait loci detection. Constitutive and response QTLs. *Theoretical and Applied Genetics*, 95: 706–713.

Morgante, M. and Salamini, F. 2003. From plant genomics to breeding practice. *Current Opinion in Biotechnology*, 14: 214–219.

Munns, R. 1993. Physiological processes limiting plant growth in saline soils: Some dogmas and hypotheses. *Plant, Cell, and Environment*, 16: 15–24.

Munns, R. 2002. Comparative physiology of salt and water stress. *Plant, Cell, and Environment*, 25: 239–250.

Munns, R. 2005. Genes and salt tolerance: Bringing them together. *New Phytologist*, 167(3): 645–663.

Munns, R. and James, R. A. 2012. Wheat grain yield on saline soils is improved by an ancestral $Na^+$ transporter gene. *National Biotechnology*, 30: 360–364.

Munns, R., Rebetzke, G. J., Husain, S., James, R. A., and Hare, R. A. 2003. Genetic control of sodium exclusion in durum wheat. *Australian Journal of Agricultural Research*, 54: 627–635.

Munns, R., Richard, A., and Jamesand, A. L. 2006. Approaches to increasing the salt tolerance of wheat and other cereals. *Journal of Experimental Botany*, 57(5): 1025–1043.

Munns, R. and Termaat, A. 1986. Whole-plant responses to salinity. *Functional Plant Biology*, 13(1): 143–160.

Munns, R. and Tester, M. 2008. Mechanisms of salinity tolerance. *Annual Review of Plant Biology*, 59: 651–681.

Neuffer, M. G., Coe, E. H., and Wessler, S. 1997. *Mutants of Maize*. New York: Cold Spring Harbor Laboratory Press.

Pakniyat, H., Powell, W. et al. 1997. AFLP variation in wild barley (*Hordeum spontaneum* C. Koch) with reference to salt tolerance and associated ecogeography. *Genome*, 40: 332–341.

Palmer, R. G. and Shoemaker, R. C. 1998. Soybean genetics. In *Soybean Institute of Field and Vegetative Crops*, M. Hrustic, M. Vidic, and D. Jackovic (eds.). Yugoslavia: Novi Sad, pp. 45–82.

Parida, A. K. and Das, A. B. 2005. Salt tolerance and salinity effects on plants: A review. *Ecotoxicology and Environmental Safety*, 60(3): 324–349.

Park, J. M., Park, C. J. et al. 2001. Overexpression of the tobacco *Tsi1* gene encoding an EREBP/AP2- type transcription factor enhances resistance against pathogen attack and osmotic stress in tobacco. *Plant Cell*, 13(5): 1035–1046.

Platten, J. D., Cotsaftis, O. et al. 2006. Nomenclature for HKT transporters, key determinants of plant salinity tolerance. *Trends Plant Science*, 11: 372–374.

Plett, D. C. and Moller, I. S. 2010. $Na^+$ transport in glycophytic plants: What we know and would like to know. *Plant Cell Environment*, 33: 612–626.

Pritchard, J. K., Stephens, M., Rosenberg, N. A., and Donnelly, P. 2000. Association mapping in structured populations. *American Journal of Human Genetics*, 67: 170–181.

Quarrie, S. A., Steed, A. et al. 2005. A high density 476 HOLLINGTON AND STEELE genetic map of hexaploid wheat (*Triticum aestivum* L.) from the cross Chinese Spring × SQ1 and its use to compare QTLs for grain yield across a range of environments. *Theoretical and Applied Genetics*, 110: 865–880.

Quesada, V., Ponce, M. R. et al. 2000. Genetic analysis of salt-tolerant mutants in *Arabidopsis thaliana*. *Genetics*, 154(1): 421–436.

Quintero, F. J., Ohta, M., Shi, H., Zhu, J. K., and Pardo, J. M. 2002. Reconstitution in yeast of the *Arabidopsis* SOS signaling pathway for $Na^+$ homeostasis. *Proceedings of the National Academy of Sciences of the United States of America*, 99(13): 9061–9066.

Ragot, M., Gay, G., Muller, J. P., and Durovray, J. 2000. Efficient selection for the adaptation to the environment through QTL mapping and manipulation in maize. In *Molecular Approaches for the Genetic Improvement of Cereals for Stable Production in Water-Limited Environments*, J. M. Ribaut and D. Poland (eds.). Mexico: CIMMYT.

Ragot, M. and Lee, M. 2007. Marker-assisted selection in maize: Current status, potential, limitations and perspectives from the private and public sectors. In *Marker-Assisted Selection, Current Status and Future Perspectives in Crops, Livestock, Forestry, and Fish*, Guimaraes, E. P. et al. (ed.). Rome, Italy: FAO.

Rahmana, A., Poustini, K., Munns, R., and James, R. A. 2010. Stomatal conductance as a screen for osmotic stress tolerance in durum wheat growing in saline soil. *Functional Plant Biology*, 37: 255–263.

Remington, D. L., Thornsberry, J. M. et al. 2001. Structure of linkage disequilibrium and phenotypic associations in the maize genome. *Proceedings of the National Academic Science of the United States of America*, 98: 11479–11484.

Ren, Z. H., Gao, J.-P. et al. 2005. A rice quantitative trait locus for salt tolerance encodes a sodium transporter. *Nature Genetics*, 37: 1141–1146.

Reyna, N. and Sneller, C. H. 2001. Evaluation of marker-assisted introgression of yield QTL alleles into adapted soybean. *Crop Science*, 41: 1317–1321.

Ribaut, J. M. and Betran, J. 1999. Single large-scale marker-assisted selection (SLSMAS). *Molecular Breeding*, 5: 531–541.

Ribaut, J. M., Edmeades, G. Perotti, E., and Hoisington, D. 2000. QTL analysis, MAS results and perspectives for drought-tolerance improvement in tropical maize. In *Molecular Approaches for the Genetic Improvement of Cereals for Stable Production in Water-Limited Environments*, Ribaut, J. M. and D. Poland (eds.). Mexico: CIMMYT.

Ribaut, J. M., Vicente, M. C., and Delannay, X. 2010. Molecular breeding in developing countries: challenges and perspectives. *Current Opinion in Plant Biology*, 13: 1–6.

Rozema, J. and Flowers, T. 2008. Ecology: Crops for a salinized world. *Science*, 322(5907): 1478–1480.

Rus, A., Baxter, I., Muthukumar, B., Gustin, J., Lahner, B., and Yakubova, E. 2006. Salt DE: Natural variants of AtHKT1 enhance $Na^+$ accumulation in two wild populations of *Arabidopsis*. *PLoS Genetics*, 2: 1964–1973.

Saleki, R., Young, P. G. et al. 1993. Mutants of *Arabidopsis thaliana* capable of germination under saline conditions. *Plant Physiology*, 101(3): 839–845.

Salvi, S., Costrini, P., Reynard, J. S., Zhao, Q., and Tuberosa, R. 2005a. Mapping major QTLs for seminal root traits and flowering time using an introgression library in maize. *Plant GEMs*, 4: 148

Salvi, S., Costrini, P., Reynard, J. S., Zhao, Q., and Tuberosa, R. 2005b. *Mapping Major QTLs for Seminal Root Traits and Flowering Time Using an Introgression Library in Maize*, Volume of Abstracts of Plant GEMs 4. Amsterdam, the Netherlands, p. 148.

Salvi, S. and Tuberosa, R. 2005. To clone or not to clone plant QTLs: Present and future challenges. *Trends Plant Science*, 10: 297–304.

Salvi, S., Tuberosa, R. et al. 2002. Toward positional cloning of Vgt1, a QTL controlling the transition from the vegetative to the reproductive phase in maize. *Plant Molecular Biology*, 48: 601–613.

Sanders, D. 2000. Plant biology: The salty tale of *Arabidopsis*. *Current Biology*, 10(13): 486–488.

Sawkins, M. C., Farmer, A. D. et al. 2004. Comparative Map and Trait Viewer (CMTV): An integrated bioinformatic tool to construct consensus maps and compare QTL and functional genomics data across genomes and experiments. *Plant Molecular Biology*, 56: 465–480.

Saxena S., Kaur, H., and Verma, P. 2013. Osmoprotectants: Potential for crop improvement under adverse conditions. In *Plant Acclimation to Environmental Stress*, N. Tuteja and S. Singh (eds.). New York: Springer.

Shi, H., Ishitani, M., Kim, C. and Zhu, J. K. 2000. The *Arabidopsis thaliana* salt tolerance gene SOS1 encodes a putative Na⁺/H⁺ antiporter. *Proceedings of the National Academy of Sciences of the United States of America*, 97(12): 6896–6901.

Shi, H. F. J., Quintero, J. M., and Zhu J. K. 2002. The putative plasma membrane Na⁺/H⁺ antiporter SOS1 controls long distance Na⁺ transport in plants. *Plant Cell*, 14(2): 465–477.

Shiozaki, N., Yamada, M. et al. 2005. Analysis of salt-stress-inducible ESTs isolated by PCR- subtraction in salt-tolerant rice. *Theoretical and Applied Genetics*, 110(7): 1177–1186.

Singh, S., Sidhu, J. S. et al. 2001. Pyramiding three bacterial blight resistance genes (xa5, xa13 and Xa21) using marker-assisted selection into indica rice cultivar PR106. *Theoretical and Applied Genetics*, 102(6): 1011–1015.

Stam, P. 1993. Construction of integrated genetic linkage maps by means of a new computer package: JoinMap. *The Plant Journal*, 3: 739–744.

Steele, K. A., Edwards, G., Zhu, J., and Witcombe, J. R. 2004. Marker-evaluated selection in rice: Shifts in allele frequency among bulks selected in contrasting agricultural environments identify genomic regions of importance to rice adaptation and breeding. *Theoretical and Applied Genetics*, 109: 1247–1260.

Steele, K. A., Price, A. H., Shashidar, H. E., and Witcombe, J. R. 2006. Marker-assisted selection to introgress rice QTLs controlling root traits into an Indian upland rice variety. *Field Crops Research*, 112: 28–221.

Steele, K. A., Virk, D. S., Kumar, R., Prasad, S. C., and Witcombe, J. R. 2007. Field evaluation of upland rice lines selected for QTLs controlling root traits. *Field Crops Research*, 101: 180–186.

Stephen, P. M. and Rita H. M. 2008. Molecular plant breeding as the foundation for 21st century. *Crop Improvement Plant Physiology*, 147: 969–977.

Stoskopf, N. C., Tomes, D. T., and Christie, B. R. 1993. *Plant Breeding Theory and Practice*. Boulder, CO: West view Press.

Stuart, J. R., Sonia, N., and Mark, T. 2014. Salt resistant crop plants. *Current Opinion in Biotechnology*, 26: 115–124.

Sunkar, R., Kapoor, A., and Zhu, J. K. 2004. Posttranscriptional induction of two Cu/Zn superoxide dismutase genes in *Arabidopsis* is mediated by down regulation of miR398 and important for oxidative stress tolerance. *Plant Cell*, 18: 2051–2065.

Suzuki, N., Koussevitzky, S., Mittler, R., and Miller, G. 2012. ROS and redox signaling in the response of plants to abiotic stress. *Plant Cell Environment*, 35: 259–270.

Suzuki, N., Miller, G., Morales, J., Shulaev V., Torres, M. A., and Mittler, R. 2011. Respiratory burst oxidases: the engines of ROS signalling. *Current Opinion in Plant Biology*, 14: 691–699.

Taji, T. and Seki, M. 2004. Comparative genomics in salt tolerance between *Arabidopsis* and *Arabidopsis*-related halophyte salt cress using *Arabidopsis* microarray. *Plant Physiology*, 135(3): 1697–709.

Tanksley, S. D. 1993. Mapping polygenes. *Annual Review of Genetics*, 27: 205–233.

Tauz, D. and Renz, M. 1984. Simple sequences are ubiquitous repetitive components of eukaryotic genomes. *Nucleic Acids Research*, 12: 4127–4138.

Tester, M. and Davenport, R. 2003. Na⁺ tolerance and Na⁺ transport in higher plants. *Annals of Botany*, 91: 503–527.

Tuberosa, R., Gill, B. S., and Quarrie, S. A. 2002. Cereal genomics: Ushering in a brave new world, *Plant Molecular Biology*, 48: 445–449.

Very, A. A. 2011. Over-expression of a Na and K permeable HKT transporter in barley improves salt tolerance. *Plant Journal*, 68: 468–479.

Virk, D. S., Singh, D. N., Prasad, S. C., Gangwal, J. S., and Witcombe, J. R. 2003. Collaborative and consultative participatory plant breeding of rice for the rainfed uplands of eastern India. *Euphytica*, 132: 95–108.

Viswanathan, C., Zhu, J., Tao, Z., and Jian, K. Z. 2007. Small RNAs: Big role in abiotic stress tolerance of plants. In *Advances in Molecular Breeding Toward Drought Salt Tolerant Crops*, A. Matthew, P. M. Jenks, S. Hasegawa, and M. Jain (eds.). Springer, Dordrecht, The Netherlands.

Vos, P., Hogers, R. et al. 1995. AFLP: A new technique for DNA fingerprinting. *Nucleic Acids Research*, 23: 4407–4414.

Walia, H., Wilson, C. et al. 2005. Comparative transcriptional profiling of two contrasting rice genotypes under salinity stress during the vegetative growth stage. *Plant Physiology*, 139(2): 822–835.

Wang, B. and Chee, P. W. 2010. Application of advanced backcross quantitative trait locus (QTL) analysis in crop improvement. *Journal of Plant Breeding and Crop Science*, 2: 221–232.

Wang, X., Jiang, G. L., Green, M., Scott, R. A., Hyten, D. L., and Cregan P. B. 2012. Quantitative trait locus analysis of saturated fatty acids in a population of recombinant inbred lines of soybean. *Molecular Breeding*, 30: 1163–1179.

Ware, D. H., Jaiswal, P. et al. 2002. Gramene, a tool for grass genomics. *Plant Physiology*, 130: 1606–1613.

Welsh, J. and McClelland, M. (1990) Fingerprinting genomes using PCR with arbitrary primers. *Nucleic Acids Research*, 18: 7231–7238.

Wenzel, G. 2006. Molecular plant breeding: Achievements in green biotechnology and future perspectives. *Applied Microbiology and Biotechnology*, 70: 642–650.

Werner, J. and Finkelstein, R. R. 1995. *Arabidopsis* mutants with reduced response to NaCl and osmotic stress. *Physiology Plant*, 93: 659–666.

Werner, J. F. 1995. *Arabidopsis* mutants with reduced response to NaCl and osmotic stress. *Physiology Plant*, 93: 659–666.

Williams, J. G. K., Kubelik, A. R., Livak, K. J., Rafalski, J. A., and Tingey, S. V. 1990. DNA polymorphisms amplified by arbitrary primers are useful as genetic markers. *Nucleic Acids Research*, 18: 6531–6535.

Wimmer, M. A., Muhling, K. H., Lauchli, A., Brown, P. H., and Goldbach, H. E. 2003. The interaction between salinity and boron toxicity affects subcellular distribution of ions and proteins in wheat leaves. *Plant, Cell, and Environment*, 26: 1267–1274.

Witcombe, J. R. and Virk, D. S. 2001. Number of crosses and population size for participatory and classical plant breeding. *Euphytica*, 122: 451–462.

Witcombe, J. R., Joshi, K. D. et al. 2005. Participatory plant breeding is better described as highly client-oriented plant breeding. Four indicators of client-orientation in plant breeding. *Experimental Agriculture*, 41: 299–319.

Xu Y. 2010. *Molecular Plant Breeding*. Wallingford, CT: CAB International.

Xu, Y. and Crouch, J. H. 2008. Marker-assisted selection in plant breeding: From publications to practice. *Crop Science*, 48: 391–407.

Xue, S., Yao, X. et al. 2011. AtHKT1;1 mediates nernstian sodium channel transport properties in *Arabidopsis* root stellar cells. *PLoS ONE*, 6: 24725.

Yang, Q., Chen, Z. Z. et al .2009. Overexpression of SOS (salt overly sensitive) genes increases salt tolerance in transgenic *Arabidopsis*. *Molecular Plant*, 2: 22–31.

Yu, J., Arbelbide, M., and Bernardo, R. 2005. Power of in silico QTL mapping from phenotypic, pedigree and marker data in a hybrid breeding program. *Theoretical and Applied Genetics*, 110: 1061–1067.

Zabeau, M. and Voss, P. 1993. Selective restriction fragment amplification: A general method for DNA fingerprinting. European Patent Application. 92402629.7 (Publ. Number 0 534 858 A1).

Zamir, D. 2001. Improving plant breeding with exotic genetic libraries. *Nature Review Genetics*, 2: 983–989.

Zhang, H. X., Hodson, J. N., Williams, J. P., and Blumwald, E. 2001. Engineering salt-tolerant *Brassica* plants: Characterization of yield and seed oil quality in transgenic plants with increased vacuolar sodium accumulation. *Proceedings of the National Academic Science of the United States of America*, 98: 12832–12836.

Zhikang, L., Fu, B. Y. et al. 2005. Genome-wide introgression lines and their use in genetic and molecular dissection of complex phenotypes in rice (*Oryza sativa* L.), *Plant Molecular Biology*, 59: 33–52.

Zhu, J. K. 2007. *Plant Salt Stress*. John Wiley & Sons Ltd.

# CHAPTER SEVENTEEN

# Mutation Breeding and Salt Stress Tolerance in Plants

*Hyun-Jee Kim, J. Grover Shannon, and Jeong-Dong Lee*

## CONTENTS

*Abstract.* Soil salinity is a major abiotic stress on field crops and is a serious problem globally. It occurs due to natural soil characteristics, inundations of salt water onto land, and use of irrigation water high in salt ions. Salinity reduces plant growth and decreases crop yield and quality and can result in plant death. Induced mutation is an important source of genetic variation in living organisms and can have a positive effect on increasing stress to saline conditions. However, mutation methods and mutagenic agents are important to target traits of interest to the breeder. Currently, mutation breeding is responsible for the release of more than 3200 varieties. Among them, 15 salt-tolerant varieties were developed almost entirely by induced ionizing radiation from gamma rays. Breeding salt-tolerant cultivars is important for producing sustainable food, fiber, biofuel, and shelter in the face of increased global salinity as time progresses. Therefore, knowledge of salt tolerance and use of mutation breeding could contribute to the development of salt-tolerant plant cultivars.

*Keywords:* mutation breeding, salinity, salt-tolerant variety.

## 17.1 INTRODUCTION

High concentrations of salts in the soil are a major abiotic stress and a serious problem for crop cultivation and production of agricultural crops (Chinnusamy et al. 2005). Saline soils are found in over 830 million ha, representing 26% of the world potential arable land or around 6% of the world's total land area (Munns 2005). Salt-affected soils are naturally present due to the accumulation of salt over long periods of time in arid and semiarid zones in more than 100 countries of the world (Leblebici et al. 2011). Wind and rain also cause accumulation of salts. Rainwater contains 6–50 mg/kg of NaCl and if rain contains 10 mg/kg of NaCl, 10 kg/ha of salt would deposit for each 100 mm of rainfall per year (Munns et al. 2008).

However, cultivated agricultural land has become saline due to clearing of natural vegetation or from irrigation with water high in salt ions. Of the approximately 230 million ha of irrigated land that produce about one-third of the world's food supply, 45 million ha are affected by salinity (Lenis et al. 2011; Munns 2005).

According to the USDA Salinity Laboratory, soils are defined as saline if the concentration of soluble salts if the $EC_e$ is $\geq 4$ dS/m (USSL 2005) where $EC_e$ is the electrical conductivity of the saturated paste extract and reflects the concentration of salts in saturated soil. NaCl is generally the most soluble and abundant salt contributing to salinity. A conductivity of 4 dS/m is equivalent to 40 mM NaCl. In regions with high salt concentrations, salinity inhibits plant growth and decreases crop yield and quality and causes plant death (Yaycili and Alikamanoglu 2012). Soil salinity also induces nutritional imbalances in plants and many salt-affected soils are waterlogged. Thus, the interaction between hypoxia and salt has a powerful depressive effect on plant growth (Barret-Lennard 2003).

Adapting crop plants to saline environments is a major challenge. To solve this problem, more tolerant varieties, developed through various breeding and biotechnology methods, are very important for successful production under saline conditions. The human population is expected to increase from 6.1 billion in 2001 to 9.3 billion by 2050 (http://www.unfpa.org/swp/2001). Thus, genetic improvements and sustainable crop production to overcome salinity are necessary. This chapter will provide a greater understanding of salt stress in plants and will present new developments and future applications in mutation breeding for salt tolerance in the face of increased salinization in the world.

## 17.2 MUTATION BREEDING

### 17.2.1 Current State of Mutation Breeding

The Food and Agriculture Organization and International Atomic Energy Agency mutant variety database (http://www-mvd.iaea.org/MVD/Default.htm), February 2014, reported there were more than 3200 mutant varieties in more than 200 crop species released. Among them approximately 3070 mutant varieties were developed by 34 countries (Table 17.1). China has released the largest number of mutant varieties with 812 since

1950, followed by Japan and India with 481 and 329 mutant varieties since the 1960s and 1950s, respectively. Among all of these mutant varieties, 15 were developed with salt tolerance. These 15 include 9 rice varieties from eight countries followed by two rapeseed and two wheat varieties from Bangladesh and China, respectively (Table 17.2). Almost all of these salt-tolerant mutant varieties were derived by ionization with gamma ray–induced mutagenesis.

### 17.2.2 Overview of Mutation Breeding

Mutation has been defined as a change in the heritable constitution of the genome and is an important source of genetic variation in living organisms, but plant breeders are interested in certain changes for a particular phenotypic trait (Roose and Williams 2007). At the molecular level, mutations can alter DNA sequence by base substitutions, insertions, deletions, or sequence rearrangement. Any of these types of change may cause a phenotypic change. Mutation breeding is a good choice when it is essential to maintain the characteristics of an existing cultivar. However, the high heterozygosity of most cultivars makes it very difficult to recover progeny with very similar characters from selection after hybridization.

Mutation breeding procedures (other than genetic engineering) do not allow the plant breeder to target specific genes or traits because mutations are random. Tissue exposed to mutagenic agents will have random mutations and often have no effect or undesirable effects. Therefore, screening of a large population is necessary to find the desired mutation. However, increasing the dosage of mutagen will not only increase the frequency of mutations but also increase plant mortality. Also, the desired mutation will often be accompanied by undesirable consequences in which the mutant line cannot be used directly as a cultivar. Thus, adapted mutation breeding methods are required to produce useful mutations without deleterious effects in plants.

The key steps in using a mutation breeding procedure are choosing a mutagenic agent, choosing a target tissue or seed for treating with a mutagenic agent, exposing the tissue or seed at a specific dosage(s) with the right protocol, propagating plants from the mutagen exposed tissue or seed, screening these plants for the desire mutation, and then evaluating selections that carry the mutation, either directly or as parents.

**TABLE 17.1**

*Varieties derived worldwide from mutation breeding (FAO-IAEA mutant variety database, Feb. 2014).*

| Rank | Country | Era | | | | | | | | Total |
|------|---------|-------|-------|-------|-------|-------|-------|-------|---------|-------|
| | | 1950s | 1960s | 1970s | 1980s | 1990s | 2000s | 2010s | Unknown | |
| 1 | China | 3 | 42 | 119 | 270 | 198 | 173 | 4 | 3 | 812 |
| 2 | Japan | 0 | 15 | 32 | 58 | 173 | 202 | 0 | 0 | 481 |
| 3 | India | 1 | 12 | 108 | 112 | 57 | 34 | 2 | 3 | 329 |
| 4 | Russian Federation | 0 | 5 | 42 | 102 | 56 | 6 | 1 | 4 | 216 |
| 5 | Netherlands | 1 | 19 | 85 | 68 | 0 | 0 | 0 | 3 | 176 |
| 6 | Germany | 3 | 18 | 32 | 86 | 21 | 11 | 0 | 0 | 171 |
| 7 | United States | 3 | 18 | 28 | 40 | 35 | 11 | 0 | 4 | 139 |
| 8 | Bulgaria | 0 | 0 | 4 | 26 | 22 | 21 | 1 | 2 | 76 |
| 9 | Viet Nam | 0 | 0 | 3 | 11 | 21 | 16 | 4 | 0 | 55 |
| 10 | Pakistan | 0 | 0 | 3 | 8 | 23 | 19 | 0 | 0 | 53 |
| 11 | Bangladesh | 0 | 0 | 4 | 3 | 18 | 18 | 1 | 0 | 44 |
| 12 | Canada | 0 | 4 | 13 | 14 | 6 | 2 | 0 | 0 | 40 |
| 13 | France | 0 | 0 | 18 | 20 | 0 | 0 | 0 | 1 | 39 |
| 14 | Italy | 0 | 3 | 7 | 23 | 2 | 0 | 0 | 0 | 35 |
| 14 | Republic of Korea | 0 | 0 | 5 | 2 | 11 | 17 | 0 | 0 | 35 |
| 15 | United Kingdom | 0 | 2 | 7 | 24 | 1 | 0 | 0 | 0 | 34 |
| 17 | Poland | 0 | 0 | 3 | 14 | 14 | 0 | 0 | 0 | 31 |
| 18 | Indonesia | 0 | 0 | 0 | 6 | 6 | 12 | 4 | 1 | 29 |
| 19 | Sweden | 3 | 7 | 9 | 6 | 0 | 0 | 0 | 1 | 26 |
| 20 | Guyana | 0 | 0 | 0 | 26 | 0 | 0 | 0 | 0 | 26 |
| 21 | Iraq | 0 | 0 | 0 | 0 | 23 | 0 | 0 | 0 | 23 |
| 22 | Belgium | 0 | 2 | 9 | 11 | 0 | 0 | 0 | 0 | 22 |
| 23 | Denmark | 0 | 0 | 3 | 17 | 1 | 0 | 0 | 0 | 21 |
| 24 | Thailand | 0 | 0 | 2 | 5 | 8 | 5 | 0 | 0 | 20 |
| 25 | Slovakia | 0 | 2 | 5 | 10 | 1 | 0 | 0 | 1 | 19 |
| 26 | Czech Republic | 0 | 1 | 8 | 7 | 2 | 0 | 0 | 0 | 18 |
| 27 | Austria | 1 | 0 | 1 | 14 | 1 | 0 | 0 | 0 | 17 |
| 28 | Mali | 0 | 0 | 0 | 0 | 14 | 1 | 0 | 0 | 15 |
| 29 | Philippines | 0 | 0 | 5 | 2 | 2 | 6 | 0 | 0 | 15 |
| 30 | Brazil | 0 | 0 | 2 | 2 | 7 | 2 | 0 | 0 | 13 |
| 31 | Cuba | 0 | 0 | 0 | 0 | 7 | 5 | 0 | 0 | 12 |
| 32 | Finland | 0 | 1 | 7 | 3 | 0 | 0 | 0 | 0 | 11 |
| 33 | Hungary | 0 | 1 | 2 | 4 | 0 | 3 | 0 | 0 | 10 |
| 34 | Ukraine | 0 | 0 | 0 | 0 | 4 | 6 | 0 | 0 | 10 |

## TABLE 17.2
*Salt-tolerant varieties derived by mutation breeding methods.*

| No | Plant species | Variety name | Parent variety | Registration year | Country | Mutagen |
|---|---|---|---|---|---|---|
| 1 | Common bean | CIAT 899 | CIAT 899 m1, CIAT 899 m2 | 2007 | Tunisia | Gr |
| 2 | Cotton | NIAB-Karishma | NIAB-86(mutant), W83–29 | 1996 | Pakistan | Gr |
| 3 | Rapeseed | Binasarisha-5 | Nap-3 | 2002 | Bangladesh | Gr |
| 4 | Rapeseed | Binasarisha-6 | Nap-3 | 2002 | Bangladesh | Gr |
| 5 | Rice | Atomita 2 | Pelita I/1 | 1983 | Indonesia | Gr |
| 6 | Rice | BINAdhan-9 | Kaloziri,Y-1281 | 2010 | Bangladesh | — |
| 7 | Rice | GINES | Jucarito −104 | 2007 | Cuba | Protons |
| 8 | Rice | Liaoyan 2 | Toyonishiki | 1992 | China | Gr |
| 9 | Rice | Mohan (= CSR-4) | IR 8 | 1983 | India | Gr |
| 10 | Rice | Niab-Irri-9 | IR-6 | 1999 | Pakistan | Fn |
| 11 | Rice | Shua 92 | Shadab | 1993 | Pakistan | Gr |
| 12 | Rice | VND 95–20 | IR64 | 1999 | Viet Nam | Gr |
| 13 | Rice | Wonhaebyeo | Donganbyeo | 2007 | Republic of Korea | Gr |
| 14 | Wheat | H6765 | Laizhou 953, 90r4−85 L//Ji87−5108 | 2004 | China | Gr |
| 15 | Wheat | Jiaxuan 1 | Maoyingafu | 1974 | China | Gr |

SOURCE: FAO/IAEA mutant varieties database. http://www-mvd.iaea.org/MVD/Default.htm.

Gr, Gamma ray; Fn, Fast neutrons.

### 17.2.3 Mutagenic Agents

Mutagenic agents are known to produce a range of mutant types. The breeder can choose from a wide range of mutagenic agents. Each agent typically induces mutations by a particular mechanism and therefore results in types of DNA alteration such as gene or point mutations and structural mutations of chromosomal rearrangements. Specific agents may be appropriate for a particular tissue or target trait. Mutagens can be classified as biological agents such as transposons and T-DNA, physical agents such as fast neutron, UV and x-ray radiation and chemical agents such as N-methyl-N-nitrosourea (MNU) and ethyl methanesulfonate (EMS).

Chemicals induce mainly point mutations and are thus adapted for producing missense and non-sense mutations. Among chemicals, EMS is the most commonly used chemical mutagen in plants (Serrat et al. 2014). Because EMS produces a large number of nonlethal point mutations, a relatively small mutant population (approximately 10,000) is sufficient to saturate the genome with mutations (Talebi et al. 2012). EMS mainly induces C-to-T substitutions resulting in C/G to T/A transitions (Krieg 1963). Among chemical mutagens, newer mutagens, like MNU, hydroxyl amine, nitrous acid, and N-ethyl-nitrosourea (ENU), are being recommended, since they induce mutations involving base substitutions, including transitions and/or transversions (Balyan et al. 2008).

Among the tools, ionizing radiations as physical mutagens were the most frequently used method to develop around 89% of the direct mutant varieties (Ahloowalia et al. 2004). There has been considerable funding for mutation breeding from national and international atomic energy agencies anxious to characterize the effects of radiation and find useful applications. Ionizing radiation can induce a wide range of mutation types including chromosome breaks and rearrangements and point mutations. In ionizing radiation, gamma rays and x-rays are the two most commonly used radiation types. Gamma rays were employed to develop 64% of the radiation-induced mutant varieties, followed by x-rays with 22% (Ahloowalia et al. 2004). Among physical mutagens, alpha particles and fast neutrons are used to generate mutations that are largely due to deletions. Although UV irradiation can be used, it generally penetrates tissues less deeply than ionizing radiation.

When using radiation, as well as other mutagens, the breeder must apply the appropriate dose and exposure time. The dose is the amount of mutagen applied per unit time, while exposure is the total absorbed radiation. The current unit of absorbed radiation energy is the Gy (Gray), while most of the older literature reports exposure in rad or Roentgen (R, which measures exposure rather than dose). One Gy equals one hundred rad.

However, it is difficult to predict the effect of a given dose and exposure on a particular tissue. Plants differ in their response to irradiation dosages due to their genotypic differences, physiological condition of the tissue, and particularly water content. Higher doses of radiation cause chromosomal damage, deceleration of the cell cycle, and delay of mitosis, which significantly affect regeneration and development of plant (Hewawasam et al. 2004; Gulsen et al. 2007). Therefore, selection of the correct dosages is very important, and breeders generally identify an appropriate dose empirically. In mutation studies, breeders typically use the $LD_{50}$, which means death of 50% of the test materials as a measure of the biological response of tissue to a particular radiation dose (Chikelu et al. 2007). Plant improvement studies have shown dosages of around 30% as effective doses ($ED_{30}$), which means an effect of 30% on the test materials is preferred (Alikamanoğlu 2002).

Transposable elements (TEs) are natural genetic elements with the ability to move from one genome location to another, either with or without duplication of the element. TEs can be used to induce mutations, because a gene mutated by a transposon insertion can be cloned using the TE as a *tag sequence*. Targeted mutation, in which mutations are induced in a specific gene or DNA sequence, is not yet efficient in plants. If this technique were successfully developed, it would certainly contribute to mutation breeding research because it would open up an alternative path to apply genomics tools, one that is likely to be less controversial than transformation.

### 17.2.4 Somatic Mutation

To improve plant tolerance to environmental stress such as salinity via mutation breeding, somatic-induced mutations are highly valuable for mutant production in vegetative plants in vitro (Saleem et al. 2005). During somatic mutation studies in tissue cultures using micropropagation techniques, $M_1V_3$, $M_1V_4$, and $M_1V_5$ generations are formed, and mutant plants with the desired characteristics can be successfully selected in vitro (Ahloowaliaand Maluszynski 2001; Hewawasam et al. 2004). Regardless of genetic differences among plants for somatic mutation, inducted radiation doses applied to node explants for in vitro tissue culture studies must be around 20 Gy. It was noted that higher doses could be lethal (Yaycili and Alikamanoglu 2012).

Saleem et al. (2005) identified moderate salt tolerance among $M_2$ in Basmati rice by using a somatic mutation method on a Murashige and Skoog (MS) medium containing 9.05 μM2, 4-D using 50 Gy of gamma ray irradiation. Even though field tests were needed in succeeding generations on the putative $M_2$ lines to prove genetic stability of salt tolerance, the combination of mutagenesis with an in vitro method was also effective in creating plant variation in a short time using in small fields to facilitate selection for the desired phenotypes.

## 17.3 MUTATION BREEDING FOR SALT TOLERANCE

Breeders often apply induced mutation methods on the best local cultivars or lines in the hope of quickly producing improved lines. However, many induced mutations have direct or indirect adverse effects. And many desirable mutations are recessive hence are difficult to track in a breeding program. According to Briggs (1978), less than 1% of all observed mutations are of value to breeders. There were many studies for salt tolerance by using mutation breeding. Among them, several successful studies are listed in Table 17.2.

### 17.3.1 Salt-Tolerant Rice Varieties

Rice (Oryza sativa L.) is one of the most important food crops in the world. Among the salt-tolerant varieties developed by mutation breeding, rice has the highest number among all crops. Rice has been conducive to successful mutation breeding methods because of its small genome size relative to other cereals, the availability of the entire genome sequence, ease of transformation and regeneration resulting in an array of different mutants (Serrat et al. 2014). Among them, "CSR10" was released in 1989 by the Central Varietal Release Committee for sodic and inland saline soils of India (Singh et al. 2010b). It was derived from a cross between the salt-tolerant variety "M40-431-24-114" and "Jaya" using in the pedigree method. The variety "M40-431-24-114" was derived from gamma ray–irradiated (10 KR), from $F_1$ seed, from the cross "CSR1/IR8." The variety "CSR10" grew well on a highly deteriorated sodic soil ($pH_2$ 9.8–10.2) and on an inland saline soil ($EC_e$ 6–10 dS/m) under a transplanted irrigated management system (Mishra et al. 1992).

In Indonesia, high yielding variety, "Atomita 2" was released in 1983 for commercial cultivation and developed from irradiation of "Pelita-1" with gamma rays (Table 17.2). It showed tolerance to salinity, resistance to brown plant hopper biotype 1, biotype 3 and had desirable palatability (Mugiono 1984).

"GINES" rice was the first variety released and developed from in vitro proton mutagenesis for salt tolerance (Table 17.2). The parent, "Jucarito-104" seeds were irradiated with different doses of protons (10, 20, 30, 40, 60, 90, 110, and 210 Gy) (Gonzalez et al. 2009). Promising lines from an $M_1V_1$ bulk population were selected in the $M_2V_1$ to $M_5V_1$ generations. The mutant line later named "GINES" had higher grain yield, grain quality, and salt tolerance than "Jucarito-104." In addition, the mutant "GINES" and the donor "Jucarito-104" showed differences in AFLP analysis indicating differences in genetic composition as a result of the induced mutation procedure applied to "Jucarito-104."

The salt-resistant rice variety "Liaoyan 2" was selected from the mutant plants from the variety "Toyonishiki" in 1988 (Table 17.2). It showed higher salt resistance (over 65%–100%) with higher yield (6.3–11.3 tons/ha), preferred grain quality (90–110 grains per panicle and 26 g for 1000-grain weight), and multiple disease resistance than the common cultivated varieties (Zhang and Xu 1992). In addition, it was more tolerant to lodging, more cold tolerant, and drought resistant and had good cooking traits compared to the other rice varieties tested.

The salt-resistant rice variety "Mohan (= CSR-4)" was released by using gamma ray–induced methods in 1983 (Table 17.2). Panda et al. (2013) studied the effect of salt stress on the pigment content and yield of different rice genotypes. The rice genotypes "CSR-4," "Canning-7," "IR-29," and "IR-64" were treated with two levels of salinity (40 and 60 mmol/L of NaCl) in seedling, tillering, and flowering stages. Salinity stress decreased chlorophyll a, chlorophyll b, total chlorophyll, and carotenoid content. The values for reduced chlorophyll value was more pronounced in "IR-29" as compared to more tolerant genotypes "CSR-4" and "Canning-7." In addition, the grain yield of different rice genotypes under salt stress condition was positively associated with total chlorophyll and carotenoid concentration. "CSR-4" and "Canning-7" performed better in grain yield as compared to "IR-29" under different levels of salinity.

"Niab-Irri-9," a productive salt-tolerant variety, was derived from "IR-6" through irradiation with fast neutrons at a dosage of 1500 rads (Table 17.2). It was evaluated as "NIAB-6" over 3 years at locations on saline soils (Cheema et al. 1999). Based on higher yield potentials, fine grain, and salt tolerance, "NIAB-6" was named and released as "Niab-Irri-9" in 1999. "NIAB-6" and two commercial varieties "KS-282" and "IR-6" were tested for the yielding ability in the moderate- ($EC_e$ dS/m = 7.0) and high-salinity ($EC_e$ dS/m = 10.0) levels in 1995–1996. At moderate- and high-salinity levels, "NIAB-6" had the highest yield (average 2.3 and 1.9 tons/ha) among the compared varieties "KS-282" (average 2.2 and 1.5 tons/ha) and "IR-6" (average 2.1 and 1.5 tons/ha), respectively. Later, "Niab-Irri-9" was evaluated with 15 genotypes during 2003–2004 on 9 locations. It showed the highest mean seed yield with 5.6 tons/ha under various environments.

After the "green revolution," "IR8" and "IR6" remained the dominant rice varieties in Sindh province, Pakistan (Mustafa et al. 1997). However, they had less consumer preference due to coarse grain. "IR8" was subjected to fast neutrons irradiation in 1972. In the $M_2$ generations, the promising plant types were selected in 1973 and later confirmed in $M_3$ and $M_4$ during 1974–1975. In subsequent generations, further selection and screening was done for desirable plant type and grain. Later, the mutant variety "Shua92" was released from "IR8" in 1993. This selected mutant showed distinctly reduced plant height, higher tiller number, longer panicles, and greater number of grains/panicle than "IR-8." In yield tests, it showed significantly higher yield with 10.2 tons/ha than commercial check varieties "IR6" (9.2 tons/ha) and "IR8" (8.9 tons/ha) during 1978–1980. During stress evaluation, "Shua92" showed salt tolerance under soils with $EC_e$ ranging from 7.1–8.0 mmho/cm with 4.6 tons/ha yield against "Pokkali" (2.2 tons/ha) as control. Consequently, "Shua92" became commercially grown cultivar with high yield, improved grain quality, and salt tolerance (Table 17.2).

The salt-tolerant rice variety "Wonhaebyeo" was developed and released through in vitro mutagenesis with 70 Gy gamma rays in 2007 (Table 17.2). In vitro culture, combined with gamma ray–induced mutations proved to be an effective way to improve salt tolerance in rice (Kim et al. 2008). The effect of gamma rays with 0, 30, 50, 70, 90, and 120 Gy on growth of the callus was investigated 40 days after an irradiation. Callus was inoculated on a N6 basal medium supplemented with 1.5% NaCl and 70 Gy, proved to be the optimum inducing for the salt-tolerant calli.

### 17.3.2 Salt-Tolerant Varieties for Other Species

The salt-tolerant cotton (*Gossypium hirsutum* L.) variety "NIAB-Karishma" was developed and released in 1996. It was derived from a cross of the salt-tolerant mutant cultivar "NIAB-86" developed from gamma ray–induced mutation breeding with an American strain "W 83–29." It is early maturing strain, with improved heat tolerance, and high yield potential (Table 17.2). "NIAB-karishma" was cultivated on 486,000 ha and brought farmers US$17 million income (Ahloowalia et al. 2004). Later, it was studied with "NIAB-86" and "K-115" during germination and early seeding growth for identifying the association between salinity and α-amylase activity (Ashraf et al. 2002). It was identified that the increase in NaCl concentration resulted in the decrease in α-amylase activity in all cultivars. Salt tolerance in "K-115" during germination, as a result of less effect on α-amylase activity.

Barley (*Hordeum vulgare* L.) has a long record as being an experimental plant for mutation research. In Sweden, a widespread research program was set up to develop useful barley mutants. By 1986 about 9000, varieties had been characterized (Lundqvists 1986). "Golden Promise" was a gamma ray–induced mutant of from the cultivar "Maythorpe." "Golden Promise" and "Maythorpe" were grown in control (25 mol/ $m^3$ NaCl) and salt stress (150 mol/ $m^3$ NaCl) conditions as described by Forster et al. (1994). There was a significant difference between the two varieties in their responses to salt stress. "Golden Promise" showed significantly less sterile spikelets/spike under salt stress than Maythorpe.

Common bean (*Phaseolus vulgaris* L.) mutants were highly tolerant to salt using gamma-irradiated clones at 700 and 800 Gy of "CIAT 899" compared to control strains (Table 17.2). This study proved that gamma irradiation was effective in changing physiological and phenotypic characteristics via random mutations (Fatnassi et al. 2011).

Two salt-tolerant rapeseed (*Brassica napus* L.) varieties, "Binasarisha-5" and "Binasarisha-6," were developed and released in 2002 using gamma ray mutation breeding (Table 17.2). "Binasarisha-5" and "Binasarisha-6" showed salt tolerance up to salt concentrations of 13 dS/m. Their maturity period ranged between 85–90 and 90–95 days, respectively, and also showed tolerance to Alternaria disease. Their maximum seed yield potential was 2.1 and 2.2 tons/ha, respectively, with an average yield was 1.4 and 1.3 tons/ha, respectively. They showed 43% and 44% of oil concentration, respectively. Seed yield of "Binasarisha-5" showed to be better than that of non-salt-tolerant varieties "Safal" and "Binasarisha-4" (Badruddin et al. 2005).

Potato (*Solanum tuberosum* L.) plant varieties are very sensitive to environmental stresses such as temperature changes, drought, and salinity due to their sparse and short root systems. When potato is grown in soil that contains 20–35 mM concentration of NaCl, it shows reductions in plant growth and yields. Potatoes show more tolerance to salinity compared to other agricultural plant such as pepper and corn; but it shows less tolerance to tomato, rice, barley, and soybean. Yaycili and Alikamanoglu (2012) identified salt-tolerant mutant potato plants were successfully created in vitro tissue cultures via gamma irradiation at around 15, 20, and 30 Gy and selected in medium containing 100 mM NaCl.

Luan et al. (2007) identified salt-tolerant mutants by using EMS in calli of sweet potato (*Ipomoea batatas* L.). Calli initiated from leaf explants were treated with 0.5% EMS for 0, 1, 1.5, 2, 2.5, and 3 h, followed by rinsing with sterile distilled water for four times. Salt-tolerant calli were subcultured on medium supplemented with 200 mM NaCl for selection of mutant cell lines. The selected calli were transferred onto somatic embryo formation medium. After propagation, mutants that showed more salt tolerance than control plants were identified.

Akyuz et al. (2013) studied four lines of salt-tolerant soybean (Glycine max L. Merr.) mutants selected from the $M_2$ generation of the gamma-irradiated "S04–05" soybean variety. The salt-tolerant mutants ($M_2$) had less protein but more lipid and cellulose contents in comparisons to control group of seeds. Salt-tolerant mutants showed increased chlorine, copper, and zinc concentrations in comparison to the control group. The increased Cu and Zn concentrations were also reported in gamma irradiated soybean seeds (Alikamanoğlu et al. 2011; Singh et al. 2010a). Soybean seeds were gamma irradiated with 150 Gy dose using Cs-137 gamma source. Salt-tolerant lines were selected based response of 14 day-old mutant seedlings to salt stress.

The mutant wheat (Triticum aestivum L.) variety "H6756" was developed by irradiation of pollen with gamma rays in 2004 (Table 17.2). It was derived from the cross "(Laizhou953 × 90r4–85L) × Ji87–5108." The anthers from $F_1$ plants were irradiated by 1.5 Gy gamma rays and cultured in a medium containing 0.4% NaCl for the callus induction. It showed high yield with an average yield of 6.2 tons/ha or 17, 3% higher than the check cultivar "DK961." In addition, it showed tolerance to salinity and drought.

## 17.4 CONCLUSIONS

Salinity is a serious problem for crop cultivation and production because it affects plant growth and decreases crop yield, quality, and plant death. To overcome this problem, developing adapted salt-tolerant varieties is a solution. Recent development of a combination of nuclear techniques with biotechnology, known as in vitro mutagenesis and manipulation, and molecular markers has been effectively introduced for the induction of new and novel types of crop varieties. Using these technologies with mutation breeding will provide solutions to increasing salinity problems worldwide. The development of salt-tolerant cultivars would contribute directly to benefit farmers

in salt-affected lands by increasing their yield and profitability. An increase in food production would benefit consumers and would improve the economy. Therefore, salt-tolerant varieties in all crops are of tremendous benefit for increased crop productivity, environmental sustainability, and food security. Use of mutation breeding has proven and will continue to be beneficial in the development of salt-tolerant crops, throughout the world.

## REFERENCES

Ahloowalia, B.S. and M. Maluszynski. 2001. Induced mutations—A new paradigm in plant breeding. Euphytica 118:167–173.

Ahloowalia, B.S., M. Maluszynski, and K. Nichterlein. 2004. Global impact of mutation-derived varieties. Euphytica 135:187–204.

Akyuz, S., T. Akyuz, O. Celik, and C. Atak. 2013. FTIR and EDXRF investigations of salt tolerant soybean mutants. J Mol Struct 1044:67–71.

Alikamanoğlu, S. 2002. Efficiency of the gamma radiation in the induction of in vitro somatic mutations. J Cell Mol Biol 1:19–24.

Alikamanoğlu, S., O. Yaycili, and A. Sen. 2011. Effect of gamma radiation on growth factors, biochemical parameters, and accumulation of trace elements in soybean plants (Glycine max L. Merrill). Biol Trace Elem Res 141:283–293.

Ashraf, M.Y., G. Sarwar, M. Ashraf, R. Afaf, and A. Sattar. 2002. Salinity induced changes in α-amylase activity during germination and early cotton seedling growth. Biol Plantarum 45:589–591.

Badruddin, M., M.M. Rhaman, N.A. Nehar, M.M. Hossain, and M.B. Hasan. 2005. Physiological characterization of mustard (Brassica sp.) genotypes for their salt tolerance. Pak J Biol Sci 8:433–438.

Barret-Lennard, G. 2003. The interaction between waterlogging and salinity in higher plants: Causes, consequences and implications. Plant Soil 253:35–54.

Balyan, H.S., N. Sreenivasulu, O. Rieralizarazu, P. Azhaguvel, and S.F. Kianian. 2008. Mutagenesis and high-throughput functional genomics in cereal crops: Current status. Adv Agron 88:357–417.

Briggs, D.E. 1978. Barley genetics. In Barley, (ed.) Briggs, D.E., Chapman and Hall Ltd., Lodon, U.K., pp. 419–441.

Cheema, A.A., M.A. Awan, and Y. Ali. 1999. Niab-Irri-9: A new salt tolerant and high yielding rice variety. Pak J Biol Sci 2:869–870.

Chikelu, M.B.A., A.F.Z.A. Rownak, M.J. Shri, B.G. Glenn, and J.Z.A. Francisco. 2007. Induced mutations for enhancing salinity tolerance in rice.

In *Advances in Molecular Breeding toward Drought and Salt Tolerant Crops*, (eds.) Jenks, M.A. et al., Springer, New York, pp. 413–454.

Chinnusamy, V., A. Jagendorf, and J.K. Zhu. 2005. Understanding and improving salt tolerance in plants. *Crop Sci* 45:437–448.

Fatnassi, I.C., S.H. Jebara, and M. Jebara. 2011. Selection of symbiotic efficient and high salt-tolerant rhizobia strains by gamma irradiation. *Ann Microbiol* 61:291–297.

Forster, B.P., H. Pakniyat, M. Macaulay, W. Matheson, M.S. Phillips, W.T.B. Thomas, and W. Powell. 1994. Variation in the leaf sodium content of *Hordeum vulgare* (barley) cultivar Maythorpe and its derived mutant cv. Golden Promise. *Heredity* 73:249–253.

Gonzalez, M.C., N. Perez., and E. Cristo. 2009. GINES: A first rice mutant obtained by proton irradiation. *Cultivos Tropicales* 30:59.

Gulsen, O., A. Uzun, H. Pala, E. Canihos, and G. Kafa. 2007. Development of seedless and Mal secco tolerant mutant lemons through budwood irradiation. *Sci Hortic* 112:184–190.

Hewawasam, W.D.C.J., D.C. Bandara, and W.M. Aberathne. 2004. New phenotypes of *Crossandra infundibuliformis* var. Danica through in vitro culture and induced mutations. *Trop Agric Res* 16:253–270.

Kim, D.S., J.Y. Song, J.B. Kim, G.J. Lee, and S.Y. Kang. 2008. A protocol for selection of radiation-induced salt tolerant rice mutants by in vitro mutagenesis. In *Breeding for Salinity Tolerant Crops*, IAEA/RCA Project RAS504, Chinese Academy of Agricultural Sciences, Beijing, China, pp. 217–223.

Krieg, D.R. 1963. Ethyl methanesulfonate-induced reversion of bacteriophage T4rII mutants. *Genetics* 48:561–580.

Leblebici, Z., A. Aksoy, and F. Duman. 2011. Influence of salinity on the growth and heavy metal accumulation capacity of *Spirodela polyrhiza* (Lemnaceae). *Turk J Biol* 35:215–220.

Lenis, J. M., M. Ellersieck, D.G. Blevins, D.A. Sleper, H.T. Nguyen, D. Dunn, J.D. Lee, and J.G. Shannon. 2011. Differences in ion accumulation and salt tolerance among Glycine accessions. *J Agron Crop Sci* 197:302–310.

Luan, Y.S., J. Zhang, X.R. Gao, and L.J. An. 2007. Mutation induced by ethyl methanesulphonate (EMS), in vitro screening for salt tolerance and plant regeneration of sweet potato (*Ipomoea batatas* L.). *Plant Cell Tissue Organ Cuit* 88:77–81.

Lundqvists, U. 1986. Barley mutants-diversity and genetics. In *Research and Results in Plant Breeding*, (ed.) Olsson, G., Svalov, LTs förlag, Stockholm, pp. 85–88.

Mishra, B., R.K. Singh, and R.K. Bhattacharya. 1992. CST 10, a newly released dwarf rice for salt affected soils. *Int Rice Res Newslett* 17:19.

Mugiono, P.S. 1984. Resistance of Indonesian mutant lines to the brown planthopper (BPH) *Nilaparvata lugens*. *Int Rice Res Newslett* 9:5.

Munns, R. 2005. Genes and salt tolerance: Bringing them together. *New Phytol* 167:645–663.

Munns, R. and M. Tester. 2008. Mechanisms of salinity tolerance. *Annu Rev Plant Biol* 59:651–681.

Mustafa, G., A.M. Soomro, A.W. Baloch, and K.A. Siddiqui. 1997. "Shua 92"—A new cultivar of rice (*Oryza sativa* L.) developed through fast neutrons irradiation. *Mut Breed Newslett* 43:35–36.

Panda, D., D.C. Ghosh, and M. Kar. 2013. Effect of salt stress on the pigment content and yield of different rice (*Oryza sativa* L.) genotypes. *Int J Bio Res Stress Manag* 4(3):431–434.

Roose, M.L. and T.E. Williams. 2007. Mutation breeding. In *Citrus Genetics, Breeding and Biotechnology*, (ed.) Khan, I.A., CAB International, Wallingford, U.K., pp. 345–352.

Saleem, M.Y., Z. Mukhtar, A.A. Cheema, and B.M. Atta. 2005. Induced mutation and in vitro techniques as a method to induce salt tolerance in Basmati rice (*Oryza sativa* L.). *Int J Environ Sci Tech* 2:141–145.

Serrat, X., R. Esteban, N. Guibourt, L. Moysset, S. Nogues, and E. Lalanne. 2014. EMS mutagenesis in mature seed-derived rice calli as a new method for rapidly obtaining TILLING mutant populations. *Plant Methods* 10:5–17.

Singh, B. and P.S. Datta. 2010a. Effect of low dose gamma irradiation on plant and grain nutrition of wheat. *Radiat Phys Chem* 79:819–825.

Singh, R.K., E. Redona, and L. Refuerzo. 2010b. Varietal improvement for abiotic stress tolerance in crop plants: Special reference to salinity in rice. In *Abiotic Stress Adaptation in Plants: Physiological, Molecular and Genomic Foundation*, (ed.) Pareek, A. et al., Springer, New York, pp. 387–415.

Talebi, A.B, A.B. Talebi, and B. Shahrokhifar. 2012. Ethyl methane sulphonate (EMS) induced mutagenesis in Malaysian rice (cv. MR219) for lethal dose determination. *Am J Plant Sci* 3:1661–1665.

USSL. 2005. George E. Brown, Jr salinity laboratory riverside, CA, USA: USDA-ARS. http://www.ussl.ars.usda.gov. (Accessed March 20, 2014.)

Yaycili, O. and S. Alikamanoglu. 2012. Induction of salt-tolerant potato (*Solanum tuberosum* L.) mutants with gamma irradiation and characterization of genetic variations via RAPD-PCR analysis. *Turk J Biol* 36:405–412.

Zhang, F.T. and L. Xu. 1992. A new salt-resistant rice variety Liaoyan 2. *J Rice Sci* 1(6):1–1.

CHAPTER EIGHTEEN

# Present Status and Future Prospects of Transgenic Approaches for Salt Tolerance in Plants/Crop Plants

*Challa Surekha, L.V. Aruna, Mohammad Anwar Hossain,*
*Shabir Hussain Wani, and Nageswara Rao Reddy Neelapu*

## CONTENTS

*Abstract.* Developing crops that are better adapted to abiotic stresses are important for food production in many parts of the world today. Multidisciplinary approaches are used to develop crops that are more tolerant to abiotic stresses, especially salinity. Prioritizing research for salinity with prospects for crop production is the immediate requirement for a sustainable future. Plants tolerate salinity by using different mechanisms; in this connection, many molecules are identified. Examining and addressing the constraints would help us in overcoming salinity. Present approaches that are used in different plants would help us in better understanding how transgenic plants are tolerating salinity. Reviewing the parameters would provide us better information of the future prospects and approaches that are to be used. Overview of the present and future methodologies would gain more insights in understanding transgenic approaches for salinity tolerance.

*Keywords:* Salt tolerance, salt tolerance mechanism, salt stress, salt sensors, transgenics for salinity

## 18.1 INTRODUCTION

Disturbance originating from the surrounding environment or from living organisms that may or can cause damage or disease is called stress. Stress originated from the surrounding environment is called abiotic stress or nonliving stress, whereas stress caused by living organisms is called biotic stresses. Abiotic stress contributes to negative impact of nonliving factors that influence the environment beyond its normal range adversely affecting the individual physiology of the organism in a significant way. Different types of abiotic stresses that a plant can experience in its lifetime are salt, water, temperature, metal, photooxidative or light, wind, and nutrient stresses (Khraiwesh et al., 2012).

Biotic stress is contributed by living organisms, like bacteria, viruses, fungi, insects, and weeds causing damage to plants. Biotic stress in plants is contributed by 8000 fungal species, 14 bacterial genera, insects, viruses, and weeds causing economic damage to crops worldwide. Fungi, bacteria, and viruses are known to cause diseases in plants, whereas insects cause physical damage to the plants or act as a vector of viruses and bacteria to transmit from infected to healthy plants. Weeds compete for space and nutrients with the desirable plants and show biotic stress by inhibiting the growth of desirable plants (Mary, 2011, Plant stresses).

Water stress is one of the most important abiotic stresses affecting plants (Osakabe et al., 2014). There are two types of water stresses, drought and flooding stress, where either of these conditions can be deadly to the plant. Plants require a certain amount of water for its survival; too little water can desiccate the plant and cause drought stress (Harb et al., 2010), whereas too much of water can make plant cells to swell and burst causing flooding stress (Bray et al., 2000; Ahmad and Prasad, 2012a,b; Striker, 2012).

Temperature stresses can also show devastating effect on a plant and its yield (Kai and Koh, 2014). There are two types of temperature stresses, low temperature or chilling stress and high temperature or heat stress (Waraich et al., 2012). Plants require optimal temperature for best growth; if the temperature is too cold for the plant, it affects the amount and rate of uptake of water and nutrients, leading to cell desiccation and starvation, which is known as cold stress or chilling stress (Miura and Furumoto, 2013; Shi et al., 2014; Kreps et al. 2002; Seki et al., 2002). Under extremely cold conditions, cell liquids can freeze outright, causing plant death, which is known as freezing stress (Beck et al., 2004; Miura and Furumoto, 2013). Hot weather can also affect plants adversely, where intense heat cause plant cell proteins to break down, a process which is called denaturation, or melt cell walls and membranes, affecting permeability of the membranes. This kind of stress is known as heat stress (Qu et al., 2013).

In farming, addition of fertilizers either in excess or in deficit can cause abiotic stress to the plant affecting balance of nutrition, which is known as nutrient stress (Tuomi et al., 1984). Wind directly damages the plant through sheer force or affects the transpiration of water through leaf stomata and cause desiccation, which is known as wind stress (Onoda and Anten, 2011). Uptake of heavy metals can occur when plants are grown in soils with heavy metal content, leading to complications in physiology and biochemical activities such as photosynthesis. This type of stress is known as metal stress (Schützendübel and Polle, 2002; Ovečka and Takáč, 2014). Photosynthetic organisms constantly face the threat of photooxidation due to fluctuating light conditions, which is known as photooxidative or light stress (Foyer et al., 1994; Havaux et al., 2009). Uptake of a high amount of salts when plants are grown in soils accumulated with salts can cause stress, subsequently leading to effect on physiology of plant and cell desiccation, which is commonly known as salt stress (Carillo et al., 2011).

In this chapter, we review salt stress, salt tolerance mechanisms in plants, salt sensors, salt sensing, and future prospects of transgenic approaches and challenges for salinity tolerance.

## 18.2 SALT STRESS

Salt-affected land in the world is nearly 800 million hectares, among them irrigated land is nearly 45 million (Hasanuzzaman et al., 2014). Saline soils are common in West and Central Asia and Australia (Dregne, 1986). India also has a good amount of saline soils, alkali soils, and coastal saline soils. Natural salinity may be due to accumulation of salts over long periods of time in arid and semiarid zones; weathering of parental rocks to release soluble salts of chlorides, sodium, magnesium, sulfates, and carbonates; and rainwater (containing 6–50 mg/kg of sodium chloride). Apart from natural salinity, land clearing or irrigation of agricultural land causes water tables to rise and concentrate the salts in the root zones (Dregne, 1986).

Salinity is a major problem for agriculture in arid and semiarid regions of the world. Adverse environments cause considerable crop yield losses with very low crop productivity. Yield losses occur due to reduction in germination, plant growth (biomass), and seed size. However, two key strategies were proposed for utilization of salt-affected lands such as reclamation of salt-affected soils and growing halophytes (Epstein, 1997). The effects of salinity include electrolyte leakage, proline accumulation, and difference in $K^+/Na^+$ ratio (Hela et al., 2012). The salt in soil water reduces the plant's ability to take up water, and this leads to slower growth of the plant, as a result of the osmotic or water-deficit effect of salinity in response to salt outside the plant. It also enters into the transpiration stream and eventually injures cells in the transpiring leaves, further reducing growth due to the salt-specific or ion excess effect of salinity inside the plant. The salt taken up by the plant concentrates in old leaves and continues transport into transpiring leaves over a long period eventually resulting in high $Na^+$ and $Cl^-$ concentrations, and the leaves die. The cause of injury is probably salt load exceeding the ability of the cells to compartmentalize salts in the vacuole. Salts then build up rapidly in the cytoplasm and inhibit enzyme activity. Alternatively, they might build up in the cell walls and dehydrate the cell (Munns, 2005).

Salt tolerance during different stages of plant development is different. Seed germination and early seedling growth have been reported to be relatively more sensitive toward salinity (Ashraf and McNeilly, 2004). In majority of the reports, salt tolerance has been assessed at the initial growth stages such as germination and seedling growth. It has been earlier reported that in most plant species, the degree of salt tolerance varies with plant development (Foolad, 2004; Ashraf et al., 2008). When salt tolerance of a crop varies with growth stage, there is a need to appraise salt tolerance at each growth stage, so as to assess overall salt tolerance of the crop.

## 18.3 SALT TOLERANCE MECHANISMS IN PLANTS

Salt tolerance is a very complex trait that is governed by numerous mechanisms at various levels. When plants are in salt stress, they either escape, avoid, tolerate, or resist. In order to maintain growth, plants have to regulate their turgor and avoid salt toxicity through osmotic adjustment and ion compartmentation. $Na^+$ exclusion, tissue tolerance or compartmentalization of $Na^+$ and $Cl^-$, osmotic tolerance, $K^+$ accumulation in cytoplasm, and $Cl^-$ tolerance are the mechanisms in plants used to tolerate salt stress (Munns and Tester, 2008). Salt sensors are initially required to sense the salt stress that is present in the surrounding environment of the plant cell. Once the salt stress is sensed, information is relayed via signaling molecules that in turn activate transcription factors, early transcription factors, and delayed responsive genes in cascade mechanism for effective salt tolerance by the plant cell. The delayed responsive genes are the functional proteins that are expressed or activated when salinity is sensed by salt sensors. They are osmolytes, water channel proteins, late embryogenesis abundant (LEA) proteins, chaperones, proteinases that remove denatured proteins, proteins that detoxify, ion channel/carrier/transporter proteins, and hormones (Ratner and Jacoby, 1976; Niu et al., 1993; Rhodes and Hanson, 1993; Vierstra, 1996; Shinozaki and Yamaguchi-Shinozaki, 2000; Polle, 2001; Zhu, 2002; Shi et al., 2003; Wang et al., 2003; Ashraf, 2004; Feller, 2004; Ashraf and Foolad, 2007; Peng et al., 2007). The regulatory proteins include both constitutive transcription factors and early responsive genes that are expressed or activated when salinity is sensed by salt sensors. They are transcription factors, protein kinases, and proteins related to phospholipid metabolism.

### 18.3.1 Compatible Organic Solutes

Overproduction of different types of compatible organic solutes is the most common form of plant responses to salt stress (Ashraf and Foolad, 2007). The most common organic solutes that play an active role in the mechanism of plant salt tolerance include proline, trehalose, sucrose, polyols, and quaternary ammonium compounds such as glycine betaine, prolinebetaine, alaninebetaine, hydroxyproline-betaine, pipecolatebetaine, and choline O-sulfate, diamines, triamines, tetraamines, and polyamines (Rhodes and Hanson, 1993) (Table 18.1).

### 18.3.2 Water Channel Proteins

Biological activities of aquaporins (AQPs) are diverse and include stomatal movement, seed germination, reproductive growth, cell elonga-tion, phloem loading and unloading, and stress responses in plants (Eisenbarth and Weig, 2005; Gao et al., 2010). Water channels, AQPs, are mem-brane proteins of 26–30 kDa in size found in all organisms. Membrane permeability is enhanced to water by AQPs in both directions of membrane by adjusting differences in osmotic pressure across a membrane (Peng et al., 2007). The probable mechanism of AQPs in salt tolerance is to reduce membrane injury, improve ion distribution, and maintain osmotic balance (Xu et al., 2014). AQPs transport glycerol, boron, and $CO_2$ along with water through membranes (Uehlein et al., 2003; Kaldenhoff and Fischer, 2006; Sade et al., 2010).

### 18.3.3 Ion Channel/Transporters/Carrier Proteins

Salt damage in plants is prevented by developing different mechanisms, either by salt uptake, trans-port, compartmentalization, or exchange ions by proton pumps, antiporters, or transporters and reduction of $Na^+$ uptake or compartmentalization of $Na^+$ into vacuoles. $Na^+/H^+$ antiporters exchange $Na^+$ or $Li^+$ for $H^+$ (Ratner and Jacoby, 1976; Niu et al., 1993; Shi et al., 2003), and regulate $Na^+$ uptake at both plasma membrane and tonoplast (Ashraf, 2004) (Table 18.2).

### 18.3.4 Enhancing Antioxidant Production

Salt stress also triggers reactive oxygen species (ROS) such as superoxide anion ($O_2^-$), hydrogen peroxide ($H_2O_2$), hydroxyl radical ($\cdot OH$), and sin-glet oxygen ($^1O_2$). ROS in plants generated under optimal growth conditions is usually low (Polle, 2001), but environmental adversaries trigger enhanced production of ROS (Karpinski et al., 2003; Laloi et al., 2004). These ROS are cytotoxic, by causing substantial oxidative damage to mem-brane lipids, proteins, and nucleic acids (McKersie

and Leshem, 1994). However, plants have their innate systems for scavenging/detoxifying the ROS, which are commonly referred to as antioxi-dants (Asada, 1999). The most common antioxi-dants found in plants are ascorbate, glutathione, α-tocopherol, lutein, morin, β-carotene, quer-cetin, kaempferol, xanthophylls, and catechins, whereas antioxidant enzymes include superoxide dismutase (SOD), catalase (CAT), ascorbate perox-idase (POD), glutathione peroxidase, and enzymes of ascorbate–glutathione cycle (Ashraf, 2009). Most of the antioxidant enzymes can substan-tially lower the levels of superoxide and hydro-gen peroxide in plants (Ali and Alqurainy, 2006). The increased expression of antioxidant enzymes decreased ROS and subsequently salt stress.

### 18.3.5 Late Embryogenesis Abundant Proteins

LEA proteins are stress proteins expressed in veg-etative tissues of plants to impart dehydration tol-erance. These LEA-type proteins are encoded by COR (cold regulated), ERD (early responsive to dehydration), KIN (cold inducible), RAB (respon-sive to abscisic acid [ABA]), and RD (responsive to dehydration) genes in different plant species (Shinozaki and Yamaguchi-Shinozaki, 2000; Zhu, 2002). Accumulation of these proteins correlated with salt stress tolerance in various species of plants suggesting a protective role under salt stress.

### 18.3.6 Chaperones

Plants activate heat-shock proteins that accumu-late upon salinity stress to play an important role of protecting cell during the salt stress. During salt stress, proteins and enzymes are usually dys-functional due to destabilization of proteins and enzymes. Molecular chaperones are known for functional stabilization of membranes and pro-teins and assist in protein refolding under salt stress conditions (Wang et al., 2003).

### 18.3.7 Proteinases That Remove Denatured Proteins

Salt stress increases inactive proteins in plants that are aggregated, denatured, or oxidatively dam-aged. Maintenance of functional conformation of a protein in a cell is necessary. Survival of a cell under stress is only possible when it can prevent aggregation of a protein, refolding of denatured

TABLE 18.1

*Osmoprotectant genes for developing transgenics.*

| Osmoprotectant | Function | Genes for developing transgenics | Source | Transgenic plants | References |
|---|---|---|---|---|---|
| Glycine betaine | Accumulation of glycine betaine protects plant from salt tolerance. | Choline oxidase Choline dehydrogenase, betaine aldehyde dehydrogenases | *Arthrobacter globiformis* *Escherichia coli* *Atriplex hortensis* | *Arabidopsis thaliana* L., *Oryza sativa* L., *Brassica juncea*, *Brassica napus* *Brassica juncea* *Brassica oleracea* *Brassica campestris* | Mohanty et al. (2002), Huang et al. (2000), Prasad et al. (2000), Bhattacharya et al. (2007), Park et al. (2005), Shirasawa et al. (2006), Ahmad et al. (2008), Duan et al. (2009), Cheng et al. (2013), Jiang et al. (2013), Nguyen et al. (2013), Lai et al. (2014) |
| Proline | Accumulation of proline protects plant from salt tolerance. | Pyrroline-5-carboxylate synthetase, pyrroline-5-carboxylate reductase | *Vigna aconitifolia* L. | *Nicotiana tabacum*, *Oryza sativa*, soya bean, citrus | Ashraf and Foolad (2007), Maggio et al. (2002), de Ronde et al. (2004), Molinari et al. (2004), Zhang et al. (2014), Surekha et al. (2014) Razavizadeh and Ehsanpour (2009), Kumar et al. (2010), Kishor et al. (1995), Sawahel and Hassan (2002) |

(Continued)

TABLE 18.1 (*Continued*)
*Osmoprotectant genes for developing transgenics.*

| Osmoprotectant | Function | Genes for developing transgenics | Source | Transgenic plants | References |
|---|---|---|---|---|---|
| Polyamines, triamines, tetramines, diamines, putrescines | Accumulation of amines protects plant from salt tolerance. | Arginine decarboxylase, ornithine decarboxylase, S-adenosylmethionine decarboxylase, spermidine synthase | | European pear L. | Liu et al. (2007), He et al. (2008) Kasukabe et al. (2004), Urano et al. (2004) |
| Mannitol | Accumulation of mannitol protects plant from salt tolerance. | Mannose-6-phosphate reductase | *Escherichia coli* | *Nicotiana tabacum* L., *Arabidopsis* L., *Triticum aestivum* L., wheat | Sickler et al. (2007), Abebe et al. (2003) Bhauso et al. (2014) |
| Trehalose | Accumulation of trehalose protects plant from salt tolerance. | Trehalose-6-phosphatase synthetase, trehalose-6-phosphate phosphatase | *Escherichia coli* | *Oryza sativa* L., *Nicotiana tabacum* L. | Serrano et al. (1999), Garg et al. (2002), Wu and Garg (2003), Penna (2003), Jun et al. (2005), Suárez et al. (2009), Guo et al. (2014), Su and Wu (2004) |

## TABLE 18.2
*Ion transporter genes for developing transgenics.*

| Gene name | Function | Source | Transgenic plants developed | References |
|---|---|---|---|---|
| Vacuolar Na⁺/H⁺ antiporter | Efflux and influx of Na⁺ ions | *Medicago sativa* *Atriplex gmelini* *Pennisetum glaucum* *Zea mays* L. *Entamoeba histolytica* *Oryza sativa* L. *Gossypium hirsutum* L. *Arabidopsis thaliana* L. *Brassica napus* L. *Hordeum brevisubulatum* | *Arabidopsis thaliana* *Oryza sativa* L. *Triticum aestivum* L., *Lolium perenne* L., *Nicotiana tabacum* L., *Lycopersicon esculentum* L., *Brassica napus* L. *Zea mays* | Ashraf and Akram (2009), Apse et al. (1999), Shi et al. (2002), Yamaguchi et al. (2003), Vasekina et al. (2004), Yu et al. (2007), Lu et al. (2014) |
| Vacuolar H⁺— pyrophosphatase | Efflux of salt ions | *Hordeum vulgare* *Beta vulgaris* L. *Nicotiana tabacum* L. *Oryza sativa* L. *Vigna radiata* L. *Cucurbita maxima* L. | *Vigna radiata* L. *Pyrus pyrifolia* L. | Nakanishi (1998), Suzuki et al. (1999), Nakanishi and Maeshima (1998) Nakanishi (1998), Suzuki et al. (1999), Gaxiola et al. (2012), Pasapula et al. (2011), Qin et al. (2013), Anjaneyulu et al. (2014), Li et al. (2014a,b), Schilling et al. (2014) |
| K⁺ channel | Efflux and influx of K⁺ ions | *Arabidopsis thaliana* L. *Sorghum bicolor* *Puccinellia tenuiflora* | *Nicotiana tabacum* L. Barley and wheat | Lebaudy et al. (2007), Bei and Laun (1998), Wang et al. (1998), Gierth et al. (2005), Takahashi et al. (2007), Obata et al. (2007), Ardie et al. (2009), Wang et al. (2014) |
| K⁺/H⁺ antiporter | Efflux and influx of K⁺ ions | *Solanum lycopersicon* L. | *Arabidopsis thaliana* L. | Venema et al. (2003), Rodriguez-Rosales et al. (2008), Zheng et al. (2013) |

proteins, and removal of nonfunctional and harmful polypeptides. Dehydrins and chaperones prevent aggregation of a protein and refolding of denatured proteins. Proteases degrade the irreversibly damaged proteins in a cell along with other proteins that are aggregated, denatured, or oxidatively damaged. The other functions of proteases are (1) selective degradation of key enzymes, (2) degradation of short-lived proteins that are involved in cell signaling, (3) recycling of aminoacids related to carbon starvation, (4) hastening of senescence under source sink regulation, and (5) protection against biotic stress (Vierstra, 1996; Feller, 2004). Thereby, proteases are very important in performing vital cell survival. Proteases are controlled or regulated by protease inhibitors.

### 18.3.8 Hormones

Phytohormones are the messengers that are synthesized at one place and are transported to the site of actions where they play an important role by regulating plant responses to stresses such as salinity. Phytohormones such as ABA, auxin, brassinosteroids (BRs), cytokinins (CKs), gibberellic acid (GA), jasmonates (JAs), salicylic acid (SA), and triazoles (TRs) are associated with salinity of plants. Phytohormones such as auxin, CK, GAs, and SA decreased in the plant tissues as salinity increased, whereas ABA and JAs increase in the plant tissues as salinity increased. Therefore, salt-induced growth reduction in plants can be mitigated by the application of plant growth regulators.

Plants exposed to salt stress induce increase in ABA concentration, which acts as a major internal signal in root-to-shoot communication in plant. ABA delays the deleterious effect of salts in plants and increases tolerance to ions by upregulating the genes involved in salt and osmotic alleviation. Thus, ABA plays an important role in salt stress.

Plants exposed to salt stress also induce increase in auxin concentration with similar action to ABA and reduce plant growth. Thus, application of plant growth regulators such as $GA_3$ decreased auxin levels in plant tissues. Auxins stimulate the transcription of auxin-responsive genes such as auxin/indoleacetic acid (*Aux/IAA*), GH3, and small auxin-up RNA (*SAUR*) gene families (Guilfoyle et al., 1993) under salt stress, but still the mechanism of action of these elements is unknown. Thus, auxins play an important role in salt stress.

CKs, the antagonists of ABA, and antagonists/synergists of auxins enhance resistance to salt (Barciszewski et al., 2000). The observed reduction in CK levels under salt stress explains the fact that CKs are the limiting factors under salt stress. Exogenous application of kinetins further support that salt stress in tomato, barley, cotton (Bozcuk, 1981), chickpea seedlings (Boucaud and Ungar, 1976), wheat seedlings (Naqvi et al., 1982), and potato plants (Abdullah and Ahmad, 1990) can be overcome. The probable role of kinetin is that it acts as a direct free radical scavenger or it may be involved in antioxidative mechanism related to the protection of purine breakdown (Chakrabarti and Mukherji, 2003). Another interesting fact is that CK receptors act as the negative regulators in ABA signaling (Tran et al., 2007). Therefore, CKs play an important role in salt stress.

GAs accumulate rapidly when exposed to salt stress (Lehmann et al., 1995), in order to alleviate deleterious effects of salinity. Application of GA in wheat, rice (Parasher and Varma, 1988; Prakash and Prathapasenan, 1990), and tomato (Maggio et al., 2010) improved growth under saline conditions. The probable role of GA can be the inhibition of free radical–induced lipid peroxidations (Choudhuri, 1988). GAs also interact with other hormones such as auxin and ABA to regulate various metabolic processes in the plants. Auxin (IAA) promotes GA biosynthesis (Wolbang et al., 2004), whereas GA application enhances the catabolism of ABA (Gonai et al., 2004). Thus, GAs play an important role in salt stress.

BRs produce a variety of chemical compounds that regulate stress responses such as salt stress. Exogenous applications of BRs were known to increase growth and yield of many economically useful plants such as cereals (Ali et al., 2008), leguminous crops (Rao et al., 2002), rapeseed plants (Hayat et al., 2000), cotton (Ramraj et al., 1997), and rice (Anuradha and Rao, 2001) under saline conditions. Though it was well established that BRs are overcoming saline conditions, the mechanism of salt tolerance is still unknown.

JAs that are involved in plant developmental processes activate plant defense mechanism in response to salinity (Cheong and Choi, 2003) with its role in plant signal transduction (Wang, 1999). JA levels were more in salt-tolerant varieties when compared to salt-sensitive varieties (Moons et al., 1997). Therefore, increased JA levels accumulated in plants may effectively protect the plants from salt stress. Exogenous application of BRs has effectively protected plants such as rice (Kang et al., 2005) and barley (Tsonev et al., 1998) from salt tolerance. Exogenous application of JA dramatically decreased sodium concentration in barley (Tsonev et al., 1998). At the same time, the application may change the hormonal balance of ABA to protect against salt stress (Kang et al., 2005).

SA is an endogenous growth regulator that also protects plants from salinity (Kaya et al., 2002). Application of SA in bean (Azooz, 2009), tomato (Tari et al., 2005), and maize (Gunes et al., 2007) alleviates salt stress in plants. The probable roles of SA to mitigate salt tolerance are accumulation of ABA and proline as in wheat (Shakirova et al., 2003); increasing the level of cell division within the apical meristem (Sakhabutdinova et al., 2003); enhancing photosynthetic rate and also maintaining the stability of membranes (El-Tayeb, 2005); decreasing $Na^+$ and $Cl^-$ accumulation (Gunes et al., 2007); increasing growth and photosynthetic rate of the plants, transpiration rate, and stomatal conductance and reducing electrolyte leakage by 32% (Stevens et al., 2006); and accumulation of both ABA and IAA (Sakhabutdinova et al., 2003). Therefore, SA is an effective plant growth regulator that mitigates the damaging effect of salt stress in plants.

TRs are plant growth regulators that protect plants against salinity (Fletcher and Hofstra, 1988). The pretreatment of seeds with TR counteracted the effect of salinity, for example, in sunflower and mung bean (Saha and Gupta, 1993). The probable role of TRs in counteracting the effect of salinity can be by enhancing the activity of free radical–scavenging systems (Kraus and Fletcher, 1994). Thus, TRs play an important role in salt stress.

### 18.3.9 Transcription Factors

Transcriptional factors regulate the salt stress–responsive gene expression in plants. Transcription factors, namely ABA-responsive element (ABRE), MYC (myelocytomatosis oncogene)/MYB (myeloblastosis oncogene), CBF (C-repeat binding factor)/DREBs (dehydration-responsive element binding protein, and NAC (NAM, ATAF and CUC), in plants are known to regulate both ABA-dependent and ABA-independent pathways and provide salt tolerance. Thus, transcription factors can be used to produce genetically engineered crops with higher tolerance to salinity (Lata et al., 2011).

### 18.3.10 Protein Kinases

Protein kinases play an important role in plants by providing salt tolerance during salt stress. Multiple-faceted roles of protein kinases such as plant hormonal signaling, inhibition of cell division, and intra- and extracellular signaling in plants have been identified in plants. The role of early salt stress inducible protein kinase *Esi47* involved in plant hormone signaling was identified in salt-tolerant wheat grass *Lophopyrum elongatum* (Shen et al., 2001). Salt stress generally inhibits cell division; ABA-induced *ICK1* in *Arabidopsis* established the link between salt stress and cell division (Zhu, 2001). Cyclin-dependent protein kinases (CDKs) have an important role in regulating cell division in plants. Wang et al. (2000) showed reduced CDK activity with increased ICK1 expression subsequently with reduced number of cells in plants when compared with their controls demonstrating that ICK1 is a CDK inhibitor and ICK1 expression inhibits cell division. Mitogen-activated protein kinase kinase (MKK) is involved in extra- and intracellular signaling in plants. Overexpression of *Oryza sativa* MKK showed enhanced salt tolerance to salt stress in rice. Cytoplasmic protein kinases in cucumber are positive regulators of salt stress responses. Overexpression of cucumber cytoplasmic protein kinases in cucumber upregulated multiple genes involved in osmotic-stress responses (Oh et al., 2014).

### 18.3.11 Protein-Related Phospholipid Metabolism

Salt stress in plants activates phospholipids to mediate early signaling events that help the plant to cope with salt stress. Early signaling events in response to plant stress are due to an increase in minor phospholipids, such as phosphoinositides and phosphatidic acid. The interacting partners are kinases, protein phosphatases, and proteins involved in cytoskeleton and membrane trafficking. Lipids affect intracellular trafficking, signaling, and cytoskeletal organization that help in enhancing salt tolerance in plants (Wang et al., 2007).

## 18.4 SALT SENSORS, SALT SENSING, SALT STRESS, AND SALT TOLERANCE

The understanding of how salt is sensed by plant cellular systems is still very limited. In theory, salt can be sensed either before or after entering the cell, or both (Zhu, 2003). A review of literature on sensing of salt by plants has helped us in identifying three different salt sensors to sense salt in the surroundings of the plant cells, and many more are to be identified even if further discoveries are still expected. An osmosensory two-component hybrid histidine kinase, nucleocytoplasmic receptors pyrabactin resistance (PYR)/PYR like (PYL)/ regulatory components of ABA receptors RCARs, and unknown receptor-like cytoplasmic kinases (RLCKs) are the few membrane proteins that sense salt and transduce the signal to induce salt tolerance.

A putative osmosensory two-component hybrid histidine kinase, ATHK1, from *Arabidopsis* is implicated in osmosensing under salt stress. ATHK1 induces expression and has the ability to complement the yeast double mutant lacking both osmosensors (*sln1_ sho1_*). Active ATHK1 may inactivate a response regulator by phosphorylation by stimulating the osmolyte biosynthesis in plants by activating a mitogen-activated protein kinases (MAPK) pathway(s) (Urao et al., 1999; Kumar and Sinha 2013) (Figure 18.1A).

Generally, $Na^+$ entry into the cytosol is through the $Na^+$ transporter HKT1. Unknown RLCKs are the membrane proteins that sense salt and transduce the signal via $Ca^{2+}$. Salt stress–elicited $Ca^{2+}$ signals are perceived by SOS3, which activates the protein kinase SOS2. Activated SOS2 phosphorylates SOS1, a plasma membrane $Na^+/H^+$ antiporter, which then transports $Na^+$ out of the cytosol. The

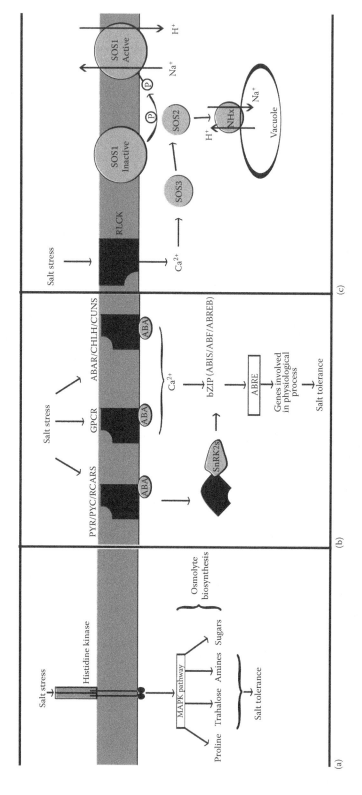

Figure 18.1. (a) Histidine kinase. (b) Receptor-like cytoplasmic kinases. (c) ABA receptors act as the sensors of salt and transduce the signal of salt stress to the plant cells.

transcript level of SOS1 is regulated by the SOS3–SOS2 kinase complex. SOS2 also activates the tonoplast Na+/H+ antiporter that sequesters Na+ into the vacuole. Na+ entry into the cytosol through the Na+ transporter HKT1 may also be restricted by SOS2. ABA-INSENSITIVE1 (ABI1) regulates the gene expression of NHX1, while ABI2 interacts with SOS2 and negatively regulates ion homeostasis either by inhibiting SOS2 kinase activity or the activities of SOS2 targets (Figure 18.1B).

ABA is accumulated in roots and transported from roots to leaves in plants. Nucleocytoplasmic receptors, PYR/PYL/RCARs, and plasma membrane–localized G protein–coupled receptors, ABA receptor (ABAR)/Mg-protoporphyrin IX chelatase subunit H (CHLH)/GENOME UNCOUPLE (GUN) 5, bind ABA. Two transporter genes, ABA exporter gene (*AtBCG25*) and ABA importer gene (*AtBCG40*), belonging to the ATP binding cassette transporter gene family transport this signal in the form of ABA accumulated in roots to leaves, thus modulating the cellular ABA signaling pathway and regulating salt stress by osmotic adjustment and/or ion compartmentation. Nucleocytoplasmic receptors PYR/PYL/RCARs bind ABA and inhibit type 2C protein phosphatases (PP2Cs) such as ABI1 and ABI2. Inactivation of PP2Cs activates accumulation of SNF1-related protein kinases (SnRKs). SnRKs regulate ABA-responsive transcription factors including ABREB/ABF and activate ABA-responsive genes and ABA-responsive physiological processes (Figure 18.1C).

## 18.5 PRESENT STATUS OF TRANSGENIC APPROACHES FOR SALINITY TOLERANCE

There are three basic approaches being used to obtain stress-tolerant varieties—screening of preexisting genotypes, conventional breeding for developing salt-tolerant lines, and generation of transgenic plants to introduce novel genes or to alter expression levels of the existing genes to affect the degree of salt stress tolerance.

### 18.5.1 Compatible Organic Solutes

The most common organic solutes are proline, trehalose, sucrose, glycine betaine, prolinebetaine, alaninebetaine, hydroxyprolinebetaine, pipecolatebetaine, diamines, triamines, tetraamines, and polyamines (Rhodes and Hanson, 1993). Salt tolerance by proline is contributed by two important

genes in proline biosynthesis, pyrroline-5-carboxylate synthetase (P5CS) and pyrroline-5-carboxylate reductase (P5CR) (Maggio et al., 2002; de Ronde et al., 2004; Molinari et al., 2004; Asharaf and Foolad, 2007). P5CS and P5CR genes were known to show salt tolerance in *Vigna aconitifolia*. P5CS129A was overexpressed in rice (Kumar et al., 2010), wheat (Vendruscolo et al., 2007), potato (Hmida-Sayari et al., 2005), tobacco (Yamachi et al., 2007; Razavizadeh and Ehsanpour, 2009; Zhang et al., 2014), *V. aconitifolia* (Hong et al., 2005), and pigeon pea (Surekha et al., 2014) to achieve salt tolerance.

Trehalose, a disaccharide, acts as a compatible solute to protect membranes and proteins by conferring salt tolerance. Trehalose is produced by trehalose-6-phosphate synthetase (TPS) and trehalose-6-phosphate phosphatase (TPP). *Escherichia coli* with TPS and TPP was known to show salt tolerance. Transgenic tobacco (*Nicotiana tabacum* L. var. SR1) plants overexpressing the *E. coli* TPS gene (*otsA*) conferred salt stress. TPS genes were overexpressed in rice, and transgenic tobacco demonstrated an increase in productivity in the presence of salt stress (Garg et al., 2002; Penna, 2003; Wu and Garg, 2003; Guo et al., 2014). TPS was overexpressed in *N. tabacum*, to achieve salt tolerance (Serrano et al., 1999; Jun et al., 2005). Transgenic alfalfa accumulated trehalose and enhanced tolerance to multiple abiotic stresses including salt tolerance (Suárez et al., 2009).

Glycine betaine is an osmoprotectant and, upon its accumulation, protects membranes and proteins by conferring salt tolerance. Choline oxidase (CodA), choline dehydrogenase (BetA), and betaine aldehyde dehydrogenases (Bet B) are the genes participating in glycine betaine. Genes in CodA, BetA, and Bet B showed salt tolerance in *E. coli*, *Arthrobacter globiformis*, and *Atriplex hortensis*. Overexpression of CodA, BetA, and Bet B genes in *N. tabacum*, *Arabidopsis thaliana* L., *O. sativa* L., *Brassica juncea* L., *Brassica napus*, *Brassica oleracea*, and *Brassica campestris* showed salt tolerance (Huang et al., 2000; Prasad et al., 2000; Mohanty et al., 2002; Park et al., 2005; Shirasawa et al., 2006; Bhattacharya et al., 2007; Ahmad et al., 2008; Duan et al., 2009; Cheng et al., 2013; Jiang et al., 2013; Nguyen et al., 2013; Lai et al., 2014).

Polyamines (Spd), triamines (Spm), tetramines, diamines, and putrescines (Put) are synthesized by arginine decarboxylase (ADC), ornithine decarboxylase, S-adenosylmethionine decarboxylase, spermidine synthase (SPDS) genes synthesize polyamines (Spd)/triamines (Spm)/tetramines/diamines/putrescines (Put) overaccumulating in plant cells to tolerate salt. Spd gene overexpressed in European pear showed tolerance to NaCl and mannitol stress (Liu et al., 2007; He et al., 2008). Spermidine synthase was also overexpressed in transgenic *A. thaliana* to enhance tolerance to multiple environmental stress such as salt stress and upregulate the expression of various stress-regulated genes (Kasukabe et al., 2004). *Arabidopsis* stress-inducible gene for ADC AtADC2 is required for the accumulation of putrescine in salt tolerance (Urano et al., 2004).

Overaccumulation of mannitol in plant cells showed salt tolerance. Overexpression of mannose-6-phosphate reductase (M6PR) in *E. coli* was known to attribute to salt tolerance. N. *tabacum* L., *Arabidopsis* L., *Triticum aestivum* L., and wheat transgenics with M6PR gene were tolerant to salt stress (Abebe et al., 2003; Sickler et al., 2007). Overexpression of bacterial *mtlD* gene conferred enhanced tolerance to salt stress and water-deficit stress in transgenic peanut (*Arachis hypogaea*) through the accumulation of mannitol (Bhauso et al., 2014).

## 18.5.2 Water Channel Proteins

AQP overexpression increased root elongation; improved hydraulic conductivity, water use efficiency, and ion distribution; and reduced membrane injury conferring salt tolerance leading to increased yield (Aharon et al., 2003; Hanba et al., 2004; Sade et al., 2010; Hu et al., 2012; Xu et al., 2014). Aharon et al. (2003) showed that overexpression of the *Arabidopsis* plasma membrane AQP (PIP) in tobacco improved plant vigor. Sade et al. (2010) showed that overexpression of the tobacco NtAQP1 in tomato (*Solanum lycopersicum*) plants improved water use efficiency, hydraulic conductivity, and yield production under salt stress. Hanba et al. (2004) showed that overexpression of the barley AQP HvPIP2;1 in rice increased $CO_2$ assimilation and internal $CO_2$ conductance in the leaves of transgenic plants. Overexpression of a wheat AQP gene, *TaAQP8*, enhances salt stress tolerance in transgenic tobacco. Overexpression of a wheat AQP gene, *TaAQP8*, in tobacco increased root elongation under salt stress, as shown by Hu et al. (2012). These attempts used AQPs to improve crop tolerance to salt stress.

## 18.5.3 Enhancing Antioxidant Production

Salt stress triggers ROS such as superoxide anion ($O_2^-$), hydrogen peroxide ($H_2O_2$), hydroxyl radical (OH), and singlet oxygen ($^1O_2$). Antioxidant enzymes such as SOD, CAT, POD, and enzymes of ascorbate–glutathione cycle (Ashraf, 2009) reduced the levels of ROS and subsequently salt stress. Transgenic plants overexpressing ROS-scavenging enzymes, such as SOD, POD, and glutathione S-transferase/glutathione peroxidase, showed increased tolerance to salt (Wang et al., 1999; Hossain et al., 2011, 2013; Hossain and Fuji, 2013). Overexpression of the tobacco NtGST/GPX gene, glutathione S-transferase, and glutathione peroxidase in transgenic tobacco plants has improved salt stresses in plants. Prashanth et al. (2008) showed that transgenic rice overexpressing SOD gene showed enhanced oxidative stress tolerance including salt stress at 150 mM NaCl. Wang et al. (2010) showed that transgenic poplar plants overexpressing TaMnSOD showed higher oxidative stress tolerance induced by salt stress. Recently, Wang et al. (2011) showed that transgenic sweet potato (*Ipomoea batatas* L. cv. Yulmi) overexpressing CuZnSOD and APX genes induce salt tolerance at different salinity levels. Higher root length, lower oxidative damage, and lower chlorophyll damage were observed in transgenics.

Genes from barley (*HvAPX1*), *Populus* peroxisomal ascorbate peroxidase (*PpAPX*), and thylakoid-bound APX (*StAPX*) in *A. thaliana* and tobacco showed tolerance to salt stress when compared with wild plants (Badawi et al., 2004; Wei-Feng et al., 2008; Xu et al., 2008; Li et al., 2009; Sun et al., 2010; Sing et al., 2013). Evidences accumulated showed that APX is having a key role in antioxidant and pathogen defense, with specialized and differentially regulated isoforms able to respond to salinity, drought, heavy metal, chilling, high light intensity, UV radiation, and pathogen-induced oxidative stress.

Monodehydroascorbate reductase (MDHAR) genes from *A. thaliana* and *Avicennia marina* were overexpressed in tobacco with significant salt tolerance (Eltayeb et al., 2007; Kavitha et al., 2010; Sultana et al., 2012; Hossain and Asada 1985). Eltayeb et al. (2007) observed MDHAR transgenics maintain photosynthetic function by limiting ROS-mediated damage. Kavitha et al. (2010) reported that transgenic plants showed an enhanced redox state of ascorbate and reduced levels of lipid peroxidation. Sultana et al. (2012) reported that transgenic rice lines showed higher seedling germination and better yield attributes such as a higher tiller number and increased 1000 grain weight.

Overexpression of dehydroascorbate reductase (DHAR) gene in plants tobacco, *Arabidopsis*, and potato (*Solanum tuberosum* L.) enhances tolerance to various oxidative stresses (Kwon et al., 2003; Eltayeb et al., 2006, 2011; Ushimaru et al., 2006; Li et al., 2012). Transgenic tobacco plants overexpressing the DHAR gene exhibited enhanced tolerance to salt in addition to $O_3$, drought, salt, and PEG-induced oxidative stresses (Eltayeb et al., 2006). Transgenic *Arabidopsis* increased ascorbic acid (AsA) content and showed tolerance to salt in addition to MV, $H_2O_2$, and low temperature–induced oxidative stresses (Kwon et al., 2003; Ushimaru et al., 2006). Eltayb et al. (2011) showed that transgenic potato (*S. tuberosum* L.) with DHAR gene (*AtDHAR1*) enhanced salinity-induced oxidative stress. Li et al. (2012) showed that transgenic tomato plants showed higher salinity and MV-induced oxidative stress tolerance.

*Suaeda salsa* GST, *Glycine soja* GST, SbGST, and PpGST overexpressed in *Arabidopsis* and tobacco enhanced salt tolerance (Ji et al., 2010; Qi et al., 2010; Jha et al., 2011; Liu et al., 2013a). Qi et al. (2010) showed that transgenic *Arabidopsis* showed higher seed germination and salt tolerance. Ji et al. (2010) observed longer root length and less growth retardation and correlated with salt and mannitol tolerance. Jha et al. (2011) showed that different stresses upregulated the expression of SbGST and tau class SbGST gene that play a vital role in abiotic stress tolerance. Liu et al. (2013a) showed that the transgenic tobacco plant overexpressing GST gene showed relatively normal growth under drought, NaCl, and cadmium (Cd) stresses.

GPX overexpressed in transgenic tobacco plants showed tolerance to higher oxidative stress caused by treatments with $H_2O_2$, MV, and environmental stress conditions, such as chilling, salinity, and drought indicating the potential physiological role of GPX in higher plants (Yoshimura et al., 2004; Gaber et al., 2006). Overexpression of CAT gene from *E. coli* (*katE*) in rice showed tolerance to salt (Moriwaki et al., 2008; Prodhan et al., 2008). Choe et al. (2013) reported that transgenic rice (*O. sativa* L. japonica cv. Ilmi) overexpressing GSH biosynthetic gene (γ-glutamylcysteine synthetase) showed higher salinity tolerance and lowered the level of ROS accumulation as compared to wild-type plants. Their findings showed that overexpression of GSH biosynthetic gene increases tolerance and germination rate in the presence of abiotic stress including salinity by improving redox homeostasis via an enhanced GSH pool.

AsA accumulation in transgenics was directly correlated with their ability to withstand abiotic stresses (Hemavathi et al., 2009, 2010a,b; Upadhyaya et al., 2011; Zhang et al., 2011; Lim et al., 2011; Lisko et al., 2013; Liu et al., 2013b). Hemavathi et al. (2009) demonstrated that transgenic potato plants (*S. tuberosum* L. cv. Taedong Valley) overexpressing strawberry D-galacturonic acid reductase (*GalUR*), a key enzyme in AsA biosynthesis, showed tolerance to abiotic stresses induced by MV, NaCl, or mannitol. Hemavathi et al. (2010a) found that transgenic potato (*S. tuberosum* L. cv. Taedong Valley) plants overexpressing L-gulono-gamma-lactone oxidase (GLOase), an essential gene for ascorbate biosynthesis, showed higher salinity tolerance. Hemavathi et al. (2010b) showed that transgenic potato plants overexpressing *GalUR* showed enhanced tolerance to various abiotic stresses. Upadhyaya et al. (2011) reported that transgenic potato (*S. tuberosum* L. cv. Taedong Valley) plants overexpressing AsA biosynthetic pathway genes (*GalUR*) showed higher salinity tolerance. Zhang et al. (2011) showed that transgenic plants overexpressing GDP-mannose 3′,5′-epimerase, an important enzyme in AsA biosynthesis pathway, showed higher abiotic stress tolerance. They proposed that improved stress tolerance was closely related to higher endogenous AsA content that increased the ability to scavenge ROS. Lim et al. (2012) showed that transgenic tomato plant (*Lycopersicon esculentum* Mill.) overexpressing the rat GLOase showed higher abiotic stress tolerance including salinity. Their finding suggests that the higher accumulation of AsA could upregulate the antioxidant system, which imparts improved tolerance against various abiotic stresses in transgenic tomato plants in comparison with wild-type plants. Liu et al. (2013b) showed that transgenic tobacco plants overexpressing L-galactono-1,4-lactone dehydrogenase (GalLDH), the last key enzyme in the AsA biosynthetic pathway, showed higher salinity tolerance. Lisko et al. (2013) reported that *Arabidopsis* lines overexpressing GalLDH gene showed tolerance to a range of various abiotic stresses like salt, cold, and heat. These attempts used antioxidants to improve crop tolerance to salt stress.

### 18.5.4 Ion Transporters

Transgenics of ion transporter genes was a success for salt tolerance. Vacuolar $Na^+/H^+$ antiporter, vacuolar $H^+$-pyrophosphatase, $K^+/H^+$ antiporter, and $K^+$ transport genes are successfully used in developing transgenics to salt tolerance. Vacuolar $Na^+/H^+$ antiporter *Medicago sativa* showed tolerance to salt in *Atriplex gmelini*, *Pennisetum glaucum*, *Zea mays* L., *Entamoeba histolytica*, *O. sativa* L., *Gossypium hirsutum* L., *A. thaliana* L., *B. napus* L., and *Hordeum brevisubulatum*. Overexpression of vacuolar $Na^+/H^+$ antiporter gene in transgenics *A. thaliana*, *O. sativa* L., *T. aestivum* L., *Lolium perenne* L., *N. tabacum* L., *L. esculentum* L., *B. napus* L., and *Z. mays* exhibited significant salt tolerance (Apse et al., 1999; Shi et al., 2002; Yamaguchi et al., 2003; Vasekina et al., 2004; Yu et al., 2007; Ashraf and Akram, 2009; Lu et al., 2014).

Vacuolar $H^+$-pyrophosphatase gene showed tolerance to salt in *Hordeum vulgare*, *Beta vulgaris* L., *N. tabacum* L., *O. sativa* L., *Vigna radiata* L., *Cucurbita maxima* L., and *Arabidopsis*. Overexpression of vacuolar $H^+$-pyrophosphatase gene in transgenics in *V. radiata* L., *Pyrus pyrifolia* L. finger millet, *T. aestivum*, and cotton exhibited significant salt tolerance and substantial crop yield (Nakanishi, 1998; Suzuki et al., 1999; Gaxiola et al., 2012; Pasapula et al., 2011; Qin et al., 2013; Anjaneyulu et al., 2014; Li et al., 2014a,b; Schilling et al., 2014).

$K^+$ transporter gene showed tolerance to salt in *A. thaliana* L. Overexpression of $K^+$ transporter gene in transgenics in *N. tabacum* L. exhibited significant salt tolerance (Bei and Laun, 1998; Wang et al., 1998; Gierth et al., 2005; Lebaudy et al., 2007; Takahashi et al., 2007; Ardie et al., 2009; Wang et al., 2014). $K^+/H^+$ antiporter gene showed tolerance to salt in *Solanum lycopersicon* L. Overexpression of $K^+/H^+$ antiporter gene in transgenics in *A. thaliana* L. exhibited significant salt tolerance (Venema et al., 2003; Rodriguez-Rosales et al., 2008; Zheng et al., 2013).

## 18.6 FUTURE PROSPECTS OF TRANSGENIC APPROACHES AND CHALLENGES FOR SALINITY TOLERANCE

Signaling molecules, transcriptional elements, hormone receptors, and LEA molecules are few prospects that also exhibit salt tolerance. More studies are required on these prospects in developing transgenics for salt tolerance. Searching for future prospects of the transgenic approaches for salinity tolerance is the next biggest challenge. Microarray, comparative genomics, comparative proteomics, protein profiling, etc., can be used to identify future prospects of salt tolerance. These genes can be harnessed for the potential property of salt tolerance.

Testing the transgenics at different stages of plant development; testing for salt tolerance for different salt ions other than $Na^+$; testing at natural field conditions, testing for generations $T_0$, $T_1$, $T_2$, $T_3$, and $T_4$; and pyramiding of multiple genes for salt tolerance are the challenges that are to be addressed.

Transgenic plants are developed and tested for salt tolerance, but salt tolerance is different in different stages of plant development. A comparative study of salt tolerance during different stages of plant development provides more insights in this regard. Studies reported that plants' response to salinity was different at different stages of plant development (Greenway and Munns, 1980). Genotypes that exhibit salt tolerance during early stages may not be tolerant at later stages. The reason can be that different genotypes contributed to salt tolerance during seed germination, vegetative growth, fruit, and yield.

Salinity is attributed to the group of salt ions ($Na^+$, $Cl^-$, and $SO_4^-$), but salt tolerance is established by evaluating with NaCl (Buckman and Brady, 1967). Earlier studies proved that the response of a crop or a crop cultivar to chloride or sulfate varies to a great extent (Curtin et al., 1993; Rogers et al., 1998; Inal, 2002). Thus, evaluating the performance of transgenic lines other than NaCl is also required.

The claims of salt tolerance in developed transgenic plants is based on the performance of transgenic lines produced and tested under controlled conditions in laboratory or greenhouse, and there are meager reports on field trials. Testing the performance of transgenic lines under natural field conditions is to be addressed. The other end of the claims on transgenics is that transgenics are salt tolerant with limited testing, that is, a maximum tolerance up to $T_0$ and $T_1$ generations. Thus, the transgenics exhibiting salt tolerance need to exhibit the same response in the next generations $T_2$, $T_3$, and $T_4$. So transgenics exhibiting the same salt tolerance for all the generations $T_0$, $T_1$, $T_2$, $T_3$, and $T_4$ would imply a stable transgenic.

Salt tolerance is not governed by a single factor; reports and studies provide us with this information. Gene pyramiding or constructs with multiple genes would be another best option to achieve salt tolerance.

## 18.7 CONCLUSION

In conclusion, achievements made to date in developing salt-tolerant plants through genetic engineering are successful in improving the yield potential and salt stress tolerance of most crops. In contrast, genetic engineering has resulted in little success in enhanced salt stress tolerance, but none of them were tested in fields. Knowledge on signaling molecules, transcriptional elements, hormone receptors, and LEA molecules would be potential candidates for salt stress tolerance in the future. Apart from these prospects for transgenics, addressing the following challenges is also required: testing the transgenics at different stages of plant development; testing for salt tolerance for different salt ions other than $Na^+$; testing in natural field conditions, testing for generations $T_0$, $T_1$, $T_2$, $T_3$, and $T_4$; and pyramiding of multiple genes for salt tolerance.

## REFERENCES

Abdullah Z, Ahmad R. 1990. Effect of pre-and post-kinetin treatments on salt tolerance of different potato cultivars growing on saline soils. *J Agron Crop Sci* 165: 94–102.

Abebe T, Guenzi AC, Martin B, Cushman JC. 2003. Tolerance of mannitol-accumulating transgenic wheat to water stress and salinity. *Plant Physiol* 131: 1748–1755.

Aharon R, Shahak Y, Wininger S, Bendov R, Kapulnik Y, Galili G. 2003. Overexpression of a plasma membrane aquaporin in transgenic tobacco improves plant vigor under favorable growth conditions but not under drought or salt stress. *Plant Cell* 15(2): 439–447.

Ahmad P, Prasad MNV. 2012a. *Abiotic Stress Responses in Plants: Metabolism, Productivity and Sustainability.* New York: Springer Science Business Media, LLC.

Ahmad P, Prasad MNV. 2012b. *Environmental Adaptations and Stress Tolerance in Plants in the Era of Climate Change.* New York: Springer Science Business Media, LLC.

Ahmad R, Kim MD, Back KH, Kim HS, Lee HS, Kwon SY, Murata N, Chung WI, Kwak SS. 2008. Stress-induced expression of choline oxidase in potato plant chloroplasts confers enhanced tolerance to oxidative, salt, and drought stresses. *Plant Cell Rep* 27: 687–698.

Ali AA, Alqurainy F. 2006. Activities of antioxidants in plants under environmental stress. In: Motohashi N, editor, *The Lutein-Prevention and Treatment for Diseases.* Kerala, India: Transworld Research Network, pp. 187–256.

Ali Q, Athar HR, Ashraf M. 2008. Modulation of growth, photosynthetic capacity and water relations in salt stressed wheat plants by exogenously applied 24-epibrassinolide. *Plant Growth Regul* 56: 107–116.

Anjaneyulu E, Reddy PS, Sunita MS, Kishor PB, Meriga B. 2014. Salt tolerance and activity of antioxidative enzymes of transgenic finger millet overexpressing a vacuolar H(+)-pyrophosphatase gene (*SbVPPase*) from *Sorghum bicolor. J Plant Physiol* 171(10): 789–798.

Anuradha S, Rao SSR. 2001. Effect of brassinosteroids on salinity stress induced inhibition of seed germination and seedling growth of rice (*Oryza sativa* L.). *Plant Growth Regul* 33: 151–153.

Apse MP, Aharon GS, Snedden WA, Blumwald E. 1999. Salt tolerance conferred by overexpression of a vacuolar $Na^+/H^+$-antiport in *Arabidopsis. Science* 285: 1256–1258.

Ardie SW, Xie L, Takahashi R, Liu S, Takano T. 2009. Cloning of a high-affinity $K^+$ transporter gene PutHKT2;1 from *Puccinellia tenuiflora* and its functional comparison with OsHKT2;1 from rice in yeast and *Arabidopsis. J Exp Bot* 60(12): 3491–3502.

Asada K. 1999. The water–water cycle in chloroplasts: Scavenging of active oxygens and dissipation of excess photons. *Annu Rev Plant Physiol Plant Mol Biol* 50: 601–639.

Ashraf M. 2004. Some important physiological selection criteria for salt tolerance in plants. *Flora* 199: 361–376.

Ashraf M. 2009. Biotechnological approach of improving plant salt tolerance using antioxidants as markers. *Biotechnol Adv* 27: 84–93.

Ashraf M, Akram NA. 2009. Improving salinity tolerance of plants through conventional breeding and genetic engineering: An analytical comparison. *Biotechnol Adv* 27: 744–752.

Ashraf M, Athar HR, Harris PJC, Kwon TR. 2008. Some prospective strategies for improving crop salt tolerance. *Adv Agron* 97: 45–110.

Ashraf M, Foolad MR. 2007. Roles of glycine betaine and proline in improving plant abiotic stress resistance. *Environ Exp Bot* 59: 206–216.

Ashraf M, McNeilly T. 2004. Salinity tolerance in *Brassica* oilseeds. *Crit Rev Plant Sci* 23(2): 157–174.

Azooz MM. 2009. Salt stress initigation by seed priming with salicylic acid in two faba bean genotypes differing in salt tolerant. *Int J Agric Biol* 11: 343–350.

Badawi GH, Kawano N, Yamauchi Y, Shimada E, Sasaki R, Kubo A, Tanaka K. 2004. Over-expression of ascorbate peroxidase in tobacco chloroplasts enhances the tolerance to salt stress and water deficit. *Physiol Plant* 121: 231–238.

Barciszewski J, Siboska G, Rattan SIS, Clark BFC. 2000. Occurrence, biosynthesis and properties of kinetin (N6-furfuryladenine). *Plant Growth Regul* 32: 257–265.

Beck EH, Heim R, Hansen J. 2004. Plant resistance to cold stress: Mechanisms and environmental signals triggering frost hardening and dehardening. *J Biosci* 29: 449–459.

Bei Q, Laun S. 1998. Functional characterization of a plant K+ channel gene in a plant cell model. *Plant J* 13(6): 857–865.

Bhattacharya A, Saini U, Sharma P, Nagar PK, Ahuja PS. 2006. Osmotin-regulated reserve accumulation and germination in genetically transformed tea somatic embryos: A step towards regulation of stress tolerance and seed recalcitrance. *Seed Sci Res* 16: 203–211.

Bhauso TD, Thankappan R, Kumar A, Mishra GP, Dobaria JR, Rajam MV. 2014. Over-expression of bacterial mtlD gene confers enhanced tolerance to salt-stress and water deficit stress in transgenic peanut (*Arachis hypogaea*) through accumulation of mannitol. *Aust J Crop Sci* 8(3): 413–421.

Boucaud J, Ungar IA. 1976. Hormonal control of germination under saline conditions of three halophyte taxa in genus *Suaeda*. *Physiol Plant* 36: 197–200.

Bozcuk S. 1981. Effect of kinetin and salinity on germination of tomato, barley and cotton seeds. *Ann Bot* 48: 81–84.

Bray EA, Bailey-Serres J, Weretilnyk E. 2000. Responses to abiotic stresses. In: Gruissem W, Buchnnan B, Jones R, editors, *Biochemistry and Molecular Biology of Plants*. Rockville, MD: American Society of Plant Physiologists, pp. 1158–1249.

Buckman HO, Brady NC. 1967. *The Nature and Properties of Soils*. New York: The MacMillan Company.

Carillo P, Grazia Annunziata M, Pontecorvo G., Fuggi A, Woodrow P. 2011. Salinity stress and salt tolerance. In: Shanker AK, Venkateswarlu B, editors, *Abiotic Stress in Plants—Mechanisms and Adaptations*. Rijeka, Croatia: InTech, pp. 21–38.

Chakrabarti N, Mukherji S. 2003. Alleviation of NaCl stress by pretreatment with phytohormones in *Vigna radiata*. *Biol Plantarum* 46: 589–594.

Cheng Y-J, Deng X-P, Kwak S-S, Chen W, Eneji AE. 2013. Enhanced tolerance of transgenic potato plants expressing choline oxidase in chloroplasts against water stress. *Bot Stud* 54: 30.

Cheong JJ, Choi YD. 2003. Methyl jasmonate as a vital substance in plants. *Trends Genetics* 19: 409–413.

Choe YH, Kim YS, Kim IS, Bae MJ, Lee EJ, Kim YH, Park HM, Yoon HS. 2013. Homologous expression of γ-glutamylcysteine synthetase increases grain yield and tolerance of transgenic rice plants to environmental stresses. *J Plant Physiol* 170(6): 610–618.

Choudhuri MA. 1988. Free radicals and leaf senescence—A review. *Plant Physiol Biochem* 15: 18–29.

Curtin D, Steppuhn H, Selles F. 1993. Plant response to sulphate and chloride salinity: Growth and ionic relations. *Soil Sci Soc Am J* 57: 1304–1310.

De Ronde W., Cress G., Krüger K., Strasser R., Van Staden J. 2004. Photosynthetic response of transgenic soybean plants, containing an *Arabidopsis* P5CR gene, during heat and drought stress. *J Plant Physiol* 161: 1211–1224.

Dregne HE. 1986. Desertification of arid lands. In: El-Baz F, Hassan MHA, editors, *Physics of Desertification*. Dordrecht, The Netherlands: Martinus, Nijhoff, pp. 4–34.

Duan X, Song Y, Yang A, Zhang J. 2009. The transgene pyramiding tobacco with betaine synthesis and heterologous expression of AtNHX1 is more tolerant to salt stress than either of the tobacco lines with betaine synthesis or AtNHX1. *Physiol Plant* 135(3): 281–295.

Eisenbarth DA, Weig AR. 2005. Dynamics of aquaporins and water relations during hypocotyl elongation in *Ricinus communis* L. seedlings. *J Exp Bot* 56: 1831–1842.

El-Tayeb MA. 2005. Response of barley grains to the interactive effect of salinity and salicylic acid. *Plant Growth Regul* 45: 215–224.

Eltayeb AE, Kawano N, Badawi G, Kaminaka H, Sanekata, T, Morishima I. 2006. Enhanced tolerance to ozone and drought stresses in transgenic tobacco overexpressing dehydroascorbate reductase in cytosol. *Physiol Plant* 127: 57–65.

Eltayeb AL, Kawano N, Badawi GH, Kaminaka H, Sanekata T, Shibahar T, Inanaga S, Tanaka K. 2007. Overexpression of monodehydroascorbate reductase in transgenic tobacco confers enhanced tolerance to ozone, salt and polyethylene glycol stresses. *Planta* 225: 1255–1264.

Eltayeb AS, Yamamoto S., Habora MEE, Yin L, Tsujimoto H, Tanak K. 2011. Transgenic potato over-expressing *Arabidopsis* cytosolic AtDHAR1 showed higher tolerance to herbicide, drought and salt stresses. *Breed Sci* 611: 3–10.

Epstein E. 1997. Genetic potentials for solving problems of soil mineral stress: Adaptation of crops to salinity. In: Wright MJ, editor, *Plant Adaptation to Mineral Stress in Problem Soils*. Ithaca, NY: Cornell University, pp. 73–123.

Feller U. 2004. Proteolysis. In: Nooden LD, editor, *Plant Cell Death Processes*. Amsterdam, the Netherlands: Elsevier Inc., pp. 107–123.

Fletcher RA, Hofstra G. 1988. Triazoles as potential plant protectants. In: Berg K, Plempel M, editors, *Sterol Biosynthesis Inhibitors: Pharmaceutical and Agrochemical Aspects*. London, U.K.: Ellis Harwood, Ltd.

Foolad MR. 2004. Recent advances in genetics of salt tolerance in tomato. *Plant Cell, Tissue Organ Cult* 76: 101–119.

Foyer CH, Lelandais M, Kunert KJ. 1994. Photooxidative stress in plants. *Physiol Plant* 92: 696–717.

Gaber A, Oshimura KY, Yamamoto T, Yabuta Y, Takeda T, Miyasaka H, Nakano Y, Shigeoka S. 2006. Glutathione peroxidase-like protein of Synechocystis PCC 6803 confers tolerance to oxidative and environmental stresses in transgenic *Arabidopsis*. *Physiol Plant* 128: 251–262.

Gao Z, He X, Zhao B, Zhou C, Liang Y, Ge R, Shen Y, Huang Z. 2010. Overexpressing a putative aquaporin gene from wheat, *TaNIP*, enhances salt tolerance in transgenic *Arabidopsis*. *Plant Cell Physiol* 51: 767–775.

Garg AK, Kim JK, Owens TG, Ranwala AP, Choi YD, Kochian LV, Wu RJ. 2002. Trehalose accumulation in rice plants confers high tolerance levels to different abiotic stresses. *Proc Natl Acad Sci USA* 99(25): 15898–15903.

Gaxiola RA, Sanchez CA, Paez-Valencia J, Ayre BG, Elser JJ. 2012. Genetic manipulation of a "vacuolar" H(+)-PPase: From salt tolerance to yield enhancement under phosphorus-deficient soils. *Plant Physiol* 159(1): 3–11.

Gierth M, Mäser P, Schroeder JI. 2005. The potassium transporter AtHAK5 functions in K(+) deprivation-induced high-affinity K(+) uptake and AKT1 K(+) channel contribution to K(+) uptake kinetics in *Arabidopsis* roots. *Plant Physiol* 137(3): 1105–1114.

Gonai T, Kawahara S, Tougou M, Satoh S, Hashiba T, Hirai N, Kawaide H, Kamiya Y, Yoshioka T. 2004. Abscisic acid in the thermo inhibition of lettuce seed germination and enhancement of its catabolism by gibberellin. *J Exp Bot* 55: 111–118.

Greenway H, Munns R. 1980. Mechanisms of salt tolerance in nonhalophytes. *Annu Rev Plant Physiol* 31: 149–190.

Guilfoyle TJ, Hagen G, Li Y, Ulmasov T, Liu Z, Strabala T, Gee MA. 1993. Auxin-regulated transcription. *Aust J Plant Physiol* 20: 489–502.

Gunes A, Inal A, Alpaslam M, Erslan F, Bagsi EG, Cicek N. 2007. Salicylic acid induced changes on some physiological parameters symptomatic for oxidative stress and mineral nutrition in maize (*Zea mays* L.) grown under salinity. *J Plant Physiol* 164: 728–736.

Guo B-T, Wang B, Weng M-L, Qiao L-X, Feng Y-B, Wang L, Zhang P-Y, Wang X-L, Sui J-M, Liu T, Duan D-L, Wang B. 2014. Expression of Porphyra yezoensis TPS gene in transgenic rice enhanced the salt tolerance. *J Plant Breed Genet* 2(1): 45–55.

Hanba YT, Shibasaka M, Hayashi Y, Hayakawa T, Kasamo K, Terashima I, Katsuhara M. 2004. Overexpression of the barley aquaporin HvPIP2;1 increases internal CO(2) conductance and CO(2) assimilation in the leaves of transgenic rice plants. *Plant Cell Physiol* 45: 521–529.

Harb A, Krishnan A, Ambavaram MM, Pereira A. 2010. Molecular and physiological analysis of drought stress in *Arabidopsis* reveals early responses leading to acclimation in plant growth. *Plant Physiol* 154(3): 1254–1271.

Hasanuzzaman M, Nahar K, Alam MM, Bhowmik PC, Hossain MA, Rahman MM, Prasad MN, Ozturk M, Fujita M. 2014. Potential use of halophytes to remediate saline soils. *Biomed Res Int*, article id: 589341.

Havaux M, Ksas B, Szewczyk A, Rumeau D, Franck F, Caffarri S, Triantaphylidès C. 2009. Vitamin B6 deficient plants display increased sensitivity to high light and photo-oxidative stress. *BMC Plant Biol* 9: 130.

Hayat S, Ahmad A, Mobin M, Hussain A, Fariduddin Q. 2000. Photosynthetic rate, growth and yield of mustard plants sprayed with 28-homobrassinolide. *Photosynthetica* 38: 469–471.

Hela M, Olfa B, Imen BS, Nawel N, Abidi W, Huang J, Hanen Z, Abdelali H, Mokhtar L, Ouerghi Z. 2012. Enhanced accumulation of root hydrogen peroxide is associated with reduced antioxidant enzymes under isoosmotic NaCl and $Na_2SO_4$ salinities. *Afr J Biotechnol* 11: 8500–8509.

Hemavathi, Upadhyaya CP, Nookaraju A, Young KE, Chun SC, Kim DH, Park SW. 2010a. Enhanced ascorbic acid accumulation in transgenic potato confers tolerance to various abiotic stresses. *Biotechnol Lett* 32: 321–330.

Hemavathi, Upadhyaya CP, Young KE, Nookaraju A, Kim HS, Heung JJ, Oh OM, Aswath CR, Chun SC, Kim DH, Park SW. 2009. Over-expression of strawberry D-galacturonic acid reductase in potato leads to accumulation of vitamin C with enhanced abiotic stress tolerance. *Plant Sci* 177: 659–667.

Hemavathi, Upadhyaya CP, Young KE, Nookaraju A, Kim HS, Heung JJ, Oh OM, Chun SC, Kim DH, Park SW. 2010b. Biochemical analysis of enhanced tolerance in transgenic potato plants overexpressing D-galacturonic acid reductase gene in response to various abiotic stresses. *Mol Breed*. 28(1): 105–115.

He ZQ, He CX, Zhang ZB, Zou ZR, Wang HS. 2007. Changes of antioxidative enzymes and cell membrane osmosis in tomato colonized by *Arbuscular mycorrhizae* under NaCl stress. *Colloid Surf B* 59: 128–133.

Hmida-Sayari A, Gargouri-Bouzid R, Bidani A, Jaoua L, Savoure A, Jaoua S. 2005. Overexpression of Δ1-pyrroline-5-carboxylate synthetase increases proline production and confers salt tolerance in transgenic potato plants. *Plant Sci* 169: 746–752.

Hong Z, Lakkineni K, Zhang Z, Pal D, Verma S. 2005. Removal of feedback inhibition of Δ1-pyrroline-5-carboxylate synthetase results in increased proline accumulation and protection of plants from osmotic stress. *Plant Physiol* 122: 1129–1136.

Hong Z, Lakkineni K, Zhang Z. Verma DPS. 2000. Removal of feedback inhibition of-pyrroline-5-carboxylate synthetase results in increased proline accumulation and protection of plants from osmotic stress. *Plant Physiol* 122: 1129–1136.

Hossain MA, Asada K. 1985. Monodehydroascorbate reductase from cucumber is a flavin adenine dinucleotide enzyme. *J Biol Chem* 260(24): 12920–12926.

Hossain MA, Fujita M. 2013. Hydrogen peroxide priming stimulates drought tolerance in mustard (*Brassica juncea* L.). *Plant Gene Trait* 4: 109–123.

Hossain MA, Hasanuzzaman M, Fujita M. 2011. Coordinate induction of antioxidant defense and glyoxalase system by exogenous proline and glycinebetaine is correlated with salt tolerance in mung bean. *Front Agric China* 5(1): 1–14.

Hossain MA, Mostofa MG, Fujita M. 2013. Heat-shock positively modulates oxidative protection of salt and drought-stressed mustard (*Brassica campestris* L.) seedling. *J Plant Sci Mol Breed* 2: 1–14.

Huang J, Hirji R, Adam L, Rozwadowski KL, Hammerlindl JK, Keller WA, Selvaraj G. 2000. Genetic engineering of glycinebetaine production toward enhancing stress tolerance in plants: Metabolic limitations. *Plant Physiol* 122: 747–756.

Hu W, Yuan Q, Wang Y, Cai R, Deng X, Wang J, Zhou S, Chen M, Chen L, Huang C, Ma Z, Yang G, He G. 2012. Overexpression of a wheat aquaporin gene, *TaAQP8*, enhances salt stress tolerance in transgenic tobacco. *Plant Cell Physiol* 53: 2127–2141.

Inal A. 2002. Growth, proline accumulation and ionic relations of tomato (*Lycopersicon esculentum* L.) as influenced by NaCl and Na$_2$SO$_4$ salinity. *Turkish J Bot* 26: 285–290.

Jha B, Sharma A, Mishra A. 2011. Expression of *SbGSTU* (tau class glutathione S-transferase) gene isolated from *Salicornia brachiata* in tobacco for salt tolerance. *Mol Biol Rep* 38(7): 4823–4832.

Jiang J, Li H, He G, Yin Y, Liu M, Liu B, Qiao G, Lin S, Xie L, Zhuo R. 2013. Over-expression of the Cod A gene by Rd29a promoter improves salt tolerance in *Nicotiana tabacum*. *Pak J Bot* 45(3): 821–827.

Ji W, Zhu Y, Li Y, Yang L, Zhao X, Cai H, Bai X. 2010. Over-expression of a glutathione S-transferase gene, *GsGST*, from wild soybean (*Glycine soja*) enhances drought and salt tolerance in transgenic tobacco. *Biotechnol Lett* 32(8): 1173–1179.

Jun S-S, Yang JY, Choi HJ, Kim N-R, Park MC, Hong Y-N. 2005. Altered physiology in trehalose-producing transgenic tobacco plants: Enhanced tolerance to drought and salinity stresses. *J Plant Biol* 48(4): 456.

Kai H, Koh I. 2014. Temperature stress in plants. In: *eLS*. Chichester, U.K.: John Wiley & Sons Ltd.

Kaldenhoff R, Fischer M. 2006. Aquaporins in plants. *Acta Physiol* 187: 169–176.

Kang DJ, Seo YJ, Lee JD, Ishii R, Kim KU, Shin DH, Park SK, Jang SW, Lee IJ. 2005. Jasmonic acid differentially affects growth, ion uptake and abscisic acid concentration in salt-tolerant and salt-sensitive rice cultivars. *J Agron Crop Sci* 191: 273–282.

Karpinski S, Gabrys H, Mateo A, Karpinska B, Mullineaux PM. 2003. Light perception in plant disease defence signalling. *Curr Opin Plant Biol* 6: 390–396.

Kasukabe Y, He L, Nada K, Misawa S, Ihara I, Tachibana S. 2004. Overexpression of spermidine synthase enhances tolerance to multiple environmental stress and upregulates the expression of various stress regulated genes in transgenic *Arabidopsis thaliana*. *Plant Cell Physiol* 45: 712–722.

Kavitha K, George S, Venkataraman G, Parida A. 2010. A salt-inducible chloroplastic monodehydroascorbate reductase from halophyte *Avicennia marina* confers salt stress tolerance on transgenic plants. *Biochimie* 92(10): 1321–1329.

Kaya C, Kirnak H, Higgs D, Saltali K. 2002. Supplementary calcium enhances plant growth and fruit yield in strawberry cultivars grown at high salinity. *Sci Hortic* 93: 65–74.

Khraiwesh B, Zhu JK, Zhu J. 2012. Role of miRNAs and siRNAs in biotic and abiotic stress responses of plants. *Biochim Biophys Acta* 1819(2): 137–148.

Kishor PBK, Hong Z, Miao CH, Hu CAA, Verma DPS. 1995. Overexpression of Δ1-pyrroline-5- carboxylate synthetase increases proline production and confers osmotolerance in transgenic plants. *Plant Physiol* 08: 1387–1394.

Kraus TE, Fletcher RA. 1994. Paclobutrazol protects wheat seedlings from heat and paraquat injury; Is detoxification of active oxygen involved? *Plant Cell Physiol* 35: 45–52.

Kreps, JA, Wu Y, Chang HS, Zhu T, Wang X, Harper JF. 2002. Transcriptome changes for *Arabidopsis* in response to salt, and cold stress. *Plant Physiol* 130: 2129–2141.

Kumar A, Sharma S, Mishra S. 2010. Influence of arbuscularmycorrhizal (AM) fungi and salinity on seedling growth, solute accumulation, and mycorrhizal dependency of *Jatrophacurcas* L. *J Plant Growth Regul* 29: 297–306.

Kumar K, Sinha AK. 2013. Overexpression of constitutively active mitogen activated protein kinase kinase 6 enhances tolerance to salt stress in rice. *Rice (NY)* 6(1): 25.

Kwon SY, Choi SM, Ahn YO, Lee HS, Lee HB, Park YM, Kwak SS. 2003. Enhanced stress-tolerance of transgenic tobacco plants expressing a human dehydroascorbate reductase gene. *J Plant Physiol* 160(4): 347–353.

Lai SJ, Lai MC, Lee RJ, Chen YH, Yen HE. 2014. Transgenic *Arabidopsis* expressing osmolyte glycine betaine synthesizing enzymes from halophilic methanogen promote tolerance to drought and salt stress. *Plant Mol Biol* 85(4–5): 429–441.

Laloi C, Apel K, Danon A. 2004. Reactive oxygen signalling: The latest news. *Curr Opin Plant Biol* 7: 323–308.

Lata Ch, Yadav A, Prasad M. 2011. Role of plant transcription factors in abiotic stress tolerance. In: Shanker A, editor, *Abiotic Stress Response in Plants - Physiological, Biochemical and Genetic Perspectives*. Rijeka, Croatia: InTech.

Lebaudy A, Véry AA, Sentenac H. 2007. K+ channel activity in plants: Genes, regulations and functions. *FEBS Lett* 581: 2357–2366.

Lehmann J, Atzorn R, Bruckner C, Reinbothe S, Leopold J, Wasternack C, Parthier B. 1995. Accumulation of jasmonate, abscisic acid, specific transcripts and proteins in osmotically stressed barley leaf segments. *Planta* 197: 156–162.

Lim JH, Park KJ, Kim BK, Jeong JW, Kim HJ. 2012. Effect of salinity stress on phenolic compounds and carotenoids in buckwheat (*Fagopyrumesculentum* M.) sprout. *Food Chem* 135: 1065–1070.

Lim MY, Pulla RK, Park JM, Ham CH, Jeong BR. 2011. Over-expression of L-gulono-γ-lactone oxidase (GLOase) gene leads to ascorbate accumulation with enhanced abiotic stress tolerance in tomato. *In Vitro Cell Dev Biol Plant* 48(5): 453–468.

Li Q, Li Y, Li C, Yu X. 2012. Enhanced ascorbic acid accumulation through over-expression of dehydroascorbate reductase confers tolerance to methyl viologen and salt stresses in tomato. *Czech J Genet Plant Breed* 48(2): 74–86.

Lisko KA, Torres R, Harris RS, Belisle M, Vaughan MM, Jullian B, Chevone BI, Mendes P, Nessler CL, Lorence A. 2013. Elevating vitamin C content via overexpression of myo-inositol oxygenase and L-gulono-1,4-lactone oxidase in *Arabidopsis* leads to enhanced biomass and tolerance to abiotic stresses. *In Vitro Cell Dev Biol Plant* 49(6): 643–655.

Liu D, Liu Y, Rao J, Wang G, Li H, Ge F, Chen C. 2013a. Overexpression of the glutathione S-transferase gene from *Pyrus pyrifolia* fruit improves tolerance to abiotic stress in transgenic tobacco plants. *Mol Breed* 47(4): 515–523.

Liu S, Cheng Y, Zhang X, Guan Q, Nishiuchi S, Hase K, Takano T. 2007. Expression of an NADP-malic enzyme gene in rice (*Oryza sativa* L.) is induced by environmental stresses; over-expression of the gene in *Arabidopsis* confers salt and osmotic stress tolerance. *Plant Mol Biol* 64: 49–58.

Liu W, An HM, Yang M. 2013b. Overexpression of *Rosa roxburghii* L-galactono-1,4-lactone dehydrogenase in tobacco plant enhances ascorbate accumulation and abiotic stress tolerance. *Acta Physiol Plant* 35(5): 1617–1624.

Li X, Guo C, Gu J, Duan W, Zhao M, Ma C, Du X, Lu W, Xiao K. 2014a. Overexpression of VP, a vacuolar H+-pyrophosphatase gene in wheat (*Triticum aestivum* L.), improves tobacco plant growth under Pi and N deprivation, high salinity, and drought. *J Exp Bot* 65(2): 683–696.

Li Y, Fan C, Xing Y, Yun P, Luo L, Yan B, Peng B, Xie W, Wang G, Li X, Xiao J, Xu C, He Y. 2014b. Chalk5 encodes a vacuolar H(+)-translocating pyrophosphatase influencing grain chalkiness in rice. *Nat Genet* 46(4): 398–404.

Li YJ, Hai RL, Du XH, Jiang XN, Lu H. 2009. Over-expression of a *Populus* peroxisomal ascorbate peroxidase (*PpAPX*) gene in tobacco plants enhances stress tolerance. *Plant Breed* 128: 404–410.

Lu W, Guo C, Li X, Duan W, Ma C, Zhao M, Gu J, Du X, Liu Z, Xiao K.. 2014. Overexpression of TaNHX3, a vacuolar Na$^+$/H$^+$ antiporter gene in wheat, enhances salt stress tolerance in tobacco by improving related physiological processes. *Plant Physiol Biochem* 76: 17–28.

Maggio A, Barbieri G, Raimondi G, De Pascale S. 2010. Contrasting effects of GA3 treatments on tomato plants exposed to increasing salinity. *J Plant Growth Regul* 29: 63–72.

Maggio A, Miyazaki S, Veronese P, Fujita T, Ibeas JI, Damsz B, Narasimhan ML, Hasegawa PM, Joly RJ, Bressan RA. 2002. Does proline accumulation play an active role in stress-induced growth reduction. *Plant J* 31: 699–712.

Mary S. 2011. Plant pathology, available from: http://www.ext.colostate.edu/mg/gardennotes/331.pdf

McKersie BD, Leshem YY. *Stress and Stress Coping in Cultivated Plants*. London, U.K.: Kluwer Academic Publishes, 1994.

Miura K, Furumoto T. 2013. Cold signaling and cold response in plants. *Int J Mol Sci* 14(3): 5312–5337.

Mohanty A, Kathuria H, Ferjani A, Sakamoto A, Mohanty P, Murata N, Tyagi AK. 2002. Transgenics of an elite indica rice variety Pusa Basmati 1 harbouring the codAgene are highly tolerant to salt stress. *Theory Appl Genet* 6: 51–57.

Molinari HBC, Marur CJ, Bespalhok-Filho JC, Kobayashi AK, Pileggi M, Júniora RPLJ, Pereirad LFP, Vieiraa LGE. 2004. Osmotic adjustment in transgenic citrus rootstock *Carrizo citrange* (*Citrus sinensis* Obs. X *Poncirustrifoliata* L. Raf.) overproducing proline. *Plant Sci* 167: 1375–1381.

Moons A, Prisen E, Bauw G, Montagu MV. 1997. Antagonistic effects of abscisic acid and jasmonates on salt-inducible transcripts in rice roots. *Plant Cell* 92: 243–259.

Moriwaki T, Yamamoto Y, Aida T, Funashi T, Shishido T, Asada M, Prodhan SH, Komanine A, Motohashi T. 2008. Overexpression of the *Escherichia coli* catalase katE, enhances tolerance to salinity stress in the transgenic indica rice cultivar, BR5. *Plant Biotechnol Rep* 2: 41–46.

Munns R, Tester M. 2008. Mechanisms of salinity tolerance. *Annu Rev Plant Biol* 59: 651–681.

Munns R. 2005. Genes and salt tolerance: Bringing them together. *New Phytol* 167(3): 645–663.

Nakanishi Y, Maeshima M. 1998. Molecular cloning of vacuolar H(+)-pyrophosphatase and its developmental expression in growing hypocotyl of mung bean. *Plant Physiol* 116(2): 589–597.

Naqvi SSM, Ansari R, Kuawada AN. 1982. Responses of salt stressed wheat seedlings to kinetin. *Plant Sci Lett* 26: 279–283.

Nguyen TX, Nguyen T, Alameldin H, Goheen B, Loescher W, Sticklen M. 2013. Transgene pyramiding of the HVA1 and mtlD in T3 maize (*Z. mays* L.) plants confers drought and salt tolerance, along with an increase in crop biomass. *Int J Agron* 10.

Niu X, Bressan RA, Hasegawa PM.1993. Halophytes up-regulate plasma membrane Hþ-ATPase gene more rapidly than glycophytes in response to salt stress. *Plant Physiol* 102: 130.

Obata T, Kitamoto HK, Nakamura A, Fukuda A, Tanaka Y. 2007. Rice shaker potassium channel OsKAT1 confers tolerance to salinity stress on yeast and rice cells. *Plant Physiol* 144(4): 1978–1985.

Oh SK, Jang HA, Lee SS, Cho HS, Lee DH, Choi D, Kwon SY. 2014. Cucumber Pti1-L is a cytoplasmic protein kinase involved in defense responses and salt tolerance. *J Plant Physiol* 171(10): 817–822.

Onoda Y, Anten NP. 2011. Challenges to understand plant responses to wind. *Plant Signal Behav* 6(7): 1057–1059.

Osakabe Y, Osakabe K, Shinozaki K, Tran LSP. 2014. Response of plants to water stress. *Front Plant Sci* 5: 86.

Ovečka M, Takáč T. 2014. Managing heavy metal toxicity stress in plants: Biological and biotechnological tools. *Biotechnol Adv* 32(1): 73–86.

Parasher A, Varma SK. 1988. Effect of pre-sowing seed soaking in gibberellic acid on growth of wheat (*Triticumaestivum* L.) under different saline conditions. *Indian J Biol Sci* 26: 473–475.

Park MY, Wu G, Gonzalez-Sulser A, Vaucheret H, Poethig RS. 2005. Nuclear processing and export of microRNAs in *Arabidopsis*. *Proc Natl Acad Sci USA* 102: 3691–3696.

Pasapula V, Shen G, Kuppu S, Paez-Valencia J, Mendoza M, Hou P, Chen J, Qiu X, Zhu L, Zhang X, Auld D, Blumwald E, Zhang H, Gaxiola R, Payton P. 2011. Expression of an *Arabidopsis* vacuolar H$^+$-pyrophosphatase gene (*AVP1*) in cotton improves drought- and salt tolerance and increases fibre yield in the field conditions. *Plant Biotechnol J* 9(1): 88–99.

Pasapula V, Shen G, Kuppu S, Paez-Valencia J, Mendoza M, Hou P, Chen J et al. 2011. Expression of an *Arabidopsis* vacuolar H $^+$ - pyrophosphatase gene (AVP1) in cotton improves drought- and salt tolerance and increases fibre yield in the field conditions. *Plant Biotechnol J* 9(1): 88–99.

Peng Y, Lin W, Cai W, Arora R. 2007. Overexpression of a *Panax ginseng* tonoplast aquaporin alters salt tolerance, drought tolerance and cold acclimation ability in transgenic *Arabidopsis* plants. *Planta* 226(3): 729–740.

Penna S. 2003. Building stress tolerance through overproducing trehalose in transgenic plants. *Trends Plant Sci* 8(8): 355–357.

Polle A. 2001. Dissecting the superoxide dismutase–ascorbate–glutathione pathway by metabolic modeling: Computer analysis as a step towards flux analysis. *Plant Physiol* 126: 445–462.

Prakash L, Prathapasenan G. 1990 NaCl and gibberellic acid induced changes in the content of auxin, the activity of cellulose and pectin lyase during leaf growth in rice (*Oryza sativa*). *Ann Bot* 365: 251–257.

Prasad KVSK, Sharmila P, Kumar PA, PardhaSaradhi P. 2000. Transformation of *Brassica juncea* (L.) Czern with a bacterial codA gene enhances its tolerance to salt stress. *Mol Breed* 6: 489–499.

Prashanth SR, Sadhasivam V, Parida A. 2008. Over expression of cytosolic copper/zinc superoxide dismutase from a mangrove plant *Avicennia marina* in indica rice var Pusa Basmati-1 confers abiotic stress tolerance. *Transgenic Res* 17(2): 281–291.

Prodhan SH, Hossain H, Nagamiya K, Komamine A, Morishima H. 2008. Improved salt tolerance and morphological variation in indica rice (*Oryza sativa* L.) transformed with a catalase gene from E. coli. *Plant Tissue Cult Biotechnol* 18: 57–63.

Qin H, Gu Q, Kuppu S, Sun L, Zhu X, Mishra N, Hu R, Shen G, Zhang J, Zhang Y, Zhu L, Zhang X, Burow M, Payton P, Zhang H. 2013. Expression of the *Arabidopsis* vacuolar $H^+$-pyrophosphatase gene *AVP1* in peanut to improve drought and salt tolerance. *Plant Biotechnol Rep* 7(3): 345–355.

Qi YC, Wang FF, Zhang H, Liu WQ. 2010. Overexpression of suadea salsa S –adenosylmethioninesynthetase gene promotes salt tolerance in transgenic tobacco. *Acta Physiol Plant* 32: 263–269.

Qu AL, Ding YF, Jiang Q, Zhu C. 2013. Molecular mechanisms of the plant heat stress response. *Biochem Biophys Res Commun* 432(2): 203–207.

Ramraj VM, Vyas BN, Godrej NB, Mistry KB, Swami BN, Singh N. 1997. Effects of 28-homobrassinolide on yields of wheat, rice, groundnut, mustard, potato and cotton. *J Agric Sci* 128: 405–413.

Rao AAR, Vardhini BV, Sujatha E, Anuradha S. 2002. Brassinosteroids a new class of phytohormones. *Curr Sci* 82: 1239–1245.

Ratner A, Jacoby B. 1976. Effect of Kþ, its counter anion, and pH on sodium efflux from barley root tips. *J Exp Bot* 27: 843–850.

Razavizadeh R, Ehsanpour AA. 2009. Effects of salt stress on proline content, expression of delta-1-pyrroline-5-carboxylate synthetase, and activities of catalase and ascorbate peroxidase in transgenic tobacco plants. *Biol Lett* 46(2): 63–75.

Rhodes D, Hanson AD. 1993. Quaternary ammonium and tertiary sulfonium compounds in higher-plants. *Annu Rev Plant Physiol Plant Mol Biol* 44: 357–384.

Rodriguez-Rosales MP, Jiang X, Gálvez FJ, Aranda MN, Cubero B, Venema K. 2008b. Overexpression of the tomato $K^+/H^+$ antiporter LeNHX2 confers salt tolerance by improving potassium compartmentalization. *New Phytol* 179(2): 366–377.

Rogers ME, Grieve CM, Shannon MC. 1998. The response of lucerne (*Medicago sativa* L.) to sodium sulphate and chloride salinity. *Plant Soil* 202: 271–280.

Sade N, Gebretsadik M, Seligmann R, Schwartz A, Wallach R, Moshelion M. 2010. The role of tobacco Aquaporin1 in improving water use efficiency, hydraulic conductivity, and yield production under salt stress. *Plant Physiol* 152: 245–254.

Saha K, Gupta K. 1993. Effect of LAB-150978-a plant growth retardant on sunflower and mungbean seedlings under salinity stress. *Indian J Plant Physiol* 36: 151–154.

Sakhabutdinova AR, Fatkhutdinova DR, Bezrukova MV, Shakirova FM. 2003. Salicylic acid prevents the damaging action of stress factors on wheat plants. *Bulg J Plant Physiol* 314–319.

Sawahel WA, Hassan AH. 2002. Generation of transgenic wheat plants producing high levels of the osmoprotectant proline. *Biotechnol Lett* 24: 721–725.

Schilling RK, Marschner P, Shavrukov Y, Berger B, Tester, M, Roy SJ, Plett DC. 2014. Expression of the *Arabidopsis* vacuolar $H^+$-pyrophosphatase gene (*AVP1*) improves the shoot biomass of transgenic barley and increases grain yield in a saline field. *Plant Biotechnol J* 12: 378–386.

Schützendübel A, Polle A. 2002. Plant responses to abiotic stresses: Heavy metal-induced oxidative stress and protection by mycorrhization. *J Exp Bot* 53(372): 1351–1365.

Seki MM, Narusaka J, Ishida T, Nanjo M, Fujita Y, Oono AM, Nakajima A, Enju T, Sakurai M, Satou K, Akiyama T, Taji K, Yamaguchi-Shinozaki P, Carninci J, Kawai Y, Hayashizaki Y, Shinozaki K. 2002.

Monitoring the expression profiles of 7000 *Arabidopsis* genes under drought, cold and high-salinity using a full-length cDNA microarray. *Plant J* 31: 279–292.

Serrano R, Culiañz-Maciá A, Moreno V. 1999. Genetic engineering of salt and drought tolerance with yeast regulatory genes. *Sci Hortic* 78: 261–269.

Shakirova FM, Sakhabutdinova AR, Bezrukova MV, Fatkhutdinova RA, Fatkhutdinova DR. 2003. Changes in the hormonal status of wheat seedlings induced by salicylic acid and salinity. *Plant Sci* 164: 317–322.

Shen W, Gómez-Cadenas A, Routly EL, Ho TH, Simmonds JA, Gulick PJ. 2001. The salt stress-inducible protein kinase gene, Esi47, from the salt-tolerant wheatgrass *Lophopyrum elongatum* is involved in plant hormone signaling. *Plant Physiol* 125(3): 1429–1441.

Shi H, Ishitani M, Kim C, Zhu JK. 2002. The *Arabidopsis thaliana* salt tolerance gene SOS1 encodes a putative Na$^+$/H$^+$ antiporter. *Proc Natl Acad Sci USA* 97: 6896–6901.

Shi H, Wu SJ, Zhu JK. 2003. Overexpression of a plasma membrane Na$^+$/H$^+$ antiporter improves salt tolerance in *Arabidopsis*. *Nat Biotechnol* 21: 81–85.

Shi H, Ye T, Zhong B, Liu X, Chan Z. 2014. Comparative proteomic and metabolomic analyses reveal mechanisms of improved cold stress tolerance in bermudagrass (*Cynodon dactylon* (L.) Pers.) by exogenous calcium. *J Integr Plant Biol* 56: 1064–1079.

Shinozaki K., Yamaguchi-Shinozaki, K. 2000. Molecular response to dehydration and low temperature: Differences and cross-talk between two stress signaling pathways. *Curr Opin Plant Biol* 3: 217–223.

Shirasawa K, Takabe T, Takabe T, Kishitani S. 2006. Accumulation of glycinebetaine in rice plants that overexpress choline monooxygenase from spinach and evaluation of their tolerance to abiotic stress. *Ann Bot* 98(3): 565–571.

Sickler CM, Edwards GE, Kiirats O, Gao Z, Loescher W. 2007. Response of mannitol-producing *Arabidopsis thaliana* to abiotic stress. *Funct Plant Biol* 34: 382–391.

Sing N, Mishra A, Jha B. 2014. Over-expression of the peroxisomal ascorbate peroxidase (SbpAPX) gene cloned from halophyte *Salicornia brachiata* confers salt and drought stress tolerance in transgenic tobacco. *Mar Biotechnol* 16(3): 321–332.

Stevens J, Senaratna T, Sivasithamparam K. 2006. Salicylic acid induces salinity tolerance in tomato (*Lycopersicon esculentum* cv. Roma): Associated changes in gas exchange, water relations and membrane stabilisation. *Plant Growth Regul* 49: 77–83.

Striker GG. 2012. Flooding stress on plants: Anatomical, morphological and physiological responses. *Botany* 1: 3–28.

Su J, Wu R. 2004. Stress-inducible synthesis of proline in transgenic rice confers faster growth under stress conditions than that with constitutive synthesis. *Plant Sci* 166: 941–948.

Sultana S, Khew CY, Morshed MM, Namasivayam P, Napis S, Ho CL. 2012. Over-expression of mono-dehydroascorbate reductase from a mangrove plant (AeMDHAR) confers salt tolerance on rice. *J Plant Physiol* 169(3): 311–318.

Sun WH, Dua MN, Sh DFU, Yang S, Meng QW. 2010. Over-expression of StAPX in tobacco improves seed germination and increases early seedling tolerance to salinity and osmotic stresses. *Plant Cell Rep* 29(8): 917–926.

Surekha Ch, Kumari KN, Aruna LV, Suneetha G, Arundhati A, Kavi Kishor PB. 2014. Expression of the *Vigna aconitifolia* P5CSF129A gene in transgenic pigeon pea enhances proline accumulation and salt tolerance. *Plant Cell, Tissue Organ Cult* 116: 27–36.

Suzuki Y, Maeshima M, Yamaki S. 1999. Molecular cloning of vacuolar H$^+$-pyrophosphatase and its expression during the development of pear fruit. *Plant Cell Physiol* 40(8): 900–904.

Suárez R, Calderón C, Iturriaga G. 2009. Enhanced tolerance to multiple abiotic stresses in transgenic alfalfa accumulating trehalose. *Crop Sci* 49: 1791–1799.

Takahashi R, Liu S, Takano T. 2007. Cloning and functional comparison of a high-affinity K$^+$ transporter gene PhaHKT1 of salt-tolerant and salt-sensitive reed plants. *J Exp Bot* 58(15–16): 4387–4395.

Tari I, Csisaal J, Szalai G, Horath F, Kiss G, Szepesi G, Szabl M, Erdei L. 2002. Acclimation of tomato plants to salinity stress after a salicylic acid pretreatment. *Acta Biol Szegediensis* 46: 55–56.

Tran LSP, Urao T, Qin F, Maruyama K, Kakimoto T, Shinozaki K, Yamaguchi-Shinozaki K. 2007. Functional analysis of AHK1/ATHK1 and cytokinin receptor histidine kinases in response to abscisic acid, drought, and salt stress in *Arabidopsis*. *Proc Natl Acad Sci USA* 104: 20623–20628.

Trueman S. Plant stresses: Abiotic and biotic stresses, available from: http://botany.about.com/od/PlantsAndTheEnvironment/a/Plant-Stresses-Abiotic-And-Biotic-Stresses.htm.

Tsonev TD, Lazova GN, Stoinova ZG, Popova LP. 1998. A possible role for jasmonic acid in adaptation of barley seedlings to salinity stress. *J Plant Growth Regul* 17: 153–159.

Tuomi J, Niemelä P, Haukioja E, Sirén S, Neuvonen S. 1984. Nutrient stress: An explanation for plant anti-herbivore responses to defoliation. *Oecologia* 61(2): 208–210.

Uehlein N, Lovisolo C, Siefritz F. 2003. Kaldenhoff: The tobacco aquaporin NtAQP1 is a membrane $CO_2$ pore with physiological functions. *Nature* 425: 734–737.

Upadhyaya CP, Venkatesh J, Gururani MA, Asnin L, Sharma K, Ajappala H, Park SW. 2011. Transgenic potato overproducing L-ascorbic acid resisted an increase in methylglyoxal under salinity stress via maintaining higher reduced glutathione level and glyoxalase enzyme activity. *Biotechnol Lett* 33: 2297–2307.

Urano K, Yoshiba Y, Nanjo T, 1to T, Yamaguchi-Shinozaki K, Shinozaki K. 2004. *Arabidopsis* stress-inducible gene for arginine decarboxylase AtADC2 is required for accumulation of putrescine in salt tolerance. *Biochem Biophys Res Commun* 313: 369–375.

Urao T, Yakubov B, Satoh R, Yamaguchi-Shinozaki K, Seki M, Hirayama T, Shinozaki K. 1999. A transmembrane hybrid stresses type histidine kinase in *Arabidopsis* functions as an osmosensor. *Plant Cell* 11: 1743–1754.

Ushimaru T, Nakagawa T, Fujioka Y, Daicho K., Naito M, Yamauchi Y, Nonaka H, Amako K, Yamawaki K, Murata N. 2006. Transgenic *Arabidopsis* plants expressing the rice dehydroascorbate reductase gene are resistant to salt stress. *J Plant Physiol* 163: 1179–1184.

Vasekina AV, Yershov PV, Reshetova OS, Tikhonova TV, Lunin VG, Trofimova MS, Babakov AV. 2004. Vacuolar $Na^+/H^+$ antiporter from barley: Identification and response to salt stress. *Biochemistry (Moscow)* 70: 100–107.

Vendruscolo ECG, Schuster I, Pileggi M, Scapim CA, Molinari HBC, Marur CJ, Viera LG. 2007. Stress-induced synthesis of proline confers tolerance to water deficit in transgenic wheat. *J Plant Physiol* 164: 1367–1376.

Venema K, Belver A, Marin-Manzano MC, Rodríguez-Rosales MP, Donaire JP. 2003. A novel intracellular $K^+/H^+$ antiporter related to $Na^+/H^+$ antiporters is important for $K^+$ ion homeostasis in plants. *J Biol Chem* 278(25): 22453–22459.

Vierstra R. 1996. Proteolysis in plants: mechanisms and functions. *Plant Mol Biol* 32: 275–302.

Wang H, Lee PD, Liu LF, Su JC. 1999. Effect of sorbitol induced osmotic stress on the changes of carbohydrate and free amino acid pools in sweet potato cell suspension cultures. *Bot Bull Acad Sinica* 40:219–225.

Wang H, Zhou Y, Gilmer S, Whitwill S, Fowke LC. 2000. Expression of the plant cyclin-dependent kinase inhibitor ICK1 affects cell division, plant growth and morphology. *Plant J* 24(5): 613–623.

Wang TB, Gassmann W, Rubio F, Schroeder JI, Glass AD. 1998. Rapid up-regulation of HKT1, a high-affinity potassium transporter gene, in roots of barley and wheat following withdrawal of potassium. *Plant Physiol* 118(2): 651–659.

Wang TT, Ren ZJ, Liu ZQ, Feng X, Guo RQ, Li BG, Li LG, Jing HC. 2014. SbHKT1;4, a member of the high-affinity potassium transporter gene family from *Sorghum bicolor*, functions to maintain optimal $Na^+/K^+$ balance under $Na^+$ stress. *J Integr Plant Biol* 56(3): 315–332.

Wang W, Vinocur B, Altman A. 2003. Plant responses to drought, salinity and extreme temperatures: Towards genetic engineering for stress tolerance. *Planta* 218: 1–14.

Wang X, Guo X, Li Q, Tang Z, Kwak S, Ma D. 2011. Studies on salt tolerance of transgenic sweet potato which harbors two genes expressing CuZn superoxide dismutase and ascorbate peroxidase with the stress-inducible SWPA2 promoter. *Plant Gene Trait* 3(2): 6–12.

Wang X, Zhang W, Li W, Mishra G. 2007. Phospholipid signaling in plant response to drought and salt stress. In: Jenks E, Matthew A, Hasegawa E, Paul M, Jain E, Mohan S, editors, *Advances in Molecular Breeding toward Drought and Salt Tolerant Crops*. Dordrecht, the Netherlands: Springer, pp. 183–192.

Wang X. 1999. The role of phospholipase D in signaling cascade. *Plant Physiol* 120: 645–651.

Wang Y, Hu J, Qin G, Cui H, Wang Q. 2012. Salicylic acid analogues with biological activity may nduce chilling tolerance of maize (*Zea mays*) seeds. *Botany* 90: 845–855.

Wang YC, Qu QZ, Li HY, Wu YJ, Wnag C, Liu GF, Yang CP. 2010. Enhanced salt tolerance of transgenic poplar plants expressing a manganese superoxide dismutase from *Tamarix androssowii*. *Mol Biol Rep* 37(2): 1119–1124.

Waraich EA, Ahmad R, Halim A, Aziz T. 2012. Alleviation of temperature stress by nutrient management in crop plants: A review. *J Soil Sci Plant Nutr* 12: 221–244.

Wei-Feng XU, Wei-Ming SHI, Ueda A, Takabe T. 2008. Mechanism of salt tolerance in transgenic *Arabidopsis thaliana* carrying a peroxisomal ascorbate peroxidase gene from barley. *Pedosphere* 18: 486–495.

Wolbang CM, Chandler PM, Smith JJ, Ross JJ. 2004. Auxin from the developing inflorescence is required for the biosynthesis of active gibberellins in barley stems. *Plant Physiol* 134: 769–776.

Wu R, Garg A. 2003. Engineering rice plants with trehalose-producing genes improves tolerance to drought, salt and low temperature, available from: http://www.seedquest.com/News/releases/2003/march/5456.htm

Xu WF, Shi WM, Ueda A, Takabe T. 2008. Mechanisms of salt tolerance in transgenic *Arabidopsis thaliana* carrying a peroxisomal ascorbate peroxidase gene from Barley. *Pedosphere* 18: 486–495.

Xu Y, Hu W, Liu J, Zhang J, Jia C, Miao H, Xu B1, Jin Z. 2014. A banana aquaporin gene, *MaPIP1;1*, is involved in tolerance to drought and salt stresses. *BMC Plant Biol* 14: 59.

Yamaguchi T, Apse MP, Shi HZ, Blumwald E. 2003. Topological analysis of a plant vacuolar $Na^+/H^+$ antiporter reveals a luminal C terminus that regulates antiporter cation selectivity. *Proc Natl Acad Sci USA* 100: 12510–12515.

Yamchi A, RastgarJazii F, Mousavi A, Karkhane A A, Renu. 2007. Proline accumulation in transgenic tobacco as a result of expression of *Arabidopsis* 1-pyrroline-5-carboxylate synthetase (P5CS) during osmotic stress. *J Plant Biochem Biotechnol* 16: 9–15.

Yoshimura K, Miyao K, Gaber A, Takeda T, Kanaboshi H, Miyasaka H, Shigeoka S. 2004. Enhancement of stress tolerance in transgenic tobacco plants overexpressing *Chlamydomonas* glutathione peroxidase in chloroplasts or cytosol. *Plant J* 37: 21–33.

Yu NJ, Huang J, Wang ZN, Zhang JS, Chen SY. 2007. An $Na^+/H^+$ antiporter gene from wheat plays an important role in stress tolerance. *J Biosci* 32: 1153–1161.

Zhang C, Liu J, Zhang Y, Cai X, Gong P, Zhang J, Wang T, Li H, Ye Z. 2011. Overexpression of *SlGMEs* leads to ascorbate accumulation with enhanced oxidative stress, cold, and salt tolerance in tomato. *Plant Cell Rep* 30(3): 389–398.

Zhang X, Tang W, Liu J, Liu Y. 2014. Co-expression of rice *OsP5CS1* and *OsP5CS2* genes in transgenic tobacco resulted in elevated proline biosynthesis and enhanced abiotic stress tolerance. *Chin J Appl Environ Biol* 20(4): 717–722.

Zheng S, Pan T, Fan L, Qiu QS. 2013. A novel *AtKEA* gene family, homolog of bacterial $K^+/H^+$ antiporters, plays potential roles in $K^+$ homeostasis and osmotic adjustment in *Arabidopsis*. *PLoS One* 8(11): e81463.

Zhu JK. 2001. Plant salt tolerance. *Trends Plant Sci* 6(2): 66–71.

Zhu JK. 2002. Salt and drought stress signal transduction in plants. *Annu Rev Plant Biol* 53: 247–273.

Zhu JK. 2003. Regulation of ion homeostasis under salt stress. *Curr Opin Plant Biol* 6: 441–445.

CHAPTER NINETEEN

# Engineering Proline Metabolism for Enhanced Plant Salt Stress Tolerance

*Vinay Kumar, Varsha Shriram, Mohammad Anwar Hossain, and P.B. Kavi Kishor*

## CONTENTS

*Abstract.* Overaccumulation of proline has been reported in numerous plant species growing in extreme environments including high soil salinity. Proline is widely regarded as a multifunctional stress-regulated molecule. Strong correlations between proline accumulation and plant osmotic and salt stress tolerance have been documented in different species. Recently, an emphasis has also been given in dissecting its role in offsetting cellular imbalances and maintaining cellular homeostasis during abiotic stresses. Several attempts have been made to increase the accumulation of this versatile proteinogenic amino acid in plants by transferring genes associated with proline metabolism and for finding the correlation between accumulation and osmotic and salt tolerance in transgenic plants. In recent times, several attempts were made to overexpress various proline biosynthetic pathway genes such as $\Delta^1$-pyrroline-5-carboxylate synthetase (P5CS), pyrroline-5-carboxylate reductase (P5CR), and ornithine-$\delta$-aminotransferase (OAT) (involved in synthesis) and to downregulate proline dehydrogenase (ProDH) and pyrroline-5-carboxylate dehydrogenase (P5CDH) (associated with proline degradation) to enhance osmotic and salinity stress tolerance. This chapter focuses on the current understandings of various aspects and the roles proline play not only in imparting osmotic and salt stress tolerance to plants but also in developmental activities.

## 19.1 INTRODUCTION

High salinity of soil and irrigation waters is a serious threat to crop production worldwide. Since plants are sessile organisms, they need to regulate their growth and development in order to respond to saline environments (Peleg and Blumwald 2011). Salinity perceptions are translated into a cascade of alterations at morphological, physiological, biochemical, and molecular levels in plants. Consequently, to maintain integrity and survival, plants evolve various defense mechanisms in response to and to cope with the salinity stress. The responses to abiotic stresses including salinity are complex and multigenic and, therefore, difficult to control and engineer abiotic stress tolerance (Chamoli and Verma 2014). One of the most effective adaptive mechanisms is the stimulated biosynthesis and accumulation of compatible osmolytes or osmoprotectants. These chemically diverse, low-molecular-weight organic metabolites figure among the most fundamental solutes in living organisms ranging from bacteria to plants and animals and are collectively known as compatible osmolytes, as they get amassed in cytoplasm and remain nontoxic even at molar concentrations. They include amino acids (proline, glutamate, glutamine, and alanine) and their derivatives (ectoine and hydroxyectoine), quaternary amines (glycine betaine, polyamines, and dimethyl sulfonioproprionate), sugars (trehalose), and polyols including sugar alcohols (mannitol, sorbitol, pinitol, glycerol, and galactinol) (Khan et al. 2009; Jewell et al. 2010). These osmolytes perform a vast array of functions including being reactive oxygen species (ROS) scavengers, cell redox balancers, osmoprotectors or osmoticums, and stabilizers of cytosolic pH, proteins, enzymes, and membranes besides being a source of carbon and nitrogen during stress conditions and recovery phases.

Among these osmolytes, proline has a distinct place owing to a plethora of roles it plays for mitigating the deleterious effects of environmental stresses including high salinity in several plant species and therefore received maximum attention of plant scientists. Proline accumulation is documented to occur under drought (Choudhary et al. 2005), salinity (Kumar et al. 2007, 2008; Kumar and Khare 2014), low temperature (Naidu et al. 1991), high-light and UV irradiation (Salama et al. 2011), heavy metals (Sharma and Dietz 2006), and oxidative stress (Yang et al. 2009) besides in response to biotic stresses (Fabro et al. 2004; Haudecoeur

et al. 2009). Strong correlations between proline accumulation and plant osmotic and salt stress tolerance have been reported on various occasions and in different species. Consequently, several attempts have been made in recent years to either increase the biosynthesis or decrease the catabolism of this versatile proteinogenic amino acid in plants by transferring proline metabolic pathway genes with an ultimate aim of developing salt stress tolerant transgenics.

## 19.2 PROLINE: A MULTIFUNCTIONAL AMINO ACID AND SIGNALING MOLECULE

Proline is a versatile amino acid with an exceptional conformational rigidity and it fulfills diverse functions in plants (Szabados and Savoure 2010). Proline is essential for primary metabolism and plays crucial roles in cellular metabolism both as a component of proteins and as a free amino acid. As an amino acid, it is a structural component of proteins with significant contributions to protein folding, structure, and stability (Lehman et al. 2010; Funck et al. 2012). Its cyclic structure and a restricted conformational flexibility consequently lead to stabilization or destabilization of secondary structures of proteins.

Though proline accumulation is one of the most prominent adaptive responses in many plants to a wide range of biotic and abiotic stresses, an increase in proline content following stress injury is generally believed beneficial for plant cells (Verbruggen and Hermans 2008; Mattioli et al. 2009). However, proline also plays important roles under nonstressed conditions, including plant development, transition from vegetative to flowering stages, plant–microbe interactions, synthesis of aromatic compounds, and programmed cell death. The major roles proline plays are shown in Figure 19.1.

Changes in free proline content also occur during the development of plants growing in nonstressed conditions, especially in reproductive organs (reviewed by Mattioli et al. 2009). There is increasing evidence that suggests that proline has regulatory functions; it controls plant growth and acts as a signaling molecule (Szabados and Savoure 2010; Hayat et al. 2012). In a recent study, Wang et al. (2014) reported that proline plays a crucial role in regulating general protein synthesis and cell cycle in maize. Szekely et al. (2008) reported that adequate proline supply is essential for embryo and plant development. There are reports of large amount of proline being accumulated in pollen grains

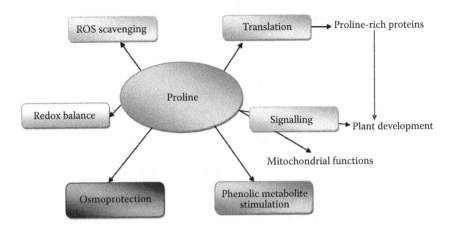

Figure 19.1. Multifunctional roles of proline in plants to mitigate deleterious effects of salinity stress as well as in protein synthesis and plant development.

(naturally desiccation-tolerant tissues) besides the fact that many flowers secrete proline-rich nectar (Lehman et al. 2010). High levels of proline in plant nectars is thought to work as attractant since several insect species prefer proline-rich nectars (Carter et al. 2006; Bertazzini et al. 2010). Chiang and Dandekar (1995) reported proline content in *Arabidopsis* reproductive tissues including florets, pollen, siliques, and seeds, as 26% of the total amino acid pool, which only accounted for 1%–3% in vegetative tissues. Even more striking variations were reported by Schwacke et al. (1999), where tomato flowers showed 60-fold higher proline content than other organs analyzed. Similarly, proline content has also been dependent on the age of the plant and of the flower and the position or part of the leaf (Verbruggen et al. 1993; Chiang and Dandekar 1995).

Proline metabolism has a significant role in biotic interactions of plants including pathogenesis and nodulation and is also suggested to influence programmed cell death in plants (Sharma and Verslues 2010; Szabados and Savoure 2010). Proline was also proposed to modulate the plant defense response to *Agrobacterium tumefaciens*. Proline accumulation has been reported in plant tumors, where it functions as a competitive antagonist of gamma-aminobutyric acid (GABA)-dependent plant defense, interfering with GABA-induced degradation of quorum sensing signal of *A. tumefaciens* (Haudecoeur et al. 2009). It indicated that proline acts as a natural antagonist of GABA signaling. Therefore, proline has a potential to promote *Agrobacterium* infection as well as horizontal transfer of the Ti plasmid. Proline has also been associated with rapid elongation of pollen tube (Schwacke et al. 1999), the elongation of

hairy roots (Trovato et al. 2001), and the elongation of primary roots of maize growing under low water potential (Spollen et al. 2008). All these studies indicate toward possible exploitation of proline by the plant cells in developmental programs involving rapid cell growth (Mattioli et al. 2009). Further, proline has also been reported to play a role in nodulation, another type of plant–microbe interaction, and nitrogen fixation (Soto et al. 1994; Curtis et al. 2004).

Proline has also shown its involvement in aroma synthesis in aromatic rice (Yoshihashi et al. 2002; Suprasanna et al. 2014). Supplementation of culture media with proline has resulted in increased aroma production in rice (Suprasanna et al. 2002). Proline has also found wide applications in cryopreservation of plant tissues, largely due to its abilities to act as a source of nitrogen and carbon and as an osmoprotectant without interfering with the normal physiology of cells (Suprasanna et al. 2014). Ogawa et al. (2012) used proline as a cryoprotectant for large-scale cryopreservation of *Arabidopsis*, *Daucus carota*, *Lotus japonicus*, *Nicotiana tabacum*, and *Oryza sativa* cells. All these studies clearly indicate that proline plays multifunctional roles in plant development and signaling, though the molecular mechanisms involved in them still need clear understanding.

## 19.3 PROLINE METABOLISM AND TRANSPORT

### 19.3.1 Proline Biosynthesis

Proline metabolism differs slightly between prokaryotes and eukaryotes (including plants). In bacteria, proline is synthesized from glutamate in three steps starting with the phosphorylation of glutamate by the enzyme γ-glutamyl kinase (γ-GK) using ATP followed by the reduction of the resultant γ-glutamyl phosphate (γ-GP) to glutamic-γ-semialdehyde (GSA) by glutamic-γ-semialdehyde dehydrogenase (GSADH) (Kavi Kishor et al. 2005; Lehmann et al. 2010). GSA then gets converted to Δ¹-pyrroline-5-carboxylate (P5C) by a reversible and spontaneous cyclization process and further gets reduced to ʟ-proline, catalyzed by Δ¹-pyrroline-5-carboxylate reductase (P5CR) using NADPH (Kavi Kishor et al. 2005; Lehmann et al. 2010). Two enzymes involved in the first steps of proline biosynthesis, γ-GK and GSADH, are conserved in bacteria (ProB and ProA) and yeast (PRO1 and PRO2), whereas in plants, both these activities are performed by a novel bifunctional enzyme (EC 2.7.2.11/1.2.1.41) called as Δ¹-pyrroline-5-carboxylate synthetase (P5CS) that requires both ATP and NADPH (Csonka 1989; Hu et al. 1992; Aral et al. 1996). *Vigna aconitifolia* P5CS has been found to have leucine zipper sequences in each of its enzymatic domains (Hu et al. 1992), which may function intramolecularly or intermolecularly to maintain tertiary structure of the enzyme or protein–protein interaction, respectively (Hu et al. 1992; Kavi Kishor et al. 2005; Suprasanna et al. 2014).

Proline biosynthesis takes place via two different precursors in plants, glutamate and/or ornithine. There are some excellent reviews for detailed proline metabolism pathways in plants including Kavi Kishor et al. (2005), Trovato et al. (2008), Lehman et al. (2010), and Suprasanna et al. (2014). The glutamate pathway is most prevalent under physiological as well as nitrogen shortage and osmotic stress conditions, whereas the ornithine pathway is thought to act particularly under supraoptimal nitrogen conditions (Delauney et al. 1993; Trovato et al. 2008). In the glutamate pathway, glutamate is converted to proline by two successive reductions catalyzed by P5CS and P5CR, respectively (Verbruggen and Hermans 2008). P5CS, which is one of most important and sought-after enzymes, catalyzes first the activation of glutamate by phosphorylation and second the reduction of the labile intermediate γ-GP into GSA, which is in equilibrium with P5C (Hu et al. 1992). The gene encoding P5CS was first isolated by means of functional complementation strategy where *V. aconitifolia* cDNA library was used for the transformation of proline-deficient *Escherichia coli* mutants (*proBA*) (Hu et al. 1992). The complementing clones harbored the 2417 bp long cDNA with

a single open reading frame that encoded a polypeptide of 73.2 kDa. The auxotrophic mutants (*proA* and *proB*) grew on proline-deficient media posttransformation with this *Vigna* cDNA. The bifunctional nature of this enzyme was confirmed when enzymatic assays were performed on the complementing moth bean protein expressed in *proBA E. coli* mutant, and it exhibited both γ-GK and GSADH activity in vitro (Hu et al. 1992). P5CS genes were later isolated from other plant species like *Arabidopsis* (Savoure et al. 1995) and rice (Igarashi et al. 1997) followed by two paralog P5CS genes—P5CS1 and P5CS2—described in different species (Fujita et al. 1998; Ginzberg et al. 1998; Armengaud et al. 2004). The activity of P5CS represents the rate-limiting step of proline biosynthesis in plants and is regulated through allosteric inhibition by the end-point proline (Hu et al. 1992; Sekine et al. 2007).

P5CR is the second and last enzyme in proline biosynthesis. Inhibitor studies on *Arabidopsis* cell cultures by Forlani et al. (2008) identified P5CR as an essential enzyme for growth. Sequence analysis has predicted a cytosolic localization for this enzyme, whereas cosedimentation of its activity with plastids was evidenced in pea and soybean (Rayapati et al. 1989; Szoke et al. 1992). However, recently, Funck et al. (2012) demonstrated an exclusive cytoplasmic localization of P5CR using P5CR-green fluorescent protein (GFP)-fluorescence analysis. Unlike P5CS, the P5CR is not considered as a rate-limiting factor in proline biosynthesis owing mainly to its lack of transcriptional regulation by stress (Szoke et al., 1992; Delauney and Verma 1993; Verbruggen et al., 1993; Hua et al., 1997; Sharma and Verslues 2010) and an observation by Szoke et al. (1992) that ectopic expression of the gene did not change proline content. However, there are some reports of plant species and stress-dependent differences since P5CR overexpression has been observed to affect proline levels in soybean (de Ronde et al. 2004). A transcriptional induction of P5CR was reported in *Arabidopsis* by heat and salinity stress but not dehydration by Hua et al. (1997). However, the increased transcript levels did not lead to increased P5CR enzyme activity or protein level. Rather, a 92 bp fragment of P5CR 5′UTR was shown to act as a regulator for controlling the mRNA stability and inhibit the protein translation of P5CR under heat or salinity stress, thus controlling protein level independently of transcript level (Hua et al. 2001). This work clearly demonstrates that abiotic stress regulation of *AtP5CR* is complex and involves the 5'UTR that acts at three levels, and also partly in opposing directions. Besides post-transcriptional regulation, its transcriptional levels are developmentally regulated with maximum expression in root tips, shoot meristem, guard cells, hydathodes, pollen grains, ovule, and developing seeds (Szoke et al. 1992; Hua et al. 1997).

The second pathway for proline production is via ornithine, and ornithine-δ-aminotransferase (OAT) catalyzes ornithine to P5C in plants (Delauney et al. 1993). While inhibition of this enzyme decreases proline content, its overexpression resulted in high proline accumulation (Yang and Kao 1999; Roosens et al. 2002). But OAT may not be associated with salt-stress-associated proline accumulation as shown by Funck et al. (2008) using the oat mutants of Arabidopsis. There is not much information on the developmental and stress-associated transcriptional regulation of this gene. Delauney et al. (1993) could not find its upregulation under stress conditions, but regulated by nitrogen content in the medium. Roosens et al. (1998) demonstrated that OAT is induced by salt stress in seedlings and in mature plants. Other reports also exist on the upregulation of OAT transcript under osmotic stress (Armengaud et al. 2004, Sharma and Verslues 2010). Recently, You et al. (2012) isolated an OAT from rice, characterized and overexpressed it in the same species. They demonstrated that both abscisic acid (ABA)-dependent and ABA-independent pathways contribute to the drought-induced expression of OsOAT. OsOAT is responsive to several of the phytohormone treatments. Further, OAT transcripts are induced by high salinity, heat, and submergence stresses (You et al. 2012). They confirmed that this OsOAT is a direct target of the stress-responsive NAC transcription factor SNAC2. Plants overexpressing it showed significantly increased proline content, tolerance to drought, and oxidative stress probably by increasing glutathione (GSH) peroxidase (GPX) activity. Thus, it appears that OsOAT confers stress tolerance by increasing the proline accumulation and ROS-scavenging capacity. Its clear role during development and stress and proline metabolism is still ambiguous, its role in abiotic stress tolerance needs to be clarified, and also not much is known about its posttranslational regulation.

## 19.3.2 Proline Catabolism

Proline catabolism is spatially separated from its biosynthetic pathway in plants and takes place in mitochondria. While proline synthesis occurs in chloroplasts and cytoplasm, its catabolism takes place in mitochondria, which takes place in two steps mediated by proline dehydrogenase (ProDH, also referred to as proline oxidase) and pyrroline-5-carboxylate dehydrogenase (P5CDH), two most important enzymes. In the first step, oxidation of proline to P5C takes place by ProDH using flavin adenine dinucleotide as a cofactor. In the second step, P5C gets converted back to glutamate by the enzyme P5CDH using $NAD^+$. ProDH is bound to the inner membrane of mitochondria. However, the enzyme kinetics of plant ProDH and post-translational regulation still remain unexplored (Verslues and Sharma 2010). The downregulation of ProDH under the conditions of low water potential, dehydration, and salinity has been reported on various occasions (Kiyosue et al. 1996; Verbruggen et al. 1996; Verslues et al. 2007; Sharma and Verslues 2010), besides its upregulation by stress release and exogenous proline (Kiyosue et al. 1996; Peng et al. 1996; Verbruggen et al, 1996; Yoshiba et al. 1997; Satoh et al. 2002; Sharma and Verslues 2010). The downregulation of ProDH expression during stress conditions is widely accepted as a control point that can promote proline accumulation under stress. Even though, there are contrasting reports of its upregulation during stress when high levels of proline were accumulating (Kaplan et al. 2007). As described earlier, P5CDH catalyzes the second step of proline oxidation, and this enzyme converts the intermediate P5C to glutamate. An interesting point to be noted here is that the plant ProDH and P5CDH differ from their bacterial counterparts, as the bacterial PutA protein incorporates activities of both these enzymes and produces proline without releasing P5C (Srivastava et al. 2010). On the other hand, the two activities are separate in plants and is therefore one of several lines of evidence suggesting a role for P5C, possibly in programmed cell death (Verslues and Sharma 2010). Both environmental signals and developmental factors regulate P5CDH in plants. Maximal expression of P5CDH has been reported in flowers (Deuschle et al. 2001). A low-water-potential-induced upregulation of P5CDH has been reported (Sharma and Verslues 2010), which is consistent with the publicly available microarray data (Genevestigator: www.genevestigator.com); similarly, P5CDH was also upregulated in response to exogenous proline (Deuschle et al. 2004).

### 19.3.3 Intra- and Interorgan Transport of Proline

Owing to the complex compartmentation of proline metabolism, where proline synthesis takes place in chloroplasts and cytosol and its catabolism in mitochondria, intracellular movement of proline is essential and important. The proline transport is regulated not only by endogenous but also by environmental signals in plants (Kavi Kishor et al. 2005). Proline not only accumulates in different plant parts but also translocates into and out of specific compartments through proline porters (Verslues and Sharma 2010). Transport assays using isolated wheat mitochondria have shown the existence of both a mitochondrial proline importer and a proline–glutamate exchanger (Di Martino et al. 2006). However, the genes encoding these activities have not been identified in plants.

Both developmental and stress-related studies have given evidence of a connection between the change in proline content and its transport (Lehman et al. 2010). Proline has been detected both in phloem and in xylem of several plant species (Bialczyk et al. 2004). There is considerable support for intercellular movement of proline, for instance, the high levels of proline in the phloem of drought-stressed plants (Lee et al. 2009). Verslues and Sharp (1999) attributed the proline deposition in the elongation zone of maize roots at low water potential as an increase in proline transport rather than its biosynthesis. Similarly, proline accumulation in maturing grapevine berries was reported not to be associated with increased P5CS expression, suggesting a contribution of transport processes (Stines et al. 1999).

Though the intracellular proline transporters (ProTs) are yet to be identified (Kavi Kishor and Sreenivasulu 2014), the patterns of intercellular transport transcription have been derived for a number of plant species (Ueda et al. 2001; Lehmann et al. 2011). The ProTs involved in its uptake across the plasma membrane have been identified in the amino acid transporter family (ATF) or amino acid/auxin permease (AAAP) family and in the amino acid polyamine choline family (Rentsch et al. 2007). The transporters that recognize proline have been identified in different subfamilies of ATF/AAAP family including the amino acid permease, the lysine histidine transporter, and the ProT family (Lehman et al. 2010). A combination of yeast complementation and sequence comparison was used to identify *Arabidopsis* ProT1, ProT2, and ProT3 (Rentsch et al. 1996; Grallath et al. 2005), and the authors have shown them to be plasma membrane localized. A salinity-stress-induced proline-specific transporter has been identified in barley root tip by Ueda et al. (2001), where authors suggested import of proline under stress. Long-distance transport of proline produced in photosynthetic tissues might be involved in buffering the cellular redox status and in its movement to the nonphotosynthetic tissues such as roots, where it may contribute to osmotic adjustment or be catabolized to support growth (Verslues and Sharma 2010). However, very little is known about the role of proline translocation besides the identification and characterization of intracellular proline transport systems, and therefore, more detailed analyses are required to reveal novel and interesting links between its metabolism and transport.

## 19.4 PROLINE ACCUMULATION AND SALT STRESS ALLEVIATION

A large body of data suggests that plants accumulate proline when they get exposed to a wide range of biotic and/or abiotic stresses including salinity as an adaptive response to these adverse conditions. However, interpretations of proline accumulation vary from its role as a useful adaptive response, helping organisms to withstand the effect of stress, to merely a consequence of stress-induced damage to the cells. To solve this debatable issue, there is need to undertake extensive studies to dissect the role of proline for combating stress tolerance and its accumulation under these circumstances. Nevertheless, a positive correlation between proline accumulation and salinity stress tolerance has been mentioned on numerous occasions, and it is a general belief that stress-induced proline overproduction is beneficial for the plant cell (Mattioli et al. 2009). Accordingly, measurement of proline accumulation is considered as an important criterion for determination of plant tolerance to salt stress and is therefore widely advocated for use as parameter of selection for salt stress tolerance (Sairam et al. 2002). Proline accumulation in stressed plant tissues is generally credited to (1) increased proline synthesis, (2) reduced degradation, (3) increased protein hydrolysis, and (4) decreased proline utilization

(Lin and Kao 2007). The level of proline accumulation varies between plant species and may even go to 100 times more than the nonstressed conditions reaching cytosol concentrations of 120–230 mM (Ashraf and Foolad 2007).

Proline plays multifunctional roles in plant stress tolerance as well as plant development, as represented in Figure 19.1. It is an excellent zwitterionic solute with high solubility (14 kg $L^{-1}$ water, Le Rudulier et al. 1984), which plays an important role in osmotic balancing and in increasing cellular osmolarity (turgor pressure) that provides turgor necessary for cell expansion under stress (Matysik et al. 2002; Yokoi et al. 2002; Sairam and Tyagi 2004). It has been proposed that it acts as a compatible osmolyte (Szabados and Savoure 2010; Kavi Kishor and Sreenivasulu 2014; Signorelli et al. 2014), and others hypothesized it might act as a nonenzymatic antioxidant (Chen and Dickman 2005). Several likely functions played by proline under salinity stress may be (1) osmotic adjustment, (2) protection of cellular structure during dehydration, (3) redox buffering, (4) storage and transfer of reductant (and nitrogen), (5) signaling molecule, and (6) ROS scavenging as suggested by Sarvesh et al. (1996) and Verslues and Sharma (2010).

An unfortunate consequence of all the abiotic stresses is the overproduction of ROS, especially singlet oxygen and free radicals, which are known to break DNA and destroy the functions of proteins and are responsible for lipid peroxidation. Interestingly, among various compatible solutes that accumulate under various abiotic stresses, proline is the only one that has been shown to protect plants against singlet oxygen and free radical–induced damages (Alia et al. 1991). Electronically, excited singlet oxygen is highly reactive and rapidly oxidizes amino acids, lipids, and DNA. Direct involvement of singlet oxygen in photobleaching of photosynthetic pigments, protein degradation, and protein cross-linking has been reported by Aro et al. (1993). Singlet oxygen also affects the structural integrity of photosystem II (PSII) membrane protein complex and destabilizes proteins and membrane. Alia et al. (2001) and Matysik et al. (2002) reported by using in vitro experiments that proline is an excellent quencher for singlet oxygen. However, recently, Signorelli et al. (2013) have shown that proline cannot quench $^1O_2$ in aqueous buffer, which presses the need to reconsider the likely role of proline in scavenging

of $^1O_2$ in plants under stress. Compared to singlet oxygen, ·OH radicals react much faster and therefore with less selectivity. Smirnoff and Cumbes (1989) for the first time have shown that proline was an effective hydroxyl radical scavenger. Very recently, Signorelli et al. (2014) have reported that proline acts as ·OH scavenger and its presence also avoided the ·OH-mediated enzyme inactivation. The authors have proposed a proline–proline cycle that could act in scavenging ·OH. In this cycle, proline captures a first ·OH by H-abstraction, followed by a second H-abstraction, which also captures another ·OH, yielding P5C, and the P5C is then recycled back to proline by the action of the P5CR/NADPH enzymatic system (Signorelli et al. 2014). In addition, binding of proline to redox-active metal ions can also protect the surrounding biological tissues from the damage by ·OH radicals as reported by Matysik et al. (2002). Owing to its action as singlet oxygen quencher and scavenger of OH· radicals, proline is able to stabilize proteins, DNA, and membranes (Alia et al. 1997; Iakobasshvili and Lapidot 1999; Sivakumar et al. 2000). Recently, Rejeb et al. (2014) reviewed the role of regulation of proline metabolism under ROS as well as its role as ROS scavenger and as a prooxidant. Accumulation of proline-rich proteins and particularly proline residues in protein are also reported to provide protection against oxidative stress (Matysik et al. 2002). In addition, proline also works as a substrate for mitochondrial respiration (Hasegawa et al. 2000) and regulation of cytosolic pH and NAD/NADH ratio and also acts as a source of carbon, nitrogen, and energy during recovery from stresses (Sairam and Tyagi 2004; Chinnusamy et al. 2005; Kavi Kishor et al. 2005). Further, there is also increasing evidence of proline performing regulatory functions besides acting as a signaling molecule.

## 19.5 IMPORTANCE OF PROLINE HOMEOSTASIS UNDER STRESS

Though proline accumulation is a common response to abiotic stress, its concentrations also go up significantly under normal physiological conditions in actively dividing cells and tissues going through senescence besides during pollen and seed maturation. But whether this occurs naturally in developmental episode or is induced by external stress needs to be given attention. Proline is known to act as a potent osmoprotectant and as a redox-buffering agent with antioxidant properties, and it may also prove to be toxic for certain tissues if it gets partially catabolized because it then induces an increase in P5C concentration, which in turn leads to apoptosis. Therefore, there is an increasing concern about the importance of proline homeostasis during development and transitional phases and stress conditions. Recently, Kavi Kishor and Sreenivasulu (2014) stressed the need to understand whether proline accumulation on its own is correlated with stress tolerance or if proline homeostasis is a more critical issue. Zhang et al. (2013) reported that proline acts as a signal of stress memory that stably gets carried to the next generation. Various authors have attributed enhanced proline content to the upregulation of P5CS coupled with downregulation of ProDH under stress conditions (Verslues et al. 2007; Sharma and Verslues 2010) and upregulation of the latter when the stress is relieved (Satoh et al. 2002). These observations revealed that the enhanced proline content arising from imposition of abiotic stresses is due not only to the activation of proline synthesis but also to the perturbation of proline homeostasis caused by the deactivation of proline catabolism that ultimately leads to the elevated levels of proline under stress (Kavi Kishor and Sreenivasulu 2014). The authors have also concluded that proline homeostasis is associated with the energy demand of young dividing cells and during the resumed growth post stress relief. Besides proline synthesis in stressed shoots and roots, it can also be transported via the phloem to the roots by ProTs as described earlier like ProT2 as phloem transporter, and thus high levels of proline accumulate under moisture stress (Lee et al. 2009). Therefore, the required proline homeostasis is perhaps attained by its porter-mediated long-distance transport. Additionally, it might also be achieved within the dividing cell via the effective coordination of proline channelling at the compartmental level (Kavi Kishor and Sreenivasulu 2014). Proline is synthesized in chloroplasts and the cytoplasm under stress conditions and subsequently gets transported to root mitochondria where it gets catabolized and generates NADPH, which presumably supports root growth (Lehmann et al. 2010; Szabados and Savoure 2010). In conclusion, proline homeostasis seems to be essential for normal plant growth and development under abiotic stresses including salinity and therefore needs to be maintained.

## 19.6 METABOLIC ENGINEERING OF PROLINE FOR ENHANCED SALT TOLERANCE

Owing to an array of roles proline plays during stress conditions, various attempts have been made to manipulate its metabolic pathways for enhanced production that in turn result in improved salt stress tolerance. Table 19.1 summarizes the examples of genetic modifications by transferring the genes associated with proline biosynthesis (P5CS1, P5CS2, and P5CR) and its degradation (ProDH). Though the proline transgenics showed better tolerance to a variety of stresses including salinity, drought, heat, cold, and heavy metals, only the transgenics generated for enhanced salinity stress tolerance has been discussed here and presented in Table 19.1. Initial attempts were made by Kavi Kishor et al. (1995) for overexpression of P5CS and generation of transgenic plants that overproduce proline. The authors overexpressed a moth bean P5CS gene in tobacco plants, and the resultant transgenics synthesized 10 to 18-fold more proline than control plants and were more salt tolerant. However, P5CS, being a rate-limiting enzyme in proline biosynthesis, is subjected to feedback inhibition by proline, and earlier reports suggested that proline accumulation in plants under stress might involve the loss of feedback regulation due to a conformational change in the P5CS protein. Therefore, Hong et al. (2000) removed the feedback inhibition using site-directed mutagenesis to replace the Phe residue at position 129 in *V. aconitifolia* P5CS with an Ala residue, and the resulting mutated enzyme (P5CS129A) was therefore no longer subjected to feedback inhibition. Consequently, the removal of this feedback inhibition resulted in two times more proline accumulation in P5CSF129A transgenics as compared to plants expressing wild-type P5CS, and this difference was further accentuated under 200 mM NaCl stress, and better protection of these plants from osmotic stress was observed (Hong et al. 2000). Later on, the same mutagenized form of this gene was used by our group (Kumar et al. 2010), and the resultant transgenic rice plants exhibited better growth performance and biomass production along with a lower level of lipid peroxidation under salt stress due to elevated proline accumulation. On the other hand, P5CS antisense *Arabidopsis* lines were hypersensitive to osmotic stress due to their impaired capacity to synthesize proline (Nanjo et al. 1999). Additionally, these antisense

## TABLE 19.1

*List of transgenics developed using proline biosynthetic pathway genes that conferred salinity and osmotic/oxidative stress tolerance.*

| Gene | Gene origin | Target plant | Enhanced tolerance and phenotype of transgenic plants | References |
|------|-------------|--------------|-------------------------------------------------------|------------|
| P5CS | *Vigna aconitifolia* | Tobacco | Increased biomass production and enhanced flower and seed development under salinity stress. | Kavi Kishor et al. (1995) |
| P5CS | *Vigna aconitifolia* | Wheat | Enhanced proline levels and salt tolerance. | Sawahel and Hassan (2002) |
| P5CS | *Vigna aconitifolia* | Carrot | Tolerance to salt stress. | Han and Hwang (2003) |
| P5CS | *Vigna aconitifolia* | Rice | Salt stress tolerance. | Anoop and Gupta (2003) |
| P5CS | *Vigna aconitifolia* | Rice | Significantly higher tolerance to drought and salt stress. | Su and Wu (2004) |
| P5CS | *Vigna aconitifolia* | *Larix leptoeuropaea* | Enhanced tolerance to cold, salinity, and freezing stresses. | Gleeson et al. (2005) |
| P5CS | *Arabidopsis* | Potato | Enhanced salt tolerance. | Hmida-Sayari et al. (2005) |
| P5CS | *Vigna aconitifolia* | Medicago | Proline essential for the maintenance of nitrogen-fixing activity under osmotic stress. | Verdoy et al. (2006) |
| P5CS | *Arabidopsis* | Tobacco | Increased proline accumulation and osmotic stress tolerance. | Yamchi et al. (2007) |
| P5CS | *Vigna aconitifolia* | Tobacco | Proline accumulation and increased enzyme activities under salinity-induced oxidative stress. | Razavizadeh and Ehsanpour (2009) |
| P5CS | *Vigna aconitifolia* | Chickpea | Enhanced proline accumulation and salt stress tolerance. | Ghanti et al. (2011) |
| P5CS | *Arabidopsis* | Tobacco | Increased proline content and salt tolerance. | Jazii et al. (2011) |
| P5CS | *Vigna aconitifolia* | Rice | Better salt stress tolerance at 200 mM NaCl. | Karthikeyan et al. (2011) |
| P5CS | Not mentioned | Olive | Enhanced tolerance to salinity stress. | Behelgardy et al.(2012) |
| P5CS | Not mentioned | *Nicotiana plumbaginifolia* | Enhanced chlorophyll content and better growth under salinity stress. | Mahboobeh and Akbar (2013) |
| P5CS | *Vigna aconitifolia* | Sugarcane | Enhanced proline content, lesser lipid peroxidation and oxidative damage, and enhanced salt tolerance. | Guerzoni et al. (2014) |
| P5CS1 | *Phaseolus vulgaris* | *Arabidopsis* | Increased plant tolerance to salt and drought stresses. | Chen et al. (2010) |
| P5CS1 | *Phaseolus vulgaris* | *Arabidopsis* | Increased tolerance to salt stress. | Chen et al. (2013) |
| P5CSF129A (mutated) | *Arabidopsis* | Tobacco | Higher proline content and enhanced resistance to osmotic stress. | Hong et al. (2000) |

(Continued)

TABLE 19.1 (*Continued*)
*List of transgenics developed using proline biosynthetic pathway genes that conferred salinity and osmotic/oxidative stress tolerance.*

| Gene | Gene origin | Target plant | Enhanced tolerance and phenotype of transgenic plants | References |
|------|-------------|--------------|-------------------------------------------------------|------------|
| P5CSF129A | *Vigna aconitifolia* | Rice | Enhanced proline, lower lipid peroxidation, and better growth performance under salt stress. | Kumar et al. (2010) |
| P5CSF129A | *Tobacco* | Tobacco | Mild proline elevation and better tolerance to abiotic stresses. | Dobra et al. (2010) |
| P5CSF129A | | Lily | Improved salt stress tolerance. | Li et al. (2013) |
| P5CSF129A | *Vigna aconitifolia* | Pigeon pea | Enhanced proline accumulation and salt tolerance. | Surekha et al. (2014) |
| P5CR | *Triticum aestivum* | *Arabidopsis* | Enhanced root growth under salt stress and decreased lipid peroxidation under salt, drought, and ABA stress. | Ma et al. (2008) |
| P5CR | *Sweet potato* | *Sweet potato* | Improved salt stress tolerance via integrating photosynthesis and activating the ROS-scavenging system. | Liu et al. (2014) |
| ProDH | *Arabidopsis* | Tobacco | Antisense plants showed increased proline content. | Kochetov et al. (2004) |
| ProDH | *Brassica oleracea italica* | Broccoli | Antisense *ProDH* resulted in low proline level. | Yang et al. (2010) |
| ProDH | *Arabidopsis* | Tobacco | Antisense suppression showed elevated proline level and tolerance to various abiotic stresses. | Ibragimova et al. (2012) |
| Glutamyl kinase, P5CS and osmotin gene | *Tobacco* | Tobacco | Transgenics showed better salt stress tolerance; however, transgenic lines containing osmotin showed better tolerance. | Sokhansanj et al. (2006) |
| OAT | *Arabidopsis* | *Nicotiana plumbaginifolia* | Overexpression increased proline levels and osmotic tolerance. | Roosens et al. (2002) |
| OAT | *Arabidopsis* | Rice | Increased proline content and improved yield under salt and drought stress. | Wu et al. (2003) |
| OAT | *Oryza sativa* | Rice | Increased proline content and tolerance to multiple abiotic stresses including oxidative stress. | You et al. (2012) |
| ProBA | *Bacillus subtilis* | *Arabidopsis* | Increased proline and enhanced osmotic tolerance. | Chen et al. (2007) |

lines also showed morphological abnormalities in epidermal and parenchymatous cells, which highlight the underlying role of proline as a major constituent of cell wall proteins (Nanjo et al. 1999). Szekely et al. (2008) reported reduced salinity stress tolerance in P5CS1 *Arabidopsis* insertion mutant, and the authors described a role in vivo for proline in ROS scavenging, a hypothesis first postulated by Smirnoff and Cumbes (1989). The manipulations with another major biosynthetic pathway gene P5CR has also been undertaken. Liu et al. (2014) observed improved salt stress tolerance of sweet potato by overexpressing P5CR gene via integrating photosynthesis and activating the ROS-scavenging system. Similarly, proline degradation has also been targeted for metabolic engineering. Yang et al. (2010) reported low proline levels in the antisense *ProDH* broccoli plants. On the other hand, antisense suppression in tobacco has shown elevated proline levels and tolerance to various abiotic stresses (Ibragimova et al. 2012).

Mani et al. (2002) reported that *ProDH* sense lines showed lower proline accumulation during osmotic stress, and these lines were more tolerant than their wild-type counterparts in the presence of exogenously supplied proline. In the second condition, the chromatography after radiolabelled proline supply showed increased glutamate degradation, suggesting proline to be a good source of energy during stress conditions.

## 19.7 EXOGENOUS APPLICATION OF PROLINE FOR ALLEVIATION OF SALT STRESS

Proline exogenously applied or produced endogenously under salt stress conditions has been found to modulate a variety of plant physiological and biochemical processes that are directly and indirectly related to salt stress adaptation reactions and stress-responsive gene expression (Hoque et al. 2008; Banu et al. 2010; Hossain et al. 2011; Nounjan et al. 2012; Shahbaz et al. 2013; Hasanuzzaman et al. 2014).

The protective role of proline was investigated in the desert plant *Pancratium maritimum* under salt stress conditions. Imposition of plants with higher salinity led to an inhibition of antioxidant enzyme activities. However, exogenous application of proline stabilizes the enzymes and induces enhanced expression of dehydrin proteins. The induction of dehydrins by proline indicates that it might be a component of regulatory processes that lead to the accumulation of dehydrins during salt stress (Khedr et al. 2003). Cuin and Shabala (2005) reported that the role of compatible solutes (proline or betaine) is not limited to conventional osmotic adjustment, but has some other regulatory functions such as maintenance of cytosolic $K^+$ homeostasis by preventing NaCl-induced $K^+$ leakage from the cell, a feature that may confer salt tolerance in many species. They found that exogenous application of proline or betaine significantly reduced the NaCl-induced $K^+$ efflux from barley roots in a dose–response manner and maintained cytosolic $K^+$ homeostasis by preventing NaCl-induced $K^+$ leakage from the cell, possibly through the enhanced activity of $H^+$-ATPase. Their finding suggests that compatible solutes played an important role in regulating ion fluxes across the plasma membrane and in modulating salt stress tolerance.

Naturally, high-proline-producing lines of niger (*Guizotia abyssinica*) exhibited higher antioxidative enzymes compared to low proline producers (Sarvesh et al. 1996). The same was later demonstrated by Hoque et al. (2007a) who studied the beneficial roles of exogenous proline against salt stress (NaCl)–induced growth reduction and on the activities of several ROS-detoxifying enzymes in cultured tobacco cells. Exogenous application of proline alleviated the salinity-induced growth reduction in tobacco (*N. tabacum*) BY-2 cells. Salt stress resulted in a profound decrease in the activities of enzymes superoxide dismutase (SOD), catalase (CAT), and peroxidase (POX), whereas addition of proline in the culture medium favorably modulates the CAT and POX activities under salt stress, suggesting that exogenous proline mitigates the damaging effects of salt stress through the modulation of antioxidant enzyme activities. Hoque et al. (2007b) studied the activities of enzymes involved in the ascorbic acid–glutathione (AsA–GSH) cycle and the levels of antioxidants in cultured tobacco cells treated with proline and exposed to salinity stress. Salinity stress led to a significant reduction in the level of nonenzymatic antioxidant contents such as AsA and reduced GSH. The activities of AsA–GSH cycle enzyme like ascorbate peroxidase (APX), monodehydroascorbic acid reductase (MDHAR), dehydroascorbate reductase (DHAR), and glutathione reductase (GR) were also found to decrease in response to salinity stress. Exogenous supplementation of proline was found to upregulate the activities of APX, DHAR, and GR. Their findings suggested that proline offers protection against salt stress by increasing the activities of enzymes involved in the antioxidant metabolism. Hoque et al. (2008) studied the regulatory roles of proline and betaine in modulating the ROS and methylglyoxal (MG) detoxification systems. Application of salt stress in tobacco BY-2-cultured cells resulted in a marked increase in protein oxidation, the reduced and oxidized glutathione (GSH/GSSG) ratio, and the activities of glutathione S-transferase (GST) and glyoxalase II (Gly II). Addition of proline or betaine in salt-stressed tobacco-cultured cells reduced the protein oxidation and increased the GSH content and the activities of GPX, GST, and Gly I. They concluded that proline acts to reduce salt-induced oxidative damage, as indicated by lower protein oxidation, through the activation of antioxidative and glyoxalase systems. In an another study, Sobahan et al. (2009) reported that addition of proline or betaine to saline medium suppressed the $Na^+$ uptake but increased the accumulation of $K^+$ that resulted to a high $K^+/Na^+$ ratio. These results suggest that exogenous proline and betaine suppressed $Na^+$-enhanced apoplastic flow to reduce $Na^+$ uptake in rice plants.

The beneficial roles of exogenous proline or betaine in inducing salinity tolerance in relation to GSH metabolism were studied by Hossain and Fujita (2010) in mung bean seedlings. The application of salinity stress resulted in a marked increase in GSH pool within 24 h. Among the GSH-utilizing and GSH-regenerating enzymes, upregulation of GPX, GST, and Gly II activities and GR and Gly I was observed after 48 h stress and 24 h treatment, respectively. Addition of proline or betaine resulted in increment of GSH pool, the GSH/GSSG ratio, and the activities of GST, GPX, GR, Gly I, and Gly II as compared to controls. Importantly, seedlings treated with salt stress alone resulted in elevated GSSG, $H_2O_2$, and MDA levels, indicating that salt stress enhances the production of ROS like $H_2O_2$ and reduces GSH with elevated MDA levels. Contrary to this, we further proved that even pretreatment of mung bean seedlings with exogenous proline and betaine improves salinity-induced oxidative stress tolerance. Imposition of salt stress (200 mM NaCl) resulted in higher oxidative damage as indicated by higher lipid peroxidation and $H_2O_2$ levels as compared to control plants due to inactivation or insufficient upregulation of ROS detoxification and glyoxalase pathway enzymes (APX, MDHAR, DHAR, CAT, GST, GPX, Gly I, and

Gly II). Importantly, betaine or proline pretreatment favorably modulated the activities of these enzymes and the GSH/GSSG ratio, rendering plants more tolerant to NaCl-induced oxidative stress. The biochemical mechanisms of proline- and betaine-induced salinity tolerance was investigated by Banu et al. (2010) in cultured tobacco cells. Salt stress resulted in an increased in $H_2O_2$, nitric oxide (NO), MG, and $O_2^{\cdot-}$ levels. Importantly, the levels of $H_2O_2$ and MG were found to decrease by proline- or betaine-supplemented salt-stressed seedlings. Exogenous proline was also found to increase GPX transcription. They suggest that proline mitigates late responses to salt stress and the production of $H_2O_2$ and MG. The reduction in $H_2O_2$ and MG level by proline could be crucial to reduce cellular damage and to improve salinity tolerance. Proline also inhibits apoptosis-like cell death.

Further, Yan et al. (2011) studied the beneficial effects of exogenously applied proline (0.2 mM) in two melon (*Cucumis melo* L.) cultivars (cv. Yuhuang and cv. Xuemei) under salt stress where biomass, chlorophyll content, photosynthetic parameters, ROS, and antioxidant enzymes activities were measured. Exogenous application of proline increased the fresh and dry weights of melon cultivars under NaCl stress; enhanced chlorophyll contents, net photosynthetic rate, and actual efficiency of photosystem II as well as enhanced the activities of SOD, POD, CAT, APX, DHAR, and GR in roots; and lowered superoxide anion radical, $H_2O_2$, and MDA levels. Proline treatment enhanced the salinity tolerance of both melon plants and alleviated the salt-mediated oxidative damage.

Although many studies have shown the beneficial effect of proline under stressful conditions, recently, it was also reported that the application of exogenous proline under nonstress conditions also improved plant growth performance and metabolic activity in two *Brassica juncea* species. Application of proline under natural conditions was found to increase plant growth, the rate of photosynthesis, and the activities of ROS-detoxifying enzymes when compared with the control seedlings (Wani et al. 2012). Recently, Nounjan et al. (2012) showed that exogenous application of proline (10 mM) to seedlings of Thai aromatic rice (cv. KDML105, salt sensitive) during salt stress and subsequent recovery induces upregulation of genes encoding the antioxidant enzymes Cu/ZnSOD, MnSOD, CytAPX, and CatC. Importantly, they also found an upregulation of proline biosynthesis genes (P5CS and P5CR) in response to exogenous proline application.

Foliar application of 10 mM proline during flowering stage resulted in increased leaf area and fruit yield in two tomato cultivars under salinity stress (Kahlaoui et al. 2013), and the mineral $Ca^{2+}$ was higher in different organs, while low accumulation of $Na^+$ occurred. However, $Cl^-$ was very low significantly in all tissues of plants of Rio Grande at the higher concentration of proline (20 mM) applied. Shahbaz et al. (2013) studied the effects of exogenous proline on growth, gas exchange characteristics, chlorophyll fluorescence, and mineral ion accumulation of two eggplant cultivars, namely L-888 and Round, grown under salinity stress (150 mM NaCl). The imposition of salt stress for 15 days resulted in reduced growth, net $CO_2$ assimilation rate ($A$), water use efficiency ($A/E$), efficiency of photosystem II (Fv/Fm), and shoot and root $K^+$ and $Ca^{2+}$ ions of both eggplant cultivars. Exogenous application of proline (10 and 20 mM) counteracted the adverse effects of salt stress on shoot fresh weight of both eggplant cultivars and $A/E$ ratio in cv. Round only.

Very recently, Hasanuzzaman et al. (2014) showed that proline or betaine improved salinity tolerance in both salt-sensitive and salt-tolerant rice cultivars. Salt stress significantly reduced the relative water and chlorophyll contents, whereas the oxidative parameters like lipid peroxidation and $H_2O_2$ content significantly increased. Importantly, exogenously applying proline or betaine improved salinity tolerance by modulating different physiological parameters as well as by enhancing ROS and MG detoxification systems.

The aforesaid examples clearly showed that apart from the multiple functions of proline, the incorporation or addition of proline exogenously can make plants more tolerant to salinity stress and salt-induced oxidative damage through increased internal proline accumulation and also by modulating different plant physiological processes including ROS and MG detoxification systems.

## 19.8 CONCLUSION AND OUTLOOK

Proline is an important osmoprotectant molecule that exhibits an array of roles during plant growth and development and under abiotic stress conditions to mitigate deleterious effects. Ever since the first report of production of transgenic plants using proline biosynthetic pathway gene to enhance salinity stress tolerance, many successful attempts have been made by the researchers to engineer salt-tolerant plants using the wild-type as well as the mutagenized version of proline metabolism genes. It is clear that proline has tremendous potential to be used as a target molecule for its optimal biosynthesis and subsequent enhanced abiotic stress tolerance via its metabolic engineering.

There is much discussion during the last 10–15 years about the multiple roles that proline plays to moderate the detrimental effects of abiotic stresses and to benefit the plants during stress and recovery phases. However, many of these proposed roles are yet to be validated at cellular or molecular levels, and more such studies should be undertaken to have a clear-cut understanding of various roles it plays in plants, particularly under adverse environmental conditions. This will also help in understanding further whether proline accumulation is itself a stress signal or it gets overproduced by plant cells to mitigate the effects of stresses. One more issue to be deciphered in detail is perhaps the proline homeostasis under normal and stressful conditions.

## ACKNOWLEDGMENTS

Vinay Kumar acknowledges the financial support from the Science and Engineering Research Board, Government of India, for a Young Scientist Project (Grant No.: SR/FT/LS-93/2011). The first author also acknowledges the use of facilities created at Modern College under Department of Science and Technology (Government of India)–Funds for Improvement of Science and Technology Infrastructure and Department of Biotechnology (Government of India) Star College Scheme. P. B. Kavikishor would like to thank Council for Scientific and Industrial Research, New Delhi, India, for awarding the Emeritus Scientist position.

## REFERENCES

Alia, P. Mohanty, and J. Matysik. 2001. Effect of proline on the production of singlet oxygen. *Amino Acids* 21:195–200.

Alia, P.P. Saradhi, and P. Mohanty. 1991. Proline enhances primary photochemical activities in isolated thylakoid membranes of *Brassica juncea* by arresting photoinhibitory damage. *Biochemistry and Biophysics Research Communications* 181:1238–1244.

Alia, P.P. Saradhi, and P. Mohanty. 1997. Involvement of proline in protecting thylakoid membranes against free radical-induced photodamage. *Journal of Photochemistry and Photobiology-B* 38:253–257.

Anoop, N. and A.K. Gupta. 2003. Transgenic indica rice cv IR-50 overexpressing *Vigna aconitifolia* delta-pyrroline-5-carboxylate synthetase cDNA shows tolerance to high salt. *Journal of Plant Biochemistry and Biotechnology* 12:109–116.

Aral, B., J.S. Schlenzig, G. Liu, and P. Kamoun. 1996. Database cloning human delta1-pyrroline-5-carboxylate synthetase (P5CS) cDNA: A bifunctional enzyme catalyzing the first two steps in proline biosynthesis. *C. R. Academy of Science III* 319:171–178.

Armengaud, P., L. Thiery, N. Buhot, G. Grenier-DeMarch, and A. Savouré. 2004. Transcriptional regulation of proline biosynthesis in *Medicago truncatula* reveals developmental and environmental specific features. *Physiologia Plantarum* 120:442–450.

Aro, E.-M., I. Virgin, and B. Anderson. 1993. Photoinhibition of photosystem II. Inactivation, protein damage and turnover. *Biochimica et Biophysica Acta* 1143:113–134.

Ashraf, M. and M.R. Foolad. 2007. Roles of glycine betaine and proline in improving plant abiotic stress tolerance. *Environmental and Experimental Botany* 59:206–216.

Banu, M.N.A., M.A. Hoque, M. Watamable-Sugimoto, M.A. Islam, M. Uraji, M. Matsuoka, Y. Nakamura, and Y. Murata. 2010. Proline and glycinebetaine ameliorated NaCl stress via scavenging of hydrogen peroxide and methylglyoxal but not superoxide or nitric oxide in tobacco cultured cells. *Bioscience, Biotechnology and Biochemistry* 74:2043–2049.

Behelgardy, M.F., N. Motamed, and F.R. Jazii. 2012. Expression of the P5CS gene in transgenic versus nontransgenic olive (*Olea europaea*) under salinity stress. *World Applied Sciences Journal* 18:580–583.

Bertazzini, M., P. Medrzycki, L. Bortolotti, L. Maistrello, and G. Forlani. 2010. Amino acid content and nectar choice by forager honeybees (*Apis mellifera* L.). *Amino Acids* 39:315–318.

Bialczyk, J., Z. Lechowski, and D. Dziga. 2004. Composition of the xylem sap of tomato seedlings cultivated on media with $HCO_3^-$ and nitrogen source as NO or $NH_4^+$. *Plant and Soil* 263:265–272.

Carter, C., S. Shafir, L. Yehonatan, R.G. Palmer, and R. Thornburg. 2006. A novel role for proline in plant floral nectars. *Die Naturwissenschaften* 93:72–79.

Chamoli, S. and A.K. Verma. 2014. Targeting of metabolic pathways for genetic engineering to combat abiotic stress tolerance in crop plants.

In *Approaches to Plant Stress and Their Management*, eds. Gaur P.K. and P. Sharma, pp. 23–37. Springer, New Delhi, India.

Chen, C.B. and M.B. Dickman. 2005. Proline suppresses apoptosis in the fungal pathogen *Colletotrichum trifolii*. *Proceedings of the National Academy of Sciences USA* 102:3459–3464.

Chen, J.B., J.W. Yang, Z.Y. Zhang, X.F. Feng, and S.M. Wang. 2013. Two P5CS genes from common bean exhibiting different tolerance to salt stress in transgenic *Arabidopsis*. *Journal of Genetics* 92:461–469.

Chen, J.B., L.Y. Zhao, and X.G. Mao. 2010. Response of PvP5CS1 transgenic *Arabidopsis* plants to drought and salt-stress. *Acta Agronomica Sinica* 36:147–153.

Chen, M., H. Wei, J. Cao, R. Liu, Y. Wang, and C. Zheng. 2007. Expression of *Bacillus subtilis proBA* genes and reduction of feedback inhibition of proline synthesis increases proline production and confers osmotolerance in transgenic *Arabidopsis*. *Journal of Biochemistry and Molecular Biology* 40:396–403.

Chiang, H.H. and A.M. Dandekar. 1995. Regulation of proline accumulation in *Arabidopsis thaliana* (L.) Heynh during development and in response to desiccation. *Plant, Cell and Environment* 18:1280–1290.

Chinnusamy, V., A. Jagendorf, and J.K. Zhu. 2005. Understanding and improving salt tolerance in plants. *Crop Science* 45:437–448.

Choudhary, N.L. R.K. Sairam, and A. Tyagi. 2005. Expression of delta1-pyrroline-5- carboxylate synthetase gene during drought in rice (*Oryza sativa* L.). *Indian Journal of Biochemistry and Biophysics* 42:366–370.

Csonka, L.N. 1989. Physiological and genetic responses of bacteria to osmotic stress. *Microbiology and Molecular Biology Reviews* 53:121–147.

Cuin, T.A. and S. Shabala. 2005. Exogenously supplied compatible solutes rapidly ameliorate NaCl-induced potassium efflux from barley roots. *Plant and Cell Physiology* 46:1924–1933.

Curtis, J., G. Shearer, and D.H. Kohl. 2004. Bacteroid proline catabolism affects N2 fixation rate of drought-stressed soybeans. *Plant Physiology* 136:3313–3318.

de Ronde, J.A., R.N. Laurie, T. Caetano, M.M. Greyling, and I. Kerepesi. 2004. Comparative study between transgenic and non-transgenic soybean lines proved transgenic lines to be more drought tolerant. *Euphytica* 138:123–132.

Delauney, A.J., C.A.A. Hu, P.B. Kavi Kishor, and D.P.S. Verma. 1993. Cloning of ornithine δ-aminotransferase cDNA from *Vigna aconitifolia* by transcomplementation in *Escherichia coli* and regulation of proline biosynthesis. *The Journal of Biological Chemistry* 268:18673–18678.

Delauney, A.J. and D.P.S. Verma. 1993. Proline biosynthesis and osmoregulation in plants. *The Plant Journal* 4:215–223.

Deuschle, K., D. Funck, G. Forlani, H. Stransky, A. Biehl, D. Leister, E. van der Graaff, R. Kunzee, and W.B. Frommer. 2004. The role of Δ¹-pyrroline-5-carboxylate dehydrogenase in proline degradation. *Plant Cell* 16:3413–3425.

Deuschle, K., D. Funck, H. Hellmann, K. Daschner, S. Binder, and W.B. Frommer. 2001. A nuclear gene encoding mitochondrial Δ¹-pyrroline-5-carboxylate dehydrogenase and its potential role in protection from proline toxicity. *The Plant Journal* 27:345–355.

Di Martino, C., R. Pizzuto, M.L. Pallotta, A. De Santis, and S. Passarella. 2006. Mitochondrial transport in proline catabolism in plants: The existence of two separate translocators in mitochondria isolated from durum wheat seedlings. *Planta* 223:1123–1133.

Dobra, J., V. Motyka, P. Dobrev, J. Malbeck, I.T. Prasil, D. Haisel, A. Gaudinova, M. Havlova, J. Gubis, and R. Vankova. 2010. Comparison of hormonal responses to heat, drought and combined stress in tobacco plants with elevated proline content. *Journal of Plant Physiology* 167:1360–1370.

Fabro, G., I. Kovacs, V. Pavet, L. Szabados, and M.E. Alvarez. 2004. Proline accumulation and gene activation are induced by plant-pathogen incompatible interactions in *Arabidopsis*. *Molecular Plant-Microbe Interactions* 17:343–350.

Forlani, G., A. Occhipinti, L. Berlicki, G. Dziedziola, A. Wieczorek, and P. Kafarski. 2008. Tailoring the structure of aminobisphosphonates to target plant P5C reductase. *Journal of Agricultural Food Chemistry* 56:3193–3199.

Fujita, T., A. Maggio, M. Garcia-Rios, R.A. Bressan, and L.N. Csonka. 1998. Comparative analysis of the regulation of expression and structures of two evolutionarily divergent genes for Δ¹-pyrroline-5-carboxylate synthetase from tomato. *Proceedings of the National Academy of Sciences USA* 118:661–674.

Funck, D., B. Stadelhofer, and W. Koch. 2008. Ornithine-delta-aminotransferase is essential for arginine catabolism but not for proline biosynthesis. *BMC Plant Biology* 8:40.

Funck, D., G. Winter, L. Baumgarten, and G. Forlani. 2012. Requirement of proline synthesis during Arabidopsis reproductive development. *BMC Plant Biology* 12:191.

Ghanti, S.K.K., K.G. Sujata, B.M.V. Kumar, N.N. Karba, K.J. Reddy, M.S. Rao, and P.B. Kavi Kishor. 2011. Heterologous expression of P5CS gene in chickpea enhances salt tolerance without affecting yield. *Biologia Plantarum* 55:634–640.

Ginzberg, I., H. Stein, Y. Kapulnik, L. Szabados, N. Strizhov, J. Schell, C. Koncz, and A. Zilberstein. 1998. Isolation and characterization of two different cDNAs of delta1-pyrroline-5-carboxylate synthase in alfalfa, transcriptionally induced upon salt stress. *Plant Molecular Biology* 38:755–764.

Gleeson, D., M.-A. Lelu-Walter, and M. Parkinson. 2005. Overproduction of proline in transgenic hybrid larch (*Larix × Leptoeuropaea* (Dengler)) cultures renders them tolerant to cold, salt and frost. *Molecular Breeding* 15:21–29.

Grallath, S., T. Weimar, A. Meyer, C. Gumy, M. Suter-Grotemeyer, J.M. Neuhaus, and D. Rentsch. 2005. The AtProT family. Compatible solute transporters with similar substrate specificity but differential expression patterns. *Plant Physiology* 137:117–126.

Guerzoni, J.T.S., N.G. Belintani, R.M.P. Moreira, A.A. Hoshimo, D.S. Domingues, J.C.B. Filho, and L.G.E. Vieira. 2014. Stress-induced Δ1-pyrroline-5-carboxylate synthetase (P5CS) gene confers tolerance to salt stress in transgenic sugarcane. *Acta Physiologiae Plantarum* doi: 10.1007/s11738–014–1579–8.

Han, K.H. and C.H. Hwang. 2003. Salt tolerance enhanced by transformation of a P5CS gene in carrot. *Journal of Plant Biotechnology* 5:149–153.

Hasanuzzaman, M., M.M. Alam, A. Rahman, M. Hasanuzzaman, K. Nahar, and M. Fujita. 2014. Exogenous proline and glycine betaine mediated up-regulation of antioxidant defense and glyoxalase systems provide better protection against salt-induced oxidative stress in two rice (*Oryza sativa* L.) varieties. *BioMed Research International*, article ID 757219.

Hasegawa, M., R. Bressan, J.K. Zhu, and H. Bhonert. 2000. Plant cellular and molecular responses to high salinity. *Annual Review of Plant Physiology* 51:463–499.

Haudecoeur, E., S. Planamente, A. Cirou, M. Tannières, B.J. Shelp, S. Moréra, and D. Faure. 2009. Proline antagonizes GABA-induced quenching of quorum-sensing in *Agrobacterium tumefaciens*. *Proceedings of the National Academy of Sciences of the United States of America* 106:14587–14592.

Hayat, S., Q. Hayat, M.N. Alyemeni, A.S. Wani, J. Pichtel, and A. Ahmad. 2012. Role of proline under changing environments: A review. *Plant Signaling and Behavior* 7:1–11.

Hmida-Sayari A., R. Gargouri-Bouzid, A. Bidani, L. Jaoua, A. Savoure, and S. Jaoua. 2005. Overexpression of Δ¹-pyrroline-5-carboxylate synthetase increases proline production and confers salt tolerance in transgenic potato plants. *Plant Science* 169:746–752.

Hong Z., K. Lakkineni, Z. Zhang, and D.P.S. Verma. 2000. Removal of feedback inhibition of pyrroline-5-carboxylate synthetase results in increased proline accumulation and protection of plants from osmotic stress. *Plant Physiology* 122:1129–1136.

Hoque, M.A., M.N.A. Banu, Y. Nakamura, Y. Shimoishi, and Y. Murata. 2008. Proline and glycinebetaine enhance antioxidant defense and methylglyoxal detoxification systems and reduce NaCl-induced damage in cultured tobacco cells. *Journal of Plant Physiology* 165:813–824.

Hoque, M.A., M.N.A. Banu, E. Okuma, K. Amako, Y. Nakamura, Y. Shimoishi, and Y. Murata. 2007b. Exogenous proline and glycinebetaine ingresses NaCl-induced ascorbate glutathione cycle enzyme activities and proline improves salt tolerance more than glycinebetaine in tobacco Bright Yellow-2 suspension-cultured cells. *Journal of Plant Physiology* 164:1457–1468.

Hoque, M.A., E. Okuma, M.N.A. Banu, Y. Nakamura, Y. Shimoishi, and Y. Murata. 2007a. Exogenous proline mitigates the detrimental effects of salt stress more than exogenous betaine by increasing antioxidant enzyme activities. *Journal of Plant Physiology* 164:553–561.

Hossain, M.A. and M. Fujita. 2010. Evidence for a role of exogenous glycinebetaine and proline in antioxidant defense and methylglyoxal detoxification systems in mungbean seedlings under salt stress. *Physiology and Molecular Biology of Plants* 16:19–29.

Hossain, M.A., M. Hasanuzzaman, and M. Fujita. 2011. Coordinate induction of antioxidant defense and glyoxalase system by exogenous proline and glycinebetaine is correlated with salt tolerance in mung bean. *Frontiers of Agriculture in China* 5:1–14.

Hu, C.A., A.J. Delauney, and D.P.S. Verma. 1992. A bifunctional D1-enzymepyrroline-5-carboxylate synthetase catalyzes the first two steps in proline biosynthesis in plants. *Proceedings of National Academy of Sciences of the United States of America* 89:9354–9358.

Hua, X.J., B. Van de Cotte, M.V. Montagu, and N. Verbruggen. 1997. Developmental regulation of pyrroline-5-carboxylate reductase gene expression in *Arabidopsis*. *Plant Physiology* 114:1215–1224.

Hua, X.J., B.V. Cotte, M.V. Montagu, and N. Verbruggen. 2001. The 5′ untranslated region of the At-P5R gene is involved in both transcriptional and post-transcriptional regulation. *The Plant Journal* 26:157–169.

Iakobashvili, R. and A. Lapidot. 1999. Low temperature cycled PCR protocol for Klenow fragment of DNA polymerase I in the presence of proline. *Nucleic Acids Research* 27:1566–1568.

Ibragimova, S.S., S.Y. Kolodyazhnaya, S.V. Gerasimova, and A.V. Kochetov. 2012. Partial suppression of gene encoding proline dehydrogenase enhances plant tolerance to various abiotic stresses. *Russian Journal of Plant Physiology* 59:88–96.

Igarashi, Y., Y. Yoshiba, Y. Sanada, and K. Yamaguchi-Shinozaki. 1997. Characterization of the gene for $\Delta^1$-pyrroline-5-carboxylate synthetase and correlation between the expression of the gene and salt tolerance in *Oryza sativa* L. *Plant Molecular Biology* 33:857–865.

Jazii, R.F., A. Yamchi, M. Hajirezaei, A.R. Abbasi, and A.A. Karkhane. 2011. Growth assessments of *Nicotiana tabaccum* cv. Xanthi transformed with *Arabidopsis thaliana* P5CS under salt stress. *African Journal of Biotechnology* 10:8539–8552.

Jewell, M.C., B.C. Campbell, and I.D. Godwin. 2010. Transgenic plants for abiotic stress resistance. In *Transgenic Crop Plants*, eds. Kole, C.C.H. Michler, A.G. Abbott, and T.C. Hall, pp. 67–132. Springer-Verlag, Berlin/Heidelberg, Germany.

Kahlaoui, B., M. Hachicha, J. Teixeira, E. Misle, F. Fidalgo, and B. Hanchi. 2013. Response of two tomato cultivars to field-applied proline and salt stress. *Journal of Stress Physiology and Biochemistry* 9:357–365.

Kaplan, F., J. Kopka, D.Y. Sung, W. Zhao, M. Popp, R. Porat, and C.L. Guy. 2007. Transcript and metabolite profiling during cold acclimation of *Arabidopsis* reveals an intricate relationship of cold-regulated gene expression with modifications in metabolite content. *The Plant Journal* 50:967–981.

Karthikeyan, A., S.K. Pandian, and M. Ramesh. 2011. Transgenic indica rice cv. ADT 43 expressing a $\Delta^1$-pyrroline-5-carboxylate synthetase (P5CS) gene from *Vigna aconitifolia* demonstrates salt tolerance. *Plant Cell Tissue and Organ Culture* 107:383–395.

Kavi Kishor, P.B., Z. Hong, G. Miao, C. Hu, and D.P.S. Verma. 1995. Over expression of $\Delta^1$-pyrroline-5-carboxylate synthetase increases proline overproduction and confers osmotolerance in transgenic plants. *Plant Physiology* 108:1387–1394.

Kavi Kishor, P.B. and N. Sreenivasulu. 2014. Is proline accumulation per se correlated with stress tolerance or is proline homeostasis a more critical issue? *Plant Cell and Environment* 37:300–311.

Kavi Kishor, P.B., S. Sangam, R.N. Amrutha, P. Sri Laxmi, K.R. Naidu, K.R.S.S. Rao, S. Rao, K.J. Reddy, P. Theriappan, and N. Sreenivasulu. 2005. Regulation of proline biosynthesis, degradation, uptake and transport in higher plants: Its implications in plant growth and abiotic stress tolerance. *Current Science* 88:424–438.

Khan, M.A., M.U. Shirazi, M.A. Khan, S.M. Mujtaba, E. Islam, and S. Mumtaz. 2009. Role of proline, K/Na ratio and chlorophyll content in salt tolerance of wheat (Triticum aestivum L.). Pakistan Journal of Botany 41:633–638.

Khedr, A.H.A., M.A. Abbas, A.A.A. Wahid, W.P. Quick, and G.M. Abogadallah. 2003. Proline induces the expression of salt-stress-responsive proteins and may improve the adaptation of Pancratium maritimum L. to salt-stress. Journal of Experimental Botany 54:2553–2562.

Kiyosue, T., Y. Yoshiba, K. Yamaguchi-Shinozaki K., and K. Shinozaki. 1996. A nuclear gene encoding mitochondrial proline dehydrogenase, an enzyme involved in proline metabolism, is upregulated by proline but downregulated by dehydration in Arabidopsis. Plant Cell 8:1323–1335.

Kochetov, A.V., S.E. Titov, Y.S. Kolodyazhnaya, M.L. Komarova, V.S. Kovel, N.N. Makarova, Y.Y. Ilyinskyi, E.A. Trifonova, and V.K. Shummy. 2004. Tobacco transformants bearing antisense suppressor of proline dehydrogenase gene are characterized by higher proline content and cytoplasm osmotic pressure. Russian Journal of Genetics 40:216–218.

Kumar, V. and T. Khare. 2014. Individual and additive effects of $Na^+$ and $Cl^-$ ions on rice under salinity stress. Archives of Agronomy and Soil Science doi: 10.1080/03650340.2014.936400.

Kumar, V., V. Shriram, N. Jawali, and M.G. Shitole. 2007. Differential response of indica rice genotypes to NaCl stress in relation to physiological and biochemical parameters. Archives of Agronomy and Soil Science 53:581–592.

Kumar, V., V. Shriram, P.B. Kavi Kishor, N. Jawali, and M.G. Shitole. 2010. Enhanced proline accumulation and salt stress tolerance of transgenic indica rice by over expressing P5CSF129A gene. Plant Biotechnology Reports 4:37–48.

Kumar, V., V. Shriram, T.D. Nikam, N. Jawali, and M.G. Shitole. 2008. Sodium chloride induced changes in mineral elements in indica rice cultivars differing in salt tolerance. Journal of Plant Nutrition 31:1999–2017.

Le Rudulier, D., A.R. Strom, A.M. Dandekar, L.T. Smith, and R.C. Valentine. 1984. Molecular biology of osmoregulation. Science 224:1064–1068.

Lee, B.R., Y.L. Jin, J.C. Avice, J.B. Cliquet, A. Ourry, and T.H. Kim. 2009. Increased proline loading to phloem and its effects on nitrogen uptake and assimilation in water-stressed white clover (Trifolium repens). New Phytologist 182:654–663.

Lehmann, S., D. Funck, L. Szabados, and D. Rentsch. 2010. Proline metabolism and transport in plant development. Amino Acids 39:949–962.

Lehmann, S., C. Gumy, E. Blatter, S. Boeffel, W. Fricke, and D. Rentsch. 2011. In planta function of compatible solute transporters of the AtProT family. Journal of Experimental Botany 62:787–796.

Li, Sh, Du, Y.-P., Wu, Zh.-Y., Huang, C.-L., Zhang, X.-H. Wang, Zh.-X., and G.-X. Jia. 2013. Excision of a selectable marker in transgenic lily (Sorbonne) using the Cre/loxP DNA excision system. Canadian Journal of Plant Science 93:903–912.

Lin, Y.C. and C.H. Kao. 2007. Proline accumulation induced by excess nickel in detached rice leaves. Biologia Plantarum 51:351–354.

Liu D., S. He, H. Zhai, L. Wang, Y. Zhao, B. Wang, R. Li, and Q. Liu. 2014. Overexpression of IbP5CR enhances salt tolerance in transgenic sweet potato. Plant Cell, Tissue and Organ Culture 117:1–16.

Ma, L., E. Zhou, L. Gao, X. Mao, R. Zhou, and J. Jia. 2008. Isolation, expression analysis and chromosomal location of P5CR gene in common wheat (Triticum aestivum L.). South African Journal of Botany 74:705–712.

Mahboobeh, R. and E.A. Akbar. 2013. Effect of salinity on growth, chlorophyll, carbohydrate and protein contents of transgenic Nicotiana plumbaginifolia over expressing P5CS gene. E3 Journal of Environmental Research and Management 4:163–170.

Mani, S., V.B. Cotte, M. Van Montagu, and N. Verbruggen. 2002. Altered levels of proline dehydrogenase cause hypersensitivity to proline and its analogs in Arabidopsis. Plant Physiology 128:73–83.

Mattioli, R., P. Costantino, and M. Trovato. 2009. Proline accumulation in plants: Not only stress. Plant Signaling and Behavior 4:1016–1018.

Matysik, J., A.B. Bhalu, and P. Mohanty. 2002. Molecular mechanisms of quenching of reactive oxygen species by proline under stress in plants. Current Science 82:525–532.

Naidu, B.P., L.G. Paleg, D. Aspinall, A.C. Jennings, and G.P. Jones. 1991. Amino acid and glycine betaine accumulation in cold-stressed wheat seedlings. Phytochemistry 30:407–409.

Nanjo, T., M. Kobayashi, Y. Yoshiba, Y. Sanada, K. Wada, H. Tsukaya, Y. Kakubari, K. Yamagushi-Shinozaki, and K. Shinozaki. 1999. Biological functions of proline in morphogenesis and osmotolerance revealed in antisense transgenic Arabidopsis thaliana. The Plant Journal 18:185–193.

Nounjan, N., P.T. Nghia, and P. Theerakulpisut. 2012. Exogenous proline and trehalose promote recovery of rice seedlings from salt-stress and differentially modulate antioxidant enzymes and expression of related genes. Journal of Plant Physiology 169:596–604.

Ogawa, Y., N. Sakurai, A. Oikawa, K. Kai, Y. Morishita, K. Mori, K. Moriya et al. 2012. High-throughput cryopreservation of plant cell cultures for functional genomics. *Plant and Cell Physiology* 53:943–952.

Peleg, Z. and A.K. Blumwald. 2011. Hormone balance and abiotic stress tolerance in crop plants. *Current opinion in Plant Biology* 14:290–295.

Peng, Z., Q. Lu, and D.P.S. Verma. 1996. Reciprocal regulation of $\Delta^1$-pyrroline-5-carboxylate synthetase and proline dehydrogenase genes controls proline levels during and after osmotic stress in plants. *Molecular and General Genetics* 253:334–341.

Rayapati, P.J., C.R. Stewart, and E. Hack. 1989. Pyrroline-5-carboxylate reductase is in pea (*Pisum sativum* L.) leaf chloroplasts. *Plant Physiology* 91:581–586.

Razavizadeh, R. and A.A. Ehsanpour. 2009. Effects of salt stress on proline content, expression of $\Delta^1$-pyrroline-5-carboxylate synthetase, and activities of catalase and ascorbate peroxidase in transgenic tobacco plants. *Biology Letters* 46:63–75.

Rejeb, K.B., C. Abdelly, and A. Savouré. 2014. How reactive oxygen species and proline face stress together. *Plant Physiology and Biochemistry* 80:278–284.

Rentsch, D., B. Hirner, E. Schmelzer, and W.B. Frommer. 1996. Salt stress induced proline transporters and salt stress-repressed broad specificity amino acid permeases identified by suppression of a yeast amino acid permease-targeting mutant. *The Plant Cell* 8:1437–1446.

Rentsch, D., S. Schmidt, and M. Tegeder. 2007. Transporters for uptake and allocation of organic nitrogen compounds in plants. *FEBS Letters* 581:2281–2289.

Roosens, N.H., F.A. Bitar, K. Loenders, G. Angenon, and M. Jacobs. 2002. Overexpression of ornithine δ-aminotransferase increases proline biosynthesis and confers osmotolerance in transgenic plants. *Molecular Breeding* 9:73–80.

Roosens, N.H.C.J., T.T. Thu, H.M. Iskandar, and M. Jacobs. 1998. Isolation of ornithine-d-aminotransferase cDNA and effect of salt stress on its expression in *Arabidopsis thaliana*. *Plant Physiology* 117:263–271.

Sairam, R.K., K.V. Rao, and G.C. Srivastava. 2002. Differential response of wheat genotypes to long term salinity stress in relation to oxidative stress, antioxidant activity and osmolytes concentration. *Plant Science* 163:1037–1046.

Sairam, R.K. and A. Tyagi. 2004. Physiology and molecular biology of salinity stress tolerance in plants. *Current Science* 86:407–421.

Salama, H.M.H., A.A.A. Watban, and A.T. Al-Fughom. 2011. Effect of ultraviolet radiation on chlorophyll, carotenoid, protein and proline contents of some annual desert plants. *Saudi Journal of Biological Sciences* 18:79–86.

Sarvesh, A., M. Anuradha, T. Pulliah, T.P. Reddy, and P.B. Kavi Kishor. 1996. Salt stress and antioxidant response in high and low proline producing cultivars of niger, *Guizotia abyssinica* (L.F) Cass. *Indian Journal of Experimental Biology* 34:252–256.

Satoh, R., K. Nakashima, M. Seki, K. Shinozaki, and K. Yamaguchi-Shinozaki. 2002. ACTCAT, a novel cis-acting element for proline- and hypoosmolarity-responsive expression of the *ProDH* gene encoding proline dehydrogenase in *Arabidopsis*. *Plant Physiology* 130:709–719.

Savouré, A., S. Jaoua, X.J. Hua, W. Ardiles, M. Van Montagu, and N. Verbruggen. 1995. Isolation and characterization, and chromosomal location of a gene encoding the $\Delta^1$-pyrroline-5-carboxylate synthetase in *Arabidopsis*. *FEBS Letters* 372:13–19.

Sawahel, W.A. and A.H. Hassan. 2002. Generation of transgenic wheat plants producing high levels of the osmoprotectant proline. *Biotechnology Letters* 24:721–725.

Schwacke, R., S. Grallath, K.E. Breitkreuz, E. Stransky, H. Stransky, W.B. Frommer, and D. Rentsch. 1999. LeProT1, a transporter for proline, glycine betaine, and gamma-amino butyric acid in tomato pollen. *Plant Cell* 11:377–392.

Sekine, T., A. Kawaguchi, Y. Hamano, and H. Takagi. 2007. Desensitization of feedback inhibition of the *Saccharomyces cerevisiae* γ-glutamyl kinase enhances proline accumulation and freezing tolerance. *Applied and Environmental Microbiology* 73:4011–4019.

Shahbaz, M., Z. Mushtaq, F. Andaz, and A. Masood. 2013. Does proline application ameliorate adverse effects of salt stress on growth, ions and photosynthetic ability of eggplant (*Solanum melongena* L.)? *Scientia Horticulture* 164:507–511.

Sharma, S. and P.E. Verslues. 2010. Mechanisms independent of abscisic acid (ABA) or proline feedback have a predominant role in transcriptional regulation of proline metabolism during low water potential and stress recovery. *Plant, Cell and Environment* 33:1838–1851.

Sharma, S.S. and K.J. Dietz. 2006. The significance of amino acids and amino acid-derived molecules in plant responses and adaptation to heavy metal stress. *Journal of Experimental Botany* 57:711–726.

Signorelli, S., J.B. Arellano, T.B. Melø, O. Borsani, and J. Monza. 2013. Proline does not quench singlet oxygen: Evidence to reconsider its protective role in plants. *Plant Physiology and Biochemistry* 64:80–83.

Signorelli, S., E.L. Coitiño, O. Borsani, and J. Monza. 2014. Molecular mechanisms for the reaction between ·OH radicals and proline: Insights on the role as reactive oxygen species scavenger in plant stress. *The Journal of Physical Chemistry-B* doi: 10.1021/jp407773u.

Sivakumar, P., P. Sharmila, and P.P. Saradhi. 2000. Proline alleviates salt stress induced enhancement in Rubisco-1,5-bisphosphate oxygenase activity. *Biochemistry and Biophysics Research Communications* 279:512–515.

Smirnoff, N. and Q.J. Cumbes. 1989. Hydroxyl radical scavenging activity of compatible solutes. *Phytochemistry* 28:1057–1060.

Sobahan, M.A., C.R. Arias, E. Okuma, Y. Shimoishi, Y. Nakamura, Y. Hirai, I.C. Mori, and Y. Murata.2009. Exogenous proline and glycinebetaine suppress apoplastic flow to reduce Na+ uptake in rice seedlings. *Bioscience, Biotechnology and Biochemistry* 73:2037–2042.

Sokhansanj, A., S.A.S. Noori, and V. Niknam. 2006. Comparison of bacterial and plant genes participating in proline biosynthesis with *Osmotin* gene, with respect to enhancing salinity tolerance of transgenic tobacco plants. *Russian Journal of Plant Physiology* 53:110–115.

Soto, M.J., A. Zorzano, F.M. Garciarodriguez, J. Mercadoblanco, I.M. Lopezlara, J. Olivares, and N. Toro. 1994. Identification of a novel *Rhizobium meliloti* nodulation efficiency NFE gene homolog of *Agrobacterium* ornithine cyclodeaminase. *Molecular Plant-Microbe Interaction* 7:703–707.

Spollen, W.G., W. Tao, B. Valliyodan, K. Chen, L.G. Hejlek, J.J. Kim, M.E. LeNoble et al. 2008. Spatial distribution of transcript changes in the maize primary root elongation zone at low water potential. *BMC Plant Biology* 8:1–32.

Srivastava, D., J.P. Schuermann, T.A. White, N. Krishnan, N. Sanyal, G.L. Hura, A.M. Tan, M.T. Henzl, D.F. Becker, and J.J. Tanner. 2010. Crystal structure of the bifunctional proline utilization A flavoenzyme from *Bradyrhizobium japonicum*. *Proceedings of the National Academy of Sciences of the United States of America* 107:2878–2883.

Stines, A.P., D.J. Naylor, P.B. Hoj, and R. van Heeswijck. 1999. Proline accumulation in developing grapevine fruit occurs independently of changes in the levels of $\Delta^1$-pyrroline-5-carboxylate synthetase mRNA or protein. *Plant Physiology* 120:923–931.

Su, J. and R. Wu. 2004. Stress-inducible synthesis of proline in transgenic rice confers faster growth under stress conditions than that with constitutive synthesis. *Plant Science* 166:941–948.

Suprasanna, P., G. Bharati, T.R. Ganapathi, and V.A. Bapat. 2002. Aroma in rice: Effects of proline supplementation and immobilization of callus cultures. *Rice Genetics Newsletter* 19:9–11.

Suprasanna, P., A.N. Rai, P. HimaKumari, S.A. Kumar, and P.B. Kavi Kishor. 2014. Modulation of proline: Implications in plant stress tolerance and development. In *Plant Adaptation to Environmental Change*, eds. Anjum, N.A., S.S. Gill, and R. Gill, pp. 68–96. CAB International.

Surekha, C., K.N. Kumari, L.V. Aruna, G. Suneetha, A. Arundhati, and P.B. Kavi Kishor. 2014. Expression of the *Vigna aconitifolia* P5CSF129A gene in transgenic pigeonpea enhances proline accumulation and salt tolerance. *Plant Cell Tissue and Organ Culture* 116:27–36.

Szabados, L. and A. Savoure. 2010. Proline: A multifunctional amino acid. *Trends in Plant Science* 15:89–97.

Szekely, G., E. Abraham, A. Cseplo, G. Rigo, L. Zsigmond, J. Csiszar, F. Ayaydin et al. 2008. Duplicated P5CS genes of *Arabidopsis* play distinct roles in stress regulation and developmental control of proline biosynthesis. *The Plant Journal* 53:11–28.

Szoke, A., G.H. Miao, Z. Hong, and D.P.S. Verma. 1992. Subcellular location of δ1-pyrroline-5-carboxylate reductase in root/nodule and leaf of soybean. *Plant Physiology* 99:1642–1649.

Trovato, M., B. Maras, F. Linhares, and P. Costantino. 2001. The plant oncogene rolD encodes a functional ornithine cyclodeaminase. *Proceedings of the National Academy of Sciences of the United States of America* 98:13449–13453.

Trovato M., R. Mattioli, and P. Costantino. 2008. Multiple roles of proline in plant stress tolerance and development. *Rendiconti Lincei* 19:325–346.

Ueda, A., W. Shi, K. Sanmiya, M. Shono, and T. Takabe. 2001. Functional analysis of salt-inducible proline transporter of barley roots. *Plant and Cell Physiology* 42:1282–1289.

Verbruggen, N. and C. Hermans. 2008. Proline accumulation in plants: A review. *Amino Acids* 35:753–759.

Verbruggen, N., X.J. Hua, M. May, and M. Van Montagu. 1996. Environmental and developmental signals modulate proline homeostasis: Evidence for a negative transcriptional regulator. *Proceedings of the National Academy of Sciences of the United States of America* 93:8787–8791.

Verbruggen, N., R. Villarroel, and M.V. Montagu. 1993. Osmoregulation of a pyrroline-5-carboxylate reductase gene in *Arabidopsis thaliana*. *Plant Physiology* 103:771–781.

Verdoy, D., T.C. de la Pena, F.J. Redondo, M.M. Lucas, and J.J. Pueyo. 2006. Transgenic *Medicago truncatula* plants that accumulate proline display nitrogen-fixing activity with enhanced tolerance to osmotic stress. *Plant, Cell and Environment* 29:1913–1923.

Verslues, P.E., Y.S. Kim, and J.K. Zhu. 2007. Altered ABA, proline and hydrogen peroxide in an *Arabidopsis* glutamate: Glyoxylate aminotransferase mutant. *Plant Molecular Biology* 64:205–217.

Verslues, P.E. and Sharma, S. 2010. Proline metabolism and its implications for plant-environment interaction. *The Arabidopsis Book American Society of Plant Biologists* 8, e014010.1199/tab.0140.

Verslues, P.E. and R.E. Sharp. 1999. Proline accumulation in maize (*Zea mays* L.) primary roots at low water potentials. II. Metabolic source of increased proline deposition in the elongation zone. *Plant Physiology* 119:1349–1360.

Wang, G., J. Zhang, G. Wang, X. Fan, X. Sun, H. Qin, N. Xu et al. 2014. Proline responding1 plays a critical role in regulating general protein synthesis and the cell cycle in maize. *Plant Cell* 26: 2582–2600.

Wani, A.S., M. Irfan, S. Hayat, and A. Ahmad. 2012. Response of two mustard (*Brassica juncea* L.) cultivars differing in photosynthetic capacity subjected to proline. *Protoplasma* 249:75–87.

Wu, L.Q., Z. Fan, L. Guo, Y. Li, W. Zhang, L.J. Qu, and Z. Chen. 2003. Overexpression of an *Arabidopsis* δ-OAT gene enhances salt and drought tolerance in transgenic rice. *Chinese Science Bulletin* 48:2594–2600.

Yamchi A., F.R. Jazii, A. Mousav, A.A. Karkhane, and Re nu. 2007. Proline accumulation in transgenic tobacco as a result of expression of *Arabidopsis* $\Delta^1$-pyrroline-5-carboxylate synthetase (p5cs) during osmotic stress. *Journal of Plant Biochemistry and Biotechnology* 16:9–15.

Yan, Z., S. Guo, S. Shu, J. Jun, and T. Tezuka. 2011. Effects of proline on photosynthesis, root reactive oxygen species (ROS) metabolism in two melon cultivars (*Cucumis melo* L.) under NaCl stress. *African Journal of Biotechnology* 10:18381–18390.

Yang, C.W. and C.H. Kao. 1999. Importance of ornithine-δ-aminotransferase to proline accumulation caused by water stress in detached rice leaves. *Plant Growth Regulation* 27:191–194.

Yang, P., L.L. Wen, C. Zhao, B. Zhao, and G. Yangdong. 2010. Cloning and functional identification of *ProDH* gene from broccoli. *Journal of Guangxi Agricultural and Biological Science* 29:206–214.

Yang, S.L., S.S. Lan, and M. Gong. 2009. Hydrogen peroxide-induced proline and metabolic pathway of its accumulation in maize seedlings. *Journal of Plant Physiology* 166:1694–1699.

Yokoi, S., R.A. Bressan, and P.M. Hasegawa. 2002. Salt stress tolerance of plants. *JIRCAS Working Report* 23, 25–33.

Yoshiba, Y., T. Kiyosue, K. Nakashima, K. Yamaguchi-Shinozaki, K., and Shinozaki, K. 1997. Regulation of levels of proline as an osmolyte in plants under water stress. *Plant and Cell Physiology* 38:1095–1102.

Yoshihashi, T., N.T.T. Huong, and H. Inatomi. 2002. Precursors of 2-acetyl-1-pyrroline, a potent flavor compound of an aromatic rice variety. *Journal of Agricultural Food Chemistry* 50:2001–2004.

You, J., H. Hu, and L. Xiong. 2012. An ornithine ι-aminotransferase gene OsOAT confers drought and oxidative stress tolerance in rice. *Plant Science* 197:59–69.

Zhang, C.Y., N.N. Wang, Y.H. Zhang, Q.Z. Feng, C.W. Yang, and B. Liu. 2013. DNA methylation involved in proline accumulation in response to osmotic stress in rice (*Oryza sativa*). *Genetics and Molecular Research* 12:1269–1277.

# CHAPTER TWENTY

# Ion Transporters

## A DECISIVE COMPONENT OF SALT STRESS TOLERANCE IN PLANTS

*Kundan Kumar and Kareem A. Mosa*

## CONTENTS

*Abstract.* Salinity stress shows negative impact on agricultural yield throughout the world. Various plants adapt themselves to salinity stress by modulating their growth and development along with diverse physiological and biochemical changes. Biochemically salt stress adaptation requires cellular ion homeostasis, and involves net intracellular toxic ion uptake and subsequent vacuolar compartmentalization without toxic ion accumulation in the cytosol. The ability of plant cells to maintain low cytosolic toxic ion concentrations is associated with the ability of selected plants, viz. halophytes, to grow in high salt concentrations. Due to the negative electrical potential inside cells, mainly $Na^+$ influx into roots occurs through ion channels or various membrane transport proteins also facilitate passive diffusion of $Na^+$ across the plasma membrane. The potentially relevant membrane transporters involved in the uptake and distribution of various ions also play an important role in mitigating salinity stress, and were identified by forward and reverse genetics, yeast complementation, *Xenopus* oocyte assays, transcriptomics, and proteomics approaches. This chapter provides an overview of ion transporters that have been assigned functions like uptake, efflux, compartmentation, and translocation of various toxic ions, that finally confers salt stress tolerance in plants. We highlight the significant contribution of different forms of ion transporter systems: present at the level of plasma membrane, transporter systems related to vacuolar compartmentation, and membrane intrinsic proteins pertaining to salinity stress tolerance. Genetic engineering applications and strategies used to improve salinity stress

tolerance by modulating the transporter system are also discussed.

*Keywords*: ion homeostasis, ion transporters, salinity, salt stress.

## 20.1 INTRODUCTION

Salinity stress in the environment is one of the most serious and major threats, limiting the worldwide agriculture productivity in the present scenario. According to FAO (2008), more than 800 million hectares of land are estimated to be salt affected, and about ~20% of the world's cultivated land and nearly half of all irrigated lands are affected by salinity. Along with the higher accumulation of salts, that is NaCl in soil, the collective presence of $Ca^{2+}$, $Mg^{2+}$, and $SO^{4-}$ in high concentrations can aggravate saline conditions (Epstein and Bloom 2005). The salt-induced oxidative stress in plants accompanying the production of ROS such as superoxide radicals ($O^{2-}$), hydrogen peroxide ($H_2O_2$), and hydroxyl radicals (OH·) has severe effects on cellular structure, metabolism and showed negative impact on their growth and development (Bartels and Sunkar 2005). High salinity induces water stress, ion toxicity, nutritional disorders, alteration of metabolic processes, membrane disorganization, reduction of cell division and expansion, and genotoxicity (Hasegawa et al. 2000; Munns 2002; Zhu 2007) and thus affects metabolism and plant survival. The efficient removal of toxic ions from photosynthetic organs can be the determining factor for the maintenance of an adequate metabolism and carbon fixation, as ion overaccumulation in the cytosol causes inhibition of protein synthesis and alters many metabolic enzymatic reactions (Flowers and Lauchli 1983; Tsunekawa et al. 2009).

Under nonsaline conditions, the cytosol of higher plant cells contains about 100 mM of both $K^+$ and $Na^+$; hence, cytosolic enzymes are optimally functional. In saline environments, cytosolic $Na^+$ and $Cl^-$ increase to more than 100 mM, these ions become cytotoxic. Major agricultural crops are unable to sustain if salt concentration exceeds 100 mM. However, halophyte plant species are able to grow and even thrive on high salt concentrations. Salt tolerance in plants is often linked to the restriction of $Na^+$ accumulation and maintenance of a high $K^+/Na^+$ ratio mostly in the

shoots (Maathuis and Amtmann 1999; Moller and Tester 2007). The rate of $Na^+$ accumulation in plant shoots is determined by the net uptake of $Na^+$ into roots and its net translocation from roots to shoots in the xylem and $Na^+$ recycling from shoots to roots in the phloem (Tester and Davenport 2003). $Na^+$ flux in the xylem involves $Na^+$ efflux from root parenchyma cells into the xylem and the recovery of $Na^+$ from the xylem. Several transporters have been identified that mediate $Na^+$ transport across the plasma membranes and tonoplasts of the cells involved in this complex system of whole plant $Na^+$ fluxes (Munns and Tester 2008).

Mechanism of salt transport across cellular membranes and over long distances affects the salinity tolerance condition in plants. The number of potential salt tolerance genes and proteins has been identified, and their function of salt tolerance is monitored by the adequate control of salt uptake at the level of roots, regulation pattern of cell influx, long distance transport control mediated by plant membrane transporters, and finally, the compartmentation at both cellular as well as tissue levels (Fowler and Colmer 2008). The ion transporter proteins have strong potential to affect plant growth in saline conditions (Maathuis 2007). The uptake, efflux, translocation, and compartmentation of toxic ions (mainly $Na^+$ and $Cl^-$) via plant transporter provide the basis of salinity tolerance in plants, and hence, potential candidates to improve crops productivity in saline soil. However, a lack of understanding regarding the molecular framework and complex interactions of membrane transport proteins has hindered further progress in developing transgenic plants with better salt tolerance strategies.

Here we have divided the chapter into three parts: first, the transporters embarked on the plasma membrane of cell including carrier type transporters viz. high-affinity potassium transporters (HKTs), salt overlay sensitive (SOS), and nonselective cation channels (NSCCs). Second, the transporters present in the tonoplast, including vacuolar $Na^+/H^+$ antiporters (NHX), cation transporters (CAX), and $H^+$ pumps, involved in

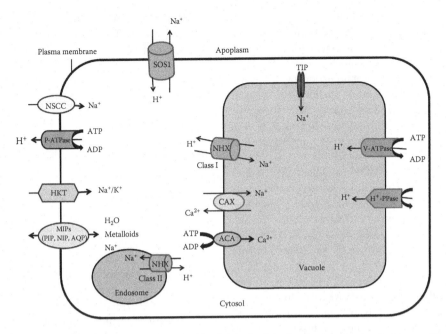

**Figure 20.1.** Diagrammatic representation of a plant cell and the subcellular localization of major plant transporters identified in the plasma membrane, vacuolar membrane, and endosomes, imparting salt stress tolerance. The transporters present in plasma membrane are $K^+/Na^+$ symporter (HKT), nonselective cation channel (NSCC), $Na^+/H^+$ antiporters (SOS1), and major intrinsic proteins (MIP) including plasma membrane intrinsic protein (PIP), nodulin intrinsic protein (NIP), and aquaporins (AQP). The transporters present in vacuolar membrane are $Na^+/H^+$ exchangers (NHX), cation antiporter (CAX), and tonoplast intrinsic protein (TIP). The endosomally localized $Na^+/H^+$ exchanger (NHX) is also represented. $Na^+$ extrusion from plant cells is powered by the electrochemical gradient generated by V-ATPases (vacuolar ATPase) and P-ATPase (plasma membrane ATPase) and $H^+$-PPase, which permits the $Na^+/H^+$ antiporters to couple the passive movement of $H^+$ inside along the electrochemical gradient and extrusion of $Na^+$ out of cytosol.

vacuolar compartmentalization. Third, the focus is on membrane intrinsic proteins (MIPs) and transportation of $Na^+$ in whole plant system involved in salt stress response. Major ion transporters, channel proteins, and pumps, working for the ion homeostasis (mainly $Na^+$) in the cell and providing salt stress tolerance in plants, present on plasma membrane, vacuolar membrane, and endosomes are represented in Figure 20.1.

## 20.2 ION TRANSPORTERS AT THE PLASMA MEMBRANE MEDIATING SALINITY STRESS TOLERANCE

Plants respond to salinity stress by different mechanisms of adaptations at the cellular level and/or the whole plant level. At the cellular level, $Na^+$ is known to be transported through various plasma membrane transporters including high-affinity potassium transporters (HKTs), SOS, and NSCCs (Figure 20.1). Studies on these transporters and the accompanying research aiming to develop salt stress tolerant plants are subsequently discussed.

## 20.2.1 High-Affinity Potassium Transporters (HKTs)

HKTs are the proteins that act as both $Na^+/K^+$ symporter or $Na^+$ uniporter (Figure 20.1), basically present in the plasma membrane of different plants including wheat, rice, and *Arabidopsis* for its role in salinity tolerance (Waters et al. 2013). Based on the phylogenetic analysis and transport selectivity, plant HKTs protein family is divided into two groups: group I HKTs are only transporting $Na^+$, and group II, HKTs are transporting $Na^+$ and $K^+$ (Corratgé-Faillie et al. 2010). It has been proposed that *AtHKT1* plays a role in sodium recirculation from shoots to roots by loading $Na^+$ from shoot into phloem and unloading it into roots (Berthomieu et al. 2003). Subsequently, *AtHKT1* was reported to be localized at the plasma membrane of xylem parenchyma cells and it has been demonstrated for its important function on salinity tolerance in *Arabidopsis* by unloading sodium from xylem vessels to xylem parenchyma cells (Sunarpi et al. 2005). Salt tolerance and reduction on shoot Na$^+$ accumulation can be achieved by manipulation of Na$^+$ transporters in specific cell types. Some reports with regard to overexpression of *Arabidopsis HKT1;1* in the root stele exhibited decreased shoot Na$^+$ accumulation by 37%–64% and showed salt tolerance phenotype as a result of Na$^+$ influx into root stele, which caused less root to shoot Na$^+$ transport. Interestingly, the transgenic plants constitutively overexpressing *HKT1;1* exhibited increased shoot Na$^+$ accumulation and showed sensitive salt phenotype (Møller et al. 2009). Barley *HKT2;1* was able to cotransport different concentrations of Na$^+$ and K$^+$ when heterologously expressed in *Xenopus* oocytes. Further, transgenic plants overexpressing *HvHKT2;1* displayed increased Na$^+$ levels in xylem sap and increased Na$^+$ translocation to leaves under 50 or 100 mM NaCl (Mian et al. 2011). *HKT1;4* isolated from a salt-tolerant durum wheat (*Triticum turgidum* L. subsp. *durum*) cultivar, Om Rabia3, exhibited high Na$^+$ selectivity when expressed in *Xenopus* oocytes (Amar et al. 2014).

## 20.2.2 Salt Overlay Sensitive (SOS1) Transporters

SOS1 transporters were identified in *Arabidopsis* (Wu et al. 1996), *Thellungiella salsuginea* (Vera-Estrella et al. 2005), rice (Martínez-Atienza et al. 2007), wheat (Xu et al. 2008), tomato (Olías et al. 2009), and other plants. AtSOS1 has been localized in the xylem parenchyma cells of leaves, stems, roots, and in the epidermal tissues of root tips, elucidated through promoter GUS analysis (Shi et al. 2002a). Further, overexpression of AtSOS1-GFP fusion proteins in *Arabidopsis* confirmed its localization on plasma membrane (Qiu et al. 2003). Functional complementation assay in yeast mutant lacking the Na$^+$-ATPases and Na$^+$/H$^+$ antiporters suggested that *Arabidopsis* plasma membrane SOS1 is transporting Na$^+$ (Shi et al. 2002a). Furthermore, AtSOS1 protein has been proposed to control the long-distance Na$^+$ transport in plants by functioning as Na$^+$/H$^+$ antiport (Shi et al. 2002a,b). SOS1 RNAi silenced lines in tomato conferred more Na$^+$ accumulation in roots and leaves but not in stem as compared to wild type plants. The less accumulation of Na$^+$ in stems suggested that SOS1 is playing a role in effluxing Na$^+$ from leaf to xylem to protect the photosynthetic organs in tomato (Olías et al. 2009). A similar strategy was followed for *Thellungiella salsuginea* using RNAi-based interference of SOS1. The SOS1 RNAi lines showed strong sensitive

phenotype and faster Na$^+$ accumulation under salt stress (Oh et al. 2009). Overexpression of SOS1 gene isolated from the extreme halophyte *Salicornia brachiate* in tobacco displayed a strong salt tolerance by effluxing the Na$^+$ outside the plasma membrane (Yadav et al. 2012).

## 20.2.3 Nonselective Cation Channels (NSCCs) Transporters

Understanding the mechanism of Na$^+$ transport from soil by the plant roots is crucial for controlling salt resistance. At high concentrations of NaCl in soil, nonselective cation channels (NSCCs) are suggested to be the main pathway for Na$^+$ transport into the roots through two classes: cyclic nucleotide-gated channels (CNGCs) and glutamate-activated channels (GLRs) (Tester and Davenport 2003). However, according to their response to changes in membrane electrical potential, NSCC candidate proteins are classified into three major families, namely, depolarization-activated NSCCs (DA-NSCCs), voltage-insensitive NSCCs (VI-NSCCs), and hyperpolarization-activated NSCCS (HA-NSCCs) (Demidchik and Maathuis 2007). The role of NSCC in catalyzing primary Na$^+$ fluxes in plants under salt stress conditions has been demonstrated in several studies (Kronzucke and Britto 2011). For example *Arabidopsis thaliana* CNGC3 has been suggested to function as Na$^+$ and K$^+$ transporter when expressed in yeast (Gobert et al. 2006). Further, under NaCl treatment, *cngc3* null mutations decreased seed germination (Gobert et al. 2006). Under salt stress conditions, rice CNGC1 exhibited more downregulation in salt-tolerant rice variety than salt-sensitive rice variety, suggesting its participation in decreasing Na influx and Na$^+$: K$^+$ ratio (Senadheera et al. 2009). *Arabidopsis thaliana* NSCC members of glutamate-activated channels, *AtGLR3;7*, *AtGLR1;1*, and *AtGLR 1;4*, showed Na$^+$ permeability when expressed in *Xenopus* oocyte (Roy et al. 2008; Tapken and Hollmann 2008).

## 20.3 ION TRANSPORTERS PRESENT ON THE TONOPLAST, MEDIATING SALINITY STRESS TOLERANCE

Plants grown in saline environment will accumulate Na$^+$ ions to some extent, due to the strong driving force for its entry. Except for some halophytic species that are able to effectively maintain very low Na$^+$ net influx, the accumulation of Na$^+$ inside vacuoles is a strategy used by many plants to survive salt stress (Hanana et al. 2007; Munns and Tester 2008). At the cellular level, vacuolar Na$^+$ accumulation will lower the amount of toxic Na$^+$ ions in the cytoplasm, and thus, the lower osmotic potential in the vacuole maintains turgor pressure and cell expansion in saline conditions. Consequently, the storage and translocation of Na$^+$ inside vacuoles in the shoot are suggested to be a key factor for sustained growth during salt stress in some plant species (Maathuis and Amtman 1999; Munns and Tester 2008). In this section, we will discuss the transporters present in vacuolar membrane viz. Na$^+$/H$^+$ antiporters (NHX), cation antiporters (CAX), autoinhibited Ca$^{2+}$-ATPase (ACA), and proton pumps such as V-H$^+$-PPase and the vacuolar H$^+$-ATPase (Figure 20.1).

## 20.3.1 Na$^+$/H$^+$ Antiporters Present in Tonoplast (NHX)

Sodium transport into vacuoles can be accomplished by the operation of tonoplast-bound NHX proteins and function as Na$^+$/H$^+$ antiporters, driven by the electrochemical gradient of protons generated by the vacuolar H$^+$-ATPase and H$^+$-PPase. The compartmentalization of Na$^+$ into vacuoles averts the deleterious effects of Na$^+$ in the cytoplasm and allows the plants to use Na$^+$ ions as an osmoticum, helping to retain the osmotic potential that drives water into the cell. Hence, NHX antiporters have major roles in the plant's response to salt stress (Rodriguez-Rosale et al. 2009). Vacuolar Na$^+$/H$^+$ antiport activity was first directly measured in tonoplast vesicles purified from red beet storage tissue (Blumwald and Poole 1985). Molecular studies and the characterization of factors influencing the antiporters activities have enabled the deciphering of their function and regulation. NHX, DNA sequences have been discovered in more than 60 plant species, including dicotyledonous, monocotyledonous angiosperms, and gymnosperms. NaCl treatment induced the expression of most of these NHX genes (Pardo et al. 2006). Plant genomes contain multiple intracellular NHX isoforms (Bassil et al. 2012) and classified into IC-NHE/NHX family (Pardo et al. 2006). According to the transport studies, the NHX catalyzes both Na$^+$/H$^+$ and K$^+$/H$^+$ exchanges (Apse et al. 2003; Leidi et al. 2010) and is involved in Na$^+$ or K$^+$ accumulation inside

vacuoles for the maintenance of cell turgor to drive cell expansion (Bassil et al. 2011; Barragan et al. 2012). On the basis of protein sequence similarity and subcellular localization, the plant NHX family is divided into two classes: class I and class II (Bassil et al. 2012). Tonoplastic NHX proteins are considered as class-I NHX antiporter and the endosomally localized NHX proteins are class-II NHX antiporters. The NHX proteins play an important role in the K$^+$ homeostasis regulation and hence play a decisive function in plant growth and development under salt stress (Pardo et al. 2006). The NHX family consists of eight isoforms in *Arabidopsis*. Most of them are intracellular (AtNHX1–6) located either to vacuole (AtNHX1-AtNHX4) or endosomal regions (AtNHX5 and AtNHX6) and an additional two more divergent members (AtNHX7/SOS1 explained earlier and AtNHX8) are present at the plasma membrane (Yokoi et al. 2002; Bassil et al. 2012). Rice transporters involved in imparting salinity stress tolerance were recently reviewed (Kumar et al. 2013a). In rice, four vacuolar (OsNHX1–4) and one endosomal Na$^+$/H$^+$ antiporter (OsNHX5) have been reported (Fukuda et al. 2011; Bassil et al. 2012). Orthologous sequences in each of the three classes (plasma membrane, vacuolar, or endosomal) are present in all sequenced genomes, suggesting that distinct functional NHX classes appeared early in evolution and might have conserved roles expressing the compartment specificity (Bassil et al. 2012). The C-terminal end regulates the cation selectivity of *At*NHX1, through the binding to a calmodulin-like protein AtCaM15 by interaction with C-terminus of AtNHX1 in a Ca$^{2+}$-and pH-dependent manner (Yamaguchi et al. 2005). By comparing different maize varieties differing in salt tolerance, it was observed that NHX transcripts were only induced in roots of a variety known to exclude Na$^+$ from the shoot (Zorb et al. 2005). Earlier reports describe that barley *Hv*NHX1 was mainly induced in roots by salt, whereas in rice, *Os*NHX1 induction was observed in shoots, suggesting that the high salt tolerance in barley is related to Na$^+$ accumulation in root cell vacuoles in order to limit transport to the shoot (Fukuda et al. 2004).

Various plant species including *Arabidopsis*, tomato, canola, cotton, rice, wheat, tobacco, and sugarbeet showed the salt tolerance by overexpression of genes encoding vacuolar NHXs, with a concomitant increase in tissue Na$^+$ level (Apse et al. 1999; Zhang and Blumwald 2001;

Zhang et al. 2001; Agarwal et al. 2013). The salt-sensitive phenotype of a yeast mutant defective for endosomal/vacuolar $Na^+/H^+$ antiporters was suppressed by the expression of vacuolar $Na^+/H^+$ antiporter NHX genes. Further, the transgenic plants overexpressing NHX displayed salt tolerance (Zhang and Blumwald 2001; Yokoi et al. 2002). This observation together with the fact that the *Arabidopsis atnhx1* mutant exhibited $Na^+$ sensitivity and significantly less vacuolar $Na^+/H^+$ antiport activity (Apse et al. 2003) strongly supported the role of vacuolar NHXs in $Na^+$ compartmentation under salinity stress. However, there has been a demonstration where the overexpression of vacuolar NHX led to accumulation of $K^+$ but not $Na^+$ (Leidi et al. 2010), but the conclusion is not clear due to certain technical problems (Smart et al. 2010; Bassil et al. 2012). The role of NHX type vacuolar antiporters as determinants of salt stress tolerance in tomato was done by comparing the expression of NHX proteins (*LeNHX1*, *LeNHX2*, *LeNHX3*, and *LeNHX4*) in salt-sensitive and salt-tolerant tomato varieties. Their results showed that during salt stress, the salt-tolerant varieties displayed enhanced expression of *LeNHX3* and *LeNHX4* (orthologues of *AtNHX1* and *AtNHX2*) with a concomitant increase in $Na^+$ in their tissues (Galvez et al. 2012). Overexpression of *OsNHX1* improved salt tolerance in rice and maize by enhancing the compartmentalization of $Na^+$ into the vacuoles (Chen et al. 2007). A new *AtNHX2*-like antiporter gene from *Ammopiptanthus mongolicus* was identified, its expression pattern under drought and salt stress was studied, and a comparative study of complementation efficiency of *AmNHX2* and *AtNHX1* using nhx1 yeast mutant and finally its functional characterization by overexpression studies in *Arabidopsis* was made (Wei et al. 2011).

The involvement of endosomal NHXs proteins in salt tolerance has also been reported in tomato (LeNHX2) and *Arabidopsis* (AtNHX5 and AtNHX6). The AtNHX5 and AtNHX6 localized in the Golgi and *trans*-Golgi network plays effective role in maintenance of endosomal pH and cation homeostasis (Rodriguez-Rosales et al. 2008; Bassil et al. 2011). The detailed biological studies are available only for the LeNHX2, the class-II NHX protein. The main function of the encoded protein is to catalyze $K^+/H^+$ exchange and affects minor $Na^+/H^+$ exchanges in proteoliposomes or plant endosomal membranes (Venema et al. 2003; Rodrıguez-Rosales et al. 2008). Indeed the mechanisms by which endosomal pH

and/or ion homeostasis affects salt tolerance are not yet clear; the possibility that excess cytosolic $Na^+$ is sequestered within vesicles that subsequently fuse to the vacuole, thus contributing to reduce high cytosolic $Na^+$ concentrations, should also be considered. Researchers cloned *TaNHX2* from bread wheat, and its expression and function were analyzed in wheat and yeast, respectively. It was suggested that *TaNHX2* might play an important role in salt tolerance of the plant through compartmentalizing $Na^+$ into vacuoles (Yu et al. 2007). Later on, it was found that wheat TaNHX2 is an endomembrane-bound protein and functions as a $K^+/H^+$ antiporter, which is involved in cellular pH regulation and potassium nutrition under normal conditions. With regard to the regulation of cellular pH and $K^+$ homeostasis, the NHX protein mediates resistance to salt stress through the process of intracellular compartmentalization of potassium, under saline conditions (Xu et al. 2013). These studies are associated with the pivotal function of the NHX family in intracellular compartmentalization of $Na^+$ or $K^+$ and salt tolerance. The SOS pathway regulated the vacuolar $Na^+/H^+$ exchange and thus its interaction with the calmodulin-like protein CaM15 in *Arabidopsis thaliana* (Qiu et al. 2004; Yamaguchi et al. 2005). AtSOS2 indirectly regulates ion transport activity in the vacuolar membrane by interacting with the vacuolar $H^+$-ATPase (Batelli et al. 2007). Multiple perspectives are essentially required to redefine the research on individual *Arabidopsis* NHX knockout mutants.

### 20.3.2 Cation Antiporters (CAX)

Calcium ($Ca^{2+}$) is an important secondary messenger in animals and plants (Bush 1995; Sanders et al. 1999; Qudeimat et al. 2008). The early events in plant signaling pathways include changes in cytoplasmic free $Ca^{2+}$, as it can accumulate to millimolar levels in the vacuole, whereas the concentrations are maintained in the submicromolar range in the cytosol (Marty 1999). Other changes also occurred including abiotic stress signaling during signaling pathways (Qudeimat et al. 2008). According to the studies by Marschner (1995), there is a complex interplay among various ions within plants, particularly between $Na^+$ and $Ca^{2+}$. For example supplemental $Ca^{2+}$ is known to mitigate the adverse effects of salinity on plant growth (Epstein 1972; Zhu 2002). However, functional redundancy and compensatory mechanisms have hindered our ability to directly assess the role of

CAX transporters in Na+ tolerance. CAXs are members of a multigene family and appear to be predominantly localized on the vacuole (Kamiya et al. 2005; Shigaki et al. 2006; Martinoia et al. 2007). Interestingly, expression of both CAX1 and CAX3 is upregulated during Ca²⁺ stress, while only CAX3 expression is significantly enhanced during Na+ stress (Hirschi 1999; Shigaki and Hirschi 2000). Zhao et al. (2008) provided insights into the diversity of CAX functions, particularly with regard to salt tolerance activity. Both CAX1 and CAX3 appeared to differentially regulate plasma membrane and tonoplast H+ pumps (Cheng et al. 2005), which in turn affects various cellular functions including H+-coupled ion transporters and pH regulation. They showed further evidence that CAX1 and CAX3 Ca²⁺ transport activities are differentially regulated by environmental stress: CAX1 by Ca²⁺ stress and CAX3 by Na+ stress. Signaling molecules maintain the H+/Na+ transport that also regulates the CAX transporters (Cheng et al. 2004). The *Arabidopsis thaliana* genome contains six CAX open reading frames and the CAX family is divided into two clusters, type IA and IB, on the basis of their amino acid sequences. The type IA constitutes CAX1, CAX3, and CAX4 whereas the CAX2, CAX5, and CAX6 are included into type IB (Manohar et al. 2011). In the rice genome, five CAXs (OsCAX1a, OsCAX1b, OsCAX1c, OsCAX2, and OsCAX3) have earlier been identified and the heterologous expression in yeast showed all rice CAXs, except OsCAX2, conferred tolerance to calcium. Differential transcript regulation patterns of OsCAX were observed in various tissues by Kamiya et al. (2009), but the role of OsCAX in salt tolerance is not known. Recently, a new gene, OsCAX4, has been annotated (Emery et al. 2012). Therefore, it is not only interesting but also important to characterize the role of all OsCAX for salt stress.

Genetically, the interplay is expressed between CAX and other transporters. Mutants in CAX1 not only exhibit 50% reduction in vacuolar H+/Ca²⁺ antiport activity, but also 40% decrease in V-ATPase activity and a 36% increase in vacuolar Ca²⁺ ATPase activity (Cheng et al. 2003). The cax1 lines also have compensatory changes in gene expression among a battery of transporters, in particular upregulation of CAX3 and CAX4. Although CAX2 and CAX3 deletions showed no alteration in H+/Ca²⁺ antiport when grown under calcium excess conditions, they also demonstrate similar reductions in V-ATPase activity (Pittman et al. 2004; Cheng et al. 2005).

While the individual CAX mutants display subtle phenotypes, stunted growth and leaf chlorosis are readily apparent when CAX1 and CAX3 are deleted simultaneously (Cheng et al. 2005). The challenge is to delineate the functional specificity of these endomembrane H+/Ca²⁺antiporters and decipher how they are integrated into various biological processes.

Plant Ca²⁺-ATPases are subgrouped into type IIA (or ECA as ER-type Ca²⁺-ATPase) and type IIB (ACA, i.e., autoinhibited Ca²⁺-ATPase) (Baxter et al. 2003). ACA contains N-terminal autoinhibitory domain that responds to Ca²⁺ signals by a Ca²⁺-induced binding of calmodulin, resulting in the activation of the Ca²⁺ pump (Hwang et al. 2000; Qudeimat et al. 2008). Eleven ACA genes were annotated in *Arabidopsis* and rice genomes. In rice, ACAs have been subdivided into four clusters based on sequence alignment and intron position (Baxter et al. 2003). AtACA4 and AtACA11 are vacuolar Ca²⁺ pumps in *Arabidopsis* (Boursiac et al. 2010). Physiological and biochemical functions of *Arabidopsis* ACAs have been studied (Bonza and Michelis 2011). The involvement of *Arabidopsis* ACA2 (Anil et al. 2008) and ACA4 (Geisler et al. 2000), and *Physcomitrella* vacuolar ACA1 (Qudeimat et al. 2008) in salt tolerance has been reported. In rice, mRNA expression profile of four putative vacuolar Ca²⁺-ATPases (OsACA4–7) was examined. The mRNA levels of OsACA4, OsACA5, and OsACA7 were relatively high and similar among sensitive cultivar IR29 and tolerant cultivar Pokkali under control conditions. Under salt stress, only OsACA4 showed a decrease in its expression in the tolerant cultivar Pokkali, followed by a significant increase (Yamada et al. 2014).

### 20.3.3 Proton Pumps

The active transport of solutes between the cytoplasm and the vacuole depends on proton gradients established by proton pumps. The electrochemical gradient of H+ generated can drive secondary transporters such as the Na+/H+ and Ca²⁺/H+antiporters (Taiz 1992). The H-gradient-driven transport of Na+ into vacuoles mediated by vacuolar Na+/H+ antiporters can not only reduce excess Na+ in the cytoplasm but also increase the osmolarity of the cells, and therefore plays an important role in the salt tolerance of plants (Zhu 2003). There are three distinct proton pumps that exhibits the proton electrochemical gradients across cell membranes: P-type ATPase, vacuolar H$^+$-pyrophosphatase (V-H$^+$-PPase), and vacuolar H$^+$-ATPase in plants. P-type ATPase pumps cytoplasmic H$^+$ across the plasma membrane into the extracellular space, whereas the V-H$^+$-PPase and the vacuolar H$^+$-ATPase acidify the vacuolar lumen to maintain a pH gradient between the cytoplasm and the vacuole (Sze et al. 1999). V-H$^+$-ATPase is a multisubunit ATP-dependent proton pump that couples the energy released upon hydrolysis of ATP to the active transport of protons from the cytoplasm to the lumen of the intracellular compartment (Jefferies et al. 2008). V-H$^+$-ATPase is composed of two distinct sectors: hydrophilic V1 domain (the peripherally associated) consists of eight different subunits (A–H) and has important role in hydrolyzing ATP to release energy. The second hydrophobic V$_0$ domain is membrane-anchored domain consisting of six different subunits, and functions in the translocation of protons across the cellular membrane (Cipriano et al. 2008; Toei et al. 2010). V-H$^+$-PPase coexists with V-H$^+$-ATPase in the vacuolar membrane, and together, they are the major components of the vacuolar membrane in plant cells (Silva and Geros 2009). The single polypeptide comprises the V-H$^+$-PPase having molecular mass of about 80 kDa, and uses pyrophosphate as its substrate (Maeshima 2000). V-H$^+$-PPases showed critical components in the regulation of cell turgor, H$^+$ electrochemical gradient, and in controlling secondary active transport of organic acids, inorganic ions, sugars, and other compounds surrounding the vacuolar membrane.

Maintaining internal water balance under osmotic stress involves high accumulation of these secondary transported solutes. Several reports have shown that upregulated expression of V-H$^+$-PPase (VP) genes can confer plant tolerance to diverse abiotic stresses, such as high salinity, drought, and inorganic phosphate (Pi) deprivation. For example the *Arabidopsis* V-H$^+$-PPase member *AVP1* can generate a H$^+$-gradient across the tonoplast (Zhen et al. 1997). Gaxiola et al. (1999) showed the heterologous overexpression of *AVP1* in yeast can restore salinity tolerance in a salt-sensitive yeast mutant. If a higher number of ions (Na$^+$) are transported across the vacuoles through the overexpression of *AVP1* in *Arabidopsis*, this significantly improves plant tolerance toward high salinity and drought conditions (Gaxiola et al.1999). The *TsVP*, a V-H$^+$-PPase

gene from *Thellungiella halophila*, overexpression exhibits a significantly greater resistance to salinity stress as compared with the wild type transgenic cotton plants (Lv et al. 2008). In wheat, a V-H$^+$-PPase gene, *TVP1*, has also been confirmed to be functional in improving salt and drought stress tolerance (Brini et al. 2007). Ectopic expression of *AVP1* and *TsVP* also exhibits improved drought tolerance in tomato (Park et al. 2005) and corn (Li et al. 2008), respectively. In addition, overexpression of *AVP1* and *TsVP* in rice, corn, tomato, and *Arabidopsis* showed improved growth under Pi deprivation as compared with the wild-type plants (Yang et al. 2007; Gaxiola et al. 2012). Overexpression of *AVP1* in cotton improved tolerance to drought and salt in greenhouse conditions, and increased fiber yield in dry land field conditions (Pasapula et al. 2011; Zhang et al. 2011). Overexpression of tonoplast H$^+$-PPase (ScVP) from *Suaeda corniculata*, a member of the Chenopodiaceae in *Arabidopsis*, exhibited much more resistance toward the higher concentrations of salt, saline-alkali stress, and water deprivation than wild type plants (Liu et al. 2011). According to the findings of Li et al. (2005), the V-H$^+$-PPase genes are the important regulators of tolerance toward the multiple abiotic stresses in plants. A considerable enhancement in root development is observed in *AVP1*-overexpressing plants, whereas an improper root development was observed in *avp1–1* loss-of-function mutants. Transgenic corn plants overexpressing *TsVP* showed higher tolerance to the high salinity, drought, and Pi deprived conditions than the wild type, reasonably associated with the improved root systems of the articulate plants (Li et al. 2008; Pei et al. 2012). These studies reveal that the enhanced vacuolar H$^+$ pumping in the transgenic plants is an effective approach in crop improvement in salt tolerance. Thus, V-H$^+$-PPase genes are important in upgrading plant tolerance to various abiotic stresses by mediating solute transport across the tonoplast and by regulating root systems.

## 20.4 MAJOR INTRINSIC PROTEINS (MIPS) AND SALT STRESS TOLERANCE

Plant MIP superfamily is a large family of channeled membranous proteins, basically grouped into four major subfamilies on the basis of their amino acid sequence similarities and subcellular localization. These are: nodulin 26-like intrinsic proteins (NIPs), plasma membrane intrinsic proteins (PIPs), tonoplast intrinsic proteins (TIPs), and small basic intrinsic proteins (SIPs) (Johanson et al. 2001). Subsequently, Danielson and Johanson (2008) also reported three other subfamilies of MIPs: hybrid intrinsic proteins (HIPs), GlpF-like intrinsic proteins (GIP), and uncategorized X intrinsic proteins (XIPs) from moss (*Physcomitrella patens*). MIPs are popularly known as the water channels (Daniels et al. 1994; Kammerloher et al. 1994). In addition, several MIPs have been reported to function as metalloid transporters (Bienert et al. 2008; Mosa et al. 2012; Kumar et al. 2013b). Furthermore, different studies have focused on exploring the role of MIPs on different abiotic stresses. With regard to salt stress, it has been reported that overexpression of SlTIP2;1 in tomato exhibited enhanced salt stress (180–200 mM NaCl) resistance (Sade et al. 2009). The overexpression of OsPIP1;1 in rice displayed resistance to salt stress consequently as a result of enhanced root hydraulic conductivity (Liu et al. 2013).

The aquaporin TIP1 was isolated from *Panax ginseng*, and TIP1-overexpressing transgenic *Arabidopsis* plants showed higher shoot Na$^+$ concentration and thus expressed salt tolerance phenotype as compared to wild type plants (Peng et al. 2007). Tomato and *Arabidopsis* plants constitutively overexpressing the tobacco stress-induced aquaporin1 (NtAQP1) showed increased water use efficiency (WUE) and enhanced biomass under salt stress (Sade et al. 2010). TaNIP, a novel NIP gene from wheat (*Triticum asetivum* L.), was reported for its role in salt tolerance signaling pathway (Gao et al. 2010) and its overexpression in *Arabidopsis* showed higher accumulation of K$^+$, Ca$^{2+}$, proline and less accumulation of Na$^+$ as compared to wild type plants. An array of genes encoding key transporters in plants has been characterized from various plant species, which may lead to improved plant tolerance against salt stresses and those discussed in this chapter are listed in Table 20.1.

## 20.5 TRANSPORTATION OF NA$^+$ IN WHOLE PLANT

Most salt tolerance is achieved through the control of ion movements into and through the plants, and Na$^+$ translocation from the root to the shoot is an important issue in providing salt stress tolerance. Halophytes actively transport Na$^+$ from root to shoot, whereas salt-sensitive glycophytes limit Na$^+$ entry into the transpirational stream to prevent Na$^+$ accumulation in the shoot (Flowers et al. 1977; Lauchli 1984). In salt-sensitive glycophytes, retranslocation of Na$^+$ from shoot to root is known and such mechanism would contribute to low shoot salt loads (Pitman 1977). Various proteins have been implicated in root shoot translocation of Na$^+$ in plants. SOS1 is the plasma membrane antiporter mainly expressed in root parenchyma cells and has a intense impact on Na$^+$ loading into the xylem sap during moderate salt stress in *Arabidopsis* (Shi et al. 2002). The functional mechanism of SOS1 is dependent on the severity of saline conditions and as in moderate salt stress condition, its expression leads to the removal of Na$^+$ from the xylem stream in plants.

The deprived functional mutations in the HKT1 gene has led to overaccumulation of Na$^+$ in shoots and leads to Na$^+$ susceptibility in *Arabidopsis*. The HKT1 is expressed mainly in leaf phloem tissues and mediates Na$^+$ loading into the phloem vessels, as studied through the RNA *in situ* hybridizations. It may also be involved in unloading of Na$^+$ from the phloem sap in roots and thus provides translocation process of Na$^+$ ions from shoot to root (Berthomieu et al. 2003).

According to the studies of Ren et al. (2005), the OsHKT1;5 is a plasma membrane Na$^+$ transporter expressed in xylem parenchyma cells that retrieves Na$^+$ from the xylem sap in rice. Rice genetic analysis revealed the crucial role of QTL, shoot K$^+$ content (SKC1), which maintain K$^+$ homeostasis and later renamed as OsHKT1;5. The heterologous expression revealed the OsHKT1;5 is a Na$^+$ transporter, and whole plant analysis indicated that it functions in the root xylem parenchyma to retrieve Na$^+$ from the xylem stream, thereby reducing Na$^+$ accumulation in the shoot (Lin et al. 2004). Salt-tolerant durum wheat landrace, line 149 showed decreased Na$^+$ transfer to shoot along with increased Na$^+$ sequestration into the leaf sheath tissue (Davenport et al. 2005). The previously mapped QTLs (Nax1 and Nax2) have been linked toward the salinity tolerance in line 149. These were found to control the two transport traits, as Nax2 locus along with the removal of Na$^+$ from the xylem in the roots also showed a Na$^+$ transporter related to OsHKT1;5 in rice (James et al. 2006; Byrt et al. 2007). Members of the H$^+$ monovalent cation exchanger family (CHX) contribute

TABLE 20.1

*List of plant transporters involved in salt tolerance identified by functional genetic analysis.*

| Gene name | Source species | Gene product | Identification method | Expressed in | References |
|---|---|---|---|---|---|
| HKT1;1 | *Arabidopsis thaliana* | Plasma membrane $Na^+$ transporter | Overexpression | *Arabidopsis* | Moller et al. (2009) |
| HKT2;1 | Barley | Plasma membrane $Na^+$ transporter | Heterologous expression | *Xenopus* oocytes | Mian et al. (2011) |
| HKT1;4 | *Triticum turgidum* | Plasma membrane $Na^+$ transporter | Heterologous expression | *Xenopus* oocytes | Amar et al. (2014) |
| SOS1 | *Arabidopsis thaliana* | Plasma membrane $Na^+/H^+$ antiporter | Yeast complementation | Yeast | Shi et al. (2002 a,b) |
| SOS1 | Tomato | Plasma membrane $Na^+/H^+$ antiporter | Knockdown (RNAi) | Tomato | Olias et al. (2009) |
| SOS1 | *Thellungiella salsuginea* | Plasma membrane $Na^+/$proton antiporter | Knockdown (RNAi) | *Thellungiela salsuginea* | Oh et al. (2009) |
| SOS1 | *Salicornia brachiate* | Plasma membrane $Na^+/H^+$ antiporter | Overexpression | Tobacco | Yadav et al. (2012) |
| CNGC3 | *Arabidopsis thaliana* | Cyclic nucleotide-gated channels | Heterologous expression | Yeast | Gobert et al. (2006) |
| CNGC1 | Rice | Cyclic nucleotide-gated channels | Transcript | Rice | Senadheera et al. (2009) |
| GLR3;7 | *Arabidopsis thaliana* | Glutamate-activated channels | Heterologous expression | *Xenopus* oocytes | Roy et al. (2008) and Tapken and Hollmann (2008) |
| GLR1;1 | *Arabidopsis thaliana* | Glutamate-activated channels | Heterologous expression | *Xenopus* oocytes | Roy et al. (2008) and Tapken and Hollmann (2008) |
| GLR 1;4 | *Arabidopsis thaliana* | Glutamate-activated channels | Heterologous expression | *Xenopus* oocytes | Roy et al. (2008) and Tapken and Hollmann (2008) |
| AtNHX1 | *Arabidopsis thaliana* | Vacuolar $Na^+/H^+$ antiporter | Overexpression, Functional complementation, knockdown | *Arabidopsis, Brassica napus*, tomato | Apse et al. (1999), Apse et al. (2003), and Zhang et al. (2001) |
| AtNHX2, AtNHX3, AtNHX4 | *Arabidopsis thaliana* | Vacuolar $Na^+/H^+$ antiporter | Transcript | *Arabidopsis* | Yokoi et al. (2002) |
| AtNHX5, AtNHX6 | *Arabidopsis thaliana* | Endosomal $K^+/H^+$ antiporter | Transcript | *Arabidopsis* | Bassil et al. (2012) |
| TaNHX2 | Wheat | Endomembrane-bound $K^+/H^+$ antiporter | Functional complementation | Wheat, yeast | Xu et al. (2013) |

(Continued)

TABLE 20.1 (Continued)

*List of plant transporters involved in salt tolerance identified by functional genetic analysis.*

| Gene name | Source species | Gene product | Identification method | Expressed in | References |
|---|---|---|---|---|---|
| OsNHX1 | Rice | Vacuolar Na$^+$/H$^+$ antiporter | Overexpression | Rice, maize | Chen et al. (2007) and Fukuda et al. (2004) |
| OsNHX2, OsNHX3, OsNHX4 | Rice | Vacuolar Na$^+$/H$^+$ antiporter | Functional complementation | Yeast | Fukuda et al. (2011) |
| OsNHX5 | Rice | Endosomal K$^+$/H$^+$ antiporter | Transcript | Yeast | Bassil et al. (2012) |
| AmNHX2 | Ammopiptanthus mongolicus | Vacuolar Na$^+$/H$^+$ antiporter | Overexpression | Arabidopsis | Wei et al. (2011) |
| LeNHX2 | Tomato | Endosomal K$^+$/H$^+$ antiporter | Overexpression, knock down | Tomato | Rodriguez-Rosales et al. (2008) |
| LeNHX1, LeNHX3, LeNHX4 | Tomato | Vacuolar Na$^+$/H$^+$ antiporter | Transcript | Tomato | Galvez et al. (2012) |
| CAX1 | Arabidopsis | Cation exchanger | Knockout | Arabidopsis | Chen et al. (2003) |
| OsCAX4 | Rice | Vacuolar cation/H$^+$ antiporter | Yeast, Transcript | Yeast | Yamada et al. (2014) |
| ACA2 | Arabidopsis | ER located Ca$^{2+}$ ATPase | Yeast complementation | Yeast | Anil et al. (2008) |
| ACA4 | Arabidopsis | Vacuolar Ca$^{2+}$ATPase | Yeast complementation | Yeast | Giesler et al. (2000) |
| OsACA4 | Rice | Calmodulin-regulated autoinhibited Ca$^{2+}$-ATPase | Transcript | Rice | Yamada et al. (2014) |
| AVP1 | Arabidopsis | Arabidopsis V-H$^+$-PPase | Yeast complementation | Yeast, Arabidopsis | Gaxiola et al. (1999) |
| TsVP | Thellungiella halophila | V-H$^+$-PPase | Overexpression | Cotton, corn | Lv et al. (2008) and Pei et al. (2012) |
| TVP1 | Wheat | V-H$^+$ PPase | Overexpression | Arabidopsis | Brini et al. (2007) |
| ScVP | Suaeda corniculata | Tonoplast H$^+$-PPase | Overexpression | Arabidopsis | Liu et al. (2011) |
| TIP2;1 | Tomato | Tonoplast intrinsic protein | Overexpression | Tomato | Sade et al. (2009) |
| PIP1;1 | Rice | Plasma membrane intrinsic protein | Overexpression | Rice | Liu et al. (2013) |
| TIP1 | Panax ginseng | Tonoplast intrinsic protein | Overexpression | Arabidopsis | Peng et al. (2007) |
| NtAQP1 | Tobacco | Aquaporin transporter | Overexpression | Arabidopsis, tomato | Sade et al. (2010) |
| NIP | Wheat | Nodulin 26-like intrinsic proteins | Overexpression | Arabidopsis | Gao et al. (2010) |

to Na$^+$ translocation. *AtCHX2;1* is mainly expressed in the root endodermis and loss of function of this gene reduced levels of Na$^+$ in the xylem sap without affecting phloem Na$^+$ concentration (Hall et al. 2006).

## 20.6 CONCLUSIONS AND FUTURE PERSPECTIVES

Soil salinity is the major environmental factor that limits agricultural productivity worldwide. The changing climatic circumstances would ultimately worsen agricultural land in terms of water unavailability and soil salinization. According to growing world population, the demand for more food will continue to rise, approximately for 9 billon people by 2050. Thus, the primary challenge of the century is the optimized food production under the limitation of water supply, and fertilizers on marginal soils. Huge amount of economic loss in the world food production is directly correlated toward the increasing problem of global land salinization; it is necessary to acquire a better understanding of the key physiological mechanisms that give salinity tolerance in crops. Salt tolerance is a quantitative trait governed by multiple genes and the success with the classical breeding programs has been low in developing salt-tolerant crops. Thus, genetic manipulation provides an alternate, yet sustainable approach toward engineering plants for salinity tolerance. Among various effective ways of gaining such knowledge by studying halophytes are the current use of models as a source of genes for engineering salt tolerance in heterologous systems. Regulatory mechanisms through genetic engineering are applied to the food crops, mostly on grasses, such as rice. In the past few years, significant efforts have been made to identify the genes encoding various ion transporters such as NHX, SOS, and HKT and successfully utilize them as genetic tools for enhancing salt tolerance of crop plants. In order to better characterize new ion transporters imparting tolerance to salt stress and give insights into the mechanisms, we need to acquire more accurate information. The development of the "omics" approaches in science, such as transcriptomic, proteomic, and metabolomic approaches, became essential to functionally characterize the transporters involved in salinity tolerance in crop plants. The progress of the different genomic sequence projects, together with the development of knockout mutants, will certainly reveal many more transporters associated with salinity stress tolerance in plants.

## ACKNOWLEDGMENTS

The work in KK's lab is supported by grants from Science & Engineering Research Board (SB/FT/LS-312/2012), Department of Science & Technology, India. KK also acknowledges Research Initiation Grant received from Birla Institute of Technology & Science, Pilani.

## REFERENCES

Agarwal, P.K., Shukla, P.S., Gupta, K., and Jha, B. 2013. Bioengineering for salinity tolerance in plants: State of art. *Molecular Biotechnology* 54:102–123.

Amar, S.B., Brini, F., Sentenac, H., Masmoudi, K., and Very, A.A. 2014. Functional characterization in *Xenopus* oocytes of Na$^+$ transport systems from durum wheat reveals diversity among two HKT1;4 transporters. *Journal of Experimental Botany* 65:213–222.

Anil, V.S., Rajkumar, P., Kumar, P., and Mathew, M.K. 2008. A plant Ca$^{2+}$ pump, ACA2, relieves salt hypersensitivity in yeast. Modulation of cytosolic calcium signature and activation of adaptive Na$^+$ homeostasis. *Journal of Biological Chemistry* 283:3497–3506.

Apse, M.P., Aharon, G.S., Snedden, A., and Blumwald, E. 1999. Salt tolerance conferred by overexpression of a vacuolar Na$^+$/H$^+$ antiport in *Arabidopsis*. *Science* 285:1256–1258.

Apse, M.P., Sottonsanto, J.B., and Blumwald, E. 2003. Vacuolar cation/H$^+$ exchange, ion homeostasis and leaf development are altered in a T-DNA insertional mutant of AtNHX1, the *Arabidopsis* vacuolar Na$^+$/H$^+$ antiporter. *The Plant Journal* 36:229–239.

Barragan, V., Leidi, E.O., Andres, Z. et al. 2012. Ion exchangers NHX1 and NHX2 mediate active potassium uptake into vacuoles to regulate cell turgor and stomatal function in *Arabidopsis*. *The Plant Cell* 24:1127–1142.

Bartels, D. and Sunkar, R. 2005. Drought and salt tolerance in plants. *Critical Review in Plant Science* 24:23–58.

Bassil, E., Coku, A., and Blumwald, E. 2012. Cellular ion homeostasis: Emerging roles of intracellular NHX Na$^+$/H$^+$ antiporters in plant growth and development. *Journal of Experimental Botany* 63:5727–5740.

Bassil, E., Ohto, M.A., Esumi, T. et al. 2011. The *Arabidopsis* intracellular Na$^+$/H$^+$ antiporters NHX5 and NHX6 are endosome associated and necessary for plant growth and development. *The Plant Cell* 23:224–239.

Batelli, G., Verslues, P.E., Agius, F. et al. 2007. SOS2 promotes salt tolerance in part by interacting with the vacuolar H$^+$-ATPase and upregulating its transport activity. *Molecular and Cellular Biology* 27:7781–7790.

Baxter, I., Tchieu, J., Sussman, M.R. et al. 2003. Genomic comparison of P-type ATPase ion pumps in *Arabidopsis* and rice. *Plant Physiology* 132:618–628.

Berthomieu, P., Conejero, G., Nublat et al. 2003. Functional analysis of AtHKT1 in *Arabidopsis* shows that Na$^+$ recirculation by the phloem is crucial for salt tolerance. *EMBO Journal* 22:2004–2014.

Bienert, G., Thorsen, M.D., Schussler, H.R. et al. 2008. A subgroup of plant aquaporins facilitate the bidirectional diffusion of As(OH)$_3$ and Sb(OH)$_3$ across membranes. *BMC Biology* 6:26.

Blumwald, E. and Poole, R. 1985. Na$^+$/H$^+$-antiport in isolated tonoplast vesicles from storage tissue of *Beta vulgaris*. *Plant Physiology* 78:163–167.

Bonza, M.C. and De-Michelis, M. 2011. The plant Ca$^{2+}$ ATPase repertoire: biochemical features and physiological functions. *Plant Biology* 13:421–430.

Boursiac, Y., Lee, S.M., Romanowsky, S. et al. 2010. Disruption of the vacuolar calcium-ATPase in *Arabidopsis* results in the activation of a salicylic acid-dependent programmed cell death pathway. *Plant Physiology* 154:1158–1171.

Brini, F., Hanin, M., Mezghani, I., Berkowitz, G.A., and Masmoudi, K. 2007. Overexpression of wheat Na$^+$/H$^+$ antiporter TaNHX1 and H$^+$-pyrophosphatase TVP1 improve salt- and drought-stress tolerance in *Arabidopsis thaliana* plants. *Journal of Experimental Botany* 58:301–308.

Bush D.S. 1995. Calcium regulation in plant cells and its role in signaling. *Annual Review of Plant Biology* 46:95–122.

Byrt, C.S., Platten, J.D. Spielmeyer, W. et al. 2007. HKT1;5 like cation transporters linked to Na$^+$ exclusion loci in wheat, Nax2 and Kna1. *Plant Physiology* 143:1918–1928.

Chen, H., An, R., Tang, J.H. et al. 2007. Overexpression of a vacuolar Na$^+$/H$^+$ antiporter gene improves salt tolerance in an upland rice. *Molecular Breeding* 19:215–225.

Cheng, N.H., Pittman, J.K., Barkla, B.J., Shigaki, T., and Hirschi, K.D. 2003. The *Arabidopsis cax1* mutant exhibits impaired ion homeostasis, development, and hormonal responses, and reveals interplay among vacuolar transporters. *Plant Cell* 15:347–364.

Cheng, N.H., Pittman, J.K., Shigaki, T. et al. 2005. Functional association of *Arabidopsis* CAX1 and CAX3 is required for normal growth and ion homeostasis. *Plant Physiology* 138:2048–2060.

Cheng, N.H., Pittman, J.K., Zhu, J.K., and Hirschi, K.D. 2004. The protein kinase SOS2 activates the *Arabidopsis* H$^+$/Ca$^{2+}$ antiporter CAX1 to integrate calcium transport and salt tolerance. *Journal of Biological Chemistry* 279:2922–2926.

Cipriano, D.J., Wang, Y.R., Bond, S., Hinton, A., and Jefferies, K.C. 2008. Structure and regulation of the vacuolar ATPases. *Biochimica et Biophysica Acta* 1777:599–604.

Corratgé-Faillie, C., Jabnoune, M., Zimmermann, S., Véry, A.A., Fizames, C., and Sentenac, H. 2010. Potassium and sodium transport in non-animal cells: the Trk/Ktr/HKT transporter family. *Cellular and Molecular Life Science* 67:2511–2532.

Daniels, M.J., Mirkov, T.E., and Chrispeels, M.J. 1994. The plasma membrane of *Arabidopsis thaliana* contains a mercury-insensitive aquaporin that is a homolog of the tonoplast water channel protein TIP. *Plant Physiology* 106:1325–1333.

Danielson, J.Å. and Johanson, U. 2008. Unexpected complexity of the aquaporin gene family in the moss *Physcomitrella patens*. *BMC Plant Biology* 8:45.

Davenport, R., James, R.A., Zakrisson-Plogander, A., Tester, M., and Munns, R. 2005. Control of sodium transport in durum wheat. *Plant Physiology* 137:807–818.

Demidchik, V. and Maathuis, F.J.M. 2007. Physiological roles of nonselective cation channels in plants: From salt stress to signalling and development. *New Phytologist* 175:387–404.

Emery, L., Whelan, S., Hirschi, K.D., and Pittman, J.K. 2012. Protein phylogenetic analysis of Ca$^{2+}$/cation antiporters and insights into their evolution in plants. *Frontiers in Plant Science* 13:1–19.

Epstein, E. 1972. *Mineral Nutrition in Plants: Principles and Perspectives*. Wiley, New York.

Epstein, E. and Bloom, A.J. 2005. *Mineral Nutrition of Plants: Principles and Perspectives*, 2nd ed. Sinauer, Sunderland, MA.

FAO (Food and Agriculture Organisation). 2008. FAO land and plant nutrition management service. http://www.fao.org/ag/agl/agll/spush.

Flowers, T.J. and Colmer, T.D. 2008. Salinity tolerance in halophytes. *New Phytologist* 179:945–963.

Flowers, T.J. and Läuchli, A. 1983. Sodium versus potassium: Substitution and compartmentation. In *Encyclopedia of Plant Physiology 15B, New Series, Inorganic Plant Nutrition*, eds. A. Läuchli and R.A. Bieleski, pp. 651–681. Springer, Berlin, Germany.

Flowers, T.J., Troke, P.F., and Yeo, A.R. 1977. The mechanism of salt tolerance in halophytes. *Annual Review of Plant Physiology* 28:89–121.

Fukuda, A., Nakamura, A., Hara, N., Toki, S., and Tanaka, Y. 2011. Molecular and functional analyses of rice NHX-type $Na^+/H^+$ antiporter genes. *Planta*, 233:175–188.

Fukuda, A., Nakaumura, A., Tagiri, A. et al. 2004. Function, intracellular localization and the importance in salt tolerance of a vacuolar $Na^+/H^+$ antiporter from rice. *Plant and Cell Physiology*. 45:146–159.

Galvez, F.J., Baghour, M., Hao, G.P., Cagnac, O., Rodriguez-Rosales, M.P., and Venema, K. 2012. Expression of LeNHX-isoforms in response to salt stress in salt sensitive and salt tolerant tomato species. *Plant Physiology and Biochemistry* 51:109–115.

Gao, Z., He, X., Zhao, B. et al. 2010. Overexpressing a putative aquaporin gene from wheat, TaNIP, enhances salt tolerance in transgenic *Arabidopsis*. *Plant Cell and Physiology* 51:767–775.

Gaxiola, R.A., Rao, R., Sherman, A., Grisafi, P., Alper, S.L., and Fink, G.R. 1999. The *Arabidopsis thaliana* proton transporters, AtNhx1 and Avp1, can function in cation detoxification in yeast. *Proceedings of the National Academy of Sciences of the United States of America* 96:1480–1485.

Gaxiola, R.A., Sanchez, C.A., Paez-Valencia, J., Ayre, B.G., and Elser, J.J. 2012. Genetic manipulation of a 'vacuolar' $H^+$-PPase: From salt tolerance to yield enhancement under phosphorus-deficient soils. *Plant Physiology* 159:3–11.

Geisler, M., Frangne, N., Gomès, E., Martinoia, E., and Palmgren, M.G. 2000. The ACA4 gene of *Arabidopsis* encodes a vacuolar membrane calcium pump that improves salt tolerance in yeast. *Plant Physiology* 124:1814–1827.

Gobert, A., Park, G., Amtmann, A., Sanders, D., and Maathuis, F.J.M. 2006. *Arabidopsis thaliana* cyclic nucleotide gated channel 3 forms a non-selective ion transporter involved in germination and cation transport. *Journal of Experimental Botany* 57:791–800.

Hall, D., Evans, A.R., Newbury, H.J., and Pritchard, J. 2006. Functional analysis of CHX21: A putative sodium transporter in *Arabidopsis*. *Journal of Experimental Botany* 57:1201–1210.

Hanana, M., Cagnac, O., Yamaguchi, T., Hamdi, S., Ghorbel, A., and Blumwald, E. 2007. A grape berry (*Vitis vinifera* L.) cation/proton antiporter is associated with berry ripening. *Plant Cell and Physiology* 48:804–811.

Hasegawa, P.M., Bressan, R.A., Zhu, J.K., and Bohnert, H.J. 2000. Plant cellular and molecular responses to high salinity. *Annual Review of Plant Physiology and Plant Molecular Biology* 51:463–499.

Hirschi, K.D. 1999. Expression of *Arabidopsis* CAX1 in tobacco: Altered calcium homeostasis and increased stress sensitivity. *Plant Cell* 11:2113–2122.

Hwang, I., Harper, J.F., Liang, F., and Sze, H. 2000. Calmodulin activation of an endoplasmic reticulum-located calcium pump involves an interaction with the N-terminal autoinhibitory domain. *Plant Physiology* 122:157–168.

James, R.A., Davenport, R.J., and Munns, R. 2006. Physiological characterization of two genes for $Na^+$ exclusion in durum wheat, Nax1 and Nax2. *Plant Physiology* 142:1537–1547.

Jefferies, K.C., Cipriano, D.J., and Forgac, M. 2008. Function, structure and regulation of the vacuolar $(H^+)$-ATPases. *Archives of Biochemistry and Biophysics* 476:33–42.

Johanson, U., Karlsson, M., Johansson, I. et al. 2001. The complete set of genes encoding major intrinsic proteins in *Arabidopsis* provides a framework for a new nomenclature for major intrinsic proteins in plants. *Plant Physiology* 126:1358–1369.

Kamiya, T., Akahori, T., and Maeshima, M. 2005. Expression profile of the genes for rice cation/$H^+$ exchanger family and functional analysis in yeast. *Plant Cell Physiology* 46:1735–1740.

Kammerloher, W., Fischer, U., Piechottka, G.P., and Schaffner, A.R. 1994. Water channels in the plant plasma membrane cloned by immunoselection from a mammalian expression system. *The Plant Journal* 6:187–199.

Kronzucker, H.J. and Britto, D.T. 2011. Sodium transport in plants: A critical review. *New Phytologists* 189:54–81.

Kumar, K., Kumar, M., Kim, S.R., Ryu, H., and Cho, Y.G. 2013a. Insights into genomics of salt stress response in rice. *Rice* 6:27.

Kumar, K., Mosa, K.A., Chikara, S., Musante, C., White, J.C., and Dhankher, O.P. 2013b. Two rice plasma membrane intrinsic proteins, OsPIP2;4 and OsPIP2;7 are involved in transport and providing tolerance to boron toxicity. *Planta* 239:187–198.

Lauchli, A. 1984. Salt exclusion: An adaptation of legumes for crops and pastures under saline condition. In *Salinity Tolerance in Plants: Strategies for Crop Improvement*, eds. R.C. Staples and G.H. Toennissen, pp. 171–187. Wiley, New York.

Leidi, E.O., Barragan, V., Rubio, L. et al. 2010. The AtNHX1 exchanger mediates potassium compartmentation in vacuoles of transgenic tomato. *The Plant Journal* 61:495–506.

Li, B., Wei, A., Song, C., Li, N., and Zhang, J. 2008. Heterologous expression of the *TsVP* gene improves the drought resistance of maize. *Plant Biotechnology Journal* 6:146–159.

Li, J., Yang, H., Peer, W.A. et al. 2005. *Arabidopsis* H$^+$-PPase AVP1 regulates auxin-mediated organ development. *Science* 310:121–125.

Lin, H.X., Zhu, M.Z., Yano, M. et al. 2004. QTLs for Na$^+$ and K$^+$ uptake of the shoots and roots controlling rice salt tolerance. *Theoretical and Applied Genetics* 143:1918–1928.

Liu, C., Fukumoto, T., Matsumoto, T. et al. 2013. Aquaporin OsPIP1;1 promotes rice salt resistance and seed germination. *Plant Physiology and Biochemistry* 63:151–158.

Liu, L., Wang, Y., Wang, N., Dong, Y.Y., Fan, X.D., Liu, X.M., Yang, J., and Li, H.Y. 2011. Cloning of a vacuolar H (+)-pyrophosphatase gene from the halophyte *Suaeda corniculata* whose heterologous overexpression improves salt, saline-alkali and drought tolerance in *Arabidopsis*. *Journal of Integrative Plant Biology* 53:731–742.

Lv, S., Zhang, K., Gao, Q. et al. 2008. Overexpression of an H$^+$-PPase gene from *Thellungiella halophila* in cotton enhances salt tolerance and improves growth and photosynthetic performance. *Plant and Cell Physiology* 49:1150–1164.

Maathuis, F.J.M. 2007. Monovalent cation transporters: Establishing a link between bioinformatic and physiology. *Plant and Soil* 301:1–15.

Maathuis, F.J.M. and Amtmann, A. 1999. Nutrition and Na$^+$ toxicity; the basis of cellular K$^+$/Na$^+$ ratios. *Annals of Botany* 84:123–133.

Maeshima, M. 2000. Vacuolar H$^+$-pyrophosphatase. *Biochimica et Biophysica Acta* 1465:37–51.

Manohar, M., Shigaki, T., and Hirschi, K.D. 2011. Plant cation/H$^+$ exchangers (CAXs): Biological functions and genetic manipulations. *Plant Biology* 13:561–569.

Marschner, H. 1995. *Mineral Nutrition of Higher Plants*. Academic, New York

Martínez Atienza, J., Jiang, X., Garciadeblas, B. et al. 2007. Conservation of the salt overly sensitive pathway in rice. *Plant Physiology* 143:1001–1012.

Martinoia, E., Maeshima, M., and Neuhaus, H.E. 2007. Vacuolar transporters and their essential role in plant metabolism. *Journal of Experimental Botany* 58:83–102.

Marty, F. 1999. Plant vacuoles. *The Plant Cell* 11:587–599.

Mian, A., Oomen, R.J., Isayenkov, S., Sentenac, H., Maathuis, F.J., and Very, A.A. 2011. Over-expression of an Na$^+$-and K$^+$-permeable HKT transporter in barley improves salt tolerance. *The Plant Journal* 68:468–479.

Møller, I.S., Gilliham, M., and Jha, D. 2009. Shoot Na$^+$ exclusion and increased salinity tolerance engineered by cell type-specific alteration of Na$^+$ transport in *Arabidopsis*. *The Plant Cell* 21:2163–2178.

Møller, I.S. and Tester, M. 2007. Salinity tolerance of *Arabidopsis*: A good model for cereals? *Trends in Plant Science* 12:534–540.

Mosa, K.A., Kumar, K., Chhikara, S. et al. 2012. Members of rice plasma membrane intrinsic protein subfamily are involved in arsenite permeability and tolerance in plants. *Transgenic Research* 21:1265–1277.

Munns, R. 2002. Comparative physiology of salt and water stress. *Plant, Cell and Environment* 25:239–250.

Munns, R. and Tester, M. 2008. Mechanism of salinity tolerance. *Annual Review of Plant Biology* 59:651–681.

Oh, D.H., Leidi, E., Zhang, Q. et al. 2009. Loss of halophytism by interference with SOS1 expression. *Plant Physiology* 151:210–222.

Olías, R., Eljakaoui, Z., Li, J. et al. 2009. The plasma membrane Na$^+$/H$^+$ antiporter SOS1 is essential for salt tolerance in tomato and affects the partitioning of Na$^+$ between plant organs. *Plant Cell and Environment* 32:904–916.

Pardo, J.M., Cubero, B., Leidi, E.O., and Quintero, F.J. 2006. Alkali cation exchangers: roles in cellular homeostasis and stress tolerance. *Journal of Experimental Botany* 57:1181–1199.

Park, S., Li, J., and Pittman, J.K. et al. 2005. Upregulation of a H$^+$-pyrophosphatase (H$^+$-PPase) as a strategy to engineer drought-resistant crop plants. *Proceedings of the National Academy of Sciences, USA* 102:18830–18835.

Pasapula, V., Shen, G., Kuppu, S. et al. 2011. Expression of an *Arabidopsis* vacuolar H$^+$-pyrophosphatase gene (*AVP1*) in cotton improves drought- and salt-tolerance and increases fiber yield in the field conditions. *Plant Biotechnology Journal* 9:88–99.

Pei, L., Wang, J., Li, K. et al. 2012. Overexpression of *Thellungiella halophila* H$^+$-pyrophosphatase gene improves low phosphate tolerance in maize. *PLoS One* 7:e43501.

Peng, Y., Lin, W., Cai, W., and Arora, R. 2007. Overexpression of a *Panax ginseng* tonoplast aquaporin alters salt tolerance, drought tolerance and cold acclimation ability in transgenic *Arabidopsis* plants. *Planta* 226:729–740.

Pitman, M.G. 1977. Ion transport into the xylem. *Annual Review of Plant Physiology* 28:71–88.

Pittman, J.K., Shigaki, T., Marshall, J.L., Morris, J.L., Cheng, N.H., and Hirschi, K.D. 2004. Functional and regulatory analysis of the *Arabidopsis* thaliana CAX2 cation transporter. *Plant Molecular Biology* 56:959–971.

Qiu, Q.S., Barkla, B.J., Vera-Estrella, R., Zhu, J.K., and Schumaker, K.S. 2003. Na+/H+ exchange activity in the plasma membrane of *Arabidopsis*. *Plant Physiology* 132:1041–1052.

Qiu, Q.S., Guo, Y., Quntero, F.J., Pardo, J.M., Schumaker, K.S., and Zhu, J.K. 2004. Regulation of vacuolar Na+/H+ exchange in *Arabidopsis thaliana* by the salt overlay sensitive (SOS) pathway. *Journal of Biological Chemistry* 279:207–215.

Qudeimat, E., Faltusz, A.M., Wheeler, G. et al. 2008. A PIIB-type Ca2+-ATPase is essential for stress adaptation in *Physcomitrella patens*. *Proceedings of the National Academy of Sciences of the United States of America* 105:19555–19560.

Ren, Z.H., Gao, J.P., Li, L.G. et al. 2005. A rice quantitative trait locus for salt tolerance encodes a sodium transporter. *Nature Genetics* 37:1141–1146.

Rodriguez-Rosales, M.P., Galvez, F.J., Huertas R. et al. 2009. Plant NHX cation/proton antiporter. *Plant Signaling and Behavior* 4:265–276.

Rodriguez-Rosales, M.P., Jiang, X.J., and Galvez, F.J. 2008. Overexpression of the tomato K+/H+ antiporters LeNHX2 confers salt tolerance by improving potassium compartmentalization. *New Phytologist* 179:366–377.

Roy, S.J., Gilliham, M., Berger, B. et al. 2008. Investigating glutamate receptor-like gene co-expression in *Arabidopsis thaliana*. *Plant, Cell and Environment* 31:861–871.

Sade, N., Gebrestasadik, M., Seligmann, R., Schwartz, A., Wallach, R., and Moshelion, M. 2010. The role of tobacco aquaporin1 in improving water use efficiency, hydraulic conductivity, and yield production, under salt stress. *Plant Physiology* 152:245–254.

Sade, N., Vinocur, B.J., Diber, A. et al. 2009. Improving plant stress tolerance and yield production: is the tonoplast aquaporin SlTIP2;2 a key to isohydric to anisohydric conversion? *New Phytologist* 181:651–661.

Sanders, D., Brownlee, C., and Harper, J.F. 1999. Communicating with calcium. *The Plant Cell* 11:691–706.

Senadheera, P., Singh, R.K., and Maathuis, F.J.M. 2009. Differentially expressed membrane transporters in rice roots may contribute to cultivar dependent salt tolerance. *Journal of Experimental Botany* 60:2553–2563.

Shi, H., Quintero, F.J., Pardo, J.M., and Zhu, J.K. 2002a. The putative plasma membrane Na+/H+ antiporter SOS1 controls long-distance Na+ transport in plants. *The Plant Cell* 14:465–477.

Shi, H., Wu, S.J., and Zhu, J.K. 2002b. Role of SOS1 as a plasma membrane Na+/H+ antiporter improves salt tolerance in *Arabidopsis*. *Nature Biotechnology* 21:81–85.

Shigaki, T. and Hirschi, K.D. 2000. Characterization of CAX-like genes in plants: implications for functional diversity. *Gene* 257:291–298.

Shigaki, T., Rees, I., Nakhleh, L., and Hirschi, K.D. 2006. Identification of three distinct phylogenetic groups of CAX cation/proton antiporters. *Journal of Molecular Evolution* 63:815–825.

Silva, P. and Gero's, H. 2009. Regulation by salt of vacuolar H+-ATPase and H+-pyrophosphatase activities and Na+/H+ exchange. *Plant Signal and Behavior* 4:718–726.

Smart, K.E., Smith, J.A.C., Kilburn, M.R., Martin, B.G.H., Hawes, C., and Grovenor, C.R.M. 2010. High resolution elemental localization in vacuolated plant cells by nanoscale secondary ion mass spectrometry. *The Plant Journal*. 63:870–879.

Sunarpi, T., Motoda, J., Kubo, M. et al. 2005. Enhanced salt tolerance mediated by AtHKT1 transporter-induced Na+ unloading from xylem vessels to xylem parenchyma cells. *The Plant Journal* 44:928–938.

Sze, H., Li, X., and Palmgren, M.G. 1999. Energization of plant cell membranes by H+-pumping ATPases: Regulation and biosynthesis. *The Plant Cell* 11:677–689.

Taiz, L. 1992. The plant vacuole. *Journal of Experimental Botany* 172:113–122.

Tapken, D. and Hollmann, M. 2008. *Arabidopsis thaliana* glutamate receptor ion channel function demonstrated by ion pore transplantation. *Journal of Molecular Biology* 383:36–48.

Tester, M. and Davenport, R.J. 2003. Na+ tolerance and Na+ transport in higher plants. *Annals of Botany* 91:503–527.

Toei, M., Saum, R., and Forgac, M. 2010. Regulation and isoform function of the VATPases. *Biochemistry* 49:4715–4723.

Tsunekawa, K., Shijuku, T., Hayashimoto, M. et al. 2009. Identification and characterization of the Na+/H+ antiporter Nhas3 from the thylakoid membrane of *Synechocystis* sp. PCC 6803. *Journal of Biological Chemistry* 284:16513–16521.

Venema, K., Belver, A., Marin-Manzano, M.C., Rodriguez-Rosales, M.P., and Donaire, J.P. 2003. A novel intracellular K+/H+ antiporter related to Na+/H+ antiporters is important for K+ ion homeostasis in plants. *Journal of Biological Chemistry* 278:22453–22459.

Vera-Estrella, R., Barkla, B.J., GarcíaRamírez, L., and Pantoja, O. 2005. Salt stress in *Thellungiella halophila* activates Na+ transport mechanisms required for salinity tolerance. *Plant Physiology* 139:1507–1517.

Waters, S., Gilliham, M., and Hrmova, M. 2013. Plant high-affinity potassium (HKT) transporters involved in salinity tolerance: Structural insights to probe differences in ion selectivity. *International Journal of Molecular Sciences* 14:7660–7680.

Wei, Q., Guo, Y.J., Cao, H.M., and Kuai, B.K. 2011. Cloning and characterization of an AtNHX2-like Na$^+$/H$^+$ antiporter gene from *Ammopiptanthus mongolicus* (Leguminosae) and its ectopic expression enhanced drought and salt tolerance in *Arabidopsis thaliana*. *Plant Cell Tissue and Organ Culture* 105:309–316.

Wu, S.J., Ding, L., and Zhu, J.K. 1996. SOS1, a genetic locus essential for salt tolerance and potassium acquisition. *The Plant Cell* 8:617–627.

Xu, H., Jiang, X., Zhan, K. et al. 2008. Functional characterization of a wheat plasma membrane Na$^+$/H$^+$ antiporter in yeast. *Archives of Biochemistry and Biophysics* 473:8–15.

Xu, Y., Zhou, Y., Hong, S. et al. 2013. Functional characterization of a wheat NHX antiporter gene TaNHX2 that encodes a K$^+$/H$^+$ exchanger. *PLoS One* 8:e78098.

Yadav, N.S., Shukla, P.S., Jha, A., Agarwal, P.K., and Jha, B. 2012. The *SbSOS1* gene from the extreme halophyte *Salicornia brachiata* enhances Na$^+$ loading in xylem and confers salt tolerance in transgenic tobacco. *BMC Plant Biol* 12:188.

Yamada, N., Theerawitaya, C., Chaum, S., Kirdmanee, C., and Takabe, T. 2014. Expression and functional analysis of putative vacuolar Ca$^{2+}$ transporters (CAXs and ACAs) in roots of salt tolerant and sensitive rice cultivars. *Protoplasma* 251:1–9.

Yamaguchi, T., Aharon, G.S., Sottosanto, J.B., and Blumwald, E. 2005. Vacuolar Na$^+$/H$^+$ antiporter cation selectivity is regulated by calmodulin from within the vacuole in a Ca$^{2+}$ and pH dependent manner. *Proceedings of the National Academy of Sciences of the United States of America* 102:16107–16111.

Yang, H., Knapp, J., Koirala, P. et al. 2007. Enhanced phosphorus nutrition in monocots and dicots overexpressing a phosphorus-responsive type I H$^+$-pyrophosphatase. *Plant Biotechnology Journal* 5:735–745.

Yokoi, S., Quintero, F.J., Cubero, B. et al. 2002. Differential expression and function of *Arabidopsis thaliana* NHX Na$^+$/H$^+$ antiporters in the salt stress response. *The Plant Journal* 30:529–539.

Yu, J.N., Huang, J., Wang, Z.N., Zhang, J.S., and Chen, S.Y. 2007. An Na$^+$/H$^+$ antiporter gene from wheat plays an important role in stress tolerance. *Journal of Biosciences* 32:1153–1161.

Zhang, H., Shen, G., Kuppu, S., Gaxiola, R., and Payton, P. 2011. Creating drought- and salt-tolerant cotton by overexpressing a vacuolar pyrophosphatase gene. *Plant Signaling and Behavior* 6:861–863.

Zhang, H.H., Hodson, J.N., Williams, J.P., and Blumwald, E. 2001. Engineering salt tolerant *Brassica* plants: Characterization of yield and seed oil quality in transgenic plants with increased vacuolar sodium accumulation. *Proceedings of the National Academy of Sciences of the United States of America* 98:12832–12836.

Zhang, H.X. and Blumwald, E. 2001. Transgenic salt-tolerant tomato plants accumulate salt in foliage but not in fruits. *Nature Biotechnology* 19:765–768.

Zhao, J., Barkla, B.J., Marshall, J., Pittman, J.K., and Hirschi, K.D. 2008. The *Arabidopsis cax3* mutants display altered salt tolerance, pH sensitivity and reduced plasma membrane H$^+$-ATPase activity. *Planta* 227:659–669.

Zhen, R.G., Kim, E.J., and Rea, P.A. 1997. The molecular and biochemical basis of pyrophosphate-energized proton translocation at the vacuolar membrane. *Advances in Botanical Research* 25:298–337.

Zhu, J.K. 2002. Salt and drought stress signal transduction in plants. *Annual Review in Plant Biology* 53:247–273.

Zhu, J.K. 2003. Regulation of ion homeostasis under salt stress. *Current Opinion in Plant Biology* 6:441–445.

Zhu, J.K. 2007. *Plant Salt Stress. In eLS*. John Wiley & Sons Ltd, Chichester, U.K.

Zorb, C., Noll, A., Karl, S., Leib, K., Yan, F., and Schubert, S. 2005. Molecular characterization of Na$^+$/H$^+$ antiporters (ZmNHX) of maize (*Zea mays* L.) and their expression under salt stress. *Journal of Plant Physiology* 162:55–66.

# CHAPTER TWENTY ONE

# Transgenic Plants for Higher Antioxidant Contents and Salt Stress Tolerance

*Nageswara Rao Reddy Neelapu, K.G.K. Deepak, and Challa Surekha*

## CONTENTS

*Abstract.* Abiotic stresses such as high salinity, drought, and extreme temperature adversely affect plant growth, crop productivity, and subsequently the death of plant. Salt stress in plants leads to stomatal closure and reduces the carbon dioxide/oxygen ratio. The reduced ratio results in the generation of high amounts of reactive oxygen species (ROS) contributing to oxidative stress. In normal aerobic conditions, ROS are effectively neutralized, but in abiotic stress conditions, there is an excess production of ROS that damages the plant. Plants have developed complex antioxidant defense systems such as ROS-scavenging enzymes and low-molecular antioxidants. Plants accumulating high amounts of antioxidant enzymes and antioxidants under salt stress were proven to maintain the plant growth and productivity. Progress has been made in developing transgenic lines that are tolerant to salt-induced oxidative stress by altering antioxidant levels in different crops. In this connection, to harness the potential of transgenic lines in counteracting stress-induced oxidative stress in abiotic conditions is reviewed.

*Keywords:* Antioxidants, Salt stress, Transgenic plants, Reactive oxygen species

## 21.1 INTRODUCTION

Abiotic factors such as high salinity, drought, extreme temperatures, excessive light, ozone exposure, UV-B irradiation, water logging, and osmotic shock show a negative impact on living organisms (Dita et al. 2006). This negative impact in a specific environment is known as stress.

Almost 71% of the earth's surface is occupied by saline water. High salinity is a common phenomenon for the arid and semiarid regions. The major cations contributing to salinity are $Na^+$, $Ca^{2+}$, $Mg^{2+}$, and $K^+$ and anions are $Cl^-$, $SO_4^{2-}$, $HCO_3^-$, $CO_3^{2-}$, and $NO_3^{2-}$. Sodium chloride is abundantly found in saline areas. There is a prediction that 30% of land will became unusable for agriculture in next 25 years because of soil salinization

(Kolodyazhnaya et al. 2009). The excess accumulation of salts in soil or water is known as salinization. Salinization can be due to natural processes such as mineral weathering, dust, precipitation, and by the movement of salt to the land surface with groundwater. Salts are carried in land from the ocean by wind and rainfall and have accumulated over long periods of time. It can also be due to artificial processes such as use of saline water for irrigation, aquaculture, salting of icy roads, and the use of potassium as fertilizer, which can form sylvite, a naturally occurring salt.

Salinity and drought are two major environmental factors that determine plant productivity, distribution, and reduced soil water potential (Groppa and Benavides 2008). Effects of salinity and drought are expressed by a series of morphological, physiological, metabolic, and molecular changes in plants. These changes adversely affect plant growth and productivity. Salinity and drought affect more than 10% of arable land. Desertification and salinization are rapidly increasing on a global scale. This results in a decrease in average crop yields by more than 50% (Bartels and Sunkar 2005; Bray et al. 2000).

Plants have developed adaptive or tolerance mechanisms to survive under salt stress (Wu et al. 2013). Initial exposure of plants to saline conditions will rapidly and temporarily decrease the growth rate (i.e., photosynthetic capacity). It is followed by a gradual recovery to a new reduced rate of growth (Munns 2002). The temporary effects are clear due to rapid and transient changes in plant water relations, that is, decrease in external water potential. Effects of high salinity become visible after several days with marked injury in older leaves. Injury is due to $Na^+$ or $Cl^-$ (or both) accumulating in the cytoplasm of leaves above toxic levels, inhibition of enzyme activity, and dehydration of cells (Flowers and Yeo 1986; Munns and Passioura 1984). The effects of salinity are more prominent in older leaves, and progressive loss of older leaves with time is the crucial issue determining the survival of plants (Munns 2002; Figure 21.1).

High concentrations of $Na^+$ damages plants in two ways: osmotic stress and toxic effect. $Na^+$ is toxic and destabilizes membranes, proteins. $Na^+$ negatively affects cellular and physiological processes such as cell division and expansion, primary and secondary metabolism, and mineral nutrient homeostasis (Munns and Tester 2008). Both $K^+$ and $Ca^{2+}$ are required to maintain the selective transport of ions across membranes and integrity of the cell membrane. Under salt stress, $Na^+$ levels increase and $Ca^{2+}$, $K^+$, and $Mg^{2+}$ levels decrease in cytosol causing cellular ion imbalance (Aleman et al. 2011). As a result, high $Na^+/K^+$ and $Na^+/Ca^{2+}$ ratios may impair the selectivity of root membranes and account for passive accumulation of $Na^+$ in the roots and shoots. High salinity causes hyperosmotic stress and ion disequilibrium, and generates secondary effects such as accumulation of ROS and growth retardation (Rishi and Sneha 2013).

ROS is a product of altered chloroplastic and mitochondrial metabolism during stress. In the absence of proper protective mechanisms, these species cause oxidative damage to different cellular components (Apel and Hirt 2004). Maintaining equilibrium between ROS and antioxidative defense determines the fate of the plant (Ajay Arora et al. 2002). The changes in equilibrium lead to sudden increase in intracellular levels of ROS posing a threat to damage cells. Damage in cells is by peroxidation of lipids, oxidation of proteins and nucleic

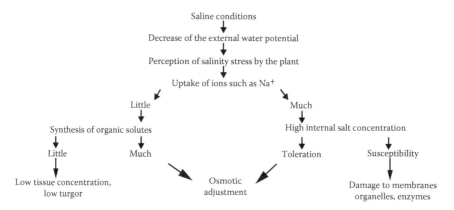

Figure 21.1. Outline of the cardinal responses of plants to saline condition.

acid damage, and enzyme inhibition. It also activates programmed cell death pathway ultimately leading to the death of cells (Maheshwari and Dubey 2009; Meriga et al. 2004; Mishra et al. 2011; Mittler 2002; Pallavi et al. 2012; Shah et al. 2001; Sharma et al., 2012b; Srivastava and Dubey 2011; Verma and Dubey 2003). In this chapter, we summarize the ROS–antioxidative system in normal conditions and in saline conditions. We also highlight the transgenic plants with increased antioxidant enzymes, which enhanced salt tolerance.

## 21.2 ROS: ANTIOXIDATIVE SYSTEM UNDER NORMAL CONDITIONS

ROS includes free radicals like superoxide anion ($O_2^-$), hydroxyl radical (OH·), as well as non-radical molecules like hydrogen peroxide ($H_2O_2$) and singlet oxygen ($^1O_2$) (Gill and Tuteja 2010; Joseph and Jini 2010; Rishi and Sneha 2013; Sharma et al. 2012b). Under normal growth conditions, ROS are produced at low level. They are produced from organelles such as chloroplasts, mitochondria, or peroxisomes with highly oxidizing metabolic activity (Gill and Tuteja 2010; Figure 21.2).

### 21.2.1 Reactive Oxygen Species

Atmospheric oxygen ($O_2$) is relatively nonreactive and harmless in its ground state. But activation of $O_2$ reverses the spin of one unpaired electron by absorbing sufficient energy or stepwise monovalent reduction of $O_2$ to form $O_2^-$, $OH^-$, $H_2O_2$, and $H_2O$ (Scandalios 2005; Figure 21.3).

Superoxide radical ($O_2^-$) is produced mostly in the thylakoid membrane-bound primary electron acceptor of photosystem I (PSI). $O_2^-$ is produced upon photoreduction of dioxygen ($O_2$) in chloroplasts during electron transport along the noncyclic pathway. The electron flow between PSI and PSII regulates the reduction state of the acceptor side of PSI. The reduction of $O_2$ to $O_2^-$ proceeds via reduced ferredoxin ($Fd_{red}$) (Ajay Arora et al. 2002; Gill and Tuteja 2010; Sharma et al. 2012b).

$$2O_2 + 2Fd_{red} \longrightarrow 2O_2^- + 2Fd_{ox}$$

The generation of $O_2^-$ triggers the formation of more reactive ROS like hydrogen peroxide ($H_2O_2$) by dismutation and singlet oxygen ($^1O_2$). Electrons from $O_2^-$ converts oxidized iron ($Fe^{3+}$) to reduced form ($Fe^{2+}$). The reduced iron ($Fe^{2+}$) in turn reduces $H_2O_2$ to form $OH^-$ (Haber–Weiss reaction) and gets reoxidized itself by $H_2O_2$ (Fenton's reaction). It is also generated by NADPH oxidases and cell wall peroxidases (Abogadallah 2010; Sagi and Fluhr 2006).

$$2O_2^- + 2H \xrightarrow{\text{SOD}} H_2O_2 + O_2$$

$$Fe^{3+} + O_2^- \longrightarrow Fe^{2+} + O_2$$

$$Fe^{2+} + H_2O_2 \longrightarrow Fe^{3+} + OH^- + OH·$$

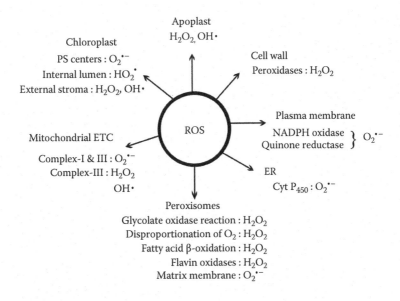

Figure 21.2. Sites of reactive oxygen species (ROS) production in plant.

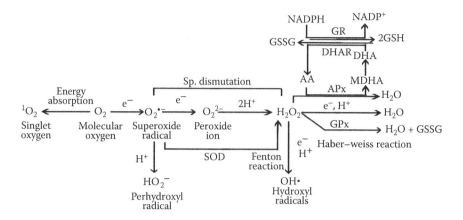

Figure 21.3. Generation of reactive oxygen species (ROS) and antioxidative mechanism in plants.

Singlet oxygen ($^1O_2$) is an unusual ROS formed due to insufficient energy dissipation during photosynthesis. Low intercellular $CO_2$ concentration in the chloroplast due to stomatal closure also favors the formation of $^1O_2$. $^1O_2$ is highly destructive oxidizing agent reacting with most of the biological molecules.

Hydrogen peroxide ($H_2O_2$) is produced by the dismutation of superoxide radicals and is moderately reactive. The majority of $H_2O_2$ is accounting from oxidation of glycolate by glycolate oxidase in peroxisomes during photorespiration (Karpinski et al. 2003; Noctor et al. 2002). $H_2O_2$ may inactivate enzymes by oxidizing thiol groups, and excess $H_2O_2$ leads to oxidative stress. $H_2O_2$ plays a dual role: at low concentrations, it acts as a signal molecule in triggering tolerance to various stresses; and at high concentrations, it leads to programmed cell death (Quan et al. 2008).

Hydroxyl radical (OH·) is the most highly reactive ROS, and the formation is dependent on both $H_2O_2$ and $O_2^-$. The Fenton reaction generates OH· from $H_2O_2$ and $O_2^-$ in the presence of suitable transitional element especially Fe.

$$Fe^{3+} + O_2^- \longrightarrow Fe^{2+} + O_2$$

$$Fe^{2+} + H_2O_2 \longrightarrow Fe^{3+} + OH^- + OH·$$

$$O_2^- + H_2O_2 \longrightarrow OH· + OH^- + O_2$$

It has a single unpaired electron that can react with oxygen in triplet ground state. OH· interacts with all biological molecules causing cellular damage (Rhoads et al. 2006).

When ROS are in low/moderate concentrations, they act as messengers in intracellular signaling cascades. These cascades mediate several plant responses such as stomatal closure (Kwak et al. 2003; Neill et al. 2002; Yan et al. 2007), programmed cell death (Bethke and Jones 2001; Mittler 2002), gravitropism (Joo et al. 2001), and acquisition of tolerance to biotic and abiotic stress (Miller et al. 2008; Torres et al. 2002; Figure 21.4). But when ROS is at high concentrations, a cell is said to be in a state of oxidative stress.

### 21.2.2 Antioxidative System

Plants possess efficient antioxidative system for scavenging ROS to protect cells from oxidative reactions. The antioxidative system comprises enzymatic and nonenzymatic antioxidants. They are found in different organelles such as chloroplasts, mitochondria, and peroxisomes (Noctor and Foyer 1998). The enzymatic antioxidative system comprises of several antioxidant enzymes such as superoxide dismutase (SOD), catalase (CAT), guaiacol peroxidase (GPX), ascorbate peroxidase (APx), monodehydroascorbate reductase (MDHAR), dehydroascorbate reductase (DHAR), and glutathione reductase (GR) (Ajay Arora et al. 2002; Fotopoulos et al. 2008; Gill and Tuteja 2010; Joseph and Jini 2010; Mhamdi et al. 2010; Miller et al. 2010; Mittler et al. 2004; Noctor and Foyer 1998; Petrov and Van Breusegems 2012; Rishi and Sneha 2013; Sharma et al. 2012b). Nonenzymatic antioxidants include ascorbate, glutathione (GSH),

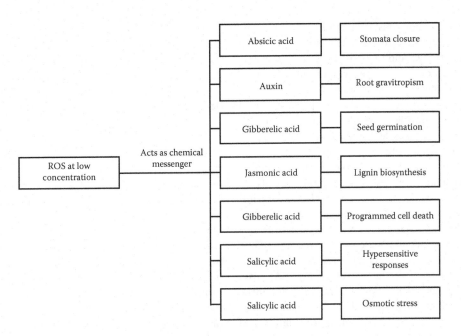

Figure 21.4. Reactive oxygen species (ROS) act as messengers in intracellular signaling and mediate several plant responses.

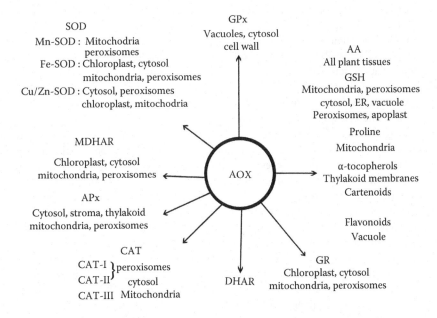

Figure 21.5. Sites of enzymatic and non-enzymatic antioxidants (AOX) production in plants.

proline, α-tocopherol, flavonoids, and carotenoids (Figure 21.5).

SOD belongs to metalloenzymes, catalyzes the dismutation of $O_2^-$ to $O_2$ and $H_2O_2$. SOD plays first line of defense against the toxic effects of elevated levels of ROS. It is present in most of the subcellular compartments that generate activated oxygen. SOD removes $O_2^-$ and, hence, decreases the risk of OH· radical formation. SOD are classified into three isoenzyme forms based on their metal cofactors: the copper/zinc (Cu/Zn-SOD), the manganese (Mn-SOD), and the iron (Fe-SOD) (Fridovich 1989; Mittler 2002; Racchi et al. 2001).

CATs are ubiquitous tetrameric heme-containing enzymes that directly dismutate two molecules of $H_2O_2$ into $H_2O$ and $O_2$. It is dispensable for ROS detoxification during stress conditions. $H_2O_2$ produced in peroxisomes is scavenged by CAT during photorespiratory oxidation, β-oxidation of fatty acids, purine catabolism, and other enzyme systems (Corpas et al. 2008; Del Rio et al. 2006; Scandalios et al. 1997).

APx plays an essential role in the control of intracellular ROS levels and protects cells in higher plants, algae, *Euglena*, and other organisms. APx scavenges $H_2O_2$ in water–water cycle and ascorbate–glutathione (ASH–GSH) cycle. It utilizes two molecules of ascorbate to reduce $H_2O_2$ to water with a concomitant generation of two molecules of monodehydroascorbate (MDHA) from chloroplasts, peroxisomes, and mitochondria. Five distinct isoenzymes of APx family are found in cytosol, stroma, thylakoid, mitochondria, and peroxisomes under stress conditions (Jimenez et al. 1997; Madhusudhan et al. 2003; Nakano and Asada 1987; Sharma and Dubey 2004; Wang et al. 1999). APx has high affinity to $H_2O_2$ than CAT making APx efficient scavengers of $H_2O_2$, which is produced within the organelles.

Glutathione peroxidase (GPx) is a heme-containing protein that prefers aromatic electron donors such as guaiacol and pyrogallol at the expense of $H_2O_2$. GPx isoenzymes are localized in vacuoles, cell wall, and cytosol (Asada 1992). GPx has a role in the lignification of cell wall, degradation of indole acetic acid, ethylene biosynthesis, and defense against biotic and abiotic stresses by consuming $H_2O_2$ (Kobayashi et al. 1996).

MDHAR is a FAD enzyme present in chloroplasts, cytosol, mitochondria, and peroxisomes. It catalyzes the regeneration of ascorbic acid (AsA) from the MDHA radical using NAD(P)H as the electron donor (Hossain and Asada 1985). The enzyme uses an organic radical (MDA) as substrate and is also capable of reducing phenoxy radicals generated by horseradish peroxidase (Sakihama et al. 2000). DHAR catalyzes the reduction of AsA using GSH as the substrate (Ushimaru et al. 1997). It plays an important role in maintaining AsA in its reduced form. It is a monomeric thiol enzyme found in dry seeds, roots, and shoots. GR is a flavoprotein oxidoreductase, potential enzyme of ASH–GSH cycle, and plays an essential role in defense system against ROS. GR, an NAD(P)H-dependent enzyme, catalyzes the

reduction of oxidized glutathione (GSSH) to GSH and maintains the high cellular GSH/GSSH ratio. It is abundantly found in chloroplasts and also in cytosol, mitochondria, and peroxisomes.

Ascorbic acid (vitamin C) is the most abundant, low-molecular-weight, and water-soluble antioxidant that has a key role in defense against oxidative stress. It is considered as a powerful antioxidant because of its ability to donate electrons in enzymatic and nonenzymatic reactions. It directly scavenges the $O_2^-$ and $OH^-$ and provides protection to membranes. It occurs in all plant tissues with the highest in photosynthetic cells and meristems.

GSH is a low-molecular-weight nonprotein thiol that plays the most important role in intracellular defense against ROS damage. It occurs in reduced form in plant tissues and is localized in cytosol, endoplasmic reticulum, vacuole, mitochondria, chloroplasts, peroxisomes, and apoplasts (Jimenez et al. 1998; Mittler and Zilinskas 1992). The balance between the GSH and the GSSG is a central component in maintaining cellular redox state. Due to its reducing power, it plays a central role in the expression of stress-responsive genes (Mullineaux and Rausch 2005), cell differentiation, cell death, and enzymatic regulation. It also regulates sulfate transport (Rausch and Wachter 2005) and conjugation of metabolites, etc. It is a potential scavenger of $^1O_2$, $H_2O_2$, and $OH^-$ (Briviba et al. 1997; Larson 1988; Noctor and Foyer 1998).

Proline acts as an osmoprotectant and plays an important role in osmotic balancing and protection and in increasing turgor necessary for cell expansion (Matysik et al. 2002; Sairam and Tyagi 2004). Proline is considered as the only osmolyte that has been shown to scavenge singlet oxygen and free radicals including hydroxyl ions. Proline acts as protein stabilizer, metal chelator, an inhibitor of lipid peroxidation, and $OH^-$ and $^1O_2$ scavenger (Ashraf and Foolad 2007; Trovato et al. 2008). It is an important molecule in redox signaling and also an effective quencher of ROS under salt stress (Alia and Pardha Saradhi 1991). It also serves as redox potential regulator and protects macromolecules such as proteins and DNA and reduces enzyme denaturation caused by heat, NaCl, and other stresses (Kumar et al. 2010; Matysik et al. 2002). Accumulation, biosynthesis, transportation, and role of proline during salinity stress have been investigated thoroughly by KaviKishor et al. (2005).

α-Tocopherol is considered as a potential scavenger of ROS and lipid radicals (Diplock et al. 1989; Hollander-Czytko 2005). Tocopherols play as antioxidants by protecting the membrane stability, scavenging $^1O_2$, and preventing the chain propagation step in lipid autooxidation. Accumulation of tocopherols has shown tolerance toward salt in different plant species (Bafeel and Ibrahim 2008; Guo et al. 2006; Munne-Bosch et al. 1999; Yamaguchi-Shinozaki and Shinozaki 1994). Carotenoids act as antioxidants and scavenge $^1O_2$ thereby inhibiting oxidative damage. They also serve as signaling molecules that influence plant development and salt stress (Li et al. 2008). Flavonoids serve as ROS scavengers by neutralizing radicals before damage takes place.

ROS production and scavenging allows dual function of ROS in plants and is thought to be under the regulation of a large network of genes (Miller et al. 2008; Mittler et al. 2004; Petrov and Van Breusegem 2012).

## 21.3 REACTIVE OXYGEN SPECIES: ANTIOXIDATIVE SYSTEM UNDER SALT STRESS CONDITION IN PLANTS

Salt tolerance in plants is a multigenic complex trait. Salt stress affects plant growth and development by imposing two stresses: osmotic stress and ion cytotoxicity affecting the activity of antioxidant enzymes leading to an increase in ROS and further cell damage (Chen et al. 2007; Hossain and Fujita 2013; Pandolfi et al. 2012). Initial stage of salt stress decreases the rate of leaf growth due to osmotic stress around the roots that decreases water absorption capacity and causes leaf cells to lose water, and therefore salt stress is also considered as hyperosmotic stress (Munns 2005). Salt stress also generates excess ROS: $O_2^-$, $OH^-$, $H_2O_2$, and $^1O_2$ by impairment of the cellular electron transport in chloroplasts and mitochondria. The water loss in leaves is transient and reduces cell elongation and cell division, which ultimately leads to slower leaf appearance and smaller size. This osmotic stress causes membrane disruption, nutrient imbalance, less scavenging of ROS due to differences in antioxidant enzymes, and finally decreased photosynthetic activity due to stomatal closure (Bartels and Sunkar 2005; Munns and Tester 2008; Rahnama et al. 2010). ROS generated due to metabolic imbalances during salt stress acts as a signal to activate acclimation and defense mechanism against oxidative stress (Davletova et al. 2005; Miller et al. 2008; Mittler et al. 2004). The equilibrium between the production of ROS and scavenging of ROS by antioxidant molecules (AOX) is lost (ROS > AOX), it results in an imbalance in intracellular ROS leading to oxidative stress that causes damage to biomolecules such as lipids, proteins, and DNA (Bose et al. 2013; Hossain and Fujita 2013; Hossain et al. 2013; Sharma et al. 2012a) (Figure 21.6).

Accumulation of salt to toxic concentrations in fully expanded leaves causes ion cytotoxicity eventually leading to leaf death. $Na^+$ and $Cl^-$ ions penetrate hydration shells and interfere with non-covalent interactions between amino acids leading to conformational changes and loss of protein function. Salt stress causes overflow, deregulation, or even disruption of electron transport chains in chloroplasts and mitochondria. Under these conditions, molecular oxygen acts as an electron acceptor giving rise to the accumulation of ROS: $O_2^-$, $OH^-$, $H_2O_2$, and $^1O_2$ harmful for cell integrity (Grob et al. 2013). ROS induced by salt stress are detoxified by antioxidant enzymes and nonenzymatic compounds.

Increased salt tolerance and activation of antioxidant enzyme relationship are demonstrated in plantago, pea, tomato, maize, sorghum, soybean, and mulberry (Azevedo-Neto et al. 2006; Cicek and Cakyrlar 2008; Costa et al. 2005; Harinasut et al. 2003; Heidari 2009; Hernandez et al. 2000; Sekmen et al. 2007). The generation of ROS and increased antioxidant activity during salt stress are reported in cotton, mulberry, wheat, tomato, rice, sugar beet, and maize (Azevedo-Neto et al. 2006; Bor et al. 2003; Desingh and Kanagaraj 2007; Harinasut et al. 2003; Mittova et al. 2004; Sairam et al. 2002; Sudhakar et al. 2001; Vaidyanathan et al. 2003). Increase in peroxidase and polyphenol oxidase activities under 50–200 mM NaCl in three *Acanthophyllum* species was observed by Meratan et al. (2008). SOD, APx, and CAT activities increased in tomato and citrus under salt tolerance (Gueta-Dahan et al. 1997; Mittova et al. 2004).

Significant increase in SOD activity under salt stress was observed in mulberry (Harinasut et al. 2003), *Cicer arietinum* (Kukreja et al. 2005), and *Lycopersicon esculentum* (Gapinska et al. 2008). Eyidogan and Oz (2005) noted the increased expression of SOD, CAT, and GR activities in *C. arietinum* leaves under salt treatment. Significant

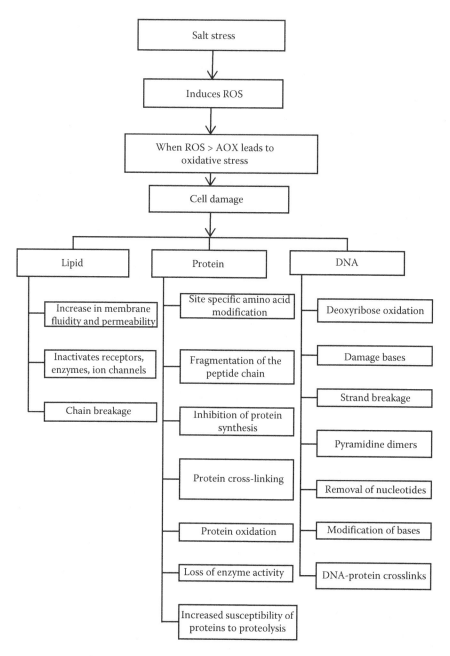

Figure 21.6. Salt stress–induced reactive oxygen species (ROS) leading to oxidative damage to lipids, proteins, and DNA.

increase in SOD and APx activity was noted under salt stress in *Anabaena doliolum* (Srivastava et al. 2005). GPx activity also increased in both leaf and root tissues of *Vigna radiata* (Panda 2001) and *O. sativa* (Koji et al. 2009) under salt stress. GR and GSH play a crucial role in determining the tolerance of a plant under various stresses.

Kukreja et al. (2005) and Srivastava et al. (2005) observed increased GR activity in *C. arietinum* roots under salt stress. Antioxidant enzymes are involved in salt tolerance and is evident by transgenic plants with reduced or increased expression of antioxidant enzymes (Willekens et al. 1995).

## 21.4 OVEREXPRESSION OF ANTIOXIDANTS IN TRANSGENIC PLANTS ENHANCING SALT TOLERANCE

Antioxidant enzymes play crucial roles in the protection of plant tissues under stress conditions including salinity (Apel and Hirt 2004; Fotopoulos et al. 2008; Joseph and Jini 2010). Some species display more salt tolerance because of increased activities of antioxidant enzymes and higher accumulation of antioxidative compounds that scavenge ROS efficiently (Apel and Hirt 2004; Bhutta 2011; Nabati et al. 2011). Overexpression of ROS-responsive regulatory genes are involved in acclimation, enhances salt tolerance, and is found to be beneficial.

SOD is nuclear encoded and targeted to its subcellular compartment by an amino-terminal targeting sequence. SOD forms are cloned in a variety of plants (Gapinska et al. 2008), and its upregulation combats oxidative stress and makes plants survive under salt stress. Overexpression of Mn-SOD in transgenic *Arabidopsis*, *L. esculentum*, and tobacco showed enhanced tolerance against salt stress (Wang et al. 2004, 2007). Overexpression of Cu/Zn-SOD from a mangrove (*Avicennia marina*) in transgenic rice was more tolerant to 150 mM NaCl stress for a period of 8 days when compared to controls (Prashanth et al. 2008). Transgenic sweet potato (*Ipomoea batatas* L. cv. Yulmi) harboring two genes Cu/Zn-SOD and APx under stress-inducible SWPA2 promoter showed significant amount of enzymes under salt stress of 100 mmol/L (Wang et al. 2012). Simultaneous expression of three antioxidant enzymes Cu/Zn-SOD, APx, and DHAR in transgenic tobacco increased 1.6–2.1 times DHAR activity and high ratio of reduced ascorbate to DHA and GSSH to reduced GSH in chloroplasts enabling plants with enhanced tolerance to multiple environmental stress (Lee et al. 2007). Transgenic *Arabidopsis* with transcription factor YAP1 from yeast showed increased activities of antioxidant enzymes: SOD, CAT, APx, GST, and GR under various NaCl concentrations (Zhao et al. 2009). Wang et al. (2010) introduced Mn-SOD gene from *Tamarix androssowii* (TaMn-SOD) into poplar and transgenic poplar, which showed enhanced SOD activity and decreased MDA content under NaCl stress.

Nagamiya et al. (2007) overexpressed CAT gene katE in japonica rice and enhanced salt tolerance up to 100 mM salt concentration. Transgenic rice with katE gene were able to grow up to 15 days in 100 mM NaCl solution at a very young stage and 7 days in 250 mM NaCl, whereas control plants died within 5 days in 100 mM NaCl (Prodhan et al. 2008). Transgenic BR5 rice plants with katE gene showed 150% (approximately) higher CAT activity when compared to non-transgenic plants and exhibited high tolerance under NaCl stress (Moriwaki et al. 2007).

Overexpression of GPx (AcGPx2) in *Arabidopsis* showed enhanced tolerance to oxidative damage caused by various abiotic stresses including salinity (Gaber et al. 2006). Badawi et al. (2004) overexpressed APx in *Nicotiana tobacum* chloroplasts, which enhanced plant salt tolerance. APx expression patterns were studied in etiolated *O. sativa* seedlings, and the expression of OsAPx8 increased in 150 and 200 mM NaCl stress (Hong et al. 2007). Overexpression of OsAPx in transgenic *Arabidopsis* enhanced salt tolerance and maintained high APx activity under NaCl stress (Lu et al. 2007). Xu et al. (2008) transformed *Arabidopsis* with pAPx gene from barley (HvAPx1), which were more tolerant to salt stress than the wild type.

Overexpression of GST and GPx activities in tobacco as Nt107 genes increased GSSG content enhance seedlings to grow under salt stress (Roxas et al. 2000). Qi et al. (2010) introduced GST gene from *Suaeda salsa* into *Arabidopsis* under the control of the CaMV35S promoter, which showed a higher level of salt tolerance. Overexpression of transcription factors such as Zat 10, Zat 12, or JERF3 enhanced the expression of ROS-scavenging genes and enhanced salt tolerance (Sakamoto et al. 2004). Overexpression of mitogen-activated kinase kinase 1 (MKK1) in *Arabidopsis* enhanced MAPK1 and ROS activity leading to increased salt tolerance (Teige et al. 2004; Xing et al. 2008). Overexpression of a vacuolar proton pyrophosphatase gene from *Sorghum bicolor* (Sb-VPPase) showed an increase in antioxidant enzymes SOD, CAT, APx, GPx, and GR by 25%–100% in transgenic finger millet under salt stress (Anjaneyulu et al. 2014).

Along with the earlier-mentioned enzymes, DHAR, MDHAR, and GR are also involved in the scavenging of superoxide radicals and $H_2O_2$ (Asada 1999). Overexpression of DHAR or MDHAR in transgenic tobacco showed higher levels of ASH and better tolerance (Ushimaru et al. 1997; Yin et al. 2010). Overexpression of MDHAR in transgenic tobacco increased the tolerance against salt and osmotic stress (Eltayeb et al. 2007). Lee and Jo (2004) introduced BcGR1 into tobacco plants isolated from Chinese cabbage, which increased tolerance to oxidative stress.

Accumulation of high amounts of proline is often related to the salt tolerance nature of plant species, and several researchers reported high proline accumulation in the salt-tolerant genotype than in their salt-sensitive counterparts including wheat (Sairam et al. 2005), sorghum (Jogeswar et al. 2006), mulberry (Kumar et al. 2003), green gram (Misra and Gupta 2005), and tobacco (Gubis et al. 2007; Yamchi et al. 2007). $\Delta$1-Pyrroline-5-carboxylate synthetase (P5CS), a rate-limiting enzyme in proline biosynthesis, is mostly used, and its overexpression in transgenics showed enhanced oxidative stress tolerance driven by drought and salt stresses. Overexpression of P5CS gene increases proline production and accumulation, which confers salt tolerance in transgenics in a number of crop plants including rice (Anoop and Gupta 2003; Karthikeyan et al. 2011; Su and Wu 2004), chickpea (Kiran Kumar Ghanti et al. 2011), wheat (Vendruscolo et al. 2007), potato (Hmida-Sayari et al. 2005), and tobacco (Yamchi et al. 2007). Overexpression of a mutated version of P5CS gene (P5CSF129A) in the overproduction of proline is seen in transgenic citrus rootstock (Molinari et al. 2004), chickpea (Bhatnagar-Mathur et al. 2009), rice (Kumar et al. 2010), and pigeon pea (Surekha et al. 2014).

## 21.5 CONCLUSIONS AND FUTURE PERSPECTIVE

The review briefly summarizes the production of ROS: $O_2^-$, $OH^-$, $H_2O_2$, and $^1O_2$ by electron transport activities of chloroplasts, mitochondria, plasma membrane, or as by-products of various metabolic pathways. Under normal growth conditions, the production of ROS is low, but salt stress leads to overproduction of ROS in plants. ROS are highly reactive and toxic to plants when the concentration exceeds in cells, ultimately leading to oxidative stress. Low concentrations of ROS in plants act as regulatory molecules that mediate several plant responses, whereas high concentration of ROS causes oxidative damage to nucleic acids, lipids, and proteins. The cells avoid oxidative damage as plants possess a complex antioxidative defense system comprising of enzymes: SOD, CAT, APx, GPx, GR, MDHAR, DHAR, and non-enzymatic components: proline, ascorbate, GSH, carotenoids, flavonoids. Overexpression of these antioxidant molecules resulted in salt tolerance in various crops due to enhanced scavenging activity. The attention is focused more on altering genes and their expression to upregulate the response to salt stress. The understanding of mechanisms that regulate multiple gene expression and the ability to transfer genes from other sources into plants will expand the ways in which plants can be made tolerant. The exploitation of cloned genes to alter the function of gene product in transgenic plants provides novel opportunities to assess their biological role in a salt stress response.

## ACKNOWLEDGMENT

The authors are grateful to GITAM University for providing necessary facilities to carry out the research work and for extending constant support in the preparation of this chapter.

## REFERENCES

Abogadallah, G.M. 2010. Antioxidative defense under salt stress. *Plant Signaling and Behavior* 5:369–374.

Ajay Arora, R., K. Sairam, and G.C. Srivastava. 2002. Oxidative stress and antioxidative system in plants. *Current Science* 82(10):1227–1238.

Aleman, F., M. Nieves Cordones, V. Martinez, and F. Rubio. 2011. Root K+ acquisition in plants: The *Arabidopsis thaliana* model. *Plant Cell Physiology* 52:1603–1612.

Alia and P. Pardha Saradhi. 1991. Proline accumulation under heavy metal stress, *Journal of Plant Physiology* 138:554–558.

Anjaneyulu, E., P.S. Reddy, M.S. Sunita, P.B.K. Kishor, and B. Meriga. 2014. Salt tolerance and activity of antioxidative enzymes of transgenic finger millet overexpressing a vacuolar H+ pyrophosphatase gene (SbVPPase) from *Sorghum bicolor*. *Journal of plant physiology* 171(10):789–798.

Anoop, N. and A.K. Gupta. 2003. Transgenic indica rice cv IR-50 overexpressing *Vigna aconitifolia* d (1) pyrroline-5-carboxylate synthetase cDNA shows tolerance to high salt. *Journal of Plant Biochemistry and Biotechnology* 12:109–116.

Apel, K. and H. Hirt. 2004. Reactive oxygen species: Metabolism, oxidative stress, and signal transduction. *Annual Review of Plant Biology* 55:373–399.

Asada, K. 1992. Ascorbate peroxidase-a hydrogen peroxide scavenging enzyme in plants. *Physiologia Plantarum* 85:235–241.

Asada, K. 1999. The water–water cycle in chloroplasts: Scavenging of active oxygen and dissipation of excess photons. *Annual Reviews in Plant Physiology and Plant Molecular Biology* 50:601–639.

Ashraf, M. and M.R. Foolad. 2007. Roles of glycine betaine and proline in improving plant abiotic stress resistance. *Environmental and Experimental Botany* 59:206–216.

Azevedo-Neto, A.D., J.T. Ptisco, J. Eneas-Filho, C.E.B. Abreu, and E. Gomez-Filho. 2006. Effect of salt stress on antioxidative enzymes and lipid peroxidation in leaves and roots of salt-tolerant and salt sensitive maize genotypes. *Environmental and Experimental Botany* 56:87–94.

Badawi, G.H., Y. Yamauchi, E. Shimada, R. Sasaki, N. Kawano, K. Tanaka, and K. Tanaka. 2004. Enhanced tolerance to salt stress and water deficit by overexpressing superoxide dismutase in tobacco (*Nicotiana tabacum*) chloroplasts. *Plant Science* 166(4):919–928.

Bafeel, S.O. and M.M. Ibrahim. 2008. Antioxidants and accumulation of α-tocopherol induce chilling tolerance in *Medicago sativa*. *International Journal of Agriculture and Biology* 10(6):593–598.

Bartels, D. and R. Sunkar. 2005. Drought and salt tolerance in plants. *Critical Reviews in Plant Science* 21:1–36.

Bethke, P.C. and R.L. Jones. 2001. Cell death of barley aleurone protoplasts is mediated by reactive oxygen species. *Plant Journal* 25(1):19–29.

Bhatnagar-Mathur, P., V. Vincent, M.J. DeviLavanya, G. Vani, and K.K. Sharma. 2009. Genetic engineering of chickpea (*Cicer arietinum* L.) with the P5CSF129A gene for osmoregulation with implications on drought tolerance. *Molecular Breeding* 23:591–606.

Bhutta, W.M. 2011. Antioxidant activity of enzymatic of two different wheat (*Triticum aestivum*. L). cultivars growing under salt stress. *Plant Soil Environment* 57:101–107.

Bor, M., F. Ozdemir, and I. Turkan. 2003. The effect of salt stress on lipid peroxidation and antioxidants in leaves of sugar beet *Beta vulgaris* L. and wild *Beta maritime* L. *Plant Science* 164:77–84.

Bose, J., A. Rodrigo-Moreno, and S. Shabala. 2013. ROS homeostasis in halophytes in the context of salinity stress tolerance. *Journal of Experimental Botany*. doi:10.1093/jxb/ert430.

Bray, E.A., J. Bailey-Serres, and E. Weretilnyk. 2000. Responses to abiotic stresses. In *Biochemistry and Molecular Biology of Plants*, Buchanan, B.B., W. Gruissem, and R.L. Jones (eds.). Rockville, MD: American Society of Plant Biologists, pp. 1158–1203.

Briviba, K., L.O. Klotz, and H. Sies. 1997. Toxic and signaling effects of photochemically or chemically generated singlet oxygen in biological systems. *Journal of Biological Chemistry* 378:1259–1265.

Chen, Z., I.I. Pottosin, T.A. Cuin, A.T. Fuglsang, M. Tester, D. Jha, I. Zepeda-Jazo, M. Zhou, M.G. Palmgren, I.A. Newman, and S. Shabala. 2007. Root plasma membrane transporters controlling $K^+/Na^+$ homeostasis in salt stressed barley. *Plant Physiology* 145:1714–1725.

Cicek, N. and H. Cakyrlar. 2008. Changes in some antioxidant enzyme activities in six soyabean cultivars in response to long-term salinity at two different temperatures. *General and Applied Plant Physiology* 34:267–280.

Corpas, F.J., J.M. Palma, L.M. Sandalio, R. Valderrama, J.B. Barroso, and L.A. del Río. 2008. Peroxisomal xanthine oxidoreductase: Characterization of the enzyme from pea (*Pisum sativum* L.) leaves. *Journal of Plant Physiology* 165(13):1319–1330.

Costa, P.H.A., A. Neto, M. Bezerra, J. Prisco, and E. Filho. 2005. Antioxidant-enzymatic system of two *Sorghum* genotypes differing in salt tolerance. *Brazilian Journal of Plant Physiology* 17:353–361.

Davletova, S., L. Rizhsky, H. Liang, Z. Shenggiang, D.J. Coutu, V. Shulev, K. Schlauch, and R. Mittler. 2005. Cytosolic ascorbate peroxidase 1 is a central component of the reactive oxygen gene network of *Arabidopsis*. *The Plant Cell* 17:268–281.

Del Rio, L.A., L.M. Scandalio, F.J. Corpas, J.M. Palma, and J.B. Barroso. 2006. Reactive oxygen species and reactive nitrogen species in peroxisomes. Production, scavenging, and role in cell signaling. *Plant Physiology* 141(2):330–335.

Desingh, R. and G. Kanagaraj. 2007. Influence of salinity stress on photosynthesis and antioxidative systems in two cotton varieties. *General Applied Plant Physiology* 33:221–234.

Diplock, T., L.J. Machlin, L. Packer, and W.A. Pryor. 1989. Vitamin E: Biochemistry and health implications. *Annals of the New York Academy of Sciences* 570:372–378.

Dita, M.A., N. Rispail, E. Prats, D. Rubiales, and K.B. Singh. 2006. Biotechnology approaches to overcome biotic and abiotic stress constraints in legumes. *Euphytica* 147:1–24.

Eltayeb, A.E., N. Kawano, G.H. Badawi, H. Kaminaka, T. Sanekata, T. Shibahara, S. Inanaga, and K. Tanaka. 2007. Overexpression of monodehydroascorbate reductase in transgenic tobacco confers enhanced tolerance to ozone, salt and polyethylene glycol stresses. *Planta* 225:1255–1264.

Eyidogan, F. and M.T. Oz. 2005. Effect of salinity on antioxidant responses of chickpea seedlings. *Acta Physiologiae Plantarum* 29:485–493.

Flowers, T.J. and A.R. Yeo. 1986. Ion relations of plant under drought and salinity. *Australian Journal of Plant Physiology* 13:75–91.

Fotopoulos, V., M.C. De Tullio, J. Barnes, and A.K. Kanellis. 2008. Altered stomatal dynamics in ascorbate oxidase over-expressing tobacco plants suggest a role for dehydroascorbate signalling. *Journal of Experimental Botany* 59(4):729–737.

Fridovich, I. 1989. Superoxide dismutases. An adaptation to a paramagnetic gas. *Journal of Biological Chemistry* 264(14):7761–7764.

Gaber, A., K. Yoshimura, T. Yamamoto, Y. Yabuta, T. Takeda, H. Miyasaka, Y. Nakano, and S. Shigeoka. 2006. Glutathione peroxidase-like protein of Synechocystis PCC 6803 confers tolerance to oxidative and environmental stresses in transgenic Arabidopsis. *Physiologia Plantarum* 128:251–262.

Gapinska, M., M. Skłodowska, and B. Gabara. 2008. Effect of short- and long-term salinity on the activities of antioxidative enzymes and lipid peroxidation I tomato roots. *Acta Physiologiae Plantarum* 30:11–18.

Gill, S.S. and N. Tuteja. 2010. Reactive oxygen species and antioxidant machinery in abiotic stress tolerance in crop plants. *Plant Physiology and Biochemistry* 48(12):909–930.

Grob, F., J. Durner, and F. Gaupels. 2013. Nitric oxide, antioxidants and prooxidants in plant defence responses. *Frontiers in Plant Science* 4:419.

Groppa, M.D. and M.P. Benavides. 2008. Polyamines and abiotic stress: Recent advances. *Amino Acids* 34(1):35–45.

Gubis, J., R. Vankova, V. Cervena, M. Dragunova, M. Hudcovicova, H. Lichtnerova, T. Dokupil, and Z. Jurekova. 2007. Transformed tobacco plants with increased tolerance to drought. *South African Journal of Botany* 73:505–511.

Gueta-Dahan, Y., Z. Yaniv, B.A. Zilinskas, and G. Ben-Hayyim. 1997. Salt and oxidative stress: Similar and specific responses and their relation to salt tolerance in citrus. *Plants* 204:460–469.

Guo, J., X. Liu, X. Li, S. Chen, Z. Jin, and G. Liu. 2006. Overexpression of VTE1 from *Arabidopsis* resulting in high vitamin E accumulation and salt stress tolerance increase in tobacco plant. *Chinese Journal of Applied and Environmental Biology* 12(4):468–471.

Harinasut, P., D. Poonsopa, K. Roengmongkol, and R. Charoensataporn. 2003. Salinity effects or antioxidant enzymes in mulberry cultivar. *Science Asia* 29:109–113.

Heidari, M. 2009. Antioxidant activity and osmolyte concentration of sorghum (*Sorghum bicolor*) and wheat (*Triticum aestivum*) genotypes under salinity stress. *Asian Journal of Plant Science* 8:240–244.

Hernandez, J.A., A. Jimenez, P. Mullineaux, and F. Sevilla. 2000. Tolerance of pea (*Pisum sativum* L.) to long term salt stress is associated with induction of antioxidant defenses. *Plant, Cell and Environment* 23:853–862.

Hmida-Sayari, A., R. Gargouri-Bouzid, A. Bidani, L. Jaoua, A. Savoure, and S. Jaoua. 2005. Overexpression of D1-pyrroline-5-carboxylate synthetase increases proline production and confers salt tolerance in transgenic potato plants. *Plant Science* 169:746–752.

Hollander-Czytko, H., J. Grabowski, I. Sandorf, K. Weckermann, and E.W. Weiler. 2005. Tocopherol content and activities of tyrosine aminotransferase and cystine lyase in *Arabidopsis* under stress conditions. *Journal of Plant Physiology* 162:767–770.

Hong, C.Y., Y.T. Hsu, Y.C. Tsai, and C.H. Kao. 2007. Expression of ascorbate peroxidase 8 in roots of rice (*Oryza sativa* L.) seedlings in response to NaCl, *Journal of Experimental Botany* 58:3273–3283.

Hossain, M.A. and K. Asada. 1985. Monodehydroascorbate reductase from cucumber is a flavin adenine dinucleotide enzyme. *Journal of Biological Chemistry* 260(24):12920–12926.

Hossain, M.A. and M. Fujita. 2013. Hydrogen peroxide priming stimulates drought tolerance in mustard (*Brassica juncea* L.). *Plant Gene and Trait* 4:109–123.

Hossain, M.A., M.G. Mostofa, and M. Fujita. 2013. Heat-shock positively modulates oxidative protection of salt and drought-stressed mustard (*Brassica campestris* L.) seedling. *Journal of Plant Science Molecular Breeding* 2:1–14.

Jimenez, A., J.A. Hernandez, L.A. del Rio, and F. Sevilla. 1997. Evidence for the presence of the ascorbate-glutathione cycle in mitochondria and peroxisomes of pea leaves. *Plant physiology* 114(1):275–284.

Jimenez, A, A. Hernandez, G. Pastori, L.A. del Rio, and F. Sevilla. 1998. Role of the ascorbate-glutathione cycle of mitochondria and peroxisomes in the senescence of pea leaves. *Plant Physiology* 118:1327–1335.

Jogeswar, G., R. Pallela, N. Jakka, P.S. Reddy, J.V. Rao, N. Sreenivasulu, and P.K. Kishor. 2006. Antioxidative response in different sorghum species under short-term salinity stress. *Acta Physiologiae Plantarum* 28(5):465–475.

Joo, J.H., Y.S. Bae, and J.S. Lee. 2001. Role of auxin-induced reactive oxygen species in root gravitropism. *Plant Physiology* 126(3):1055–1060.

Joseph, B. and D. Jini. 2010. Insight into the role of antioxidant enzymes for salt tolerance in plants. *International Journal of Botany* 6(4):456–464.

Karpinski, S., H. Gabrys, A. Mateo., B. Karpinska, and P.M. Mullineaux. 2003. Light perception in plant disease defence signaling. *Current Opinion in Plant Biology* 6:390–396.

Karthikeyan, A., P. ShumugiahKarutha, and M. Ramesh. 2011. Transgenic indica rice cv. ADT 43 expressing a D1-pyrroline-5-carboxylate synthetase (P5CS) gene from *Vignaa conitifolia* demonstrates salt tolerance. *Plant Cell Tissue Organ Culture* 107:383–395.

KaviKishor, P.B., S. Sangam, R.N. Amrutha, P.S. Laxmi, K.R. Naidu, K.R.S.S. Rao., S. Rao, K.J. Reddy, P. Theriappan, and N. Sreenivasulu. 2005. Regulation of proline biosynthesis, degradation, uptake and transport in higher plants: Its implications in plant growth and abiotic stress tolerance. *Current Science* 88:424–438.

Kiran Kumar Ghanti, S., K.G. Sujata, B.M. Vijay Kumar, N. Nataraja Karba, K. Janardhan Reddy, M. Srinath Rao, and P.B. KaviKishor. 2011. Heterologous expression of P5CS gene in chickpea enhances salt tolerance without affecting yield. *Biologia Plantarum* 55(4):634–640.

Kobayashi, K., Y. Kumazawa, K. Miwa, and S. Yamanaka. 1996. ε-(γ-Glutamyl) lysine cross-links of spore coat proteins and transglutaminase activity in *Bacillus subtilis*. *FEMS Microbiology Letters* 144(2–3):157–160.

Koji, Y., M. Shiro, K. Michio, T. Mitsutaka, and M. Hiroshi. 2009. Antioxidant capacity and damages caused by salinity stress in apical and basal regions of rice leaf. *Plant Production Science* 12:319–326.

Kolodyazhnaya, Y.S., N.K. Kutsokon, B.A. Levenko, O.S. Syutikova, D.B. Rakhmetov, and A.V. Kochetov. 2009. Transgenic plants tolerant to abiotic stresses. *Cytology and Genetics* 43:132–149.

Kukreja, S., A.S. Nandval, N. Kumar, S.K. Sharma, S.K. Sharma, V. Unvi, and P.K. Sharma. 2005. Plant water status, $H_2O_2$ scavenging enzymes, ethylene evolution and membrane integrity of *Cicer arietinum* roots as affected by salinity. *Biologia Plantarum* 49:305–308.

Kumar, S.G., A. Reddy, and C. Sudhakar. 2003. NaCl effects on proline metabolism in two high yielding genotypes of mulberry (*Morus alba* L.) with contrasting salt tolerance. *Plant Science* 165:1245–1251.

Kumar, V., V. Shriram, P.B. KaviKishor, N. Jawali, and M.G. Shitole. 2010. Enhanced proline accumulation and salt stress tolerance of transgenic indica rice by over-expressing P5CSF129A gene. *Plant Biotechnology Report* 4:37–48.

Kwak, J.M., I.C. Mori, and Z.M. Pei. 2003. NADPH oxidase Atrboh D and Atrboh F genes function in ROS-dependent ABA signaling in *Arabidopsis*. *EMBO Journal* 22(11):2623–2633.

Larson, R.A. 1988. The antioxidants of higher plants. *Phytochemistry* 27:969–978.

Lee, H. and J. Jo. 2004. Increased tolerance to methyl viologen by transgenic tobacco plants that overexpress the cytosolic glutathione reductase gene from *Brassica campestris*. *Journal of Plant Biology* 47:111–116.

Lee, Y.-P., S.-H. Kim, J.-W. Bang, H.-S. Lee, S.-S. Kwak, and S.-Y. Kwon. 2007. Enhanced tolerance to oxidative stress in transgenic tobacco plants expressing three antioxidant enzymes in chloroplasts. *Plant Cell Reports* 26:591–598.

Li, R., J. Vallabhaneni, T. Yu, T. Rocheford, and E.T. Wurtzel. 2008. The maize phytoene synthase gene family: Overlapping roles for carotenogenesis in endosperm, photomorphogenesis, and thermal stress tolerance. *Plant Physiology* 147(3):1334–1346.

Lu, Z., D. Liu, and S. Liu. 2007. Two rice cytosolic ascorbate peroxidases differentially improve salt tolerance in transgenic *Arabidopsis*. *Plant Cell Reports* 26:1909–1917.

Madhusudhan, R., T. Ishikawa, Y. Sawa, S. Shigeoka, and H. Shibata. 2003. Characterization of an ascorbate peroxidase in plastids of tobacco BY-2 cells. *Physiologia Plantarum* 117(4):550–557.

Maheshwari, R. and R.S. Dubey. 2009. Nickel-induced oxidative stress and the role of antioxidant defence in rice seedlings. *Plant Growth Regulation* 59(1):37–49.

Matysik, J., Alia, B. Bhalu, and P. Mohanty. 2002. Molecular mechanisms of quenching of reactive oxygen species by proline under stress in plants. *Current Science* 82:525–532.

Meratan, A.A., S.M. Ghaffari, and V. Niknam. 2008. Effects of salinity on growth, proteins and antioxidant enzymes in three *Acanthophyllum* species of different ploidy levels. *JUST* 33:1–8.

Meriga, B., B.K. Reddy, K.R. Rao, L.A. Reddy, and P.B.K. Kishor. 2004. Aluminium-induced production of oxygen radicals, lipid peroxidation and DNA damage in seedlings of rice (*Oryza sativa*). *Journal of Plant Physiology* 161:63–68.

Mhamdi, A., G. Queval, S. Chaouch, S. Vanderauwera, F. Van Breusegem, and G. Noctor. 2010. Catalase function in plants: A focus on *Arabidopsis* mutants as stress-mimic models. *Journal of Experimental Botany* 61:4197–4220.

Miller, G., V. Shulaev, and R. Mittler. 2008. Reactive oxygen signaling and abiotic stress. *Physiologia Plantarum* 133(3):481–489.

Miller, G., N. Sujuki, S. Ciftci-Yilmaz, and R. Mittler. 2010. Reactive oxygen species homeostasis and signalling during drought and salinity. *Plant Cell and Environment* 33:453–467.

Mishra, S., A.B. Jha, and R.S. Dubey. 2011. Arsenite treatment induces oxidative stress, upregulates antioxidant system, and causes phytochelatin synthesis in rice seedlings. *Protoplasma* 248(3):565–577.

Misra, N. and A.K. Gupta. 2005. Effect of salt stress on proline metabolism in two high yielding genotypes of green gram. *Plant Science* 169:331–339.

Mittler, R. 2002. Oxidative stress, antioxidants and stress tolerance. *Trends in Plant Science* 7(9):405–410.

Mittler, R., S. Vanderauwera, M. Gollery, and F. Van Breusegem. 2004. Reactive oxygen gene network of plants. *Trends in Plant Science* 9:490–498.

Mittler, R. and B.A. Zilinskas. 1992. Molecular cloning and characterization of a gene encoding pea cytosolic ascorbate peroxidase. *Journal of Biological Chemistry* 267(30):21802–21807.

Mittova, V., M. Guy, M. Tal, and M. Volokita. 2004. Salinity upregulates the antioxidative system in root mitochondria and peroxisomes of the wild salt-tolerant tomato species *Lycopersicon pennellii*. *Journal of Experimental Botany* 55:1105–1113.

Molinari, H.B.C., C.J. Marur, J.C.B. Filho, A.K. Kobayashi, M. Pileggi, R.P.L. Junior, L.F.P. Pereira, and L.G.E. Vieira. 2004. Osmotic adjustment in transgenic citrus rootstock Carrizo citrange (*Citrus sinensis* Osb. X *Poncirus trifoliate* L. Raf.) overproducing proline. *Plant Science* 167:1375–1381.

Moriwaki, T., Y. Yamamoto, T. Aida, T. Funashi, T. Shishido, M. Asada, S.H. Prodhan, A. Komanine, and T. Motohashi. 2007. Overexpression of the *Escherichia coli* catalase katE, enhances tolerance to salinity stress in the transgenic indica rice cultivar, BR5. *Plant Biotechnol Rep* 2:41–46.

Mullineaux, P.M. and T. Rausch. 2005. Glutathione, photosynthesis and the redox regulation of stress responsive gene expression. *Photosynthetic Research* 86:459–474.

Munné-Bosch, S., K. Schwarz, and L. Alegre. 1999. Enhanced formation of α-tocopherol and highly oxidized a betane diterpenes in water stressed rosemary plants. *Plant Physiology* 121:1047–1052.

Munns, R. 2002. Comparative physiology of salt and water stress. *Plant, Cell and Environment* 25:239–250.

Munns, R. 2005. Genes and salt tolerance: Bringing them together. *New Phytologist* 167(3):645–663.

Munns, R. and J.B. Passioura. 1984. Effect of prolonged exposure to NaCl on the osmotic pressure of leaf xylem sap from intact, transpiring barley plants. *Australian Journal of Plant Physiology* 11:497–507.

Munns, R. and M. Tester. 2008. Mechanisms of salinity tolerance. *Annual Review of Plant Biology* 59:651–681.

Nabati, J., M. Kafi, A. Nezami, P.R. Moghaddam, A. Masoumi, and M.Z. Mehrjerdi. 2011. Effect of salinity on biomass production and activities of some key enzymatic antioxidants in kochia (*Kochia scoparia*). *Pakistan of Journal of Botany* 43:539–548.

Nagamiya, K., T. Motohashi, K. Nakao, S.H. Prodhan, E. Hattori, S. Hirose, K. Ozawa, Y. Ohkawa, T. Takabe, T. Takabe, and A. Komamine. 2007. Enhancement of salt tolerance in transgenic rice expressing an *Escherichia coli* catalase gene, katE. *Plant Biotechnology Report* 1:49–55.

Nakano, Y. and K. Asada. 1987. Purification of ascorbate peroxidase in spinach chloroplasts; its inactivation in ascorbate—Depleted medium and reactivation by monodehydroascorbate radical. *Plant and Cell Physiology* 28(1):131–140.

Neill, S., R. Desikan, and J. Hancock. 2002. Hydrogen peroxide signaling. *Current Opinion in Plant Biology* 5(5):388–395.

Noctor, G. and C.H. Foyer. 1998. Ascorbate and glutathione: Keeping active oxygen under control. *Annual Review of Plant Biology* 49:249–279.

Noctor, G., S. Veljovic-Jovanovic, S. Driscoll, L. Novitskaya, and C.H. Foyer. 2002. Drought and oxidative load in the leaves of C3 plants: A predominant role for photorespiration. *Annals of Botany* 89:841–850.

Panda, S.K. 2001. Response of green gram seeds under salinity stress. *Indian Journal of Plant Physiology* 6:438–440.

Pandolfi, C., S. Mancuso, and S. Shabala. 2012. Physiology of acclimation to salinity stress in pea (*Pisum sativum*). *Environmental and Experimental Botany* 84:44–51.

Petrov, V.D. and F. Van Breusegem. 2012. Hydrogen peroxide-a central hub for information flow in plant cells. *Physiology* 141:336–340.

Prashanth, S.R., V. Sadhasivam, and A. Parida. 2008. Over expression of cytosolic copper/zinc superoxide dismutase from a mangrove plant *Avicennia marina* in indica rice varPusa Basmati-1 confers abiotic stress tolerance. *Transgenic Research* 17(2):281–291.

Prodhan, S.H., A. Hossain, K. Nagamiya, A. Komamine, and H. Morishima. 2008. Mishima. Improved salt tolerance and morphological variation in indica rice (*Oryza sativa* L.) transformed with a catalase gene from *E. coli*. *Plant Tissue Culture and Biotechnolgy* 18:57–63.

Qi, Y.C., W.Q. Liu, L.Y. Qiu, S.M. Zhang., L. Ma, and H. Zhang. 2010. Overexpression of glutathione S-transfer gene increases salt tolerance of *Arabidopsis*. *Russian Journal of Plant Physiology* 57:233–240.

Quan, L.J., B. Zhang, W.W. Shi, and H.Y. Li 2008. Hydrogen peroxide in plants: A versatile molecule of the reactive oxygen species network. *Journal of Integrated Plant Biology* 50:2–18.

Racchi, M.L., F. Bagnoli, I. Balla, and S. Danti. 2001. Differential activity of catalase and superoxide dismutase in seedlings and in vitro micropropagated oak (*Quercus robur* L.). *Plant Cell Reports* 20(2):169–174.

Rahnama, A., R.A. James, K. Poustini, and R. Munns. 2010. Stomatal conductance as a screen for osmotic stress tolerance in durum wheat growing in saline soil. *Functional Plant Biology* 37(3):255–263.

Rausch, T. and A. Wachter. 2005. Sulfur metabolism: A versatile platform for launching defence operations. *Trends in Plant Science* 10:503–509.

Rhoads, D.M., A.L. Umbach, C.C. Subbaiah, and J.N. Siedow. 2006. Mitochondrial reactive oxygen species. Contribution to oxidative stress and inter organellar signaling *Plant Physiology* 141:357–366.

Rishi, A. and S. Sneha. 2013. Antioxidative defense against reactive oxygen species in plants under salt stress. *International Journal on Current Research* 5:1622–1627.

Roxas, V.P., S.A. Lodhi, D.K. Garrett, J.R. Mahan, and R.D. Allen. 2000. Stress tolerance in transgenic tobacco seedlings that overexpress glutathione S-transferase/glutathione peroxidase. *Plant Cell Physiology* 41:1229–1234.

Sagi, M. and R. Fluhr. 2006. Production of reactive oxygen species by plant NADPH oxidases. *Plant Physiology* 141:336–340.

Sairam, R.K., K.V. Rao, and G.C. Srivastava. 2002. Differential response of wheat genotypes to long term salinity stress in relation to oxidative stress, antioxidant activity and osmolyte concentration. *Plant Science* 163:1037–1046.

Sairam, R.K, G.C. Srivastava, S. Agarwal, and R.C. Meena. 2005. Differences in response to salinity stress in tolerant and susceptible wheat genotypes. *Biologia Plantarum* 49:85–91.

Sairam, R.K. and A. Tyagi. 2004. Physiology and molecular biology of salinity stress tolerance in plants. *Current Science* 86:407–421.

Sakamoto, H., K. Maruyama, Y. Sakuma, T. Meshi, M. Iwabuchi, K. Shinozaki, and K. Yamaguchi-Shinozaki. 2004. *Arabidopsis* Cys2/His2-type zinc-finger proteins function as transcription repressors under drought, cold, and high-salinity stress conditions. *Plant Physiology* 136:2734–2746.

Sakihama, Y., J.I Mano, S. Sano, K. Asada, and H. Yamasaki. 2000. Reduction of phenoxyl radicals mediated by monodehydroascorbate reductase. *Biochemical and Biophysical Research Communications* 279(3):949–954.

Scandalios, G., L. Guan, and A.N. Polidoros. 1997. Catalases in plants: Gene structure, properties, regulation and expression. In *Oxidative Stress and the Molecular Biology of Antioxidants Defenses*, Scandalios, J.G. (ed.). New York: Cold Spring Harbor Laboratory Press, pp. 343–406.

Scandalios, J.G. 2005. Oxidative stress: Molecular perception and transduction of signals triggering antioxidant gene defenses. *Brazilian Journal of Medical and Biological Research* 38(7):995–1014.

Sekmen, A.H., I. Turkana, and S. Takiob. 2007. Differential responses of antioxidative enzymes and lipid peroxidation to salt stress in salt-tolerance *Plantago maritime* and salt-sensitive *Plantago media*. *Physiologia Plantarum* 131:399–411.

Shah, K., R.G. Kumar, S. Verma, and R.S. Dubey. 2001. Effect of cadmium on lipid peroxidation, superoxide anion generation and activities of antioxidant enzymes in growing rice seedling. *Plant Science* 161(6):1135–1144.

Sharma, P. and R.S. Dubey. 2004. Ascorbate peroxidase from rice seedlings: Properties of enzyme isoforms, effects of stresses and protective roles of osmolytes. *Plant Science* 167(3):541–550.

Sharma, P. and R.S. Dubey. 2005. Drought induces oxidative stress and enhances the activities of antioxidant enzymes in growing rice seedlings. *Plant Growth Regulation* 46(3):209–221.

Sharma, P., A.B. Jha, R.S. Dubey, and M. Pessarakli. 2012a. Reactive oxygen species, oxidative damage, signalling during drought and salinity. *Plant Cell and Environment* 33:453–467.

Sharma, P., A.B. Jha, R.S. Dubey, and M. Pessarakli. 2012b. Reactive oxygen species, oxidative damage, and antioxidative defense mechanism in plants under stressful conditions. *Journal of Botany* 1–26. doi:10.1155/2012/217037.

Srivastava, A.K., P. Bhargava, and L.C. Rai. 2005. Salinity and copper-induced oxidative damage and changes in antioxidative defense system of *Anabaena doliolum*, W. *Journal of Microbiology and Biotechnology* 22:1291–1298.

Srivastava, S. and R.S. Dubey. 2011. Manganese-excess induces oxidative stress, lowers the pool of antioxidants and elevates activities of key antioxidative enzymes in rice seedlings. *Plant Growth Regulation* 64:1–16.

Su, J. and R. Wu. 2004. Stress-inducible synthesis of proline in transgenic rice confers faster growth under stress conditions than that with constitutive synthesis. *Plant Science* 166:941–948.

Sudhakar, C., A. Lakshmi, and S. Giridarakumar. 2001. Changes in the antioxidant enzyme efficacy in two high yielding genotypes of mulberry (*Morus alba* L.) under NaCl salinity. *Plant Science* 161:613–619.

Surekha, Ch., K. NirmalaKumari, L.V. Aruna, G. Suneetha, A. Arundhati, and P.B. Kavi Kishore. 2014. Expression of the *Vigna aconitifolia P5CSF129A* gene in transgenic pigeonpea enhances proline accumulation and salt tolerance. *Plant Cell Tissue Organ Culture: Journal of Biotechnology* 116:27–36. doi:10.1007/s11240–013–0378-z.

Teige, M., E. Scheikl, T. Eulgem, R. Doczi, K. Ichimura, K. Shinozaki, J.L. Dangl, and H. Hirt. 2004. The MKK2 pathway mediates cold and salt stress signaling in *Arabidopsis*. *Molecular Cell* 15:141–152.

Torres, M.A., J.L. Dangl, and J.D.G. Jones. 2002. *Arabidopsis* gp91phox homologues Atrbohd and Atrbohf are required for accumulation of reactive oxygen intermediates in the plant defense response. *Proceedings of the National Academy of Sciences of the United States of America* 99(1):517–522.

Trovato, M., R. Mattioli, and P. Costantino. 2008. Multiple roles of proline in plant stress tolerance and development. *Rendiconti Lincei* 19:325–346.

Ushimaru, T., Y. Maki, K. Koshiba, K. Asada, and H. Tsuji. 1997. Induction of enzymes involved in the ascorbate-dependent antioxidative system, namely, ascorbate peroxidase, monodehydroascorbate reductase and dehydroascorbate reductase, after exposure to air of rice (*Oryza sativa*) seedlings germinated under water. *Plant and Cell Physiology* 38(5):541–549.

Vaidyanathan, H., P. Sivakumar, R. Chakrabarsty, and G. Thomas. 2003. Scavenging of reactive oxygen species in NaCl-stresses rice (*Oryza sativa* L.)-differential response in salt-tolerant and sensitive areas. *Plant Science* 165:1411–1418.

Vendruscolo, E.C., I. Schuster, M. Pileggi, C.A. Scapim, H.B. Molinari, C.J. Marur, and L.G. Vieira. 2007. Stress-induced synthesis of proline confers tolerance to water deficit in transgenic wheat. *Journal of Plant Physiology* 164:1367–1376.

Verma, S. and R.S. Dubey. 2003. Lead toxicity induces lipid peroxidation and alters the activities of antioxidant enzymes in growing rice plants. *Plant Science* 164(4):645–655.

Wang, J., H. Zhang, and R.D. Allen. 1999. Overexpression of an *Arabidopsis* peroxisomal ascorbate peroxidase gene in tobacco increases protection against oxidative stress. *Plant and Cell Physiology* 40(7):725–732.

Wang, S., D. Liang, C. Li, Y. Hao, F. Ma, and H. Shu. 2012. Influence of drought stress on the cellular ultrastructure and antioxidant system in leaves of drought-tolerant and drought-sensitive apple rootstocks. *Plant Physiology and Biochemistry* 51:81–89.

Wang, Y.C., G.Z. Qu, H.Y. Li, Y.J. Wu, C. Wang, G.F. Liu, and C.P. Yang. 2010. Enhanced salt tolerance of transgenic poplar plants expressing a manganese superoxide dismutase from *Tamarix androssowii*. *Molecular Biology Reports* 37(2):1119–1124.

Wang, Y., M. Wisniewski, R. Meilan, S.L. Uratsu, M.G. Cui, A. Dandekar, L. Fuchigami. 2007. Ectopic expression of Mn-SOD in *Lycopersicon esculentum* leads to enhanced tolerance to salt and oxidative stress. *Journal of Applied Horticulture* 9:3–8.

Wang, Y., Y. Ying, J. Chen, and X. Wang. 2004. Transgenic *Arabidopsis* overexpressing Mn-SOD enhanced salt-tolerance. *Plant Science* 167(4):671–677.

Willekens, H., M. van Montagu, and W. van Camp. 1995. Catalase in plants. *Molecular Breeding* 1: 207–228.

Wu, D., S. Cai, M. Chen, L. Ye, Z. Chen, H. Zhang, F. Wu, and G. Zhang. 2013. Tissue metabolic responses to salt stress in wild and cultivated barley. *Public Library of Science One* 8(1):e55431.

Xing, Y., W. Jia, and J. Zhang. 2008. At MKK1 mediates ABA-induced CAT1 expression and $H_2O_2$ production via AtMPK6-coupled signaling in *Arabidopsis*. *The Plant Journal* 54:440–451.

Xu, W.F., W.M. Shi, A. Ueda, and T. Takabe. 2008. Mechanisms of salt tolerance in transgenic *Arabidopsis thaliana* carrying a peroxisomal ascorbate peroxidase gene from Barley. *Pedosphere* 18(4):486–495.

Yamaguchi-Shinozaki, K. and K. Shinozaki. 1994. A novel cis-acting element in an *Arabidopsis* gene is involved in responsiveness to drought, low-temperature, or high-salt stress. *Plant Cell* 6:251–264.

Yamchi, A., F.R. Jazii, A. Mousav, and A.A. Karkhane. 2007. Proline accumulation in transgenic tobacco as a result of expression of *Arabidopsis* D1-pyrroline-5-carboxylate synthetase (P5CS) during osmotic stress. *Journal of Plant Biochemistry and Biotechnology* 16:9–15.

Yan, J., N. Tsuichihara, T. Etoh, and S. Iwai. 2007. Reactive oxygen species and nitric oxide are involved in ABA inhibition of stomatal opening. *Plant, Cell and Environment* 30(10):1320–1325.

Yin, L., S. Wang, A.E. Eltayeb, M.I. Uddin, Y. Yamamoto, W. Tsuji, Y. Takeuchi, and K. Tanaka. 2010. Overexpression of dehydroascorbate reductase, but not monodehydroascorbate reductase, confers tolerance to aluminium stress in transgenic tobacco. *Planta* 231:609–621.

Zhao, J.S., Guo, H. Zhang., and Y. Zhao. 2009. Expression of yeast YAP1 in transgenic *Arabidopsis* results in increased salt tolerance. *Journal of Plant Biology* 52:56–64.

# CHAPTER TWENTY-TWO

# Salinity Tolerance in Plants

## INSIGHTS FROM TRANSCRIPTOMIC STUDIES

*Mohammad Rashed Hossain, Laura Vickers, Garima Sharma,*
*Teresa Livermore, Jeremy Pritchard, and Brian V. Ford-Lloyd*

## CONTENTS

*Abstract.* Salinity stress poses a great threat to agricultural productivity worldwide as it can severely affect crop yields. Understanding the physiological and molecular aspects of the tolerance mechanism at whole-plant, tissue, cellular, and molecular levels along with the complex network of genes, proteins, and metabolites involved in this process is of paramount importance. This chapter reviews the transcripts identified by microarray studies in the recent past that are involved in various cellular and molecular processes such as response to stimulus, transcription and translation factor activity, SNAP receptor, and chaperone binding activity along with the possibilities of identifying novel genes from noncrop sources using these technologies. A major issue of crop, growth stage, tissue, and stress-specific profiling of transcripts and integrating this huge amount of microarray data thus generated is also discussed. Light is shed on the current trends in identifying candidate genes and the potential of bioinformatic integration of the data generated by microarrays with the pregenomic and various other high-throughput postgenomic technologies.

*Keywords:* microarray, salinity tolerance, QTLs, transcriptomics, transcription factors, multiomic databases, and system biology

## 22.1 INTRODUCTION

Climate change–induced events along with inappropriate irrigation practices are increasing soil salinity and are responsible for loss in crop production. Research in the last few decades has made significant advances in understanding the physiology and molecular mechanism of salinity tolerance especially on how salt is harmful to plants and what mechanisms they employ to overcome harmful effects of salt stress (Flowers et al. 2000; Yu et al. 2012).

Pregenomic studies on natural and recombinant genetic diversity have led to the identification of several quantitative trait loci (QTL) whose fine mapping had helped in further identification of a few key genes for salinity tolerance in various crop species (Collins et al. 2008). However, the major breakthroughs started to come with the advent of postgenomic high-throughput technologies like microarrays (Langridge and Fleury 2011). Being the pioneer postgenomic technology, microarrays have fully exploited the whole-genome sequences to reveal the stress-responsive transcripts that have helped to change our understanding of stress tolerance in terms of number of genes involved, as it was readily revealed that hundreds of genes are involved in different aspects of tolerance mechanisms at the whole-plant, physiological, biochemical, cellular, and molecular levels (Hirayama and Shinozaki 2010).

More sophisticated next-generation sequence-based technologies along with other omic technologies are now being used to identify the stress-responsive genes (Lister et al. 2009; Cabello et al. 2014). As transcriptomics is the gateway to explore the genomic information followed by proteomics and metabolomics, linking transcriptomics with other omics can lead to the identification of candidates for salinity tolerance. The recent rise in systems biology has further shown that the mass of microarray data thus generated so far would be of great use in the future (Fukushima et al. 2009; Cramer et al. 2011).

In this chapter, we summarize the advances made in the pregenomic era in identifying salinity-tolerant QTL and the main studies on salinity tolerance using microarray technologies in terms of the genes identified by microarray studies in various plant species ranging from cultivated crops to wild relatives, mangroves, and halophytes. Also discussed are the potentialities and challenges in integrating those genes with

other omic technologies that can be helpful in our overall understanding of salinity tolerance mechanisms.

## 22.2 NATURAL ALLELIC VARIATION FOR SALT TOLERANCE: ADVANCES FROM PREGENOMIC ERA

Over the last few decades, attempts have been made to identify QTL linked with the component physiological traits of salinity tolerance in numerous populations of various crop species using the natural and recombinant allelic variation for salinity tolerance (Flowers et al. 2010). For $Na^+$ or $K^+$ concentration in shoots and roots of rice, Lin et al. (2004) identified eight QTL from F2 and F3 populations from a cross between Nonabokra and Koshihikari. One of these is QTL SKC1 (shoot $K^+$ concentration), located on chromosome 1, which accounts for 40.1% of phenotypic variance for shoot $K^+$ concentration as revealed by fine mapping using fixed recombinant progeny testing (Ren et al. 2005). The SKC1 QTL encodes a high-affinity $K^+$ transporter (HKT) family ion antiporter OsHKT1;5 in rice that mediates $Na^+$ reabsorption. The ortholog of the OsHKT1;5 gene in bread wheat is TaHKT1;5-D, a candidate for a $K^+/Na^+$ discrimination gene 1 (Kna1) on chromosome arm 4DL, TmHKT1;5-A (HKT8), a candidate for a $Na^+$ exclusion gene 2 (Nax2) on chromosome arm 5AL, and TmHKT7-A2, a candidate for Nax1 on chromosome arm 2AL (Byrt et al. 2007). Other major QTL identified in rice are Saltol on chromosome 1 for ion uptake in salt-tolerant cultivar Pokkali (and also in other rice varieties) accounting for 64%–80% of the phenotypic variation (Bonilla et al. 2002), QNa (chromosome 1) for high $Na^+$ uptake, QNa:K (chromosome 4) for $Na^+/K^+$ ratio, etc. (Flowers et al. 2000). For root $Na^+/K^+$ ratio, Ming-zhe et al. (2005) identified two QTL on chromosomes 2 and 6, while Sabouri et al. (2009) identified several QTL on all but chromosome 9 in rice and three QTL for ion exchange on chromosomes 3 and 10. Of the 13 QTL identified by Wang et al. (2007a) on chromosomes 1, 2, 5, 6, 7, and 12, the QTL qSC1b accounted for 45% of the total phenotypic variability. On chromosome 4, Koyama et al. (2001) detected 10 QTL and Lin et al. (2004) identified 3 QTL for seedling survival days under elevated salt stress.

In barley, Xue et al. (2009) identified a number QTL for various traits such as shoot $Na^+$, $K^+$, and $Na^+/K^+$ ratio. In sunflower, Lexer et al. (2003) detected 10 QTL for ion uptake and later several candidate genes were found to be linked with those QTL as revealed by studies based on expressed sequence tag (EST) and single-nucleotide polymorphism (SNP) mapping strategy (Lexer et al. 2004). In durum wheat (*Triticum turgidum* L.), Na$^+$ exclusion was reported to be linked to *Nax1*, a locus that promotes Na retention in the leaf sheath, and *Nax2*, a locus that harbors the Na$^+$ transporters HKT7 and HKT8, respectively (James et al. 2006). Frary et al. (2010) identified a number of stress inducible QTL for the production of various antioxidants in *Solanum pennelli*. Similarly, QTL for the production of various osmoprotectants such as proline were identified on chromosomes 2 and 4 of barley under salt stress (Siahsar and Narouei 2010). The accuracy of QTL identification is already well established (Price 2006) and has led to the identification of a number of candidate genes (Bansal et al. 2014).

## 22.3 IDENTIFICATION OF STRESS-RESPONSIVE TRANSCRIPTS: REVOLUTIONARY ASPECTS OF MICROARRAYS

One of the revolutionary advantages of microarrays is that they straightway utilize the whole genome to identify the transcripts that encode for stress-responsive genes upon exposure to stresses and allowed researchers to identify many genes that are involved in different processes of stress adaptation on a holistic basis. However, the identification of hundreds of genes poses a challenge to functionally categorize those genes in a way to explain their roles in stress adaptation. The improvement in gene ontology enrichment analyses has made the task easier. This section attempts to discuss some of the transcripts identified by microarray experiments that are involved in key biological processes and molecular functions with special emphasis on model species such as rice and *Arabidopsis*. However, the transcripts related to signaling and transporters are not discussed in this section due to limitation in space and will be published as a separate book chapter soon.

### 22.3.1 *Transcripts Involved in "Response to Stimulus"*

"Response to stimulus" is an important gene ontology term (GO:0050896) under biological processes (GO:0008150), which include response to biotic, abiotic, endogenous, external, and chemical stimuli. In a microarray study using Agilent 4 × 44 K rice microarray with rice varieties Pokkali, FL478, Hassawi, and Nonabokra, Hossain (2014) identified around 56 transcripts that were involved in "response to stimulus." Liu et al. (2013) identified 59 transcripts in *Arabidopsis thaliana* ecotype Columbia (Col-0), Rasmussen et al. (2013) identified 286 transcripts in 10 *Arabidopsis* ecotypes, and Gao et al. (2013) identified a number of transcripts in barley. The list of selected transcripts is shown in Table 22.1.

The "late embryogenesis abundant protein (LEA)" encoding transcript Os01g0159600 and AT1G55450 was highly upregulated in rice and *Arabidopsis*, respectively. Wang et al. (2007b) identified constitutive and induced expression of 34 *OsLEA* genes in rice under several stress conditions. Microarray and reverse transcription polymerase chain reaction (RT-PCR) techniques have identified plasma membrane intrinsic proteins (PIPs) such as *OsPIP2;4*, *OsPIP2;5*, *OsPIP2;7*, *OsTIP1;2*, *OsPIP1;1*, and *OsTIP4;1* in rice, while AT3G61430, AT2G45960, AT1G01620, and AT3G53420 in *Arabidopsis* (Rasmussen et al. 2013). Microarray studies have identified a number of transcripts involved in salt overly sensitive (SOS) pathway that plays important roles in regulating Na$^+$ transport. For example, Rasmussen et al. (2013) identified the transcripts AT2G01980, AT5G35410, AT4G30960, AT1G06040, AT1G10570, etc., in *Arabidopsis* (Table 22.1). Transcripts Os01g0959200 and Os01g0959100 encoding an "abscisic stress–ripening protein 1 (OzAsr1)" were upregulated in rice genotype Pokkali (>3 fold) and Nonabokra (1.72 fold) (Hossain 2014).

Plant γ-thionins or defensins, as they are commonly known, are believed to play important roles in defense against pests and pathogens. The transcript Os03g0130300 encoding Cp-thionin was identified as salt stress–responsive in rice (Table 22.1). A number of transcripts encoding dehydrins (e.g., Os11g0451700 [dehydrin 9]), heat shock proteins (e.g., Os03g0745000, Os09g0482600, and Os03g0188100), universal stress family proteins (Usp, e.g., Os01g0511100 and Os10g0437500), and peroxidases (e.g., Os04g0465100 and Os01g0963200) were also identified by microarray studies in rice (Table 22.1) and all are believed to play roles in abiotic stress adaptation in plants.

TABLE 22.1
*Lists of selected transcripts involved in the biological process "response to stimulus" (GO:0050896) under salt stress in rice and Arabidopsis.*

| Locus ID | Annotation |
|---|---|
| In rice (Hossain 2014) | |
| Os01g0159600 | Embryonic abundant protein 1 |
| Os01g0959200 | Abscisic stress–ripening protein 1 |
| Os04g0465100 | Haem peroxidase, plant/fungal/bacterial family protein |
| Os04g0549600 | Heat shock protein DnaJ family protein |
| Os10g0542900 | Chitinase (EC 3.2.1.14) (Fragment) |
| Os03g0130300 | Cp-thionin |
| Os07g0694700 | Ascorbate peroxidase (EC 1.11.1.11) |
| Os02g0643800 | Auxin-responsive SAUR protein family protein |
| Os11g0451700 | Dehydrin 9 |
| Os02g0669100 | Dehydrin COR410 (cold-induced COR410 protein) |
| Os11g0454200 | Dehydrin RAB 16B |
| Os11g0453900 | Dehydrin RAB 16D |
| Os01g0702500 | Dehydrin RAB25 |
| Os05g0349800 | Embryonic abundant protein 1 |
| Os04g0493400 | Endochitinase A precursor (EC 3.2.1.14) (seed chitinase A) |
| Os03g0745000 | Heat shock factor (HSF)–type, DNA-binding domain containing protein |
| Os09g0482600 | Heat shock protein 82 |
| Os06g0553100 | Heat shock transcription factor 29 (fragment). |
| Os01g0571300 | Heat shock transcription factor 31 (fragment) |
| Os10g0345100 | Multi-antimicrobial extrusion protein MatE family protein |
| Os01g0963200 | Peroxidase BP 1 precursor |
| Os03g0390200 | Protein kinase 3 |
| Os12g0626500 | Seed maturation protein domain containing protein |
| Os06g0341300 | Seed maturation protein domain containing protein |
| Os02g0782500 | Small heat stress protein class CIII |
| Os06g0517700 | Thionin Osthi1 |
| Os01g0511100 | Universal stress protein (Usp) family protein |
| In *Arabidopsis* (Liu et al. 2013; Rasmussen et al. 2013) | |
| AT1G55450 | Embryo-abundant protein related |
| AT3G61430 | Pip1a (plasma membrane intrinsic protein 1a) |
| AT2G45960 | Pip1b (named plasma membrane intrinsic protein 1b) |
| AT1G01620 | Pip1c (plasma membrane intrinsic protein 1c) |
| AT3G53420 | Pip2a (plasma membrane intrinsic protein 2a) |
| AT4G32150 | Atvamp711 (*A. thaliana* vesicle-associated membrane protein 711) |
| AT5G11150 | Atvamp713 (vesicle-associated membrane protein 713) |
| AT2G01980 | Sos1 (salt overly sensitive 1) |
| AT5G35410 | Sos2 (salt overly sensitive 2) |

(Continued)

TABLE 22.1 (*Continued*)

Lists of selected transcripts involved in the biological process "response to stimulus" (GO:0050896)
under salt stress in rice and Arabidopsis.

| Locus ID | Annotation |
| --- | --- |
| AT4G30960 | Sip3 (sos3-interacting protein 3) |
| AT1G06040 | Sto (salt tolerance) |
| AT1G10570 | Ots2 (overly tolerant to salt 2) |
| AT1G13930 | Involved in response to salt stress |
| AT1G32230 | Rcd1 (radical-induced cell death 1) |
| AT1G35515 | Hos10 (high response to osmotic stress 10) |
| AT2G17840 | Erd7 (early response to dehydration 7) |
| AT2G33380 | Rd20 (responsive to desiccation 20) |
| AT3G51920 | Cam9 (calmodulin 9) |
| AT3G55530 | Sdir1 (salt- and drought-induced ring finger 1) |
| AT5G08620 | Strs2 (stress response suppressor 2) |
| AT1G27730 | Stz (salt tolerance zinc finger) |

### 22.3.2 Transcripts Encoding Transcription Factors

The rice genome is known to have 2777 genes encoding proteins of 93 different transcription factor families. Kawasaki et al. (2001) identified 17 transcription factors (TFs) in the root of rice cultivar Pokkali at the vegetative stage, Chao et al. (2005) identified 32 TFs in the rice cultivar Nonabokra, and a number of TFs was identified in other microarray studies on rice upon exposure to salt stress (Walia et al. 2005; Senadheera et al. 2009). In a recent study with eight rice genotypes, Hossain (2014) identified 81 positively expressed TFs for the trait shoot Na/K using significance analysis of microarrays (SAM). Jiang and Ramachandran (2010) has shown the differential expression of 319 TF genes in either Nipponbare or Pokkali, of which one-third showed similar patterns of expression under salt stress suggesting an expression divergence of TF genes between these two genotypes. The *Arabidopsis* genome is known to possess at least 1819 predicted TFs belonging to 56 protein families (Riechmann 2000). Jiang and Ramachandran (2010) identified 289 upregulated and 139 downregulated TFs in *Arabidopsis* upon exposure to salt stress; Liu et al. (2013) identified 33 TFs in 4-week-old 150 mM NaCl-stressed *A. thaliana* ecotype Columbia (Col-0) seedlings, and Rasmussen et al. (2013) identified 72 TFs in 10 *Arabidopsis* ecotypes under multiple stresses.

A number of micro-RNAs (miRNAs) encoding genes for TFs were also identified in *Arabidopsis* using microarrays (Liu et al. 2008). In common wheat, Kawaura et al. (2008) identified 109 TFs that belong to four major protein families such as ethylene-responsive element-binding protein, EREBP (21), MYB (42), NAC (23), and WRKY (23), and Schreiber et al. (2009) identified a number of TFs in Triticeae. Using microarrays, a number of abiotic stress–induced TFs have also been identified in barley (Talamè et al. 2007) and maize (Ding et al. 2009). Transcription factors that are induced by different stresses including salt stress in different plants are summarized in some recent reviews (Agarwal and Jha 2010; Atkinson and Urwin 2012), and the overexpression of some of those TFs has shown to increase tolerance to salt and other abiotic stresses (Table 22.2), with substantial cross talk being shown to exist between different abiotic stress responses in different crop species.

### 22.3.3 Transcripts Involved in Translation Factor Activity

The regulation of translation facilitating the selective synthesis of required proteins is one of the versatile strategies plants have evolved to cope with environmental stresses. Hossain (2014) identified 36 positively and 26 negatively expressed transcripts that are involved in translation factor activity (GO:0008135) while analyzing the gene expression and shoot Na/K data of 120 mM NaCl-stressed seedlings of eight rice genotypes using

TABLE 22.2
Transcription factor genes whose overexpression has shown to increase tolerance to salt and other abiotic stresses in different plants.

| Plant | Gene | Increased tolerance to | References |
|---|---|---|---|
| Rice | OsMYB54 | Salt | James and Dow (2012) |
| | OsMYB3R-2 | Salt, drought, cold | Dai et al. (2007) |
| | OsDREB1A | Salt and cold | Dubouzet et al. (2003) |
| | OsDREB1F | Salt, drought | Wang et al. (2008) |
| | OsDREB2A | Salt, drought | Dubouzet et al. (2003) |
| | OsEREB67 | Salt | James and Dow (2012) |
| | OsNAC | Salt, drought | Song et al. (2011) |
| | SNAC1 | Salt, osmotic stress | Yokotani et al. (2009) |
| | SNAC2/rice | Salt, cold | Hu et al. (2008) |
| | OsNAC5 | Salt, drought, cold | Takasaki et al. (2010) and Song et al. (2011) |
| | OsNAC6 | Salt, drought, cold | Ohnishi et al. (2005) and Nakashima et al. (2007) |
| | ONAC10 | Salt, drought, cold | Jeong et al. (2010) |
| | ONAC045 | Salt, drought | Zheng et al. (2009) |
| | OsNAC063 | Salt, osmotic stress | Yokotani et al. (2009) |
| | OsWRKY01 OsWRKY02 | Drought | Berri et al. (2009) and Ramamoorthy et al. (2008) |
| | OsWRKY07 OsWRKY45 | Salt, drought | Ramamoorthy et al. (2008) |
| | OsWRKY05 OsWRKY43 | Salt, drought, osmotic stress | |
| | OsOrphan19 OsbHLH17 OsLUX | Salt | James and Dow (2012) |
| Wheat | TaMYB2A | Salt and drought | Mao et al. (2011) |
| | TaPIMP1(MYB) | Salt and drought | Abu Qamar et al. (2009) |
| | TaNAC69 | Salt, dehydration | Xue et al. (2011) |
| | AIDFa | Salt, dehydration | Xu et al. (2007) |
| | TaDREB1 | Salt, drought, cold | Shen et al. (2003) |
| | TaWRKY2 | Salt and drought | Niu et al. (2012) |
| | TaWRKY19 | Salt and drought | Niu et al. (2012) |
| Barley | DREB1 | Salt | Xu et al. (2009) |
| | HsDREB1A | Salt, dehydration | Zhang et al. (2011) |
| Soybean | GmMYB76 | Salt | Liao et al. (2008) |
| | GmMYB92 GmMYB177 | Salt, cold | Liao et al. (2008) |
| | GmNAC1 GmNAC20 | Salt, cold | Hao et al. (2011a) |
| | GmNAC11 | Salt | Hao et al. (2011a) |
| | GmDREB2 | Salt, drought, cold | Chen et al. (2007) |
| | GmDREBa GmDREBb | Salt, drought, cold | Li et al. (2005) |

(Continued)

TABLE 22.2 (Continued)

Transcription factor genes whose overexpression has shown to increase tolerance to salt and other abiotic stresses in different plants.

| Plant | Gene | Increased tolerance to | References |
|-------|------|------------------------|------------|
| | GmDREBc | Salt, drought | Li et al. (2005) |
| | ERF3 | Salt, dehydration | Zhang et al. (2009) |
| | GmbZIP44 | Salt, freezing | Liao et al. (2008) |
| | GmbZIP62 GmbZIP78 GmbZIP132 | | |
| | GmWRKY13 | Salt, mannitol | Zhou et al. (2008) |
| | GmWRKY54 | Salt, drought | Zhou et al. (2008) |
| *Arabidopsis* | AtMYBC1 | Salt, drought, ABA | Zhai et al. (2010) |
| | AtMYB2 | Salt, cold, drought, ABA | Abe et al. (1997) |
| | AtMYB15 AtMYB41 | Salt, drought, ABA | Ding et al. (2009) |
| | AtMYB44 | Salt, drought, cold, ABA | Jung et al. (2008) |
| | AtMYC2 | Salt and drought | Abe et al. (1997) |
| | AtNAC2 | Salt, drought | He et al. (2005), Liu et al. (2011) |
| | AtNAC019 | Salt and drought | Tran et al. (2004) |
| | AtNAC055 AtNAC072 | | |
| | AtDREB2A | Salt, heat, dehydration | Sakuma et al. (2006) |
| | AtWRKY25 AtWRKY33 | Salt, oxidative stress, ABA | Jiang and Deyholos (2009) |

SAMs (Table 22.3). Under salt stress, the positively expressed transcripts are Os02g0146600 (*eIF4A*), Os03g0566800 (*eIF4A-3*), Os05g0566500 (*eIF-3 zeta*), Os07g0124500 (*eIF3 p110*), Os07g0555200, Os02g0101100 (*eIF3 family protein*), Os07g0555200 (*eIF4G*), Os05g0592600 (*eIF2 family protein*), etc. (Table 22.2). The transcript eukaryotic initiation factor, *eIF4A* (Os02g0146600), which is considered as the gold standard for DEAD box helicases, has a role beyond translation and is reported to be induced by biotic stress in *Glycine max* cv. Kent (Öktem et al. 2008) and its ortholog conferred tolerance to salt stress in tobacco (Mishra et al. 2005). In addition, *eIF1A* and *eIF3* also play an important role in the regulation of translation initiation under abiotic stress in plants (Echevarría et al. 2013), and the overexpression of sugar beet *eIF1A* is shown to increase salinity tolerance in *Arabidopsis* (Rausell et al. 2003). The positively expressed transcripts encoding the genes for *eIF4G* and *eIF2* were also known to play important roles in regulating translation initiation under stressed condition (Clemens 2001; Lageix et al. 2008). Besides

the positively expressed transcripts, a number of other transcripts encoding translation elongation factors, such as Os03g0177400 and Os03g0178000 (EF-1 α), Os11g0116400 (EF-P), Os12g0541500 (EF-Ts), Os07g0614500 (EF-1-*beta*), and Os02g0220500 and Os02g0220600 (EF-1-*gamma*), and translation initiation factors, for example Os02g0300700 (EIF-1A), Os07g0681000 (*eIF-2-beta*) (P38), Os01g0120800 (*eIF-3 theta*), Os07g0167000 (*eIF-3 p48*), Os01g0970400 (*eIF4E-1*), Os05g0107700 (TFIIA-*gamma*), and Os07g0639800 (IF6 family protein), were also found to be negatively expressed in rice (Table 22.2) indicating their putative role in stress adaptation.

### 22.3.4 Transcripts Involved in "SNAP Receptor" and "Chaperone Binding" Activity

SNAP receptor activity (GO:0005484) is regulated by a superfamily of proteins known as soluble N-ethylmaleimide-sensitive factor attachment protein receptors (SNAREs) that act as a marker to identify a membrane and selectively interact with

## TABLE 22.3
*Lists to selected transcripts involved in the molecular functions under salt stress in rice as determined by SAM analysis.*

| Name | Annotation | Expression |
|------|-----------|------------|
| (a) Translation factor activity (GO:0008135) | | + |
| Os07g0662500 | Elongation factor 1-beta' (EF-1-beta') | + |
| Os06g0571400 | Elongation factor 1-gamma (EF-1-gamma) (eEF-1B gamma) | + |
| Os01g0742200 | Elongation factor EF-2 (fragment) | + |
| Os02g0146600 | Eukaryotic initiation factor 4A (eIF4A) (eIF-4A) | + |
| Os03g0566800 | Eukaryotic initiation factor 4A-3 (eIF4A-3) (eIF-4A-3) | + |
| Os05g0566500 | Eukaryotic translation initiation factor 3 subunit 7 (eIF-3 zeta) (eIF3d) | + |
| Os07g0124500 | Eukaryotic translation initiation factor 3 subunit 8 (eIF3 p110) (eIF3c) | + |
| Os07g0555200 | Eukaryotic translation initiation factor 4G | + |
| Os07g0597000 | Eukaryotic translation initiation factor 5A (eIF-5A) | + |
| Os03g0758800 Os12g0507200 | Eukaryotic translation initiation factor 5A-2 (eIF-5A) (eIF-4D) | + |
| Os05g0592600 | Initiation factor 2 family protein | + |
| Os02g0101100 | Initiation factor 3 family protein | + |
| Os05g0575300 | Translation initiation factor IF-2, chloroplast precursor (PvIF2cp) | + |
| Os02g0557600 Os05g0498400 | Translation initiation factor SUI1 family protein | + |
| Os03g0177400 | EF-1 alpha | − |
| Os03g0178000 | EF-1 alpha | − |
| Os11g0116400 | Elongation factor P (EF-P) | − |
| Os12g0541500 | Elongation factor Ts (EF-Ts) | − |
| Os07g0614500 | Elongation factor 1-beta (EF-1-beta) | − |
| Os02g0220500 | Elongation factor 1-gamma (EF-1-gamma) (eEF-1B gamma) | − |
| Os02g0220600 | Elongation factor 1-gamma (EF-1-gamma) (eEF-1B gamma) | − |
| Os02g0300700 | Eukaryotic translation initiation factor 1A (EIF-1A) (EIF-4C) | − |
| Os07g0681000 | Eukaryotic translation initiation factor 2 beta subunit (eIF-2-beta) (P38) | − |
| Os01g0120800 | Eukaryotic translation initiation factor 3 subunit 10 (eIF-3 theta) | − |
| Os07g0167000 | Eukaryotic translation initiation factor 3 subunit 6 (eIF-3 p48) (eIF3e) | − |
| Os01g0970400 | Eukaryotic translation initiation factor 4E-1 (eIF4E-1) (eIF-4E-1) | − |
| Os12g0607100 | Histone-lysine N-methyltransferase, H3 lysine-9 specific (EC 2.1.1.43) | − |
| Os02g0794400 | Initiation factor 3 family protein | − |
| Os05g0107700 | Transcription initiation factor IIA gamma chain (TFIIA-gamma) | − |
| Os07g0639800 | Translation initiation factor IF6 family protein | − |
| Os02g0641800 | RNA helicase | − |
| (b) SNAP receptor activity (GO:0005484) | | |
| Os02g0820700 | Bet1-like SNARE 1-1 (AtBET11) (Bet1/Sft1-like SNARE 14a) (AtBS14a) | − |

(Continued)

TABLE 22.3 (Continued)

*Lists to selected transcripts involved in the molecular functions under salt stress in rice as determined by SAM analysis.*

| Name | Annotation | Expression |
|------|------------|------------|
| Os08g0563300 | Bet1-like SNARE 1-1 (AtBET11) (Bet1/Sft1-like SNARE 14a) (AtBS14a) | − |
| Os02g0119400 | Syntaxin 52 (AtSYP52) | − |
| Os07g0164300 | Syntaxin, N-terminal domain containing protein | − |
| Os01g0254900 | Syntaxin | − |
| Os06g0168500 | Syntaxin-like protein (fragment) | − |
| (c) Chaperone activity (GO:0051087) | | |
| Os08g0338700 | Chaperone GrpE type 2 | − |
| Os04g0431100 | GrpE protein family protein | − |
| Os08g0464000 | Protein of unknown function DUF704 domain containing protein | − |
| Os09g0284400 | GrpE protein family protein | − |
| Os12g0456200 | Diaphanous protein homologue 2 | − |

NOTE: + and − indicate significant positive and negative expressions, respectively as determined by SAM analysis (Hossain 2014).

SNAREs on other membrane surfaces to mediate membrane fusion, providing a continuous flux of membranes via transport vesicles. This vesicle traffic is believed to be involved in cell homeostasis, growth, and development of plants (Kim and Brandizzi 2012). Using SAM analysis, Hossain (2014) identified six negatively expressed transcripts that encode genes involved in SNAP receptor activity (Table 22.2), of which the bet-like SNARE-AtBS14a (Os02g0820700 and Os08g0563300) is reported to control cell growth in Arabidopsis. In poplar (Populus yunnanensis) the transcript estExt_fgenesh4_pg.C_290237 (AT4G17730) encoding SYP23; SNAPs were upregulated in males but downregulated in females under salinity stress (Jiang et al. 2012). The syntaxin identified is AtSYP52 (encoded by Os02g0119400) and was very recently described to act as a t-SNARE when distributed in membrane TGN/PVC and plays a putative inhibitory role when present on the tonoplast in Arabidopsis (Benedictis et al. 2013). Another syntaxin, OSM1/SYP61, was also reported to be involved in osmotic stress tolerance in Arabidopsis (Zhu 2001). Suppression of the expression of Arabidopsis vesicle-SNARE gene AtVAMP7C inhibited the fusion of $H_2O_2$-containing vesicles with tonoplast and increased salt tolerance (Leshem et al. 2006).

Chaperones are proteins involved in noncovalent folding or unfolding of other proteins and are believed to be expressed in response to stresses. Hossain (2014) identified six transcripts such as Os08g0338700 (GrpE type 2), Os04g0431100, and Os09g0284400 (GrpE protein family protein) that are involved in chaperone binding activity (GO:0051087) in rice genotypes under salt stress (Table 22.3). The transcript EST F443 (GenBank: BU671999) encoding a copper chaperone was reported to be repressed by salt and cold stresses in rice (Cui et al. 2005) and sunflower (Fernandez et al. 2008). Overexpression of genes such as AtCSP1 (At4g36020), AtCSP2 (At4g38680), and AtCSP3 (At2g17870) that are involved in the RNA chaperone activity was reported to be induced by cold stress (Kim et al. 2009), while overexpression of NADPH–thioredoxin reductase type C resulted in enhanced tolerance to heat stress in Arabidopsis (Chae et al. 2013) and to drought stress in tobacco (Sarwat), and the cytosolic chaperonin-containing TCP-1α (CCTα) homologue showed increased tolerance to salt in the mangrove plant Bruguiera sexangula (Yamada et al. 2002).

### 22.3.5 Proteins of Unknown Function

The high-throughput studies such as microarray and proteomics aiming at identifying stress-responsive genes and proteins usually identify a large number of genes and proteins whose function is yet to be known (Garg et al. 2013). But the stress responsiveness of these proteins of unknown function (PUFs) signifies the fact that these PUFs may have vital roles in stress adaptation. The lack of similarity with well-characterized sequences (genes and proteins) further increases the interest in these groups of genes, a fact that points toward the possibility of these genes having unique functions and roles in stress tolerance. With the progress in genome sequencing and functional annotation, the specific roles of these genes could be established, which may lead to the discovery of novel candidates and new alternate pathways (Luhua et al. 2008). A number of techniques have already been proposed to identify the functions of these PUFs, for example identification of different kinds of PUFs based on known and predicted features, for example, protein domains and identification of the natural ligands of PUFs by the crystallographic screening of the binding of a metabolite library using metabolite cocktail screening, followed by a focused search in the metabolic space that can link those proteins to a specific biological context (Meier et al. 2013; Shahbaaz et al. 2013). The bulk of PUFs identified by microarrays or proteomic studies may thus be of huge importance in further broadening our understanding of abiotic stress tolerance of crop plants once their functional roles can be revealed.

## 22.4 SOURCE OF NOVEL GENES: HALOPHYTES, MANGROVES, AND CROP WILD RELATIVES

The commonality of the mechanisms of salt tolerance in plants with the diverse mechanisms employed by halophytes points toward the possibility of the plants becoming adapted to extreme conditions (Colmer et al. 2006; Flowers et al. 2010). The sequence similarity (almost 90%) of the salt cress (Thellungiella halophila) with Arabidopsis has made it a model halophyte for studying salinity tolerance (Zhu 2001). Using comparative cDNA microarrays, higher constitutive expression of stress-responsive genes (such as SOS1, SOS2, NHX1, and HKT1) having essential roles in Na+ excretion, compartmentation, and diffusion was observed in T. halophila compared to A. thaliana (Taji et al. 2010). Zhang et al. (2008) generated a cDNA library of salt-treated Thellungiella seedlings and identified 7% ESTs as Thellungiella uniquely indicating the possibility of discovering novel salt stress–related genes. Using semiquantitative RT-PCR, Edelist et al. (2009) observed the constitutive under- or overexpression of K and Ca transport–related genes (homologues of KT1, KT2, ECA1) in the halophyte Helianthus

paradoxus. Using whole transcriptome sequencing, Fan et al. (2013) identified a number of differentially expressed transcripts at different time points in root and shoot development of another halophyte, Salicornia europaea under salt stress. The transfer of these novel genes may be useful in enhancing the level of tolerance to salt stress in crop plants. For example the transfer of a vacuolar H+-ATPase subunit c1 (SaVHAc1) gene from the halophytic grass Spartina alterniflora Loisel has shown enhanced tolerance to salt stress in transgenic rice (Baisakh et al. 2012).

A number of genes that were unknown previously have been identified in various mangrove species (Rajalakshmi and Parida 2012). Mehta et al. (2005) identified 30% novel cDNAs from EST sequencing of 1841 clones of an Avicennia marina leaf library under salt stress, and Miyama and Hanagata (2007) identified 38% novel genes from a combined library of 26,400 sequenced cDNAs from the mangrove species, Bruguiera gymnorrhiza. Ezawa and Tada (2008) identified 44 putative salt tolerance genes in B. gymnorrhiza under salt stress (500 mM NaCl) using microarray expression profiling. From the identified genes in a later experiment, two genes (homologues of Bg70 and cyc02) showed enhanced tolerance to salt stress in transgenic Arabidopsis. Using suppression subtractive hybridization, Wong et al. (2007) identified 42 novel salt-induced cDNAs in the root of a mangrove species Bruguiera cylindrica. Using the Illumina sequencing platform, a number of salt stress–responsive genes was identified in a semimangrove plant, Millettia pinnata (Huang et al. 2012), and in a highly salt-tolerant mangrove species, Sonneratia alba (Chen et al. 2011). Miyama and Hanagata (2007) identified six major coexpression clusters in Burma mangrove, and salt stress–responsive transcripts encoding novel genes were also identified from olive using whole transcriptome profiling (Zhang et al. 2014).

Besides the halophytes and mangroves, wild relatives of crop plants can also provide novel genes for salinity tolerance (Atwell et al. 2014). For example a novel ion transporter gene GmCHX1 was identified in wild soybean (Qi et al. 2014). Salinity tolerance of Oryza coarctata (Porteresia coarctata), a wild species of rice, is well documented, which has less similarity with rice genome compared to other wild relatives of rice (Flowers et al. 1990). Very recently, the transcriptome of this genotype was sequenced using the Illumina platform, and more than 4000 genes including 118 TFs were found to be differentially regulated under salt stress, of which the function of a significant number of genes is unknown and is believed to be species specific (Garg et al. 2013). Salt tolerance traits from O. *coarctata* were successfully introgressed into the variety IR56 that opens up the avenue to develop supersalt-tolerant rice in the future (Sengupta et al. 2010). Using the Affymetrix Tomato Genome Array, a gene encoding salicylic acid–binding protein 2 was identified in wild halophytic tomato species, *Solanum pimpinellifolium* "PI365967" (Sun et al. 2010). The novel genes thus identified from halophytes, mangroves, and the wild relatives of crop plants using microarray studies can reveal novel adaptive mechanisms and can further enhance our understanding of salinity tolerance in crops.

## 22.5 UTILIZING THE GAINS MADE BY MICROARRAY TRANSCRIPTOMICS: THE WAY FORWARD

The past decade has seen a rapid rise in the generation of gene expression data using microarrays in various plants under different biotic and abiotic stress conditions (Hazen et al. 2003). However, with the progress in technology and the affordability of whole genome sequencing, the world has arguably started to see the beginning of the end of microarrays as more and more sophisticated techniques are emerging and starting to be employed to reach to appropriate candidate genes (Lister et al. 2009; Cabello et al. 2014). Nevertheless, the huge datasets that have been generated by microarrays and that are deposited in public databases (e.g., Gene Expression Omnibus of NCBI, Array Express of EMBL-EBI) can still be utilized in enhancing our understanding of stress tolerance (Deyholos 2010).

Combined analysis of microarray and QTL data was successfully used for different traits in crops and has led to the identification of a number of abiotic stress–related transcripts. For example, Hossain (2014) identified 25 salt stress–responsive transcripts (e.g., zinc finger, $C_3HC_4$ type [RING finger], mitochondrial import inner membrane translocase) that are located within the salt stress–related QTL *Saltol* and *qSKC1* in chromosome 1 of rice. Pandit et al. (2010) identified two genes, an integral transmembrane protein DUF6 and a cation chloride cotransporter within the QTL intervals

that were mapped on chromosomes 1, 8, and 12 for the salt ion concentrations in rice. Wang et al. (2005) identified 16 candidate genes within the QTL region responsible for drought tolerance in rice. Besides, another potential use of these datasets can be the identification of the molecular signatures in a way to predict the tolerance status of genotypes whose tolerance to a particular stress is unknown. Identification of such prediction was successful with human cancer cell data in predicting the prostate cancer features and in identifying molecular signatures that characterize two different stages of rheumatoid arthritis (Sha et al. 2004; Trevino et al. 2011).

Owing to the limitation of resources in most of the cases, microarray experiments have been conducted on a limited scale while studying the transcriptome of certain genotypes of a particular species from a given tissue at a certain growth stage under a given stress level. Such an experiment on its own cannot reveal the overall stress responses of that species at different growth stages and at various tissue levels under different stress conditions. For example, there was much discrepancy in the gene expression of rice genotypes between root and shoot tissues (Cotsaftis et al. 2011), and between vegetative growth stage and panicle initiation stages under salinity stress (Walia et al. 2007).

But now that substantial numbers of datasets are available online, the challenges will be the effective integration of datasets, preferably by minimizing the variation between multiplatform data, for a particular species to determine the stress-responsive genes. This can better profile the transcriptome of a particular species and can provide a global overview of the stress-responsive genes that are tissue (and/or cell), growth stage, and stress-level specific. These can then be integrated with other "omic" technologies such as proteomics, metabolomics, epigenomics, and phenomics using a systems biology approach (as shown in Figure 22.1) to precisely identify the appropriate candidates. For example, the identification of the network of genes, such as the signaling genes that activate a certain set of transcription factors that regulate the activity of further downstream genes that confer the tolerance to a particular stress, can narrow down our efforts in targeted manipulation of the appropriate candidates using biotechnological and breeding approaches to develop stress-tolerant crops.

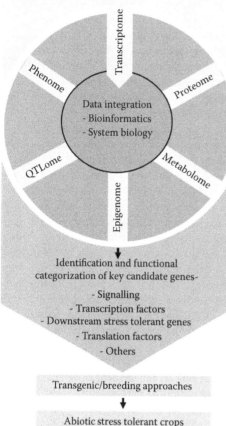

Stress inducible genes from different-

- Species, e.g., rice, wheat, *Arabidopsis* etc.
- Growth stages, e.g., seedling, vegetative stage ect.
- Tissues, e.g., root, shoot and leaves etc.
- Stress levels, e.g., low, moderate and high etc.
- Microarray platforms, e.g., Affymetrix, Agilent etc.

Transcriptome

Phenome

Proteome

Data integration
- Bioinformatics
- System biology

QTLome

Metabolome

Epigenome

Identification and functional
categorization of key candidate genes-

- Signalling
- Transcription factors
- Downstream stress tolerant genes
- Translation factors
- Others

Transgenic/breeding approaches

Abiotic stress tolerant crops

Figure 22.1. Simplified overview of systems biology approach showing the possible integration of datasets generated by microarrays with other omic techniques and phenomics in identifying the appropriate candidates for abiotic stress tolerance.

## 22.6 IDENTIFYING THE CANDIDATE GENES: CURRENT TRENDS IN PLANT GENOMICS

The recent technological advancement has offered novel techniques for identifying the candidate genes. Cell-specific expression profiling of the whole genome is one such example. Even though nearly every cell of a multicellular organism contains the same genome, there is likely to be variation in transcriptional activity in different cells, an avenue that has started to catch plant biologists' attention. Knowledge on the transcriptomic

variation in different types of cells such as root hair, epidermal, cortical, and vascular bundle cells may further enhance our understanding of the plant's abiotic stress adaptation strategies (Long 2011; Rogers et al. 2012).

With next-generation sequencing being more affordable, the transcript abundance by sequencing the entire set of mRNAs using RNA-seq has gained much popularity as it is suitable for nonmodel organisms and can provide accurate measurement of gene expression including the features such as allele-specific expression, novel transcribed regions, and a comprehensive capability to capture alternative splicing (Jain 2012; Bolger et al. 2014). RNA-seq has been used to identify salt stress–responsive transcripts from the wild halophytic rice, *P. coarctata*, wild soybean, common bean, poplar, barley, petunia, and mangrove species *S. alba* (Villarino et al. 2014; Zhang et al. 2014).

The availability of genome sequences has allowed the comparison and identification of sequence variation, that is SNPs within many plant genetic resources. For example, using ESTs, Leonforte et al. (2013) identified the SNP markers that are linked with the genomic regions containing candidate genes associated with salt stress tolerance in field pea (*Pisum sativum* L.). Similarly, SNPs linked with salt and drought tolerance in wheat (Mondini et al. 2012), maize (Hao et al. 2011b), *Arabidopsis* (Hao et al. 2008), etc., were also discovered that can be used to develop functional markers for marker-assisted selection-based crop breeding programs. Targeted sequencing of the genes identified through microarray and other high-throughput studies can further reveal the SNPs in the important biotic and abiotic stress–related genes in plants (Kumar et al. 2010). Genome-wide association studies (GWASs), another approach that uses the genome-wide markers in association with phenotypic variation across a large genetic population, can be used to identify the candidate loci for the trait of interest related to biotic and abiotic stress tolerance (Bolger et al. 2014). For example, GWAS was used to identify QTL related to 107 phenotypes including tolerance to salt in *Arabidopsis* (Atwell et al. 2010).

The small noncoding RNAs such as miRNAs and small interfering RNAs are also believed to play vital roles in regulating transcriptional gene expression and posttranscriptional gene silencing by influencing DNA methylation, mRNA levels, chromatin remodeling, histone modification, translational repression, etc., under various biotic and abiotic stresses (Trindade et al. 2010). A number of novel miRNAs have been identified using either conventional cloning or small RNA-seq analysis to be involved in plant biotic and abiotic stress responses (Khraiwesh et al. 2012). The epigenetics or the posttranslational modifications such as various histone modifications or DNA methylation processes of the stress-responsive genes may also play vital roles in plant adaptation to stresses (Chinnusamy and Zhu 2009). The recent progress in bioinformatics and the technological advancement have allowed the whole genome or selected gene-based detection of chromatin conditions (e.g., ChIP) or modified bases in DNA (e.g., bisulfite method) that can further enhance our understanding of the role of stress-responsive genes in stress adaptation (Hirayama and Shinozaki 2010).

## 22.7 INTEGRATING MULTIOMIC DATASETS: A SYSTEMS BIOLOGY PERSPECTIVE

Molecular plant breeding aiming at improving a plant's ability to withstand environmental stresses requires the precise point of intervention or selection of targets whose modification can offer the greatest genetic gain (Deyholos 2010). To reach those appropriate candidates, it is essential to understand the intertwining network of the genes and their isoforms that can have similar mechanisms of action across species (Platten et al. 2013) and the dynamics of the proteins and metabolic pathways they are involved in along with the possible complexity in epigenetics that can be linked back to the biotic and abiotic stress physiology and phenomics (Golldack et al. 2011; Duque et al. 2013). This has led to the development of the systems biology approach, which aims at integrating various omic datasets such as genomics, transcriptomics, proteomics, phosphoproteomics, secretomics, interactomics, metabolomics, regulomics, fluxomics, ionomics, physiomics, and phenomics (Tripathi et al. 2014).

The rapid progress in research on various omic technologies has generated large numbers of datasets, but the integration of these datasets in a functioning systems context is still in its infancy (Palsson and Zengler 2010). However, various networking and meta-analytical applications have shown some promising breakthroughs that are

yet to be developed as an effective tool to identify the novel candidates from multiomic datasets (Mochida and Shinozaki 2011). For example, using *Arabidopsis* microarray data, Less and Galili (2008) proposed a novel regulatory program of combined transcriptional and posttranslational controls that modulate fluxes of amino acid metabolism in response to abiotic stresses. Wang et al. (2013) also constructed a regulatory network of rice salt-tolerant genes using novel computational systems biology methods. Integration of metabolomic and transcriptomic data has also been used effectively to confirm inferences about stress-related gene expression (Urano et al. 2009). Using pathway analysis of transcriptome and metabolome data, Janz et al. (2010) identified evolutionary adaptation of stress-tolerant mechanisms in poplar species. Carrera et al. (2009) further extended network analysis using combined genome-wide expression data and reverse-engineering network modeling to identify the genomic signatures for stress tolerance. Combining multiomic datasets is still challenging, which mainly arises from the diverse nature of omic data that are available from many data sources (Deshmukh et al. 2014). But some algorithmic and computational model–based approaches have been proposed to overcome such limitations (Ge et al. 2003; Yizhak et al. 2010). Systems biology can reveal the holistic model of plants' stress responses that will enable us to selectively target the novel candidates to develop plant varieties with strengthened fitness to ever-changing climates.

## 22.8 CONCLUSION

Microarray analysis is one of the first tools of the postgenomic era that has helped us to explore transcriptomic information as a gateway to entering into the whole genomic information. Even though the world has probably started to see the beginning of the end of microarrays, this technique has successfully led to the identification of numerous stress-responsive genes under various stresses in many crop species. Inclusion of microarray data along with other omic technologies such as proteomics, metabolomics, epigenetics, and phenomics using modern computational and systems biology approaches will lead to the identification of novel points of intervention or the key genetical determinants of stress tolerance whose modifications through advanced breeding and biotechnological approaches will be helpful in developing crop varieties that can be better adapted to the stressed environments to ensure global food security.

## REFERENCES

Abe H, K Shinozaki, T Urao, T Iwasaki, D Hosokawa et al. 1997. Role of *Arabidopsis* MYC and MYB homologs in drought- and abscisic acid-regulated gene expression. *Plant Cell* 9:1859.

Abu Qamar S, H Luo, K Laluk, MV Mickelbart, and T Mengiste. 2009. Crosstalk between biotic and abiotic stress responses in tomato is mediated by the AIM1 transcription factor. *The Plant Journal* 58:347–360.

Agarwal PK and B Jha. 2010. Transcription factors in plants and ABA dependent and independent abiotic stress signalling. *Biologia Plantarum* 54(2):201–212.

Atkinson NJ and PE Urwin. 2012. The interaction of plant biotic and abiotic stresses: From genes to the field. *Journal of Experimental Botany* 63(10):3523–3543.

Atwell BJ, H Wang, and AP Scafaro. 2014. Could abiotic stress tolerance in wild relatives of rice be used to improve *Oryza sativa*? *Plant Science* 215–216:48–58.

Atwell S, YS Huang, BJ Vilhjálmsson, G Willems, M Horton et al. 2010. Genome-wide association study of 107 phenotypes in *Arabidopsis thaliana* inbred lines. *Nature* 465(7298):627–631.

Baisakh N, MV RamanaRao, K Rajasekaran, P Subudhi, J Janda et al. 2012. Enhanced salt stress tolerance of rice plants expressing a vacuolar H⁺ -ATPase subunit c1 (*savhac1*) gene from the halophyte grass *Spartina alterniflora* Löisel. *Plant Biotechnology Journal* 10(4):453–464.

Bansal KC, SK Lenka, and TK Mondal. 2014. Genomic resources for breeding crops with enhanced abiotic stress tolerance. *Plant Breeding* 133(1):1–11.

Benedictis M, G Bleve, M Faraco, E Stigliano, F Grieco et al. 2013. *AtSYP51/52* functions diverge in the post-golgi traffic and differently affect vacuolar sorting. *Molecular Plant* 6:916–930.

Berri S, P Abbruscato, O Faivre-Rampant, AC Brasileiro, I Fumasoni et al. 2009. Characterization of WRKY co-regulatory networks in rice and *Arabidopsis*. *BMC Plant Biology* 9(1):120.

Bolger ME, B Weisshaar, U Scholz, N Stein, B Usadel et al. 2014. Plant genome sequencing-applications for crop improvement. *Current Opinion in Biotechnology* 26:31–37.

Bonilla P, J Dvorak, D Mackill, K Deal, and G Gregorio. 2002. RFLP and SSLP mapping of salinity tolerance genes in chromosome 1 of rice (*Oryza sativa* L.) using recombinant inbred lines. *Philippines Agricultural Science* 85:68–76.

Byrt CS, JD Platten, W Spielmeyer, RA James, ES Lagudah et al. 2007. HKT1;5-like cation transporters linked to Na⁺ exclusion loci in wheat, *Nax2* and *Kna1*. *Plant Physiology* 143:1918–1928.

Cabello JV, AF Lodeyro, and MD Zurbriggen. 2014. Novel perspectives for the engineering of abiotic stress tolerance in plants. *Current Opinion in Biotechnology* 26:62–70.

Carrera J, G Rodrigo, A Jaramillo, and SF Elena. 2009. Reverse-engineering the *Arabidopsis thaliana* transcriptional network under changing environmental conditions. *Genome Biology* 10:R96.

Chae HB, JC Moon, MR Shin, YH Chi, YJ Jung et al. 2013. Thioredoxin reductase type C (NTRC) orchestrates enhanced thermotolerance to *Arabidopsis* by its redox-dependent holdase chaperone function. *Molecular Plant* 6:323–336.

Chao DY, YH Luo, M Shi, D Luo, and HX Lin. 2005. Salt-responsive genes in rice revealed by cDNA microarray analysis. *Cell Research* 15(10):796–810.

Chen M, QY Wang, XG Cheng, ZS Xu, LC Li et al. 2007. GmDREB2, a soybean DRE-binding transcription factor, conferred drought and high-salt tolerance in transgenic plants. *Biochemical and Biophysical Research Communications* 353:299–305.

Chen S, R Zhou, Y Huang, M Zhang, G Yang et al. 2011. Transcriptome sequencing of a highly salt tolerant mangrove species *Sonneratia alba* using Illumina platform. *Marine Genomics* 4(2):129–136.

Chinnusamy V and JK Zhu. 2009. Epigenetic regulation of stress responses in plants. *Current Opinion in Plant Biology* 12:133–139.

Clemens MJ. 2001. Translational regulation in cell stress and apoptosis. Roles of the *eIF4E* binding proteins. *Journal of Cellular & Molecular Medicine* 5:221–239.

Collins NC, F Tardieu, and R Tuberosa. 2008. Quantitative trait loci and crop performance under abiotic stress: Where do we stand? *Plant Physiology* 147(2):469–486.

Colmer TD, TJ Flowers, and R Munns. 2006. Use of wild relatives to improve salt tolerance in wheat. *Journal of Experimental Botany* 57(5):1059–1078.

Cotsaftis O, D Plett, AAT Johnson, H Walia, C Wilson et al. 2011. Root-specific transcript profiling of contrasting rice genotypes in response to salinity stress. *Molecular Plant* 4(1):25–41.

Cramer GR, K Urano, S Delrot, M Pezzotti, and K Shinozaki. 2011. Effects of abiotic stress on plants: A systems biology perspective. *BMC Plant Biology* 11(1):163.

Cui S, F Huang, J Wang, X Ma, Y Cheng, and J Liu. 2005. A proteomic analysis of cold stress response in rice seedlings. *Proteomics* 5:3162–3172.

Dai X, Y Xu, Q Ma, W Xu, T Wang et al. 2007. Overexpression of an R1R2R3 MYB gene OsMYB3R-2, increases tolerance to freezing, drought, salt stress in transgenic *Arabidopsis*. *Plant Physiology* 143:1739–1751.

Deshmukh R, H Sonah, G Patil, W Chen, S Prince et al. 2014. Integrating omic approaches for abiotic stress tolerance in soybean. *Frontiers in Plant Science* 5:244.

Deyholos MK. 2010. Making the most of drought and salinity transcriptomics. *Plant, Cell & Environment* 33(4):648–654.

Ding D, L Zhang, H Wang, Z Liu, Z Zhang et al. 2009. Differential expression of miRNAs in response to salt stress in maize roots. *Annals of Botany* 103(1):29–38.

Dubouzet JG, Y Sakuma, Y Ito, M Kasuga, M Dubouzet et al. 2003. OsDREB genes in rice, *Oryza sativa* L., encode transcription activators that function in drought-, high-salt- and cold-responsive gene expression. *Plant Journal* 33:751–763.

Duque AS, AM Almeida, AB Silva, JM Silva, AP Farinha et al. 2013. Abiotic stress responses in plants: Unraveling the complexity of genes and networks to survive. In K Vahdoti and C Leslie, editors, *Abiotic Stress-Plant Responses and Applications in Agriculture*, InTech, Rijeka, Croatia, pp. 49–101.

Echevarría ZS, E Yángüez, NF Bautista, AC Sanz, A Ferrando, and M Castellano. 2013. Regulation of translation initiation under biotic and abiotic stresses. *International Journal of Molecular Sciences* 14(3):4670–4683.

Edelist C, X Raffoux, M Falque, C Dillmann, D Sicard et al. 2009. Differential expression of candidate salt-tolerance genes in the halophyte *Helianthus paradoxus* and its glycophyte progenitors H. *annuus* and H. *petiolaris*. *American Journal of Botany* 96(10):1830–1838.

Ezawa S and Y Tada. 2008. Identification of salt tolerance genes from the mangrove plant *Bruguiera gymnorrhiza* using *Agrobacterium* functional screening. *Plant Science* 176:272–278.

Fan P, L Nie, P Jiang, J Feng, S Lv et al. 2013. Transcriptome analysis of *Salicornia europaea* under saline conditions revealed the adaptive primary metabolic pathways as early events to facilitate salt adaptation. *PloS One* 8(11):e80595.

Fernandez P, JD Rienzo, L Fernandez, HE Hopp, N Paniego et al. 2008. Transcriptomic identification of candidate genes involved in sunflower responses to chilling and salt stresses based on cDNA microarray analysis. *BMC Plant Biology* 8:11.

Flowers TJ, SA Flowers, MA Hajibagheri, and AR Yeo. 1990. Salt tolerance in the halophytic wild rice, *Porteresia coarctata* Tateoka. *New Phytologist* 114:675–684.

Flowers TJ, ML Koyama, SA Flowers, C Sudhakar, KP Singh, and AR Yeo. 2000. QTL: Their place in engineering tolerance of rice to salinity. *Journal of Experimental Botany* 51(342):99–106.

Flowers TJ, HK Galal, and L Bromham. 2010. Evolution of halophytes: Multiple origins of salt tolerance in land plants. *Functional Plant Biology* 37(7):604.

Frary A, G Deniz, K Davut, Ö Bilal, P Hasan et al. 2010. Salt tolerance in *Solanum pennellii*: Antioxidant response and related QTL. *BMC Plant Biology* 10:58.

Fukushima A, M Kusano, H Redestig, M Arita, and K Saito. 2009. Integrated omics approaches in plant systems biology. *Current Opinion in Chemical Biology* 13(5–6):532–538.

Gao R, K Duan, G Guo, Z Du, Z Chen et al. 2013. Comparative transcriptional profiling of two contrasting barley genotypes under salinity stress during the seedling stage. *International Journal of Genomics* 852–972.

Garg R, M Verma, S Agrawal, R Shankar, M Majee, and M Jain. 2013. Deep transcriptome sequencing of wild halophyte rice, *Porteresia coarctata*, provides novel insights into the salinity and submergence tolerance factors. *DNA Research*, 1–16.

Ge H, AJM Walhout, and M Vidal. 2003. Integrating "omic" information: A bridge between genomics and systems biology. *Trends in Genetics* 19:551–560.

Golldack D, I Lüking, and O Yang. 2011. Plant tolerance to drought and salinity: Stress regulating transcription factors and their functional significance in the cellular transcriptional network. *Plant Cell Reports* 30(8):1383–1391.

Hao GP, XH Zhang, YQ Wang, ZY Wu, and CL Huang. 2008. Nucleotide variation in the NCED3 region of *Arabidopsis thaliana* and its association study with abscisic acid content under drought stress. *Journal of Integrative Plant Biology* 51:175–183.

Hao YJ, W Wei, QX Song, HW Chen, YQ Zhang et al. 2011a. Soybean NAC transcription factors promote abiotic stress tolerance and lateral root formation in transgenic plants. *The Plant Journal* 68:302–313.

Hao Z, X Li, C Xie, J Weng, M Li et al. 2011b. Identification of functional genetic variations underlying drought tolerance in maize using SNP markers. *Journal of Integrative Plant Biology* 53:641–652.

Hazen SP, Y Wu, and JA Kreps. 2003. Gene expression profiling of plant responses to abiotic stress. *Functional & Integrative Genomics* 3(3):105–111.

He XJ, RL Mu, WH Cao, ZG Zhang, and JS Zhang. 2005. AtNAC2, a transcription factor downstream of ethylene and auxin signaling pathways, is involved in salt stress response and lateral root development. *Plant Journal* 44:903–916.

Hirayama T and K Shinozaki. 2010. Research on plant abiotic stress responses in the post-genome era: Past, present and future. *The Plant Journal* 61(6):1041–1052.

Hossain MR. 2014. Salinity tolerance and transcriptomics in rice. PhD thesis, School of Biosciences, The University of Birmingham, U.K., http://etheses.bham.ac.uk/5092/

Hu H, J You, Y Fang, X Zhu, Z Qi, and L Xiong. 2008. Characterization of a transcription factor gene *SNAC2* conferring cold and salt tolerance in rice. *Plant Molecular Biology* 67:169–181.

Huang J, X Lu, H Yan, S Chen, W Zhang et al. 2012. Transcriptome characterization and sequencing-based identification of salt-responsive genes in *Millettia pinnata*, a semi-mangrove plant. *DNA Research* 19(2):195–207.

Jain M. 2012. Next-generation sequencing technologies for gene expression profiling in plants. *Briefings in Functional Genomics* 11:63–70.

James RA, RJ Davenport, and R Munns. 2006. Physiological characterization of two genes for Na⁺ exclusion in durum wheat, *Nax1* and *Nax2*. *Plant Physiology* 142:1537–1547.

James M and S Dow. 2012. Transcription factors important in the regulation of salinity tolerance. PhD thesis, University of Adelaide, Adelaide SA, Australia. Available from: https://digital.library.adelaide.edu.au/dspace/handle/2440/77095.

Janz D, K Behnke, JP Schnitzler, B Kanawati, PS Kopplin, and A Polle. 2010. Pathway analysis of the transcriptome and metabolome of salt sensitive and tolerant poplar species reveals evolutionary adaptation of stress tolerance mechanisms. *BMC Plant Biology* 10(1):150.

Jeong JS, YS Kim, KH Baek, H Jung, SH Ha et al. 2010. Root-specific expression of *OsNAC10* improves drought tolerance and grain yield in rice under field drought conditions. *Plant Physiology* 153:185–197.

Jiang H, S Peng, S Zhang, X Li, H Korpelainen, and C Li. 2012. Transcriptional profiling analysis in *Populus yunnanensis* provides insights into molecular mechanisms of sexual differences in salinity tolerance. *Journal of Experimental Botany* 63(10):3709–3726.

Jiang SY and S Ramachandran. 2010. Assigning biological functions to rice genes by genome annotation, expression analysis and mutagenesis. *Biotechnology Letters* 32(12):1753–1763.

Jung C, JS Seo, SW Han, YJ Koo, CH Kim et al. 2008. Overexpression of *AtMYB44* enhances stomatal closure to confer abiotic stress tolerance in transgenic *Arabidopsis*. *Plant Physiology* 146:623–635.

Kawasaki S, C Borchert, M Deyholos, H Wang, S Brazille et al. 2001. Gene expression profiles during the initial phase of salt stress in rice. *The Plant Cell* 13:889–905.

Kawaura K, K Mochida, and Y Ogihara. 2008. Genome-wide analysis for identification of salt-responsive genes in common wheat. *Functional & Integrative Genomics* 8(3):277–286.

Khraiwesh B, J Zhu, and J Zhu. 2012. Role of miRNAs and siRNAs in biotic and abiotic stress responses of plants. *Biochimica et Biophysica Acta* 18–19:137–148.

Kim SJ and F. Brandizzi 2012. News and views into the SNARE complexity in *Arabidopsis*. *Frontier in Plant Science* 3:28.

Kim MH, K Sasaki, and R Imai. 2009. Cold shock domain protein 3 regulates freezing tolerance in *Arabidopsis thaliana*. *Journal of Biological Chemistry* 284:23454–23460.

Koyama ML, A Levesley, RMD Koebner, TJ Flowers, and AR Yeo. 2001. Quantitative trait loci for component physiological traits determining salt tolerance in rice. *Plant Physiology* 125:406–422.

Kumar GR, K Sakthivel, RM Sundaram, CN Neeraja, SM Balachandran et al. 2010. Allele mining in crops: Prospects and potentials. *Biotechnology Advances* 28:451–461.

Lageix S, E Lanet, MNP Pelissier, MC Espagnol, C Robaglia et al. 2008. *Arabidopsis* eIF2alpha kinase GCN2 is essential for growth in stress conditions and is activated by wounding. *BMC Plant Biology* 8:134.

Langridge P and D Fleury. 2011. Making the most of "omics" for crop breeding. *Trends in Biotechnology* 29(1):33–40.

Leonforte A, S Sudheesh, NOI Cogan, PA Salisbury and ME Nicolas et al. 2013. SNP marker discovery, linkage map construction and identification of QTLs for enhanced salinity tolerance in field pea (*Pisum sativum* L.). *BMC Plant Biology* 13(1):161.

Leshem Y, NM Book, O Cagnac, G Ronen, Y Nishri et al. 2006. Suppression of *Arabidopsis* vesicle-SNARE expression inhibited fusion of $H_2O_2$-containing vesicles with tonoplast and increased salt tolerance. *Proceedings of the National Academy of Sciences of the United States of America* 103(47):18008–18013.

Less H and G Galili. 2008. Principal transcriptional programs regulating plant amino acid metabolism in response to abiotic stresses. *Plant Physiology* 147:316–330.

Lexer C, ME Welch, JL Durphy, and LH Rieseberg. 2003. Natural selection for salt tolerance quantitative trait loci (QTLs) in wild sunflower hybrids: Implications for the origin of *Helianthus paradoxus*, a diploid hybrid species. *Molecular Ecology* 12:1225–1235.

Lexer C, B Heinze, R Alia, and LH Rieseberg. 2004. Hybrid zones as a tool for identifying adaptive genetic variation in outbreeding forest trees: Lessons from wild annual sunflowers (*Helianthus* spp.). *Forest Ecology Management* 197:49–64.

Li XP, AG Tian, GZ Luo, ZZ Gong, J Zhang, and SY Chen. 2005. Soybean DRE-binding transcription factors that are responsive to abiotic stresses. *Theoretical and Applied Genetics* 110:1355–1362.

Liao Y, H Zou, W Wei, YJ Hao, AG Tian et al. 2008. Soybean GmbZIP44, GmbZIP62 and GmbZIP78 genes function as negative regulator of ABA signaling and confer salt and freezing tolerance in transgenic *Arabidopsis*. *Planta* 228:225–240.

Lin HX, MZ Zhu, M Yano, JP Gao, ZW Liang et al. 2004. QTLs for Na$^+$ and K$^+$ uptake of the shoots and roots controlling rice salt tolerance. *Theoretical and Applied Genetics* 108(2):253–260.

Lister R, BD Gregory, and JR Ecker. 2009. Next is now: New technologies for sequencing of genomes, transcriptomes, and beyond. *Current Opinion in Plant Biology* 12(2):107–118.

Liu HH, X Tian, YJ Li, CA Wu, and CC Zheng. 2008. Microarray-based analysis of stress-regulated microRNAs in *Arabidopsis thaliana*. *RNA* 14(5):836–843.

Liu X, L Hong, XY Li, Y Yao, B Hu, and L Li. 2011. Improved drought and salt tolerance in transgenic *Arabidopsis* overexpressing a NAC transcriptional factor from *Arachis hypogaea*. *Bioscience, Biotechnology and Biochemistry* 3:443–450.

Liu Y, X Ji, L Zheng, X Nie, and Y Wang. 2013. Microarray analysis of transcriptional responses to abscisic acid and salt stress in *Arabidopsis thaliana*. *International Journal of Molecular Sciences* 14(5):9979–9998.

Long TA. 2011. Many needles in a haystack: Cell-type specific abiotic stress responses. *Current Opinion in Plant Biology* 14:325–331.

Luhua S, SC Yilmaz, J Harper, J Cushman, and R Mittler. 2008. Enhanced tolerance to oxidative stress in transgenic *Arabidopsis* plants expressing proteins of unknown function. *Plant Physiology* 148: 280–292.

Mao X, D Jia, A Li, H Zhang, S Tian et al. 2011. Transgenic expression of *TaMYB2A* confers enhanced tolerance to multiple abiotic stresses in *Arabidopsis*. *Functional & Integrative Genomics* 11:445–465.

Mehta PA, K Sivaprakash, M Parani, G Venkataraman, and AK Parida. 2005. Generation and analysis of expressed sequence tags from the salt tolerant mangrove species *Avicennia marina* (Forsk) Vierh. *Theoretical and Applied Genetics* 110:416–424.

Meier M, RV Sit, and SR Quake. 2013. Proteome-wide protein interaction measurements of bacterial proteins of unknown function. *Proceedings of the National Academy of Sciences of the United States of America* 110(2):477–482.

Ming-zhe Y, W Jian-fei, C Hong-you, Z Hu-qu, and Z Hong-sheng. 2005. Inheritance and QTL mapping of salt tolerance in rice. *Rice Science* 12(1):25–32.

Mishra SN, XH Pham, SK Sopory, and N Tuteja. 2005. Pea DNA helicase 45 overexpression in tobacco confers high salinity tolerance without affecting yield. *Proceedings of the National Academy of Sciences of the United States of America* 102(39):509–514.

Miyama M and N Hanagata. 2007. Microarray analysis of 7029 gene expression patterns in Burma mangrove under high-salinity stress. *Plant Science* 172:948–995.

Mochida K and K Shinozaki. 2011. Advances in omics and bioinformatics tools for systems analyses of plant functions. *Plant Cell Physiology* 52:2017–2038.

Mondini L, M Nachit, E Porceddu, and MA Pagnotta. 2012. Identification of SNP mutations in DREB1, HKT1, and *WRKY1* genes involved in drought and salt stress tolerance in durum wheat (*Triticum turgidum L. var durum*). *Omics* 16:178–187.

Nakashima K, LSP Tran, DV Nguyen, M Fujita, K Maruyama et al. 2007. Functional analysis of a NAC-type transcription factor OsNAC6 involved in abiotic and biotic stress-responsive gene expression in rice. *Plant Journal* 51:617–630.

Niu CF, W Wei, QY Zhou, AG Tian, YJ Hao et al. 2012. Wheat WRKY genes *TaWRKY2* and *TaWRKY19* regulate abiotic stress tolerance in transgenic *Arabidopsis* plants. *Plant Cell and Environment* 35(6):1156–1170.

Ohnishi T, S Sugahara, T Yamada, K Kikuchi, Y Yoshiba et al. 2005. *OsNAC6*, a member of the NAC gene family, is induced by various stresses in rice. *Genes & Genetic Systems* 80:135–139.

Öktem HA, F Eyidogan, F Selçuk, M Tufan, JA Teixeira et al. 2008. Revealing response of plants to biotic and abiotic stresses with microarray technology. *Genes, Genomes and Genomics* 2(1):14–48.

Palsson B and K Zengler. 2010. The challenges of integrating multi-omic data sets. *Nature Chemical Biology* 6:787–789.

Pandit A, V Rai, S Bal, S Sinha, V Kumar et al. 2010. Combining QTL mapping and transcriptome profiling of bulked RILs for identification of functional polymorphism for salt tolerance genes in rice (*Oryza sativa* L.). *Molecular Genetics and Genomics* 284:121–136.

Platten JD, JA Egdane, and AM Ismail. 2013. Salinity tolerance, Na$^+$ exclusion and allele mining of HKT1;5 in *Oryza sativa* and *O. glaberrima*: Many sources, many genes, one mechanism? *BMC Plant Biology* 13:32.

Price AH. 2006. Believe it or not, QTLs are accurate. *Trends in Plant Science* 11:213–216.

Qi X, M-W Li, M Xie, X Liu, M Ni et al. 2014. Identification of a novel salt tolerance gene in wild soybean by whole-genome sequencing. *Nature Communications* 5:4340.

Rajalakshmi S and A Parida. 2012. Halophytes as a source of genes for abiotic stress tolerance. *Journal of Plant Biochemistry and Biotechnology* 21(1):63–67.

Ramamoorthy R, SY Jiang, N Kumar, PN Venkatesh, and S Ramachandran. 2008. A comprehensive transcriptional profiling of the *WRKY* gene family in rice under various abiotic and phytohormone treatments. *Plant Cell Physiology* 49(6):865–879.

Rasmussen S, P Barah, MCS Rodriguez, S Bressendorff, P Friis et al. 2013. Transcriptome responses to combinations of stresses in *Arabidopsis*. *Plant Physiology* 161(4):1783–1794.

Rausell A, R Kanhonou, L Yenush, R Serrano, and R Ros. 2003. The translation initiation factor eIF1A is an important determinant in the tolerance to NaCl stress in yeast and plants. *Plant Journal* 34:257–267.

Ren ZH, JP Gao, LG Li, XL Cai, W Huang et al. 2005. A rice quantitative trait locus for salt tolerance encodes a sodium transporter. *Nature Genetics* 37(10):1141–1146.

Riechmann JL. 2000. *Arabidopsis* transcription factors: Genome-wide comparative analysis among eukaryotes. *Science* 290(5499):2105–2110.

Rogers ED, T Jackson, T Moussaieff, A Aharoni, and PN Benfey. 2012. Cell type-specific transcriptional profiling: Implications for metabolite profiling. *Plant Journal* 70:5–17.

Sabouri H, AM Rezai, A Moumeni, A Kavousi, M Katouzi, and A Sabouri. 2009. QTLs mapping of physiological traits related to salt tolerance in young rice seedlings. *Biologia Plantarum* 53(4):657–662.

Sakuma Y, K Maruyama, Y Osakabe, F Qin, M Seki et al. 2006. Functional analysis of an *Arabidopsis* transcription factor, DREB2A, involved in drought-responsive gene expression. *Plant Cell* 18:1292–1309.

Schreiber AW, T Sutton, RA Caldo, E Kalashyan, B Lovell, et al. 2009. Comparative transcriptomics in the *Triticeae*. *BMC Genomics* 10(January):285.

Senadheera P, RK Singh, and FJM Maathuis. 2009. Differentially expressed membrane transporters in rice roots may contribute to cultivar dependent salt tolerance. *Journal of Experimental Botany* 60(9):2553–2563.

Sengupta A and Majumder L. 2010. *Porteresia coarctata* (Roxb.) Tateoka, a wild rice: a potential model for studying salt-stress biology in rice. *Plant Cell and Environment* 33:526–542.

Sha N, M Vannucci, MG Tadesse, PJ Brown, I Dragoni, et al. 2004. Bayesian variable selection in multinomial probit models to identify molecular signatures of disease stage. *Biometrics* 60(3):812–819.

Shahbaaz M, MI Hassan, and F Ahmad. 2013. Functional annotation of conserved hypothetical proteins from *Haemophilus influenzae* Rd KW20. *PloS One* 8(12):e84263.

Shen YG, WK Zhang, DQ Yan, BX Du, JS Zhang et al. 2003. Characterization of a DRE-binding transcription factor from a halophyte *Atriplex hortensis*. *Theoretical & Applied Genetics* 107:155–161.

Siahsar BA and M Narouei. 2010. Mapping QTLs of physiological traits associated with salt tolerance in 'Steptoe'×'Morex' doubled haploid lines of barley at seedling stage. *International Journal of Food, Agriculture and Environment* 8(2):751–759.

Song SY, Y Chen, J Chen, XY Dai, and WH Zhang. 2011. Physiological mechanisms underlying OsNAC5-dependent tolerance of rice plants to abiotic stress. *Planta* 234:331–345.

Sun W, X Xu, H Zhu, A Liu, L Liu et al. 2010. Comparative transcriptomic profiling of a salt-tolerant wild tomato species and a salt-sensitive tomato cultivar. *Plant & Cell Physiology* 51(6):997–1006.

Taji T, K Komatsu, T Katori, Y Kawasaki, Y Sakata et al. 2010. Comparative genomic analysis of 1047 completely sequenced cDNAs from an *Arabidopsis*-related model halophyte, *Thellungiella halophila*. *BMC Plant Biology* 10:261.

Takasaki H, K Maruyama, S Kidokoro, Y Ito, Y Fujita et al. 2010. The abiotic stress-responsive NAC-type transcription factor OsNAC5 regulates stress-inducible genes and stress tolerance in rice. *Molecular Genetics & Genomics* 284:173–183.

Talamè V, NZ Ozturk, HJ Bohnert, and R Tuberosa. 2007. Barley transcript profiles under dehydration shock and drought stress treatments: A comparative analysis. *Journal of Experimental Botany* 58(2): 229–240.

Tran LSP, K Nakashima, Y Sakuma, SD Simpson, Y Fujita et al. 2004. Isolation and functional analysis of *Arabidopsis* stress-inducible NAC transcription factors that bind to a drought-responsive *cis*-element in the early responsive to dehydration stress 1 promoter. *Plant Cell* 16:2481–2498.

Trevino V, MG Tadesse, M Vannucci, F Al-Shahrour, P Antczak et al. 2011. Analysis of normal-tumour tissue interaction in tumours: Prediction of prostate cancer features from the molecular profile of adjacent normal cells. *PloS One* 6(3):e16492.

Trindade I, C Capitão, T Dalmay, MP Fevereiro, DM Santos et al. 2010. miR398 and miR408 are up-regulated in response to water deficit in *Medicago truncatula*. *Planta* 231(3):705–716.

Tripathi P, RC Rabara, and PJ Rushton. 2014. A systems biology perspective on the role of WRKY transcription factors in drought responses in plants. *Planta* 239(2):255–266.

Urano K, Y Kurihara, M Seki, and K Shinozaki. 2009. "Omics" analyses of regulatory networks in plant abiotic stress responses. *Current Opinion in Plant Biology* 13(2):132–138.

Villarino GH, A Bombarely, JJ Giovannoni, MJ Scanlon, and NS Mattson. 2014. Transcriptomic analysis of *Petunia hybrida* in response to salt stress using high throughput RNA sequencing. *PloS One* 9(4):e94651.

Walia H, C Wilson, P Condamine, X Liu, AM Ismail et al. 2005. Comparative transcriptional profiling of two contrasting rice genotypes under salinity stress during the vegetative growth stage. *Plant Physiology* 139(2):822–835.

Walia H, C Wilson, L Zeng, AM Ismail, P Condamine, and TJ Close. 2007. Genome-wide transcriptional analysis of salinity stressed japonica and indica rice genotypes during panicle initiation stage. *Plant Molecular Biology* 63:609–623.

Wang B, T Lan, and WR Wu. 2007a. Mapping of QTLs for Na+ content in rice seedlings under salt stress. *Chinese Journal of Rice Science* 21(6):585–590.

Wang J, L Chen, Y Wang, J Zhang, Y Liang, and D Xu. 2013. A computational systems biology study for understanding salt tolerance mechanism in rice. *PloS One* 8(6):e64929.

Wang Q, Y Guan, Y Wu, H Chen, F Chen, and C Chu. 2008. Overexpression of a rice OsDREB1F gene increases salt, drought, and low temperature tolerance in both *Arabidopsis* and rice. *Plant Molecular Biology* 67:589–602.

Wang XS, J Zhu, L Mansueto, and R Bruskiewich. 2005. Identification of candidate genes for drought stress tolerance in rice by the integration of a genetic (QTL) map with the rice genome physical map. *Journal of Zhejiang University Science B*5(5):382–388.

Wang XS, HB Zhu, GL Jin, HL Liu, WR Wu, and J Zhu. 2007b. Genome-scale identification and analysis of LEA genes in rice (*Oryza sativa* L.). *Plant Science* 172:414–420.

Wong YY, CL Ho, PD Nguyen, SS Teo, JA Harikrishna et al. 2007. Isolation of salinity tolerant genes from the mangrove plant, *Bruguiera cylindrica* by using suppression subtractive hybridization (SSH) and bacterial functional screening. *Aquatic Botany* 86:117–122.

Xu ZS, LQ Xia, M Chen, XG Cheng, RY Zhang et al. 2007. Isolation and molecular characterization of the *Triticum aestivum* L. ethylene-responsive factor 1 (TaERF1) that increases multiple stress tolerance. *Plant Molecular Biology* 65:719–732.

Xu ZS, ZY Ni, ZY Li, LC Li, M Chen, DY Gao et al. 2009. Isolation and functional characterization of HvDREB1, a gene encoding a dehydration-responsive element binding protein in *Hordeum vulgare*. *Journal of Plant Research* 122:121–130.

Xue D, Y Huang, X Zhang, K Wei, S Westcott et al. 2009. Identification of QTLs associated with salinity tolerance at late growth stage in barley. *Euphytica* 169:187–196.

Xue GP, HM Way, T Richardson, J Drenth, PA Joyce et al. 2011. Overexpression of *TaNAC69* leads to enhanced transcript levels of stress up-regulated genes and dehydration tolerance in bread wheat. *Molecular Plant* 4:697–712.

Yamada A, M Sekiguchi, T Mimura, and Y Ozeki. 2002. The role of plant CCTα in salt- and osmotic-stress tolerance. *Plant and Cell Physiology* 43:1043–1048.

Yizhak K, T Benyamini, W Liebermeister, E Ruppin, and T Shlomi. 2010. Integrating quantitative proteomics and metabolomics with a genome-scale metabolic network model. *Bioinformatics* 26:255–260.

Yokotani N., T Ichikawa, Y Kondou, M Matsui, H Hirochika et al. 2009. Tolerance to various environmental stresses conferred by the salt-responsive rice gene *ONAC063* in transgenic *Arabidopsis*. *Planta* 229:1065–1075.

Yu S, W Wang, and B Wang. 2012. Recent progress of salinity tolerance research in plants. *Genetika* 48(5):590–598.

Zhai H, X Bai, YM Zhu, Y Li, H Cai et al. 2010. A single-repeat R3-MYB transcription factor MYBC1 negatively regulates freezing tolerance in *Arabidopsis*. *Biochemical and Biophysical Research Communications* 394(4):1018–1023.

Zhang G, M Chen, L Li, Z Xu, X Chen et al. 2009. Overexpression of the soybean GmERF3 gene, an AP2/ERF type transcription factor for increased tolerances to salt, drought, and diseases in transgenic tobacco. *Journal of Experimental Botany* 60:3781–3796.

Zhang J, J Feng, J Lu, Y Yang, X Zhang et al. 2014. Transcriptome differences between two sister desert poplar species under salt stress. *BMC Genomics* 15(1):337.

Zhang L, D Xi, S Li, Z Gao, S Zhao et al. 2011. A cotton group CMAP kinase gene, GhMPK2, positively regulates salt and drought tolerance in tobacco. *Plant Molecular Biology* 77(1–2).

Zhang Y, J Lai, S Sun, Y Li, Y Liu et al. 2008. Comparison analysis of transcripts from the halophyte *Thellungiella halophila*. *Journal of Integrative Plant Biology* 50(10):1327–1335.

Zheng X, B Chen, G Lu, and B Han. 2009. Overexpression of a NAC transcription factor enhances rice drought and salt tolerance. *Biochemical and Biophysical Research Communications* 379:985–989.

Zhou QY, AG Tian, HF Zou, ZM Xie, G Lei et al. 2008. Soybean WRKY-type transcription factor genes, *GmWRKY13*, *GmWRKY21*, and *GmWRKY54*, confer differential tolerance to abiotic stresses in transgenic *Arabidopsis* plants. *Plant Biotechnology Journal* 6(5):486–503.

Zhu JK. 2001. Plant salt tolerance. *Trends in Plant Science* 6:66–71.

# INDEX